HANDBOOK OF
Chemical Mass Transport in the Environment

HANDBOOK OF
Chemical Mass Transport in the Environment

Edited by
Louis J. Thibodeaux
Donald Mackay

CRC Press
Taylor & Francis Group
Boca Raton London New York

CRC Press is an imprint of the
Taylor & Francis Group, an **informa** business

Symbols used to create cover art courtesy of the Integration and Application Network (ian.umces.edu/symbols/), University of Maryland Center for Environmental Science.

First published in paperback 2024

First published 2011
by CRC Press
2385 NW Executive Center Drive, Suite 320, Boca Raton FL 33431

and by CRC Press
4 Park Square, Milton Park, Abingdon, Oxon, OX14 4RN

CRC Press is an imprint of Taylor & Francis Group, LLC

© 2011, 2024 by Taylor and Francis Group, LLC

Library of Congress Cataloging-in-Publication Data

Handbook of estimation methods for chemical mass transport in the environment /
    editors, Louis J. Thibodeaux, Donald Mackay.
        p. cm.
    Includes bibliographical references and index.
    ISBN 978-1-4200-4755-4 (hardcover : alk. paper)
    1. Chemical kinetics. 2. Mass transfer. 3. Partition coefficient (Chemistry) 4.
Estimation theory. 5. Environmental chemistry. I. Thibodeaux, Louis J. II. Mackay,
Donald, 1936-

QD502.H357 2011
628.5'2--dc22   2010029280

ISBN: 978-1-032-91806-8 (pbk)
ISBN: 978-1-420-04755-4 (hbk)
ISBN: 978-0-429-13883-6 (ebk)

DOI: 10.1201/b10262

Visit the Taylor & Francis Web site at
http://www.taylorandfrancis.com

and the CRC Press Web site at
http://www.crcpress.com

Contents

Preface ix
Acknowledgments xiii
Editors xv
Contributors xvii

Chapter 1   Introduction: Chemical Mobility in the Environment 1

Louis J. Thibodeaux and Donald Mackay

Chapter 2   Mass Transport Fundamentals from the Environmental
Perspective 15

Louis J. Thibodeaux

Chapter 3   The Fugacity Approach to Mass Transport and MTCs 43

Donald Mackay

Chapter 4   Flux Equations for Mass Transport Processes across Interfaces 51

Louis J. Thibodeaux

Chapter 5   Estimating Molecular Diffusivities in Environmental Media 71

Justin E. Birdwell and Thomas R. Marrero

Chapter 6   Deposition from the Atmosphere to Water and Soils with Aerosol
Particles and Precipitation 103

Matthew MacLeod, Martin Scheringer, Christian Götz,
Konrad Hungerbühler, Cliff I. Davidson, and
Thomas M. Holsen

Chapter 7   Mass Transfer between the Atmosphere and Plant
Canopy Systems 137

Michael S. McLachlan

Chapter 8   Mass Transfer within Surface Soils 159

Thomas E. McKone, Shannon L. Bartelt-Hunt, Mira S. Olson,
and Fred D. Tillman

Chapter 9 Air–Water Mass Transfer Coefficients 213

John S. Gulliver

Chapter 10 Deposition and Resuspension of Particles and the Associated
Chemical Transport across the Sediment–Water Interface 253

*Joseph V. DePinto, Richard D. McCulloch, Todd M. Redder,
John R. Wolfe, and Timothy J. Dekker*

Chapter 11 Advective Porewater Flux and Chemical Transport
in Bed Sediment...301

Louis J. Thibodeaux, John R.Wolfe, and Timothy J. Dekker

Chapter 12 Diffusive Chemical Transport across Water and Sediment
Boundary Layers 321

Louis J. Thibodeaux, Justin E. Birdwell, and Danny D. Reible

Chapter 13 Bioturbation and Other Sorbed-Phase Transport Processes in
Surface Soils and Sediments.......................................359

Louis J. Thibodeaux, Gerald Matisoff, and Danny D. Reible

Chapter 14 Mass Transport from Soil to Plants.................................389

*William Doucette, Erik M. Dettenmaier, Bruce Bugbee,
and Donald Mackay*

Chapter 15 Dispersion and Mass Transfer in Groundwater Near-Surface
Geologic Formations 413

Tissa H. Illangasekare, Christophe C. Frippiat, and Radek Fučík

Chapter 16 Dust Resuspension and Chemical Mass Transport from Soil to
Atmosphere 453

Cheryl L. McKenna-Neuman

Chapter 17 Deposition of Dissolved and Particulate-Bound Chemicals from
the Surface Ocean 495

Rainer Lohmann and Jordi Dachs

Chapter 18 Chemical Exchange between Snow and the Atmosphere 513

Torsten Meyer and Frank Wania

Chapter 19   Chemical Dynamics in Urban Areas 531

Miriam L. Diamond and Louis J. Thibodeaux

Chapter 20   Mixing in the Atmosphere and Surface Waters with
Application to Compartmental Box Models 565

*Ellen Bentzen, Matthew MacLeod, Brendan Hickie, Bojan Gasic,
Konrad Hungerbühler, and Donald Mackay*

Chapter 21   Environmental Mass Transfer State-of-the-Art, Deficiencies,
and Future Directions 589

Justin E. Birdwell, Louis J. Thibodeaux, and Donald Mackay

Index...595

Preface

THREE PILLARS OF CHEMICAL FATE IN THE ENVIRONMENT

The body of knowledge comprising the environmental science of chemicals rests on three fundamental pillars. All have their theoretical roots in physical chemistry and are essential components for understanding chemical fate.

From the kinetic theory of gases, we know that the mean speed of molecules is proportional to the square root of its absolute temperature. Millions of intermolecular collisions occur per second and are proportional to the concentration of molecules. In a gas of nonuniform composition, this random interaction results in molecular diffusion until the composition is uniform throughout. Similar diffusive transport processes occur in liquids and solids as well. Diffusion is thus one aspect of the *transport pillar* and is the subject of this handbook. It comprises rates of chemical transport within and between media. We assert that this topic has been the "poor relation" in environmental science.

The absence of chemical equilibrium between media phases is conveniently expressed by unequal fugacities. This condition dramatically influences the concentration gradients in the vicinity of interfaces and the chemical transport rate or flux. This forms the *thermodynamic pillar*, commonly termed chemical equilibrium partitioning between the environmental media. A huge literature exists on partition coefficients, and estimation methods are well developed at least for conventional contaminants.

The *reaction pillar* is chemical mechanisms and kinetics involving electron rearrangements in both abiotic and biotic systems. It is of practical importance to predict how quickly a reaction mixture will attain its equilibrium state. The processes may involve molecule rearrangements, molecule destruction, ionic species transformations, etc. All these mechanisms alter and influence chemical concentration gradients as well. Extensive literature can be found especially on atmospheric reactions, although much information is also available on aqueous and solid phase reactions.

A BRIEF HISTORY ON THE ORIGIN OF THE HANDBOOK

The textbook entitled *Environmental Chemodynamics* (Thibodeaux, 1996) was the first unified attempt to address multimedia chemical mass transfer in natural systems such as air, water, soil, sediments, etc. Arguably, mass transfer is the key chemodynamic process responsible for the presence of many anthropogenic chemicals worldwide as trace contaminants and these chemicals achieving pollutant levels at many locales. Mass transport coefficients (MTCs) are the kinetic parameters that quantify chemical movement rates within and between these natural media. Coefficient estimation methods based upon theoretically sound transport mechanisms and experimental measurements are essential inputs to all environment mathematical models designed to quantify chemical fate in the environment. These models vary significantly in type and level of mathematical sophistication. One genre is the multimedia chemical evaluative typified by the Mackay fugacity models (Mackay, 2001). Earlier versions include the U.S. Environmental Protection Agency's EXAMS; the European Union's model EUSES is a more recent entry. Another genre is the single- or dual-media models such as those designed for chemical fate studies in flowing streams, lakes, terrestrial watersheds, agricultural soils, air pollutant dispersion, groundwater, etc. The last genre includes single-media vignette models that are site-specific and locale-focused. These include chemical fate models for landfills, waste treatment soils, aquatic surface impoundments, contaminated land, industrial manufacturing sites, bed sediment sites, etc. All these models require MTCs. The input values needed range in their certainty levels from roughly order-of-magnitude estimates to more accurate numerical estimates aimed at making the most exacting of biota exposure predictions.

Besides serving as a general introduction to the subject of environmental mass transfer, a key objective of this handbook is to provide model users with mass transport coefficient estimation methods for all genres. This is a first-time attempt at compiling information on numerical estimates and correlating equations in a unified fashion in a single document. The idea for a handbook originated in 1995 in an international workshop, which was attended by the editors, held to evaluate the performance of five multimedia chemical fate evaluative models in current use worldwide (Cohen et al., 1995). The initial comparison of model-predicted numerical results on concentration, residence times, and certain other response times revealed some unexpected large discrepancies, which proved to be from two sources. The most important differences arose because intermedia transport data were selected independently by each modeler. Upon adjusting the MTCs, the models gave essentially identical numerical results. We, the editors of this handbook, realized the importance of the outcome of this serendipitous workshop and made tentative plans to address the issue of estimating environmental MTCs. Other demands in our professional lives got in the way for over a decade. In 2006 we decided the time was right and the process of creating the handbook began.

With the needs of these model-users in mind as the target audience, the topics for chapters were decided upon and general guidelines for their contents established. Recognized experts in the various MTC subject areas were identified and invited to contribute as authors. The individuals accepting the offer were charged with gathering and organizing the appropriate technical matter.

Why Buy This Book and What Can the User Expect?

In this handbook we have sought to present a unique and comprehensive account of the current state of the science in this wide area. Four chapters by the editors set out the underlying principles. Sixteen chapters written by 38 selected experts discuss specific transport processes in a consistent, user-friendly format. Transport data and correlations are provided along with examples of calculation procedures. The last chapter offers our personal glimpse of future needs and emerging priorities.

Our hope is that this handbook will be of value to students, teachers, researchers, and practitioners and serve to strengthen the transport pillar so important to the base of environmental science.

LITERATURE CITED

Cowen, C.E., et al. 1995. *The Multimedia Fate Model*, Society of Environmental Toxicology and Chemistry Press, Pensacola, Florida.

Mackay, D. 2001. *Multimedia Environmental Models*, 2nd ed., Lewis Publishers, Boca Raton, Florida.

Thibodeaux, L.J. 1996. *Environmental Chemodynamics*, 2nd ed., Wiley, New York.

Acknowledgments

Many of our professional colleagues contributed to this handbook. These contributions include technical materials, reviews, and ideas. Early on, at the conceptual stage, many voiced the need for such a document and encouraged us to do it exclaiming, "who else is more qualified?" Many of these people are chapter contributors and we gratefully acknowledge their contribution.

Both institutions, Louisiana State University (LSU) and Trent University, provided the supportive and sustaining academic atmosphere that encourages the diffusion of knowledge by writing books. Specifically, LSU is acknowledged for funding my sabbatical leave between January and May 2006. The time was used to travel, so that Don and I could meet and agree on a conceptual plan. Additional traveling was done to recruit contributors. In this regard we thank Professor K.T. Valsaraj, chair of the Cain Department of Chemical Engineering, and Professor Zaki Bassouni, dean of the College of Engineering, for supporting the sabbatical specifically, and the handbook creation activity in general. At Trent University, we sincerely thank Professor David A. Ellis, director of The Canadian Centre for Environmental Modelling and Chemistry and Eva Webster, the associate director, for their efforts in supporting this activity.

Special thanks to Dr. Justin E. Birdwell. As a postdoctoral student he provided invaluable leadership on a host of conceptual and technical issues concerned with creating the handbook. Late in the handbook writing process he assumed the leadership role in preparing two book chapters. In addition, he also assisted with the production and review of the manuscript.

Thanks to Melanie McCandless for organizing the various chapter styles and formats received from the various contributors into a homogeneous document. Thanks to Danny Fontenot and Darla Dao for word processing duties on the chapters I authored and other handbook-related tasks. The editorial and production staff of Taylor & Francis Group LLC, have been very understanding and helpful through this four-year project. It began in December 2006 and there were many delays along the way and missed deadlines. Special thanks to Jill Jurgensen, Hillary Rowe, and Rachael Panthier. Also thanks to S.M. Syed of Techset Composition for overseeing the production of this book.

They knew the drill, the slog and sacrifices involved when they approved our doing this project that seemed never ending. We lovingly acknowledge Ness Mackay and Joyce Thibodeaux for their enduring support.

Editors

Louis J. Thibodeaux is the Jesse Coates Professor in the Cain Department of Chemical Engineering at Louisiana State University (LSU), Baton Rouge, LA. He was born in Louisiana and obtained all his degrees from LSU: the BS in 1962 in petroleum chemical engineering. He worked for a few years in the nuclear and hazardous waste field for chemical and paper companies. In 1968 he joined the University of Arkansas and worked there for 17 years. In 1984 he returned to LSU as the director of the U.S. Environmental Protection Agency funded Center for Hazardous Waste Research. After 11 years he returned to fulltime teaching and research. Recently he served as the chair of geology and geophysics for two years.

As a graduate student he held a fellowship from the National Council for Air and Stream Improvement (pulp & paper industry). While working on chemical fate in rivers, his career interest turned to the dynamics of chemical behavior in geospheres. For 40+ years his research has focused on processes that drive the mobility of both anthropogenic chemicals and geochemicals in the natural environment. The first edition of the textbook titled *Environmental Chemodynamics* was published in 1979 by John Wiley; the second edition appeared in 1996. Both received worldwide recognition and formed the basis of college- and university-level environmental courses in diverse academic units such as chemistry, chemical engineering, civil-environmental engineering, environmental science, geosciences, and public health.

Professor Thibodeaux has published approximately 200 articles and has authored or contributed to 33 books. Awards received in recognition of his environmental work include the Louisiana Engineering Society Professionalism Award, the Cecil Environmental Award of the American Institute of Chemical Engineers, and the Pohland Medal from the Association of Environmental Engineering and Science Professors.

Donald Mackay was born and educated in Glasgow, Scotland. He graduated in chemical engineering in 1961 and worked for some years in the petrochemical industry. In 1967 he joined the University of Toronto, and taught and researched there for 28 years in the Department of Chemical Engineering and Applied Chemistry and in the Institute for Environmental Studies. In 1995 he moved to Trent University in Peterborough, Ontario, where he became director of the Canadian Environmental Modelling Centre. He is now professor emeritus at both the University of Toronto and Trent University and is director emeritus of the Canadian Environmental Modelling Centre.

His interests have included the fate and effects of oil spills, especially under cold climate and arctic conditions, environmentally relevant physical chemical properties of organic chemicals and the development and validation of models of chemical fate in the environment. He has introduced the concept of fugacity to environmental modeling using "Mackay-type" models and has a particular interest in human exposure to chemical substances, bioaccumulation processes, that is, uptake of chemicals by a variety of organisms, and chemical transport to and fate in cold climates.

He has authored over 600 publications including authoring or co-authoring 12 books. He has received a number of awards recognizing his work including the Honda Prize for Ecotoxicology, the Order of Ontario, and the Order of Canada.

Contributors

Shannon L. Bartelt-Hunt
Department of Civil
  Engineering
University of Nebraska-Lincoln
Lincoln, Nebraska

Ellen Bentzen
Environmental and Resources
  Studies Program
Trent University
Peterborough, Ontario, Canada

Justin E. Birdwell
Energy Resources Team
U.S. Geological Survey
Denver, Colorado

Bruce Bugbee
Crop Physiology Laboratory
Utah State University
Logan, Utah

Jordi Dachs
Department of Environmental
  Chemistry
Consejo Superior de Investigacones
  Cientificas
Barcelona, Spain

Cliff I. Davidson
Department of Civil and
  Environmental Engineering
Carnegie Mellon University
Pittsburgh, Pennsylvania

Tim Dekker
Water Sciences and Engineering
LimnoTech. Inc.
Ann Arbor, Michigan

Joseph V. DePinto
Water Sciences and Engineering
LimnoTech, Inc.
Ann Arbor, Michigan

Erik M. Dettenmaier
Utah Water Research Laboratory
Utah State University
Logan, Utah

Miriam L. Diamond
Department of Geography
University of Toronto
Toronto, Ontario, Canada

William Doucette
Utah Water Research Laboratory
Utah State University
Logan, Utah

Christophe C. Frippiat
Department of Civil and
  Environmental Engineering
Université catholique
  de Louvain, BE
Louvain-la-Neuve, Belgium

Radek Fučík
Department of Mathematics
Czech Technical University
  in Prague
Prague, Czech Republic

Bojan Gasic
Institute for Chemical and
  Bioengineering
Swiss Federal Institute of
  Technology (ETH Zurich)
Zurich, Switzerland

Christian Götz
Institute for Chemical and
    Bioengineering
Swiss Federal Institute of Technology
    (ETH Zurich)
Zurich, Switzerland

John S. Gulliver
Department of Civil Engineering
University of Minnesota
Minneapolis, Minnesota

Brendan Hickie
Environmental and Resources
    Studies Program
Trent University
Peterborough, Ontario, Canada

Thomas M. Holsen
Department of Civil and
    Environmental Engineering
Clarkson University
Potsdam, New York

Konrad Hungerbühler
Institute for Chemical
    and Bioengineering
Swiss Federal Institute of
    Technology (ETH Zurich)
Zurich, Switzerland

Tissa H. Illangasekare
Department of Civil Engineering
Colorado School of Mines
Golden, Colorado

Rainer Lohmann
Graduate School of Oceanography
University of Rhode Island
Narragansett, Rhode Island

Donald Mackay
Canadian Environmental
    Modelling Centre

Trent University
Peterborough, Ontario, Canada

Matt MacLeod
Institute for Chemical
    and Bioengineering
Swiss Federal Institute of
    Technology (ETH Zurich)
Zurich, Switzerland

Thomas R. Marrero
Department of Chemical
    Engineering
University of Missouri-Columbia
Columbia, Missouri

Gerald Matisoff
Department of Geological Sciences
Case Western Reserve University
Cleveland, Ohio

Richard D. McCulloch
Water Sciences and Engineering
LimnoTech, Inc.
Ann Arbor, Michigan

Cheryl L. McKenna-Neuman
Department of Geography
Trent University
Peterborough, Ontario, Canada

Thomas E. McKone
Department of Indoor
    Environment
Lawrence Berkeley National
    Laboratory
Berkeley, California

Michael S. McLachlan
Institute of Applied Environmental
    Research
Stockholm University
Stockholm, Sweden

Torsten Meyer
Department of Chemical Engineering
    and Applied Chemistry
and
Department of Physical and
    Environmental Sciences
University of Toronto, Scarborough
Toronto, Ontario, Canada

Maria S. Olsen
Department of Environmental
    Engineering
Drexel University
Philadelphia, Pennsylvania

Todd Redder
Water Sciences and Engineering
LimnoTech, Inc.
Ann Arbor, Michigan

Danny D. Reible
Department of Civil and
    Environmental Engineering
University of Texas
Austin, Texas

Martin Scheringer
Institute for Chemical and
    Bioengineering
Swiss Federal Institute of Technology
    (ETH Zurich)
Zurich, Switzerland

Louis J. Thibodeaux
Department of Chemical
    Engineering
Louisiana State University
Baton Rouge, Louisiana

Fred D. Tillman
U.S. Geological Survey
    Arizona Water Science Center
Tucson, Arizona

Frank Wania
Department of Chemical Engineering
    and Applied Chemistry
and
Department of Physical and
    Environmental Sciences
University of Toronto, Scarborough
Toronto, Ontario, Canada

John R. Wolfe
Water Sciences and Engineering
LimnoTech, Inc.
Ann Arbor, Michigan

1 Introduction
Chemical Mobility in the Environment

Louis J. Thibodeaux and Donald Mackay

CONTENTS

1.1 Aims of this Handbook 1
    1.1.1 Chemical Mobility 1
    1.1.2 The Purpose of this Handbook 2
    1.1.3 Overview and Summary of Handbook Contents 3
1.2 The Earth's Environmental Compartments 5
    1.2.1 Manufacturing and Earth Science 5
1.3 Flux: The Basic Chemical Mobility Concept 5
    1.3.1 Chemical Flux Defined 5
    1.3.2 Diffusive and Advective Transport Processes 6
1.4 Mass Transport and Mass Conservation: A Brief History 7
1.5 Specimen Chemicals 9
    1.5.1 The Central Role of Chemical Properties 9
    1.5.2 Chemical Phase Partitioning Overview 12
1.6 Conclusion 13
Literature Cited 13

1.1 AIMS OF THIS HANDBOOK

1.1.1 CHEMICAL MOBILITY

The subject of this handbook is the transport and hence mobility of chemicals in the environment. This is a broad subject and its theoretical foundations are derived from many scientific and engineering disciplines. The concept is applied in almost all areas of science and engineering. It is an interdisciplinary subject and not highly advanced for applications to chemicals in the environment. Chemical transport processes combine a diversity of biological, chemical, and physical mechanisms that drive the movement of molecules and particles in the multimedia compartments of the Earth's surface. Whereas geoscientists focus primarily on the natural substances, environmental chemists and engineers focus primarily on the anthropogenic substances. Both groups must deal with nature on its own terms, a task that is very unlike

1

dealing with chemicals in human controlled and engineered environments. Transport is the key chemodynamic process driving the migration of anthropogenic and natural chemicals throughout all compartments or phases of the Earth's environment. Transport across phase boundaries results in the chemicals deposited in the adjoining phases and is manifest as enhanced thermodynamic partitioning, biota uptake, and reactions of decay and daughter product formation. The phases include air, water, soil, plants, fish, mammals, and all other organisms.

Although many users of this book are familiar with the subject of chemical molecular diffusion transport, it may be new to others or they may have limited knowledge about the broader subject area of chemical transport and its fundamentals. There are numerous rapid transport mechanisms compared to the slow process of molecular diffusion. These rapid processes efficiently move chemicals within and across the various media compartments of the Earth. We have thus sought to present the subject of chemical mobility in a handbook-type format so as to make the material applied rather than theoretical, while being useful and accessible and relevant to a broad range of users. Our aim is to document and reflect the present "state of the science."

It is now universally accepted that chemicals of commerce and those that may be formed inadvertently by processes such as combustion should be subjected to evaluation for their possible adverse effects on humans, the environment, and its various ecosystems. Earth surface processes are continually active with atmospheric, oceanic, and terrestrial media forces that foster chemical mobilization with long-range chemical transport within continental land masses, across the oceans, and on a global scale between the hemispheres. Monitoring data from remote locations provide evidence of this transport and these assertions are confirmed by the theoretical results of a variety of multimedia chemical fate and transport models. At the local level, the other geographic extreme, chemical sources are more intense and the pathways shorter and the impacts are therefore more severe. Anthropogenic substances have been mobilized and now exist in every nook, cranny, and recess of the physical environment and within many biological species.

Chemical exposure must be assessed both from a long-range perspective as well as at the local and at all geographic scales in between. Models applied at these various scales are very different and require different numerical values of the transport coefficients. Some users need only ballpark average values for their estimation models whereas other users need very site-specific transport coefficients and the corresponding estimating algorithms that can capture the significant forcing functions, which reflect the variations in environmental conditions in both time and position. A key objective of this handbook is to provide sufficient information so as to be of service to a diversity of users. To this end, the intent is to provide a broad base of environmental transport information that reflects the current state of knowledge that is supported by laboratory or field data. It is up to the user to access this information, evaluate it, and then translate it for needs of the particular predictive model or application.

1.1.2 THE PURPOSE OF THIS HANDBOOK

The primary purpose of this handbook is to present the current state of the science concerning the estimation of these mass transfer velocities or coefficients. The material covers several chemical transport topics but it is not a textbook on the fundamentals of mass transfer. However, it presents and explains, without derivation, the fundamental equation of chemical transport (the species continuity equation), introduces the concept of chemical flux, gives the most relevant environmental flux equations, presents numerical data on the kinetic parameters in the flux equations, and provides the algorithms appropriate for use in the estimation of these kinetic transport parameters.

In many respects, the format of the handbook is modeled on a companion *Handbook of Property Estimation Methods for Chemicals* edited by Mackay and Boethling (2000) that focuses on partitioning and reactive processes. As editors, we have been fortunate in enlisting the efforts of an impressive number of authors who have contributed to the many chapters. We have sought to assist the reader by ensuring a degree of consistency in the material presented. Each chapter consists of a similar number of subsections containing a specific type of information arranged in an order convenient to the user. We hope this arrangement will be useful to environmental scientists, engineers, educators, and regulators as well as applied geoscientists: in short, all those seeking a fuller quantitative understanding of chemical mobility in the environment.

1.1.3 OVERVIEW AND SUMMARY OF HANDBOOK CONTENTS

The handbook consists of 21 chapters prepared by 38 contributing authors. The chapters fall within two broad undesignated categories. The first category consists of four chapters, specifically Chapters 1 through 4, authored by the editors. Apart from this introductory chapter (Chapter 1), the other three chapters are on mass transfer topics. Chapter 2 is on the subject of mass transport fundamentals from an environmental perspective. It presents, without derivation, the species continuity equation obtained from a Lavoisier species mass balance performed on an arbitrary medium or compartment. It is used to define the concept of chemical fate, and goes on to cover steady-state flux equations and simple models for environmental mass transfer coefficients and their estimation. Chapter 3 presents the fugacity approach to environmental mass transfer. This approach has enjoyed worldwide adoption in the field of multimedia modeling and has evolved as a practical and theoretically sound complement and alternative to the conventional approach to the subject as illustrated by application to a simple problem. Chapter 4 is concerned with the individual mass transport processes and the appropriate flux equations required for a quantitative expression. As editors, we identified 41 individual environmental chemical transport processes that we believe to be most significant environmentally. These are defined and listed in Table 4.1. Numerous individual processes, both diffusive and advective in nature, actively transport chemicals across media interfaces. Procedures for combining these to account for the total flux are presented.

The 41 individual chemical transport processes provide a realistic description of mass transport chemodynamics in natural systems. Based on the limited number of environmental mass transfer experts, the subset of willing contributors, and the subject area divisions, these 41 processes were assigned to 16 chapters. At the outset, the editors established a generic chapter outline for use by the contributing authors, specifying each of the chapter subsections; however, the contributors were free to choose the material to present. The generic chapter outline approach was adopted so the material in the respective chapters had a high degree of presentation uniformity, was positioned in similar locations, and maintained a consistency of style. This consistency of chapter layout and organization should make the handbook user-friendly.

The generic outline used in Chapters 5 through 20 is as follows. The chapter subsections are as follows: (1) the introduction that defined the specific process or processes covered in the chapter, (2) a detailed qualitative description of the processes, (3) a presentation of the theoretical underpinnings of the processes using mathematical formulations *without* any derivation from first principles, (4) a description of field or laboratory measurements used to obtain the process transport parameters, (5) data tables and empirical algorithms for obtaining numerical estimates, (6) a guide for users who are familiar with the process fundamentals and seeking a direct and easy pathway to obtain the numerical results needed, and (7) some computed results using example problems consisting of case studies and/or student-type exercises that showed worked-through solutions giving calculation procedures and numerical answers. Still some chapters deviated from this generic chapter outline. In these few cases, the processes of concern had not reached a stage of scientific development appropriate for presenting the required minimum level of theoretical underpinning and/or empirical data, information, evidence, so as to complete the appropriate subsections.

These 16 chapters focus on the 41 individual transport processes. A brief summary of each chapter follows.

Chapter 5 deals with molecular diffusion. Numerical data and estimation methods for obtaining chemical diffusivities in gases, liquids, solids, and porous media are given. Chapter 6 outlines various deposition processes by which chemicals are transported from the atmosphere to earthen surfaces and water bodies. Chapter 7 is similar but covers the important characteristic of a plant canopy lying between the atmosphere and the soil. Chapter 8 addresses chemical transport processes on both the soil side and the air-side of the air–soil interface. Chapter 9 targets the transport processes on either side of the air–water interface. Chapter 10 discusses particle deposition to and re-suspension from the sediment water surface and the associated chemical loadings. Chapter 11 explains the advective pore water-driven flux and chemical transport in bed sediments. Chapter 12 covers the diffusive-type chemical transport processes on both sides of the sediment–water interface. Chapter 13 describes the so-called bioturbation process that occurs in the surface layers of soils and sediments that move particles and pore water, thus enhancing chemical transport in these layers. Chapter 14 addresses the uptake and transport of chemicals within plants. Chapter 15 is concerned with estimating chemical dispersion in near-surface groundwater flows and the dissolution of nonaqueous liquids within porous media. Chapter 16 covers the suspension of dust and related particles into the atmosphere by surface winds as well as the related chemical transport process. Chapter 17 is concerned with the deposition of dissolved and particulate-bound chemicals from the ocean surface to greater depths. Chapter 18 studies the chemical exchange between snow, ice, and the atmosphere. Chapter 19 addresses chemical dynamics and transport between the atmosphere and urban surfaces. Chapter 20 is on the diffusive-like mixing processes within the lower stratified atmospheric layers as well as the analogous stratified layers in water bodies. In the final chapter (Chapter 21) we, as editors, reflect on the state of this science, on emerging and unfinished topics, and suggest paths forward to enhance our understanding of chemical mobility in the natural environment.

1.2 THE EARTH'S ENVIRONMENTAL COMPARTMENTS

1.2.1 MANUFACTURING AND EARTH SCIENCE

Environmental chemodynamics as a subject exists at the interface between the fields of engineered chemical manufacture, materials and energy processing, and the earth sciences. Particles, molecules, and elements that enter the complex environment of the Earth's systems are primarily at the mercy of the forces of nature. Once released, humans and their machines have limited control on their subsequent chemodynamic behavior. However, prior to substance release, humans do have the technology, which allows us to forecast to some degree on the chemical/particle destiny, fate, and impact on biota. By the use of computerized chemical fate and transport (CF&T) models, the location end-points in space, media concentrations levels, fluxes, lifetimes, and biotic loadings can be estimated within a range of uncertainty, depending upon the model used. These models are at an early stage of development. Enormous challenges await chemical/particle data collection efforts in all the environmental media and the evaluation vis-à-vis model predictions so as to assess the present state their development. The outcome of this comparison will likely yield models capable of making more realistic quantitative predictions, thus reducing the envelope of uncertainty. This model evolution process will follow the advances made in understanding more about the chemodynamic and geoscience processes within Earth's surface systems and the advances made in machine computation power. Until that time, many simpler tools are available and appropriate for use.

A very useful tool that is currently available which allows for visualizing, understanding, and simulating the behavior of chemicals in our complex environment is to view it as comprising a number of connected compartments such as air, water, and soil. Mathematical simulations using computer or "vignette" models are facilitated if it can be assumed that concentrations are homogeneous within the compartment. This leads to "multimedia" or "multibox" models that can give an approximate description of chemical distribution in a complex environment. In some cases of heterogeneity in the compartment, it can be subdivided to achieve greater simulation fidelity or spatially variable concentration expressions can be sought. Examples of subcompartments include the layers in the atmosphere, the epilimnion and hypolimnion in water, biota masses in water, terrestrial plants, and layers of soil or sediments. The material in this handbook is also applicable to transport between and within such defined compartments or media.

1.3 FLUX: THE BASIC CHEMICAL MOBILITY CONCEPT

1.3.1 CHEMICAL FLUX DEFINED

Flux provides the transition concept and theoretical basis for combining the kinetics of chemical movement with their thermodynamic drive toward achieving equilibrium (i.e., equal fugacity) both across and within environmental compartments

(Thibodeaux, 1996; Thibodeaux and Mackay, 2007). Flux facilitates the use of the Lavoisier mass balance that is the basis of all CF&T models. It is quantified as the mass rate through a unit of plane area positioned perpendicular to the movement direction. Its dimensions are mass/time/length squared; the SI unit of flux is $kg/s/m^2$. In mathematical terms, it is a vector quantity having both a direction of movement and a numerical magnitude. As a theoretical concept, it has evolved fully through the science of transport phenomena. It is a tool of chemical transport that has received wide applicability (Bird et al., 2007; Thibodeaux, 1996). Its power is manifest in its generality; it accommodates all types of theoretical and empirical algorithms that are need to mathematically express the numerous chemical transport mechanisms encountered in natural systems. The concept and related aspects is fully developed and discussed in Chapters 2 through 4. A brief introduction follows in the next section.

1.3.2 DIFFUSIVE AND ADVECTIVE TRANSPORT PROCESSES

Mathematical equations for flux must be formulated to describe movement due to chemical, physical, biological, or combined mechanisms. There are two distinct processes by which chemicals migrate within and between environmental compartments. In both cases, the chemical flux equation is the product of two basic parts. One is a kinetic parameter with velocity units and generally referred to as a mass-transfer coefficient (MTC). The other is a chemical concentration or fugacity term.

Diffusive processes result from random mixing. If a concentration gradient (i.e., fugacity gradient) or difference exists within a phase, mixing will tend to preferentially move chemical mass from the higher fugacity state to the lower, thus attempting to eradicate the difference. The net result is transport from regions of high to low concentration. If a difference in chemical potential or fugacity exists between adjoining phases, mixing will again result in net transport as long as a concentration difference between phases still exists. The difference in fugacity only sets up the potential for the possibility of chemical movement. Once it exists, the rate of chemical movement is controlled by the numerical value of the MTC. In diffusion, the degree of phase mixing between the two levels of fugacity determines the value of the MTC. Without an open pathway for molecules, no mass transfer will occur even if a difference in fugacity potential exists. The diffusive flux is thus driven by the product of the concentration differences within a phase or by fugacity differences between phases and the magnitude of the MTC. If either is zero, no flux or chemical movement occurs.

As luck would have it, the natural systems that constitute the Earth's surface are generally replete with mass transport mechanisms that promote efficient chemical movement within and across environmental compartments. Examples are air–water exchange of dissolved gases and sediment bed to water column exchange of hydrophobic chemicals from pore water. A counterexample of efficient flux would be a chemical "trapped" or sealed at high-fugacity potential inside the pores of an ideal, monolithic, granitic, or impermeable natural solid material. Although a potential exists, virtually no pathways are open for transport, the MTC is zero, and therefore the chemical flux from the monolith is zero as well. Such ideas drive the ongoing search for isolation and long-term storage of nuclear wastes.

Therefore, the key information required to quantify chemical mobility are the numerical values of the MTCs, concentrations in the phases and the equilibrium partition coefficient, or correspondingly the respective fugacities. The direction of the diffusive flux can be inferred from the fugacity gradient or difference; diffusion is directed from high to low fugacity.

Advective processes have been defined as directed movement of a chemical by virtue of its presence in a medium (often a particle) that is moving or flowing from location to location. Examples include the deposition of chemical sorbed to aerosols. As the particles settle, the bound chemical fraction is along for the ride. It arrives and contacts the water, soil, snow/ice, or vegetation surface. A similar advective process occurs in the aquatic media as sediment particles are resuspended from the surface of the bed. The biotic ingestion of chemical in food is an advective process. Other examples are transport in flowing air or water. The key information required is the flow rate of the carrier phase and the chemical concentration in it. In this mode of chemical transport, the velocity (i.e., volumetric flux of the medium in $m^3/m^2/s$) of the flowing media is in effect a type of MTC. The velocity and the product of chemical concentration or fugacity in the flowing stream yield the flux. Its direction is dictated by the direction of the flowing medium.

Hydrophobic substances are of particular environmental interest because they tend to bioaccumulate and they often volatilize from aqueous solution because their low solubility can result in high air–water partition coefficients. They partition into organic phases such as foliage, organic carbon, and non-lipid organic present in biota. These phases can thus act as advective "carriers" of hydrophobic substances. A fundamental quantitative understanding of the biogeochemistry of organic carbon is thus essential for determining transport rates of chemicals in aquatic and marine systems and for assessing the fate of chemicals such as pesticides in soils. Correspondingly, there is a need for an understanding of the dynamics of lipids and nonlipid organic matter in uptake and clearance in organisms and food webs.

Often, advective and diffusive process act together, in series, in parallel, or in opposition. Solute and colloid bound chemicals diffuse in the open pores of bed-sediment driven by concentration gradients. Depending on the relative positions of the hydraulic heads in ground water compared to the water body, a positive or negative Darcian velocity exists in the bed, resulting in an in-or-out advective chemical transport superimposed on the diffusive transport. Both transport processes must be appropriately combined for assessing the overall chemical mobility (see Section 4.4).

1.4 MASS TRANSPORT AND MASS CONSERVATION: A BRIEF HISTORY

The ultimate objective of this handbook is to enable the user to obtain numerical estimates of mass transport parameters for inclusion in quantitative models describing chemical fate. All quantitative models that address chemical fate derive their origin from the Lavoisier mass balance. On the subject of calcinations of metals (i.e., oxidation) he wrote: "I realize that the property of increasing weight by calcinations, which is simply a slower combustion, was not particular to metals, as has been thought, but that it was a general law in nature to which a large number of solid and liquid bodies could be subjected" (Poirier, 1998). Repeating experiments of others, in addition to his own, accounting for the mass changes with reaction vessels and mass flows entering and exiting, allowed him to make definitive statements about otherwise subjective chemical reactions. It earned him the title of the "father of modern chemistry." While Greek philosophers had speculated that mass was conserved, Lavoisier proved it with precise measurements. The species continuity equation is the current manifestation of Lavoisier's general law of the conservation of mass. It is the basis of chemical stoichiometry and thus of the modern chemical industry.

The earliest quantitative description of mass transport was by Fick, who in 1855 established the relationship between mass flux, and concentration difference in which the proportionality constant is diffusivity or diffusion coefficient. This steady-state expression, now known as Fick's first law, can be written as

$$N = AD\Delta C/\Delta Y \tag{1.1}$$

where N is the mass rate (g/h), A is the area normal to the direction of flux (m^2), and ΔC is the concentration difference or "driving force" (g/m^3) that applies over the diffusion path ΔY (m). At small path length, the group $\Delta C/\Delta Y$ becomes dC/dY, the concentration gradient. The proportionality constant D (m^2/h) is the diffusion coefficient and depends on the average velocity with which the molecules move along the pathway. Because diffusion is essentially a random mixing process, the net velocity depends on the distance ΔY, and is slower over greater distances. The ratio $D/\Delta Y$ with units of velocity (m/h) is the actual average net velocity of the diffusing molecules and is termed a mass transfer velocity or mass transfer coefficient.

Fick's flux equation is similar in concept to the heat flux equation suggested by Fourier in 1822 but the "driving force" for transport of heat is temperature difference or temperature gradient and the proportionality constant is thermal conductivity. Both are analogous to Newton's law of momentum transport in which the driving force is velocity difference or gradient, and the proportionality constant is the fluid viscosity. Ohm's law is also similar in concept. Quantifying rates of mass transport is critical in designing and simulating industrial chemical processes such as distillation, gas absorption, and solvent extraction. Accordingly, chemical engineers have devoted considerable efforts to measuring and correlating mass transfer coefficients in a variety of geometries and to establishing the mathematical basis of "transport phenomena" in which the similarities between heat, mass, and momentum transfer are exploited (Bird et al., 2007). The first application of mass transfer to environmental systems was done in 1925 to river reaeration (oxygen uptake) by civil engineers who were concerned about oxygen depletion in rivers impacted by oxygen-demanding sewage (Streeter and Phelps, 1925).

Two formidable and highly mathematical treatises have been invaluable to those seeking to simulate mass transfer processes. The text on heat transmission by Carslaw and Jaeger (1959) addresses many of the diffusion situations encountered in mass transfer. The texts by Crank on the *Mathematics of Diffusion* (1975) and Crank et al. (1981) on *Diffusion Processes in Environmental Systems* address a wealth of diffusion problems. Many of the more useful equations have been conveniently compiled by Choy and Reible (2000).

The mathematics of mass diffusion within a single phase has thus been well established. Diffusion between phases such as air and water was not fully understood until 1923 when Whitman proposed the "two-film theory" in which transfer is expressed using two mass transfer coefficients in series, one for each phase (Whitman, 1923). This concept has been rediscovered in pharmacology and probably in other areas and is now more correctly termed the "two-resistance theory."

The concept of the mass transfer coefficient as a diffusivity divided by a diffusion path length is conceptually simple but in environmental systems such as air–water exchange it is now clear that conditions are more dynamic and transient in nature. The pioneering work of Higbie (1935) and Danckwerts (1951) on surface renewal and penetration theory has resulted in a marked improvement in mass transfer correlations and especially on the relationship between diffusivities and mass transfer coefficients.

In 1908 while seeking to measure rate constants for chemical reactions in gases, Max Bodenstein and Karl Wolgast (1908) and subsequently Irving Langmuir (1908) clarified the difference between reactors that operated under unmixed flow and "well mixed" conditions. This resulted in what are now known as the continuous stirred tank reactor (CSTR) and plug flow reactor (PFR). These concepts have been exploited and developed by engineers undertaking chemical reactor designs and simulations. They have been widely adopted by environmental scientists when simulating chemical fate and transport processes. The CSTR concept forms the basis of the compartment concept used in many multimedia models.

In recent decades, there has been a remarkable increase in research on environmental processes, resulting in a number of texts that address environmental mass transfer. Notable ones are listed in Table 1.1.

Two fields in which a considerable and specialized literature on chemical transport has developed are atmospheric dispersion and the hydrodynamics of groundwater. We do not address these topics, except peripherally, because of the abundant existing literature. Our focus is primarily on transport between adjoining environmental compartments rather than on distribution within a single compartment. As editors, we have sought in this handbook to summarize and consolidate the available information on environmental mass transfer and we gratefully acknowledge the contributions of these and other authors.

1.5   SPECIMEN CHEMICALS

1.5.1   The Central Role of Chemical Properties

Example calculations and case studies of MTC applications are presented in each chapter using a wide selection of chemicals. A list of specimen chemicals and their properties appear in Table 1.2. These have been chosen to represent the wide range of properties and phase partition coefficients possessed by environmental chemical contaminants. Many contributing authors have used these in the examples. Chemical properties influence fluxes in a very complex and nonlinear fashion. This is especially significant when there are resistances in series as occurs during air–water and sediment–water exchange. Selected example calculations using a range of chemical

TABLE 1.1
Environmental Mass Transfer Books

1. Boeker, E. and Grondelle, R. van. 1995. *Environmental Physics*, 2nd ed. John Wiley & Sons, New York.
2. Brutsaert, W. and Jirka, G.H., Eds. 1984. *Gas Transfer at Water Surfaces*. D. Reidel Publishing Company, Dordrecht, Holland.
3. Campbell, G.S. 1997. *An Introduction to Environmental Biophysics*. Springer-Verlag, New York.
4. Choy, B. and D.D. Reible. 2000. *Diffusion Models of Environmental Transport*. Lewis Publishers, Boca Raton, FL.
5. Clark, M.M. 1996. *Transport Modeling for Environmental Engineers and Scientists*. John Wiley & Sons, New York.
6. Cohen, Y., Ed. 1986. *Pollutants in a Multimedia Environment*, Plenum Press, New York.
7. Csanady, G.T. 1973. *Turbulent Diffusion in the Environment*. D. Reidel Publishing Company, Dordrecht, Holland.
8. de Vries, D.A. and N.H. Afgan. 1975. *Heat and Mass Transfer in the Biosphere—1. Transfer Processes in Plant Environment*. Scripta Book Company, Washington, DC.
9. DiToro, D.M. 2001. *Sediment Flux Modeling*. John Wiley & Sons, New York.
10. Friedlander, S.K. 1977. *Smoke, Dust and Haze—Fundamentals of Aerosol Behavior*. John Wiley & Sons, New York.
11. Grathwohl, P. 1998. *Diffusion in Natural Porous Media: Contaminant Transport, Sorption/Desorption and Dissolution Kinetics*. Kluwer Academic, London.
12. Gulliver, J.S. 2007. *Introduction to Chemical Transport in the Environment*. Cambridge University Press, Cambridge.
13. Haque, R. and V.H. Freed, Eds. 1975. *Environmental Dynamics of Pesticides*. Plenum Press, New York.
14. Haque, R., Ed. 1980. *Dynamic, Exposure and Hazard Assessment of Toxic Chemicals*. Ann Arbor Science, Ann Arbor, MI.
15. Hemond, H.F. and Fechner, E.J. 1994. *Chemical Fate and Transport in the Environment*. Academic Press, San Diego, CA.
16. Klecka, G., Boething, B., Franklin, J., et.al., Eds. 2000. *Evaluation of Persistence and Long-Range Transport of Organic Chemicals in the Environment*. SETAC Press, Pensacola, FL.
17. Kullenberg, G., Ed. 1982. *Pollutant Transfer and Transport in the SEA*, Vols. I and II. CRC Press, Boca Raton, FL.
18. Lerman, A. 1979. *Geochemical Processes Water and Sediment Environments*. John Wiley & Sons, New York.
19. Lick, W. 2009. *Sediment and Contaminant Transport in Surface Water*, CRC Press, Boca Raton, FL.
20. Lipnick, R.L., R.P. Mason, M.L. Phillips, and C.U. Pittman, Jr. Eds. 2001. *Chemicals in the Environment—Fate, Impacts and Remediation*. ACS Symposium Series 806. American Chemical Society, Washington, DC.
21. Logan, B. 1999. *Environmental Transport Processes*. John Wiley & Sons, New York.
22. Lyman, W.J., Reehl, W.F., and Rosenblatt, D.H. 1977. *Handbook of Chemical Property Estimation Methods: Environmental Behavior of Organic Compounds*. McGraw-Hill, New York.

TABLE 1.1   (continued)
Environmental Mass Transfer Books

23. Mackay, D. 2001. *Multimedia Environmental Models: The Fugacity Approach*, 2nd ed. Lewis Publishers, Boca Raton, FL.
24. Monteith, J.L. and Unsworth, M.H. 1990. *Principles of Environmental Physics*, 2nd ed. Butterworth/Heinemann, Oxford, UK.
25. Neely, W.B. 1980. *Chemicals in the Environment—Distribution, Transport, Fate, Analysis*, Marcel Dekker, New York.
26. Neely, W.B. and G.E. Blau, Eds. 1985. *Environmental Exposure from Chemicals*, Volume I and II. CRC Press, Boca Raton, FL.
27. Okubo, A. and Levin, S.A. 2001. *Diffusion and Ecological Problems: Modern Perspectives*. Springer, New York.
28. Reible, D.D. 1998. *Fundamentals of Environmental Engineering*, Lewis Publishers, Boca Raton, FL.
29. Samiullah, Y. 1990. *Prediction of the Environmental Fate of Chemicals*. Elsevier Science, London.
30. Saxena, J. and F. Fisher, Eds. 1981. *Hazard Assessment of Chemicals—Current Developments*, Part I. Saxena, J., Editor. 1983. Part II. Academic Press, New York.
31. Scharzenbach, R.P., Gschwend, P.M., and Imboden, D.M. 1993. *Environmental Organic Chemistry*. John Wiley & Sons, New York.
32. Schnoor, J.L., Ed. 1992. *Fate of Pesticides & Chemicals in the Environment*. John Wiley & Sons, New York.
33. Schnoor, J.L. 1996. *Environmental Modeling: Fate and Transport of Pollutants in Water, Air, and Soil*. John Wiley & Sons, New York.
34. Seinfeld, J.H. and Pandis, S.N. 1998. *Atmopheric Chemistry and Physics: From Air Pollution to Climate Change*. John Wiley & Sons, New York.
35. Suffet, I.H. Ed. 1977. *Fate of Pollutants in the Air and Water Environments*, Parts I & II. John Wiley & Sons, New York.
36. Swann, R.L. and A. Eschenroeder, Eds. 1983. Fate of chemicals in the environment— Compartmental and multimedia model predictions. ACS Symposium Series 225. American Chemical Society, Washington, DC.
37. Thibodeaux, L.J. 1996. *Environmental Chemodynamics: Movement of Chemicals in Air, Water, and Soil*. 2nd ed. John Wiley & Sons, New York.
38. Tinsley, I.J. 1979. *Chemical Concepts in Pollutant Behavior*, 2nd ed. 2004. John Wiley & Sons, New York.
39. Trapp, S. and M. Matthies. 1998. *Chemodynamics and Environmental Modeling—An Introduction*. Springer, Berlin.
40. Valsaraj, K.T. 2009. *Elements of Environmental Engineering—Thermodynamics and Kinetics*. CRC Press, Boca Raton, FL.

properties can demonstrate this point. Although partitioning does not regulate the magnitude of the individual phase MTCs or mobility coefficients, it does influence chemical flux by determining the relative concentrations in adjoining phases near the interface. The next section is a brief overview of the subject of chemical phase partitioning.

TABLE 1.2

Properties of Specimen Chemicals at 25°C from Mackay et al. (2006)

Chemical	Benzene	Hexachloro-benzene	Lindane	B(a)P	Nickel	Copper
Molecular formula	C_6H_6	C_6Cl_6	$C_6H_6Cl_6$	$C_{20}H_{12}$	Ni	Cu
Molar mass (g/mol)	78.1	284.8	290.8	252.3	58.71	63.54
Vapor pressure (Pa)	12700	0.0023	0.00374	7.0×10^{-7}	0	0
Solubility (g/m^3)	1780	0.0050	7.3	0.0038	–	–
Henry's law constant (Pa m^3/mol)	557	131	0.149	0.0465	0	0
K_{AW}	0.225	0.0529	6.0×10^{-5}	1.88×10^{-5}	0	0
log K_{AW}	−0.65	−1.28	−4.22	−4.73	–	–
log K_{OW}	2.13	5.50	3.70	6.04	–	–
log K_{OA}	2.78	6.78	7.92	10.77	–	–
log K_{OC}	1.67	5.04	3.24	5.58	∼2.0	∼5.0

Note: Lindane is γ-hexachlorocyclohexane, B(a)P is benzo(a)pyrene. The organic carbon–water partition coefficient K_{OC} is only an estimate and can vary considerably. For organics, it is estimated as $0.35 \cdot K_{OW}$ and for metals typical empirical vales are used.

1.5.2 CHEMICAL PHASE PARTITIONING OVERVIEW

It is generally accepted that for organic substances there are three useful and fundamental physicochemical partition coefficients, K_{AW} for air–water, K_{OW} for octanol–water, and K_{OA} for octanol–air. K_{OA} can be estimated as K_{OW}/K_{AW}. Alternatively, log K_{OA} can be estimated as log K_{OW} − log K_{AW}. Other partition coefficients are estimated from K_{OW}, notably K_{OC}, the organic carbon (OC)–water partition coefficient, K_{OM}, the organic matter (OM)–water partition coefficient, and K_{LW}, the lipid–water partition coefficient.

Convenient, simple correlations are

$$K_{OC} = 0.35 K_{OW} \quad \text{(Seth et al., 1999)} \tag{1.2}$$

$$K_{OM} = 0.20 K_{OW} \quad \text{(implying that OC is 56\% of OM)} \tag{1.3}$$

$$K_{LW} = K_{OW} \quad \text{(implying that octanol and lipids are equivalent solvents)} \tag{1.4}$$

whereas K_{AW}, K_{OW}, and K_{OA} are dimensionless ratios of concentrations in units such as g/m^3, K_{OC} and K_{OM} generally have dimensions of L/kg or reciprocal density. This arises from the use of, for example, mg/kg for the concentration in the organic carbon (OC) and organic matter (OM) and mg/L for the concentration in water. The coefficients of 0.35 and 0.20 thus have dimensions of L/kg. If partitioning is primarily applied to the organic carbon or organic matter, the bulk phase partition coefficient (generally designated K_P or K_D, L/kg) can be estimated as (OC) K_{OC} or (OM) K_{OM} where OC and OM are the mass fractions of organic carbon and organic matter in the dry solid phase. Organic matter is often assumed to be 56% organic carbon, thus K_{OM} is $0.56 K_{OC}$.

The air–water partition coefficient K_{AW} can be measured directly or estimated from the solubility S (g/m^3), the molar mass M (g/mol), the vapor pressure P (Pa), the gas constant R of 8.314 Pa m^3/mol and the absolute temperature T (K).

$$K_{AW} = H/RT = PM/SRT \tag{1.5}$$

where H the Henry's law constant is PM/S (Pa m^3/mol). It is noteworthy that the solubility and vapor pressure must refer to the same solid or liquid state of the chemical and the same temperature. For water-miscible chemicals such as acetone, solubility is an inappropriate quantity and K_{AW} must be determined empirically. For gases such as propane, it is important to ensure that the solubility S is determined at the vapor pressure P. Often, S is reported at atmospheric pressure. In this case, P is atmospheric pressure of 101,325 Pa. Care should be taken when interpreting air–water partition coefficients as units can vary and in many atmospheric studies H or K_{WA} are actually water–air partition coefficients, that is the reciprocal of K_{AW}.

The subject of chemical phase partitioning continues to grow and there are three chapters in this handbook that contain specific information on chemical phase partitioning (see Chapter 6, Section 6.2.2 for gas-to-particle partitioning; Chapter 14, Section 14.3 for plant-to-water partitioning; and Chapter 18, Section 18.2 for snow-to-air partitioning).

The properties of the specimen chemicals are listed in Table 1.2, and the values are taken from Mackay et al. (2006). These partition coefficients are temperature-dependent, thus mass transfer rates are also usually temperature-dependent.

1.6 CONCLUSION

In this introductory chapter, we have sought to outline some of the incentives, major concepts, history, and practices adopted when addressing mass transport in the environment. In the following chapters, we describe in more detail the mathematical approaches used to express these transfer processes.

LITERATURE CITED

Bird, R.B., Stewart, W.E., and Lightfoot, E.N. 2007. *Transport Phenomena*, 2nd ed., John Wiley & Sons, New York, NY.

Bodenstein, M. and Wolgast K. 1908. Reaktionsgeschwindigkeit in Strömenden Gasen. *Z. Phys. Chem.* 61:422–436.

Carslaw, H.S. and Jaeger, J.C. 1959. *Conduction of Heat in Solids*, 2nd ed. Oxford University Press, Clarendon.

Choy, B. and Reible, D.D. 2000. *Diffusion Models of Environmental Transport*. CRC Press, Boca Raton, FL.

Crank, J. 1975. *The Mathematics of Diffusion*, 2nd ed. Oxford University Press, Clarendon.

Crank, J., McFarlane, N.R., Newby, J.C., Paterson, G.D., and Pedley, J.B. 1981. *Diffusion Processes in Environmental Systems*. Macmillan, London.

Danckwerts, P.V. 1951. Significance of liquid-film coefficients in gas absorption. *Indust. Eng. Chem.* 43:1460–1467.

Higbie, R. 1935. The rate of absorption of a pure gas into a still liquid during short periods of exposure. *Trans. Am. Inst. Chem. Engr.* 31:365–388.

Langmuir, I. 1908. The velocity of reactions in gases moving through heated vessels and the effect on convection and diffusion. *J. Am. Chem. Soc.* 30:1742–1754.

Mackay, D. and Boethling, R.S. (Eds.) 2000. *Handbook of Property Estimation Methods for Chemicals: Environmental and Health Sciences.* CRC Press, Boca Raton, FL.

Mackay, D., Shiu, W.Y., Ma, K.C., and Lee, S.C. 2006. *Handbook of Physical Chemical Properties and Environmental fate for Organic Chemicals*, 2nd ed., 4 volumes, CRC Press, Boca Raton, FL. Also available in CD-ROM format.

Poirier, J.-P. 1998. *Lavoisier: Chemist, Biologist, Economist.* University of Pennsylvania Press, Philadelphia, PA.

Seth, R., Mackay, D., and Muncke, J. 1999. Estimating the organic carbon partition coefficient and its variability for hydrophobic chemicals. *Environ. Sci. Technol.* 33:2390–2394.

Streeter, H.W. and Phelps, E.B. 1925. A study of the pollution and natural purification of the Ohio River. U.S. Public Health Service Bulletin 146. Washington, DC.

Thibodeaux, L.J. 1996. *Environmental Chemodynamics*, 2nd ed. John Wiley & Sons, New York, NY.

Thibodeaux, L.J. and Mackay, D. 2007. The Importance of Chemical Mass-transport Coefficients in Environmental and Geo-Chemical Models. Society Environmental Toxicology and Chemistry, SETAC GLOBE. July–August, pp. 29–31.

Whitman, W.G. 1923. The two-film theory of gas absorption. *Chem. Metal. Eng.* 29:146–150.

2 Mass Transport Fundamentals from the Environmental Perspective

Louis J. Thibodeaux

CONTENTS

2.1 Introduction 16
2.2 The Chemical Species Continuity Equation 17
    2.2.1 The Lavosier Mass Balance 17
    2.2.2 Chemical Fate: Defined for an Arbitrary Compartment 17
2.3 Chemical Flux 19
2.4 Steady-State Flux Equations 20
    2.4.1 Fick's First Law of Molecular Diffusion 20
    2.4.2 Diffusion across Interface Films and Finite Fluid Layers 20
    2.4.3 Advective Transport of Solutes in Fluid Media 21
    2.4.4 Advective Transport of Particles within a Fluid 21
    2.4.5 Time-Smoothed Fluctuating Advective Turbulent Flux 21
    2.4.6 Fick's Law Analogy for Turbulent Diffusion 21
    2.4.7 Turbulent Diffusion across Finite Layers 22
    2.4.8 Convective Mass Transfer 22
2.5 Convective Mass Transfer Coefficients and Their Models 23
    2.5.1 Film Theory 23
    2.5.2 Penetration Theory 24
    2.5.3 Surface Renewal Theory 25
    2.5.4 Hanratty's Approach 26
    2.5.5 Boundary Layer Theory 26
    2.5.6 Analogy Theories 27
    2.5.7 Natural Convection Models 32
2.6 General Data Correlations for MTCs 34
    2.6.1 Forced Convection and Natural (FREE) Convection 34
    2.6.2 Mixed Forced and Natural Convection Mass Transfer 35
2.7 Unsteady-State Vignette Chemodynamic Models 36
    2.7.1 Diffusion in a Semi-Infinite Solid Media 37
    2.7.2 Diffusion and Advection in a Semi-Infinite Solid Media 38

2.8   A User Guide to Chapter 2 38
2.9   Example Problems 39
        2.9.1   A Water-Side MTC Estimate for Sea-Surface Waves Interacting
                 at Rocky Shoreline 39
        2.9.2   Estimate the Air-Side MTC for a Large and Flat Plowed Field 40
        2.9.3   Stream Bed Water-Side MTC 40
        2.9.4   Urban Surface Air-Side MTC 41
Literature Cited 42
Further Reading 42

2.1   INTRODUCTION

The subject of this book is environmental mass transport coefficients. Generally speaking, it is a highly advanced and broad subject with theoretical developments arising from many of the science and engineering disciplines. Anthropogenic chemicals and particles are continually being introduced into Earth's natural environment. Mass transport is the chemical mobility process that results in their appearance throughout all the phases, which includes air, water, soil, plants, and other biological forms. Although many users of this book are familiar with the subject, others are new to it or have limited knowledge and acquaintance with it. The objective of this chapter is to provide a brief but formal overview of the general subject of chemical transport from an environmental perspective.

The ultimate objective of the user of this book is to obtain numerical estimates of mass-transport parameters for inclusion in a quantitative model for projecting chemical fate. The chapter begins with a precise definition of chemical fate in the environmental context. All quantitative models that address chemical fate derive their origin from the Lavoisier mass balance.

The species continuity equation (CE) is the current manifestation of Lavoisier's general law of the conservation of mass. It is presented in the first section in the context of chemical fate in the environment. This is followed by introducing the chemical flux concept. It captures the idea of chemical mobility and provides the mathematical means needed for its inclusion in the mass balance. In multimedia natural systems, chemical mobility is the result of numerous and complex physical, chemical, and biological mechanisms. Precisely defined and science-based flux expressions are needed to quantify the individual transport mechanisms. Several basic and useful flux equations are presented in the next section; these represent the current state for applying the most significant environmental chemical transport processes. The material presented defines and summarizes the basic types of transport parameters appearing throughout the various chapters in the book.

The fundamentals portion of this chapter concludes with a presentation of generic models and correlations useful for estimating convective MTCs at the interface regions of simple geometric shapes and fluid flow situations. Typically these formulations will yield first-order numerical estimates in cases where directly applicable, process-specific studies and data correlations are lacking. Oftentimes, the user will need to be creative in applying one or more of these generic correlations to the specific circumstances at environmental interfaces involved.

The remainder of the chapter contains several, practical, mathematical, vignette chemodynamic models. These are simple but useful chemical fate models for obtaining numerical results when the dominant or controlling transport process is known. Those selected for presentation have stood the test of time, having proved useful to the authors on numerous occasions.

2.2 THE CHEMICAL SPECIES CONTINUITY EQUATION

2.2.1 THE LAVOSIER MASS BALANCE

The conservation of mass, performed by applying the Lavosier species mass balance to chemicals in the natural media, is the basic concept underlying environmental chemodynamics. The species CE is the result and is a good context in which to present the various types of chemical mass transport processes needed for environmental chemical modeling and chemical fate analysis. For constant physical properties; first-order reaction, and dilute solutions in any media the CE in vector notation is

$$\frac{\partial C}{\partial t} + \bar{v} \cdot \overline{\nabla} C = D^{(\ell)} \nabla^2 C - \overline{\nabla} \cdot j^{(t)} - kC, \tag{2.1}$$

where C in mg/m^3 is concentration, t in seconds is time, $\bar{v}$ in m/s is the velocity vector, $D^{(\ell)}$ in m^2/s is molecular diffusivity, $j^{(t)}$ in mg/m^2 s is a general flux vector representing turbulent flow and other mass-transfer processes, and k in s^{-1} is the first-order reaction rate constant. This chemical species CE is the starting point for using the Lavoisier mass balance to environmental compartments. Similar balances apply to particles as well; these include aerosols, aquasols, nanoparticles, and so on.

2.2.2 CHEMICAL FATE: DEFINED FOR AN ARBITRARY COMPARTMENT

The phrase environmental chemical fate and transport is in common use. Although not precisely defined, it generally refers to the eventual demise and/or final location of anthropogenic chemicals in nature following their introduction and movement within and across the various media surface and compartments. Fate in this use is often construed to mean reactive degradation, "ultimate" and complete elimination from the nature or some similar final ending. Nevertheless, the usage and implication is that chemical fate and transport are two different concepts that can be disconnected. Its use in this way can be misinterpreted or misrepresented; however, use of the CE produces a precise definition of fate and clarifies the situation.

At this juncture, a conceptual application of Equation 2.1 for an arbitrary environmental compartment space will provide a convenient definition and description of the commonly used term "chemical fate." In addition, it will clearly distinguish the role of transport in relation to chemical fate. Once the geometric dimensions and content of the environmental compartment is precisely defined, the CE can be applied to its chemical mass. As written, Equation 2.1 is applicable to a vanishingly small differential space of volume $dV = dx \cdot dy \cdot dz$ where x, y, and z are the symbols for lengths in three-dimensional space. A focus on $\partial C / \partial t$, the accumulation term, provides a means of defining chemical fate. Simply interpreted, it is the chemical concentration change per unit time within dV. In other words, the accumulation term is the mathematical equivalent to the proverbial "canary in the mine" with regard to characterizing chemical fate in the environmental compartment. Clearly, an inspection of Equation 2.1 demonstrates the left-hand side, which is the fate or response term, must equal the right-hand side, which is the causative group of terms. Furthermore, it is apparent that the transport terms are not the only causative terms.

The four causative terms in Equation 2.1 account for the advection, diffusion, and turbulent transport processes, respectively, these being the mobility terms, and reactive decay, the final term in the equation. Used in a quantitative modeling context, each of the terms will have a numerical value with either a positive $(+)$ or a negative $(-)$ sign or a zero value. The sum total will also be a number with an appropriate sign, or zero, and equal to $\partial C / \partial t$. It is a numerical concentration rate term that can be positive, negative, or zero, indicating respectively increasing, decreasing, or a steady-state chemical build-up behavior in the compartment with increasing time. Taken together, these four categories of processes regulate the environmental fate of a chemical in a differential segment of media such as air, water, and so on. Furthermore, the relative magnitudes of the individual terms signify for the user the important processes that control or dominate chemical fate in the compartment. Typically in many practical situations, only one or two processes dominate chemical fate at a particular locale. The primary focus of this handbook is concerned with the three transport processes. Chemical reaction within the media phases is covered elsewhere; see, for example, Boethling and Mackay (1997), Stumm and Morgan (1996), Seinfeld and Pandis (1998), and Soil Science Society of America (1989).

This CE is always the starting point used to examine the environmental fate of chemicals and has numerous environmental chemodynamic modeling applications. Such balances applies to each media phase, however most environmental modeling application do not require the simultaneous solution of three or more partial differential equations such as Equation 2.1. Based on the expertise of the user, realistic simplifying assumptions plus eliminating the insignificant processes reduce the set of equations substantially. Limitations on computing power and limited knowledge about some processes and the availability of only sparse environmental data may require modeling compromises that further reduce the allowable number of state variables and model parameters, including the mass transport ones.

Compartmental or media box models offer an alternative practical approach. They are derived from applying integrated forms of the CE. These involve volume and area integrals over the boxes. The volume integrals sum the mass accumulation and reaction terms while the area integrals direct the flux terms to account for the movement of chemicals between the boxes. Typically, the result is a set of linear ordinary differential equations capable of mathematically mimicking many of the key dynamic and other features of the chemodynamics in natural systems. This handbook provides the mass transport parameters needed for both model types.

2.3   CHEMICAL FLUX

Chemical mass balances use flux, the definitive term for chemical mobility, in their formulation (Bird et al., 2002). It simplifies, clarifies, and unifies the derivation procedure needed in this complex subject. Therefore, fluxes are imbedded in the CE. Decomposing Equation 2.1, as is done in the following paragraph, reveals the basic flux terms and provides an entrée to the ones used in chemodynamic modeling and the mass-transport processes covered in this book. By assuming steady-state, constant properties, no reaction, dilute solution (i.e., <5%) and focusing only on the z-dimension Equation 2.1 can be written as:

$$\frac{d}{dz}(V_z C) = \frac{d}{dz}\left(D^\ell \frac{dC}{dz}\right) - \frac{d}{dz}(j^{(t)}) \tag{2.2}$$

Integrating this differential equation over z using the indefinite integral formulation introduces an arbitrary constant. We define it as the total flux n_z with unit mg/m^2 s. Performing the integration yields

$$n_z = -D^{(\ell)}\frac{dC}{dz} + V_z C + j^{(t)} \tag{2.3}$$

This is the one-dimension (i.e., 1-D) expression for the total chemical flux and the definitive expression of chemical mobility. It has units of mass rate through an x–y plane surface area perpendicular to the z-direction. Used as is this 1-D form the flux expresses the total mass rate of chemical moving through the atmosphere–soil interface plane, for example.

Each of the three fluxes in Equation 2.3 is a vector quantity, having both magnitude and direction; together they captured the three general categories of flux encountered in mass transport. The first category refers to all the molecular and Brownian diffusion processes that quantify the movement of solute molecules and particles (i.e., aquasols, aerosols, nanoparticles and its clusters, etc.). The superscript (ℓ) denotes the fluid medium must be in laminar flow or stagnant. As written, the second category refers to all those processes where chemical transport occurs by the bulk movement of the media phase. These include the wind and water currents, media pore-water velocity, for example. The velocity, V_z in m/s, is alternately viewed as the "volumetric flux" of the media phase, in mg/m^2 s, and imparts the vector nature to the group. By extension, this group also represents other flux processes such as particle deposition from air or water, dust, or sediment resuspension, and so on. The last flux term in Equation 2.3 category is a general representation used for quantifying various fluid turbulence-related chemical transport processes and is denoted by the superscript (t). These $j^{(t)}$ fluxes, in mg/m^2 s, include the vertical and horizontal turbulent or eddy diffusion (i.e., mixing processes) of chemicals in surface water, groundwater, and the atmosphere as well as a lesser known biological-driven turbulence processes in surface soils and sediments such as bioturbation.

The above theory is presented in a narrative form, primarily. However, it contains enough mathematical structure so as to give the user a perspective of the quantitative nature of the fundamental relationships involved. The above presentation develops the flux concept. The resulting flux equations provide the basis for further defining the individual transport parameters and related MTCs, which are the primary subjects of this handbook. Those users needing additional information on the subject of transport theory in the natural environment should consult Thibodeaux (1996); those interested in the general subject of transport fundamentals should consult Bird et al. (2002) or similar works of the subject of mass transport phenomena.

2.4   STEADY-STATE FLUX EQUATIONS

The section expands upon the flux concept, which resulted in Equation 2.3 above. Its application within and across media interfaces focuses the user to consider all the individual mechanisms that are operative and may result in chemical transport. Several steady-state flux equation types have emerged over the years as being appropriate and practical for quantifying known transport mechanisms. These are used throughout the book. Specific chapters are devoted to estimating these transport parameters. The most significant ones are presented in the following sections.

2.4.1   Fick's First Law of Molecular Diffusion

This classical flux equation is

$$n = -D^{(\ell)} \frac{dC}{dz} \tag{2.4}$$

where $D^{(\ell)}$ is the molecular diffusivity of the chemical species in the fluid or solid media in SI units m^2/s. It has environmental applications for diffusive transport within stagnant fluid layers or fluids in laminar flow, which include water, air, and in porous sediment, soil, or geologic media, and so on. The equation is applicable for solutes as well as particles undergoing Brownian motion. See Chapter 5 for the estimation of molecular and Brownian diffusivities.

2.4.2   Diffusion across Interface Films and Finite Fluid Layers

The integrated form of Equation 2.4 across a length Δz is

$$n = \frac{D^{(\ell)}}{\Delta z} \Delta C \tag{2.5}$$

where ΔC is the concentration difference across the film or layer. This well-known flux equation is the basis for the "film" theory of mass transfer. When applied to transport across films of known thickness Δz, either stagnant or in laminar flow, numerical values of the flux can be estimated. However, due to theoretical problems concerning the existence of these films in many situations, the approach has been marginalized for use in estimating flux except for very special cases. For example, these include applications to porous soil and sediment surface layers where molecular or Brownian diffusion is almost certain to exist in the fluid-filled spaces.

2.4.3 ADVECTIVE TRANSPORT OF SOLUTES IN FLUID MEDIA

The flux equation is

$$n = V_z C, \tag{2.6}$$

where V_z is the fluid velocity (i.e., of the wind or water stream) in units of m/s or the volumetric flux in units m^3/m^2 s. It also applies to porous media where the Darcian velocity is used for either soil gas or water. Being a vector V_z can be $(+)$ or $(-)$ and gives the flux its direction. The subject of this advective transport process is covered in detail in Chapter 11.

2.4.4 ADVECTIVE TRANSPORT OF PARTICLES WITHIN A FLUID

The flux equation is

$$n = V_p C_p \rho_p, \tag{2.7}$$

where the "p" subscript denotes particle. The term V_p is the solid of liquid particle flux within the media, m^3 particle/m^2 area/s, and C_p is the chemical concentration or mass loading on the particles, mg/kg, and ρ_p is the particle bulk density, kg/m^3. Thus, advective flux formulation has been used for dry desorption of chemical from the atmosphere to soil, water and plant surfaces, atmospheric wet deposition as well, chemical-laden particles in water depositing on the sediment bed surface, sediment resuspension, deep ocean particle settling, snow, and ice forms are the main examples. In the case of solid particles accumulating on a bed surface it is termed the sedimentation velocity. The equation finds many applications (see Chapters 6, 7, 10, 16–19).

2.4.5 TIME-SMOOTHED FLUCTUATING ADVECTIVE TURBULENT FLUX

Time-smoothing the CE upon applying both time-varying velocities and concentration yields the equation

$$n = \overline{V_z' C'} \tag{2.8}$$

where the primer denotes the fluctuation velocity and concentration components and the over-bar denotes the average values of the product. In reality it finds limited use in chemodynamic modeling as written but has found some application in the field measurement of chemical fluxes. These have been applied to vertical fluxes in the atmospheric boundary layer, primarily. Commencing with Equation 2.8 and using the Prandtl mixing-length theory yields a more practical flux equation. However, these in turn require the time-smoothed velocities and concentration profiles obtained from measurements (Bird et al., 2002; Thibodeaux, 1996).

2.4.6 FICK'S LAW ANALOGY FOR TURBULENT DIFFUSION

An alternative to Equation 2.8 for chemical transport in the presence of turbulent fluid mixing is the following empirical expression. By analogy with Fick's first law of diffusion the following turbulent flux is defined

$$n = -D^{(t)} \frac{d\bar{C}}{dz} \tag{2.9}$$

This is the defining equation for the turbulent diffusivity $D^{(t)}$, in m^2/s, called the eddy diffusivity. Within a single fluid, it is highly variable and depends on position, direction, and the nature of the flow field. Nevertheless, by selecting appropriate, average numerical values, it enjoys some use in chemodynamic model dispersion applications such as horizontal dispersion of chemicals in surface waters, vertical dispersion in the water column, dispersion in the atmosphere and in groundwater and biodiffusions in soils and sediment (see Chapters 13, 15, and 20).

2.4.7 TURBULENT DIFFUSION ACROSS FINITE LAYERS

The integrated form of Equation 2.9 for a layer of thickness Δz, in meters, is

$$n = \frac{D^{(t)}}{\Delta z} \Delta \bar{C}. \tag{2.10}$$

This flux equation finds application across finite fluid layers and porous media layers. It has found use for chemical transport in stratified fluid bodies where the compartments are characterized by average eddy diffusion coefficients, and in biodiffusion of particles and solutes in surface soils and sediment. The latter are termed bioturbation and pedbioturbation respectively (see Chapters 13, 15, and 20).

2.4.8 CONVECTIVE MASS TRANSFER

The fluid dynamic structure within the boundary layers adjacent to natural solid surfaces such as soil, sediment, snow, and ice, is complex. Typically, the flows fields have both laminar and turbulent regions. The flows magnitudes and directions respond according to the angle of incidence to the surface and the overall shape of the object as well as thermal-induced fluid density differences (i.e., stratification) and so on. All these factors operate into shaping the mass transfer boundary layer which controls the chemical flux. Ironically, the traditional approach to handling such complex flow situations has been to use a simple flux equation. The so-called convective mass flux equation is

$$n = k_c \Delta C, \tag{2.11}$$

where k_c is called the convective MTC with velocity units of m/s. The concentration difference reflects the concentration in the fluid at the interface and the bulk average fluid concentration (i.e., cup mixing) at a distance removed from the interface.

Borrowed from heat-transport science the convective flux equation enjoys the simple, linear, and theoretically correct concentration functionality. However, the MTC has one adjustable parameter that must reflect all the complex interacting factors mentioned above, which affect the magnitude of mass transfer in the boundary layer. So created k_c is essentially a fudge factor! Not satisfied with this lowly theoretical status scientists and engineers are constantly active elevating its rank by providing it with various theoretical model underpinnings. This has been a good effort because some very significant and useful models have been developed and these appear in the next section. The conceptual models yield simple algebraic equations for estimating k_c from easily obtained independent variables and provide the suggested parameters needed in correlating k_c data.

2.5   CONVECTIVE MASS TRANSFER COEFFICIENTS AND THEIR MODELS

Experimental measurements have advanced our understanding of fluid turbulence including the flow interactions at fluid–solid and fluid–fluid interfaces. Computational fluid dynamic (CFD) software models are able to numerically solve the fundamental equations of fluid flow, heat transfer, and mass transfer verifying and extending these measurements. Without the availability of computational horsepower, the earlier researchers resorted to creating conceptual models approximation the perceived physical structure of turbulence and laminar element in the flow fields adjacent to these interfaces. The conceptualizations were expressed in appropriate mathematical terms, based on interpretations of the available measured velocity profiles and velocity fluctuation, mechanistic flux relationships, known boundary conditions, and intuition. CFD results, to a large degree, have confirmed the key results of these earlier conceptual models. Arguably, CFD numerical results consisting of detailed profiles and copious numerical minutia supersede those of the conceptual models developed for flux estimates at interfaces.

Nevertheless conceptual models remain useful for the user acquiring a basic mechanistic understanding of chemical transport at solid–fluid interfaces. Typically, the final results were simple formulations for estimating the MTCs. As the state of the science in fluid dynamic and mass transport knowledge progressed over the last century, these earlier and crude conceptual models were replaced by more sophisticated ones. The following section contains an overview of all. Taken together, they provide unique qualitative and quantitative descriptive insights into the turbulent/laminar flow structures that regulate chemical transport at these interfaces.

2.5.1   FILM THEORY

The stagnant film theory was developed by Nernst in 1904. Because no flow can exist at the solid interface, a stagnant or laminar film of fluid (i.e., hydraulic film) of thickness, S, was assumed to exist, and beyond this film the fluid was assumed to be turbulent.

For mass transfer the effective film (i.e., concentration film) thickness is δ_c, in meters, and the chemical flux is regulated by molecular diffusion. Being highly turbulent, the fluid beyond is rapidly mixed and providing no resistance to mass transfer.

Based on these assumptions the final conceptual model equation is

$$k_c = \frac{D}{\delta_c},$$
(2.12)

where D is the diffusivity, m^2/s, and k_c is the MTC in m/s. The theory suggests that the fluid-side MTC is proportional to D^1, which is a testable hypothesis. Oftentimes, experimental results reflect an exponent dependence less than unity. The D can be a gas-phase or a liquid-phase diffusivity, thus yielding the corresponding gas-side or liquid-side MTC. In addition, both δ_c values are highly variable parameters and one numerical constant does not apply to all situations. Later theoretical developments including boundary layer theory indicate that the momentum transport δ and δ_c typically have different magnitudes. This suggests that the mass-transfer thickness height extends into the turbulent portion of the flow field, which is impossible; therefore, an unfortunate conceptual glitch exists in the film model. Applying the film theory to fluid–fluid interfaces such as the gas–liquid is problematic, because it is hard to argue that a zero velocity exists at the interface or elsewhere in the film. Therefore, molecular diffusion across a stagnant film as a conceptual model for explaining how k_c works is problematic. However, it remains a good pedagogical tool. The net result is the film theory likely applies to some (few) real-world situations. It has great pedagogical use highlighting the fact that mass transfer is controlled by molecular diffusion processes in thin fluid layer near the interface. It cannot be used for making calculations estimating k_c because S_c is usually unknown.

2.5.2 PENETRATION THEORY

The penetration theory for mass transfer is attributed to Higbie (1935). A permanent, stagnant, or laminar film is not required. The notion that fluids in turbulent flow contain "eddies" gained favor based on experimental advances in observing the complex fluid dynamic flow patterns near solid surfaces. These eddies were visualized as finite volume elements of fluid possessing rotational velocity component within, in addition the whole ensemble of eddies has a directed average velocity in the flow field. Their birth place is in the fluid layers near the wall, and their size increases with distance from the wall. They disintegrate in size with time, resulting in smaller ones that in turn disintegrate, and so on. Therefore, by a constant birthing process a steady-state size distribution of succeeding small-to-large eddies exist in the flowing fluid. Being mobile and randomly directed all sizes eventually crash into or otherwise encounter the solid interface. This constitutes the qualitative description of the conceptual model in regard to the fluid dynamics.

Assuming the solid surface is made of a substance with solubility, C_i, in mg/m^3, is a convenient notion to understand the mass transfer. The bulk fluid far removed has a lesser concentration, C_∞, same as does each fluid eddies that approaches the interface. All eddies that encounter the interface reside in direct contact for a time period, $\bar{\tau}$, in seconds. While in contact for this short time-period, diffusive mass transport occurs from the solid surface into the adjoining fluid driven by the $[C_i - C_\infty]$ concentration differences. The entering chemical mass is viewed to be penetrating into the eddy, hence the origin of the term "penetration theory" for mass transfer. Depending on eddy size and its kinetic energy level, the contact time-periods are highly variable. Most have very short contact time-periods; only a few can reside for very long time-periods. Nevertheless, while in contact, mass is transferred by and unsteady-state process conventionally termed Fick's second law of diffusion. Higbie assumed that one average contact time-period $\bar{\tau}$ was sufficient to characterize the eddy behavior. From the semi-infinite slab boundary condition analytical solution of Ficks' second law the following equation for the MTC results:

$$k_c = 2\sqrt{D/\pi\bar{\tau}}, \tag{2.13}$$

where $\bar{\tau}$ is the average eddy contact time in seconds.

The above theoretical equation is the key result of the penetration theory. Note that the MTC is proportional to $D^{1/2}$ power. Such dependences have been observed in rapidly mixed fluid mass-transfer experiments. The assumption of a single average eddy-to-solid surface contact time is problematic because a distribution of contact times is more likely. Similar to the one unknown parameter in the film model, the penetration theory MTC model contains $\bar{\tau}$. No single constant value of $\bar{\tau}$ is sufficient for all solid–fluid or fluid–fluid contact situation. Unlike the film theory, the penetration theory can be visualized as the mass transport mechanisms at fluid–fluid interfaces. Here eddies can position themselves to exchange chemical mass by diffusion from both sides of the interface. Also eddies in liquids, such as water, as well as eddies in gases, such as air, can be visualized on opposite sides of the interface plane exchanging chemical mass in either direction. This implies the use of the appropriate gas or liquid diffusivity for D in Equation 2.13 yield the corresponding gas-side or liquid-side MTC. The hard part is selecting the appropriate $\bar{\tau}$ in real-world applications (see Example 2.9.1).

2.5.3   SURFACE RENEWAL THEORY

Whereas Higbie assumed a constant renewal time, Dankwert's (1951) extension of the penetration theory employs a wide spectrum of eddy contact times. These eddy-like fluid packets are assumed to remain in contact with the interface for variable times from zero to infinity. He assumed a surface-age distribution function, which skews the contact times to small values; it is

$$\phi = se^{-st}, \tag{2.14a}$$

where ϕ represents the contact time probability distribution. The appropriate Fick's second law diffusion model is applied, which yields the following relationship for the MTC:

$$k_c = \sqrt{Ds}, \tag{2.14b}$$

where s, in s^{-1}, is the fractional renewal rate. This value is a constant, no matter what the time of contact, t, all are renewed at the same rate, s. The D can be a gas-phase or liquid-phase diffusivity, thus yielding the corresponding gas-side or liquid-side MTC.

Just as with the penetration theory $k_c \sim D^{1/2}$. The s is that one hard-to-find parameter; no one numerical value fits all mass transfer situations. Since values of s are not generally available, its appearance here presents the same problem as does S_c for the film model and τ for the Higbie penetration model. If s is set equal $1/\tau$, the fractional exposure time, Equations 2.13 and 2.14b give nearly equal numerical estimates of the MTC. The penetration theory value is only 1.13 times larger.

2.5.4 Hanratty's Approach

Progress in fluid transport turbulence measurements has led Hanratty and co-workers (1970, 1986) to develop a conceptual model that focuses on the time (t) and position (z) functionality of the vertical velocity profile, $w(z,t)$, in the bulk fluid beyond the layers at the fluid–solid and fluid–fluid interface. It is assumed that $w(z,t)$ could be expressed as the product of two separate functions, one of time (t) only and the other of distance (z) only. Each function contained turbulent eddy attributes representative of measurements in the flow field above the surface. The time-varying function, $\beta(t)$, known as Hannatty's β, in s^{-1}, represents the driving turbulence outside the concentration boundary layer and $g(z)$, the positions varying function represents the stretching, shrinking, and overall dissipation of the turbulent eddies as they approach the interface (Gulliver, 2007).

The net result is a physical representation of turbulence in conceptual terms that is more realistic than those of the film, penetration, or surface renewal models. As with these models, this one also depends upon measurements; in this case, it is $\beta(t)$. The end result of the analysis yields a correlation for the MTC at a solid interface, which is a function of $\beta(t)$:

$$k_c = U * F[\beta(t)]/Sc^{7/10}, \tag{2.15}$$

where $U*$ is the friction velocity (m/s), and Sc is the Schmidt number. This leads to $k_c \sim D^{7/10}$, which is yet another function of the diffusivity. For the free interface, or the fluid–fluid interface, with its corresponding $g(t)$, the result is

$$k_c = \sqrt{(\beta(t)^2)^{1/2} \, D/2}. \tag{2.16}$$

As can be appreciated by comparison the results for a solid surface are quite different from that for a fluid–fluid interface. The latter, in its $D^{1/2}$ functionality, is consistent with both the penetration and surface renewal theories. It appears that the approach of Hanratty et al. is based on a more realistic qualitative and quantitative characterization of the turbulence in a flowing fluid stream than the simpler ones.

2.5.5 Boundary Layer Theory

A flat plate of negligible thickness submerged in a tangential flowing fluid stream with approach velocity, V_∞, creates a slower moving boundary layer because of the "no slip" velocity condition at the wall. Transport to and from this idealized submerged object has been studied extensively yielding significant theoretical insight into the process. Based on the CE of species A in the fluid mixture concentration gradients and composition profiles can be determined numerically (Bird et al., 2002). These, in turn, lead to algorithms for estimating k_c. The direct application of the result is problematic for many environmental geophysical flow situations. However, they do provide useful first-order approximations for some. The main problem is the length of the flat plate. The velocity profile develops with distance down the plate and the thickness of the boundary layer grows as well. Few realistic natural environment fluids moving over and around large solid objects meet these conditions. The average MTC for a flat object of length L, meters, in laminar flow is

$$k_c = 0.664\frac{D}{L}\text{Re}_L^{1/2}\text{Sc}^{1/3}, \qquad (2.17)$$

where flow is directed tangential to the surface with velocity V_∞ in m/s. In addition the $\text{Re}_L(=V_\infty L/\nu)$, the Reynold's number must be less than $1 \times \text{E}5$. Such laminar situations are rare for objects of finite size submerged in flowing air or water streams. Extension of the BL analysis turbulent BL flows using empirical velocity and concentration profiles yields

$$k_c = .0365\frac{D}{L}\text{Re}_L^{0.8}\text{Sc}^{1/3}, \qquad (2.18)$$

where $\text{Re}_L > 1\text{E}5$. These are useful and have been verified for mass transport from area sources of length L situated on flat surfaces experiencing tangential flows (i.e., fluid flow parallel to the surface plane). The last results suggest $k_c \sim D^{2/3}$.

2.5.6 ANALOGY THEORIES

The analogy theories derive their name from the similarity of the transport process mechanisms and the mathematical descriptions of the phenomena. For the purposes of this section, the mass and momentum flux relationships are needed. Each is quantified by a simple equation using gradients of concentration and velocity and the respective "coefficients of diffusivity" as constants.

The analogy theory boundary layers adjacent to a flat solid surface are assumed to be fully developed of constant thickness for a fluid flowing parallel to the surface. The k_c in both Equations 2.17 and 2.18 decrease with increasing L; this is a direct result of an increasing boundary layer thickness for both laminar and turbulent flow. Therefore the analogy theories provide a realistic scenario for mass transfer at large, flat earthen surfaces with a well-developed, steady-state velocity, and concentration profile. Such real-world MT situations include pesticide evaporation from agricultural fields and soluble chemicals absorbing onto bed sediments of a flowing aquatic stream, and the dry deposition of gases and vapors from the atmosphere to water or soil surfaces.

The theoretical models proposed by Prandtl and von Karman view the flowing fluid as composed of successive layers. Beginning at the solid surface these include: a laminar layer, a "buffer layer" and a turbulent flowing fluid zone beyond. Steady-state momentum and species mass fluxes are coupled mathematically and used with empirical expressions for the layer thicknesses. In combination the development yields a MTC, k_c (m/s), which is a function of the skin friction, C_f, of the flat surface:

$$k_c = \frac{v_\infty C_f/2}{1 + 5\sqrt{C_f/2}\left\{Sc - 1 + \ln\left[1 + \frac{5}{6}(Sc - 1)\right]\right\}}. \qquad (2.19)$$

Here, v_∞ (m/s) is the fluid velocity well away from the surface, C_f is its Fanning friction factor (dimensionless), and Sc is the chemical Schmidt number, $Sc = v/D_M$. For gases with $Sc = 1$ the denominator in Equation 2.19 is unity. For liquids with large Schmidt numbers $k_c \sim Sc^{-1}$.

Using an experimental data base, Chilton and Colburn (1933, 1934) produced an analogy algorithm for the MTC that was similar but algebraically simple:

$$k_c = v_\infty C_f/2Sc^{2/3}. \qquad (2.20)$$

This equation is exact for flat plates and satisfactory for other geometric shapes provided no form drag exists. The Chilton–Coburn analogy is valid for gases and liquids within the range $0.6 < Sc < 2500$ and is based on data collected in both laminar and turbulent flow regimes. This largely empirical result contains a Schmidt number function of the $-2/3$ power.

Both the theoretical and empirical results of the analogy theories relate one transport coefficient to another. In this case, they are k_c, the MTC, and C_f, the coefficient of skin friction or the momentum transfer coefficient. This is a fortuitous outcome, because as will be seen in the next few paragraphs much information is available on the coefficients of friction, C_f, for fluid moving over solid surfaces in natural systems.

The skin friction coefficient at the atmosphere soil surface. This section provides means for obtaining numerical estimating the coefficient of skin friction on the air side of large, flat terrestrial or water surfaces during neutral stability and windy conditions. The C_f is then used with the appropriate analogy theory algorithm to obtain a numerical estimate of the air-side MTC.

The well-known logarithmic velocity profile law follows from Prandtl mixing length theory. It applies well to the constant flux surface layer and has been verified numerous times from measurements taken both in the laboratory and in the field for neutral and near-neutral stability atmospheric surface layers. The result is the following relationship for the skin friction:

$$C_f = 2k^2/[\ln(y/y_0)]^2, \qquad (2.21)$$

where $k = 0.41$, the von Karman constant and y, in m, is the elevation above the flat surface, and y_0(m) is the surface roughness parameter. Typical values of C_f and y_0 appear in Table 2.1. The friction coefficients increase with increasing surface roughness, however, for these rather flat and smooth terrain types the C_f values vary less than $5\times$.

TABLE 2.1
Aerodynamic Skin Friction Coefficients for Various Types of Terrains

Terrain Type or Description	Y_0(m)	C_f^*
Ice, mud flats	2E − 5	0.0019
Calm open sea	1E − 4	0.0024
Snow-covered flat or rolling ground	1E − 4	0.0024
Desert (flat)	5E − 4	0.0033
Off-sea wind in coastal water areas	9E − 4	0.0037
Natural snow surface (farm land)	2E − 3	0.0044
Fairly level grass (∼3 cm) plains	6E − 3	0.0058
Fairly level uncut grass plains	2E − 2	0.0083

Source: From Arya, S.P., *Introduction to Micrometeorology*. Academic Press, San Diego, CA, 1988. With permission.
C_f^* from Equation 2.21, obtained with assumed $y = 10$ m, this being elevation commonly used for wind velocity measurement.

In general, an increasing MTC with increasing surface roughness may be problematic in some cases. However, very small surface roughness has been observed to increase MTCs in some cases (Dawson and Trass, 1972). In most cases, the opposite is true. In fluid dynamics, a surface is considered aerodynamically smooth if the small-scale surface protrusions or irregularities allow the formation of laminar sublayer in which they are completely submerged. Otherwise if the irregularities are large enough to present the formation of a viscous sublayer the surface is aerodynamically rough. For atmospheric flows, the sublayer thickness is usually ∼1 mm or less while most surface protrusions are generally larger than 1 mm. The analogy theory equations presented above are appropriate for use with aerodynamically smooth surfaces and may be appropriate for some in the transitional roughness regime, which is neither smooth nor completely rough. Clearly the presence of grasses, hedges, crops, tree groves, and so on are large protrusions in the boundary layer, which increase C_f while dampening the fluid velocities within the "canopies." As a limiting condition the flow may become stagnant in very dense canopies. In such cases, the effective MTC may be dramatically reduced due to the existence of a larger diffusive path length. An effort was made in selecting y_0 values for Table 2.1 that did not reflect the presence of a "crop canopy," the factors controlling mass transfer for this situation are presented in Chapter 7.

The empirical parameter y_0 in the logarithmic wind-profile equation is a constant, which is termed the roughness length. It is interpreted to be the height y_0 above the $y = o$ plane at which the wind velocity is zero. When interpreted this way, it is a height within the surface roughness elements, which is the origin of the momentum flux. Since perfectly smooth surfaces are generally absent in such natural settings y_0 undoubtedly reflects skin friction as well as some form of drag. Monteith and Unsworth (1990) note that the apparent sources of water vapor at the soil–air interface will, in general, be found at a lower level, say at y'_0, which is smaller than y_0. This gives rise to an addition resistance for water vapor transport, which requires the re-formulation of Equation 2.21:

$$C'_f/2 = 1 / \left\{ \left[B^{-1} + \ln(y/y_0)/k \right] \ln(y/y_0)/k \right\},$$ (2.22)

where $B^{-1} = \ln(y_0/y'_0)/k$ is the parameter, which a number of workers have used to analyze processes of chemical exchange at rough surfaces. Values of the additional resistance are seldom determined directly, but are usually estimated using measurements of both profiles and Equation 2.22. If Equation 2.21 is used then $\sqrt{C_f/2} = k/\ln(y/y_0)$ and Equation 2.22 can be recast as

$$C'_f/2 = 1 / \left\{ \left[B^{-1} + \sqrt{2/C_f} \right] \sqrt{2/C_f} \right\}$$ (2.22a)

Table 2.1 contains numerical values of relevant observed skin friction coefficients, C_f. The size of the additional resistance for real and model surfaces has been reviewed by Montieth and Unsworth (1990) and is summarized as follows. For water vapor transfer to rigid rough surfaces the following expression applies:

$$B^{-1} = 7.3 \, Re^{0.25} \, Sc^{0.5} - 5.0.$$ (2.23)

It fits observations well and may be used for ploughed fields. The Reynolds number $Re* \equiv U * y_0/v$ where $U*$ the friction velocity, which can be estimated from $U_* = U_\infty C_f/2$. Equation 2.23 is valid for $1 < Re* < 1E4$. The expression is also applicable to the diffusive transport of particles in the size range where impaction and sedimentation are unimportant, this is normally assumed to be those within the radius $= 0.5 \, \mu m$.

The skin friction coefficient at the aquatic-sediment surface. This section provides means for obtaining numerical estimates of the coefficient of skin friction on the water-side of lengthy and flat soil or sediment surfaces positioned beneath a flowing water stream or bottom water marine current. The flowing water situations include surface streams, rivers, estuaries, shallow marine flows, flooded soils, flood plains, and so on. The C_f is then used with the appropriate analogy theory algorithm to obtain a numerical estimate of the water-side MTC.

From a theoretical perspective, our understanding of the flow field above this aquatic surface tracks that presented above for the atmospheric boundary layer. For the neutral-stability class of turbulent flows the logarithmic velocity profile, the constant flux layer assumption and so on, apply as well. Although Equation 2.21 is valid for use in estimating C_f less measurement on y_0, the bottom roughness parameters are available in aquatic environments for producing summary results as shown in Table 2.1. In the absence of these site-specific y_0 values, an alternative approach is used to estimate C_f for hydraulic flows; it is presented next.

It is well known both theoretical and experimentally in the fluid dynamic field that the skin friction coefficient is independent of the Reynold's number once it becomes very large. Such is the usual case for open channel flows. For this very common situation, Manning proposed the resistance equation:

$$v = R^{2/3}S^{1/2}/n, \tag{2.24}$$

where v is average velocity (m/s), R is hydraulic radius of the channel (m), S is the energy slope, and n is Manning's roughness coefficient (s/m$^{1/3}$). This coefficient is independent of stream velocity and reflects only the roughness of the flow channel bottom and sides. A comprehensive database has been compiled containing numerical values of n for various types of flow channels (Yang, 2003).

Defining the skin friction factor C_f, for open channel flow and using Equation 2.24 yields

$$C_f = 8gn^2/R^{1/3}, \tag{2.25}$$

where g is the gravitational constant equal to 9.81 m/s^2. The hydraulic radius is the channel cross-sectional area divided by its welted parameter; for shallow and wide channels $R = h(m)$, the channel depth. Clearly, Equation 2.25 indicates that as bottom roughness increases so does C_f and therefore so will the MTC in using either analogy Equation 2.19 or 2.20. As noted above for the aerodynamic boundary layers case, this functionality taken to the limit is problematic. In the present case of very rough bottom surfaces made of gravel, cobbles, and boulders but including trees, weeds, and aquatic vegetation, the counterargument is that MTC should decrease due to the bottom features dampening the stream flow velocities near the MT interface and reducing the effective coefficient values. For hydraulically smooth bottom surfaces made of fine grain material, Arya (1988) offers Strickler's result for Manning's n as equal to $d^{1/16}/21.1$ for uniform sand of diameter d(m). Table 2.2 contains a summary of Manning's roughness coefficients appropriate for use in estimating C_f. This selection of

---

TABLE 2.2
Values of Manning's Roughness Coefficient n (s/m$^{1/3}$)
for Aquatic Streams

Type of Channel and Description	n
Excavated earth, straight, uniform, clean	0.018
Same but after weathering	0.022
Excavated earth, winding, sluggish, no vegetation	0.025
Same but grass, some weeds	0.030
Minor stream less 30 m width, clean and straight	0.030
Same but winding, some pools	0.040
Flood plains, pasture, no bush, short grass	0.030
Same but cultivated areas with no crops	0.030
Major streams greater than 30 m width, no boulders or brush	0.025
Same with irregular and rough section	0.035

Source: Adapted from tables presented by Yang, C.T. 2003. *Sediment Transport Theory and Practice*, Krieger Publishing, Malabar, FL.

---

values represents relative smooth bottom surfaces for which assumptions supporting the analogy theory algorithms (i.e., Equations 2.19 and 2.20) are likely met. The user is cautioned in using n values larger than those appearing in Table 2.2 for estimating MTC for lengthy flat hydraulic surfaces in that very large and unrealistic numerical values may result. Also see Chapter 12 (Table 12.4) for additional n values.

2.5.7 NATURAL CONVECTION MODELS

The above theoretical approaches apply estimating MTCs for the forced convection of fluid flowing parallel to a surface. Typically, the fluid forcing process is external to the fluid body. In the case of natural or "free" convection fluid motion occurs because of gravitational forces (i.e., $g = 9.81\ \mathrm{m/s^2}$) acting upon fluid density differences within (i.e., internal to) regions of the fluid. Temperature differences across fluid boundary layers are a major factor enhancing chemical mass transport in these locales. Concentration differences may be present as well, and these produce density differences that also drive internal fluid motion (i.e., free convection).

The phenomenon of free convection results in nature, primarily from the fact that when the fluid is heated, the density (usually) decreases; the warmer fluid portions move upward. This process is dramatically evident in rural areas on sunny days with low to no-wind when the soil surface is significantly hotter than the air above. The air at the soil surface becomes heated and rises vertically, producing velocity updrafts that carry the chemical vapor and the fine aerosol particles, laden with adsorbed chemical fractions, upward into the atmospheric boundary layer. When accompanied by lateral surface winds, the combined processes produce a very turbulent boundary layer and numerically large MTCs. This section will outline the major aspects of the theory of natural convection using elementary free convection concepts. Details are presented in Chapter 10 of *Transport Phenomena* (Bird et al., 2002).

Thermal-free convection heat transfer. The basic parameter that regulates a fluid response to temperature variations is the thermal coefficient of volume expansion at constant pressure it is defined as

$$\beta \equiv \frac{1}{\rho}\left(\frac{\partial \rho}{\partial T}\right)_p, \tag{2.26}$$

where ρ is the density in $\mathrm{kg/m^3}$, T is temperature, $°\mathrm{K}$, and β has dimensions $°\mathrm{K}^{-1}$. For air at 27°C the numerical value is $\beta = 0.72\mathrm{E}-4/°\mathrm{K}$ and for water $\beta = 10.3\mathrm{E}-4/°\mathrm{K}$.

Middleman (1998) presents the natural convection generated velocity profile in a fluid between two vertical parallel plates separated by distance w, meters, across which a temperature gradient $\Delta T/w$, in $°\mathrm{K/m}$, exist. The fluid region near the hot plate experiences an upward flow and that adjacent to the cold plate, a downward flow. The maximum velocity in each region is

$$v_{mx} = .032\frac{v}{w}\left[\frac{g\beta w^3}{v^2}\Delta T\right], \tag{2.27}$$

where v, in m^2/s, is the average fluid kinematic viscosity. The group of parameters enclosed within the brackets is the thermal Grashoff number, symbol Gr. It is a dimensionless number ΔT is the absolute value temperature difference between surface and fluid. It is evident from this simple result that fluid movement is driven by the thermal gradients within and the magnitude is in large part influenced by Gr. It in effect plays the equivalent role Re does in forced convection.

A more realistic and useful, but nevertheless simple, natural connection geometric shape is a single vertical plate with one face in contact with the fluid. It approximates the side of a building, a vertical cliff, steep mountain side or a similar feature in nature. An analysis of the heat flux for this situation has resulted in the following equation for the heat-transfer coefficient, h in $W/m^2 s$:

$$h = \frac{ck}{L}(\text{Gr Pr})^{1/4}, \tag{2.28}$$

where k, in W/m^2 s, is the thermal conductivity of the fluid and Pr, the dimensionless fluid Prandtl number equal to $\rho C_p v/k$, with heat capacity, C_p, in $J/kg \cdot K°$. The L (in m) variable is the vertical length of the "plate." The appropriate form of the thermal Grashoff number is given above. The ΔT is the absolute value of the temperature difference between the "plate" surface and the fluid far removed from the surface. For laminar fluid flow, when Gr Pr $< 10^9$, $c = 0.52$ for air and 0.62 for liquid water; these values based on experimental measurements (Bird et al., 2002). Representative numerical values of Pr at ambient temperatures are 0.73 for air and 10 for water.

Concentration difference driven free convection mass transfer. Analogous to the previous development of thermal density driven free convection, density variations also occur near fluid–solid interfaces as a result chemical concentration difference. These density imbalances generate fluid motion that enhances chemical transport is a more subtle conduction process. For example, at the aquatic boundary layer, chemical concentrations in the interface fluid may be very different from those in the water column above. This in turn produces a water density difference, ΔC_A (kg/m^3), where C_A is the concentration in water. Similar to the thermal coefficient of volume expansion given by Equation 2.26; it is termed the mass coefficient of volume expansion defined as

$$\zeta \cong \frac{|\Delta C_A|}{\rho}. \tag{2.29}$$

The corresponding Grashoff number (dimensionless) is

$$\text{Gr}_A \equiv \frac{L^3 g |\Delta C_A|}{v^2 \rho}, \tag{2.29a}$$

where L (m) is the length of the chemical source or sink, g is the gravitational constant, and v (m^2/s) is the kinematics viscosity of the fluid.

Data correlations of mass transfer coefficient employ empirical equations similar to the theoretical one, Equation 2.28 above. These use of the Sherwood number (dimensionless), Sh $\equiv k_c L/D$, as the dependent variables:

$$\text{Sh} = a(\text{Gr}_A \text{Sc})^b, \tag{2.30}$$

where Sc is the Schmidt number (dimensionless) and defined as v/D. The a and b constants are empirical data fitting parameters. See Chapter 12, Equations 12.14 and 12.15 for specific examples applicable to the aquatic benthic boundary layer.

2.6   GENERAL DATA CORRELATIONS FOR MTCs

2.6.1   FORCED CONVECTION AND NATURAL (FREE) CONVECTION

In the previous Sections 2.5.1 through 2.5.7, theoretical models describing and formulas for estimating MTCs are presented. These have developed over several decades. They are based on fluid dynamic mixing and diffusion arguments describing the mass transport process at fluid–fluid and solid–fluid interfaces. Simple model equations are given for each and can be used to estimate MTCs. However, they provide only approximate values. They apply to very hypothetical or idealized situations where the key independent parameters are oftentimes roughly related to the real independent variables. These theoretical models do identify many of the independent variables needed for characterizing k_c in the form of dimensional groups. Typically these are the Reynolds (Re), Schmidt (Sc), Prandtl (Pr), Grashoff (Gr), and so on. The definitions of each appear in Table 2.3. Ultimate usefulness of these dimensionless groups is producing correlations derived from experimental laboratory measurements on k_c

---

TABLE 2.3
Sherwood and Nusselt Correlations for Transfer Coefficients: Plates of Length L

Forced Convection Flow Parallel to Surface:
    Laminar   Equation 2.17
    Turbulent   Equation 2.18

Natural Convection at Surface:

    Horizontal hot plate facing up or cold plate facing down

$$1E5 < Ra < 2E7 \quad Nu = 0.54\,Ra^{1/4} \tag{2.31}$$
$$2E7 < Ra < 3E10 \quad Nu = 0.14\,Ra^{1/3} \tag{2.32}$$

    Horizontal hot plate facing down or cold plate facing up

$$3E5 < Ra < 1E10 \quad Nu = 0.27\,Ra^{1/4} \tag{2.33}$$

Vertical Plate:

$$Nu = [.825 + .387Ra^{1/6}/\{1 + (.492/Pr)^{9/16}\}^{8/27}]^2 \tag{2.34}$$

Ra over 13 order of magnitudes; applies to all fluids.

     To Convert Nu to Sh use: $Sh = Nu(Sc/Pr)^{1/3}$ A Chilton–Colburn analogy.

Definition of dimensionless groups: $Sh = k_c L/D$, $Nu = hL/k$, $Sc = v/D$, $Pr = v\rho Cp/k$, $Re = LV/v$, $Ra = PrGr$, $Gr = L^3 g \Delta \rho/\rho v^2$.

Source: From Welty et al. 2001. *Fundamentals of Momentum, Heat and Mass Transfer*, 4th ed. Wiley, New York, Chapters 20 and 28. With permission.

TABLE 2.4

Nusselt Correlations for Transfer Coefficients: Cylinders and Spheres

Forced Convection, Cylinders:

Flow parallel to long axis, length, L

Use plate Equations 2.17 or 2.18, Table 2.3

Flow perpendicular to long axis, diameter L.

$$\text{Nu} = 0.30 + \frac{0.62\,\text{Re}^{1/2}\,\text{Pr}^{1/3}}{\left[1+(0.4/\text{Pr})^{2/3}\right]^{1/4}}\left[1 + \left(\frac{\text{Re}}{2.82\text{E}5}\right)^{5/8}\right]^{4/5} \qquad (2.35)$$

Natural Convection, Cylinder:

Vertical cylinder, length L

Use plate Equation 2.34 Table 2.3

Horizontal cylinder, diameter L

$$\text{Nu} = \left[0.06 + 0.387\,\text{Ra}^{1/6}\bigg/\left\{1 + (0.559/\text{Pr})^{9/16}\right\}^{8/27}\right]^2 \qquad (2.36)$$

$$1\text{E} - 5 < \text{Ra} < 1\text{E}12$$

Forced Convection, Sphere Diameter L:

$$\text{Nu} = 2 + 0.4\,\text{Re}^{1/2} + 0.06\,\text{Re}^{2/3})\text{Pr}^{0.4}\,(\mu_\infty/\mu_s)^{1/4} \qquad (2.37)$$

$$3.5 < \text{Re} < 7.6\text{E}4, 0.71 < \text{Pr} < 350, \mu_\infty/\mu_s < 3.2$$

Natural Convection, Sphere Diameter L:

$$\text{Nu} = 2 + 0.43\,\text{Ra}^{1/4} \qquad (2.38)$$

$$1 < \text{Ra} < 1\text{E}5, \text{Pr} \sim 1, \text{strictly}$$

$$\mu = \text{viscosity, s at surface, } \infty \text{ at distance}$$

Definitions of dimensionless groups, see Table 2.3.

Source: From Welty et al. 2001. *Fundamentals of Momentum, Heat and Mass Transfer*, 4th ed. Wiley, New York, Chapters 20 and 28. With permission.

---

and other transport coefficients. Many such useful empirical correlations appear in Tables 2.3 and 2.4.

Altogether 10 such formulas appear in these two tables. They represent forced and natural convection fluid flow conditions associated with simple shapes such as flat plates, single cylinders, and single spheres. Several of the correlations in the tables are based on thermal measurements containing the Nusselt (Nu) number and can be used as approximations of k_c by converting it into the equivalent Sherwood (Sh) number. For the forced convection ones the Chilton–Colburn analogy is used for this conversion, it is given in Table 2.3. In the absence of specific field data representing mass-transfer coefficients in natural environmental situations, there correlations can be used as first-order estimates (see Example 2.9.4).

2.6.2 MIXED FORCED AND NATURAL CONVECTION MASS TRANSFER

The data correlations presented in the previous section include both forced convection and free convection processes. In nature as well as engineering applications, there are important mass transfer situations that require a combination of forced and free. The subject appears not to be fully developed but some information is available. What follows is summarized from Bird et al. (2002, see specifically pages 698, 699, and 445).

Additivity of Grashof numbers. For the combined mixed free-convection heat and mass transfer the procedure is to replace the Gr_T or Gr_A with $Gr_T + Gr_A$, the sum. For example, Equation 2.31 in Table 2.3 becomes

$$Sh = 0.54 \ (Gr_T + Gr_A)^{1/4} Pr^{1/4}. \tag{2.39}$$

The limitation on this procedure is $Sc_A \simeq Pr$ and this is frequently the case for transport processes in air. Since the thermal Grashoff number is normally much larger, neglecting the interaction would greatly underestimate the MTC.

Free convection heat transfer as a source of forced convection mass transfer. It has been demonstrated on numerous occasions that the Chilton–Colburn analogy appearing in Table 2.3 is applicable for converting a forced-convection Nusselt number to a forced-convection Sherwood number as a means of converting the imbedded HTC into its equivalent MTC. In the present situation, the thermal buoyant forces provide the momentum source, which in effect provides the forced-convective flow that drives the mass transfer process. In addition, $Gr_T \gg Gr_A$ and $Sc > Pr$. For this case the alternative equation is

$$Sh = Nu \ (Sc/Pr)^{1/12}, \tag{2.40}$$

which in effect combines the 1/4-power of free convection with the 1/3-power of forced convection as $((Sc/Pr)^{1/4})^{1/3}$. Equation 2.40 is applicable to water boundary layers because the Schmidt number, Sc, is much larger than the Prandtl number Pr.

Mixed free and forced convection. In this mass transfer situation, a free convection process driven either by thermal and/or concentration $\Delta\rho$ differences is present in the boundary layer formed by forced convection flow parallel to the surface. The situation of pesticide evaporation from a hot soil surface into a light wind moving across is a good example. An empirical combination rule adapted from the heat transfer is proposed; it is

$$Sh^{total} = \left[\left(Sh^{free} \right)^3 + \left(Sh^{forced} \right)^3 \right]^{1/3} \tag{2.41}$$

for the total Sherwood number. Bird et al. (2002) note that this rule appears to hold reasonably well for all geometries and situations, provided only that the forced and free convection have the same primary flow direction.

2.7   UNSTEADY-STATE VIGNETTE CHEMODYNAMIC MODELS

This section contains time-tested, useful theoretical equations of chemodynamic processes in a solid porous media zone adjacent to a fluid. All cases presented assume a semi-infinite solid geometry and transient chemical concentration behavior in the y-dimension. Although a seemingly absurd shape, it realistically represents earthen layers beneath the atmospheric boundary layer or water column relevant to the chemo-dynamics of environmental pollutants. The vignette models are analytical solutions to Fick's second law and the advection–diffusion equation that include reactive decay and interphase transport. The Choy and Reible (2000) compilation of environmental transport models was the primary literature source. The selected models fall into two transport categories: diffusion only and advection–diffusion.

2.7.1 DIFFUSION IN A SEMI-INFINITE SOLID MEDIA

The typical application of these models involves transient chemodynamics within the solid yielding equations for the concentration profiles and the flux to the adjoining fluid media. The applications involve chemicals in soils and sediments for idealized scenarios of uniform concentration, $C_o(g/m^3)$ at $t = 0$, from the interface at $y = 0$ downward in the y-direction to infinity. Within the solid chemical transport is by Fickian diffusion with coefficient $D_e(m^2/s)$. At times, $t > 0$, the surface concentration is assumed zero causing chemical quantities to move from the solid into the adjoining fluid phase.

Diffusion with reactive decay. For this case, the time-varying concentration profile in the solid is given by

$$C = C_o \cdot \exp(-k_1 t) \cdot \mathrm{erf}\sqrt{y^2 R_f / 4 D_e t}, \tag{2.42}$$

where R_f is the so-called retardation factor and k_1 (s^{-1}) the first-order decay constant. The R_f is defined as the ratio of the total chemical mass per unit volume solid to the mobile chemical mass. The flux from the solid to the fluid is

$$n = C_o \exp(-k_1 t) \sqrt{D_e R_f / \pi t}. \tag{2.43}$$

Both of the above equations apply when $k_1 = 0$, which is no reactive decay.

Diffusion with reactive decay and mass transfer to the adjoining fluid phase. This case extends the previous one by including a MTC, k_a (m/s), on the fluid side to capture this transport resistance. The concentration profile equation for the same idealized initial solid phase conditions is

$$C = C_o \left[\mathrm{erf}\sqrt{\beta} + \exp(\sigma + \alpha)\mathrm{erfc}\left(\sqrt{\beta} + \sqrt{\alpha}\right) \right] \exp(-k_1 t) \tag{2.44}$$

and the flux is

$$n = k_a C_o \exp(\alpha - k_1 t)\mathrm{erfc}\sqrt{\alpha}, \tag{2.45}$$

where

$$\alpha \equiv \frac{k_a^2 t}{D_e R_f}, \quad \beta \equiv \frac{R_f y^2}{4 D_e t}, \quad \text{and} \quad \sigma \equiv \frac{k_a y}{D_e}.$$

The above results are valid for the case $k_1 = 0$.

2.7.2 Diffusion and Advection in a Semi-Infinite Solid Media

The typical application of these models is similar to those presented above except that porewater flow is characterized by a Darcian velocity v (m/s) so that advective transport is superimposed on the diffusive transport. The solid is infinite in extent and the processes are transient with transport occurring in the y-direction.

Bed of uniform initial concentration subjected to fluid inflow of constant concentration. The initial concentration is C_i (g/m^3) and the inflow concentration is C_o; both are constant. The concentration profile is

$$C = C_i + \left(\frac{C_o - C_i}{2}\right)\left[\text{erfc}\left(\frac{R_f y - vt}{\sqrt{4R_f t}}\right) + \exp\left(\frac{vy}{D}\right)\text{erfc}\left(\frac{R_f y + vt}{\sqrt{4R_f t}}\right)\right]. \quad (2.46)$$

The various parameters have been defined above.

A solid region capped with a uniform layer of different uniform concentration. The initial concentration in the capped layer of thickness y_1 is C_1, and the underneath layer is at concentration C_2. The concentration at the fluid-capped layer interface is maintained at C_o to start the process diffusion–advective process. The concentration profile is

$$C = C_2 + (C_1 - C_2)\cdot\frac{\varphi}{2} + (C_o - C_1)\cdot\frac{\phi}{2}, \quad (2.47)$$

where

$$\varphi = \text{erf}\left[\frac{R_f(y - y_1) - vt}{\sqrt{4D_e R_f t}}\right] + \exp\left(\frac{vy}{D_e}\right)\text{erfc}\left[\frac{R_f(y + y_1) + vt}{\sqrt{4D_e R_f t}}\right] \quad (2.48)$$

and

$$\phi = \text{erfc}\left[\frac{R_f y - vt}{4D_e R_f t}\right] + \exp\left(\frac{vy}{D_e}\right)\text{erfc}\left[\frac{R_f y + vt}{\sqrt{4D_e R_f t}}\right]. \quad (2.49)$$

This final case applies to a polluted sediment or soil of uniform concentration upon which a fresh layer of similar but pollutant-free material was suddenly placed to form a cap to retard the pollutant flux process.

2.8 A USER GUIDE TO CHAPTER 2

Sections 2.1 through 2.4 contain fundamental concepts and definitions, primarily. It is presented in textbook style. Detailed qualitative descriptions of environmental chemodynamic processes rather than rigorous derivations are used to present the material. It is suggested reading and studying for those new to the subject of mass transfer. A few of the resulting equations are useful in obtaining approximate numerical estimates of MTCs.

Section 2.5 contains equations and data that are more useful for estimating individual phase MTCs at environmental interfaces. They are some formulas that users may choose to apply to any number of transport situations. At first, these formulas may not appear to be explicitly obvious in their potential uses but are generally applicable to various interface chemical transport situations. Applications are subject to the ingenuity of the individual user. However, there are several equations and tables in Section 2.5.6 on analogy theories and Section 2.5.7 on natural convection, which have clear and explicit applications for estimating particular MTCs. Tables 2.3 and 2.4 in Section 2.6 contain ten empirical correlations for use in estimating MTCs in fluids adjacents to solid surfaces. The chapter ends with Section 2.7 containing a few time-tested vignette chemodynamic models.

2.9   EXAMPLE PROBLEMS

2.9.1   A WATER-SIDE MTC ESTIMATE FOR SEA-SURFACE WAVES INTERACTING AT ROCKY SHORELINE

Estimate the enhanced oxygen reaeration MTC in the surface waters occupying an area several meters in width parallel to the shoreline. Here the wind-generated water waves encounter the shoreline generating a highly turbulent zone.

Solution: The wave interacting with the shore will result in surface water currents being deflected and reversed. This will create an area near shore with enhanced turbulence as the opposing currents converge and plunge below the surface. One can envision a surface renewal action occurring each time a wave arrives and encounters the shore. The surface renewal model result, Equation 2.14b, contains a parameter which generally characterizes the fractional surface renewal rate. The parameter s (s^{-1}) is assumed to be equal to the inverse wave time period. With assumed wave travel time periods crest-to-crest of 1–5 s the s values are 1 to $0.2\,\text{s}^{-1}$. The molecular diffusivity of O_2 in water is $2.4\text{E} - 5\,\text{cm}^2/\text{s}$; the MTCs can be estimated using Equation 2.14b:

$$k_c \sqrt{2.4\text{E} - 5\frac{\text{cm}^2}{\text{s}} \left|\frac{0.2}{s}\right| \cdot 3600\frac{s}{n}} = 7.9\,\text{cm/h}$$

and 18 cm/h, respectively.

The turbulence level in the shoreline mixing zone is assumed to be superimposed onto the background level existing offshore. As a first approximation, the MTCs are summed. A value of 20 cm/h average is commonly used for the CO_2 MTC in the open ocean. With a CO_2 diffusivity equal to $1.8\text{E} - 5\,\text{cm}^2/\text{s}$, the O_2 equivalent is

$$k_c = 20 \left(\frac{2.4\text{E} - 5}{1.8\text{E} - 5}\right)^{1/2} = 23\,\text{cm/h}.$$

This calculation assumes $k_c \sim D^{1/2}$ for a fluid–fluid interface. So, it appears that the shoreline interaction may provide a 34–78% enhancement in the near-shore area compared to the sea-surface background value.

2.9.2 Estimate the Air-Side MTC for a Large and Flat Plowed Field

For a 20 km square swath of SW Louisiana prairie farmland, estimate the air-side MTC for chloropyritos with molecular diffusivity, $0.046 \, cm^2/s$ in air at 25°C, Schmidt number equals 3.4 and widespeed is 5 m/s.

Solution: The analogy theory equations provide an approach to estimating MTC for the case of a fully developed and steady-state boundary layer. The traditional turbulent BL theory correlation (Equation 2.18) suggests that the MTC decreases with increasing fetch as $L^{-0.2}$. The Chilton–Colburn analogy result, Equation 2.20 will be used, it is the simplest and based on experimental data. From Table 2.1, $C_f = 0.0044$ and $y_o = 2E - 3m$ is assumed appropriate for farmland, however a 30% variation about this C_f is arguable, based on values of similar terrain types in the table. Equation 2.22a is the alternative choice for estimation C_f. Using Equation 2.23 to obtain β^{-1} yields 21.1 and a $C_f' = 0.0022$ results. The MTC computation is

$$k_c = \frac{v_\infty C_f'}{2Sc^{2/3}} = \frac{5(0.0022)}{2(3.4)^{2/3}} = 2.43E - 3 \, m/s.$$

Therefore the MTC is 880 cm/h with range [610–1140].

The turbulent BL theory result is Equation 2.18. With $L = 20,000 \, m$ the $Re_L = 6.39E9$, the MTC computation is:

$$k_c = 0.0365 \left(\frac{4.6E - 6\frac{m^2}{s}}{2E4m} \right) (6.39E9)^{0.8}(3.4)^{1/3} = 8.82E - 4 \, m/s,$$

and the MTC is 320 cm/h.

It appears that the analogy theory estimate of 880 cm/h is approximately $3\times$ higher than the turbulent boundary layer result. However, both values fall within the range of reported field observations; see Thibodeaux (1996, p. 390).

2.9.3 Stream Bed Water-Side MTC

Use the von Karman analogy result Equation 2.19 to estimate the water-side coefficient for phenol in the Mississippi River. Velocity $= 1.2 \, m/s$ and depth $= 7.62 \, m$. Phenol diffusivity in water is $0.84E - 5 \, cm^2/s$ and $Sc = 1200$ at 25°C.

Solution: It is assumed that a Manning's roughness coefficient $n = 0.025$ in Table 2.2 applies to the Mississippi River. Equation 2.25 is appropriate for estimating the friction factor, applying it yields:

$$C_f = \frac{8gn^2}{R^{1/3}} = \frac{8(9.81)(0.025)^2}{(7.62)^{1/3}}, \quad C_f = 0.0249.$$

Entering the appropriate numerical values into the denominator of Equation 2.19 yields a value of equal to 674. The k_c computation is:

$$k_c = 1.2\frac{m}{s} \frac{(0.0249)}{2(674)} = 2.22E - 5 \, m/s \, (7.98 \, cm/h).$$

Assuming the Manning's roughness may vary to maximum of 0.030 and minimum of 0.020 the k_c range is [5.1 to 11.5 cm/h]. For stream velocity of 0.12 m/s, the range falls to 0.5 to 1.2 cm/h. Field data are on the lower end of these ranges, values of 0.21–1.6 cm/h for rivers have been reported (Thibodeaux, 2005). Recalculations using the Chilton–Colburn analogy results in a k_c six times larger. More appropriate estimating methods are given in Section 12.2 of Chapter 12.

2.9.4   URBAN SURFACE AIR-SIDE MTC

The built environment is constructed of various solid materials such as brick, concrete, glass, steel, iron, asphalt, wood, and so on. Such exposed exterior surfaces accumulate organic films to which chemicals, metals, spores, dust, and so on. accumulate (Diamond and Hodge, 2007). Estimate the air-side MTC for a structure 50 m in length, 5 m/s wind speed, 45°C surface temperature, and 20°C air temperature. Assume the chemical is chloropyritos.

Solution: (a) For windy condition, assume the urban structure is a building 50 m high and 50 m diameter. An appropriate correlation is Equation 2.35 in Table 2.4. It models the building as a vertical cylinder with flow perpendicular to the cylinder axis.

Properties: Molecular diffusivity, 0.046 cm^2/s; Schmidt number, 3.1, Pr No. (20°C) = 0.702 and 0.699 at 45°C; air viscosity 15.1E − 6 and 17.4E − 6 m^2/s at 20° and 45°C, respectively. Calculated results: Re_L = 1.06E7 at 20°C. For Equation 2.35, Nu = 1.16E4.

The Nusselt to Sherwood number conversion: $Sh = Nu(Sc/Pr)^{1/3}$ = 2.64E4.

The MTC calculation is: $k_c = ShD/L$ = 2.64E4 (0.046cm^2/s)/5000 cm for k_c = 0.243 cm/s or 870 cm/h. At 45°C, k_c = 770 cm/h.

Solution: (b) For windless conditions, assume thermal natural convection between air at 20°C and building surface temperature 45°C. The appropriate heat-transfer correlation is Equation 2.34 in Table 2.3. Calculated results: the Grashoff number is the group in the brackets in Equation 2.27. With $\beta = 0.722E − 4/K°$ and $Gr = L^3 g\beta\Delta T/v_i^2$

$$Gr = (50\,m)^3 \frac{9.81}{s^2} \left(\frac{0.722E − 4}{K°}\right)(45 − 20)K°/(0.174E − 4\,m^2/s)^2, \quad Gr = 7.31\,E\,12$$

Ra = PrGr = 5.13E12. Equation 2.24 yields the Nusselt number Nu = 1880.

The Nusselt-to-Sherwood conversion is: $Sh = Nu(Sc/Pr)^{1/3}$ with Sh = 3089 the MTC is:

$$k_c = 3089\frac{(0.046\,cm^2/s)}{5000\,cm} = 0.0248\,cm/s = 89.3\,cm/h.$$

Summary: Equation 2.41 is appropriate for combining the free and forced convection MTCs. Since the D and L in the Sherwood number, $Sh = k_cL/D$ are constant for both MTCs, Equation 2.41 becomes

$$k_c = \left[(89.3)^3 + (870)^3\right] = 870cm/h.$$

In this windy case it appears that the forced convection MTC dominates the air-side MTC in the urban setting.

LITERATURE CITED

Arya, S.P. 1988. *Introduction to Micrometeorology*. Academic Press, San Diego, CA, Chapter 10.

Bird, R.E., Stewart, E.W., and Lightfoot, E.N. 2002. *Transport Phenomena*, 2nd ed. Wiley, New York.

Boethling, R.S. and Mackay, D. 2000. *Handbook of Property Estimation Methods for Chemicals*. Lewis Publishers, Boca Raton, FL.

Campbell, J.A. and Hanratty, T.J. 1982. *J. Am. Inst. Chem. Eng.*, 28, 988.

Chilton, T.H. and Colburn, A.P. 1934. *Ind. Eng. Chem.*, 26, 1183.

Choy, B. and Reible, D.D. 2000. *Diffusion Models of Environmental Transport*, Lewis Publishers, Boca Raton, FL.

Colburn, A.P. 1933. *Trans. AIChE*, 29, 174–210.

Danckwerts, P.V. 1951. *Ind. Eng. Chem.*, 23(6), 1460.

Dawson, D.A. and Tass, O. 1972. *Int. J. Heat Mass Transfer*, 15, 1317–1336.

Diamond, M. and Hodge, E. 2007. *Environ. Sci. Technol.* 41(11), 3796–3805.

Gulliver, J.S. 2007. *Introduction to Chemical Transport in the Environment*. Cambridge University Press, NY.

Higbie, R. 1935. *Am. Inst. Chem. Eng. J.*, 31, 365.

McCready, M.A., Vassiliadou, E., and Hanratty, T.J. 1986. *Am. Instit. Chem. Eng. J.* 32(7), 1108.

Monteith, J.L. and Unsworth, M.H. 1990. *Principles of Environmental Physics*, 2nd ed. Butterworth and Heinemann, Oxford, UK, Chapter 15.

Nernst, W. 1904. *Zietschvif fur Physikalische Chemi*. 47, 52.

Seinfeld, J.H. and Pandis, S.N. 1998. *Atmospheric Chemistry and Physics*. Wiley, New York, NY.

Sikar, T.T. and Hanratty, T.J. 1970. *J. Fluid Mech.* 44, 589.

Soil Science Society of America 1989. In B.L. Sawney and K. Brown, eds. *Reaction and Movement of Organic Chemicals in Soils*. SSSA Inc., Madison, WI.

Stumm, W. and Morgan, J.L. 1996. *Aquatic Chemistry*, 3rd ed. Wiley, New York.

Thibodeaux, L.J. 1996. *Environmental Chemodynamics*, 2nd ed. Wiley, New York, Chapter 3.

Welty, J.R., Wicks, C.E., Wilson, R.E., and Rorrer, G.L. 2001. *Fundamentals of Momentum, Heat and Mass Transfer*, 4th ed. Wiley, New York.

Yang, C.T. 2003. *Sediment Transport Theory and Practice*, Krieger Publishing Co., Malabar, FL, Chapter 3.

FURTHER READING

Boeker, E. and van Grondelle, R. 1999. *Environmental Physics*, 2nd ed. Wiley, Chickester, West Sussex, UK.

Choy, B. and Reible, D.D. 2000. *Diffusion Models of Environmental Transport*. Lewis Publishers, Boca Raton, FL.

Mackay, D. 2001. *Multimedia Environmental Models—The Fugacity Approach*, 2nd ed. Lewis Publishers, Boca Raton, FL.

Monteith, J.L. and Unsworth, M.H. 1990. *Principles of Environmental Physics*, 2nd ed. Butterworth-Heinemann, Oxford, UK.

Okubo, A. and Lavin, S.A. 2001. *Diffusion and Ecological Problems: Modern Perspectives*, 2nd ed. Springer, New York.

Thibodeaux, L.J. 1996. *Environmental Chemodynamics*, 2nd ed. Wiley, New York, Chapter 3.

3 The Fugacity Approach to Mass Transport and MTCs

Donald Mackay

CONTENTS

3.1 Introduction 43
3.2 Equilibrium Partitioning 44
3.3 Flux Parameters or D Values 45
3.4 Inherent Advantages of the Fugacity Approach 46
    3.4.1 Fugacity: The Term 46
    3.4.2 Fugacity: Simplicity in Formulating Interphase Rate Expressions ... 46
3.5 Characteristic Times for Interphase Transport 47
3.6 Illustration of a Simple Fugacity Mass Balance Model 48
Literature Cited 50

3.1 INTRODUCTION

The conventional formulation of the mass transfer process rate equation expresses the diffusive driving force in terms of concentration gradients or differences. For interphase diffusion, this driving force is adjusted using the equilibrium partition coefficient K_{ij} as shown below. An alternative approach is to express the flux equations in terms of fugacity instead of concentration. The driving force then becomes simply the fugacity gradient or difference. The chemical transfer rate N(mol/h) for both approaches is given by

$$N = Ak(C_j - C_i/K_{ij}) = D(f_j - f_i). \tag{3.1}$$

The theoretical and practical relationships between the mass transfer coefficient k (m/h), the area A (m^2), the concentration C (mol/m^3), the fugacity transport coefficient D (mol/Pa/h), and the fugacity of the chemical f (Pa) is the subject of this chapter. The use of fugacity is a particularly useful concept for numerous reasons.

In this chapter, we present the theoretical basis for the fugacity approach by developing its equilibrium and flux parameters in relation to the conventional approach. We then discuss and illustrate the inherent advantages of using the fugacity approach. We also define and discuss the use of characteristic times as an aid to interpreting the results of fugacity-type multicompartment models. The last part contains an example problem of a fugacity-based multimedia model application, the result being identical to a conventional concentration-rate constant calculation. The reader is referred to the text by Mackay (2001) for further details.

3.2 EQUILIBRIUM PARTITIONING

Using air–water partitioning as an example, the air–water partition coefficient K_{AW} is defined as C_A/C_W at equilibrium where C_A is the concentration in air (mol m^{-3} or g m^{-3}) and C_W is the concentration in water in identical units. K_{AW} is thus dimensionless. For other pairs of phases, it is often more convenient to use different units in the two phases, in which case the partition coefficient has dimensions. For example for sediment–water, C_S may be expressed in mg kg^{-1} and C_W in mg L^{-1} thus K_{SW} has units of L kg^{-1}. It can be converted into dimensionless form by multiplying by the sediment density in units of kg L^{-1}.

When using equilibrium partition coefficients in the fugacity formulation, it is necessary to convert all these coefficients into their dimensionless form as a ratio of amount/volume in both phases.

The K_{AW} and K_{SW} partition coefficients given above are a function of the properties of the chemical and the phase composition in *two* phases. In the fugacity approach, each phase is treated individually using a quantity termed the fugacity capacity or Z_i value, which expresses the affinity of the chemical for that phase. The relationship between fugacity f_i (Pa) and C_i (mol m^{-3}) is

$$C_i = Z_i \cdot f_i, \tag{3.2}$$

where Z_i is the fugacity capacity of the particular phase; it has units of mol m^{-3} Pa^{-1}. In many respects, Z_i is analogous to a heat capacity of a phase that relates concentration of heat to temperature.

The definitions of the Z_is start with the one for the air phase, Z_A. The ideal gas law is used to obtain it as follows:

$$C = \frac{n}{V} = \frac{P}{(RT)} = \frac{f}{(RT)} = Z_A f \tag{3.3}$$

hence Z_A is $1/(RT)$, R being the gas constant 8.314 Pa m^3 mol^{-1} K^{-1}, n is moles of chemical, V is the volume (m^3), and T is the absolute temperature (K). Under assumed ideal conditions, the partial pressure P is equated to the fugacity (Pa). If K_{AW} is known Z_W, the water phase Z, can then also be obtained as follows:

$$K_{AW} = \frac{C_A}{C_W} = \frac{Z_A f}{(Z_W f)} = \frac{Z_A}{Z_W}. \tag{3.4}$$

TABLE 3.1
List of Partition Coefficients Dependent on Z_A or Z_W and K_{iA} or K_{iW}

Partition Coefficient		Definition
Air–water	$K_{AW} = Z_A/Z_W$	
Octanol–water	$K_{OW} = Z_O/Z_W$	Z_O is Z value in octanol, mol m^{-3} Pa^{-1}
Octanol–air	$K_{OA} = Z_O/Z_A$	
Aerosol–air	$K_{QA} = Z_Q \cdot K_{OA} = Z_Q/Z_A$ where	Z_Q is Z value in aerosols, mol m^{-3} Pa^{-1}
	$Z_Q = \varphi_Q \cdot Z_O \cdot (\rho/1000)$	φ_Q is organic matter mass fraction in aerosols
		ρ is aerosol density, kg m^{-3}
Soil (earth)–air	$K_{EA} = Z_E/Z_A$	Z_E is Z value in soil (earth), mol m^{-3} Pa^{-1}
Soil (earth)–water	$K_{EW} = Z_E/Z_W$	
Sediment–water	$K_{SW} = Z_S/Z_W$	Z_S is Z value in sediment, mol m^{-3} Pa^{-1}
Fish–water	$K_{FW} = \varphi_F \cdot K_{OW} = \varphi_F \cdot Z_O/Z_W$	φ_F is the lipid fraction in fish

Source: From Mackay, D. 2001. *Multimedia Environmental Models: The Fugacity Approach*, 2nd ed. Lewis Publishers, Boca Raton, FL, pp. 1–261. With permission.

Finally,

$$Z_W = \frac{Z_A}{K_{AW}} = \frac{1}{H}. \tag{3.5}$$

Since K_{AW} is P/C_W, the ratio of chemical vapor pressure to water solubility, it can also be expressed as $H/(RT)$ where H is the Henry's law constant or with units of Pa m^3 mol^{-1}.

All Z_i values in the remaining phases can be calculated from Z_A or Z_W in a similar fashion using the appropriate bi-phasic partition coefficients. In all cases K_{ij} is Z_i/Z_j. Table 3.1 contains the relationships needed in performing these calculations for the most common environmental phases; the Z_i values are defined within the table.

3.3   FLUX PARAMETERS OR *D* VALUES

The general rate equation in the fugacity formalism for a chemical moving *from* a particular phase is given by

$$N = Df, \tag{3.6}$$

where f is the fugacity in the source phase and D is a rate parameter. The D values relationships needed depend on the type of mass transfer process involved and the associated form of the mathematical rate expression. The most common formulations for the coefficients and related quantities are shown in Table 3.2.

TABLE 3.2
***D* Value Definitions**

Process	*D* Value (mol Pa^{-1} h^{-1})	Definition
Diffusion expressed by a diffusivity	$B \cdot A \cdot Z / \Delta Y$	B is diffusivity, m^2 h^{-1} A is area, m^2 ΔY is path length, m
Diffusion expressed by a mass transfer coefficient	$k_M \cdot A \cdot Z$	k_M is mass transfer coefficient, m h^{-1}
Advection	$G \cdot Z$	G is advective flow rate, m^3 h^{-1}
Reaction	$k_R \cdot V \cdot Z$	k_R is the reaction or degradation rate constant, h^{-1} V is volume, m^3
Phase growth dilution	$(dV/dt)Z$	

Note: In all cases Z (mol m^{-3} Pa^{-1}) applies to the source phase.

3.4 INHERENT ADVANTAGES OF THE FUGACITY APPROACH

3.4.1 FUGACITY: THE TERM

This state variable has several attributes that make it preferred over concentration for environmental modeling. The pressure concept is a familiar one to most scientists and engineers. The "concentration" level in all phases is quantified by this one variable that has pressure units of Pascals (Pa). Typically, the values for chemicals in the various media are usually very low numerically, being much less than the standard, sea level, atmospheric pressure; 1 atmosphere is 1.01E5 Pa. All f values are either zero or positive; a zero value for a phase or media indicates the absence of the chemical. Unlike chemical concentrations in the various phases, which cannot be compared directly, their fugacities can. By comparing the numerical values of chemical fugacity, in Pascals, for each phase, it is clear to the user what the relative levels of "concentration" are. The numerical differences in f indicate the relative magnitudes of chemical presence and if the f values in adjoining phases are equal, they are in chemical equilibrium. This latter attribute is particularly useful for assessing interphase mass transfer.

3.4.2 FUGACITY: SIMPLICITY IN FORMULATING INTERPHASE RATE EXPRESSIONS

An advantage of this approach is that if the chemical is leaving a phase by a variety of parallel mechanisms such as diffusion, advection, and reaction, the respective D values can be added to give a D value expressing the total rate of total losses. The relative importance of the loss processes then becomes immediately apparent and the algebraic or differential equations become simpler to manipulate. This is particularly valuable when there is two-way reversible diffusion, for example between air and water. The volatilization rate from the water is $D_V \cdot f_W$ and the absorption rate from the atmosphere is $D_V \cdot f_A$ thus the net volatilization flux is $D_V(f_W - f_A)$. When writing the mass balance equations for the water phase, $D_V \cdot f_W$ is a loss process and $D_V \cdot f_A$ is treated as an entirely separate input process.

When a chemical experiences two diffusion processes in series as occurs during air–water exchange used earlier as an example, the respective D values add reciprocally, for example,

$$\frac{1}{D_V} = \frac{1}{D_A} + \frac{1}{D_W}, \tag{3.7}$$

where D_V is the overall parameter, D_A is the air-side parameter, and D_W is the water-side parameter. The appropriate rate equation is Equation 3.1.

For the conventional approach to mass transfer the rate expression is:

$$N = A \cdot k_V \left(C_W - \frac{C_A}{K_{AW}} \right), \tag{3.8}$$

where the overall MTC k_V is obtained by the familiar two-resistance theory relationship:

$$\frac{1}{k_V} = \frac{1}{k_W} + \frac{1}{(k_A K_{AW})}. \tag{3.9}$$

Here k_W and k_A are the individual water and air mass transfer coefficients, in units of m/h. The relationships connecting the fugacity mass-transport parameters to the conventional are: D_A is $k_A \cdot A \cdot Z_A$, D_W is $k_W \cdot A \cdot Z_W$, K_{AW} is Z_A/Z_W, C_A is $Z_A \cdot f_A$, and C_W is $Z_W \cdot f_W$.

These are for the air–water chemical evaporation example and obtained from Tables 3.1 and 3.2. Thus, the algebraically equivalent rate of volatilization is

$$N = D_V(f_W - f_A). \tag{3.10}$$

In this case the D values, which are conductivities, add reciprocally since the $1/D$ quantities are effectively resistances. The relative resistances become immediately apparent. In more complex situations, there may be several series and parallel resistances, an example being volatilization from soil in which there is diffusion in both soil air (pore air) and soil water (pore water) followed by an air boundary layer resistance. Later an example is given involving multiple transport processes and illustrates the simplicity and transparency of the fugacity mass balance equations.

3.5 CHARACTERISTIC TIMES FOR INTERPHASE TRANSPORT

When chemical is present in a compartment and is subject to loss processes, it is useful to calculate the time that is required for a defined fraction of the chemical mass in the compartment to be lost by that process. A half-life or half-time is commonly used to characterize rates of reaction or radioactive decay. More fundamental is the characteristic time, which is the reciprocal of the rate constant of loss. The half-time is thus 0.693 or ln 2 of the (longer) characteristic time. For example, a characteristic time of 100 h corresponds to a halftime of 69.3 h. The characteristic time is thus the time to achieve 63.2% depletion with 36.8% remaining. It can be regarded as approximately a "two thirds" time.

Characteristic times can be calculated for individual process as VZ/D or for combined parallel processes as $VZ/\Sigma D$ where V and Z apply to the source compartment and D to the rate of movement. Alternatively, if a mass balance is available, they can be calculated as the amount of chemical in the compartment (VZf) divided by the appropriate rate of transport or reaction (Df) or ($f\Sigma D$).

The calculation of characteristic times aids in the interpretation of multimedia mass balance for environmental compartments by conveying how long it will take for the compartment or phase to adjust as a result of each rate process. Obviously, the shorter characteristic time processes are the most important with respect to chemical fate. The following simple example demonstrates the use of fugacity, Z values, D values, and characteristic times when interpreting the results of mass balances.

3.6 ILLUSTRATION OF A SIMPLE FUGACITY MASS BALANCE MODEL

To illustrate the use of Z and D values in a simple multimedia model, we present below a steady-state mass balance for an air–water–sediment system representing a small lake with inflow and outflow. It is an application of the quantitative water air sediment interaction (QWASI) model that is available from the Web site www.trentu.ca/cemc. The chemical is similar in properties to a volatile hydrocarbon such as benzene. Table 3.3 lists the lake properties, the chemical input rates in the inflowing water, its properties as partition coefficients, Z values and D values for all the transport and reaction rates.

The steady-state input–output mass balance equations are simple and transparent, namely

For the water compartment: $I + A + D_{SW} \cdot f_S = f_W(D_O + D_V + D_{RW} + D_{WS})$,
For the sediment compartment: $D_{WS} \cdot f_W = f_S(D_{RS} + D_{SW} + D_B)$.

Here the input rates are I by inflow and A by atmospheric deposition. Solving the two equations simultaneously for the chemical fugacity in water f_W, then substituting the numbers yields a value of 0.1 Pa. In the sediment f_S is 0.05 Pa. The mass balance diagram (Figure 3.1) shows the rates at steady state and the D values. The concentrations are then fZ or 0.001 mol m^{-3} in the water and 0.05 mol m^{-3} in the sediment with corresponding amounts of 10^4 and 2000 mol in the respective compartments.

The relative importance of the various chemical rates with regard to the chemical fate in the aquatic system can be inferred from the magnitudes of the D values. In this case, the water to sediment transport is relatively unimportant since water outflow (50%), degradation in water (25%), and volatilization (15%) account for 90% of the loss from the water compartment. The sediment fugacity is half that in the water, largely because of the relatively high rate of degradation in the sediment. Complete

TABLE 3.3
Selected Input Data for the Mass Balance Example

Water volume	10^7 m^3
Sediment volume	4×10^4 m^3
Water outflow rate	10^4 m^3 h^{-1}
Sediment burial rate	5 m^3 h^{-1}
Z Values (mol m^{-3} Pa^{-1})	
Air	4×10^{-4}
Water	0.01
Sediment	1.0
Chemical inflow rate I	15 mol h^{-1}
Chemical atmospheric deposition rate A	4 mol h^{-1}
D Values (mol Pa^{-1} h^{-1})	
Water outflow, D_O	100
Volatilization, D_V	30
Water-to-sediment transport, D_{WS}	20
Sediment-to-water transport, D_{SW}	20
Sediment burial, D_B	5
Degradation in water, D_{RW}	50
Degradation in sediment, D_{RS}	15

mass balances are thus established for water and sediment as shown in Figure 3.1 and the rates of chemical movement can be examined and checked for consistency.

The characteristic time of the chemical in water can be calculated as $VZ/\Sigma D$ or from $VZ f_W / f_W \Sigma D$ as 10,000/20 h or 500 h or 21 days where the amount in the water is 10,000 mol and the total rate of loss is 20 mol h^{-1}. For the sediment, the characteristic

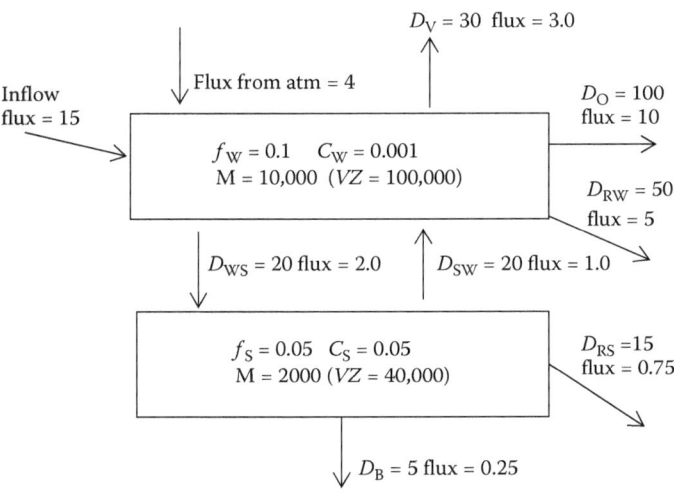

FIGURE 3.1 Illustrative mass balance diagram showing the steady-state fluxes (mol h^{-1}), fugacities (Pa), concentrations (mol m^{-3}), amounts (mol), and VZ products (mol Pa^{-1}).

time is 1000 h or 42 days. For the system as a whole, it is 631 h or 26 days. These times can be interpreted as the approximate times for each compartment to build up in concentration 63% towards steady state or to fall to 37% of the steady-state value if emissions ceased. The times for each loss process to deplete the quantity in the water or sediment, if acting alone, can be calculated as VZ/D. For example, for outflow from water, this is 1000 h and for volatilization it is 3333 h.

The dynamic or unsteady-state differential equations can be readily derived and solved either analytically or numerically. The solution gives the time response of the concentrations, amounts, fugacities, and fluxes as a function of the initial condition and the varying input rates. As the number of compartments increase and the mass balance equations become more numerous and lengthy, the simplicity and transparency of these equations becomes increasingly advantageous and interpretation of the results is facilitated. The relative equilibrium status of compartments is obtainable directly from the fugacities. Often, compartment concentrations differ by orders of magnitude (a factor of 50 in the illustration) but the fugacities are closer in magnitude (a factor of 2 in the illustration) thus the "reasonableness" of the fugacities is more readily checked and errors arising from misuse of unit conversions can be identified more easily. Phases that are in intimate contact, such as water and suspended particles or air and aerosol particles, have short characteristic times for approaching equilibrium or equifugacity and the resulting fugacities are approximately equal. Indeed it is often assumed that equilibrium applies, in which case the phases can be "lumped" into a single bulk compartment and there is no need to obtain the interphase D values or mass transfer coefficients.

For substances that do not partition into a phase (e.g., ionic species into air), the Z value is zero and a "division by zero" issue can arise when solving the mass balance equations. This can be circumvented by using aquivalence (essentially f/H) as the equilibrium criterion or activity (concentration in water/solubility in water or equivalently fugacity/vapor pressure). Indeed, when examining fugacities, it is desirable to calculate the activity to ensure that subsaturation conditions prevail, that is, all fugacities are less than the liquid or subcooled liquid vapor pressure.

LITERATURE CITED

Mackay, D. 2001. *Multimedia Environmental Models: The Fugacity Approach*, 2nd ed. Lewis Publishers, Boca Raton, FL, pp. 1–261.

4 Flux Equations for Mass Transport Processes across Interfaces

Louis J. Thibodeaux

CONTENTS

4.1 Introduction 51
4.2 An Overview of the Individual Processes 52
4.3 Flux Equations for Individual Processes 54
    4.3.1 Diffusive-Type Fluxes 54
    4.3.2 Advective-Type Fluxes 58
    4.3.3 Closure 59
4.4 Environmental Models and Interface Fluxes 60
    4.4.1 The "Interface Compartment" Concept 60
    4.4.2 Differential Equation Model Interface Fluxes 61
    4.4.3 Compartment Box Model Diffusive-Type Fluxes 63
    4.4.4 Compartmental Box Models: Advective Plus Diffusive Transport
        Processes 65
    4.4.5 Closure 67
4.5 A Step-by-Step Procedure for Obtaining MTCs and Flux Equations 67
Acknowledgment 69
Literature Cited 69
Additional Reading Material 69

4.1 INTRODUCTION

The subject of environmental mass transport is a broad and complex. Many individual chemodynamic processes occur both within the natural media compartments and across the natural interfaces and they impact the magnitude of chemical movement rates. With several transport processes occurring simultaneously in series and in parallel, certain procedures need to be followed for their proper accounting of all. The objective of this chapter is to make the user generally aware of the most significant individual processes and the necessary procedures for combining these to obtain the overall chemical mass flux. The subject is about the fundamentals of mass transport, the rate equations, and the transport coefficients for arriving at chemical flux at and across the interfaces that separate the natural media compartments. It uses a chemical flux framework to introduce the types of individual processes, the types of mass transfer coefficients, and procedures for combining them. It also provides details for incorporating MTCs and flux expressions into the two most common environmental mathematical models. To this end, applications for differential equation models as well as compartmental box models will be demonstrated.

Following this introduction, an overview of all the individual processes will be given in a tabular listing. This will be followed by a section giving and describing the various types of theoretical flux expressions. This in turn will be followed by a section on the "interface compartment"; it is a theoretical concept that uses chemical flux continuity across interfaces to connect the adjoining media or compartments. It will be developed and applied to several realistic environmental chemodynamic scenarios for the user. Finally a generalized step-by-step procedure will be given. It serves as a guide the user for arriving at the appropriate flux expression and MTCs for quantifying chemical transport rates across environmental interfaces or between environmental compartments.

4.2 AN OVERVIEW OF THE INDIVIDUAL PROCESSES

A total of 41 individual processes have been identified as being most significant in order to realistically describe the mass transfer chemodynamics in natural systems. The list appears in Table 4.1 and all are covered in Chapters 5 through 20 where detailed descriptions and data are presented. The processes are divided into four broad categories: (1) the air–water interface, (2) the water–sediment interface, (3) the soil–air interface, and (4) intramedia processes. Water, as used in Table 4.1, includes ice and snow. Some are well known and have been studied in detail while others have not. Specifically these include advection and dispersion processes within the atmosphere, water bodies, and the groundwater associated with soils and sediments. Although plants and the built urban environment are "separate phases" they are grouped with the air phase.

The subject of chemical fate and transport in groundwaters within deep and varied subterranean geologic formations is not covered in this handbook. However, dispersion and mass-transport processes within near-surface (approximately 0 to 3 m) porous media, is covered in Chapter 15. Figure 4.1 illustrates many of the individual processes listed in Table 4.1 in the general context of their locales.

As an aid to the user, Table 4.1 serves as a comprehensive check-off list choosing suspected-to-be significant MT processes. In addition, those thought-to-be insignificant should be evaluated as well. The model applications include site-specific situations as well as the local, regional, continental, and global scale multicompartmental models. The user is directed to the HDBK table of contents for location of the chapter and section where detailed descriptions of particular MT processes and data are presented.

TABLE 4.1
A List of Individual Mass Transport Processes

1. Air–Water Interface Chemical Exchange Processes

Air-Side Processes

AW1 Dry removal deposition velocity for gases and vapors from the air
AW2 Dry removal deposition velocity for particles with associated sorbed chemicals from air
AW3 Wet removal velocity of gases and vapors from air
AW4 Wet removal velocity of particles and associated sorbed chemicals from air
AW5 Gas and vapor phase MTC from chemicals moving from water interface to air
AW6 Transport of chemicals in air moving to the surface of snow and ice

Water-Side Processes

AW7 Aqueous phase MTC for solutes and particles
AW8 Transport of chemicals within ice and snow packs

2. Water–Sediment Interface Chemical Exchange Processes

Water-Side Processes

WS1 Solute transport MTC on the water-side of interface
WS2 Particle re-suspension velocity from the bed surface
WS3 Particle deposition velocity onto the bed surface

Sediment-Side Processes

WS4 Groundwater velocity in bi-directional seepage across the sediment bed
WS5 Surface bed-form induced in-bed pore-water velocities (i.e., hyporheic flows)
WS6 Solute molecular diffusion coefficient in the pore-waters of the bed
WS7 Brownian diffusion coefficient for aqua sols (i.e., DOC, colloids, etc.) in the bed pore-waters
WS8 Bioturbation driven bio-diffusion coefficient for particle and pore-water transport within the bed
WS9 Particle settling and deposition into the deep ocean

3. Soil–Air Interface Chemical Exchange Processes

Air-Side Processes w/o Canopy

SA1 Dry deposition velocity for gases and chemical vapors from air-to-the soil and ground surfaces
SA2 Dry deposition velocity for particles and associated sorbed chemicals from air to earthen surfaces
SA3 Wet deposition velocity for gases and chemical vapors from air to earthen surfaces
SA4 Wet deposition velocity for particles and associated sorbed chemicals from air to earthen surfaces
SA5 Air-side MTC for gases and vapor from soil and ground surfaces into the lower atmospheric
  boundary layer
SA6 Dust re-suspension from earthen surfaces

Air-Side Canopy Processes

SA7 Dry deposition velocities and MTCs for gases and vapors moving between plant canopy surfaces
  and the lower atmospheric boundary layer
SA8 Dry deposition velocities for particles and associated chemicals from air to the plant canopy
  surfaces
SA9 Air-side MTC for gases and vapors from plant canopy to sub-canopy air
SA10 Air-side MTC for gases and vapors to surface within the built urban canopy

continued

TABLE 4.1   (continued)
A List of Individual Mass Transport Processes

Soil-Side Processes

SA11   Molecular transport coefficients of gases and vapors in the soil pore–air
SA12   Molecular transport coefficients for solutes in the soil pore–water
SA13   Bioturbation driven biodiffusion coefficient for sorbed-phase chemicals in upper soil layers
SA14   Mass transfer from air–plant interface into and within plants
SA15   Mass-transfer from soil interface into and within plants
SA16   Advective transport of gases and vapor in soil pore–air
SA17   Advective transport of solutes in soil pore-water

4. Intraphase Chemical Exchange Processes

Dispersion in the Fluid Media

IT1   The lower atmospheric boundary layer with diel characteristics (thermal density stratification)
IT2   The stratified aquatic system (thermal and salinity stratification)
IT3   The surface waters of lakes, estuaries, and oceans
IT4   Flowing streams
IT5   The lower atmospheric boundary layer

Transport in Porous Media

IT6 Dispersivity in soils and the up-most groundwaters of geologic formations
IT7 MTCs in pore-waters adjacent to particles and non-aqueous liquids

4.3   FLUX EQUATIONS FOR INDIVIDUAL PROCESSES

4.3.1   Diffusive-Type Fluxes

This subject of chemical flux appears in Section 2.3. Equation 2.3 is the starting point for addressing the various types of flux relationships needed for environmental chemodynamic mass transfer applications. This subsection will address the diffusive-type fluxes; it is the first group on the right side of Equation 2.3. Within this category, there are two subtypes: the concentration gradient and the concentration difference. Due to the large number of mass transport processes, the number of flux equations is large as well. As an introduction a few specific processes listed in Table 4.1 will be used. For the diffusive-type fluxes, those applicable at the water–sediment (WS) interface will be used.

Process WS6 is for soluble chemical diffusion in bed-sediment porewater. The concentration gradient flux n, in mg/m$^2 \cdot$ s, is

$$n = -D_s \left. \frac{dC_{wi}}{dz} \right|_{z=0} \tag{4.1}$$

where D_s, in m^2/s, is the effective diffusion coefficient in the sediment water filled pore-spaces. The coefficient is a scalar quantity with no implied direction of chemical movement. The dC_{wi}/dz, in mg/m^4, term is the concentration gradient at the sediment

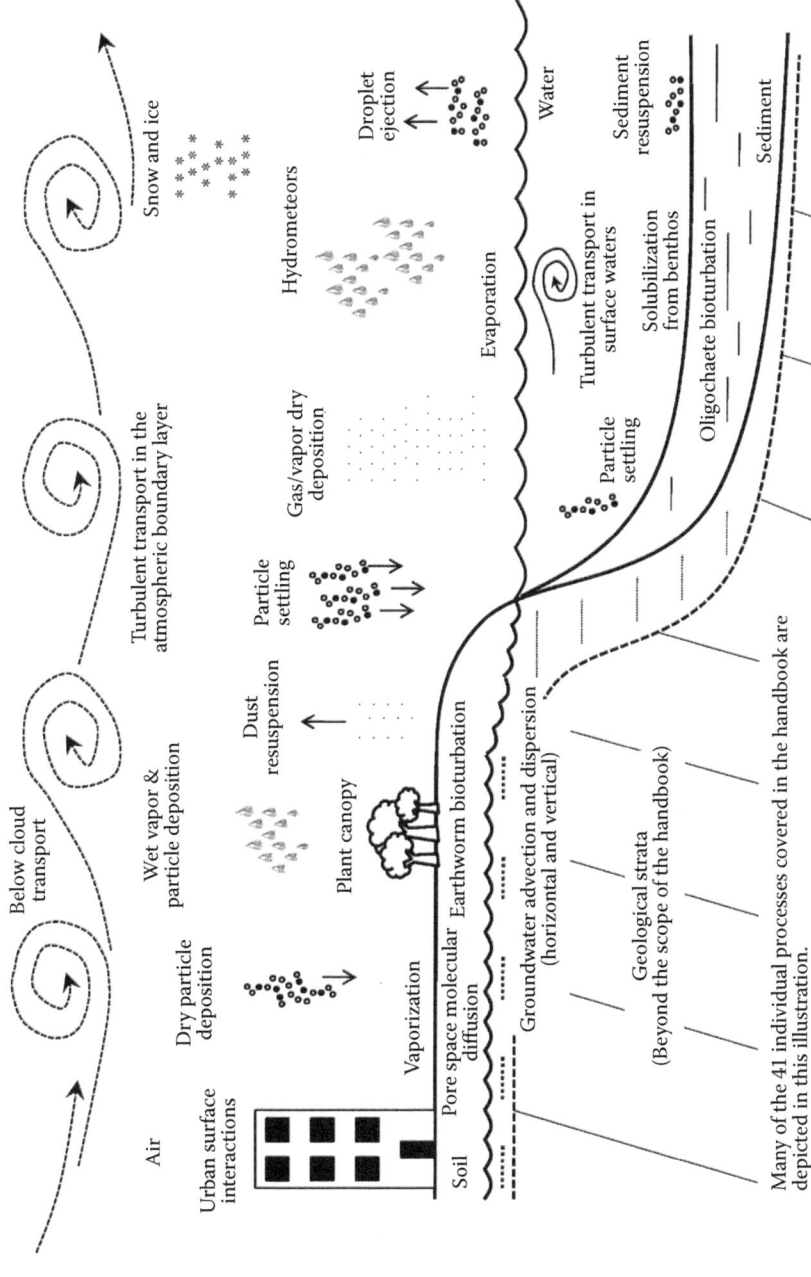

FIGURE 4.1 An illustration of several individual mass-transfer processes.

water interface (i) located at position $z = 0$; it is a vector quantity directed upward away from interface in the positive z-direction.

The applicable mathematical form of the flux equation is model dependent. For example, the first type model consists of differential equations (DEs). They are developed to yield concentration profiles in the sediment layers as well as the flux. These DEs typically use Equation 4.1 as a boundary condition. The solutions to these DEs require one or more of the following boundary condition categories: the Dirichlet condition, the Neuman condition, or "a third condition." The first two types are the most common; these require mathematical functions containing gradients of the dependent variable (i.e., C_w) as well as functions of the dependent variable itself. For these diffusive-type fluxes, the transport parameter is a diffusion coefficient such as D_s. Several other transport parameters are commonly used and represent diffusion in air and the biodiffusion or bioturbation of soil/sediment particles.

In the case of environmental compartmental box models, the second type model, a concentration difference equation is used. The procedure for obtaining the flux equations is to integrate the concentration over some specified distance, Δz in meters, adjacent to the media interface that represents a real or a characteristic boundary-layer thickness. When the integration process is complete, a concentration difference, Δc in mg/m^3, results. The Δz may be explicitly known as a measurable depth, h in metres, such as associated with the sediment bioturbation depth, for example. Or it may represent a hypothetical stagnant fluid layer thickness. In the latter case, characterizing the length may be too problematic by actual measurement, theoretical approximation, or otherwise. As a result, the practice is to incorporate h within the diffusive-type transport parameter and simply create a new quantity called the individual mass-transfer coefficient (MTC). These coefficients are commonly given lower case k notation with velocity (m/s) units. The generality of the k-notation has grown and is now used to replace the ratios $D^{(1)}/h$ and $D^{(t)}/h$ as well as others; all are referred to as MTCs. As an example, returning to Process WS6 introduced above and performing the integration as described above yields

$$n = \frac{D_s}{h}(C_{wh} - C_{wi}) \qquad (4.2)$$

where, h is measured in meters. In this case, the MTC is the transport parameter represented by a porous sediment-side surface layer through which chemical diffusion is occurring. Its length is usually apparent from concentrations profile data. C_{wh} is the porewater concentration at $z = h$. The concentration difference in parenthesis when divided by h is in effect the average concentration gradient and regulates the flux magnitude and gives it direction.

Next, process WS1 soluble chemical transport on the water side of the interface will now be covered. This is an integrated product as well but the diffusive film thickness of water is unknown. The concentration difference flux equation is

$$n = k'_w(C_{wi} - C_{w\infty}) \qquad (4.3)$$

where k'_w in m/s is the water-side MTC; it is a scalar quantity. The $C_{w\infty}$ is the concentration in the water column far removed (i.e., $z = \infty$) from the interface. Additional

TABLE 4.2

Compatible Transfer Rate Constants (m/s) and Media Concentrations (kg/m³)

	Media Concentrations		
Interface and transport media	C_a	C_w	$C_s \rho_s$
A/W, water side solute	$k_w/(H_c)$	k_w	—
A/W, water side vapor	k_a	$k_a(H_c)$	—
A/So, air side vapor	k_a	—	$k_a/(K_{sa})$
A/So, soil side vapor	D_a/h	—	$D_a/h(K_{sa})$
A/So, soil side particle bioturbation	$(\rho_s K_{sa})D_b/h$	—	D_b/h
Sd/W, water side solute	—	k'_w	$k'_w/(K_d)$
Sd/W, sediment side solute	—	D_s/h	$D_s/h(K_d)$
Sd/W, sediment side particle bioturbation	—	$(\rho_s K_d)D_b/h$	D_b/h
Sd/W, water side particle resuspension	—	$(\rho_s K_d)v_r$	v_r
A/So, soil side particle resuspension	$(\rho_s K_{sa})v_r$	—	v_r
A/W, air side hydrometeor deposition	$v_w/(H_c)$	v_w	—
A/W, air side particle deposition	$(\rho_s K_{sa})v_d$	—	—
Sd/W, water side particle setting	—	$(\rho_s K_d)v_s$	v_d
A/So, soil side solute	—	D_s/h	$D_s/h(K_d)$
Sd/W, sediment side colloids in porewater	—	$(C_{sw}K_c)D_s/h$	$(C_{sw})D_s/h$

individual process MTCs much like Equations 4.2 and 4.3 are presented in Table 4.2 as entries under the C_w column.

Of the 41 listed in Table 4.1 the 16 most common mass transport processes representing the air, water, and soil and sediment media appear in Table 4.2. The media of prime concern often dictate the most convenient phase concentration used in the flux equation. For example, water quality models usually have C_w as the state variable and therefore the flux expression must have the appropriate MTC group based on C_w and these appear in the center column of Table 4.2. Aquatic bed sediment models usually have C_s, the chemical loading on the bed solids, as the state variable. The MTC groups in the right column are used. All the MTC groups in Table 4.2 contain a basic transport parameter that reflects molecule, element, or particle mobility. Both diffusive and advective types appear in the table. These are termed the individual phase MTCs with SI units of m/s. Examples of each type in Table 4.2 include k'_w for water solute transport and v_s for sediment particle deposition (i.e., setting).

In addition, many of the MTC groups in the table also contain equilibrium-type, chemical partition coefficients as modifiers. These modifiers are not kinetic parameters but they can and do significantly amplify or attenuate the magnitude of the basic transport parameter. For example, a water quality model that uses C_a in the flux expression needs the Henry's law constant multiplier, H_c, as the appropriate modifier applied to k_w as is shown in column one of Table 4.2. Therefore, this table is an aid to users formulating the flux equations for multimedia models. The primary focus of this handbook is onto estimating the various basic transport parameters listed. These include k_w, k_a, D_a, D_b, k'_w, v_r, D_s, v_s, and so on. The uncertainty in the numerical estimate of these basic transport parameters are a key factor as well. It becomes a particularly important issue because the base coefficient uncertainties are amplified or attenuated by the numerical magnitudes of the chemical-specific partition coefficients.

Estimating the uncertainties in the basic transport parameters is presently very problematic. Nevertheless, it is addressed to some degree in the respective chapters. Obtaining estimates of the partition coefficients and the associated uncertainties is beyond the scope of this book; see Boethling and Mackay (2000). See Chapter 1, Section 1.5.2 for a brief summary of general phase partition coefficients. However, some key partition coefficient relations have appeared in the literature recently and these have been provided in the appropriate chapters. Specifically see Chapter 6, Section 6.2.2 on gas-to-particle partitioning Chapter 14, Section 14.3 on plants-to-water partitioning; and Chapter 18, Section 18.2 on snow/ice-to-air partitioning.

4.3.2 ADVECTIVE-TYPE FLUXES

Advective-type fluxes will be discussed in this section. These are in the middle group of Equation 2.2 in Section 2.3. The equations used as examples will be for some of the processes at the air–water (AW2) and soil–air (SA2) interfaces. An advective-type chemical flux is

$$n = v_d \rho_p C_s \tag{4.4}$$

where v_d, in $m^3/m^2 \cdot s$, is the particle deposition flux (i.e., m/s); it is a vector directed from air-to-the-soil surface. The ρ_p, in kg/m^3, is the particle dry mass per air volume and C_s, in mg/kg, is the chemical loading on the particle mass. The mechanism of chemical transport is not a concentration gradient–driven diffusion process. It is driven by the movement of the bulk phase; in this example, it is the solid particles moving downward (i.e., settling in air) through the atmospheric boundary layer.

Process AW3 and SA3-wet deposition removal of vapor phase chemicals from air to water or soil follows. The flux equation is:

$$n = v_w C_w \tag{4.5}$$

where v_w, in $m^3/m^2 \cdot s$, is the vector component is the water (i.e., hydrometeor) flux or precipitation rate in m/s. C_w, in mg/m^3, is the chemical concentration in the falling hydrometers. The individual process transport parameters in Equations 4.4 and 4.5 are recorded in Table 4.2 as entries under the $C_s \rho_s$ and C_w columns, respectively.

The overall advective flux equation combining dry and wet deposition is defined using the vapor concentration in the air, C_a, in mg/m^3. This air concentration measurement is typically performed to assess the chemical mass source strength in the atmosphere. The equation is:

$$n = v C_a \tag{4.6}$$

where v, in m/s, is the total combined deposition velocity. Using the particle-to-air chemical partition coefficient K_{pa}, in L/kg, and the Henrys' constant H_c, in m^3/m^3, the total is

$$v = v_d \rho_p K_{pa} + v_w/H_c \tag{4.7}$$

To obtain this result, the individual conductances are summed since both deposition processes occur in parallel. Using the phase partition coefficients K_{pa} and H_c assumes the concentrations on the particles and the water droplets are in chemical equilibrium with the surrounding air and this state has been achieved prior to arriving at the water surface.

Processes SA2 and SA3 are the air-to-soil advective deposition processes identical to the air-to-water processes presented above. Process SA6-dust resuspension from the soil can be included as well, its flux equation is

$$n = v_r \rho_s C_s \tag{4.8}$$

where v_r, in m/s, directed upward, is the dust resuspension velocity, ρ_s, in kg/m^3, is dry surface dust particle density and C_s, in mg/kg, is the chemical loading on the surface soil dust. In this case, the overall air-to-soil deposition velocity represents the three mechanisms of particle deposition, rain deposition, and particle resuspension:

$$v = v_d \rho_p K_{pa} + v_w/H_c - v_r \rho_s K_{sa} \tag{4.9}$$

The negative sign indicates that the resuspension velocity is in the opposite direction of both deposition velocities. The dust-to-air partition coefficient is similar to that for the airborne solid particles. The dust resuspension rate parameter defined by Equation 4.8 appears in Table 4.2 in the $\rho_s C_s$ column. All three of the advective rate parameters that were presented in this section were converted to their equivalent air concentration forms by using the appropriate equilibrium or phase partition coefficient. The final result appears in Equation 4.9. These converted forms are also presented individually in Table 4.2 and appear under the C_a column.

4.3.3 CLOSURE

It is obvious to the user at this juncture that the subject of environmental chemical fate models enjoys many individual mass transfer processes. Besides this, the flux equations used for the various individual processes are often based on different concentrations such as C_a, C_w, C_s, and so on. Since concentration is a state variable in all EC models, the transport coefficients and concentrations must be compatible. Several concentrations are used because the easily measured ones are the logical mass-action rate drivers for these first-order kinetic mechanisms. Unfortunately, the result is a diverse set of flux equations containing various mechanism-oriented rate parameters and three or more media concentrations. Complications arise because the individual process parameters are based on a specific concentration or concentration difference. As argued in Chapter 3, the fugacity approach is much simpler. Conversions to an alternative but equivalent media chemical concentration are performed using the appropriate thermodynamic equilibrium statement or equivalent phase partition coefficients. The process was demonstrated above in obtaining the overall deposition velocity; Equation 4.9. In this regard, the key purpose of Table 4.2 is to provide the user with the appropriate transport rate constant compatible with the concentration chosen to express the flux. For each interface, there are two choices of concentration and either is appropriate for use. The table serves as an aid for converting transport rate constants based on one concentration to another. For example, $C_a k_w/(H_c)$ is equal to $C_w k_w$.

4.4 ENVIRONMENTAL MODELS AND INTERFACE FLUXES

The subject of mathematical models and the appropriate flux equations was introduced in the previous section. Although a few examples were given, a general approach for accommodating individual flux equations of various types into the overall accounting for the combined processes was not explicitly developed. The objective of this section is to introduce and develop the "interface compartment" concept as a consistent approach of combined the various flux expressions as they apply to differential equations, and compartmental-box models. As was done in the previous section the development will include the application of the concept to several realistic interface chemical transport scenarios.

4.4.1 THE "INTERFACE COMPARTMENT" CONCEPT

In order to be realistic, environmental models must contain multiple phases, and are often referred to as "multiphasic," or multimedia compartment models. Being so requires interfaces that separate the phases and media. These interfaces can be real or idealized (i.e., imaginary). Two-dimensional interface planes are assumed to exist between the air–water, water–sediment, and soil–air phases or media. In reality, chemical transport across the air–water interphase plane involves a true phase change. The watery interface plane at the water-bed sediment junction and airy interface plane at the soil–air junction are only separated by imaginary planer surfaces. Nevertheless, due to the dramatic changes that typically occur within the fluid and the associated media fluid dynamics on either interface side, different transport processes occur on opposite sides, typically. Therefore it is also practical to define an interfacial compartment for such imaginary and idealized interface situations. The interfacial compartment concept for multimedia, interphase chemical transport is based on the following ideas:

a. A two-dimensional surface (i.e., interface plane) containing no mass separates the compartments.
b. The chemical flux direction is perpendicular to the surface plane.
c. The net entering and departing fluxes are equal therefore steady state is implied.
d. Hypothetical concentrations are assumed to exist in the interface plane.
e. If the adjoining media are separate phases chemical equilibrium is assumed to exist and if not the concentrations on either side of the plane are equal.

In other words, if the adjoining media compartments consist of separate phases (i.e., air–water), the concentrations in the adjoining media phases on either side of the interface are assumed to be in thermodynamic equilibrium. If they are not separate phases, such as the atmosphere against a porous surface soil planar interface where the air pathway is continuous between compartments, then the interface concentrations in the air and pore–air on either side of the interface are equal. The same idea applies between water and the adjoining porous sediment surface.

The interface compartment is not a media compartment in the normal sense of the word; it contains no mass since it is a two-dimensional surface. Through this surface, the chemical species is moving and in a direction perpendicular to the plane area. Without mass, the interface plane cannot accumulate materials so a transient condition in the compartment cannot occur, therefore the chemical flux through the compartment must be at steady-state. However, chemical concentrations are assumed to exist with their normal mass per volume units such as mg/m^3. Obviously this concentration is a hypothetical term; it is not a measurable quantity. Nevertheless, as will be demonstrated, the interface concentration is the parameter needed for connecting the adjoining media compartments.

The essence of the interface compartment concept has existed in the chemical engineering literature for decades (see Cussler, 1997; Bird et al., 2002). The idea originated and evolved through the efforts of Nernst, Whitman, and Lewis over the 20-year time-period 1904–1924. One current chemical engineering application is for gas-to-liquid mass transfer where it appears as a step in the derivation of the overall MTC, relating it to the individual phase MTCs (resistance-in-series concept). Another use is estimating concentrations at the gas–liquid interface.

The presentation given in this section generalizes the earlier concept, which has been used solely for the interphase molecular diffusions process. The term "interface compartment" is the name applied. It broadens the concept and thus applies to all transport mechanisms occurring at or across adjoining multimedia compartments and is particularly useful in linking both diffusive-type and advective-type flux expressions. The key result is a simple algebraic relationship for the interface compartment concentration, which in turn yields a single flux equation containing the individual MTCs and the bulk chemical concentrations in the adjoining media compartments. When the media compartments are adjoining true phases, such as gas–liquid, and involve only diffusive-type as a special case, the concept yields the usual relationship for the overall MTC. The following two sections will demonstrate the utility of the "interface compartment" by using several of the transport processes listed in Table 4.1.

4.4.2 DIFFERENTIAL EQUATION MODEL INTERFACE FLUXES

Scenario 1. Chemical solubilization across the WS interface. Fate models consisting of one or more differential equation require appropriate boundary conditions that will account for the mass-transfer resistance occurring on both sides of the interface. This MT situation was covered in the previous section under diffusive-type individual process where the flux equations appear; see Equations 4.1 and 4.2. Figure 4.2 depicts the interface situation for this mass-transfer scenario. The positive z-direction is from the sediment into the water column. S, I, and W denote positions of defined water concentrations in the sediment bed (S), the interface (I), and water (W), respectively. A tortuous dashed line tipped with an arrowhead traces the chemical flux during its molecular diffusion process pathway from within the porous bed, to interface and through the benthic boundary layer finally arriving in the turbulent water column.

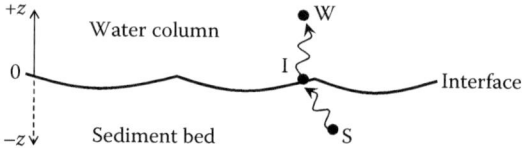

FIGURE 4.2 Chemical solubilization across the sediment–water interface.

A pore–water concentration gradient is assumed to exist between the bed and the interface layer; it is not shown in the figure. Processes WS1 and WS6 listed in Table 4.1 are depicted in this chemical transport scenario.

Applying the interface compartment concept is a two-step procedure. First, write a flux balance to sum the rate of all processes moving chemical through the interface. The approach is to write a balance of those entering being equal to those departing. In this case, it is

$$n = -D_s \frac{dC_{wi}}{dz} = k'_s[C_{wi} - C_{w\infty}] \qquad (4.10)$$

where C_{wi}, in mg/m^3, is the interface porewater chemical concentration. The rate of chemical movement departs from the sediment side of the interface; concentration in the porewater decreases toward the interface and the negative sign counters the negative concentration gradient yielding a positive departing flux. On the water-side of the interface, the chemical rate is away from the interface plane into the water column as driven by the concentration difference shown in the brackets, in mg/m^3. The second step in the procedure is to apply the equilibrium condition at the interface. Since the water phase is continuous through the interface plane a continuous smooth curve represents the pore-water concentration. So, the chemical concentration is equal whether the plane is approached from either the $+z$ direction or the $-z$ direction.

The connection needed relating concentrations in the adjoining phases is Equation 4.10. It provides the so-called third type BC since it contains both the interface concentration and the concentration gradient at $z = 0$. With known values of D_s, k'_s, and $C_{w\infty}$, it is the mathematical result typically needed for this type of mathematical model.

A realistic extension to Scenario 1 for riverine as well as some estuarine environments is to include the process of particle re-suspension; see process WS2 in Table 4.1. The solid particles originating from the bed have interface concentration loadings of C_{si}, in mg/kg. Re-suspension is expressed as an advective-type flux for the chemical departing the interface and entering the water column. Equation 4.10 reconstituted now appears as

$$n = -D_s \frac{dC_{wi}}{dz} = k'_s(C_{wi} - C_{w\infty}) + v_r(\rho_s K_d)C_{wi} \qquad (4.11)$$

where the use of the sediment-to-water equilibrium partition coefficient K_d, in m^3/kg, converts the solid particle chemical loading ratio to an equivalent concentrations in water. All three rate parameters in Equation 4.11 appear in Table 4.2. The purpose of this extended version of scenario 1 was to demonstrate two parallel processes on the water-side of the interface and the application of the equilibrium condition at the interface.

4.4.3 COMPARTMENT BOX MODEL DIFFUSIVE-TYPE FLUXES

The diffusive gradient is integrated over a specified film or layer thickness to accommodate the compartmental box models; the resulting flux equations contain a concentration difference (see Section 4.3). What follows are some chemodynamic mass transfer scenarios using concentration difference flux equations for demonstrating the interface compartment concept used with compartmental-box models.

Scenario 2. Chemical volatilization from water follows. Processes AW1 and AW7 are listed in Table 4.1. The chemical pathway is from solution in water with turbulent transport to the interface where vaporization occurs followed by turbulent transport away and into the bulk air beyond. The first step in applying the AW interface compartment concept is to sum the entering and departing fluxes. This is

$$n = k_w(C_w - C_{wi}) = k_a(C_{ai} - C_a) \qquad (4.12)$$

where k_w and k_a, in m/s, are the individual MTCs for water and air, respectively, taken from Table 4.2. The second step in the procedure is to apply the phase chemical equilibrium at the interface. In this case Henrys' law applies:

$$C_{ai} = H_c C_{wi} \qquad (4.13)$$

A simultaneous solution of Equations 4.12 and 4.13 will yield for the concentration on the water-side of the interface:

$$C_{wi} = (C_w k_w + C_a k_a)/(H_c k_a + k_w) \qquad (4.14)$$

Using this result and Equation 4.13 yields the concentration on the air side of the interface. Knowing either interface concentration allows the flux n, in $mg/m^2 \cdot s$, to be obtained from Equation 4.12. With this approach, Equation 4.14 provides a means for connecting the adjoining compartments mathematically by specifying the hypothetical interface concentrations.

An alternative means of connecting adjoining compartments is to use a flux equation containing an overall MTC. This approach commences by defining the overall flux equation. This scenario is the classical chemical engineering application for chemical transfer across a gas–liquid interface (Bird et al., 2002); it is followed here. The flux equation is defined by the following equation:

$$n = K_w[C_w - C_a/H_c] \qquad (4.15)$$

where K_w, in m/s, is the overall MTC. Substituting Equation 4.14 into 4.12 and doing the algebra eliminating C_{wi} yields an expression for the K_w; it is

$$1/K_w = 1/k_w + 1/k_a H_c \qquad (4.16)$$

This result is similar to Ohm's law and is commonly referred to as the resistance-in-series law in mass transfer. Each resistance is the inverse of a conductance. The individual MTCs on the right side of Equation 4.16 appear in the center column under C_w in Table 4.2. Use of the two equations provides a means for obtaining the chemical flux across the interface plane. In doing so, it provides an alternative mathematical means of connecting the adjoining phase without explicitly obtaining the interface concentrations.

The above alternative scenario development for liquid-to-gas phase chemical transport has a long history of successful application in the area of chemical separation in the chemical process industries (Bird et al., 2002). It gives support to the assumptions behind the interface compartment concept. However, the resistance-in-series algorithm shown in Equation 4.16 will work correctly only if the flux equations are of the diffusive concentration-difference-type.

Scenario 3. Chemical vapor transport from air into the soil column follows. The transport pathways are depicted in Figure 4.3. Vapor phase transport from position (A) to position (I) occurs by process SA1 in Table 4.1. From position (I) to position (S), the transport is by two parallel processes. One process SA11 and the other is SA13; the former is vapor diffusion in air-filled pore spaces in the soil column and the latter is bioturbation of soil particles containing quantities of sorbed chemicals by earthworms primarily. The latter two processes are depicted in Figure 4.3. The soil column transport depth is h in meters.

Application of the interface compartment concept follows. The fluxes of chemical to and from the interface by the respective rate equations are as follows:

$$n = k_a(C_a - C_{ai}) = \frac{D_a}{h}(C_{ai} - C_{ah}) + \frac{D_b \rho_s}{h}(C_{si} - C_{sh}) \qquad (4.17)$$

where D_a and D_b in units m²/s, are the air and biodiffusion coefficients respectively. The C_{ah} is pore air concentration at depth h, mg/m³, and C_{sh} is soil particle concentration, mg/kg. The latter reflects a solid particle diffusion process. The air phase is continuous through the interface connecting adjoining compartments however, an equilibrium relationship between air and soil solid exists at the compartment interface; it is

$$C_{si} = K_{sa}C_{ai} \qquad (4.18)$$

where K_{sa}, in L/kg, is the soil-to-air partition coefficient. At this point, it is possible to solve Equations 4.17 and 4.18 simultaneously for C_{ai} in terms of C_a and C_{sh}, the

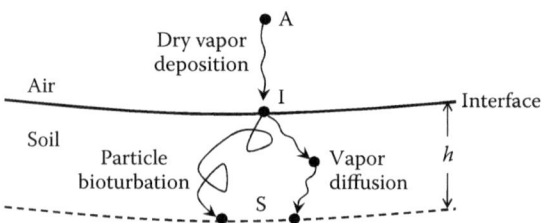

FIGURE 4.3　Chemical transport across the soil–air interface.

resulting relationship is linear and contains all the transport parameters involved. The overall flux expression is defined as

$$n = K_a(C_a - C_{sh}/\rho_s K_{sa})$$ (4.19)

where K_a, in m/s, is the overall MTC. A simultaneous solution of Equations 4.17 through 4.19 yields for K_a:

$$\frac{1}{K_a} = \frac{1}{k_a} + \frac{1}{\left[\dfrac{D_a}{h} + (\rho_s K_{sa})\dfrac{D_b}{h}\right]}$$ (4.20)

This is a classical application of the Ohm's law, the conductance-in-parallel rule is used for obtaining the overall conductance for the two, in-parallel, soil-side processes and the resistance-in-series rule is used for combining the air-side with the overall soil-side. The various individual MTCs appearing in Equation 4.20 are listed under column C_a in Table 4.2.

There are obvious advantages to being able to obtain simple relationships for the overall MTC, such as Equations 4.16 and 4.20, as demonstrated above in Scenarios 2 and 3. The equations are simple and formulating the overall resistance created by selecting the combination of in-series and in-parallel individual process parameters from Table 4.2 is a straightforward procedure. However, as will be demonstrated in the next scenario the procedure for obtaining an overall MTC using the classical rules is not universally applicable to all situations, particularly to those interface transport situations involving a complex set of processes requiring the use of a combination of diffusive and advective flux relationships.

4.4.4 COMPARTMENTAL BOX MODELS: ADVECTIVE PLUS DIFFUSIVE TRANSPORT PROCESSES

Multimedia environmental models using the fugacity and conventional approaches are based on a number of phases or compartments which are in contact. An interface plane either real or imaginary separates the adjoining compartments. The Lavoisier mass balance is applied to each compartment. In doing so, the fluxes across the interface planes provide the mathematical connection needed between compartments. First-order rate equations using linear source concentrations are convenient because they make the equations algebraically simple and enable the user to compare numerical rates at which chemicals move by the various mechanisms (Mackay, 2001). Mackay and coworkers developed a technique using first order rate equations for expressing both the advective and diffusive processes operative across the interface plane between compartments. It is based on the electrical analog of resistance-in-series and conductances-in-parallel which simplifies writing and using the mass balance equations and an alternative to using the interface compartment concept.

Scenario 4. A modification of Scenario 3 follows. This example demonstrates the use of the interface compartment concept for obtaining the overall flux expression connecting adjoining media compartments when a combination of advective and diffusive individual flux equations are required for the transport process. Where the air-side originally contained the single process of dry deposition of vapor, SA1 in Table 4.1, the additional processes of atmospheric particle dry deposition, SA2, is added. Using the "interface compartment" concept, the net sum of chemical flux from the air to the interface equals the net sum departing the interface plane to deeper locations within the soil column. Therefore the interface compartment flux balance yields n, in mg/m^2·s at steady-state is

$$n = k_a(C_a - C_{ai}) + v_d \rho_p C_p = \frac{D_a}{h}(C_{ai} - C_{ah}) + \frac{D_b}{h}\rho_s(C_{si} - C_{sh}) \qquad (4.21)$$

This balance is Equation 4.17 with an additional term. An advective flux accompanies the diffusive flux on the air-side. Vapor diffusion and particle biodiffusion processes represent the fluxes on the soil side. The appropriate phase equilibrium relationships are

$$C_p = K_{Pa}C_a \quad \text{and} \quad C_s = K_{sa}C_a \qquad (4.22a \text{ and } b)$$

for solid particles and surface soil particles-to-air partitioning, respectively. The latter applies for both C_{ai} and C_{ah}. Inserting these into Equation 4.21 yields

$$k_a(C_a - C_{ai}) + v_d \rho_p K_{pa} C_a = \left(\frac{D_a}{h} + \frac{D_b \rho_s K_{sa}}{h}\right) C_{ai} + \left[\frac{D_a}{h} + \frac{D_b \rho_s K_{sa}}{h}\right] C_{ah}$$
$$(4.23)$$

The various individual MTCs used appear in Table 4.2. The equation contains C_a and C_{ah}, the bulk phase concentrations in the adjoining phases plus the interface concentration C_{ai}. Equation 4.23 is solved for the interface concentration, the result is

$$C_{ai} = \left(\frac{k_a + v_d \rho_p K_{pa}}{k_a + \dfrac{D_a}{h} + \dfrac{D_b \rho_s K_{sa}}{h}}\right) C_a + \left(\frac{\dfrac{D_a}{h} + \dfrac{D_b \rho_s K_{sa}}{h}}{k_a + \dfrac{D_a}{h} + \dfrac{D_b \rho_s K_{sa}}{h}}\right) C_{ah} \qquad (4.24)$$

This equation for C_{ai} shows it depends on all the transport parameters, both diffusive and advective. It is then substituted into the right side of Equation 4.21 and gives

$$n = k_a \left\{ \frac{\dfrac{D_a}{h} + \dfrac{(D_b \rho_s K_{sa})}{h}}{k_a + \dfrac{D_a}{h} + \dfrac{(D_b \rho_s K_{sa})}{h}} \right\} \left[C_a\left(1 + \frac{v_d \rho_p K_{pa}}{k_a}\right) - C_{ah} \right] \qquad (4.25)$$

This is the flux expression for the combined processes. The transport parts of Equation 4.25 cannot be put in the simple format of a single overall MTC such Equations 4.16 and 4.20. In addition, it cannot be constructed by applying the Ohm's law rules of adding the appropriate conductance and resistances in series. Although the air-phase concentration, C_a and the equivalent soil-side, air-phase concentration, C_{ah}, are present they appear in an atypical form. The complex algebra results because the equations required for expressing the diffusive and advective processes are mathematically dissimilar.

A final note. The interface compartment concept has application for the next generation of multimedia compartment box models and other models as well. Presently several of these models use versions of Ohm's law as noted above as well as other procedures. The numerical results produced appear to be reasonable and provide good approximations apparently. However, without the additional mass balance provided by applying the interface compartment concept, the advective transport processes fail to impact the magnitude of the interface concentration. This influences the flux and finally the media mass concentrations. Comparative model studies using the present-day approaches of combining interface fluxes and the IC model approach need to be performed and the results evaluated. Such studies may aid the development of a more realistic and appropriate approaches for connecting chemical flux between multimedia environmental compartments.

4.4.5 Closure

Most users have an application for MTCs that in one way or another involve an environmental model at some mathematical level. The model may be as simple as one steady-state flux equation requiring a single value of the MTC representing a single simple and individual process. However, the user is typically concerned with a more complex application, which involves mathematical models of coupled mass balances across several media and numerous coupled individual transport processes. The previous section contains material appropriate to these types of applications. The emphasis was on both the differential equation models and the compartmental box models. Since the connecting interface and their adjoining boundary layers involved the interphase mass transport resistances, the subject received special attention. The interface compartment balance was defined and presented as the basic concept upon which to obtain the appropriate flux expressions and associated MTCs at the media interfaces. Procedures were demonstrated using it for both differential equation and compartmental box models. Specific mass transfer scenarios were used as examples and these also serve as guides for users to follow in developing results for alternative scenarios. In the next section, an attempt is made to organize the various specific procedures developed above and to present a generalized step-by-step approach for obtaining the appropriate interface fluxes expressions and the appropriate MTCs for the combined mass transport processes required in environmental fate models.

4.5 A STEP-BY-STEP PROCEDURE FOR OBTAINING MTCs AND FLUX EQUATIONS

1. Choose an interface separating two adjoining media. In most applications, there are three choices, AW, WS, and SA. Additional interfaces include ice–air, urban surfaces, plant–air, and so on.
2. Review Table 4 .1 and select the individual processes likely to be presented on either side. Do not prejudge the significance of any individual process; be inclusive rather than exclusive. Table 4.1 is divided into four sections. The

first three apply to processes at the air–water (AW), water–sediment (WS), and soil–air (SA) interfaces respectively and are relevant to this chapter.

3. If your model involved one or more differential equations go to step 4 otherwise go to step 8.

4. Hand draw an interface diagram showing the adjoining media and illustrate the locations, types of individual processes and expected chemical transport pathway directions. Many of the individual processes are illustrated in Figure 4.1. However, see Figures 4.2 and 4.3, and uses these as a guide for developing the interface pathway flux diagram. Label each phase or media using A for air, W for water, Sd for sediment, So for soil, and others as needed.

5. Use the interface compartment balance to derive the differential equation boundary conditions for the adjoining phases or media's. Flux equations are used in the balance. The general algorithm for using the interface compartment balance is:

$$n = \begin{bmatrix} \text{Sum of individual processes} \\ \text{fluxes entering the interface} \\ \text{plane} \end{bmatrix} = \begin{bmatrix} \text{Sum of individual processes} \\ \text{fluxes exiting the interface} \\ \text{plane} \end{bmatrix}$$

Each flux has units of $mg/m^2 \cdot s$. The individual processes flux equations may contain both diffusive and advective fluxes. Mimic the examples presented in Section 4.4.2 above. Flux expressions with concentration gradients and effective diffusion coefficients are normal for the soil and sediment side of these interfaces. For the air and water side of interfaces, concentration difference flux equations are normally used for the diffusive processes. Table 4.2 contains a representative summary of MTCs and equivalent parameters. As a part of these MTC parameters, the Table 4.2 entries also contain several effective diffusion coefficients that are needed to accompany the concentration gradient in the flux equations.

6. Apply the equilibrium condition for interfaces concentrations as needed.

7. The result at this step will be an equality containing at least one concentration gradient and other functions containing concentrations. Typically it will be the so-called "third type" boundary condition common to differential equations. Consult the appropriate chapters for estimating numerical values of the mass transport parameters in the equality. This is the last step for differential equation model application.

8. For adjoining compartments, hand draw an interface diagram showing the adjoining phases or media and illustrate the locations, types of individual processes and the expected chemical transport pathways. Many of the individual processes are illustrated in Figure 4.1. However, use Figures 4.2 and 4.3 serve as a guide for entering both the diffusive and advective types. Label each phase or media using A for air, W for water Sd for sediment, So for soil and others as needed.

9. Use the interface diagram and apply the interface compartment concept. The flux n is equal to the sum of individual processes moving A to the plane

which is set equal to those moving away from the plane. Equation 4.23 is a good example to follow.

10. Solve the resulting equation for the interface concentration in terms of the concentrations in the adjoining bulk compartments.

11. Using either expression for n in step 9, substitute the interface concentration and solve for the flux. Equation 4.24 is an example. This result provides the flux expression needed to connect the adjoining compartments.

12. Consult the appropriate chapters for estimating numerical values of the MTCs required. This is the last step for the compartment box model application.

Ultimately the user of this handbook will exit the document with a set of flux equations and numerical values of their associated MTCs for use in a specific environmental fate model for making computations of chemical movement rates. The applications are for chemical transport across environmental interfaces and the associated boundary layers rather than intramedia transport within the bulk media phases. For this case, see Chapter 20.

ACKNOWLEDGMENT

Thanks to Danny Fontenot and Darla Dao for the word processing and Justin Birdwell for editing and artwork.

LITERATURE CITED

R.B. Bird, W.E. Stewart, and E.N. Lightfoot. 2002. *Transport Phenomena*, 2nd ed. John Wiley, New York, Section 22.4.

R.S. Boethling and D. Mackay. 2000. *Handbook of Property Estimation Methods for Chemicals*. Lewis Publishers, New York.

E.L. Cussler. 1997. *Diffusion-Mass Transfer in Fluid Systems*, 2nd ed. Cambridge University Press, Cambridge, UK, Section 8.2.2.

D. Mackay. 2001. *Multimedia Environmental Models—The Fugacity Approach*. Lewis Publishers, Chelsea, Michigan, Section 7.11.

ADDITIONAL READING MATERIAL

L.J. Thibodeaux. 1996. *Environmental Chemodynamics*, 2nd ed. John Wiley, New York.

5 Estimating Molecular Diffusivities in Environmental Media

Justin E. Birdwell and Thomas R. Marrero

CONTENTS

5.1 Introduction 71
5.2 Estimating Binary Molecular Diffusivities 73
    5.2.1 Estimating Diffusion Coefficients in Air 73
    5.2.2 Estimating Diffusion Coefficients in Aqueous Systems 80
5.3 Diffusion Coefficients in Other Environmentally Relevant Media 86
    5.3.1 Diffusion Coefficients in Porous Media 86
    5.3.2 Effective Diffusivities in Snow and Ice 88
    5.3.3 Diffusion Coefficients in NAPLs 89
    5.3.4 Diffusion through Solid Mineral Lattices 89
5.4 Summary and User's Guide 91
5.5 Illustrative Calculations 91
References 100

5.1 INTRODUCTION

Molecular diffusion is an inherent and ubiquitous mass transport process by which chemical species move within and between environment phases. All environmental mass transfer coefficients, in one way or another, reflect the molecular diffusivity of the chemical species in the environmental "solvents" (i.e., air, water, organic matter, nonaqueous liquids, solids, etc.). It is therefore the fundamental transport parameter for molecular mass transport. It is important to keep in mind that diffusion is generally only used to describe molecular motion in the absence of mechanical mixing or advection. The material in this chapter provides a brief, theoretical introduction to diffusivities and follows this with estimation techniques, developed using empirical data, for chemicals in the significant environmental media compartments. The chapter concludes with example calculations.

Molecular diffusion can be defined as the movement of chemical compounds, driven by their kinetic energy (gas) or chemical potential (liquid, solid), from one point within a phase to another or across a phase boundary. Diffusion is more rapid in air (~four orders of magnitude) than in water due to the higher density of liquids, which leads to a reduced mean free path between diffusing molecules and those in the surrounding media as well as intensified intermolecular interactions between diffusing molecules and those in the bulk. Diffusion through air and water imbibed in inter- and intraparticle pore spaces of sediment beds and soil columns are covered in Chapters 8 and 12, respectively, and are briefly reviewed. True solid phase diffusion is of little relevance for environmental transport processes (unless they occur over geological timescales) and is only briefly covered for the case of diffusion along grain boundaries. In this chapter, the methods described deal primarily with neutral molecules and aqueous ions. The concentrations of diffusing species considered are limited to trace amounts, as defined in Chapter 1, therefore the effect of these compounds on the relevant properties of air and water are assumed to be negligible.

Diffusion coefficients are proportionality constants used to describe the motion of a particular compound in a phase or medium with a particular composition. The ratio of the mass (or molecular) flux to the concentration gradient in a system defines the diffusion coefficient, or diffusivity. Binary molecular diffusivities apply to species present in gaseous and liquid mixtures consisting of only two different components. In the context of environmental transport, air and surface waters (as well as fluids imbibed in soil or sediment) are treated as homogeneous phases, despite the fact that both represent mixtures with a wide range of components. For air, this simplification is reasonable in most cases in the lower atmosphere, since the composition of air is fairly constant and its major constituents are all apolar (excluding water vapor and CO_2), diatomic or simple gases. In the case of natural surface waters, aqueous diffusivities applied to fresh water are generally the same as those for pure water; these values are also fairly accurate for chemical species diffusing through seawater.

Diffusivities are scalar quantities. Chemical species diffuse from regions of high concentration to regions of relatively lower concentration. Therefore, the net direction of molecular movement is dependent on the concentration gradient, or more precisely the chemical potential. Chemical potential is driven by differences in the mass or molar distribution of dilute chemical species in a bulk phase, but can also be due to gradients in temperature, pressure, or external force fields (e.g., electric, magnetic). In this chapter, only concentration gradient-driven diffusion will be considered. Diffusive transport within environmental media may be driven by concentration gradients within a single phase, across phase boundaries or both. Methods for estimating diffusivities across phase boundaries are covered in Chapter 4.

The purpose of this chapter is to provide the reader with robust, easily applied methods for estimating diffusivities within the dominant natural phases (air, water, solid matrix, oil, etc.) for a range of environmentally and geochemically relevant compounds. The types of compounds considered include atomic (Hg) and diatomic (O_2, Cl_2) elements, simple molecules (H_2S, CO_2, NO), organic contaminants (poly-nuclear aromatic hydrocarbons, polychlorinated biphenyls, chlorinated solvents), as well as dissolved cationic (Pb^{2+}, Cr^{3+}) and anionic (Br^-, SO_4^{2-}) species. Examples showing the application of estimation methods are provided and the results of different methods are compared to measured values from the literature. When available, estimates regarding the accuracy of theoretical and empirical methods are included. Experimental methods for determining diffusivities in air, water, porous media, and other relevant phases have been reviewed extensively elsewhere (Cussler, 1984; Marrero and Mason, 1972) and will not be included here.

5.2 ESTIMATING BINARY MOLECULAR DIFFUSIVITIES

Kinetic theory indicates that molecular diffusion can be described as the random movement of atoms or molecules. Molecular motion appears random when small numbers of molecules are considered; this randomness does not contradict the observation of concentration gradient-driven movement of chemical species when very large numbers of molecules are considered. Truly random molecular diffusion will only be observed in a completely homogeneous phase. Molecules move because they have kinetic energy. The kinetic energy a molecule possesses is primarily a function of the mass of the molecule (molecular weight) and the prevailing temperature within the bulk medium. Because a diffusing molecule will interact with the components of the phase in which it resides, diffusivities are also dependent on the properties of the bulk phase, particularly viscosity. This is not a major consideration for gases, which are not particularly viscous under environmentally relevant temperatures and pressures, but can be important when dealing with liquid water. This difference in dependence on the composition of the bulk is due to the differences in density between gases and liquids; a molecule diffusing in a gas encounters other molecules much less frequently than one diffusing through a liquid.

Theoretical models exist that can be used to estimate diffusivities in gaseous and liquid media. Models of gaseous diffusion are based on the kinetic theory of gases and for ambient pressures are quite accurate, but for liquids theoretical approaches are less useful due to the need to account for intermolecular interactions between the diffusing species and bulk phase. The methods for estimating diffusivities in air and water are based on theoretical arguments, empirical functions based on measured data, or a combination of theoretical formulae with empirically determined correction factors and coefficients.

5.2.1 ESTIMATING DIFFUSION COEFFICIENTS IN AIR

The most common theoretical approach for describing molecular diffusion in gases is application of the kinetic theory of gases. Gas molecules are treated as hard spheres that interact only through completely elastic collisions. The velocities of gaseous molecules are determined by how much kinetic energy they possess, which is proportional to the square root of the temperature in the bulk phase. The velocity is also inversely proportional to the square root of the mass of the individual molecules, described by the molar mass or molecular weight (M_i, g mol^{-1}). Diffusion of molecules through gases is limited by the mean free path (λ, cm) between the diffusing species and the molecules of the bulk. This value can be estimated from the sizes of the compound of interest and the surrounding molecules (in the case of air, the average of the dominant constituents, diatomic oxygen and nitrogen, is used).

Derivations of the equations for estimates of gaseous molecular diffusivities from the kinetic theory of gases are available elsewhere (Cussler, 1984). Here a basic overview of the concepts is given to provide a review of the framework on which other, semiempirical formulae are based. As previously stated, diffusivity for a chemical species in air is a function of the molecular velocity and mean free path. The mean free path is a function of the temperature, pressure, and the sizes of the molecules involved. The velocity is a function of the temperature and molecular weight. The resulting equation for the theoretical approach is

$$D_{A1} = \frac{2 \left(\dfrac{RT}{\pi} \right)^{3/2}}{3NM_A^{-1/2} r_{crit}^2 P},$$ (5.1)

where D_{A1} = diffusivity of A in air (subscript 1 denotes air), $cm^2 s^{-1}$; R = the gas constant, 8.206×10^{-5} atm m^3 mol^{-1} K^{-1}; T = absolute temperature, K; N = Avogadro's number, 6.02×10^{23} mol^{-1}; M_A = the molecular weight of A, g mol^{-1}; r_{crit} = the critical distance for collisions of A with air, cm; and P = pressure, atm.

It should be noted that Equation 5.1 is consistent in terms of units, a hallmark of properly derived theoretical formulas, and therefore can be applied with any set of consistent units.

Application of the theoretical model requires only knowledge of the conditions in the bulk phase (temperature and pressure) and basic information on the compound of interest (molecular weight and size). However, the theoretical approach often does not produce diffusivity estimates in good agreement with measured values. This is because the assumption that gaseous molecules can be treated like hard spheres interacting only through physical collisions is insufficient. However, the form of the theoretical model provides a useful framework for methods that include terms that account for attractive and repulsive forces between gaseous molecules. The Leonard-Jones potential terms included in the method of Hirschfelder et al. (1949) were an early attempt to augment the theoretical model. The model has the form of the Chapman–Enskog equation (Chapman and Cowling, 1952), which comes from a solution of the Boltzmann equation

$$D_{A1} = \frac{0.001858 \, T^{3/2} \left(\dfrac{1}{M_A} + \dfrac{1}{M_1} \right)^{1/2}}{P \sigma_{A1}^2 \Omega_D},$$ (5.2)

where M_1 = the molecular weight of air, 28.97 g mol^{-1}; σ_{A1}^2 = the Leonard-Jones collision diameter of the binary system, Å; and Ω_D = the collision integral of molecular diffusion, dimensionless.

The collision diameter for a binary mixture can be calculated using the values for each pure component

$$\sigma_{A1} = \frac{\sigma_A + \sigma_1}{2},$$ (5.3)

where σ_A = the collision diameter of A, Å and σ_1 = the collision diameter of air, 3.617 Å.

In the absence of measured values for a particular chemical, the collision diameter can be estimated from the molecular volume at the chemical's boiling point using the following empirical relationship:

$$\sigma_A = 1.18(V_A^{(b)})^{1/3}, \tag{5.4}$$

where $\overline{V}_A^{(b)}$ = the molecular volume of A at the boiling point under atmospheric pressure (cm^3 mol^{-1}).

The collision diameter can also be calculated from critical point properties

$$\sigma_A = 2.44\left(\frac{T_A^{(c)}}{P_A^{(c)}}\right)^{1/3}, \tag{5.5}$$

where $P_A^{(c)}$ = the critical pressure of A, atm and $T_A^{(c)}$ = the critical temperature of A, K.

The collision integral term accounts for the effects of intermolecular interactions on the diffusion process and is a function of the reduced temperature

$$T^* = \frac{RT}{N\varepsilon_{A1}}, \tag{5.6}$$

where T^* = the reduced temperature, K and $N\varepsilon_{A1}$ = interaction energy between air and A, ergs (1 erg = 10^{-7} J = 9.48×10^{-11} BTU).

The Leonard-Jones interaction energy can be calculated from pure component values for apolar molecule pairs using the following relationship:

$$\varepsilon_{A1} = \sqrt{\varepsilon_A \varepsilon_1}, \tag{5.7}$$

where ε_A = interaction energy between molecules of A, ergs and ε_1 = interaction energy between molecules of air, 1.34×10^{-14} ergs.

As with the collision diameter, the interaction energies of apolar compounds can be estimated from boiling and critical point properties:

$$\varepsilon_A = \frac{1.15\, T_A^{(b)} R}{N}, \tag{5.8}$$

$$\varepsilon_A = \frac{0.77\, T_A^{(c)} R}{N}, \tag{5.9}$$

where $T_A^{(b)}$ = the boiling point of A at atmospheric pressure, K.

Once the reduced temperature has been determined, the collision integral can be read off Figure 5.1 or calculated using the equation shown in the upper portion of the plot.

The previous equations for determining the Leonard-Jones potential terms (τ_A, ε_A, and Ω_D) are for apolar compounds only. For polar chemicals, the previously

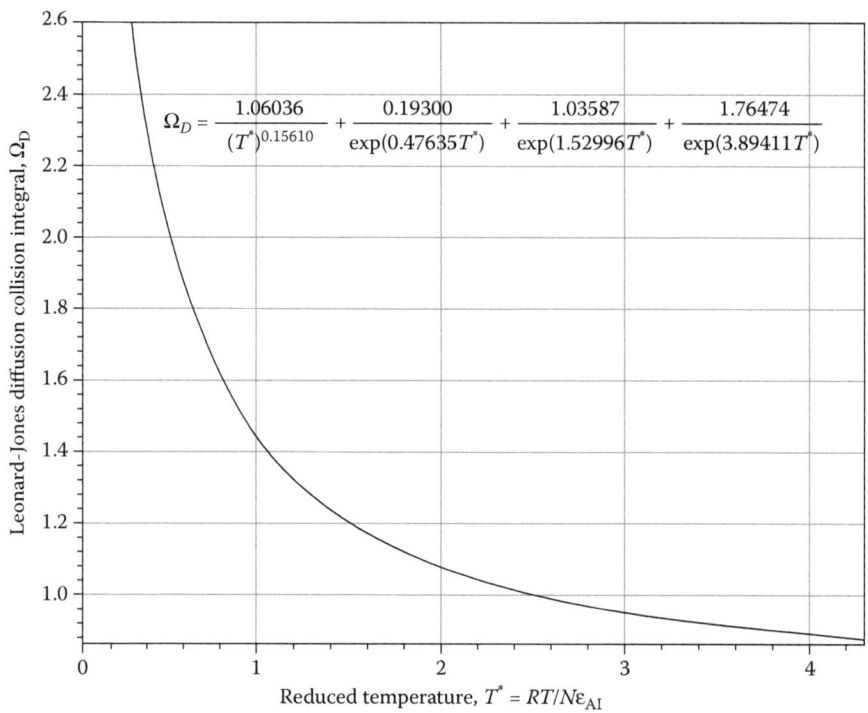

FIGURE 5.1   Plot of the Leonard-Jones reduced temperature and diffusion collision integral.

described correlations will not produce accurate estimates of diffusivity coefficients in air. Instead, the following set of equations are used to obtain collision diameters, interaction energies, and diffusion integrals (Brokaw, 1969) that are then used with the Chapman–Enskog equation (Equation 5.2)

$$\Omega_{D,polar} = \Omega_D + \frac{0.196\,\delta_{A1}^2}{T^*}, \tag{5.10}$$

$$\delta_{A1} = \sqrt{\delta_A \delta_1}, \tag{5.11}$$

$$\delta_A = \frac{1940\mu_A^2}{\overline{V}_A^{(b)} T_b}, \tag{5.12}$$

$$\varepsilon_{A,polar} = \frac{1.18\left(1 + 1.3\delta_A^2\right)RT_b}{N}, \tag{5.13}$$

$$\sigma_{A,polar} = \left(\frac{1.585\,\overline{V}_A^{(b)}}{1 + 1.3\,\delta_A^2}\right), \tag{5.14}$$

where δ_{A1} = the polarity parameter of A in air, dimensionless; δ_A = the polarity parameter of A, dimensionless; δ_1 = the polarity parameter of air ~ 0.822 (using

$\overline{V}_1^{(b)} = 29.9 \, \text{g mol}^{-1}$, $T_b = 79 \, \text{K}$); $\mu_1 = 1.0006$ Debyes with Equation 5.12); and μ_A = the dipole moment of A, debyes.

Semiempirical correlations for estimating chemical diffusivities in air include those of and Wilke and Lee (1955) and Fuller et al. (1966). The Wilke–Lee model uses many of the same Leonard-Jones potential terms described above, but includes an empirically determined correction factor in the numerator

$$D_{A1} = \frac{\left(3.03 - 0.98 M_{A1}^{1/2}\right) T^{3/2} M_{A1}^{1/2}}{\sigma_{A1}^2 \Omega_D P}, \tag{5.15}$$

where P = atmospheric pressure, kPa and M_{A1} = a function of the molecular weights of A and air,

$$M_{A1} = \frac{1}{M_A} + \frac{1}{M_1} \tag{5.16}$$

and all other terms have been defined. The equation is not dimensionally consistent due to the correction term (in parentheses in the numerator); therefore in order to get diffusivities with units $\text{cm}^2 \, \text{s}^{-1}$, the other parameters must have the following units: molecular weight in g mol^{-1}, temperature in K, pressure in kiloPascals, and the collision diameter in Å^2.

The method of Fuller et al. (1966) is a correlation of binary diffusivities based on the average molar volumes of the diffusing species. This dependence on molar volume eliminates the collision diameter term. The Fuller et al. equation also lumps the explicit (3/2 power in the theoretical treatments) and implicit (from the collision integral) temperature dependence of the diffusion coefficient. The final equation has the form

$$D_{A1} = \frac{10^{-3} T^{7/4} M_{A1}^{1/2}}{P(\overline{V}_A^{1/3} + \overline{V}_1^{1/3})^2}, \tag{5.17}$$

where $\overline{V}_1$ = the molar volume of air = $19.7 \, \text{cm}^3 \, \text{mol}^{-1}$ at 25°C and 1 atm, and all other terms have been defined. If the molar volume of species A is not readily available, it can be estimated from the molecular formula using the data in Table 5.1. By summing the volume contribution of each constituent atom and in some cases taking into account the functional group in which they reside (aromatic or heterocyclic rings), a reasonable estimate of the molar volume for many simple inorganic and organic molecules can be obtained.

In addition to the theoretical and semiempirical methods for estimating molecular diffusivities in air, several empirical methods, based on regressions of measured diffusivities and other parameters, like molecular weight and molar volume, are available. These methods provide easily calculated estimates when the property data required to apply the more rigorous semiempirical methods are not available. Molecular weight is an excellent parameter for empirical correlations because it is easily determined from the chemical formula and is an important parameter in the theoretical model. Diffusion coefficients in air can be expected to have power-law dependence on molecular

TABLE 5.1
Diffusion Volumes in the Gas Phase for Atomic Constituents and Simple Organic and Inorganic Molecules

Element	Volume Contribution (cm^3 mol^{-1})
Carbon	15.9
Hydrogen	2.31
Nitrogen	4.54
Oxygen	6.11
Sulfur	22.9
Fluorine	14.7
Chlorine	21.0
Bromine	21.9
Iodine	29.8
Aromatic carbon or carbon in heterocyclic ring	12.9

Molecule	Volume (cm^3 mol^{-1})
N_2	18.5
O_2	16.3
Air	19.7
Ar	16.2
CO	18.0
CO_2	26.7
N_2O	35.9
NH_3	20.7
H_2O	13.1
SO_2	41.8

Source: Adapted from Fuller, E.N., Ensley, K., and Giddings, J.C. 1969. *Industrial and Engineering Chemistry* **73**, 3679–3685.

weight of –0.5 based on theory. Figure 5.2 shows a data set consisting of measured diffusivities for a range of environmentally relevant compounds in air plotted against molecular weight. Nonlinear regression was applied and a power-law fit was obtained ($R^2 = 0.895$) for 35 compounds with molecular weights ranging from 17 (ammonia) to 354 (DDT). The estimating equation appears in lower left-hand corner of Figure 5.2.

A similar correlation can be obtained by relating diffusivities (cm^2 s^{-1}) with molar volume (cm^3 mol^{-1}); one such correlation is available in Schwarzenbach et al. (2002):

$$D_{A1} = 2.35 \cdot V_A^{-0.73}. \tag{5.18}$$

Molecular diffusivities in air can also be estimated for compounds when measured values are not readily available using a referencing method in which the diffusivity of a known (preferably a chemical similar to the unknown compound) is used to infer the diffusivity of the unknown species. This method is based on the theoretical

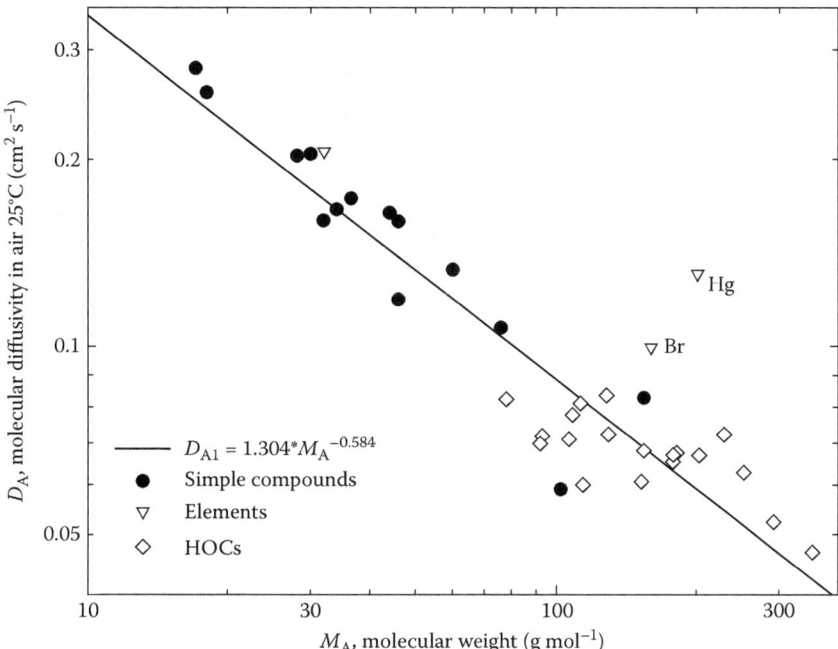

FIGURE 5.2 Correlation of molecular diffusivities in air at 25°C with molecular weight. Hg and Br shown as outliers (not included in correlation). (Values compiled by Thibodeaux, L.J. 1996. *Environmental Chemodynamics: Movement of Chemicals in Air, Water, and Soil*, Wiley, New York and from measurements by Gustafson, K.E. and Dickhut, R.M. 1994b. *Journal of Chemical and Engineering Data* **39**,286–289.)

relationship of diffusivity to molecular weight

$$D_{A1} = D_{B1} \left(\frac{M_A}{M_B} \right)^{-1/2}, \qquad (5.19)$$

where D_{B1} = the known diffusivity of compound B, cm^2 s^{-1} and M_B = the molecular weight of compound B, g mol^{-1}.

This approach may provide more accurate estimates for particular types of chemicals than the previous empirical correlation with molecular weight, provided that a measured diffusivity is available for at least one compound of the particular type of chemical.

Temperature adjustments over the range of environmentally relevant conditions ($-10°C$ to $45°C$) can be made using 25°C diffusivities as a reference and adjusting using a power-law correction. According to theory, diffusivity is proportional to absolute temperature (K) to the 3/2 power; however, this does not take into account the temperature dependence of the intermolecular interactions experienced by gaseous molecules. The Fuller et al. semiempirical approach indicates that the dependence is described by a 7/4 power law, which is very similar to the dependence indicated by

Marrero and Mason (1972) of 1.724. To correct for temperature, use the following relationship:

$$D_{A1}(T_2) = D_{A1}(T_R)\left(\frac{T_2}{T_R}\right)^{7/4}, \tag{5.20}$$

where $D_{A1}(T) =$ the diffusivity of A in air at temperature T, cm^2 s^{-1}; $T_2 =$ the temperature of interest, K; and $T_R =$ the reference temperature, K.

For instance, when diffusivities are needed for gaseous mixtures other than air or for air mixed with another gas where the new component accounts for more than 10% of the total volume (or mole fraction >0.1), a method described by Fairbanks and Wilke (1950) is available for estimating diffusivities for species of interest in the mixture. The method uses a weighted average approach. The component A-free mole fractions for each component of the mixture and the binary diffusivities for component A in the individual components of the mixture are used to calculate the diffusivity of A in the new mixture according to the following equations:

$$D_{A\text{-mixture}} = \frac{1}{\dfrac{y_2'}{D_{A-2}} + \dfrac{y_3'}{D_{A-3}} + \ldots + \dfrac{y_n'}{D_{A-n}}}, \tag{5.21}$$

$$y_2' = \frac{y_2}{y_2 + y_3 + \ldots + y_n}, \tag{5.22}$$

where $D_{A\text{-mixture}} =$ the diffusivity of A in the mixture, cm^2 s^{-1}; $y_2', y_3', y_n' =$ the component A-free mole fraction of 2, 3, $\ldots, n$ in the mixture; $D_{A-2}, D_{A-3}, D_{A-n} =$ the binary diffusivities for A in component 2, 3, $\ldots, n$, cm^2 s^{-1}; and $y_2, y_3, y_n =$ the mole fraction of 2, 3, $\ldots, n$ in the mixture including A.

It should be noted that in nearly all cases relevant to environmental transport of dilute chemicals in air, the difference between the component A-free and overall mole fractions will be negligible.

5.2.2 Estimating Diffusion Coefficients in Aqueous Systems

Unlike molecular diffusion in gases, there is no well-developed theoretical approach for estimating diffusion coefficients in liquids. Two rough theories for estimating the order of magnitude of liquid diffusivities for dilute, nonionic species are the hydrodynamic theory and the Eyring theory. The hydrodynamic theory is derived from arguments similar to those posed in the kinetic theory of gases–liquid molecules have a velocity and the molecules are under the influence of some combination of forces exerted on them as they move through the bulk liquid phase. In the kinetic theory of gases, the mass of the molecule limits its movement; in hydrodynamic theory, it is the viscosity of the bulk liquid that resists diffusion. The most commonly used derivation from hydrodynamic theory is the Stokes–Einstein equation

$$D_{A2} = \frac{RT}{6\pi N \eta_2 r_A}, \tag{5.23}$$

where D_{A2} = the diffusivity of A in water (subscript 2 indicates water), cm^2 s^{-1}; η_2 = the viscosity of water, kg m^{-1} s^{-1}; and r_A = the radius of A, Å.

This formula works well for solute molecules that are significantly larger than the solvent ($r_2 \sim 1.0$ Å) and are roughly spherical, with deviations from measured values of $\sim$20%. It is also useful for estimating the diffusivity of colloids. The Eyring theoretical approach treats diffusion as a rate process where movement from one point to another requires transition through an activated state. It yields the same dependence on temperature and viscosity as the Stokes–Einstein result from hydrodynamic theory.

As with diffusion in air, semiempirical relationships between the properties of diffusing species and water can be used to obtain estimates of molecular diffusivity. One of the most commonly used approaches for estimating diffusion coefficients for nonionic species in liquids at dilute concentrations is that of Wilke and Chang (1955). This method incorporates the dependence on temperature and viscosity the theoretical derivations obtained with a solvent association parameter and explicit dependence on the molar volume of the diffusing species

$$D_{A2} = \frac{7.4 \times 10^{-8} (\Phi_2 M_2)^{1/2} T}{\left(\overline{V}_A^{(b)} \right)^{0.6} \eta_2}, \tag{5.24}$$

where D_{A2} = the diffusivity of A in water, cm^2 s^{-1}; Φ_2 = the association constant for water, 2.26, dimensionless; M_2 = the molecular weight of water, 18.0 g mol^{-1}; T = the system temperature, K; $\overline{V}_A^{(b)}$ = the molar volume of A at its boiling point (1 atm), cm^3 mol^{-1}; and η_2 = the viscosity of water, cP (1 kg m^{-1} s^{-1} = 1 Pa s = 1000 cP).

The relationship described in Equation 5.24 is not dimensionally consistent; therefore accurate estimates of aqueous diffusivity will only be obtained if the values of the parameters used have the specified units. The Wilke–Chang correlation provides estimates with average errors of $\sim$10%.

Another semiempirical correlation for estimating diffusivities was developed based on the Wilke–Chang equation by Othmer and Thakar (1953) with further modifications by Hayduk and Laudie (1974) for nonionic compounds in water:

$$D_{A2} = \frac{13.26 \times 10^{-5}}{\left(\overline{V}_A^{(b)} \right)^{0.589} \eta_2^{1.14}}, \tag{5.25}$$

where all terms are the same (and have the same units) as those in the Wilke–Chang equation. Small changes were made to the parameter values (Φ_2 was originally determined to be 2.6; the more accurate value 2.26 is indicated above) and to the power-law dependencies on molar volume and viscosity. The advantage of the Hayduk–Laudie method over that of Wilke–Chang is that it was specifically developed for estimating diffusivities in water. The average error anticipated for this correlation is <6%. Estimates of molar volume for molecules at their normal boiling point can be determined by summing the volume contribution of constituent atoms using the data listed in Table 5.2.

Another modification of the Wilke–Chang equation was proposed by Scheibel (1954) that eliminates the association constant and has been found to be useful for

TABLE 5.2

Diffusion Volume Contribution of Atomic Constituents of Simple Molecules in Liquid Phase

Element	Volume Contribution (cm^3 mol^{-1})
Bromine	27.0
Carbon	14.8
Chlorine	21.6
Hydrogen	3.7
Iodine	37.0
Nitrogen, double bond	15.6
Nitrogen, primary amines	10.5
Nitrogen, secondary amines	12.0
Oxygen (except for cases below)	7.4
Oxygen, in methyl esters	9.1
Oxygen, in methyl ethers	9.9
Oxygen in higher esters and ethers	11.0
Oxygen, in acids	12.0
Sulfur	25.6

Corrections for Ring Structures

Three-membered ring (ethylene oxide)	−6.0
Four-membered ring (cyclobutane)	−8.5
Five-membered ring (furan)	−11.5
Six-membered ring (benzene, pyridine)	−15.0

Source: Adapted from Le Bas, G. 1915. *The Molecular Volumes of Liquid Chemical Compounds*, Longmans, Green & Company, Ltd., London, NY.

estimating diffusivities in organic solvents, lending it to applications involving water and nonaqueous phase liquids (NAPLs) in the environment. The Scheibel formula is

$$D_{A2} = \frac{K\,T}{\left(\overline{V}_A^{(b)}\right)^{1/3}\eta_2},$$
(5.26)

where

$$K = 8.2 \times 10^{-8}\left[1 + \left(\frac{3\overline{V}_2^{(b)}}{\overline{V}_A^{(b)}}\right)^{2/3}\right],$$
(5.27)

$\overline{V}_2^{(b)}$ = the molar volume of water at its normal boiling point, $18.9\,cm^3\,mol^{-1}$.

Equation 5.26 is further simplified in instances where the molar volume of A is similar to that of the solvent; for benzene if $\overline{V}_A^{(b)} < 2\overline{V}_{benzene}^{(b)}$, $K = 18.9 \times 10^{-8}$ and for other organic solvents, if $\overline{V}_A^{(b)} < 2.5\,\overline{V}_{solvent}^{(b)}$, $K = 17.5 \times 10^{-8}$. The Scheibel method typically yields estimates with errors of $<20\%$ for diffusivities in organic solvents.

All of the previous theoretical and semiempirical methods described are for use with nonionic species in water. However, the diffusion of cations and anions in surface and pore waters is also a very important phenomenon. For salts, the dimensionally consistent Nernst equation has the following form:

$$D_{A2} = \frac{nRT}{\left(\dfrac{1}{\lambda_+^0} + \dfrac{1}{\lambda_-^0}\right)\mathfrak{I}^2}, \tag{5.28}$$

where $n = 2$ for univalent salts; $\lambda_+^0 =$ the limiting cationic conductance, A cm^{-2} (V cm^{-1}) g-equivalent cm^{-3}; $\lambda_-^0 =$ the limiting anionic conductance, A cm^{-2} (V cm^{-1}) g-equivalent cm^{-3}; and $\mathfrak{I} =$ Faraday's constant, 96,500 Coulombs g-equivalent^{-1}.

For polyvalent ions, the n term in the numerator of Equation 5.28 is replaced by

$$n = \frac{1}{x^+} + \frac{1}{x^-}, \tag{5.29}$$

where $x^+ =$ the valence number of the cation and $x^- =$ the valence number of the anion.

Table 5.3 contains limiting ionic conductances and valence charges for several common cations and anions.

Measured diffusion coefficients for chemicals in water can be correlated to molecular weight and molar volume in the same way that diffusivities in air are correlated

TABLE 5.3
Limiting Ionic Conductances in Water at 298 K

Cations	λ_+^0 [A cm^{-2} (V cm^{-1}) g-eq.cm^{-3}]	Anions	λ_-^0 [A cm^{-2} (V cm^{-1}) g-eq. cm^{-3}]
Ba^{2+}	127.2	Br^-	78.1
Ca^{2+}	119.0	$CH_3CO_2^-$	40.9
Cs^+	77.2	Cl^-	76.4
Cu^{2+}	107.2	ClO_4^-	67.3
H^+	349.6	CO_3^{2-}	138.6
K^+	73.5	$(CO_2)_2^{2-}$	148.2
Li^+	38.7	F^-	55.4
Mg^{2+}	106.0	$[Fe(CN)_6]^{3-}$	302.7
Na^+	50.1	$[Fe(CN)_6]^{4-}$	442.0
$[N(C_2H_5)_4]^+$	32.6	HCO_2^-	54.6
$[N(CH_3)_4]^+$	44.9	I^-	76.8
NH_4^+	73.5	NO_3^-	71.5
Rb^+	77.8	OH^-	199.1
Sr^{2+}	118.9	SO_4^{2-}	160.0
Zn^{2+}	105.6		

Source: Adapted from Atkins, P. 1998. *Physical Chemistry*, W.H. Freeman and Company, New York, NY.

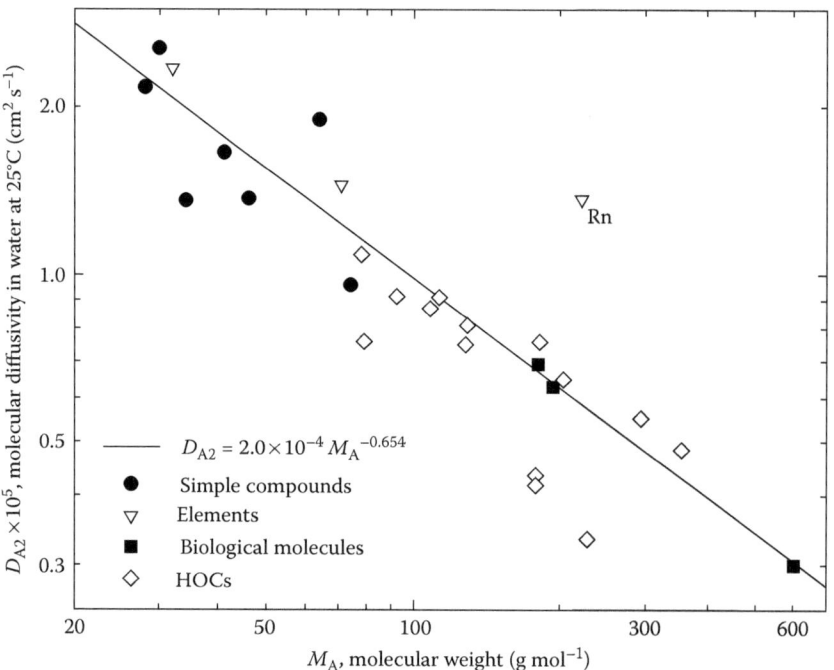

FIGURE 5.3 Correlation of molecular diffusivities in water at 25°C with molecular weight. Rn shown as an outlier (not included in correlation). (Values compiled by Thibodeaux, L.J. 1996. *Environmental Chemodynamics: Movement of Chemicals in Air, Water, and Soil*, Wiley, New York and measurements by Gustafson, K.E. and Dickhut, R.M. 1994a. *Journal of Chemical and Engineering Data* **39**, 281–285.)

to these parameters. Figure 5.3 shows a collection of measured aqueous diffusivities at 25°C plotted against the molecular weight of each compound, with a power-law regression line ($R^2 = 0.853$).

Using the molecular weight correlation shown in Figure 5.3, aqueous diffusivities for ionic species (cations and anions) can be estimated. First, calculate the diffusivity using the molecular weight of the ion, then multiply that value by the following correction factor, which is a function of the ionic species molecular weight and valence number (charge)

$$F_{ion,2} = 0.12 \left(\frac{M_{ion}}{x_{ion}} \right)^{0.54}, \qquad (5.30)$$

where $F_{ion,2}$ = the correction factor for the aqueous diffusivity molecular weight correlation; M_{ion} = the molecular weight of the ionic species, g mol^{-1}; and x_{ion} = the absolute value of the ion valence number, dimensionless.

As with diffusion coefficients in air, a referencing method can be used to estimate the diffusivities in water using a reference compound when measured values for the compound of interest are not available

$$D_{A2} = D_{B2} \left(\frac{M_A}{M_B} \right)^{-3/5}, \qquad (5.31)$$

where D_{B2} = the diffusivity of the reference compound in water, $cm^2 \ s^{-1}$.

This function corresponds to Equation 5.19 for diffusion coefficients in air. A similar relationship exists for aqueous diffusivity and molar volume, with the same form as Equation 5.31, but the ratio of molar volumes is taken to the -0.589 power instead of $-3/5$.

The temperature dependence of aqueous diffusion coefficients is related to both the increased molecular motion of the solute at higher temperatures and a reduction in the viscosity of water. Using the following equation, diffusion coefficients can be estimated using a known diffusivity for the compound of interest at some reference temperature that is within $\pm 40°$ of the desired temperature:

$$D_{A2}(T_2) = D_{A2}(T_R) \left(\frac{T_2}{T_R} \right) \left[\frac{\eta_2(T_R)}{\eta_2(T_2)} \right], \qquad (5.32)$$

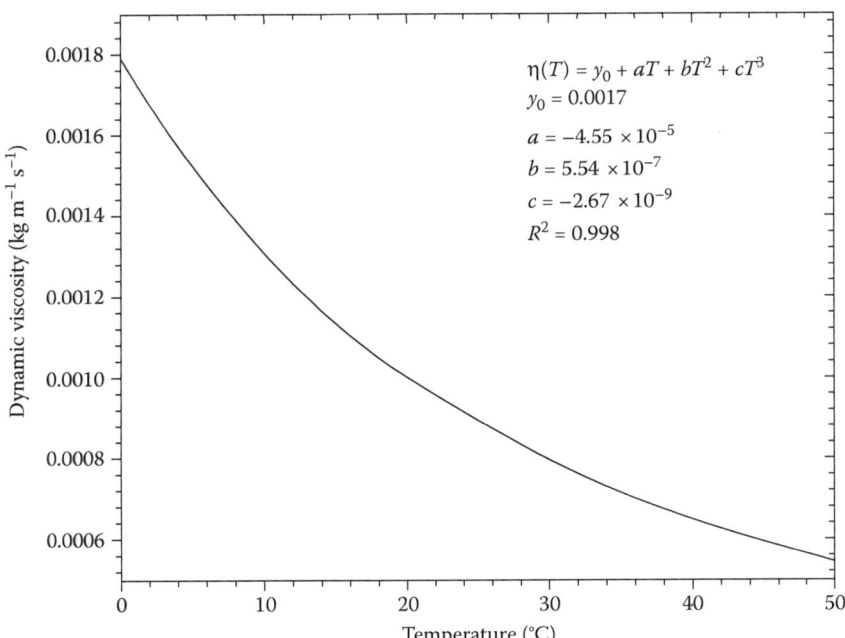

FIGURE 5.4 Dynamic viscosity of water as a function of temperature. (Data from Perry, R.H. and Green, D.W. 1997. *Perry's Chemical Engineers Handbook*, 7th Ed., McGraw-Hill, Columbus, OH.)

where $D_{A2}(T)$ = the aqueous diffusivity at temperature T, cm^2 s^{-1}; T_R = the reference temperature for known diffusivity, K; T_2 = the temperature of interest, K and $\eta_2(T)$ = the dynamic viscosity of water at temperature T, kg m^{-1} s^{-1}.

Dynamic viscosity values can be read off of Figure 5.4 or can be calculated using the polynomial equation shown in the upper right hand corner of the figure can be used to calculate values directly from the temperature ($^\circ$C).

5.3 DIFFUSION COEFFICIENTS IN OTHER ENVIRONMENTALLY RELEVANT MEDIA

In addition to air and water, there are many other environmentally relevant phases in which diffusion and diffusion-related processes are important for understanding and forecasting chemical transport. Many of these processes will be covered in depth in subsequent chapters (diffusion in sediment beds, Chapter 12; biodiffusion, Chapter 13; dispersion in groundwater, Chapter 15) and will only be touched on here. Some of the media considered here are natural materials (soil, sediment, ice, snow), while others are of anthropogenic origin (oil slicks, organic liquids accumulated in soils and sediment).

5.3.1 DIFFUSION COEFFICIENTS IN POROUS MEDIA

Soil and sediment grain packs contain significant void space between particles (30–90% of total volume), which is generally occupied by air and water. Porous media in which essentially all voids are occupied by water are called saturated, while those containing both air and water are unsaturated. Bed sediments are saturated at all times under normal conditions, while surface soils experience a dynamic range of moisture conditions depending on rainfall, relative humidity, and the depth of the water table. The moisture content (fraction of void space occupied by water) of soils in regions adjacent to water bodies or wells can also be affected by the movement of water in these formations. Diffusion coefficients for transport through porous media are determined by the fluid or fluids present within and the structure of the media itself. In saturated soil or sediment, diffusion occurs through the aqueous phase and is inhibited by physical and chemical interactions with the particles. In unsaturated porous media, the amount of moisture present is an important factor, since diffusivities in air are 10,000 times larger than those in water. This means that in unsaturated porous media the water can be viewed as hindering transport.

Diffusion coefficients for porous media are generally referred to as *effective diffusivities*, since the actual molecular diffusion process occurs in the fluid phase and interactions with the porous medium inhibits the chemical movement. There are both physical and chemical factors that go into estimating effective diffusivities. The physical effects are twofold. First, some fraction of the porous media is solid, limiting the volume through which fluid phase diffusion can occur. This is quantified by the porosity, which is defined as the ratio of the volume of void space to the total volume. Second, the connectivity between pore spaces in soil and sediment grain packs (as well as other porous media) are circuitous and lengthen the distance a molecule must travel to traverse the material. This lengthening of the diffusion path is quantified by the tortuosity of the media, which is defined as the ratio of the distance an ion or molecule travels around particles to the distance between the starting and ending points in the absence of particles. Chemical interactions between diffusing species with porous media are mostly attributed to adsorption or absorption of the compound to mineral or organic surfaces of particles. This type of interaction is akin to chromatographic processes and the strength of the interaction(s) can be described by a solid–fluid (air or water) partition coefficient. Chemical interactions can also included reactions of the solute with other dissolved or surface-bound chemical species.

For saturated porous media, the effective diffusivity can be estimated using simple arguments to quantify the physical and chemical resistances that reduce the aqueous diffusivity. The following equations can be used provided that some basic properties of the soil or sediment and compound of interest are available:

$$D_{\text{eff}} = \frac{D_{A2}}{R_P R_C}, \tag{5.33}$$

where D_{eff} = the effective diffusivity of A in the porous medium, $cm^2 \ s^{-1}$; R_P = the physical resistance factor, dimensionless; and R_C = the chemical resistance factor, dimensionless.

The physical resistance factor can be estimated from the square of the tortuosity

$$R_P = \theta_3^2, \tag{5.34}$$

where θ_3 = the tortuosity of the porous medium (subscript 3 indicates soil or sediment phase), $cm \ cm^{-1}$, which can be estimated from the porosity using a number of different relationships depending on the type of porous media (Shen and Chen, 2007). The following equations are recommended in the absence of detailed information on the specific porous medium of interest:

$$\theta_3^2 = \phi_3^{-1/3}, \tag{5.35}$$

$$\theta_3^2 = 1 - \ln(\phi_3^2), \tag{5.36}$$

where ϕ_3 = the porosity (total void space) of the porous medium, $cm^3 \ cm^{-3}$.

Equation 5.35 provides a general relationship between porosity and tortuosity while Equation 5.36 is particularly useful for fine-grained unlithified sediments (Boudreau, 1996).

The chemical resistance factor is defined as

$$R_C = 1 + \frac{\rho_3}{\varphi_3} K_{23}, \tag{5.37}$$

where ρ_3 = the bulk density of the soil or sediment, $g \ cm^{-3}$ and K_{23} = the solid–water partition coefficient, $cm^3 \ g^{-1}$. It applies only to the application of unsteady-state (i.e., transient) diffusion. For steady-state diffusion, $R_C = 1$. Solid–water partition coefficients are chemical specific and can be measured using simple batch experiments. For hydrophobic organic compounds (HOCs), like polynuclear aromatic hydrocarbons or polychlorinated biphenyls, soil, or sediment organic matter dominate the chemical interaction with particles, therefore the organic carbon content of the soil or sediment is an important parameter for estimating diffusivities of these compounds in porous media. Sorption to soil or sediment organic carbon can be described by the organic carbon–water partition coefficient, which is related to K_{23} and the organic carbon content by the following relationship:

$$K_{23} = \varphi_{OC} K_{OC}, \tag{5.38}$$

where ϕ_{OC} = is the fraction organic carbon of the soil or sediment, g g^{-1} and K_{OC} = the organic carbon-water partition coefficient, cm^3 g^{-1}. K_{OC} values can be found in the literature for a wide range of compounds or estimated from aqueous solubility or octanol–water partition coefficients (K_{OW}). It must be reemphasized that the chemical resistance factor applies to chemodynamic models that contain time as an independent variable. Estimating K_{OC} and related partition coefficients is beyond the scope of this handbook; however, the reader is referred to Chiou (2002) and Boethling and Mackay (2000) for values and estimation methods for these parameters.

In unsaturated soils, effective diffusivities can be estimated using the following empirical relationship originally developed by Millington and Quirk (1961):

$$D_{\text{eff}} = D_{A1} \frac{\varphi_1^{10/3}}{\varphi_3^{2}}, \tag{5.39}$$

where ϕ_1 = the fraction of void space filled with air, cm^3 cm^{-3}.

Application of this model should be limited to chemical species that do no interact strongly with the water or solid phases.

5.3.2 Effective Diffusivities in Snow and Ice

During the winter seasons in northern latitudes and in perennially ice-covered regions near the poles, ice and snow represent major phases for chemical accumulation. Measurements of diffusivities in snow and ice are limited, so empirical models are particularly useful. Two semiempirical effective diffusivity models were outlined by Herbert et al. (2006) for snow, but both were found to severely underpredict measured diffusivities for several semivolatile organic contaminants in new and aged snow. A simple empirical correlation was used to fit the measured values

$$D_{A4} = D_{A1} \varphi_4^{m}, \tag{5.40}$$

where D_{A4} = the diffusivity of A in snow or ice (subscript 4 indicates ice/snow), cm^2 s^{-1}; ϕ_4 = the snow porosity, cm^3 cm^{-3}; and m = an experimentally determined exponent, -0.072 ± 0.025.

The snow porosity can be estimated using the following equation:

$$\varphi_4 = 1 - \frac{\rho_{\text{snow}}}{\rho_{\text{ice}}}, \tag{5.41}$$

where ρ_{snow} = the density of snow, kg m^{-3} and ρ_{ice} = the density of ice, ~917 kg m^{-3}.

Interestingly, the exponent for Equation 5.40 was not found to vary significantly between fresh and aged snow. See Chapter 18 for more information on transport processes in ice and snow.

5.3.3 DIFFUSION COEFFICIENTS IN NAPLs

When large spills of nonaqueous phase liquids (NAPLs), primarily hydrocarbons and chlorinated solvents, occur onto soils and sediments, it can lead to the formation of a persistent organic phase within the natural solid–fluid matrix. NAPLs migrate down through porous media under the influence of gravity. Diffusion of organic contaminants through NAPLs is slower than through water imbibed in soil or sediments due to the higher viscosity of these fluids. Diffusion through NAPLs present in soil or sediment is also affected by the physical resistances previously described for porous media.

In addition to the Scheibel method discussed previously (Equations 5.26 and 5.27), another semiempirical model for estimating diffusion coefficients in organic liquids developed by Hayduk and Minhas (1982) is available. Though this model has similar functionality to the other liquid diffusivity estimation methods described above, this formula is specifically for diffusion of chemical species in paraffin solutions

$$D_{\text{A-NAPL}} = \frac{13.3 \times 10^{-8} T^{1.47} \eta_{\text{NAPL}}^{K}}{(\overline{V}_{A}^{(b)})^{0.71}}, \qquad (5.42)$$

where $D_{\text{A-NAPL}}$ = the diffusivity of A in NAPL, cm^2 s^{-1} and η_{NAPL} = the viscosity of NAPL, cP, and the exponent on the NAPL viscosity term can be calculated using the following equation:

$$K = \frac{10.2}{\overline{V}_{A}^{(b)}} - 0.791. \qquad (5.43)$$

Typical viscosities for light (benzene) and dense (diesel fuel) NAPLs are 0.65 and 3.60 cP, respectively at 25°C. Materials like creosote and coal tar can have viscosities as high as 20 cP at ambient conditions and may not be well represented by this method. Equation 5.42 can be used along with the porous media formulae outlined above (Equations 5.33 and 5.39) to estimate effective diffusivities of soil or sediment when NAPLs are present.

5.3.4 DIFFUSION THROUGH SOLID MINERAL LATTICES

Aside from pore fluid diffusion, geochemically relevant processes in solid media are driven by diffusion along grain boundaries and through crystal lattices. However, at ambient temperatures and pressures, diffusion of even the smallest elements and molecules in solid minerals is exceedingly slow. Properties of diffusing ionic species that affect the rate of diffusion include the size, charge, and polarizability of the ion. Diffusion through solid minerals is of particular interest in the field of nuclear waste disposal, since uranium (and related compounds and ions), plutonium, and other radioactive species must be stored for long periods of time, often within granite and basalt formations or those composed of welded tuffs (e.g., Yucca Mountain).

Deep within the Earth, where temperatures and pressures are higher than those at surface, diffusion coefficients within solid media are significantly higher, but are still more than 10 orders of magnitude smaller than diffusivities in water. For typical geological timescales, diffusion through mineral lattices is only a significant bulk transport process at temperatures exceeding 900°C ($\sim$100 km beneath the Earth's surface).

Diffusivities in solid media have an Arrhenius-like temperature dependence, as shown below:

$$D_{A5}(T) = D_{A5}^0 \exp\left(-\frac{Q}{RT}\right), \tag{5.44}$$

where D_{A5} = the diffusivity of A in the mineral lattice (subscript 5) at temperature T, cm^2 s^{-1}; D_{A5}^0 = the pre-exponential factor of A in the mineral lattice, cm^2 s^{-1}; and Q = the activation energy of diffusion in the mineral lattice, kJ mol^{-1}.

Diffusion activation energies range from 100 to 300 kJ mol^{-1} in solid media. Table 5.4 lists preexponential and activation energy values for several diffusing species in some common minerals.

TABLE 5.4
Diffusion Parameters for Silicate Minerals

Mineral	Diffusing Species	D_0 (cm^2 s^{-1})	Q (kJ mol^{-1})	Temperature Range (K)
Quartz[a]	Sodium (Na)	3.8×10^{-6}	102.4	673–1273
Quartz[a]	Potassium (K)	1.8×10^{-5}	132.5	573–773
Quartz[a]	Lithium (Li)	6.9×10^{-7}	85.7	573–773
Quartz[a]	Oxygen (^{18}O)	1.1×10^{-14}	195	1143–1453
Kaolin[a]	Calcium (^{45}Ca)	3.71×10^{-2}	292.6	1473–1673
Illite[a]	Calcium (^{45}Ca)	0.12	292.6	1473–1673
Biotite[a]	Argon (Ar)	3.42×10^{-5}	167	NR
Biotite[a]	Barium (Ba)	4.4×10^{-7}	41.4	293–395
Biotite[a]	Calcium (Ca)	9.4×10^{-8}	44.7	293–395
Biotite[a]	Strontium (Sr)	1.1×10^{-7}	39.7	293–395
Plagioclase feldspar (albite)[b]	Lithium (Li)	1.6×10^{-4}	146	473–1073
Plagioclase feldspar (albite)[b]	Potassium (K)	4.0×10^{-4}	296	873–1273
Plagioclase feldspar (albite)[b]	Rubidium (Rb)	1.6×10^{-6}	283	1003–1273

[a] Compiled by Freer, R. 1981. *Contributions to Mineralogy and petrology*, **76**, 440–454.
[b] Giletti, B.J. and Shanahan, T.M. 1997. *Chemical Geology* **139**, 3–20.

The distance a diffusing ion, atom, or molecule moves within a specified period of time can be estimated using the following equation:

$$X = \sqrt{2D_{A5}t},\tag{5.45}$$

where X = the penetration depth of A within the mineral lattice, cm and t = the time A has been diffusing through the mineral lattice, s.

As an example of how slow diffusion through mineral lattices is, consider diffusion of lithium through albite feldspar at 600°C ($D_{A5} = 2.7 \times 10^{-9}$ cm^2 s^{-1}); it would take over 58,000 years for a lithium ion to move 1 m through the bulk mineral.

5.4 SUMMARY AND USER'S GUIDE

The preceding discussion provides the user with an assortment of estimation methods for obtaining reasonable values of molecular diffusivities in air, water, organic liquids, and other environmentally and geochemically relevant media. Methods presented for estimating effective diffusivities in porous media represent extensions of the parameters determined for fluids to situations where diffusion occurs through a fluid imbibed within solid material. The presentation of theoretical, semiempirical, and purely empirical formulae gives the user the ability to choose an estimation method based on the available information about the chemical(s) of interest and the conditions within the medium.

Molecular diffusion coefficients for environmental media can be easily and accurately estimated using semiempirical methods. For diffusion in air, the method of Fuller et al. (1966) in the form of Equation 5.17 can be used with good accuracy and requires only temperature and pressure information (for typical environmental conditions, 25°C and 1 atm are sufficient) and the molecular weight and molar volume of the chemical of interest (which can be estimated using Table 5.1 and the chemical formula). Diffusivities in water can be estimated using the method of Hayduk and Laudie (1974), which as shown in Equation 5.25 requires only the molar volume of the chemical of interest and the viscosity of water at the prevailing temperature (see Figure 5.5). Empirical formulas based on molecular weight can be used to obtain estimates of chemical diffusivities in both air and water with excellent results (see Figures 5.2 and 5.4). Measured values or those estimated using the purely empirical molecular weight relationships described above (which apply to 25°C only) can be adjusted for use at other temperatures using Equation 5.20 for air and Equation 5.32 for water. For all other environmental media, diffusivity estimating methods described in this chapter represent only a cursory review of what is available in the literature and the reader is encouraged to review more in depth examinations of transport processes in some of these types of media in subsequent chapters in this handbook.

5.5 ILLUSTRATIVE CALCULATIONS

A set of illustrative calculations of molecular diffusivities using many of the methods described in the chapter are provided in the following section. For both air and water, diffusivities estimated using several different methods are compared to measured values. The compound used in most examples is naphthalene ($C_{10}H_8$, $M_A = 128.17\,\text{g}\,\text{mol}^{-1}$), a common household fumigant that is an important contaminant in all major environmental phases and is considered a possible carcinogen in humans.

Example 5.1: Diffusivity of Naphthalene in Air at 25°C

Estimate molecular diffusivities ($\text{cm}^2\,\text{s}^{-1}$) for naphthalene in air using the following methods described in Chapter 5: (a) kinetic theory of gases, (b) Chapman–Enskog, (c) Wilke–Lee, (d) Fuller et al. (e) the molecular weight correlation (Figure 5.2), and (f) the molar volume correlation. Compare the results to the measured value, $D_{A1} = 0.084 \pm 0.010\,\text{cm}^2\,\text{s}^{-1}$ (Gustafson and Dickhut, 1994b).

SOLUTION

(a) Calculate the molar volume of naphthalene from the chemical formula ($C_{10}H_8$) using the data in Table 5.1 to estimate the radius, then calculate the critical collision distance (r_{crit}) (assume the radius for air is ~1 Å):

Volume contribution for aromatic carbon (v_{C-arom}): 12.9 $\text{cm}^3\,\text{mol}^{-1}$ atom^{-1}.

Volume contribution for hydrogen (v_H): 2.31 $\text{cm}^3\,\text{mol}^{-1}$ atom^{-1}.

$$\overline{V}_A = 10 v_{C-Arom} + 8 v_H = 10\left(12.9\text{cm}^3/\text{mol}\right) + 8\left(2.31\text{cm}^3/\text{mol}\right)$$

$$= 147.48\text{cm}^3/\text{mol}.$$

Assuming that naphthalene can be approximated as a spherical molecule, r_{crit} can be estimated:

$$r_{crit} = r_1 + \left(\frac{3(\overline{V}_A)}{4\pi N}\right)^{1/3} = 10^{-10}\,\text{m} + \left[\frac{3\left(147.48\text{cm}^3/\text{mol}\right)\left(1\,\text{m}^3/10^6\,\text{cm}^3\right)}{4\pi(6.02 \times 10^{23}\,\text{mol}^{-1})}\right]^{1/3}$$

$$= 4.88 \times 10^{-10}\,\text{m} = 4.88\,\text{Å}.$$

Now calculate the diffusivity of naphthalene in air using Equation 5.1:

$$D_{A1} = \frac{2\left(\frac{RT}{\pi}\right)^{3/2}}{3NM_A^{1/2}r_{crit}^2 P}$$

$$= \frac{2\left[\dfrac{\left(8.314\text{Pa}\,\text{m}^3/\text{mol}\cdot\text{K}\right)(298\text{K})}{\pi}\right]^{3/2}\left(10,000\,\text{cm}^2/\text{m}^2\right)}{3(6.02 \times 10^{23}\,\text{mol}^{-1})\left(128.17\text{g/mol}\right)^{1/2}(4.88 \times 10^{-10}\,\text{m})^2(101,325\,\text{Pa})}$$

$$= 0.001\text{cm}^2/\text{s}.$$

(b) Calculate the collision diameter for naphthalene using Equation 5.4 or 5.5, then calculate the binary collision diameter for air–naphthalene using Equation 5.3 ($\sigma_1 = 3.617\,\text{Å}$):

$$\sigma_A = 1.18 \left(V_A^{(b)}\right)^{1/3} = 1.18 \left(147.48\,\text{cm}^3/\text{mol}\right)^{1/3} = 6.23\,\text{Å},$$

$$\sigma_A = 2.44 \left(\frac{T_A^{(c)}}{P_A^{(c)}}\right)^{1/3} = 2.44\ (748\,\text{K}/40.46\,\text{atm})^{1/3} = 6.45\,\text{Å},$$

$$\sigma_{A1} = \frac{\sigma_A + \sigma_1}{2} = \frac{6.45\,\text{Å} + 3.617\,\text{Å}}{2} = 5.03\,\text{Å}.$$

To determine the collision integral using Figure 5.1, the reduced temperature must be calculated (Equation 5.6), which requires the determination of the interaction energies for naphthalene (Equation 5.8 or 5.9) and the binary air–naphthalene system (Equation 5.7, $\varepsilon_1 = 1.34 \times 10^{-14}$ ergs):

$$\varepsilon_A = \frac{1.15\,T_A^{(b)}R}{N} = \frac{1.15\,(490\,\text{K}) \left(8.314 \times 10^7\,\text{ergs/mol\,K}\right)}{6.02 \times 10^{23}\,\text{mol}^{-1}} = 7.78 \times 10^{-14}\,\text{ergs},$$

$$\varepsilon_A = \frac{0.77\,T_A^{(c)}R}{N} = \frac{0.77\,(748\,\text{K}) \left(8.314 \times 10^7\,\text{ergs/mol\,K}\right)}{6.02 \times 10^{23}\,\text{mol}^{-1}} = 7.95 \times 10^{-14}\,\text{ergs},$$

$$\varepsilon_{A1} = \sqrt{\varepsilon_A \varepsilon_1} = \sqrt{(1.34 \times 10^{-14}\,\text{ergs})(7.95 \times 10^{-14}\,\text{ergs})} = 3.26 \times 10^{-14}\,\text{ergs},$$

$$T^* = \frac{RT}{N\varepsilon_{A1}} = \frac{\left(8.314 \times 10^7\,\text{ergs/mol} \cdot \text{K}\right)(298\,\text{K})}{(6.02 \times 10^{23}\,\text{mol}^{-1})(3.26 \times 10^{-14}\,\text{ergs})} = 1.26\ \text{here.}$$

The diffusion collision integral, $\Omega_D = 1.3$ (read from Figure 5.1). The diffusivity can then be calculated using Equation 5.2:

$$D_{A1} = \frac{0.001858\,T^{3/2} \left(\dfrac{1}{M_A} + \dfrac{1}{M_1}\right)^{1/2}}{P\sigma_{A1}^2\Omega_D}$$

$$= \frac{0.001858\,(298\,\text{K}) \left(\dfrac{1}{128.17\,\text{g/mol}} + \dfrac{1}{28.97\,\text{g/mol}}\right)^{1/2}}{(1\,\text{atm})(5.03\,\text{Å})^2(1.3)} = 0.060\,\text{cm}^2/\text{s}.$$

(c) Calculate the value of the molecular weight correction term (Equation 5.16) and use the previously determined binary collision diameter and collision integral to calculate the diffusivity:

$$M_{A1} = \frac{1}{M_A} + \frac{1}{M_1} = \frac{1}{128.17\,\text{g/mol}} + \frac{1}{28.97\,\text{g/mol}} = 0.042\,\text{mol/g},$$

$$D_{A1} = \frac{\left(3.03 - 0.98M_{A1}^{1/2}\right)\,T^{3/2}M_{A1}^{1/2}}{\sigma_{A1}^2\,\Omega_D P}$$

$$= \frac{\left[3.03 - 0.98\left(0.042\,\text{mol/g}\right)^{1/2}\right]\,(298\,\text{K})^{3/2}\left(0.042\,\text{mol/g}\right)^{1/2}}{(5.03\,\text{Å})^2(1.3)(1.013 \times 10^3\,\text{kPa})}$$

$$= 0.095\,\text{cm}^2/\text{s}.$$

(d) Using the molecular volume calculated above, the diffusivity can be determined using Equation 5.17:

$$D_{A1} = \frac{10^{-3}\,T^{7/4}M_{A1}^{1/2}}{P\left(\overline{V}_A^{1/3} + \overline{V}_1^{1/3}\right)^2}$$

$$= \frac{10^{-3}(298\,\text{K})^{7/4}\left(0.042\,\text{mol/g}\right)^{1/2}}{(1\,\text{atm})\left[(147.48\,\text{cm}^3/\text{mol})^{1/3} + (19.7\,\text{cm}^3/\text{mol})^{1/3}\right]^2} = 0.069\,\text{cm}^2/\text{s}.$$

(e) Using the correlation shown in Figure 5.2, the molecular weight can be used to estimate the diffusivity:

$$D_{A1} = 1.304 \cdot M_A^{-0.584} = 1.304 \cdot \left(128.17\,\text{g/mol}\right)^{-0.584} = 0.077\,\text{cm}^2/\text{s}.$$

(f) Using the parameters described in Equation 5.18:

$$D_{A1} = 2.35 \cdot \overline{V}_A^{-0.73} = 2.35 \cdot \left(147.48\,\text{cm}^3/\text{mol}\right)^{-0.73} = 0.061\,\text{cm}^2/\text{s}.$$

Comparison of estimated values:

Method	D_{A1} (cm^2 s^{-1})	Error
Kinetic theory of gases	0.001	—
Chapman–Enskog	0.060	−29%
Wilke–Lee	0.095	13%
Fuller et al.	0.069	−18%
Molecular weight correlation	0.077	−8%
Molar volume correlation	0.061	−27%
Measured value	0.084 ± 0.010	

Example 5.2: Diffusivity of Naphthalene in Water at 25°C

Estimate molecular diffusivities (cm^2 s^{-1}) for naphthalene in water using the following methods described in Chapter 5: (a) Stokes–Einstein, (b) Wilke–Chang, (c) Hayduk–Laudie, (d) Scheibel, and (e) the molecular weight correlation (Figure 5.4). Compare the results to the measured value, $D_{A2} = 7.64 \times 10^{-6}$ cm^2 s^{-1} (Gustafson and Dickhut, 1994a).

SOLUTION

(a) To calculate the diffusivity, the viscosity of water (Figure 5.5, $\eta_2 = 0.0009$ kg m^{-1} s^{-1} = 0.0009 Pa s) and the molecular radius are required. The radius of naphthalene can be estimated using the molar volume calculated in Example 5.1 ($V_A = 147.48$ cm^3 mol^{-1}):

$$r_A = \left(\frac{3\overline{V}_A}{4\pi N}\right)^{1/3} = \left[\frac{3 \cdot \left(147.48\,\text{cm}^3/\text{mol}\right)\left(10^6\,\text{m}^3/\text{cm}^3\right)}{4\pi(6.02 \times 10^{23}\,\text{mol}^{-1})}\right]^{1/3}$$

$$= 3.88 \times 10^{-10}\,\text{m} = 3.88\,\text{Å}.$$

Now applying Equation 5.23:

$$D_{A2} = \frac{RT}{6\pi N\eta_2 r_A} = \frac{\left(8.314\,\text{Pa}\cdot\text{m}^3/\text{mol}\cdot\text{K}\right)(298\,\text{K})\left(10^4\text{cm}^2/\text{m}^2\right)}{6\pi(6.02 \times 10^{23}\,\text{mol}^{-1})(0.0009\,\text{Pa}\cdot\text{s})(3.88 \times 10^{-10}\,\text{m})}$$

$$= 6.25 \times 10^{-6}\,\text{cm}^2/\text{s}.$$

(b) From the molar volume at the normal boiling point (atmospheric pressure) and viscosity data, the diffusivity can be calculated using the Wilke–Chang equation (Equation 5.24). The molar volume can be calculated from the chemical formula using the data in Table 5.2:

$$\overline{V}_A = 10v_C + 8v_H - C_{Arom}$$

$$= 10\left(14.8\,\text{cm}^3/\text{mol}\right) + 8\left(3.70\,\text{cm}^3/\text{mol}\right) - 15\,\text{cm}^3/\text{mol} = 162.2\,\text{cm}^3/\text{mol},$$

$$D_{A2} = \frac{7.4 \times 10^{-8}\,(\Phi_2 M_2)^{1/2}\,T}{(\overline{V}_A^{(b)})^{0.6}\eta_2}$$

$$= \frac{7.4 \times 10^{-8}\left[(2.26)\left(18\,\text{g/mol}\right)\right]^{1/2}(298\,\text{K})}{\left(162.2\,\text{cm}^3/\text{mol}\right)^{0.6}(0.9\,\text{cP})}$$

$$= 7.38 \times 10^{-6}\,\text{cm}^2/\text{s}$$

(c) As in part b, all parameters required to calculate the diffusivity are available; applying Equation 5.25:

$$D_{A2} = \frac{13.26 \times 10^{-5}}{\left(\overline{V}_A^{(b)}\right)^{0.589}\eta_2^{1.14}} = \frac{13.26 \times 10^{-5}}{\left(162.2\,\text{cm}^3/\text{mol}\right)^{0.589}(0.9\,\text{cP})^{1.14}} = 7.46 \times 10^{-6}\,\text{cm}^2/\text{s}.$$

(d) Diffusivity is estimated using Equations 5.26 and 5.27:

$$K = 8.2 \times 10^{-8} \left[1 + \left(\frac{3\overline{V}_2^{(b)}}{V_A^{(b)}} \right)^{2/3} \right] = 8.2 \times 10^{-8} \left[1 + \left(\frac{3\left(18.9\,\text{cm}^3/\text{mol}\right)}{(162.2\,\text{g/mol})} \right)^{2/3} \right]$$

$$= 1.23 \times 10^{-7},$$

$$D_{A2} = \frac{K\,T}{\left(V_A^{(b)}\right)^{1/3} \eta_2} = \frac{(1.23 \times 10^{-7})\,(298\,\text{K})}{(162.2\,\text{cm}^3/\text{mol})^{1/3}\,(0.9\,\text{cP})} = 7.47 \times 10^{-6}\,\text{cm}^2/\text{s}.$$

(e) Using the correlation shown in Figure 5.4, the molecular weight can be used to estimate the aqueous diffusivity:

$$D_{A2} = 2 \times 10^{-4} \cdot M_A^{-0.654}$$

$$= 2 \times 10^{-4} \cdot (128.17\,\text{g/mol})^{-0.654}$$

$$= 8.37 \times 10^{-6}\,\text{cm}^2/\text{s}$$

Comparison of estimated values:

Method	$D_{A2} \times 10^6$ (cm^2 s^{-1})	Error
Stokes–Einstein	6.25	−18%
Wilke–Chang	7.38	−3%
Hayduk–Laudie	7.46	−2%
Scheibel	7.47	−2%
Molecular weight correlation	8.37	−10%
Measured value	7.64 ± 0.08	

Example 5.3: Diffusivity of gypsum in water at 25°C

Estimate the molecular diffusivity (cm^2 s^{-1}) of (a) calcium sulfate salt using the Nernst equation and (b) the cationic (Ca^{2+}) and anionic (SO_4^{2-}) components of calcium sulfate in water using the molecular weight correlation shown in Figure 5.4 and the correction factor (Equation 5.30) for ionic species in water. Compare the results to the measured values for the individual ions, $D_{Ca^{-2}} = 7.93 \times 10^{-6}$ cm^2 s^{-1} and $D_{SO_4^{-2}} = 10.7 \times 10^{-6}$ cm^2 s^{-1} (Thibodeaux, 1996).

SOLUTION

(a) Look up values for the limiting cationic and anionic conductances of calcium and sulfate (Table 5.3). Then calculate the n term using Equation 5.29 and diffusivity can be calculated using Equation 5.28:

$$n = \frac{1}{x^+} + \frac{1}{x^-} = \frac{1}{2} + \frac{1}{2} = 1,$$

$$D_{A2} = \frac{nRT}{\left(\dfrac{1}{\lambda_+^0} + \dfrac{1}{\lambda_-^0}\right)\Im^2} = \frac{(8.314\,\text{J/mol}\cdot\text{K})\,(298\,\text{K})}{\left(\dfrac{\text{cm}^6}{119A\cdot V\cdot g_{eq}} + \dfrac{\text{cm}^6}{160A\cdot V\cdot g_{eq}}\right)\left(96\,500\dfrac{C}{g_{eq}}\right)^2}$$

$$= 18.20 \times 10^{-5}\,\text{cm}^2/\text{s}.$$

(b) Determine the ion correction factors (Equation 5.30) then use the molecular weight correlation in Figure 5.4 to estimate the diffusivity for each ion:

$$F_{Ca^{-2}} = 0.12\left(\frac{M_{ion}}{x_{ion}}\right)^{0.54} = 0.12\left(\frac{40.1\,\text{g/mol}}{2}\right)^{0.54} = 0.606,$$

$$D_{Ca^{-2}} = 2 \times 10^{-4} \cdot F_{Ca^{-2}} M_A^{-0.654} = 2 \times 10^{-4} \cdot 0.606 \cdot (40.1\,\text{g/mol})^{-0.654}$$

$$= 11.40 \times 10^{-6}\,\text{cm}^2/\text{s},$$

$$F_{SO_4^{-2}} = 0.12\left(\frac{M_{ion}}{x_{ion}}\right)^{0.54} = 0.12\left(\frac{96.1\,\text{g/mol}}{2}\right)^{0.54} = 0.971,$$

$$D_{SO_4^{-2}} = 2 \times 10^{-4} \cdot F_{SO_4^{-2}} \cdot M_A^{-0.654}$$

$$= 2 \times 10^{-4} \cdot 0.971 \cdot (96.1\,\text{g/mol})^{-0.654} = 10.50 \times 10^{-6}\,\text{cm}^2/\text{s}.$$

The estimated value for the diffusivity of Ca^{2+} in water is 30% higher than the measured value and the estimated value for SO_4^{2-} was 2% lower.

Example 5.4: Effective Diffusivity of Naphthalene in Saturated Sediment

Estimate the effective diffusivity of naphthalene in saturated sediment collected from Campus Lake on the LSU campus in Baton Rouge using the following properties of the material and the saturated porous media method described in Chapter 5.

Diffusivity of naphthalene in water $(D_{A2}) = 7.64 \times 10^{-6}\,\text{cm}^2\,\text{s}^{-1}$
Moisture content $(MC) = 0.65\,\text{g}\,\text{g}^{-1}$
Fraction organic carbon $(f_{OC}) = 0.06\,\text{g}\,\text{g}^{-1}$
Bulk density $(\rho_3) = 1.58\,\text{g}\,\text{cm}^{-3}$
Organic carbon–water partition coefficient $(K_{OC}) = 1000\,\text{cm}^3\,\text{g}^{-1}$

SOLUTION

The moisture content can be used to estimate the porosity of the sediment bed using the density of water $(1.0\,\text{g}\,\text{cm}^{-3})$ and the mineral component of the sediment $(\sim 2.65\,\text{g}\,\text{cm}^{-3})$

$$\varphi_3 = \frac{\dfrac{MC}{\rho_1}}{\dfrac{MC}{\rho_1} + \dfrac{(1-MC)}{\rho_{min}}} = \frac{\dfrac{0.65\,\text{g/g}}{1.0\,\text{g/cm}^3}}{\dfrac{0.65\,\text{g/g}}{1.0\,\text{g/cm}^3} + \dfrac{(1-0.65\,\text{gg})}{2.65\,\text{g/cm}^3}} = 0.83.$$

By substituting Equations 5.34, 5.36, 5.37, and 5.38 into Equation 5.33, the effective diffusivity in the sediment can be estimated:

$$D_{eff} = \frac{D_{A2}}{R_P R_C} = \frac{D_{A2}}{\left(\theta_3^2\right)\left(1 + \frac{\rho_3}{\varphi_3}K_{23}\right)} = \frac{D_{A2}}{\left[1 - \ln(\varphi_3^2)\right]\left[1 + \frac{\rho_3}{\varphi_3}(\varphi_{OC}K_{OC})\right]},$$

$$D_{eff} = \frac{7.64 \times 10^{-6}\,cm^2/s}{[1 - \ln(0.83^2)]\left[1 + \dfrac{1.58\,g/cm^3}{0.83}(0.06 \cdot 1000\,cm^3/g)\right]} = 4.83 \times 10^{-8}\,cm^2/s.$$

Example 5.5: Effective Diffusivity of Naphthalene in Unsaturated Soil

Estimate the effective diffusivity of naphthalene in a sandy soil using the given properties and the Millington–Quirk method described in Chapter 5:

> Diffusivity of naphthalene in air $(D_{A1}) = 0.084\,cm^2\,s^{-1}$
> Total porosity $(\varphi_3) = 0.4\,cm^3\,cm^{-3}$
> Fraction of void space occupied by air $(\varphi_1) = 0.9\,cm^3\,cm^{-3}$
> Bulk density $(\rho_3) = 1.67\,g\,cm^{-3}$

SOLUTION

Applying Equation 5.39, the effective diffusivity through is estimated to be

$$D_{eff} = D_{A1}\frac{\varphi_1^{10/3}}{\varphi_3^2} = 0.084\,cm^2/s\frac{(0.9 \cdot 0.4)^{10/3}}{(0.4)^2} = 0.049\,cm^2/s.$$

Example 5.6: Diffusivity of Naphthalene in Hydraulic Oil

Estimate molecular diffusivities $(cm^2\,s^{-1})$ for naphthalene in hydraulic oil $(\eta = 3.09\,cP)$ using the following methods described in Chapter 5: (a) Scheibel and (b) Hayduk–Minhas. Compare the results to the measured value, $D_{A-NAPL} = 5.40 \times 10^{-6}\,cm^2\,s^{-1}$ (Timothy et al., 1973).

SOLUTION

(a) To calculate the diffusivity using the Scheibel method, the K parameter must be calculated or assumed (using the rules of thumb presented). For hydraulic oil, it is assumed that the criteria for using $K = 17.5 \times 10^{-8}$ $\left(\overline{V}_A^{(b)} < 2.5\,\overline{V}_{solvent}^{(b)}\right)$ are met and Equation 5.26 can be applied:

$$D_{A\text{-}NAPL} = \frac{K\,T}{\left(\overline{V}_A^{(b)}\right)^{1/3}\eta_{NAPL}} = \frac{(17.5 \times 10^{-8})\,(298\,K)}{(147.48\,cm^3/mol)^{1/3}\,(3.09\,cP)} = 3.19 \times 10^{-6}\,cm^2/s.$$

(b) To calculate the diffusivity of naphthalene in hydraulic oil using the Hayduk–Minhas method, the exponent K in Equation 5.42 must be determined using Equation 5.43:

$$K = \frac{10.2}{V_A^{(b)}} - 0.791 = \frac{10.2}{(147.48\,cm^3/mol)} - 0.791 = -0.722,$$

$$D_{A-NAPL} = \frac{13.3 \times 10^{-8} T^{1.47} \eta_{NAPL}^K}{\left(V_A^{(b)}\right)^{0.71}} = \frac{13.3 \times 10^{-8}(298\,K)^{1.47}(3.09\,cP)^{-0.722}}{(147.48\,cm^3/mol)^{0.71}}$$

$$= 7.34 \times 10^{-6}\,cm^2/s.$$

The Scheibel method underestimates the diffusivity by 41%, while the Hayduk–Minhas method over predicts the measured value by 36%.

Example 5.7

Using the temperature adjustment parameters in the text (Equations 5.20 and 5.32 for air and water, respectively), estimate the diffusivity of naphthalene at (a) 0°C and (b) 40°C in air and water using the measured values for 25°C from Gustafson and Dickhut (1994a,b) as reference values.

$$D_{A1}\,(25°C) = 0.084\,cm^2\,s^{-1},$$
$$D_{A2}\,(25°C) = 7.64 \times 10^{-6}\,cm^2\,s^{-1}.$$

SOLUTION

(a) First, convert all temperatures into Kelvin by adding 273.15 to each value. To estimate the diffusivity of naphthalene in air at 0°C (273.15 K) and 40°C (313.15 K), use Equation 5.20

$$D_{A1}(T_2 = 273\,K) = D_{A1}(298\,K)\left(\frac{T_2}{298\,K}\right)^{7/4} = 0.084\,cm^2/s\left(\frac{273\,K}{298\,K}\right)^{7/4}$$

$$= 0.072\,cm^2/s,$$

$$D_{A1}(T_2 = 313\,K) = 0.084\,cm^2/s\left(\frac{313\,K}{298\,K}\right)^{7/4} = 0.092\,cm^2/s.$$

The measured value for the diffusivity of naphthalene in air at 40°C is 0.09 cm^2 s^{-1} (2% error).

(b) To estimate the diffusivity of naphthalene in water at 0°C and 40°C, the viscosity of water at both temperatures must be determined using Figure 5.5: $\eta(0°C) = 1.7\,cP$; $\eta(40°C) = 0.6\,cP$. Then Equation 5.32 can be applied

$$D_{A2}(T_2 = 273\,K) = D_{A2}(T_R)\left(\frac{T_2}{T_R}\right)\left[\frac{\eta_2(T_R)}{\eta_2(T_2)}\right]$$

$$= 7.64 \times 10^{-6}\,cm^2/s\left(\frac{273\,K}{298\,K}\right)\left[\frac{0.9\,cP}{1.7\,cP}\right] = 3.71 \times 10^{-6}\,cm^2/s,$$

$$D_{A2}(T_2 = 313\,\text{K}) = 7.64 \times 10^{-6}\,\text{cm}^2/\text{s} \left(\frac{313\,\text{K}}{298\,\text{K}}\right) \left[\frac{0.6\,\text{cP}}{0.9\,\text{cP}}\right] = 12.04 \times 10^{-6}\,\text{cm}^2/\text{s}.$$

The measured value for the diffusivity of naphthalene in water at 40°C is $10.60 \times 10^{-6}\,\text{cm}^2\,\text{s}^{-1}$ (14% error).

REFERENCES

Atkins, P. 1998. *Physical Chemistry*, W.H. Freeman and Company, New York, NY.

Boethling, R.S. and Mackay, D. 2000. *Handbook of Property Estimation Methods for Chemicals: Environmental and Health Sciences*, CRC Press, Boca Raton, FL.

Boudreau, B.P. 1996. The diffusive tortuosity of fine-grained unlithified sediments, *Geochimica et Cosmochimica Acta* **60**, 3139–3142.

Brokaw, R.S. 1969. Predicting transport properties of dilute gases, *Industrial and Engineering Chemistry Process Design and Development* **8**, 240–253.

Chapman, S. and Cowling, T.G. 1952. *The Mathematical Theory of Nonuniform Gases*, Cambridge University Press, New York, NY.

Chiou, C. 2003. *Partition and Adsorption of Organic Contaminants in Environmental Systems*, Wiley, New York, NY.

Cussler, E.L. 1984. *Mass Transfer in Fluid Systems*, Cambridge University Press, New York, NY.

Fairbanks, D.F. and Wilke, C.R. 1950. Diffusion coefficients in multicomponent gas mixtures, *Industrial and Engineering Chemistry* **42**, 471–475.

Freer, R. 1981. Diffusion in silicate minerals and glasses: A data digest and guide to the literature, *Contributions to Mineralogy and Petrology* **76**, 440–454.

Fuller, E.N., Schettler, P.D., and Giddings, J.C. 1966. A new method for prediction of binary gas-phase diffusion coefficients, *Industrial and Engineering Chemistry* **58**, 19–27.

Fuller, E.N., Ensley, K., and Giddings, J.C. 1969. Diffusion of halogenated hydrocarbons in helium. The effect of structure on collision cross sections, *Industrial and Engineering Chemistry* **73**, 3679–3685.

Giletti, B.J. and Shanahan, T.M. 1997. Alkali diffusion in plagioclase feldspar, *Chemical Geology* **139**, 3–20.

Gustafson, K.E. and Dickhut, R.M. 1994a. Molecular diffusivity of polycyclic aromatic hydrocarbons in aqueous solution, *Journal of Chemical and Engineering Data* **39**, 281–285.

Gustafson, K.E. and Dickhut, R.M. 1994b. Molecular diffusivity of polycyclic aromatic hydrocarbons in air, *Journal of Chemical and Engineering Data* **39**, 286–289.

Hayduk, W. and Laudie, H. 1974. Prediction of diffusion coefficients for nonelectrolytes in dilute aqueous solutions, *American Institute of Chemical Engineers Journal* **20**, 611–615.

Hayduk, W. and Minhas, B.S. 1982. Correlations for prediction of molecular diffusivities in liquids, *Canadian Journal of Chemical Engineering* **60**, 295–299.

Herbert, B.M.J., Halsall, C.J., Jones, K.C., and Kallenborn, R. 2006. Field investigation into the diffusion of semi-volatile organic compounds into fresh and aged snow, *Atmospheric Environment* **40**, 1385–1393.

Hirschfelder, J.O., Bird, R.B., and Spotz, E.L. 1949. The transport properties of gases and gaseous mixtures. II, *Chemical Reviews* **44**, 205–231.

Le Bas, G. 1915. *The Molecular Volumes of Liquid Chemical Compounds*, Longmans, Green & Company, Ltd., London, UK.

Marrero, T.R. and Mason, E.A. 1972. Gaseous diffusion coefficient, *Journal of Physical and Chemical Reference Data* **1**, 1–118.

Millington, R.J. and Quirk, J.M. 1961. Permeability of porous solids, *Transactions of the Faraday Society* **57**, 1200–1207.

Othmer, D.F. and Thakar, M.S. 1953. Correlating diffusion coefficients in liquids, *Industrial and Engineering Chemistry* **45**, 589–593.

Perry, R.H. and Green, D.W. 1997. *Perry's Chemical Engineers' Handbook*, 7th Ed., McGraw-Hill, Columbus, OH.

Scheibel, E.G. 1954. Correspondence, *Industrial and Engineering Chemistry* **46**, 2007–2008.

Schwarzenbach, R.P., Gschwend, P.M., and Imboden, D.M. 2002. *Environmental Organic Chemistry*, Wiley, New York, NY.

Shen, L. and Chen, Z. 2007. Critical review of the impact of tortuosity on diffusion, *Chemical Engineering Science* **62**, 3748–3755.

Thibodeaux, L.J. 1996. *Environmental Chemodynamics: Movement of Chemicals in Air, Water, and Soil*, Wiley, New York, NY.

Timothy, G., Hiss, E.L., and Cussler, E.L. 1973. Diffusion in high viscosity liquids, *American Institute of Chemical Engineers Journal* **19**, 698–703.

Welty, J., Wicks, C.E., Wilson, R.E., and Rorrer, G.L. 2001. *Fundamentals of Momentum, Heat, and Mass Transfer*, Wiley, New York, NY.

Wilke, C.R. and Lee, C.Y. 1955. Estimation of diffusion coefficients for gases and vapors, *Industrial and Engineering Chemistry* **47**, 1253–1257.

Wilke, C.R. and Chang, P. 1955. Correlations of diffusion coefficients in dilute solutions, *American Institute of Chemical Engineers Journal* **1**, 264–270.

6 Deposition from the Atmosphere to Water and Soils with Aerosol Particles and Precipitation

Matthew MacLeod, Martin Scheringer, Christian Götz, Konrad Hungerbühler, Cliff I. Davidson, and Thomas M. Holsen

CONTENTS

6.1 Introduction 104
6.2 The Transport Processes 104
    6.2.1 Overview of Dry and Wet Deposition Processes 104
    6.2.2 Methods for Estimating Gas/Particle Partitioning
         for Organic Chemicals 106
         6.2.2.1 One-Parameter Correlations 107
         6.2.2.2 Two-Parameter Correlations 109
         6.2.2.3 Polyparameter Linear Free Energy Relationships 109
         6.2.2.4 Summary of Methods for Estimating Gas/Particle
                Partitioning of Organic Chemicals 110
6.3 Transport Theory 111
    6.3.1 Dry Particle Deposition 111
    6.3.2 Wet Particle and Wet Gaseous Deposition..........................112
    6.3.3 Wet Deposition on Snow 114
6.4 Transport Parameters and Correlations.....................................116
    6.4.1 Dry Particle Deposition 116
    6.4.2 Wet Particle and Wet Gaseous Deposition..........................121
6.5 Example Calculations 122
    6.5.1 Gas/Particle Partitioning..122
    6.5.2 Dry Deposition 124
    6.5.3 Wet Deposition 125
    6.5.4 Look-Up Tables for Mass Transfer Coefficients 127
6.6 Summary and Conclusion..128

Literature Cited 131
Further Reading 135

6.1 INTRODUCTION

Chemical contaminants in the atmosphere can be deposited to surfaces in association with aerosol particles or falling rain and snow. These are advective transport processes, since the chemical moves in association with aerosol particles, raindrops, or snowflakes. This chapter describes methods for estimating chemical fluxes associated with deposition of aerosol particles and precipitation, and provides recommended values for mass transfer coefficients for a range of environmental conditions. In this chapter we do not consider transport of gaseous species in the atmosphere and adjacent surfaces. These convective transport processes, termed dry deposition of gases, are covered in Chapter 2, Section 2.5.6 and Chapter 7, Section 7.3. exchange between air and plants in Chapter 7, air and water in Chapter 9, and air and snow in Chapter 18.

The deposition processes described in this chapter can be separated into two categories: dry and wet. *Dry deposition* refers to chemical deposited to the surface in association with depositing particles. Here, we deal primarily with deposition to "flat" surfaces with low surface roughness, such as bare soils and water. Chemical transport from the atmosphere to plants and within the canopy is covered in Chapter 7. Particle dry deposition velocities to vegetation may be considerably higher than deposition to flat surfaces. *Wet deposition* refers to chemical deposited to the surface in association with rain or snow. Rain and snow scavenge both particle-associated and gas-phase pollutants, thus we can further differentiate *wet particle deposition* and *wet gaseous deposition*. By their nature, wet deposition processes are episodic.

In this chapter, we first provide a qualitative description of the processes of wet and dry deposition. Since the rates of these processes are controlled to a large extent by the partitioning of contaminants between the gas phase and aerosols in the atmosphere, we review some common methods for estimating the relevant partition coefficient. We then present theory and data to support quantitative modeling of dry deposition, wet deposition during continuous rain, and wet deposition assuming episodic rain events within a steady-state model. We briefly address differences between wet deposition with snow as distinct from rain. Finally, we present a summary of recommended ranges of mass transfer coefficients to describe these processes, and present example calculations for hypothetical chemical property ranges that are representative of the universe of chemicals.

6.2 THE TRANSPORT PROCESSES

6.2.1 OVERVIEW OF DRY AND WET DEPOSITION PROCESSES

Figure 6.1 illustrates the pathways and partitioning that are components of the dry and wet deposition processes. Ignoring diffusion, a prerequisite for either dry or wet deposition to remove a chemical from the atmosphere is that the chemical must partition to either aerosol particles or falling precipitation. Usually this is described

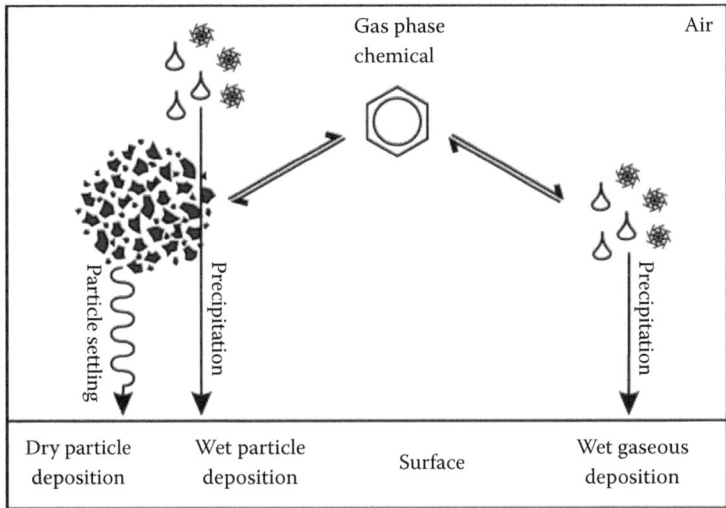

FIGURE 6.1 Pathways for dry particle deposition, wet particle deposition, and wet gaseous deposition of chemical contaminants from the atmosphere to the surface. Diffusive deposition processes are not dealt with in this chapter, and are not shown.

using a model that assumes equilibrium partitioning between the gas phase and aerosol particles, rain, or snow, indicated in Figure 6.1 by the double-headed arrows. Wet and dry deposition by advection are generally not important processes for volatile chemicals that are virtually entirely in the gas phase at all environmental temperatures.

Rates of both wet deposition and dry particle deposition can be estimated from mass transfer coefficients. The mass transfer coefficient for wet deposition is directly proportional to the rainfall rate (r, m/h, or equivalently $m^3/m^2/h$). Similarly, the dry deposition velocity of particles (U_D, m/h) determines the mass transfer coefficient for dry particle deposition. Both r and U_D can be converted into equivalent volumetric flow rates since wet deposition and dry particle deposition are advective processes.

Wet gaseous deposition is important for chemicals that are normally present in the gas phase but can be scavenged by rain drops and/or snowflakes, that is, chemicals that have large water–air and/or ice–air partition coefficients. Polar semivolatile chemicals such as many modern pesticides are examples of chemicals that are efficiently removed from the atmosphere by wet gaseous deposition.

Dry particle deposition and wet particle deposition are important processes for pollutants that are strongly associated with aerosols in the atmosphere. For example, heavy metals released from natural and anthropogenic sources are completely associated with aerosol particles in the atmosphere. Organic chemicals with low vapor pressure are also strongly associated with aerosol particles in the atmosphere, and are thus subject to dry and wet particle deposition. Because of the importance of partitioning between aerosols and the gas phase to the process of wet and dry particle deposition, we briefly review estimation methods below.

6.2.2 METHODS FOR ESTIMATING GAS/PARTICLE PARTITIONING FOR ORGANIC CHEMICALS

Methods for estimating the degree of sorption of chemicals to aerosols have been reviewed by Bidleman and Harner [1], Wania et al. [2], and Cousins and Mackay [3]. Here we provide a brief summary of methods reviewed by these authors, and some additional details of methods that have emerged since the publication of these reviews.

The distribution of chemicals between the gas phase and aerosol particles has most commonly been quantified by K_P (m^3 air/μg aerosol),

$$K_P = \frac{C_P^*}{C_A}, \tag{6.1}$$

where C_P^* is the particle-associated concentration (mol/μg aerosol) and C_A is the gas-phase concentration (mol/m^3 air). The chemical quantity unit (mol) cancels and can be substituted by a mass quantity such as grams or milligrams. Vapor–particle partitioning is usually determined from field studies using a high-volume air sampler equipped with a filter to capture aerosol particles followed in-line by a vapor trap to collect gas-phase chemical. The amount of chemical contaminant captured on the filter and in the vapor trap is determined analytically, and K_P is calculated as

$$K_P = \frac{\left(\dfrac{C_F}{TSP}\right)}{C_A}, \tag{6.2}$$

where C_F (mol/m^3 air) is the amount of chemical on the filter per unit volume of air, and TSP is the total suspended particle concentration (μg aerosol/m^3 air).

It is conceptually important to distinguish between *ab*sorption of chemical into the bulk particle and *ad*sorption of chemical to the surface of the particles. If absorption is the dominant sorption process and the aerosol particles are assumed to be homogeneous, the sorption capacity of aerosol particles is proportional to their volume. In this case, K_P can be converted directly into a volumetric partition coefficient, K_{PA} (m^3 air/m^3 aerosol particles), with knowledge of the density of aerosol particles (ρ_p, kg/m^3)

$$K_{PA} = K_P \rho_P \times 10^9, \tag{6.3}$$

where the factor 10^9 converts micrograms of aerosol into kilograms. K_{PA} is a convenient measure of gas/particle partitioning since it is C_P (mol/m^3 aerosols)/ C_A (mol/m^3 air), and it is thus directly analogous to other volumetric partition coefficients such as K_{AW} and K_{OW}. Assuming a density of aerosol particles of 2000 kg/m^3 [1,2], Equation 6.3 becomes $K_{PA} = K_P \times 2 \times 10^{12}$.

Alternatively, if adsorption dominates total sorption, K_P is proportional to the total particle surface area and it can be expressed as a surface area-based partition coefficient, K_{SA} (m^3 air/m^2 aerosol particle area):

$$K_{SA} = \frac{K_P \rho_P \times 10^9}{\left(\dfrac{SA}{V}\right)}, \qquad (6.4)$$

where SA/V is the surface area-to-volume ratio of the aerosol particles (m^2 particles/m^3 particles). If we assume aerosol particles to be spherical, the ratio of surface area-to-volume is $3/r$, where r is the particle radius. For an average particle radius of $2\,\mu m$, SA/V is $3/2 \times 10^6$ m^2 particles/m^3 particles. Thus, for our generic aerosol, Equation 6.4 becomes $K_{SA} = K_P \times 4/3 \times 10^6$.

Despite the important conceptual difference, it is often difficult to distinguish between absorption and adsorption as the dominant mechanism, particularly for data gathered in the field. Therefore it is useful to be able to convert between K_{PA} and K_{SA}, which are related by the surface area-to-volume ratio,

$$K_{PA} = \frac{SA}{V} \times K_{SA}. \qquad (6.5)$$

Assuming spherical particles with a radius of $2\,\mu m$, Equation 6.5 is $K_{PA} = 3/2 \times 10^6 \times K_{SA}$. Thus, for a generic aerosol particle scenario, K_{PA} can be estimated as $K_P \times 2 \times 10^{12}$ regardless of whether an absorptive or adsorptive mechanism is dominant. And, with knowledge of average particle radius and aerosol particle density, K_P, K_{PA}, and K_{SA} can be interconverted. For most mass balance modeling applications, the most convenient form is K_{PA}, and we use it exclusively for the remainder of this chapter.

The fraction of particle-associated chemical in bulk air Φ, (dimensionless), can be calculated from K_{PA} and the volume fraction of aerosol particles in air,

$$\phi = \frac{K_{PA}}{K_{PA} + \dfrac{V_A}{V_P}}, \qquad (6.6)$$

where V_A and V_P are the volumes of air and aerosols, respectively.

6.2.2.1 One-Parameter Correlations

Pankow proposed two models of vapor/particle partitioning based on surface adsorption [4] and absorption [5]. Both models predict inverse relationships between K_{PA} and the liquid-state saturation vapor pressure (P_L, Pa) that have the form [2]:

$$K_{PA} = \frac{x}{P_L}. \qquad (6.7)$$

In the adsorption model, the parameter x is a function of available surface area and the enthalpy of sorption. In the absorption model, x is proportional to the amount of available organic matter and inversely proportional to the activity coefficient of chemical dissolved in organic matter. Based on data for PAH sorption to urban aerosols,

Mackay [6] has suggested $x = 6 \times 10^6$ Pa as a reasonable estimate that allows K_{PA} to be estimated from Equation 6.7.

Finizio et al. [7] provided an alternative interpretation of the Pankow absorption model [5] in terms of the octanol/air partition coefficient (K_{OA}, m^3 air/m^3 octanol)

$$K_{PA} = BK_{OA}. \tag{6.8}$$

The coefficient B(m^3 octanol/m^3 aerosol particles) is directly proportional to the fraction of available organic matter in aerosols (f_{OM}), the ratio of densities of aerosol particles and octanol (ρ_P/ρ_O), the ratio of molecular weights of octanol and organic matter (M_O/M_{OM}), and the ratio of activity coefficients of the chemical in octanol and organic matter (γ_O/γ_{OM}),

$$B = f_{OM} \frac{\rho_P}{\rho_O} \frac{M_O}{M_{OM}} \frac{\gamma_O}{\gamma_{OM}}. \tag{6.9}$$

Bidleman and Harner [1] suggested a generic value of $B = 0.5$ based on $\rho_O = 840 \, \text{kg/m}^3$, and the assumptions that $f_{OM} = 0.5$, $\rho_P = 2000 \, \text{kg/m}^3$, $(M_O/M_{OM}) = 1$ and $(\gamma_O/\gamma_{OM}) = 1$.

Götz et al. [8] reviewed the composition of aerosol particles in different representative environmental scenarios and found that the fine fraction ($<2.5 \, \mu\text{m}$ diameter) has f_{OM} that is typically 3.5 times higher than the coarse fraction (diameter between 2.5 and 10 μm). This difference in f_{OM} is attributable to different sources of the fine and coarse fractions. Götz et al. [8] also recommend $(M_O/M_{OM}) = 0.26$ based on measurements of the molecular weight of aerosol organic matter by Kalberer et al. [9]. Adopting this recommendation with the other assumptions above implies $B = 0.13$ is a reasonable generic value that can be used to estimate K_{PA} from K_{OA}. Further assuming that the total aerosol volume is equally distributed between the fine and coarse fractions implies B for the fine fraction is 0.20 and B for the coarse fraction is 0.057.

Xiao and Wania [10] compared vapor pressure and octanol/air partition coefficient as descriptors of partitioning between air and organic matter. Using a data set for over 200 nonpolar, nonionizing organic chemicals they demonstrated empirically that K_{OA} and P_L are approximately inversely proportional to one another,

$$P_L (\text{Pa}) \approx \frac{10^{6.7}}{K_{OA}}. \tag{6.10}$$

They conclude that K_{OA} and P_L are both equally well suited for describing partitioning of nonpolar organic chemicals between the gas phase and aerosol organic matter.

By combining Equations 6.7 and 6.10, we observe that assuming $x = 6 \times 10^6$ in Equation 6.7 is equivalent to assuming $B = 1.2$ in Equation 6.8. This result is not conceptually satisfying since it implies that bulk aerosols have a sorption capacity for nonpolar organic chemicals that exceeds that of octanol, which is unlikely. We therefore recommend $x = 6 \times 10^5$ for predicting K_{PA} from vapor pressure using Equation 6.7, and $B = 0.13$ for estimating K_{PA} from K_{OA} using Equation 6.8. These recommendations should provide estimates of K_{PA} that are approximately consistent whether they are based on vapor pressures or octanol/air partition coefficients.

6.2.2.2 Two-Parameter Correlations

Data from field studies of partitioning of semivolatile chemicals between aerosols and air are often empirically fitted using two-parameter equations based on P_L or K_{OA},

$$\log K_{PA} = m \log P_L + b, \qquad (6.11)$$

$$\log K_{PA} = m' \log K_{OA} + b', \qquad (6.12)$$

where m, m', b, and b' are fitting parameters specific to the set of data under consideration. Equation 6.11 is equivalent to the one-parameter approach (Equation 6.7) when $m = -1$ and $b = \log x$. Likewise, when $m' = 1$ and $b' = \log B$, Equation 6.12 reduces to the one-parameter Equation 6.8. Bidleman and Harner [1] provide a summary table of regressions of $\log K_P$ versus $\log P_L$ from field studies, and Xiao and Wania [10] provide references to studies using both P_L and K_{OA}. Generally, the absolute value of m and m' determined from field studies ranges between 0.6 and 1.

6.2.2.3 Polyparameter Linear Free Energy Relationships

In a critical review of gas/solid and gas/liquid partitioning of organic chemicals, Goss and Schwarzenbach [11] call attention to the potential for error in applying one- or two-parameter correlations to predict gas/particle partitioning. A one-parameter correlation with P_L assumes that activity coefficients of all sorbates in aerosol particles are equal and a one-parameter correlation with K_{OA} assumes that organic matter absorption is the dominant sorption process in aerosols and that the ratio of sorbate activities in octanol and aerosol organic matter are equal for all possible sorbates. However, compound-specific interactions between the sorbate molecules and aerosols can render these assumptions invalid, and this is especially possible for chemicals with polar functional groups. Deviations from these assumptions lead to slopes in Equations 6.11 and 6.12 with absolute values different from 1, and would imply errors in the one-parameter correlations. In addition, Goss and Schwarzenbach [11] point out that two-parameter correlations are only strictly valid within a group of structurally related substances because of the potential for different types of specific interactions to apply for different compound classes.

As an alternative, Goss and coworkers have advocated modeling partitioning between the gas phase and aerosol particles using polyparameter linear free energy relationships (pp-LFERs) [12]. A pp-LFER for absorption into the bulk particle phase has the general form

$$\log K_{PA} = a_{bulk} \log K_{HXA} + b_{bulk} \Sigma\beta + c_{bulk} \Sigma\alpha + d_{bulk} V_m + \text{const}_{bulk}. \qquad (6.13)$$

Each of the multiplicative terms in Equation 6.13 conceptually represents a type of possible interaction between the chemical and the aerosol particle bulk phase. These are van der Waals interactions ($a_{bulk} \log K_{HXA}$), electron donor–acceptor interactions ($b_{bulk} \Sigma\beta$ and $c_{bulk} \Sigma\alpha$) and cavity formation in the sorbent ($d_{bulk} V_m$). The constant, const_{bulk}, depends on the dimensions of $\log K_{PA}$. Log K_{HXA} is the decadic logarithm of the hexadecane/air partition coefficient, $\Sigma\beta$ and $\Sigma\alpha$ are chemical descriptors that characterize electron donor and electron acceptor ability, and V_m is the

McGowan molar volume of the chemical. These four chemical properties are examples of Abraham chemical descriptors, which are named after Michael Abraham, who pioneered this approach [13]. The parameters a_{bulk}, b_{bulk}, c_{bulk}, and d_{bulk} are empirically derived descriptors that are properties of the aerosol under consideration. These aerosol property descriptors are temperature-dependent and also dependent on relative humidity when adsorption is the dominant sorption mechanism. Polyparameter linear free energy relationships have been developed to describe absorptive partitioning between air and bulk aerosol particles [14,15], as well as adsorption to aerosol components such as mineral surfaces and salts [16,17] and to snow [18].

Götz et al. [8] compared a gas/particle partitioning model based on the pp-LFER approach with the K_{OA} approach with the goal of defining the limits of applicability of the K_{OA} model. They concluded that for nonpolar chemicals and cases where sorption is dominated by absorption into organic matter, the two model approaches are highly correlated and either is appropriate. However, for polar chemicals sorbing to low-organic matter aerosols at low humidity, such as in arctic or desert regions, the two approaches differed by up to several orders of magnitude and the pp-LFER approach is preferable.

6.2.2.4 Summary of Methods for Estimating Gas/Particle Partitioning of Organic Chemicals

Table 6.1 summarizes recommended methods for estimating the dimensionless aerosol particle/air partition coefficient, K_{PA}, from vapor pressure, K_{OA} or using pp-LFERs.

TABLE 6.1
Recommended Methods for Estimating K_{PA}

One-Parameter Correlations

$K_{PA} = 6 \times 10^{-5}/P_L$ (Pa)	Screening-level estimation of partitioning of polar and
$K_{PA} = 0.13 K_{OA}$	nonpolar substances in generic evaluative
Fine fraction $\quad K_{PA} = 0.20 K_{OA}$	environments.
Coarse fraction $\quad K_{PA} = 0.057 K_{OA}$	

Two-Parameter Correlations

$\log K_{PA} = m \log P_L + b$	Calculation of site-specific partitioning of polar and
$\log K_{PA} = m' \log K_{OA} + b'$	nonpolar substances where absorption by organic matter dominates. Fitting parameters can be applied within a substance class.

Polyparameter Linear Free Energy Relationships

$\log K_{PA} = a_{bulk} \log K_{HXA} + b_{bulk} \Sigma\beta$ $\quad + c_{bulk} \Sigma\alpha + d_{bulk} V_m$ $\quad + const_{bulk}$	General calculation of site-specific partitioning for polar and nonpolar substances and absorptive or adsorptive mechanisms.

6.3   TRANSPORT THEORY

A detailed and comprehensive theoretical description of dry particle deposition and wet deposition processes can be found in Seinfeld and Pandis [19], and treatments with a specific focus on trace contaminants are provided in the texts by Mackay [6], Thibodeaux [20], and Scheringer [21]. In this section, we review this transport theory and place special emphasis on a method to approximate the effect of intermittent rainfall events in a steady-state model.

6.3.1   DRY PARTICLE DEPOSITION

The flux of dry particle-associated chemical from the atmosphere to the surface (N_D, mol/h) can be calculated from the estimated dry deposition velocity of particles (U_D, m/h);

$$N_D = U_D A C_P V_P / V_A, \tag{6.14}$$

where A (m^2) is the horizontal area that receives the deposition, C_P (mol/m^3 particles) is the concentration of chemical in aerosol particles and V_P/V_A is the volume fraction of particles in air. In general cases it is more convenient to calculate N_D from the total chemical concentration in bulk air, C_B, where $C_B = C_A + C_P V_P / V_A$. Then,

$$N_D = U_D \phi A C_B. \tag{6.15}$$

The group $U_D \phi$ can be interpreted as a chemical- and environment-specific mass transfer coefficient for dry particle deposition, k_D (m/h).

Selecting accurate and representative estimates of U_D is a challenge because it bundles several competing processes. Figure 6.2 illustrates a conceptual resistance model for dry deposition of small and large particles. Small, low-density particles are mixed in the atmosphere like gases and do not fall under the influence of gravity. For these particles, U_D can be calculated as

$$U_D = \frac{1}{1/k_A + 1/k_B}, \tag{6.16}$$

where k_A (m/h) is the mass transfer coefficient for aerodynamic transport in the constant flux layer and k_B (m/h) is the mass transfer coefficient for transport through the stagnant air layer above the depositional surface by random Brownian motion. At a given time and location, k_A and k_B are determined by wind speed, atmospheric stability, and roughness height of the depositional surface. Windy conditions and a rough depositional surface increase k_A and k_B and thus increase U_D of small particles [19]. Additional information about k_A and k_B are presented in Chapter 7, where they are referred to as their reciprocal resistances ($R_A = 1/k_A$ and $R_B = 1/k_B$). For large and dense aerosol particles, dry deposition under the influence of gravity dominates and U_D is approximately equal to the gravitational settling velocity of particles through viscous air (U_S, m/h) [19].

In this chapter, we focus on deposition to surfaces with low surface roughness, such as bare soils and water. Dry deposition of fine particles to grassland or forested

Atmosphere

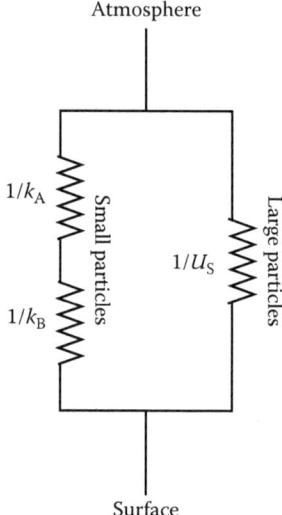

Surface

FIGURE 6.2 Conceptual resistance model of resistances that control the rate of dry particle deposition.

areas can be much higher because of the high surface roughness of the receiving surface, and is discussed in Chapter 7.

6.3.2 WET PARTICLE AND WET GASEOUS DEPOSITION

Chemical flux due to wet particle deposition (N_{WP}, mol/h) can be calculated as

$$N_{WP} = U_R Q A C_P V_P / V_A, \tag{6.17}$$

where U_R is the rain rate (m/h) and Q (dimensionless) is the scavenging ratio, representing the volume of air that is efficiently scavenged of aerosol particles by a unit volume of rain. This can also be expressed in terms of the total concentration in bulk air, C_B,

$$N_{WP} = U_R Q \phi A C_B. \tag{6.18}$$

The group $U_R Q \phi$ in Equation 6.18 can be interpreted as a chemical- and environment-specific mass transfer coefficient for wet particle deposition, k_{WP} (m/h).

Wet gaseous deposition (N_{WG}, mol/h) can be calculated as

$$N_{WG} = U_R A C_R, \tag{6.19}$$

where C_R (mol/m^3 rain) is the concentration of chemical in falling rain. Assuming that immediately after rain starts the gas-phase concentration of chemical (C_A, mol/m^3 air) becomes distributed between the gas phase and raindrops, we obtain

$$C_A V_A = C_{A,R} V_A + C_R V_R, \tag{6.20}$$

where $C_{A,R}$ (mol/m^3 air) is the concentration in the gas phase during rainfall. Rearranging Equation 6.20 yields

$$\frac{C_A}{C_R} = \frac{C_{A,R}}{C_R} + \frac{V_R}{V_A}. \tag{6.21}$$

If we assume equilibrium partitioning between the gas phase and raindrops during the rain event, $C_{A,R}/C_R = K_{AW}$, and

$$C_R = \frac{C_A}{K_{AW} + \dfrac{V_R}{V_A}} = \frac{C_B(1 - \phi)}{K_{AW} + \dfrac{V_R}{V_A}}. \tag{6.22}$$

Thus the concentration in rain and the wet gaseous deposition flux is dependent on K_{AW} and the volume fraction of raindrops in air during rainfall [22]. The term V_R/V_A in Equation 6.22 acts as a limit on the chemical concentration that can develop in raindrops for very hydrophilic substances with very low K_{AW}. It has a typical value of 6×10^{-8} m^3 water/m^3 air [22]. This term has often been omitted from calculations of wet gaseous deposition [6,21], which leads to overestimation of C_R and N_{WG} for hydrophilic compounds with log K_{AW} less than ~7.5.

Equation 6.19 can be expressed in terms of total concentration in bulk air as

$$N_{WG} = \frac{U_R(1 - \phi)}{K_{AW} + \dfrac{V_R}{V_A}} A C_B. \tag{6.23}$$

Thus the group $(U_R(1 - \phi))/(K_{AW} + V_R/V_A)$ in Equation 6.23 can be interpreted as a chemical- and environment-specific mass transfer coefficient for wet gaseous deposition, k_{WG} (m/h).

The equations above are formulated to calculate time-averaged wet particle and wet gaseous deposition flux over several rain events. This approach is particularly useful for calculating steady-state solutions to mass balance chemical fate models, since the time-variant rate of deposition with rainfall is represented as a continuous process, that is, the model assumes a constant light drizzle of rain. However, the effects of intermittent rain events on average chemical concentrations in air over several rain events are not well described for all chemicals by assuming constant rainfall [22,23].

Jolliet and Hauschild [22] point out that the intermittent nature of rainfall events imposes a limit on the timescale that chemicals can be removed from the atmosphere by wet particle and wet gaseous deposition. Consider a hypothetical chemical that is efficiently removed from the atmosphere by rain. Figure 6.3 illustrates the atmospheric residence time due to wet deposition for such a chemical over a rainfall cycle that includes a dry period of duration t_{dry} and a wet period of duration t_{wet}. If the chemical is released to the atmosphere immediately following the end of a rain event, its residence time in the atmosphere is t_{dry}. If it is released during a rain event, its residence time is approximately zero. The average atmospheric residence time for a chemical released during the dry period is $t_{dry}/2$. Over the long term,

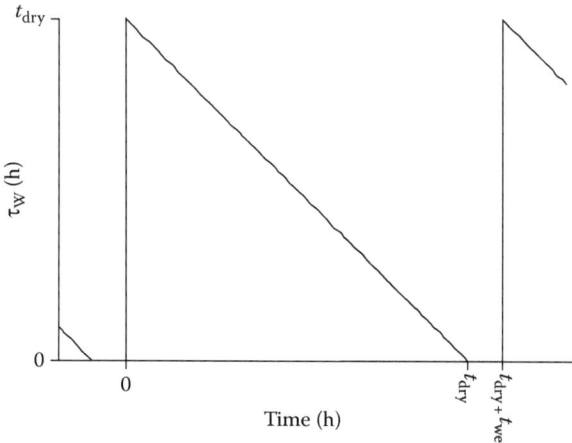

FIGURE 6.3 Atmospheric residence time due to wet deposition processes of a hypothetical chemical that is efficiently removed from the atmosphere by rainfall (τ_W, h) during a rain event cycle including a dry period (t_{dry}) and a wet period (t_{wet}).

the average atmospheric residence time over several cycles of dry and wet periods, $\bar{\tau}_{W,MIN}$ (h), is

$$\bar{\tau}_{W,MIN} = \frac{t_{dry}}{2} \times \frac{t_{dry}}{t_{dry} + t_{wet}}. \tag{6.24}$$

The subscript "MIN" denotes that $\bar{\tau}_{W,MIN}$ is the minimum possible long-term average atmospheric residence time due to wet deposition processes for chemicals that are efficiently scavenged by rain. The maximum possible long-term average flux of chemical to the surface due to wet deposition processes is

$$N_{W,MAX} = \frac{2(t_{dry} + t_{wet})}{t_{dry}^2} hAC_B, \tag{6.25}$$

where h (m) is the mixing height of the atmosphere, which is typically set at 1000 m, representing the planetary boundary layer. The term $2h(t_{dry} + t_{wet})/t_{dry}^2$ can be interpreted as an environment-specific maximum limit on the total mass transfer coefficient for wet deposition ($k_{W,MAX}$, m/h). When the group ($k_{WP} + k_{WG}$), is calculated assuming continuous rain, is greater than $k_{W,MAX}$, $k_{W,MAX}$ should be used to estimate the time-averaged wet deposition flux over several rain events.

6.3.3 Wet Deposition on Snow

If the temperature is below 0°C, snow instead of rain causes "wet" deposition of chemicals. Like rain, snow is an advective process that transfers both gaseous and particle-bound chemicals from the air to the ground, and it is desirable to be able to estimate the mass transfer coefficients for these processes. The most important factors determining the rate of wet deposition by snow are the snow–air partition coefficient of the chemical [18], the specific surface area of the snowflakes [24], and temperature, which influences both the snow–air partition coefficient and the specific surface area [25,26].

The mechanism of sorption of chemicals to snow is different from their dissolution in rain droplets. Equilibrium partitioning between the gas phase and snowflakes can be described using a partition coefficient between snow surface and air, K_{SA}. Recently, Roth et al. [18] have measured K_{SA} to snow for a set of 57 organic chemicals and derived the following pp-LFER describing sorption of organic chemicals:

$$\log K_{SA}(-6.8°C) = 0.639 \cdot \log K_{HXA} + 3.53 \cdot \sum \alpha + 3.38 \cdot \sum \beta - 6.85. \quad (6.26)$$

As described above in Section 6.2.2, K_{SA} has units of m^3 air/m^2 snow surface. The volumetric snowflake–air partition coefficient, $K_{FA}(m^3$ air/m^3 snow) is analogous to K_{PA}, and can be obtained from Equation 6.5, where the surface area-to-volume ratio of snow is calculated as the product of the specific surface area of the snowflakes (SSA_F in m^2 snow surface/kg snow), and the density of the snowflakes (ρ_F, in kg snow/m^3snow). Typical values for SSA_F and ρ_F are 100 m^2/kg [24] and 920 kg/m^3.

It is important to note that the extent of partitioning to snowflakes is temperature-dependent and will be higher at lower temperatures. Goss and Schwarzenbach [27], and Roth et al. [28] describe methods for estimating the temperature dependence of partitioning to snow.

From the snowflake–air partition coefficient (K_{FA}) and the precipitation rate for snow, U_F (m/h), the chemical flux due to wet deposition with snowflakes can be derived:

$$N_{WG,F} = \frac{U_F(1 - \phi)}{(K_{FA})^{-1} + \dfrac{V_F}{V_A}} \cdot AC_B, \quad (6.27)$$

where V_F/V_A is the volume fraction of snowflakes in air during a snowfall event. The group $(U_F(1 - \phi))/((K_{FA})^{-1} + V_F/V_A)$ in Equation 6.27 can be interpreted as a chemical- and environment-specific mass transfer coefficient for wet gaseous deposition on snowflakes, $k_{WG,F}$ (m/h).

Wet gaseous deposition by snow can be very efficient if the temperature is low. In particular, large molecules with a high affinity to snow because of high $\Sigma\beta$ values, such as α-HCH, are efficiently deposited with falling snow. Compounds with very low vapor, pressure such as high molecular weight PAHs and decabromodiphenylether, have high K_{FA} but also have high octanol–air partition coefficients, and are therefore mainly bound to atmospheric aerosol particles [25,26].

Wet particle deposition by snow can be calculated using an equation analogous to Equation 6.18,

$$N_{WP,F} = U_F Q\phi AC_B. \quad (6.28)$$

As a first approximation, the same scavenging ratio (Q) can be assumed to apply to both snow as rain in generic models.

6.4 TRANSPORT PARAMETERS AND CORRELATIONS

6.4.1 DRY PARTICLE DEPOSITION

Figure 6.4 summarizes the range of possible particle dry deposition velocities (U_D) as a function of particle diameter (the gray area and the primary vertical axis). This figure is adapted from information presented in Seinfeld and Pandis [19] on deposition velocity determined in a wind tunnel and on generic aerosol distributions. The range of U_D shown in gray in Figure 6.4 corresponds approximately to surface roughness heights in the range 0.1–3 cm and particle densities between 1000 and 4000 kg/m^3.

Generally speaking, the uncertainty range for particles smaller than 0.02–0.03 μm is a factor of 10× whereas for larger particles it is a factor of 5×. So, it is evident from Figure 6.4 that any generic, single value selected for U_D will be subject to large uncertainties due to variability in particle size, surface roughness, and particle density. Table 6.2 summarizes recommended values for f_{OM} and V_P/V_A for different environmental scenarios, based on data presented in Figure 6.4. The value for the generic aerosol scenario is from Bidleman and Harner [1], and values for other scenarios are estimated from data presented by Putaud et al. [29] and Götz et al. [8].

It is of interest to ask what deposition velocities might exist for smooth, flat surfaces, such as soil or water surfaces, or artificial surfaces such as asphalt, under a range of wind conditions. Such values may useful in defining lower limits to deposition from the ambient atmosphere. Measurements and models for dry deposition to rougher surfaces such as vegetation are discussed in Chapter 7.

For particles with diameter greater than several micrometers in light winds, deposition to smooth surfaces will be dominated by gravity; values of the sedimentation velocity U_S for particles of various sizes depositing under the influence of gravity are given by Seinfeld and Pandis [19]. For particles smaller than this size range, or for particles depositing to smooth, flat surfaces in strong winds, turbulent inertial deposition will increase the deposition velocity.

Data on particle deposition to smooth surfaces were obtained in the wind tunnel experiments by Sehmel [30] and Sehmel and Hodgson [31] for a wide range of particle sizes. For example, Figure 6.5 shows U_D versus particle diameter d_p for a roughness height of 0.001 cm at friction velocities of 720 m/h (light winds) and 3600 m/h (strong winds) for unit density particles. The line corresponding to the sedimentation velocity is also shown for comparison. Note that the curve for 720 m/h joins the sedimentation line for d_p greater than about 15 μm. The curve for 3600 m/h joins the sedimentation line for d_p greater than about 60 μm.

Figure 6.5 can be used with size distributions of chemical species from the literature to estimate a lower limit to the dry deposition velocity for that species on rougher natural surfaces. If the molar airborne concentration of a chemical in size range i is given by C_{Pi} (mol/m^3 particles) and the volume fraction of particles of size range i in air is given by V_{Pi}/V_A, then the overall average deposition velocity $U_{D-average}$ is given by

$$U_{D-average} = \frac{\sum_{i=1}^{n} U_{Di} C_{Pi} \dfrac{V_{Pi}}{V_A}}{\sum_{i=1}^{n} C_{Pi} \dfrac{V_{Pi}}{V_A}}. \tag{6.29}$$

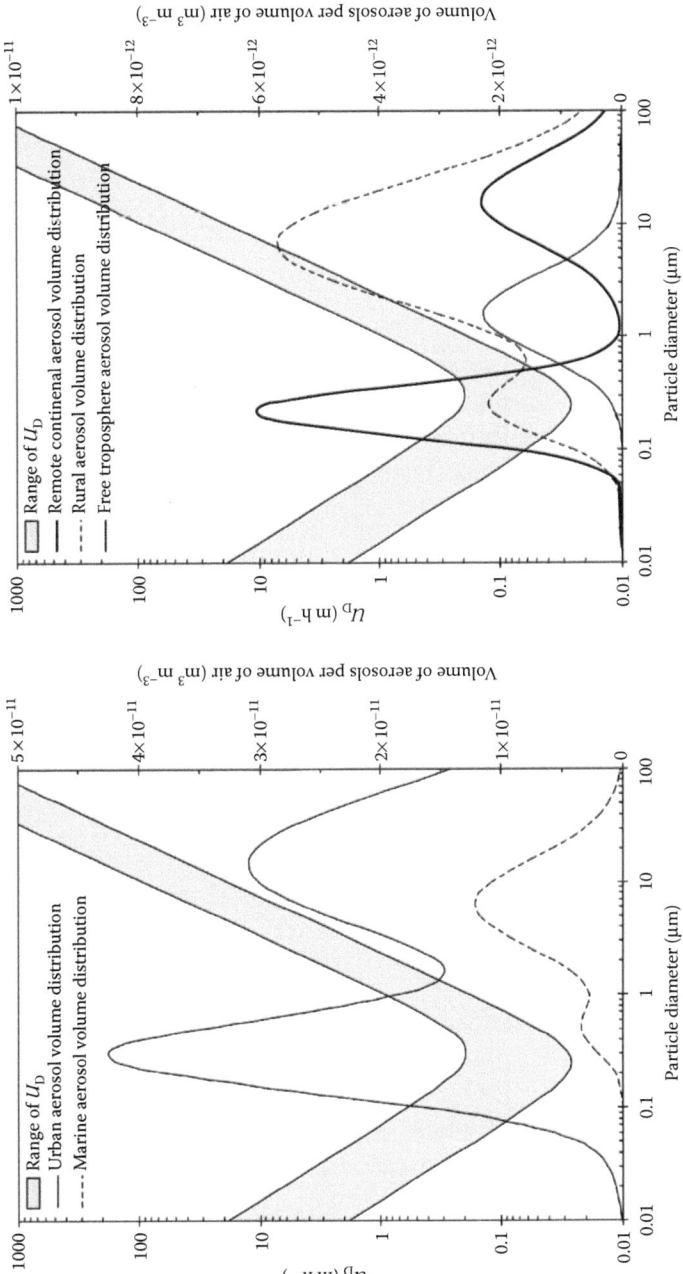

FIGURE 6.4 Particle dry deposition velocity (U_D, primary vertical axis) as a function of particle diameter compared with representative volume distributions of aerosol particles (secondary vertical axis).

TABLE 6.2
Recommended Values of Aerosol Particle Volume Fraction in Air (V_P/V_A, dimensionless) and Volume Fraction of Organic Matter in Aerosols (f_{OM}, dimensionless) for Different Environmental Scenarios

Aerosol Scenario		U_D (m/h)	V_P/V_A	f_{OM}
Generic		4.6	2×10^{-11}	0.5
Free troposphere		1	2×10^{-11}	0.2
Urban	Fine fraction	0.2	4×10^{-10}	0.3
	Coarse fraction	20	5×10^{-10}	0.1
Rural	Fine fraction	0.2	2×10^{-11}	0.3
	Coarse fraction	20	8×10^{-11}	0.1
Remote continental	Fine fraction	0.2	4×10^{-11}	0.3
	Coarse fraction	20	2×10^{-11}	0.1
Marine	Fine fraction	0.2	2×10^{-11}	0.1
	Coarse fraction	10	1×10^{-10}	0.05

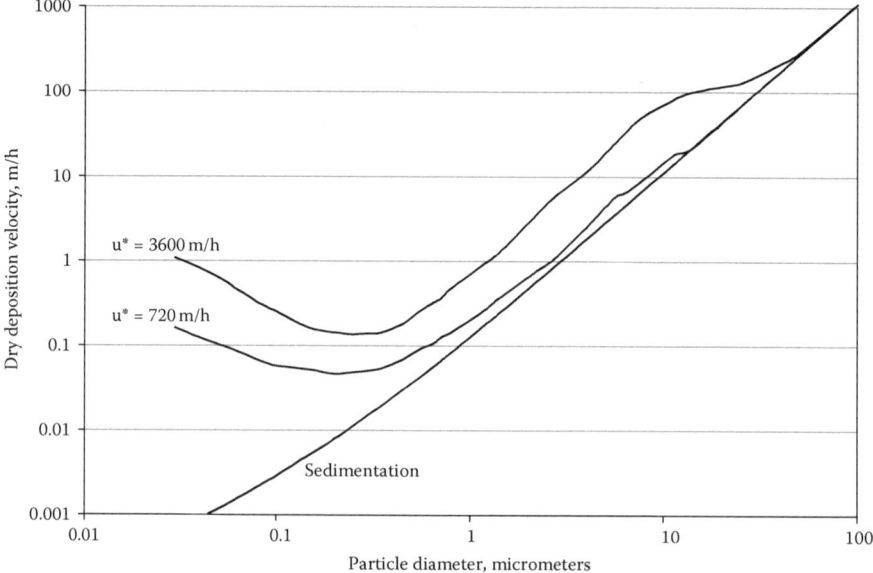

FIGURE 6.5 Dry deposition velocity of unit-density particles on a smooth, flat surface at friction velocities of $u^* = 720$ m/h and 3600 m/h. The line for sedimentation by gravity is also shown for comparison. Calculations for this graph are taken from the model of Sehmel and Hodgson [31].

TABLE 6.3

Calculated Dry Deposition Velocities for a Smooth, Flat Surface Based on Size Distribution Data in the Literature for Pb, Mn, and Ca

Metal	N	Average Diameter (μm)	Geom. Mean Concentration (ng/m^3)	U_{D-ave} (m/h)	
Pb	41	0.55	310	3.5	Sedimentation
				3.6	$u_* = 720$ m/h
				12	$u_* = 3600$ m/h
Mn	32	2.06	20	7.9	Sedimentation
				8.3	$u_* = 720$ m/h
				24	$u_* = 3600$ m/h
Ca	20	4.64	330	13	Sedimentation
				15	$u_* = 720$ m/h
				41	$u_* = 3600$ m/h

We have used size distribution data from the review of Milford and Davidson [32] for three trace metals with Equation 6.29 to calculate $U_{D-average}$ for these metals. Results are shown in Table 6.3. Note that N refers to the number of separate size distributions reported in the literature for each metal. Pb is primarily submicron, and is known to be a result of anthropogenic combustion processes including use of leaded gasoline, metallurgical processing, coal combustion, and numerous other sources. The small values of $U_{D-average}$ reflect the relatively small particle sizes associated with Pb. Note that $U_{D-average}$ is only slightly greater for a friction velocity of 720 m/h compared with sedimentation due to weak turbulent inertial deposition on the surface. However, the value of $U_{D-average}$ is considerably greater for a friction velocity of 3600 m/h due to much greater turbulent delivery to the surface that accompanies higher wind speeds.

The original calculations show that most of the mass flux of Pb is due to those particles in the uppermost size ranges. Milford and Davidson [32] assumed a maximum aerodynamic diameter of 40 μm; the estimated values of $U_{D-average}$ are sensitive to this assumed maximum.

Mn is a trace metal of intermediate sizes, with particles that originate both from anthropogenic and natural sources. The greater values of $U_{D-average}$ compared with Pb reflect the larger particle sizes.

Ca is mainly derived from the Earth's crust, with airborne particles resulting mainly from abrasion of rock and soil. The particle sizes of Ca are much larger than either Pb or Mn, and the deposition velocities are correspondingly greater. Table 6.4 summarizes typical $U_{D-average}$ values. Curves of U_D versus particle diameter for a smooth surface can be used as a rough estimate of deposition to a calm water surface as long as wave action is negligible. For a lake or ocean surface with significant waves, particle deposition will be enhanced. This is due to increased air turbulence from the wave motion as well as greater inertial impaction as particles are captured by the wave. Zufall et al. [33] have modeled waves with a variety of ratios of $2a/\lambda$ where

TABLE 6.4
Typical Overall Deposition Velocities

Pollutant	Overall U_D, m/h (Typical Range)	Comments	References
SO_4^{2-}	3–100		[37–39]
NO_3^-	3–70		[37,38,40]
NH_4^+	3–70		[37]
Cl^-	40–200		[37]
Primarily crustal metals (Al, Ca, Mg, K …)	40–100 70–400	Nonurban Urban	[37,41–45] [37,42–45]
Primarily anthropogenic metals (Pb, Cu, Zn, Ni, Cd …)	3–40 40–250	Nonurban Urban	[37,42–45] [37,42–45]
Particle associated semivolatile organics (PCBs, PAHs …)	3–70 20–200	Nonurban Urban	[42,46–49] [46,49–53]
TSP	20–200 40–400	Nonurban Urban	[40] [40]

$2a$ = peak-to-trough amplitude and λ = length of the wave. For $2a/\lambda = 0.03$, deposition of 1 μm diameter particles is enhanced by a factor of 50% over the corresponding deposition to a smooth, flat surface. For $2a/\lambda = 0.01$, deposition is enhanced by a factor of 100%.

If the particles are hygroscopic, deposition to a water surface may also be enhanced by particle growth in the humid boundary layer just over the surface. For example, ammonium sulfate particles will grow by water uptake and thus have a greater deposition velocity compared with nonhygroscopic particles. However, particle transport through the humid boundary layer may be faster than hygroscopic growth so that the effects of growth are limited.

Ammonium sulfate particles are normally dry crystalline solids as relative humidity increases up to a value of about 80%. If the humidity exceeds that value, the particle will uptake water and grow, transitioning from a solid to a solution droplet. But if humidity then falls below 80%, the droplet will not dry out but rather exist as a supersaturated solution as long as the humidity stays above 36%. Liquid droplets existing in such a supersaturated state are *metastable*, and such particles behave differently than particles that have never deliquesced even though they are identical in all other respects.

Figure 6.6 shows a graph of U_D versus particle diameter d_p for ammonium sulfate for several conditions. Curve A refers to the hypothetical case where there is no uptake of water and therefore no increase in U_D. This is the base case. Curve B shows the maximum increase in deposition velocity that can be expected, if hygroscopic growth

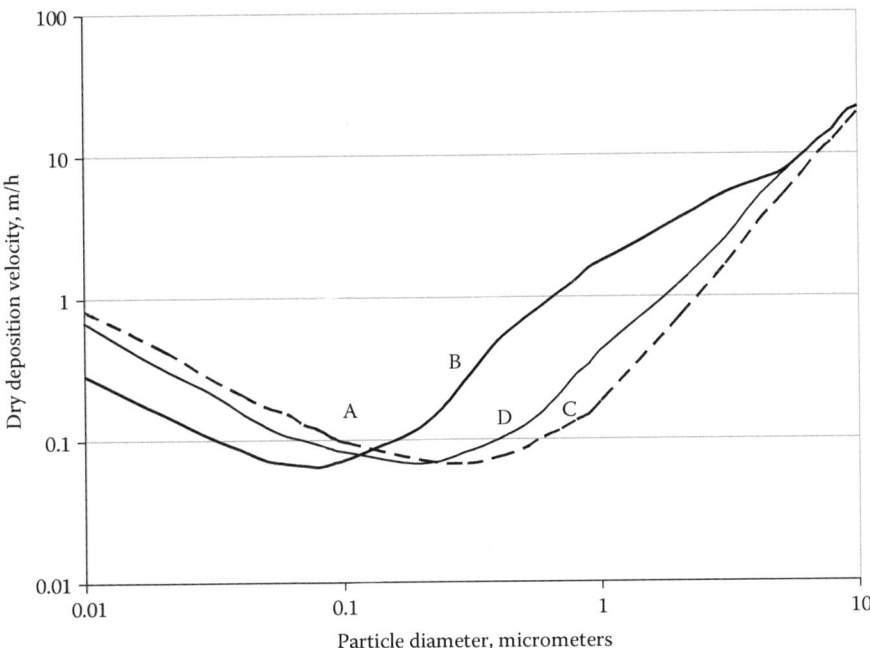

FIGURE 6.6 Dry deposition velocity of ammonium sulfate particles to a smooth water surface for a windspeed of 2 m/s, $T = 20°C$, relative humidity at 1 m = 75%, and initial particle
height = 1 m. Curve A is the baseline case without any hygroscopic growth, while Curve B is
for growth to equilibrium. Curve C is for partial growth of metastable liquid particles. Curve D
is for partial growth of particles that deliquesce from solid crystals to become liquid. [Adapted
from Zufall, M.J. et al., *Environ. Sci. Technol.*, 1998, **32**(5), 584–590.]

is so rapid that the particle reaches equilibrium well within the humidity boundary
layer. Curve C shows the deposition velocity for metastable particles that are still in
a solution phase. These particles have much slower growth since they are already in
a liquid state; a realistic humidity profile has been taken from Avery [34]. Finally,
Curve D shows ammonium sulfate particles that deliquesce from a solid crystalline
form to a liquid droplet taking into account the humidity profile.

The large values of U_D for the equilibrium case in this figure, say for 1 μm,
compared with the cases that account for time needed for growth suggest that one
cannot assume equilibrium has been achieved. Rather it is important to account for
timescales of growth when dealing with hygroscopic particles. The third and fourth
curves are probably the most realistic when estimating ammonium sulfate deposition
to an ambient water surface.

6.4.2 WET PARTICLE AND WET GASEOUS DEPOSITION

The long-term global average precipitation rate (U_R) is 9.7×10^{-5} m/h, equivalent to
0.85 m/year, and this value has been recommended as a generic value for chemical fate
modeling [6,21]. However, rain rate is highly variable geographically and seasonally,

TABLE 6.5

Recommended Generic Yearly Average Values of Rain Rate (U_R, m/h), Time between Rain Events (t_{dry}, h) and Duration of Rain Events (t_{wet}, h) for Selected Environmental Scenarios

Environmental Scenarios	U_R (m/h)	(m/year)	t_{dry} (h)	t_{wet} (h)
Generic global average	9.7×10^{-5}	0.85	120	12
Desert	1×10^{-6}	8.8×10^{-3}	720	12
Continental	1×10^{-4}	0.88	120	12
Tropical	3×10^{-4}	2.6	48	12

therefore locally relevant data should be used whenever possible. The International Research Institute for Climate and Society at Columbia University provides historical precipitation rate and precipitation frequency (t_{dry}) data through their Web site, http://iri.ldeo.columbia.edu/. We have compiled a selection of these data in Table 6.5 as recommended values for generic regional assessments. In the absence of data for a specific location, we recommend assuming that the interval between rain events (t_{dry}) is 5 days (120 h) and that the duration of rain events (t_{wet}) is 12 h [22]. Using these generic values the minimum possible long-term average atmospheric residence time due to wet deposition processes ($\bar{\tau}_{W,MIN}$, see Equation 6.24) is 55 h.

A scavenging ratio (Q) of 2×10^5 is recommended by Mackay [6] as a typical generic value. This corresponds to a $1\,mm^3$ raindrop falling from a height of 200 m and intercepting all particles in its path, or falling from 1000 m and intercepting 20% of particles. Table 6.6 summarizes measured values of Q found in the literature. Seinfeld and Pandis [19] review factors that control the collision efficiency between raindrops and aerosol particles. By analogy to dry deposition, large aerosol particles are efficiently collected by falling raindrops by inertial impaction and small aerosol particles are efficiently collected as a result of Brownian diffusion into the raindrop. However, particles in the $0.1–1\,\mu m$ size range are too large to have appreciable Brownian diffusivity, and too small to be efficiently collected by impaction.

6.5 EXAMPLE CALCULATIONS

Below we present sample calculations of the important parameters discussed in this chapter for selected organic chemicals with different properties and nickel, which is representative of involatile substances that have no vapor pressure. The chemical property data used in this section are all taken from the handbook of Mackay et al. [35].

6.5.1 GAS/PARTICLE PARTITIONING

a. Given that the logarithm of the octanol–air partition coefficient (log K_{OA}) of benzene is 2.78, estimate the volumetric gas–particle partition coefficient (K_{PA}) and the fraction associated with aerosols (ϕ) under typical generic environmental conditions.

TABLE 6.6
Typical Values of Scavenging Ratio, Q

Pollutant	Scavenging Ratios, Volume Basis $(m/V)_{rain}/(m/V)_{air}$, Typical range or value[a]	Comments	References
SO_4^{2-}	0.1–17	Rain	[54–56]
	0.02–6	Snow	[55,56]
NO_3^-	0.3–1.4	Rain	[54]
	1–3	Snow	[56]
		(total nitrate)	
NH_4^+	0.1–3	Rain	[56]
	1	Snow	[56]
Primarily crustal metals (Al, Ca, Mg, Mn, K ...)	0.1–4.5	Rain	[54,57]
Primarily anthropogenic metals (Pb, Cu, Zn, Ni, Cd ...)	0.02–0.5	Rain	[54,55,57]
Particle associated semivolatile organics (PCBs, PAHs ...)	0.001–0.6	Rain	[47,54,57,58]
	0.1–50	Snow	[58,59]
Gas-associated semivolatile organics (PCBs, PAHs ...)	0.01–1	Rain	[54,59]
	0.01–0.3	Snow	[58]
Total PCBs (gas + particle)	0.001–2	Rain	[55,58,60]
	0.05–10	Snow	[55,58]
Total PAHs (gas + particle)	0.001–300	Rain	[55,58]
	0.01–50	Snow	[55,58]

[a] Values have been multiplied by 10^{-6}.

Using Equation 6.8 and the constant recommended by Götz et al. [8], $K_{PA} = 0.13 K_{OA}$. The octanol–air partition coefficient of benzene is $10^{2.78} = 600$, therefore, K_{PA} for benzene is 78 m^3 air/m^3 aerosol particles.

Using Equation 6.6 and assuming the volume fraction of aerosols in air is 2×10^{-11} m^3 aerosol particles/m^3 air (Table 6.2), $\phi = K_{PA}/(K_{PA} + 2 \times 10^{11})$. Substituting $K_{PA} = 78$ m^3 air/m^3 aerosol particles for benzene, $\phi = 3.9 \times 10^{-10}$ which is approximately zero. Therefore, benzene is expected to be virtually entirely in the gas phase in the atmosphere.

b. Calculate K_{PA} and ϕ from log K_{OA} for hexachlorobenzene (HCB), Lindane, benzo(a)pyrene (B(a)P), and nickel using the same assumptions.

Results are summarized below in Table 6.7. Note that since nickel has a negligible vapor pressure it does not partition to air and K_{OA} and K_{PA} are essentially infinite.

TABLE 6.7
Log Octanol/Air Partition Coefficient (log K_{OA}) and Calculated Volumetric Particle/Air Partition Coefficient (K_{PA}) and Fraction Associated with Aerosol Particles (ϕ) under a Generic Scenario for Five Example Chemicals

Substance	log K_{OA}	K_{PA}	ϕ
Benzene	2.78	78	3.9×10^{-10}
Hexachlorobenzene	6.78	7.8×10^5	3.9×10^{-6}
Lindane	7.92	1.1×10^7	5.4×10^{-5}
Benzo(a)pyrene	10.77	7.7×10^9	0.037
Nickel	(Large)	(Large)	1

From the form of Equation 6.6 it is clear that ϕ will be low for chemicals that have K_{PA} lower than V_A/V_P, which has a typical value of about 2×10^{11}. Benzene, HCB, and Lindane all have very low ϕ and can be expected to be found predominantly in the gas phase under most environmental conditions. B(a)P is estimated to be 4% on particles under this generic scenario, and ϕ may be considerably higher at low temperatures, or when aerosol particle concentrations are higher, for example in an urban area. Nickel can be expected to be 100% particle-associated in the atmosphere under all conditions.

6.5.2 DRY DEPOSITION

a. Calculate the mass transfer coefficient for dry deposition (k_D, m/h) of benzene, HCB, Lindane, B(a)P, and nickel for the generic environmental scenario, then calculate the time required for half of the bulk atmospheric burden of each chemical to be removed by dry deposition.

From Equation 6.15, $k_D = U_D\phi$. For the generic scenario, $U_D = 4.6$ m/h (Table 6.2). Using this value along with ϕ of 3.9×10^{-10} for benzene, we calculate k_D for benzene as 1.8×10^{-9} m/h, and k_D for the other four substances as HCB: 1.8×10^{-5} m/h, Lindane: 2.5×10^{-4} m/h, B(a)P: 0.17 m/h, and nickel: 4.6 m/h. The half time for removal of a chemical from the atmosphere by dry deposition is ($\ln 2 \times h/k_D$) where h is the mixing height of the atmosphere and is assumed to be 1000 m, representing the planetary boundary layer. Substituting values for k_D for each chemical yields the following times for half of the atmospheric burden to be removed by dry deposition; benzene: 3.9×10^{11} h, or about 44 million years; HCB: 3.9×10^7 h, or about 4 thousand years; Lindane: 2.8×10^6 h or about 320 years; B(a)P: 4.1×10^3 h or about 6 months; nickel: 1.5×10^2 h, or about 1 week.

It is apparent from these calculations that dry deposition is not a very efficient process in the generic scenario except for chemicals that are highly sorbed to aerosols. However, it can be important under specific conditions, especially when chemicals

TABLE 6.8

Equations and Parameter Values for Calculating Wet Deposition for the Example Chemicals

Process	Equation	Parameter Values
Wet particle deposition	$k_{WP} = U_R Q \phi$	$U_R = 9.7 \times 10^{-5}$ m/h
Wet gaseous deposition	$k_{WG} = \dfrac{U_R(1 - \phi)}{(K_{AW} + V_R/V_A)}$	$Q = 5 \times 10^4$
		$V_R/V_A = 6 \times 10^{-8}$
Maximum wet deposition mass transfer coefficient	$k_{W,MAX} = \dfrac{2h(t_{dry} + t_{wet})}{t_{dry}^2}$	$h = 1000$ m
		$t_{dry} = 120$ h
		$t_{wet} = 12$ h

are associated with coarse particles, or in cases where the receiving surface has a high roughness height and is efficient at scavenging fine particles.

6.5.3 Wet Deposition

Given the air/water partition coefficient (K_{AW}), calculate mass transfer coefficients and half-times for removal by wet particle and wet gaseous deposition of the example chemicals for a generic environmental scenario assuming constant drizzling rain. Compare the sum of these mass transfer coefficients to the maximum possible long-term average mass transfer coefficient for wet deposition given the intermittent nature of rainfall.

Equations and generic environmental parameters required to calculate the mass transfer coefficients for wet deposition are summarized from the text in Table 6.8.

Table 6.9 provides the logarithm of the air–water partition coefficient for each example chemical and values for the calculated mass transfer coefficients for wet deposition in units of m/h.

TABLE 6.9

Logarithm of the Air/Water Partition Coefficient (log K_{AW}) for the Example Chemicals and Mass Transfer Coefficients for Wet Deposition in Units of m/h

Substance	log K_{AW}	k_{WP}	k_{WG}	$k_{WP} + k_{WG}$	$k_{W,MAX}$
Benzene	−0.65	1.9×10^{-9}	4.3×10^{-4}	4.3×10^{-4}	18.3
Hexachlorobenzene	−1.28	1.9×10^{-5}	1.9×10^{-3}	1.9×10^{-3}	18.3
Lindane	−4.22	2.6×10^{-4}	1.6	1.6	18.3
Benzo(a)pyrene	−4.73	0.18	5	5.2	18.3
Nickel	n/a	4.9	0	4.9	18.3

As shown in Table 6.9, $(k_{WP} + k_{WG})$ is less than $k_{W,MAX}$ for all of the example chemicals. Therefore, the continuous rain assumption is not producing unrealistically high wet deposition rates for these chemicals. The half-times for total wet deposition of the substances are benzene, 1.6×10^6 h or about 180 years; HCB, 3.7×10^5 h or about 40 years; Lindane, 431 h or 18 days; B(a)P, 134 h or 5.6 days; and nickel, 143 h or 6 days. Benzene and HCB are not efficiently removed from the atmosphere by wet deposition. Lindane and B(a)P are removed by wet gaseous deposition, and nickel by wet particle deposition. The 6-day half-life of nickel in the atmosphere represents the half-life for removal of particles by wet deposition in the generic scenario. This value is sensitive to the selected values for the rain rate and the scavenging ratio. The value of the maximum mass transfer coefficient for wet deposition corresponds to a minimum half-life in air of 38 h or 1.6 days.

TABLE 6.10

Fraction of Particle-Associated Chemical in the Atmosphere (ϕ)

					$\log K_{OW}$					
$\log K_{AW}$	-1	0	1	2	3	4	5	6	7	8
2	0	0	0	0	0	0	0	0	0	0
1	0	0	0	0	0	0	0	0	0	0
0	0	0	0	0	0	0	0	0	0	0
-1	0	0	0	0	0	0	0	0	0	0
-2	0	0	0	0	0	0	0	0	0	0.03
-3	0	0	0	0	0	0	0	0	0.03	0.21
-4	0	0	0	0	0	0	0	0.03	0.21	0.72
-5	0	0	0	0	0	0	0.03	0.21	0.72	0.96
-6	0	0	0	0	0	0.03	0.21	0.72	0.96	1.00
-7	0	0	0	0	0.03	0.21	0.72	0.96	1.00	1.00
-8	0	0	0	0.03	0.21	0.72	0.96	1.00	1.00	1.00
-9	0	0	0.03	0.21	0.72	0.96	1.00	1.00	1.00	1.00
-10	0	0.03	0.21	0.72	0.96	1.00	1.00	1.00	1.00	1.00
-11	0.03	0.21	0.72	0.96	1.00	1.00	1.00	1.00	1.00	1.00

Note: Φ calculated as a function of $\log K_{OW}$ and $\log K_{AW}$ using Equation 6.6 and assuming $K_{PA} = 0.13 K_{OA}$ and $V_P/V_A = 2 \times 10^{-11}$.

6.5.4 LOOK-UP TABLES FOR MASS TRANSFER COEFFICIENTS

The mass transfer coefficients described in this chapter are functions of chemical partitioning properties and environmental conditions. For nonionizing organic chemicals, the range of possible partitioning in the environment can be represented by octanol–water partition coefficients between 10^{-1} and 10^8 and air–water partition coefficients between 10^{-11} and 10^2 [36]. To a first approximation, $K_{OA} = K_{OW}/K_{AW}$, thus the relevant range of partitioning encompasses a range of K_{OA} from 10^{19} to 10^{-1}.

Tables 6.10 through 6.15 contain results for example calculations of Φ and the mass transfer coefficients k_D, k_{WP}, and k_{WG} calculated using our recommended values for a generic environmental scenario for the range of plausible partitioning properties of organic chemicals defined above. Also included is a look-up table for $k_{W,TOT}$,

TABLE 6.11

Mass Transfer Coefficient for Dry Particle Deposition (K_D, m/h)

log K_{AW} \ log K_{OW}	−1	0	1	2	3	4	5	6	7	8
2	0	0	0	0	0	0	0	0	0	0
1	0	0	0	0	0	0	0	0	0	0
0	0	0	0	0	0	0	0	0	0	0
−1	0	0	0	0	0	0	0	0	0	0.01
−2	0	0	0	0	0	0	0	0	0.01	0.12
−3	0	0	0	0	0	0	0	0.01	0.12	0.95
−4	0	0	0	0	0	0	0.01	0.12	0.95	3.32
−5	0	0	0	0	0	0.01	0.12	0.95	3.32	4.43
−6	0	0	0	0	0.01	0.12	0.95	3.32	4.43	4.58
−7	0	0	0	0.01	0.12	0.95	3.32	4.43	4.58	4.60
−8	0	0	0.01	0.12	0.95	3.32	4.43	4.58	4.60	4.60
−9	0	0.01	0.12	0.95	3.32	4.43	4.58	4.60	4.60	4.60
−10	0.01	0.12	0.95	3.32	4.43	4.58	4.60	4.60	4.60	4.60
−11	0.12	0.95	3.32	4.43	4.58	4.60	4.60	4.60	4.60	4.60

Note: Calculated as $U_D\phi$, assuming $U_D = 4.6$ m/h.

TABLE 6.12

Mass Transfer Coefficient for Wet Particle Deposition (k_{WP}, m/h)

		$\log K_{OW}$									
		−1	0	1	2	3	4	5	6	7	8
	2	0	0	0	0	0	0	0	0	0	0
	1	0	0	0	0	0	0	0	0	0	0
	0	0	0	0	0	0	0	0	0	0	0
	−1	0	0	0	0	0	0	0	0	0	0.01
	−2	0	0	0	0	0	0	0	0	0.01	0.12
	−3	0	0	0	0	0	0	0	0.01	0.12	1.00
$\log K_{AW}$	−4	0	0	0	0	0	0	0.01	0.12	1.00	3.50
	−5	0	0	0	0	0	0.01	0.12	1.00	3.50	4.67
	−6	0	0	0	0	0.01	0.12	1.00	3.50	4.67	4.83
	−7	0	0	0	0.01	0.12	1.00	3.50	4.67	4.83	4.85
	−8	0	0	0.01	0.12	1.00	3.50	4.67	4.83	4.85	4.85
	−9	0	0.01	0.12	1.00	3.50	4.67	4.83	4.85	4.85	4.85
	−10	0.01	0.12	1.00	3.50	4.67	4.83	4.85	4.85	4.85	4.85
	−11	0.12	1.00	3.50	4.67	4.83	4.85	4.85	4.85	4.85	4.85

Note: Calculated as $U_R Q \phi$, where ϕ is from Table 6.10 and assuming $U_R = 9.7 \times 10^{-5}$ m/h and $Q = 5 \times 10^4$.

which represents the total mass transfer coefficient for wet deposition considering the maximum possible rate due to the intermittent nature of rainfall events, and for k_{TOT}, the total mass transfer coefficient for all aerosol-associated deposition processes simultaneously. Values calculated above for the example chemicals can be cross checked in the look-up tables using the log K_{AW} values in Table 6.9 and the following log K_{OW} values: benzene, 2.13; HCB, 5.50; Lindane: 3.70; B(*a*)P, 6.04; nickel, large.

6.6 SUMMARY AND CONCLUSION

In this chapter we have reviewed the theory of dry particle and wet deposition of chemical contaminants from the atmosphere to surfaces, and we have provided

TABLE 6.13

Mass Transfer Coefficient for Wet Gaseous Deposition (k_{WG}, m/h)

	$\log K_{OW}$									
$\log K_{AW}$	−1	0	1	2	3	4	5	6	7	8
2	0	0	0	0	0	0	0	0	0	0
1	0	0	0	0	0	0	0	0	0	0
0	0	0	0	0	0	0	0	0	0	0
−1	0	0	0	0	0	0	0	0	0	0
−2	0.01	0.01	0.01	0.01	0.01	0.01	0.01	0.01	0.01	0.01
−3	0.10	0.10	0.10	0.10	0.10	0.10	0.10	0.10	0.09	0.08
−4	0.97	0.97	0.97	0.97	0.97	0.97	0.97	0.94	0.77	0.27
−5	9.64	9.64	9.64	9.64	9.64	9.62	9.40	7.65	2.68	0.36
−6	91.5	91.5	91.5	91.5	91.3	89.2	72.6	25.4	3.39	0.35
−7	606	606	606	605	591	481	168	22.5	2.32	0.23
−8	1386	1385	1382	1351	1100	385	51.3	5.31	0.53	0.05
−9	1590	1586	1550	1262	442	58.9	6.09	0.61	0.06	0.01
−10	1610	1573	1281	448	59.8	6.18	0.62	0.06	0.01	0
−11	1575	1283	449	59.9	6.19	0.62	0.06	0.01	0	0

Note: Calculated as $(U_R(1-\phi)/K_{AW} + V_R/V_A)$, where ϕ is from Table 6.10 and assuming $U_R = 9.7 \times 10^{-5}$ m/h and $V_R/V_A = 6 \times 10^{-8}$ m^3 water/m^3 air.

recommended values to estimate rates of deposition under generic environmental scenarios. Accurately estimating the rates of these processes is a considerable challenge because they depend on both (1) the extent of partitioning of chemical between the gas phase and aerosol particles or precipitation and (2) the rate of deposition of the aerosol particles and precipitation. We stress that there is considerable uncertainty in estimation equations to describe gas–particle partitioning, and in estimates of dry deposition velocities and scavenging ratios. In addition, there is a high degree of variability in aerosol size distribution and composition, which impacts both partitioning and deposition velocities.

Considering these uncertainties and variability, it is clear that estimates of the rates of dry particle and wet deposition based on the generic parameters suggested here

TABLE 6.14

Total Maximum Mass Transfer Coefficient for Wet Gaseous Deposition ($k_{W,TOT}$, m/h)

					log K_{OW}					
log K_{AW}	−1	0	1	2	3	4	5	6	7	8
2	0	0	0	0	0	0	0	0	0	0
1	0	0	0	0	0	0	0	0	0	0
0	0	0	0	0	0	0	0	0	0	0
−1	0	0	0	0	0	0	0	0	0	0.01
−2	0.01	0.01	0.01	0.01	0.01	0.01	0.01	0.01	0.02	0.13
−3	0.10	0.10	0.10	0.10	0.10	0.10	0.10	0.11	0.22	1.08
−4	0.97	0.97	0.97	0.97	0.97	0.97	0.98	1.07	1.77	3.77
−5	9.64	9.64	9.64	9.64	9.64	9.63	9.52	8.65	6.18	5.03
−6	18.3	18.3	18.3	18.3	18.3	18.3	18.3	18.3	8.06	5.18
−7	18.3	18.3	18.3	18.3	18.3	18.3	18.3	18.3	7.15	5.08
−8	18.3	18.3	18.3	18.3	18.3	18.3	18.3	10.1	5.38	4.90
−9	18.3	18.3	18.3	18.3	18.3	18.3	10.9	5.46	4.91	4.86
−10	18.3	18.3	18.3	18.3	18.3	11.0	5.47	4.91	4.86	4.85
−11	18.3	18.3	18.3	18.3	11.0	5.47	4.91	4.86	4.85	4.85

Note: Calculated as the minimum of ($k_{WP} + k_{WG}$) from Tables 6.11 and 6.12 and $2h(t_{dry} + t_{wet})/t_{dry}^2$, assuming $h = 1000$ m, $t_{dry} = 120$ h, and $t_{wet} = 12$ h.

could easily be in error by an order of magnitude or more when compared to measurements from field studies. Site- and event-specific data or regression equations should be used in preference to generic values to estimate the rates of dry particle and wet deposition whenever data are available. Experimental work in several areas is desirable to further evaluate and refine the models and parameter values recommended here. First, more measurements of gas-particle partitioning of organic chemicals in a variety of different aerosols under laboratory or field conditions are needed. Experiments that determine the concentration and composition of the aerosol in addition to the chemical concentrations in the gas phase and on particles are particularly valuable; however, this type of study has only rarely been conducted. Second,

TABLE 6.15
Total Mass Transfer Coefficient for Dry and Wet Deposition (k_{TOT}, m/h)

		log K_{OW}								
log K_{AW}	−1	0	1	2	3	4	5	6	7	8
2	0	0	0	0	0	0	0	0	0	0
1	0	0	0	0	0	0	0	0	0	0
0	0	0	0	0	0	0	0	0	0	0
−1	0	0	0	0	0	0	0	0	0	0.03
−2	0.01	0.01	0.01	0.01	0.01	0.01	0.01	0.01	0.03	0.25
−3	0.10	0.10	0.10	0.10	0.10	0.10	0.10	0.12	0.33	2.03
−4	0.97	0.97	0.97	0.97	0.97	0.97	0.99	1.18	2.72	7.09
−5	9.64	9.64	9.64	9.64	9.64	9.64	9.64	9.60	9.50	9.46
−6	18.3	18.3	18.3	18.3	18.3	18.4	19.3	21.7	12.5	9.76
−7	18.3	18.3	18.3	18.3	18.4	19.3	21.7	22.8	11.7	9.68
−8	18.3	18.3	18.3	18.4	19.3	21.7	22.8	14.7	9.98	9.50
−9	18.3	18.3	18.4	19.3	21.7	22.8	15.5	10.1	9.51	9.46
−10	18.3	18.4	19.3	21.7	22.8	15.6	10.1	9.51	9.46	9.45
−11	18.4	19.3	21.7	22.8	15.6	10.1	9.51	9.46	9.45	9.45

Note: Calculated as the sum of k_D from Table 6.11 and $k_{W,TOT}$ from Table 6.14.

additional field measurements of deposition mass fluxes of chemicals are required to reduce uncertainty in the parameter values recommended here and to evaluate their variability.

LITERATURE CITED

1. Bidleman, T.F. and T. Harner, Sorption to aerosols, in *Handbook of Property Estimation Methods for Chemicals*, R.S. Boethling and D. Mackay, Eds. 2000, CRC Press, Boca Raton, FL, pp. 233–260.
2. Wania, F., J. Axelman, and D. Broman, A review of processes involved in the exchange of persistent organic pollutants across the air–sea interface. *Environmental Pollution*, 1998. **102**(1): 3–23.

3. Cousins, I.T. and D. Mackay, Gas–particle partitioning of organic compounds and its interpretation using relative solubilities. *Environmental Science and Technology*, 2001. **35**(4): 643–647.

4. Pankow, J.F., Review and comparative-analysis of the theories on partitioning between the gas and aerosol particulate phases in the atmosphere. *Atmospheric Environment*, 1987. **21**(11): 2275–2283.

5. Pankow, J.F., An absorption-model of gas-particle partitioning of organic-compounds in the atmosphere. *Atmospheric Environment*, 1994. **28**(2): 185–188.

6. Mackay, D., *Multimedia Environmental Models: The Fugacity Approach.* 2001, Lewis Publishers, Boca Raton, FL, p. 261.

7. Finizio, A., et al., Octanol–air partition coefficient as a predictor of partitioning of semi-volatile organic chemicals to aerosols. *Atmospheric Environment*, 1997. **31**(15): 2289–2296.

8. Götz, C.W., et al., Alternative approaches for modeling gas–particle partitioning of semivolatile organic chemicals: Model development and comparison. *Environmental Science and Technology*, 2007. **41**(4): 1272–1278.

9. Kalberer, M., et al., Identification of polymers as major components of atmospheric organic aerosols. *Science*, 2004. **303**(5664): 1659–1662.

10. Xiao, H. and F. Wania, Is vapor pressure or the octanol–air partition coefficient a better descriptor of the partitioning between gas phase and organic matter? *Atmospheric Environment*, 2003. **37**(20): 2867–2878.

11. Goss, K.U. and R.P. Schwarzenbach, Gas/solid and gas/liquid partitioning of organic compounds: Critical evaluation of the interpretation of equilibrium constants. *Environmental Science & Technology*, 1998. **32**(14): 2025–2032.

12. Goss, K.U., Conceptual model for the adsorption of organic compounds from the gas phase to liquid and solid surfaces. *Environmental Science and Technology*, 1997. **31**(12): 3600–3605.

13. Abraham, M.H., A. Ibrahim, and A.M. Zissimos, Determination of sets of solute descriptors from chromatographic measurements. *Journal of Chromatography A*, 2004. **1037**(1–2): 29–47.

14. Roth, C.M., K.U. Goss, and R.P. Schwarzenbach, Sorption of a diverse set of organic vapors to diesel soot and road tunnel aerosols. *Environmental Science and Technology*, 2005. **39**(17): 6632–6637.

15. Roth, C.M., K.U. Goss, and R.P. Schwarzenbach, Sorption of a diverse set of organic vapors to urban aerosols. *Environmental Science & Technology*, 2005. **39**(17): 6638–6643.

16. Goss, K.U. and R.P. Schwarzenbach, Adsorption of a diverse set of organic vapors on quartz, $CaCO_3$, and alpha-Al_2O_3 at different relative humidities. *Journal of Colloid and Interface Science*, 2002. **252**(1): 31–41.

17. Goss, K.U. and S.J. Eisenreich, Adsorption of VOCs from the gas phase to different minerals and a mineral mixture. *Environmental Science and Technology*, 1996. **30**(7): 2135–2142.

18. Roth, C.M., K.U. Goss, and R.P. Schwarzenbach, Sorption of diverse organic vapors to snow. *Environmental Science and Technology*, 2004. **38**(15): 4078–4084.

19. Seinfeld, J.H. and S.N. Pandis, *Atmospheric Chemistry and Physics: From Air Pollution to Climate Change.* 1998, Wiley, New York, p. 1326.

20. Thibodeaux, L.J., *Enviornmental Chemodynamics: Movement of Chemicals in Air, Water, and Soil.* 2nd ed. 1996, Wiley, New York, p. 624.

21. Scheringer, M., *Persistence and Spatial Range of Environmental Chemicals: New Ethical and Scientific Concepts for Risk Assessment*. 2002, Wiley-VCH Berlag, Weinheim, p. 308.

22. Jolliet, O. and M. Hauschild, Modeling the influence of intermittent rain events on long-term fate and transport of organic air pollutants. *Environmental Science & Technology*, 2005. **39**(12): 4513–4522.

23. Hertwich, E.G., Intermittent rainfall in dynamic multimedia fate modeling. *Environmental Science and Technology*, 2001. **35**(5): 936–940.

24. Legagneux, L., A. Cabanes, and F. Domine, Measurement of the specific surface area of 176 snow samples using methane adsorption at 77 K. *Journal of Geophysical Research-Atmospheres*, 2002. **107**(D17).

25. Lei, Y.D. and F. Wania, Is rain or snow a more efficient scavenger of organic chemicals? *Atmospheric Environment*, 2004. **38**(22): 3557–3571.

26. Stocker, J., et al., Modeling the effect of snow and ice on the global environmental fate and long-range transport potential of semi-volatile organic compounds. *Environmental Science & Technology,* 2007. **41**(17): 6192–6198.

27. Goss, K.U. and R.P. Schwarzenbach, Empirical prediction of heats of vaporization and heats of adsorption of organic compounds. *Environmental Science & Technology*, 1999. **33**(19): 3390–3393.

28. Roth, C.M., K.U. Goss, and R.P. Schwarzenbach, Adsorption of a diverse set of organic vapors on the bulk water surface. *Journal of Colloid and Interface Science*, 2002. **252**(1): 21–30.

29. Putaud, J.P., et al., European aerosol phenomenology-2: Chemical characteristics of particulate matter at kerbside, urban, rural and background sites in Europe. *Atmospheric Environment*, 2004. **38**(16): 2579–2595.

30. Sehmel, G.A., Particle diffusivities and deposition velocities over a horizontal smooth surface. *Journal of Colloid and Interface Science*, 1971. **37**: 891–906.

31. Sehmel, G.A. and W.H. Hodgson, *A Model for Predicting Dry Deposition of Particles and Gases to Environmental Surfaces*. 1978. Battelle Pacific Northwest Laboratories. Report, Richland, WA.

32. Milford, J.B. and C.I. Davidson, The sizes of particulate trace elements in the atmosphere– a review. *Journal of the Air Pollution Control Association*, 1985. **35**: 1249–1260.

33. Zufall, M.J., et al., Dry deposition of particles to wave surfaces: I. Mathematical modeling. *Atmospheric Environment*, 1999. **33**(26): 4273–4281.

34. Avery, K.R., Literature Search for Atmospheric Humidity Profile Models from the Sea Surface to 1000 Meters. 1972, NOAA Technical Memorandum EDS NODC-1: Silver Spring, MD.

35. Mackay, D., W.Y. Shiu, and K.C. Ma, *Handbook of Physical-Chemical Properties and Environmental Fate for Organic Chemicals*, 2nd ed. 2006, CRC Press, Boca Raton, FL.

36. Fenner, K., et al., Comparing estimates of persistence and long-range transport potential among multimedia models. *Environmental Science and Technology*, 2005. **39**(7): 1932–1942.

37. Davidson, C.I. and Y.L. Wu, Dry deposition of particles and vapors, in *Acidic Precipitation Volume 3*, Lindberg, A.L. and S.A. Norton, Eds. 1990. Springer-Verlag, New York, pp. 103–216.

38. Ruijgrok, W., H. Tieben, and P. Eisinga, The dry deposition of particles to a forest canopy: A comparison of model and experimental results. *Atmospheric Environment*, 1997. **31**(3): 399–415.

39. Brook, J.R., et al., Description and evaluation of a model of deposition velocities for routine estimates of dry deposition over North America. Part II: Review of past measurements and model results. *Atmospheric Environment*, 1999. **33**(30): 5053–5070.

40. Yang, H.H., L.T. Hsieh, and S.K. Cheng, Determination of atmospheric nitrate particulate size distribution and dry deposition velocity for three distinct areas. *Chemosphere*, 2005. **60**(10): 1447–1453.

41. Davidson, C.I., et al., Airborne trace elements in Great Smoky Mountains, Olympic, and Glacier National Parks. *Environmental Science and Technology*, 1985. **19**(1): 27–35.

42. Yi, S., et al., Overall elemental dry deposition velocities measured around Lake Michigan. *Atmospheric Environment*, 2001. **35**(6): 1133–1140.

43. Fang, G.C., et al., Overall dry deposition velocities of trace elements measured at harbor and traffic site in central Taiwan. *Chemosphere*, 2007. **67**(5): 966–974.

44. Fang, G.C., et al., Ambient air particulates, metallic elements, dry deposition and concentrations at Taichung Airport, Taiwan. *Atmospheric Research*, 2007. **84**(3): 280–289.

45. Zufall, M.J., et al., Airborne concentrations and dry deposition fluxes of particulate species to surrogate surfaces deployed in Southern Lake Michigan. *Environmental Science and Technology*, 1998. **32**(11): 1623–1628.

46. Vardar, N., M. Odabasi, and T. Holsen, Particulate dry deposition and overall deposition velocities of polycyclic aromatic hydrocarbons. *Journal of Environmental Engineering–ASCE*, 2002. **128**(3): 269–274.

47. Sharma, M. and E. McBean, Atmospheric PAH deposition: Deposition velocities and washout ratios. *Journal of Environmental Engineering–ASCE*, 2002. **128**(2): 186–195.

48. McVeety, B.D. and R.A. Hites, Atmospheric deposition of PAHs to water surfaces: A mass balance approach. *Atmospheric Environment Part a—General Topics*, 1988. **22**(2): 511–536.

49. Franz, T., S. Eisenreich, and T. Holsen, Dry deposition of particulate polychlorinated biphenyls and polycyclic aromatic hydrocarbons to Lake Michigan. *Environmental Science and Technology*, 1998. **32**(23): 3681–3688.

50. Tasdemir, Y., et al., Dry deposition fluxes and velocities of polychlorinated biphenyls (PCBs) associated with particles. *Atmospheric Environment*, 2004. **38**(16): 2447–2456.

51. Bozlaker, A., M. Odabasi, and A. Muezzinoglu, Dry deposition and soil-air gas exchange of polychlorinated biphenyls (PCBs) in an industrial area. *Environmental Pollution*, 2008. **156**(3): 784–793.

52. Sheu, H.L., et al., Dry deposition of polycyclic aromatic hydrocarbons in ambient air. *Journal of Environmental Engineering*, 1996. **112**(12): 1101–1109.

53. Odabasi, M., et al., Measurement of dry deposition and air-water exchange of polycyclic aromatic hydrocarbons with the water surface sampler. *Environmental Science and Technology*, 1999. **33**(3): 426–434.

54. Poster, D. and J. Baker, Mechanisms of atmospheric wet deposition of chemical contaminants, in *Atmospheric Deposition of Contaminants to the Great Lakes and Coastal Waters: Proceedings from a Session at SETAC's 15th Annual Meeting*, J. Baker, Ed. 1997, SETAC Press, Pensacola, FL.

55. Franz, T., D. Gregor, and S. Eisenreich, Snow deposition of atmospheric semivolatile organic chemicals, in *Atmospheric Deposition of Contaminants to the Great Lakes and Coastal Waters*, J. Baker, Ed. 1997, SETAC Press, Pensacola, FL.

56. Kasper-Giebl, A., M.F. Kalina, and H. Puxbaum, Scavenging ratios for sulfate, ammonium and nitrate determined at Mt. Sonnblick (3106 m asl). *Atmospheric Environment*, 1999. **33**(6): 895–906.

57. Schnoor, J., *Environmental Modeling*. 1996. Wiley Interscience, New York.

58. Franz, T.P. and S.J. Eisenreich, Snow scavenging of polychlorinated biphenyls and polycyclic aromatic hydrocarbons in Minnesota. *Environmental Science and Technology*, 1998. **32**(12): 1771–1778.

59. Wania, F., D. Mackay, and J.T. Hoff, The importance of snow scavenging of polychlorinated biphenyl and polycyclic aromatic hydrocarbon vapors. *Environmental Science and Technology*, 1999. **33**(1): 195–197.

60. Agrell, C., et al., PCB congeners in precipitation, wash out ratios and depositional fluxes within the Baltic Sea region, Europe. *Atmospheric Environment*, 2002. **36**(2): 371–383.

61. Zufall, M.J., M.H. Bergin, and C.I. Davidson, Effects of non-equilibrium hygroscopic growth of $(NH_4)_2SO_4$ on dry deposition to water surfaces. *Environmental Science and Technology*, 1998. **32**(5): 584–590.

FURTHER READING

Boethling, R.S. and D. Mackay, Eds. *Handbook of Property Estimation Methods for Chemicals*, 2000, CRC Press, Boca Raton, FL.

Schnoor, J., *Environmental Modeling*. 1996. Wiley Interscience, New York.

Seinfeld, J.H. and S.N. Pandis, *Atmospheric Chemistry and Physics: From Air Pollution to Climate Change*. 1998. Wiley, New York, p. 1326.

Thibodeaux, L.J., *Environmental Chemodynamics: Movement of Chemicals in Air, Water, and Soil*. 2nd ed. 1996. Wiley, New York, p. 624.

7 Mass Transfer between the Atmosphere and Plant Canopy Systems

Michael S. McLachlan

CONTENTS

7.1 Introduction 137
7.2 Transport Processes 138
7.3 Transport Theory 143
    7.3.1 Dry Gaseous Deposition 144
    7.3.2 Dry Aerosol-Bound Deposition 147
    7.3.3 Measurement of Transport Parameters.................................149
7.4 Transport Parameters 150
7.5 Example Calculations 153
    7.5.1 What are the Values of k_g and $k_{p(D\&I)}$ for Deposition
           to a Grassy Coastal Plain in Mid-Summer, with Wind Speeds
           of on Average $8\,\mathrm{m\,s^{-1}}$ 153
    7.5.2 Calculate k_g for Deposition to a Mature 25 m High Deciduous
           Canopy and a Mature 25 m High Coniferous Canopy in Bayreuth
           and Compare Them to the Values in Table 7.2 155
7.6 Summary and Conclusions..156
Acknowledgment 156
Literature Cited 157

7.1 INTRODUCTION

Plant canopies can play an important role in the environmental fate of organic chemicals. The exchange of chemicals between the atmosphere and plant canopies can be much more rapid than the exchange between the atmosphere and soil or water surfaces. Consequently, plant canopies can "filter" chemicals out of the atmosphere and eventually transfer them to the underlying soil where they may be sequestered for longer periods of time. Vegetation also has a large capacity to reversibly store many chemicals. This, combined with the rapid exchange kinetics, allows plant canopies to serve as chemical buffers for the atmosphere, accumulating chemicals during periods when concentrations in the atmosphere are high (for instance, following a chemical accident), and releasing them back when the concentrations in the atmosphere drop again. In addition, the canopy may be an important site for transformation of a chemical, for instance via photochemical reactions on leaf surfaces or biodegradation within the foliage.

Chemical concentrations in a plant can be important in themselves. Plants stand at the base of most terrestrial food webs, and hence chemical accumulation in plants is a fundamental component of bioaccumulation to all trophic levels in terrestrial ecosystems. In other cases, the phytotoxic effects of a chemical may be an issue that requires an understanding of its uptake into plants from the atmosphere. Another related question is phytoremediation, the use of plants to detoxify soil. For some chemicals, phytoremediation can be accomplished by transport of chemicals from the roots into the foliage, and from there to the atmosphere. Clearly, there are a multitude of questions that require an understanding of organic chemical transport to, from, and within plant canopies. These issues are also discussed by Doucette et al. in Chapter 14 with a focus on soils as the chemical source.

Net transport of chemical to and within the canopy system is the result of the combined effects of a multitude of transport processes. These include advective processes such as wet deposition and dry deposition of particles from the atmosphere to the canopy, and from the canopy to the forest floor, as well as diffuse processes such as gaseous exchange between the atmosphere and the canopy, and between the subcanopy air and the forest floor. The relative importance of these processes depends on the chemical source situation, the physical chemical properties of the chemical, as well as the characteristics of the environment.

This chapter begins with a definition of the different transfer processes involved in chemical transport in the atmosphere–canopy–soil surface system. A qualitative description of each process is followed by an example of how the relevance of the different processes changes with the physical chemical properties of the chemical. Then, a theoretical framework is presented for the two processes for which this is available, namely dry gaseous deposition and dry particle-bound deposition. This is accompanied by a description of the measurement methods available to quantify these processes. The last section is devoted to summarizing the available correlations and presenting several example calculations of mass transfer coefficients.

7.2 TRANSPORT PROCESSES

The transport processes involved in the exchange of organic chemicals between the atmosphere, plant canopies, and the soil surface are illustrated in Figure 7.1. Transfer from the atmosphere to the canopy can occur by wet deposition in the form of rain, snow or fog, dry deposition of chemical associated with atmospheric aerosols, or dry deposition of gases. Transfer from the canopy to the atmosphere can occur via suspension of chemical associated with particles or volatilization. Transfer from the canopy to surface soil can occur via sedimentation, whereby again both wet and dry processes contribute. Resuspension of soil particles can result in transfer of chemical to the canopy foliage. Finally, chemicals can volatilize from the canopy to the subcanopy air, whereby some fraction may sorb to the soil (with the remainder being advected either vertically or horizontally out of the canopy), or they may volatilize

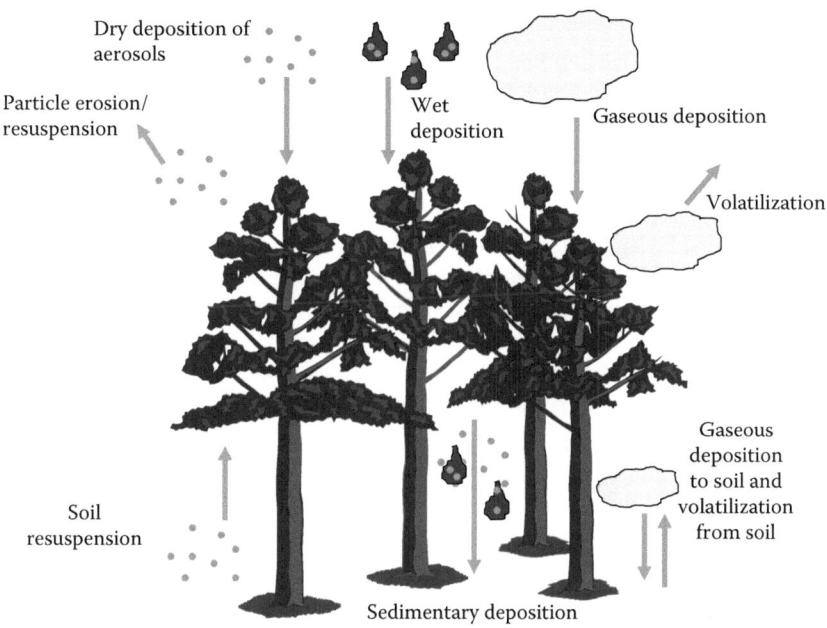

Dry deposition of aerosols

Particle erosion/resuspension

Wet deposition

Gaseous deposition

Volatilization

Soil resuspension

Gaseous deposition to soil and volatilization from soil

Sedimentary deposition

FIGURE 7.1   Chemical transfer process between the atmosphere, canopy, and soil.

from the soil and sorb to the canopy (or be advected out). Most of these processes are advective in nature, with the exception of volatilization and gaseous absorption, which are diffusive. In the following, the primary factors influencing these processes are discussed. Transfer via the roots between the soil and the canopy is discussed in Chapter 14.

Wet deposition of chemical can occur dissolved in, adsorbed to the surface of, and associated with aerosols entrapped in the precipitation. Which of these processes is most important depends on the partitioning properties of the chemical between air, water, aerosol organic matter, and surfaces. A low Henry's law constant favors deposition in dissolved form, a high aerosol organic matter/air partition coefficient favors deposition with entrapped aerosols, and a high surface adsorption coefficient favors deposition adsorbed to the surface of the raindrops or snow/ice crystals. For volatile chemicals, all of the above partitioning properties are unfavorable and wet deposition is generally not a relevant process. The properties of the precipitation, for example, temperature, the quantity and properties of the entrained aerosols, and the specific surface area, also play a role. MacLeod et al. in Chapter 6 provide a more detailed treatment of these interactions and deposition with particles and precipitation to water and soils. Generally, wet deposition to plant canopies can be modeled in the same manner as wet deposition to water and soils. Orographic fog is one notable exception. It can occur in uneven terrain (e.g., mountains) where the meteorological conditions result in moving fog banks or clouds. Plant canopies can be efficient at scavenging the fog/cloud droplets and the chemicals associated with them.

Dry deposition of chemical associated with aerosols occurs when a chemical is associated with particles that deposit on canopy surfaces. The relevance of this process compared to the competing processes shown in Figure 7.1 is determined by the partitioning of the chemical between the gas phase and aerosols, and by the precipitation rate. A description of the factors governing gas/aerosol partitioning is given in Chapter 6. For dry deposition via aerosols to be important compared with gaseous deposition, a significant fraction (typically more than half) of the chemical in the atmosphere must be associated with aerosols. For it to be important compared to wet deposition, the climate must generally be dry. In temperate climates, wet deposition is usually greater due to the efficient scavenging of aerosols by rain and snow. An exception can be situations in which a large fraction of the chemical is associated with large particles, which favors dry deposition as discussed below. Nevertheless, although it may not be the most important deposition process, dry deposition via aerosols is frequently significant. More insight into the factors governing this process and how it can be modeled is given in Section 7.3.

Gaseous deposition from the atmosphere to the canopy and volatilization from the canopy to the atmosphere are in effect the same process: diffusive gaseous exchange. It is the dominant form of deposition to plant canopies for a wide range of volatile organic chemicals. Furthermore, it is the much faster gaseous exchange to plant canopies compared with water or soil that renders plant canopies important in the environmental fate of many organic contaminants. In contrast to the other processes in this chapter, gaseous exchange is governed by a thermodynamic gradient in chemical potential or fugacity and must be expressed as the product of that gradient, a mass transfer coefficient, and a surface area. This is dealt with in more detail in Section 7.3.

It is not always appreciated that particles are transferred from plant canopies to the atmospheric boundary layer. For instance, it has been estimated that 8–38% of the extractable organic material in aerosols from remote arid areas have their origin in the erosion of plant cuticles, the wax layers that cover leaf surfaces (Mazurek et al., 1991). These particles will contain organic contaminants that have been deposited on the leaves. This is not a process that has been extensively studied in connection with organic contaminant fate. It is generally assumed that the chemical flux is small compared with the reverse process, namely dry deposition of chemical associated with aerosols. There may nevertheless be situations where it needs to be considered, for instance in the dissipation of chemical from a canopy following a contamination event.

Turning to exchange between the canopy and the underlying soil, sedimentation processes are of primary importance. This includes deposition via precipitation, whereby, as outlined above, organic chemicals can be transported dissolved in the water, associated with particles trapped in the precipitation, or sorbed to the surface of the water/ice. The chemical fluxes below the canopy are, however, much more difficult to estimate than the wet deposition to the canopy, due to the interactions between the precipitation and the canopy. When a raindrop hits a leaf, some of the chemical dissolved in the rain may sorb to the leaf, but on the other hand, chemical already present on the leaf may dissolve in the rain. Similarly, particles in a raindrop may be retained on the leaf surface (for instance by leaf hairs), or particles which had already accumulated on the leaf surface (for instance via dry deposition) may be washed off. Furthermore, some portion of the rain may fall through the canopy, while some portion will also be retained in the canopy and eventually evaporate. During some rain events, the flux of the chemical in the precipitation can be higher below the canopy than above, while in other cases, it may be much lower (Kaupp, 1996). This depends on a multitude of factors: chemical (e.g., the partitioning properties of the chemicals), botanical (e.g., canopy density, leaf surface roughness), meteorological (e.g., precipitation quantity and intensity), and historical (e.g., how long had it not rained, prehistory of chemical accumulation in the canopy). This makes estimating the mass transfer coefficients difficult, and the approach taken will vary depending on the situation. For instance, for a water-soluble chemical for which rain is the primary source to the canopy, it may be appropriate to estimate the canopy to soil transfer over the long term using an average canopy interception factor of, for example, 80%.

Dry sedimentation is also an important process for transferring chemical from the canopy to the underlying soil. The most obvious example is leaf fall, whereby chemical associated with the leaves is deposited on the canopy floor. The more general term is litter fall, which refers to all debris that falls out of the canopy. The mass transfer coefficient for leaf fall is usually easily estimated from the foliage biomass and the seasonality of leaf fall. However, for the overall litter fall this may be more complex. One component of litter fall that is easily overlooked is the deposition of wax droplets and other components of the leaf cuticle. It has been shown that this is the dominant flux of several classes of persistent lipophilic organic contaminants beneath a spruce canopy in Germany (Horstmann et al., 1997). Cuticle erosion varies with the plant species as well as both seasonally and between years, so that there is no simple way of estimating this process. Again, different approaches are needed depending on the situation being modeled. As an example, when predicting the deposition over a time scale of years it may be appropriate to make a steady state assumption and set the total annual dry and wet sedimentation equal to the net accumulation (total atmospheric deposition plus net input from the roots minus volatilization, resuspension, and degradation) in the canopy over the year.

Soil particle resuspension is one process that can transfer chemicals from soil to the plant canopy. This can occur via wind or via rain splash. Wind driven resuspension is favored by a very open or low plant canopy that allows high wind speeds at ground level, and by dry conditions with low vegetation cover. Hence, while this process can generally be neglected in tropical and temperate forests, it may be important in the steppe, tundra, for small islands of vegetation (e.g., hedges, windbreaks), and for some agricultural crops. Transfer via rain splash is favored by bare soil and vegetation very close to the ground. It can be important for the uptake of some chemicals in agricultural crops, particularly when they are growing on contaminated soil. These processes are difficult to capture with a theoretical modeling approach. In those cases when they are important, site-specific measurements form the best basis for model estimates. Before such measurements are initiated, a modeling exercise should be undertaken using conservative assumptions to assess the potential importance of the soil resuspension pathways compared to other transfer processes in the soil–canopy– atmosphere system.

Finally, chemical can exchange between soil and foliage via gaseous transport, whereby the atmosphere will also be involved as some portion of the chemical in the subcanopy air will have originated in the atmosphere, having not been intercepted during passage through the canopy, just as some portion of the chemical in the subcanopy air will also be advected by convective mixing back into the atmosphere. Generally speaking, gaseous transport to the soil is much slower than gaseous transport to the canopy since the higher wind speeds above the canopy, its higher surface roughness, and its greater surface area all contribute to higher mass transfer coefficients. However, it is conceivable that the reverse process, gaseous transport from the soil to the canopy and the atmosphere, is an important flux in the soil–canopy–atmosphere system under some conditions. This process will be particularly favored when there is a strong gradient in chemical potential from the soil to the atmosphere (i.e., a strong tendency for the chemicals to want to volatilize from soil to air) and when the soil/air partition coefficient is low enough that a significant fraction of the chemical in soil can actually partition into the gas phase. Such a situation could develop for instance as a result of chemical application directly to the soil, or for an ecosystem where a persistent chemical has accumulated in the soil as a result of long-term atmospheric exposure, but where the atmospheric levels are currently low. Air–soil mass transfer coefficients of gases as discussed in Chapter 2 can provide an initial estimate of the magnitude of this process.

The interactions between the different transport pathways and how their relative importance changes with the partitioning properties of the chemical is nicely illustrated by the concept of the forest filter effect. The forest filter factor (F) is defined as the ratio of the net atmospheric deposition of a persistent chemical to a forest canopy to its net deposition to a bare soil (McLachlan and Horstmann, 1998). This quotient is plotted as a function of the octanol:air partition coefficient K_{OA} (a surrogate for the vegetation:air partition coefficient) and the air:water partition coefficient in Figure 7.2, which illustrates F for a deciduous forest in central Europe. A value of $F > 1$ implies that the forest is more efficient at filtering the chemical out of the atmosphere than a bare soil. For chemicals with low K_{AW} values, $F \approx 1$, that is, there is no forest filter effect. These chemicals are highly water soluble and are deposited primarily dissolved in rain. Since there is usually no difference in wet deposition to a forest or a bare soil, there is no filter effect. For chemicals with low K_{OA} values, there is also no forest filter effect. For these chemicals gaseous exchange is the dominant transport process, and the mass transfer coefficient to the forest is much higher than to the bare soil. However, the canopy has a limited capacity to take up the chemical due to the low vegetation:air partition coefficient, and thus the net gaseous deposition to the forest is low. When both K_{OA} and K_{AW} are high, F is about 3. Chemicals in this partitioning property range are deposited primarily associated with particles. Although dry particle deposition of smaller particles is much greater to the forest, some of the particle-associated flux derives from wet deposition and, to a smaller extent, sedimentation of large particles, and these fluxes are the same to the forest and the bare soil. However, for chemicals with intermediate K_{OA} values and high K_{AW} values, there is a more pronounced forest filter effect, with F reaching a maximum of 10.5. Chemicals in this range are also deposited primarily as gases, but in this case the high vegetation:air partition coefficients implies that the canopy can accumulate the chemicals, and hence the much higher gaseous deposition velocities to the canopy actually do result in an enhanced net deposition to the forest.

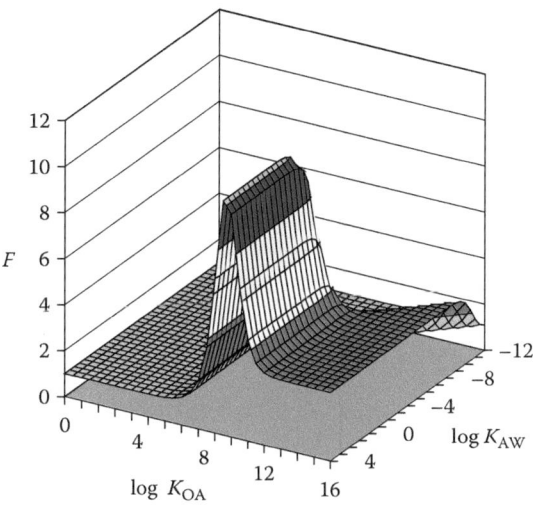

FIGURE 7.2 Illustration of the forest filter factor (F) as a function of the log octanol/air partition coefficient (log K_{OA}) and the log air/water partition coefficient (K_{AW}). F is defined as the quotient of the net deposition flux of a chemical to a forest and the net deposition flux to a nonvegetated soil. The figure was determined based on parameters for a deciduous forest in Germany. (Adapted from McLachlan, M.S., and M. Horstmann. 1998. *Environmental Science & Technology*, 32: 413–420.)

Having now outlined the different mass transfer processes in the soil–canopy–atmosphere system and briefly explored their interactions, the next section will examine two of the most important processes in more detail, namely dry deposition with aerosols and diffusive gaseous exchange.

7.3 TRANSPORT THEORY

Dry gaseous deposition and dry deposition of chemical associated with atmospheric aerosols can be described using a mass transfer coefficient according to the equation

$$N_y = k_y A C_{Ay}, \tag{7.1a}$$

where N is the flux (mol h^{-1}), k is the mass transfer coefficient (m h^{-1}), A is the surface area over which the deposition occurs (m^2), C_A is the concentration of the chemical in the appropriate phase in the air (i.e., gas phase or associated with aerosols, mol m^{-3} air), and y refers to the deposition process, either gaseous ($y = g$) or associated with aerosols ($y = p$). Volatilization, the diffusive gaseous transport from the canopy to the atmosphere, is modeled using a similar equation:

$$N_V = k_g A \frac{C_V}{K_{VA}}. \tag{7.1b}$$

Note that concentration in the gas phase from Equation 7.1a has been replaced by the quotient of the concentration in the vegetation C_V (mol m^{-3}) and the vegetation/air partition coefficient K_{VA} ($m^3\,m^{-3}$), which is equivalent to the concentration that would be present in air if it was in equilibrium with the leaf. The "driving force" for transfer is thus $(C_A - C_V/K_{VA})$.

When modeling deposition to canopies, care must be taken to ensure that the mass transfer coefficient k and the surface area A are defined with respect to the same plane. Values of k may be defined either for deposition through the vertical plane above the canopy or through the plane parallel to the surface of the individual leaves. The value of A must be selected to be consistent with this. In the following we briefly examine the theoretical principles that determine k_g and k_p.

7.3.1 DRY GASEOUS DEPOSITION

A three resistance model has been developed to model the deposition of gases to vegetation canopies. These three resistances, which are aligned in series, describe the turbulent transfer from the atmospheric boundary layer into the canopy (the bulk aerodynamic resistance R_a), the transfer from the air in the canopy to the surface of the vegetation (the quasilaminar sublayer resistance R_b), and the transfer from the surface of the vegetation to reservoir or sink for the chemical in the vegetation (the canopy resistance R_c). These resistances are related to the mass transfer coefficient k_g according to

$$k_g = \frac{1}{R_a + R_b + R_c}, \tag{7.2}$$

where the resistances all have units of reciprocal velocity ($h\,m^{-1}$). For a good overview of the three resistance model, the reader is referred to Hicks et al. (1987).

The bulk aerodynamic resistance R_a is the resistance to transfer of momentum from the atmosphere to the canopy. It can be viewed as the resistance to turbulent diffusion of the chemical between the atmosphere and the canopy. This resistance is governed by the wind speed (i.e., a higher wind speed creates higher turbulence, lowering the resistance), by the "roughness" of the canopy (i.e., a rougher canopy creates more turbulence, lowering the resistance), and by the atmospheric stability (a stable atmosphere dampens turbulence, increasing the resistance). Under conditions of neutral atmospheric stability, R_a can be approximated by (Monteith and Unsworth, 1990):

$$R_a = \frac{\ln[(z - d)/z_0]}{ku_*}, \tag{7.3}$$

where u_* is the friction velocity, defined as

$$u_* = \frac{\kappa u}{\ln[(z - d)/z_0]}. \tag{7.4}$$

Here u is the wind speed ($m\,h^{-1}$), z is the height above the ground at which u is measured (m), d is the displacement height (m), z_0 is the surface roughness of the canopy (m), and κ is the von Karman constant, which has a value of 0.4. The displacement height, which is the height in the canopy at which the wind speed is theoretically zero, can be estimated by

$$d = 0.75 \, h, \tag{7.5}$$

where h is the height of the canopy. Rearranging yields

$$R_a = \frac{\{\ln[(z - d)/z_0]\}^2}{\kappa^2 u}, \tag{7.6}$$

which gives the interesting result that R_a is linearly proportional to $1/u$. Typical values of z_0 range from 0.02 m for closely cropped grassland to 0.05 m for rangeland, 0.1 m for agricultural crops, and 1 m for deciduous and coniferous forests (Wesley and Lesht, 1989). Table 7.1 lists z_0 values for a range of vegetation types and other natural surfaces. In contrast to water surfaces, the roughness height z_0 is largely independent of the friction velocity u_* for most plant canopies.

The quasilaminar sublayer resistance R_b describes the excess resistance for the transfer of matter from the atmosphere to the surfaces of the vegetation, that is, the difference between the resistance for matter and the resistance for momentum. It is primarily associated with molecular diffusion through quasi laminar boundary layers. Several parameterizations for R_b have been developed, but that employed by Brook et al. (1999), which like Equations 7.3 and 7.6 is valid for conditions of neutral atmospheric stability, is particularly easy to apply:

$$R_b = \frac{2}{\kappa u_*} Sc^{2/3}, \tag{7.7}$$

where Sc is the Schmidt number, which is defined by

$$Sc = \frac{\nu}{D}. \tag{7.8}$$

Here ν is the kinematic viscosity ($m^2 \, h^{-1}$) and D is the molecular diffusion coefficient ($m^2 \, h^{-1}$). Substituting Equations 7.4 and 7.8 into Equation 7.7 yields

$$R_b = \frac{2 \ln[(z - d)/z_0]}{\kappa^2 u} \left(\frac{\nu}{D}\right)^{2/3}. \tag{7.9}$$

This relationship illustrates that R_b, like R_a, is inversely proportional to the wind speed. Intuitively, one would expect higher winds to result in thinner laminar boundary layers around the canopy surfaces, and hence more rapid diffusive transport. Furthermore, R_b is inversely related to the molecular diffusion coefficient, in contrast to R_a, which is independent of the properties of the chemical. Finally, R_b is inversely related to the roughness height; an increased roughness height increases the turbulence in the canopy, which also results in thinner laminar boundary layers.

The canopy resistance R_c depends on the location of the reservoirs/sinks for the chemical in the vegetation, as well as on the properties of the chemical. Chemicals

TABLE 7.1
Typical Roughness Heights (z_0) for Different Vegetation Canopies and Other Natural Surfaces

Canopy Type	Surface Roughness z_0 (m)
Evergreen needleleaf trees	0.9
Evergreen broadleaf trees	2.0
Deciduous needleleaf trees	0.4–0.9
Deciduous broadleaf trees	0.4–1.0
Tropical broadleaf trees	2.5
Drought deciduous trees	0.6
Evergreen broadleaf shrubs	0.2
Deciduous shrubs	0.05–0.2
Thorn	0.2
Short grass and forbs	0.04
Long grass	0.02–0.1
Crops	0.02–0.1
Rice	0.02–0.1
Sugar	0.02–0.1
Maize	0.02–0.1
Cotton	0.02–0.1
Irrigated crops	0.05
Tundra	0.03
Swamp	0.1
Desert	0.04
Mixed wood forests	0.6–0.9
Transitional forests	0.6–0.9

Source: Adapted from Zhang, L., J.R. Brook, and R. Vet. 2003. *Atmospheric Chemistry and Physics*, 3: 2067–2082.

can enter the vegetation either by moving through the stomatal openings or onto and through the cuticle, the wax-impregnated layer of the polymer cutin that coats the leaves. Consequently, R_c is frequently modeled as the sum of a cuticular resistance and a stomatal resistance acting in parallel. For many lipophilic chemicals, the cuticle itself is an important sink in the vegetation, and this can greatly reduce the cuticular resistance and R_c. It has been suggested that the cuticular pathway is generally dominant for organic chemicals and that the stomatal pathway can be neglected (Riederer 1990). In this case the canopy resistance can be defined as

$$R_c = \frac{y_{Cu}}{K_{Cu/A}D_{Cu}}, \tag{7.10}$$

where D_{Cu} is the diffusion coefficient of the chemical in the cuticle/vegetation ($m^2\,h^{-1}$), y_{Cu} is the diffusion path length between the vegetation surface and the sink/reservoir of the chemical (m), and $K_{Cu/A}$ is the cuticle/air partition coefficient

($m^3 \, m^{-3}$). Experimental evidence indicates that D_{Cu} is not inversely proportional $K_{Cu/A}$. This suggests that R_c should be dependent on $K_{Cu/A}$. The fact that field measurements of k_g for lipophilic chemicals have shown no dependence on $K_{Cu/A}$ has been interpreted as evidence that R_c is small compared to R_a and R_b, and that it can be neglected (Horstmann and McLachlan, 1998; McLachlan, 1999; Su and Wania, 2005). However, this conclusion remains controversial in the literature (Barber et al., 2004b; Kerstiens, 2006); there may be situations where R_c cannot be neglected, for instance when R_a and R_b are particularly low (e.g., high wind speed, high surface roughness, high instability) and/or for plant–chemical combinations with high values of R_c.

Summarizing the above, assuming R_c is negligible, k_g can generally be defined by

$$
\begin{aligned}
k_g &= \left[\frac{\ln[(z-d)/z_0]}{\kappa u_*} + \frac{2}{\kappa u_*} Sc^{2/3} \right]^{-1} \\
&= \left[\frac{\{\ln[(z-d)/z_0]\}^2}{\kappa^2 u} + \frac{2 \ln[(z-d)/z_0]}{\kappa^2 u} \left(\frac{\nu}{D} \right)^{2/3} \right]^{-1}
\end{aligned}
\tag{7.11}
$$

7.3.2 Dry Aerosol-Bound Deposition

Dry deposition of chemical associated with atmospheric aerosols to a plant canopy is in many ways similar to the corresponding deposition pathway to water and soil as discussed in Chapter 6. Atmospheric aerosols can be deposited via Brownian diffusion (dominant for small aerosols ($<\sim 0.1 \, \mu m$) and characterized by intermediate deposition velocities, see Figure 6.4), impaction or interception (medium size aerosols ($<\sim 2 \, \mu m$), low deposition velocities), or sedimentation (large aerosols ($>\sim 2 \, \mu m$), high deposition velocities).

Which of these deposition processes dominates depends on the relative concentration of the chemical in different aerosol sizes classes. Many chemicals are able to partition freely between the gas phase and aerosols; they will have the highest affinity for the aerosol size class with the highest sorption capacity (e.g., organic carbon content). Other chemicals may have been entrapped in the aerosol during its formation (e.g., PAHs in combustion aerosols). Some portion of these chemicals will not be able to freely partition between the different aerosol sizes in the atmosphere; their concentration in different aerosol size classes will be determined by the size of the aerosols they were entrapped in and the coagulation history of the aerosols. A third possibility is that a chemical is present in its pure form, for instance as a result of emissions of aerosols during production of the chemical or erosion of aerosols from goods and products. In this case, the chemical concentration in different aerosol size classes will also be determined by the specific aerosol formation process.

The second factor determining whether diffusion, impaction/interception, or sedimentation dominates is the relative abundance of the aerosol size classes in the atmosphere (see Figure 6.4). This is determined by the atmospheric half-lives of the aerosols and distance from the sources of aerosol and chemical emissions. Large aerosols have the shortest half-lives as they are deposited most rapidly, whereas medium-size aerosols have the longest half-lives. Consequently, sedimentation may be the dominant dry deposition pathway close to sources of large aerosols, whereas in more remote regions it is frequently diffusion or impaction/interception due to the paucity of large aerosols. Note that local particle resuspension processes may cause a significant aerosol sedimentation flux in remote areas, but that this flux can be viewed as a local vertical cycling phenomenon rather than as lateral contaminant translocation via long-range atmospheric transport.

The relative importance of the deposition mechanisms influences the effect of plant canopies on dry deposition of chemicals via aerosols. While the presence of foliage does not influence sedimentation (i.e., the deposition velocity is the same to a vegetated and a nonvegetated surface), aerosol deposition via impaction/interception is generally greater to plant canopies than it is to water or soil. This is a consequence of the greater aerodynamic roughness of the plant canopy, which results in more intense turbulence, reducing the thickness of laminar boundary layers on the surfaces and accentuating aerosol impaction/interception. Dry deposition of aerosols is also influenced by the "stickiness" of the canopy surfaces, that is, their ability to retain deposited aerosols.

The Brownian diffusion deposition mechanism is governed by the same principles as the deposition of gases (see Section 7.3.1). It can be treated using the same three resistance model consisting of aerodynamic, quasilaminar sublayer, and canopy resistances (R_a, R_b, and R_c, see Equation 7.2). If the properties of the aerosols carrying the chemical do not change during the transport process, R_a and R_b can be formulated in the same manner as for gases, whereby the aerosol size must be taken into consideration in the Schmidt number correction of the latter (van Jaarsveld et al., 1997). The canopy resistance R_c accounts for the "stickiness" of the canopy surfaces for the aerosols.

Particle impaction/interception may also be described with the three resistance model. The aerodynamic resistance (R_a) is the same as for gases or Brownian diffusion. The quasilaminar sublayer resistance (R_b) and canopy resistance (R_c) are, however, complex functions of both aerosol size and the properties of the canopy surfaces. Since particle size-specific information is typically not available for organic contaminants, it is expedient to combine R_b and R_c for both Brownian diffusion and impaction/interception into a single resistance R_{ds}.

$$k_{p(D+I)} = \frac{1}{R_a + R_{ds}}. \tag{7.12}$$

As a first approximation, under neutral atmospheric conditions R_{ds} is inversely proportional to the friction velocity:

$$R_{ds} = \frac{B}{u_*}, \tag{7.13}$$

where B is a proportionality constant that is specific for the canopy (Brook et al., 1999), whereby a value of 500 is used for short canopies. Su and Wania (2005) suggest values for B of 94 and 1321 for deciduous and coniferous canopies, respectively, based on measurements of the deposition of organic contaminants. A more sophisticated approach to calculating R_{ds} for tall canopies can be found in Brook et al. (1999). Substituting in Equation 7.4, one obtains

$$R_{ds} = \frac{B \ln[(z-d)/z_0]}{ku}.$$ (7.14)

This illustrates that R_{ds} is also inversely proportional to the wind speed.

Summarizing

$$k_{p(D+I)} = \left[\frac{\ln[(z-d)/z_0]}{\kappa u_*} + \frac{500}{u_*} \right]^{-1}.$$ (7.15)

As mentioned above, the mass transfer coefficient for particle sedimentation, $k_{p(S)}$, is no different for vegetated surfaces than it is for nonvegetated surfaces or water, and it can be treated in the same manner. For further insight into the physics of particle sedimentation, the reader is referred to the text by Seinfeld and Pandis (1998).

7.3.3 MEASUREMENT OF TRANSPORT PARAMETERS

The theory presented in the preceding sections has relied heavily on the study of the deposition of inorganic species. There have, however, been a few field measurements of the mass transfer coefficients of organic contaminants to plant canopies.

One measurement approach is to quantify the accumulation of organic chemicals in a canopy and to relate this to the levels of the chemicals in the air above the canopy. Dividing the rate of accumulation by the surface area and the atmospheric concentration yields the mass transfer coefficient. Two major challenges in such measurements are the attribution of chemical accumulation in the canopy to specific deposition processes and accounting for loss of chemical from the canopy. As outlined above, there are several possibilities for a chemical to enter or leave the canopy. To ameliorate these difficulties, it has proven beneficial to focus on persistent chemicals and to study simultaneously a selection of chemicals with a broad range of partitioning properties. The relationships between accumulation behavior and partitioning properties can be used to identify the dominant chemical transfer processes (Horstmann and McLachlan, 1998; Böhme et al., 1999; Su et al., 2007). Furthermore, measuring deposition under a canopy and comparing it with the bulk deposition measured outside of the canopy allows one to separate the contributions of wet and dry sedimentary deposition from gaseous deposition and dry deposition of aerosols via diffusion and impaction/interception. This is, however, impractical in short canopies, which explains why most measurements to grass and agricultural crops are based on the measured accumulation of the chemical in the canopy foliage. In addition to the limitation just mentioned, this method has a further weakness in that it does not capture the atmospheric deposition which is subsequently lost via litter fall or wash-off to the canopy floor. Hence mass transfer coefficients determined in this manner must be evaluated carefully.

A second approach is to measure the vertical gradient in the concentrations of chemicals in the atmosphere above a canopy. This can be related to the vertical flux of chemical into the canopy by the eddy diffusivity, which in turn is determined from the vertical gradient in temperature and wind speed (Monteith and Unsworth, 1990). This method can only be applied over relatively short periods, and thus it cannot be used for many organic chemicals, since long sampling times are required to quantify concentrations in air. It also places high demands on the precision of the analytical methodology. However, as analytical techniques improve, this technique will be useful.

Relaxed eddy accumulation is another promising method. It uses a high-speed sensor to measure the vertical component of the air velocity above a canopy. The sensor controls a valve on an air sampling inlet which directs the sampled air to one of two chemical sampling trains, depending on whether the air flux is upward or downward. The difference in the chemical quantity collected in the two sampling trains can then be translated into the net flux into or out of the canopy during the sampling period (for an application to aerosol deposition see Grönholm et al., 2007). This technique requires very sensitive analytical methods, which has so far hampered its application to semivolatile organic contaminants.

7.4 TRANSPORT PARAMETERS

Measured values for the mass transfer coefficients for gaseous deposition k_g and dry deposition associated with aerosols via diffusion and impaction/interception $k_{p(D\&I)}$ are summarized in Tables 7.2 through 7.4. The data for were separated into two tables, those determined from measurements of deposition under the canopy and including correction for wet and dry sedimentary deposition (Table 7.3), and those based on measurements of accumulation in the foliage only (Table 7.4). All values are based on measurements with a seasonal or annual time scale, and hence they represent long-term average mass transfer coefficients. Note the much higher values of k_g to forests compared to grass, which reflects the greater surface roughness of the former (see Section 7.3.3). Note also that the mass transfer coefficients are higher for gases than for particles. This suggests that R_{ds}, and not R_a, is the resistance limiting dry deposition of aerosols to canopies.

In extrapolating these values to other environmental situations, one has two choices: one can scale the values using the dependencies on environmental variables outlined in Section 7.3, or one can calculate the values using the equations given in Section 7.3 together with estimates of the necessary meteorological and canopy properties (e.g., surface roughness, wind speed, canopy stickiness). An example of each approach is given in Section 7.5, and a good example of scaling on a global scale is found in the Supporting Information of Su and Wania (2005). When the modeling goal is well represented by the conditions and assumptions used to generate the data in the tables, it is anticipated that the first method will give the better result. Otherwise, calculating using the equations in Section 7.3 may be the method of choice. Each method contains considerable uncertainty, as there are too few empirical data available on the influence of meteorological and canopy variables on the values of k_g and $k_{p(D\&I)}$. However, the methods do allow their estimation to within an order of magnitude.

TABLE 7.2

Measured Mass Transfer Coefficients for Gaseous Deposition (k_g) of Organic Contaminants to Various Plant Canopies

Canopy	Location	Period	Chemicals	k_g (m h^{-1})	Comments	Source
Mature grass	Bayreuth, Germany	16 weeks, summer, early autumn	PCBs, PAHs, PCDD/Fs	5	Based on net accumulation in foliage	Böhme et al., 1999
Pasture grass	Bayreuth and Magdeburg, Germany	Various time periods during growth	PCBs	8	Based on net accumulation in foliage	Welsch-Pausch, 1998
Pasture grass	Bayreuth and Magdeburg, Germany	Various time periods during growth	PCDDs	6	Based on net accumulation in foliage	Welsch-Pausch, 1998
Pasture grass	Bayreuth and Magdeburg, Germany	Various time periods during growth	PCDFs	4	Based on net accumulation in foliage	Welsch-Pausch, 1998
Corn	Bayreuth, Germany	Full growing period, 15 weeks	PCBs, PAHs, PCDD/Fs	6	Based on net accumulation in leaves only, not stems, fruit	Calculated from Böhme et al., 1999
Deciduous forest (mixed oak and beech)	Bayreuth, Germany	Foliage period (May–November)	PCBs, PAHs, PCDD/Fs	130		Horstmann and McLachlan, 1998
Coniferous forest (spruce)	Bayreuth, Germany	Full year	PCBs, PAHs, PCDD/Fs	28		Horstmann and McLachlan, 1998
Deciduous forest (mixed maple and aspen)	Southern Ontario, Canada	Foliage period (May–November)	PCBs, PAHs, PBDEs	97		Su et al., 2007

TABLE 7.3
Measured Mass Transfer Coefficients for Dry Deposition Associated with Aerosols Via Diffusion and Impaction/Interception $k_{p(D\&I)}$ of Organic Contaminants to Various Plant Canopies

Canopy	Location	Period	Chemicals	$k_{p(D\&I)}$ (m h^{-1})	Comments	Source
Deciduous forest (mixed oak and beech)	Bayreuth, Germany	Foliage period (May–November)	PCDD/Fs	7		Horstmann and McLachlan, 1998
Deciduous forest (mixed oak and beech)	Bayreuth, Germany	Foliage period (May–November)	PAHs	26		Horstmann and McLachlan, 1998
Coniferous forest (spruce)	Bayreuth, Germany	Full year	PAHs	2		Horstmann and McLachlan, 1998
Deciduous forest (mixed maple and aspen)	Southern Ontario, Canada	Foliage period (May–November)	PBDEs	29		Su et al., 2007
Deciduous forest (mixed maple and aspen)	Southern Ontario, Canada	Foliage period (May–November)	PAHs	4		Su et al., 2007

TABLE 7.4
Measured Mass Transfer Coefficients for Net Dry and Wet Deposition Associated with Aerosols $k_{W\&D}$ of Organic Contaminants to Various Plant Canopies (Based on Accumulation in Foliage)

Canopy	Location	Period	Chemicals	$k_{W\&D}$ (m h^{-1})	Comments	Source
Pasture grass	Bayreuth and Magdeburg, Germany	Various time periods during growth	PCDD/Fs	3		Welsch-Pausch, 1998
Corn	Bayreuth, Germany	Second half of growing period (August & September)	PAHs	0.8	Based on uptake in leaves only	Kaupp, 1996
Corn	Bayreuth, Germany	Second half of growing period (August & September)	PCDD/Fs	2	Based on uptake in leaves only	Kaupp, 1996

The physical chemical properties of the chemical being deposited can also affect k_g and $k_{p(D\&I)}$. For gaseous deposition, the bulk aerodynamic resistance R_a is independent of the properties of the chemical, but the quasilaminar sublayer resistance R_b is dependent on the Schmidt number (Sc), the quotient of the chemical's kinematic viscosity in air and its molecular diffusion coefficient in air (Equation 7.7). If R_b is the controlling resistance, then in a rigorous analysis one would correct R_b for the differences in Sc between the chemical of interest and the chemical on which the measured values are based. However, for most organic contaminants the range in Sc is small, and the error introduced by assuming a constant value is small compared to the uncertainty in the available measured values of k_g and their extrapolation to different environmental conditions. For deposition associated with aerosols, the variability of $k_{p(D\&I)}$ amongst different chemicals can be expected to be linked to the properties of the aerosols (particularly size and "stickiness") that the chemicals are associated with. Different values of $k_{p(D\&I)}$ have been measured for different chemicals (see Table 7.3) and attributed to differences in particle size distribution (Horstmann and McLachlan, 1998; Su et al., 2007).

The situation may arise that one needs to quantify the mass transfer to different plant species within the same canopy. Again, data are limited, but a study of the accumulation of PCBs, PAHs, and PCDD/Fs in eight different herb and grass species of widely ranging morphology growing in the same pasture canopy indicated that the average k_g (in this case defined through the plane parallel to the leaf) over the growing period ranged over a factor of ~ 2.5 (Böhme et al., 1999). This suggests that the variability in k_g to different species within a canopy is considerably less than the variability between canopies. For an extensive discussion of the influence plant properties on the uptake of organic contaminants, see Barber et al. (2004b).

7.5   EXAMPLE CALCULATIONS

7.5.1   What are the Values of k_g and $k_{p(D\&I)}$ for Deposition to a Grassy Coastal Plain in Mid-Summer, with Wind Speeds of on Average $8\,m\,s^{-1}$

Method A: Scaling measured data

From Table 7.1, select a value of k_g of 8 m h^{-1} for deposition of PCBs to grassland in Bayreuth (the lower values of k_g for PCDD/Fs to the same canopy may be due to photodegradation on the leaf surface). Bayreuth is an inland station. According to the meteorological records the average wind speed in this region during summer is ~ 2 m s^{-1}. Referring to Equations 7.6 and 7.9, both R_a and R_b are inversely proportional to the wind speed, and hence k_g is linearly proportional to the wind speed. Scaling

$$k_{g(8\,m/s)} = k_{g(2\,m/s)} \times \frac{8}{2} = 32\,m\,h^{-1}.$$

For $k_{p(D\&I)}$ we select a value of 3 m h^{-1} for deposition of PCDD/Fs to grassland in Bayreuth from Table 7.3, remembering that this value is uncertain because it also includes the effects of wet deposition and aerosol sedimentation (possible positive bias), but on the other hand is based on net accumulation in the foliage, not gross deposition to the canopy (possible negative bias). Referring to Equation 7.15, $k_{p(D\&I)}$ is also linearly proportional to the wind speed, so we can correct from the conditions in Bayreuth to the conditions at our coastal station in the same manner.

$$k_{p(D\&I)(8\,m/s)} = k_{p(D\&I)(2\,m/s)} \times \frac{8}{2} = 12\,m\,h^{-1}.$$

Method B: Calculation from the equations

Assume that the wind speed was measured at 5 m above the ground, and that the terrain surrounding the meteorological station has a surface roughness typical for that of the coastal plain. Assuming that the grass is ~0.5 m high, $(z - d)$ can be approximated by $(5 - 0.5)$ m. Referring to Table 7.3, select a typical roughness height of 0.04 m (for short grass and forbs, also in range for long grass). Substituting into Equation 7.4

$$u_* = \frac{\kappa u}{\ln[(z - d)/z_0]} = \frac{(0.4)(8)}{\ln[(5 - 0.375)/0.04]} = 0.67\,m\,s^{-1}.$$

Using $v = 0.0000142\,m^2\,s^{-1}$ for dry air at 10°C and choosing a "typical" value of $D = 0.000005\,m^2\,s^{-1}$ for organic contaminants in air yields $Sc = 2.8$ (Equation 7.8). Substituting into Equation 7.11 yields

$$k_g = \left[\frac{\ln[(z - d)/z_0]}{\kappa u_*} + \frac{2}{\kappa u_*} Sc^{2/3} \right]^{-1}$$

$$= \left[\frac{\ln[(5 - 0.375)/0.04]}{(0.4)(0.67)} + \frac{2}{(0.4)(0.67)} (2.8)^{2/3} \right]^{-1} = 0.031\,m\,s^{-1} = 111\,m\,h^{-1}.$$

Substituting into Equation 7.15 yields

$$k_{p(D+I)} = \left[\frac{\ln[(z - d)/z_0]}{\kappa u_*} + \frac{500}{u_*} \right]^{-1} = \left[\frac{\ln[(5 - 0.375)/0.04]}{(0.4)(0.71)} + \frac{500}{0.67} \right]^{-1}$$

$$= 0.0013\,m\,s^{-1} = 4.7\,m\,h^{-1}.$$

Comparing the results, Method B yields a value for k_g that is 3.5 times higher than the value from Method A. Possible explanations include an underestimation by Method A, for instance because the measured data were based on net, not gross, accumulation in the grass, or an overestimation by Method B, for instance due to the assumption of neutral atmospheric stability, or because the canopy resistance R_c was not negligible in this case.

In contrast, Method B yields a value for $k_{p(D\&I)}$ that is lower than the Method A result by a factor of 2.5. This may be attributable to the fact that the measured value of $k_{p(D\&I)}$ in Method A included the contribution from wet deposition and aerosol sedimentation. On the other hand, the constant (500) in Equation 7.15 was determined from data for the deposition of SO_2 aerosols, which may have a lower "stickiness" than the organic aerosols of interest here.

7.5.2   CALCULATE k_g FOR DEPOSITION TO A MATURE 25 m HIGH DECIDUOUS CANOPY AND A MATURE 25 m HIGH CONIFEROUS CANOPY IN BAYREUTH AND COMPARE THEM TO THE VALUES IN TABLE 7.2

First we need to estimate z, the height above the canopy where the wind speed is $2\,m\,s^{-1}$ (because of the different surface roughness at the meteorological station and for the forest canopies, the wind profile with height will differ). To do this, we first calculate the wind speed at an arbitrary height far above the meteorological station (say 1000 m). Then we use this to estimate the height above the canopy where the wind speed would drop to $2\,m\,s^{-1}$. First, apply Equation 7.4 to the wind profile at the meteorological station, and use the fact that the friction velocity of a surface is independent of height.

$$\left\langle \frac{\kappa u}{\ln[(z-d)/z_0]} \right\rangle_{1000\,m} = \left\langle \frac{\kappa u}{\ln[(z-d)/z_0]} \right\rangle_{5\,m}.$$

Assume that the surface roughness at the meteorological station is 0.04 m, the displacement height is 0.375 m and that the wind speed is being measured at 5 m (as above). Solving for $u(1000\,m)$ gives

$$u(1000\,m) = \left\langle \frac{\ln[(z-d)/z_0]_{1000\,m}}{\ln[(z-d)/z_0]_{5\,m}} u(5\,m) \right\rangle_{5\,m}$$

$$= \frac{\ln[(1000-0.375)/0.04]}{\ln[(5-0.375)/0.04]} 2 = 4.3\,m\,h^{-1}.$$

Choose a roughness height of 0.9 m for the canopies from Table 7.1 (deciduous broadleaf and evergreen needleleaf), calculate d from the canopy height (Equation 7.5), and rearrange the equation above to solve for the unknown height (z) above the canopy where the wind speed is $2\,m\,h^{-1}$.

$$\ln[(z-d)/z_0] = \frac{u}{u(1000\,m)} \ln[(z-d)/z_0]_{1000\,m}$$

$$= \frac{2}{4.3} \ln[(1000-18.75)/(0.9)]; \quad z = 43\,m.$$

Note that the wind speed increases with height more slowly above the forest canopies than at the meteorological station as a result of the greater surface roughness of the canopies.

Now, use these values of z, d, and z_0 to calculate u_* and k_g as above:

$$u_* = \frac{\kappa u}{\ln[(z-d)/z_0]} = \frac{(0.4)(2)}{\ln[(42.7-18.75)/(0.9)]} = 0.37\,m\,s^{-1},$$

$$k_g = \left[\frac{\ln[(z-d)/z_0]}{\kappa u_*} + \frac{2}{\kappa u_*} Sc^{2/3} \right]^{-1}$$

$$= \left[\frac{\ln[(42.7 - 18.75)/(0.9)]}{(0.4)(0.37)} + \frac{2}{(0.4)(0.37)}(2.8)^{2/3} \right]^{-1}$$

$$= 0.020 \, \mathrm{m \, s^{-1}} = 73 \, \mathrm{m \, h^{-1}}.$$

The calculated value lies between the values measured for the coniferous canopy $(28 \, \mathrm{m \, h^{-1}})$ and the deciduous canopy $(130 \, \mathrm{m \, h^{-1}})$ in Bayreuth. This indicates that the equations give a reasonable estimate of k_g to tree canopies, but that they do not capture the canopy-specific differences in the mass transfer coefficients.

7.6 SUMMARY AND CONCLUSIONS

This chapter has defined the major processes that contribute to transfer of organic chemicals between the atmosphere and plant canopy systems, reviewed the theoretical underpinnings of the two most important processes, and summarized the available measurements of the mass transfer parameters. It was illustrated that forest canopies can play an important part in the environmental fate of organic contaminants, particularly for those with intermediate K_{OA} values and high K_{AW} values, in addition to the more obvious consequences of chemical accumulation in plants for phytotoxicity and bioaccumulation in terrestrial food chains. The methods presented here allow order of magnitude calculations of the most important chemical fluxes and illustrate how they will be influenced by chemical and environmental properties.

Despite the valuable insight that the existing literature offers, it is also clear that the empirical basis for estimating the gaseous and aerosol-bound mass transfer coefficients of organic contaminants from the atmosphere to the canopy is very thin. Only a handful of measurements have been conducted, and these are limited to a small selection of chemicals, canopy types, and climatic conditions. More measurements are required if more reliable estimation methods are to be obtained. The canopy accumulation method described in Section 7.3.3 has proven to be valuable for studying mass transfer over periods of weeks to months, while micrometeorological methods hold promise for quantifying fluxes of organic contaminants over much shorter time scales.

A comprehensive understanding of chemical transfer between the atmosphere and canopy systems will also require more research into the other transport processes involved. Volatilization from forest soils could become particularly relevant for the fate of persistent organic pollutants as atmospheric concentrations decrease. It will also be necessary to refine our knowledge of the partition coefficients controlling gaseous exchange in the canopy/atmosphere system. The greatest deficits here are for vegetation/air partitioning and it temperature dependence, for which more measurements are urgently needed.

ACKNOWLEDGMENT

Daniel Partridge is thanked for offering valuable comments to a draft version of this chapter.

LITERATURE CITED

Barber, J.L., G.O. Thomas, R. Bailey, G. Kerstiens, and K.C. Jones. 2004a. Exchange of polychlorinated biphenyls (PCBs) and polychlorinated naphthalenes (PCNs) between air and a mixed pasture sward. *Environmental Science & Technology*, 38: 3892–3900.

Barber, J.L., G.O. Thomas, G. Kerstiens, and K.C. Jones. 2004b. Current issues and uncertainties in the measurement and modeling of air-vegetation exchange and within-plant processing of POPs. *Environmental Pollution*, 128: 99–138.

Böhme, F., K. Welsch-Pausch, and M.S. McLachlan. 1999. Uptake of airborne semivolatile organic compounds in agricultural plants: Field measurements of interspecies variability. *Environmental Science & Technology*, 33: 1805–1813.

Brook, J.R., L. Zhang, F. Di-Giovanni, and J. Padro. 1999. Description and evaluation of a model of deposition velocities for routine estimates of air pollutant dry deposition over North America. Part I: model development. *Atmospheric Environment*, 33: 5037–5051.

Grönholm, T., P.P. Aalto, V. Hiltunen, Ü. Rannik, J. Rinne, L. Laakso, S. Hyvönen, T. Vesala, and M. Kulmala. 2007. Measurements of aerosol particle dry deposition velocity using the relaxed eddy accumulation technique. *Tellus*, 59B: 381–386.

Hicks, B.B., D.D. Baldocchi, T.P. Meyers, R.P. Hosker Jr., and D.R. Matt. 1987. A preliminary multiple resistance routine for deriving dry deposition velocities from measured quantities. *Water, Air, and Soil Pollution*, 36: 311–330.

Horstmann, M. and M.S. McLachlan. 1998. Atmospheric deposition of semivolatile organic compounds to two forest canopies. *Atmospheric Environment*, 32: 1799–1809.

Horstmann, M., U. Bopp, and M.S. McLachlan. 1997. Comparison of the bulk deposition of PCDD/Fs in a spruce forest and an adjacent clearing. *Chemosphere*, 34: 1245–1254.

Kaupp, H. 1996. Atmosphärische Eintragswege und Verhalten von polychlorierten Dibenzo-p-dioxinen und –furanen sowie polyzyklischen Aromaten in einem Maisbestand. Doctoral dissertation, University of Bayreuth.

Kerstiens, G. 2006. Parameterization, comparison, and validation of models quantifying relative change of cuticular permeability with physicochemical properties of diffusants. *Journal of Experimental Botany*, 57: 2525–2533.

Mazurek, M.A., G.R. Cass, and B.R.T. Simoneit. 1991. Biological input to visibility-reducing aerosol particles in the remote arid southwestern United States. *Environmental Science & Technology*, 25: 684–694.

McLachlan, M.S. 1999. Framework for the interpretation of measurements of SOCs in plants. *Environmental Science & Technology*, 33: 1799–1804.

McLachlan, M.S. and M. Horstmann. 1998. Forests as filter of airborne organic pollutants: A model. *Environmental Science & Technology*, 32: 413–420.

Monteith, J.L. and M.H. Unsworth. 1990. *Principles of Environmental Physics*, 2nd ed. Edward Arnold, London.

Riederer, M. 1990. Estimating partitioning and transport of organic chemicals in the foliage/atmosphere system: Discussion of a fugacity based model. *Environmental Science & Technology*, 24: 829–837.

Seinfeld, J. H. and S.N. Pandis. 1998. *Atmospheric Chemistry and Physics: From Air Pollution to Climate Change*. Wiley, New York.

Su, Y. and F. Wania. 2005. Does the forest filter effect prevent semivolatile organic compounds from reaching the Arctic? *Environmental Science & Technology*, 39: 7185–7193.

Su, Y., F. Wania, T. Harner, and Y.D. Lei. 2007. Deposition of polybrominated diphenyl ethers, polychlorinated biphenyls, and polycyclic aromatic hydrocarbons to a boreal deciduous forest. *Environmental Science & Technology*, 41: 534–540.

van Jaarsveld, J.A., W.A.J. van Pul, and F.A.A.M. de Leeuw. 1997. Modelling transport and deposition of persistent organic pollutants in the European region. *Atmospheric Environment*, 31: 1011–1024.

Welsch-Pausch, K. 1998. Atmosphärische Deposition polychlorierte Dibenzo-*p*-dioxine und Dibenzofurane auf Futterpflanzen. Doctoral dissertation, University of Bayreuth.

Wesely, M.L. and B.B. Lesht. 1989. Comparison of RADM dry deposition algorithms with a site-specific method for inferring dry deposition. *Water, Air, and Soil Pollution*, 44: 273–293.

Zhang, L., J.R. Brook, and R. Vet. 2003. A revised parameterization for gaseous dry deposition in air-quality models. *Atmospheric Chemistry and Physics*, 3: 2067–2082.

8 Mass Transfer within Surface Soils

Thomas E. McKone, Shannon L. Bartelt-Hunt, Mira S. Olson, and Fred D. Tillman

CONTENTS

Part 1   A Multilayer Box Approach for Use in Fugacity
        Multimedia Fate Models
        Thomas E. McKone

8.1   Introduction to Part 1 160
8.2   Sources of Contamination in Soils 162
8.3   Composition of Soils 163
8.4   Transport Processes in the Soil Column.......................................164
8.5   Transformation Processes in Soil 166
8.6   Soil Mass Transfer: Theory and Equations.....................................167
    8.6.1   Background..167
    8.6.2   Conceptual Model..168
    8.6.3   Mass Balance Equations 168
    8.6.4   Normalized Gradient of Soil Concentration 170
8.7   Soil Transport Parameters...171
8.8   Soil–Air Mass Transfer Rates...175
    8.8.1   Two-Compartment (Air–Soil) Model
          Assuming a Well-Mixed Soil 176
    8.8.2   Two-Compartment (Air–Soil) Model with a Concentration
          Gradient in the Soil 179
    8.8.3   Discussion of the Two Approaches to a Single
          Soil Compartment..180
    8.8.4   Multiple Soil Layers with Concentration Gradients.................182
8.9   Summary..184
Acknowledgments 185
Literature Cited 186

Part 2   Advective–Diffusive Transport Processes for
        Chemicals in Surface Soils
        Shannon L. Bartelt-Hunt, Mira S. Olson, and Fred D. Tillman

8.10 Introduction to Part 2...187

8.11 Continuity Equation 187
    8.11.1 Lavoisier Mass Balance and the Advective–Diffusive
        Equation 187
    8.11.2 Overview of Closely Related Surface Soil
        Chemodynamic Processes 188
8.12 Advective Chemical Transport in the Mobile Phases of Soil................188
    8.12.1 Water-Saturated Soils: Darcy's Law 188
    8.12.2 Water-Unsaturated Soils: Richards' Law and Capillary Rise........188
    8.12.3 Nonaqueous Phase Liquids 190
    8.12.4 Soil Gas Movement 193
        8.12.4.1 Barometric Pumping 194
        8.12.4.2 Soil Vapor Intrusion 195
        8.12.4.3 Landfill Gas Generation.....................................196
8.13 Diffusive Chemical Transport in Mobile Phases...........................197
    8.13.1 Diffusive Transport in the Aqueous Phase: Saturated
        and Unsaturated Conditions 197
    8.13.2 Diffusive Transport by Particle Bio-Diffusion.......................197
    8.13.3 Diffusive Transport in the Vapor Phase 198
    8.13.4 The Dispersion Process in Saturated Porous Media 201
8.14 Contaminant Migration by Leaching: Mass-Transfer between Mobile and
    Immobile Phases 202
    8.14.1 NAPL-to-Porewater Transport in Porous Media 202
    8.14.2 Solute-to-Solid Phase Transport in Porous Media...................203
    8.14.3 Vapor-to-Solid Phase Transport in Porous Media 204
8.15 Case Study: Soil Gas Movement and Contaminant Flux under Natural
    Conditions 204
    8.15.1 Effect of Moisture Content and Predictive Relationship
        on Estimates of Effective Diffusion Coefficient 206
Literature Cited 207
Additional Reading 210

PART 1   A MULTILAYER BOX APPROACH FOR USE IN FUGACITY MULTIMEDIA FATE MODELS

Thomas E. McKone

8.1   INTRODUCTION TO PART 1

Contaminants in soil can impact human health and the environment through a complex web of interactions. Soils exist where the atmosphere, hydrosphere, geosphere, and biosphere converge. Soil is the thin outer zone of the Earth's crust that supports rooted plants and is the product of climate and living organisms acting on rock. A true soil is a mixture of air, water, mineral, and organic components. The relative proportions of these components determine the value of the soil for agricultural and for other human uses. These proportions also determine, to a large extent, how a substance added to soil is transported and/or transformed within the soil (Sposito, 2008). In mass-balance models, soil compartments play a major role, functioning both as reservoirs and as the principal media for transport among air, vegetation, surface water, deeper soil, and ground water (Mackay, 2001). Quantifying the mass transport of chemicals within soil and between soil and atmosphere is important for understanding the role soil plays in controlling fate, transport, and exposure to multimedia pollutants.

Soils are characteristically heterogeneous. A trench dug into soil typically reveals several horizontal layers having different colors and textures. As illustrated in Figure 8.1, these multiple layers are often divided into three major horizons:

1. A horizon, which encompasses the root zone and contains a high concentration of organic matter.
2. B horizon, which is unsaturated, lies below the roots of most plants, and contains a much lower organic carbon content.
3. C horizon, which is the unsaturated zone of weathered parent rock consisting of bedrock, alluvial material, glacial material, and/or soil of an earlier geological period.

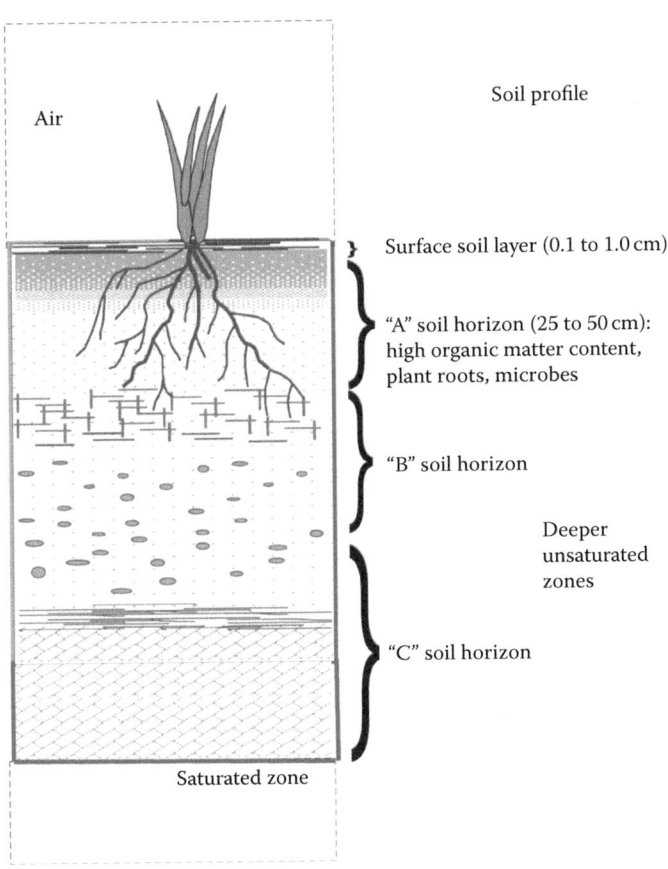

FIGURE 8.1 An illustration of the "horizons" in a typical soil column. (Reproduced from McKone, T. E. 2005. *Encyclopedia of Toxicology*, 2nd ed. Philip Wexler, ed., Elsevier, Oxford, pp. 489–495.)

Below these three horizons lies the saturated zone, a zone that encompasses the area below ground surface in which all interconnected openings within the geologic media are completely filled with water. Similarly to the unsaturated zone with three major horizons, the saturated zone can be further divided into other zones based on hydraulic and geologic conditions. Wetland soils are a special and important class in which near-saturation conditions exist most of the time.

When a contaminant is added to or formed in a soil column, there are several mechanisms by which it can be dispersed, transported out of the soil column to other parts of the environment, destroyed, or transformed into some other species. Thus, to evaluate or manage any contaminant introduced to the soil column, one must determine whether and how that substance will (1) remain or accumulate within the soil column, (2) be transported by dispersion or advection within the soil column, (3) be physically, chemically, or biologically transformed within the soil (i.e., by hydrolysis, oxidation, etc.), or (4) be transported out of the soil column to another part of the environment through a cross-media transfer (i.e., volatilization, runoff, ground water infiltration, etc.). These competing processes impact the fate of physical, chemical, or biological contaminants found in soils. In order to capture these mechanisms in mass transfer models, we must develop mass-transfer coefficients (MTCs) specific to soil layers. That is the goal of this chapter. The reader is referred to other chapters in this handbook that address related transport processes, namely Chapter 13 on bioturbation, Chapter 15 on transport in near-surface geological formations, and Chapter 17 on soil resuspention.

This chapter addresses the following issues: the nature of soil pollution, composition of soil, transport processes and transport parameters in soil, transformation processes in soil, mass-balance models, and MTCs in soils. We show that to address vertical heterogeneity in soils it is necessary to define a characteristic scaling depth and use this to establish process-based expressions for soil MTCs. The scaling depth in soil and the corresponding MTCs depend strongly on (1) the composition of the soil and physical state of the soil, (2) the chemical and physical properties of the substance of interest, and (3) transformation rates in soil. Our particular focus is on approaches for constructing soil-transport algorithms and soil-transport parameters for incorporation within multimedia fate models. We show how MTCs can be developed to construct a simple two-compartment air–soil system. We then demonstrate how a multilayer box model approach for soil-mass balance converges to the exact analytical solution for concentration and mass balance. Finally, we demonstrate and evaluate the performance of the algorithms in a model with applications to the specimen chemicals benzene, hexachlorobenzene, lindane (gamma-hexachlorocyclohexane), benzo(a)pyrene, nickel, and copper.

8.2　SOURCES OF CONTAMINATION IN SOILS

Soil contamination is found throughout the world and can be traced to local, regional, and global pollution sources. This pollution is often the result of human and natural activities that involve the direct emission of contaminants to soil through waste disposal practices, pesticide applications, leaking tanks and pipelines, irrigation with contaminated water, and disposal of sewage and industrial sludge. Soil contamination also results from the deposition and exchange of contaminants from the atmosphere.

Metal species and radionuclides released from combustion processes or from volcanoes, and persistent organic pollutants (POPs) migrate globally in the atmosphere and result in low levels of soil contamination through deposition from the atmosphere. Some sources of contamination, such as local high concentrations of toxic elements, the natural production of radon in soils, and the growth of toxic organisms are not external but internal to the soil. Pesticide use and the disposal of radioactive, biological, and chemical wastes can lead to much higher but localized levels of soil contamination.

Contaminant releases to soil are normally quantified in terms of the mass of substance per unit area per unit time or per release episode. For example, pesticide applications to agricultural fields can range from under $1 \, \text{kg ha}^{-1}$ to over $20 \, \text{kg ha}^{-1}$ per application. Organic contaminants with low water solubility, when introduced to the soil, will migrate to the organic carbon phase of the soil where they can be retained for relatively long periods. Some metal species can also accumulate and persist in soil if their soil chemistry favors the binding of these contaminants into the mineral phase. A large fraction of the sewage sludge produced in many regions of the world is used as soil amendments often after treatment to reduce the content of harmful microorganisms. Sewage sludge contains contaminants and pathogens that are discharged to the sewer system from homes, businesses, industries, and streets (National Research Council, 2002).

As is discussed in more detail in Chapters 6 and 7, contaminants in the atmosphere can be transferred to soil either directly through dry deposition, wet deposition, and vapor partitioning or indirectly through deposition to plants with subsequent transfer to soil. Dry deposition is the process by which particulate matter settles out of the atmosphere and onto soil and plant surfaces. Contaminants that are attached to these particles will be transferred to soil through this deposition process. Atmospheric contaminants on particles are also washed out of the air to soil with rain or snow in the wet deposition of the particles. Contaminants dissolved in the gas phase of air and not bound to particles can also be transferred to soil through a combination of wet deposition and chemical partitioning. Gaseous contaminants with low air–water or K_{AW} partition coefficients are readily washed out during rain and snow by wet deposition. In addition, contaminants with high octanol–air (K_{OA}) and low K_{AW} partition coefficients are transferred preferentially from air to soil through partitioning that involves chemical diffusion from "solution" in air to solution in the soil water. Similarly, hydrophobic contaminants that are sparingly soluble in water but highly lipid soluble can be carried from air to soil through partitioning into the organic phases of soil. In this process, the contaminants diffuse from solution in air directly to the organic phase of soil. Finally, as discussed in Chapter 7, contaminants in air can be transferred from air to vegetation surfaces by dry deposition, wet deposition, and by partitioning into the lipid and water phases of plants. When the plants decay, lose leaves, or are mowed; residual contamination is transferred to soil. This litterfall is a primary source of the organic matter in soil.

8.3 COMPOSITION OF SOILS

Soils are composed of three major phases—gases, liquids, and solids. The volume fraction of soil that is gas (air) ranges from 10% in clay soils, to 25% in sandy soils and typically decreases with increasing depth (Jury et al., 1983; Brady and Weil, 2004). The soil solution is mostly water but also includes dissolved minerals, nutrients, and organic matter such as fulvic acid. The volume fraction of soil that is liquid ranges from 10% typical for sandy soils to 40% typical for clay soils (Jury et al., 1983; Brady and Weil, 2004). The fraction of solid material in soil accounts for some 50–80% by volume and 75–90% by mass (Brady and Weil, 2004). Soil solids include both mineral (i.e., the parent rock) and organic components, including humic acids and decaying matter. The organic phase of soil is often characterized by its organic-carbon content. While the mineral component of soil is in the range of 70–90% by mass, the organic-carbon content of soil ranges from a fraction of 1% by mass for desert and/or sandy soils to as much as 5% by mass for clay soils and even as high as 10% by mass for peat bogs (Jury et al., 1983; Brady and Weil, 2004).

8.4 TRANSPORT PROCESSES IN THE SOIL COLUMN

Chemicals move through soil by advection in the liquid phase through water transport; diffusion in the gas phase, and to some extent in the liquid phase; bioturbation caused by soil-dwelling organisms such as worms; and erosion near the soil surface (Roth and Jury, 1993; Anderson et al., 1991; Thibodeaux, 1996; Mullerlemans and Vandorp, 1996). The partitioning of chemicals among the components of soils (gas, liquid, mineral, and organic) strongly impacts the rates of transport and transformation in and among soil compartments. The key conceptual issue for multimedia models is how soils function as systems to store, destroy, and transport substances within the soil column. Of particular interest is the way these factors combine to determine the structure and performance of MTCs for soil in environmental mass-balance models. Table 8.1 summarizes processes by which contaminants are transferred to, from, and within soils.

Studies of radioactive fallout in agricultural land-management units reveal that, in the absence of tilling, particles deposited from the atmosphere accumulate in and are resuspended from a thin ground- or surface-soil layer with a thickness in the range 0.1–1 cm at the top of the A soil horizon (Whicker and Kirchner, 1987).

TABLE 8.1
Processes by Which Contaminants are Transferred to and from Soils Layers

Gains	Losses
Direct application	Volatilization to air
Deposition from air	Resuspension of soil particles
Washout from air by rainfall	Mass transfer (diffusion and advection) downward
Dry deposition of air particles	to deeper soil layers and to ground water
Mass transfer (diffusion and advection) upward	Transfers to vegetation
from deeper soil layers and from ground water	Soil solution runoff
	Erosion (mineral runoff) to surface water
	Chemical, physical, and biological transformation

The ground-surface-soil layer has a lower water content and higher gas content than underlying layers. Contaminants in this surface-soil layer are transported horizontally by mechanical runoff and soil-solution runoff to nearby surface waters. Surface-soil contaminants are susceptible to wind erosion (as discussed in detail in Chapter 16), volatilization, photolysis, biodegradation, and transfer to plant surfaces by rainsplash. These contaminants are transferred to and from the root-zone soil by diffusion and leaching.

The roots of most plants are confined within the first meter of soil depth. In agricultural lands, the depth of plowing is 15–25 cm. Contaminants in the A horizon below the surface layer, that is, in the root-zone soil, are transported upward by diffusion, volatilization, root uptake, and capillary motion of water; transported downward by diffusion and leaching; and transformed chemically primarily by biodegradation or hydrolysis. The presence of clay and organic matter in the root-zone layer serves to retain water, resulting in a higher water content. In addition, the diffusion depth, which is the depth below which a contaminant is unlikely to escape by diffusion, is on the order of a meter or less for all but the most volatile contaminants (Jury et al., 1990).

The deeper unsaturated soil (generally below 1 m depth) includes the soil layers below the root zone and above the saturated zone, where all pore spaces are filled with water. The soil in this layer typically has a lower organic carbon content and lower porosity than the root-zone soil. Contaminants in this layer move downward to the ground-water zone primarily by capillary motion of water and leaching.

The partitioning of chemicals among the components of soils (gas, liquid, mineral, and organic) strongly impacts the rates of transport and transformation in and among soil compartments. This is illustrated in Figure 8.2. Also affecting the rate of transport and transformation are climate and landform properties, which include temperatures of air and soil, rainfall rates, soil properties (bulk density, porosity), and variability of these properties within soil. Finally as discussed in Chapter 13, there is bioturbation—the mixing of mass within the soil column by detritivores such as worms and burrowing animals such as shrews, moles, and mice.

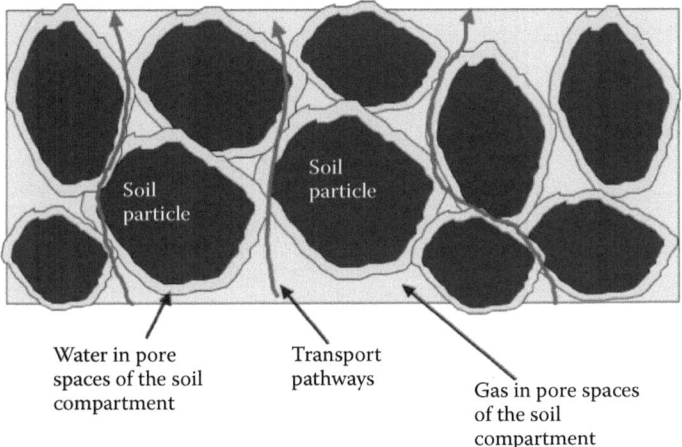

FIGURE 8.2 Microscopic view of partitioning and transport in soil layers.

8.5 TRANSFORMATION PROCESSES IN SOIL

The transformation of toxic substances in soil can have a profound effect on their potential for transport and accumulation at different soil depths. The rate of transformation processes impact the effective penetration depth of contaminants in soil, which in turn determines the length scale needed to define the soil compartment dimensions in mass transfer models. Transformation processes in soil include chemical conversions such as photolysis, hydrolysis, and oxidation/reduction; biological processes such as microbial transformations; and physical processes such as radioactive decay.

Most organic contaminants are capable of undergoing photolytic decomposition. Although the atmosphere attenuates solar radiation before it reaches the Earth's surface, solar radiation is generally sufficient to break bonds in many compounds at this surface. Phototransformation in soil impacts only those contaminants on the soil surface. However, in agricultural lands that are tilled, contaminants in the tilling horizon ($\sim$15–20 cm) can be brought to the surface where photo-transformation can occur. Photo-transformations can result in relatively short half-lives (e.g., hours to days) for contaminants such as pesticides that are applied directly to crops or soil surfaces.

Hydrolytic transformation of organic chemicals can be a significant destructive process for compounds that are present in the aqueous phase of soils. Hydrolysis is most important for chemicals that have functional groups (e.g., amides, esters, carbamates, organophosphates), which can be rapidly altered (e.g., minutes to days) in the presence of water. Conversely, hydrolytic degradation of compounds that contain stable constituents (e.g., halogenated compounds such as carbon tetrachloride) can have half-lives of several hundred years. Because hydrolytic reactions are driven by the availability of hydrogen and hydroxide ions, the pH of the soil can dramatically affect the rate of hydrolysis for any given compound.

Many inorganic and organic chemicals can undergo oxidation or reduction reactions in the soil. An indicator of a compound's ability to be oxidized or reduced is provided by its oxidation potential (EO), which is the voltage at which it is transformed to its reduced state. A similar measure of a soil's ability to reduce a compound is provided by the redox potential (pE), which is a measure of electron activity. Redox potentials are relatively high and positive in oxidized environments (e.g., surface waters), and low and negative in reduced environments (e.g., aquatic sediments and the subsurface soil layers). These environmental conditions are especially important for inorganic chemicals that are rarely present in their elemental form in the environment. Arsenic, for example, exists primarily in its oxidized form (arsenate) in the atmosphere and in surface waters and in its reduced form (arsenite) in sediments.

Because of their broad range of enzymatic capabilities, microorganisms are capable of destroying other microorganisms and transforming many inorganic and organic compounds. These transformations can result in the partial degradation of a compound (e.g., conversion of trinitrotoluene to dinitrotoluene), mineralization (i.e., complete transformation to carbon dioxide and water), or synthesis of a stable product (e.g., formation of methyl arsenicals from arsenate). While these processes generally result in the detoxification of the parent compound, toxic products may also be formed. For example, the microbial metabolism of aromatic amines can result in the formation of toxic by-products.

Radioactive decay applies only to radioactive elements, which have unstable nuclei and emit atomic radiation as part of a process of attaining stability, resulting in the formation of another chemical element, which may be stable or may be radioactive resulting in further decay.

8.6   SOIL MASS TRANSFER: THEORY AND EQUATIONS

In this section, we describe and illustrate an approach for constructing soil-transport parameters, algorithms, and MTCs. The resulting MTCs account for diffusion in gas and liquid components: advection in gas, liquid, or solid phases and multiple transformation processes. They also provide an explicit quantification of the characteristic soil penetration depth. We apply transport algorithms to and develop MTCs for both a simple one-layer soil model and more detailed and presumably more realistic multilayer models.

8.6.1   Background

There are significant variations in both the complexity and structure of transport models applied to the soil compartment. Among the differences are the types of soils considered, the number of layers used for each soil type, and how the depth of the soil compartment is selected. Models developed for assessing the behavior of contaminants in soils can be categorized in terms of the transport/transformation processes being modeled. Partitioning models such as the fugacity models of Mackay (1979, 2001) and Mackay and Paterson (1981, 1982) describe the distribution of a contaminant among the liquid, solid, and organic phases of soils. Jury et al. (1983) have developed an analytical screening model that can be used to calculate the extent to which contaminants buried in soil evaporate to the atmosphere or infiltrate down to deeper soil layers. For radioactive-fallout deposition on agricultural lands, Whicker and Kirchner (1987) have developed a model that includes three soil layers, surface soil (0–0.1 cm), intermediate soil (0.1–25 cm), and deep soil (>25 cm).

Because experimental and theoretical evidence shows a large variation in the depth to which different chemicals penetrate, multimedia model developers have acknowledged that each specific chemical substance requires a different soil depth to scale its transport into soil (Cowan et al., 1995; McKone and Bennett, 2003). When a model is used for groundwater or soil protection or when the source resides below the soil–atmosphere interface, detailed soil concentration profiles are vitally important. This often requires detailed numerical simulation models. But when the modeler's primary goal is to assess multimedia transport, then capturing the magnitude of chemical transport between air and soil is as important or (in some cases) more important than capturing the concentration profile within the soil. Examples include efforts to address the role of air/soil exchange in assessments of persistence and long-range transport in the atmosphere. Multimedia models require simple but reliable mass-exchange algorithms between the air compartment and soil.

8.6.2 Conceptual Model

Contaminant penetration in soil and the gaseous exchange between surface soil and the atmosphere are influenced by chemical-specific rates of dispersion/diffusion, advection, and transformation. The relative magnitude of these processes depends strongly on the relative partitioning among the mobile (liquid and gas) and sorbed phases of the soil. The partitioning of a chemical between the three soil phases is defined by any two among the three primary partition properties: the octanol–water partition coefficient (K_{OW}), the octanol–air partition coefficient (K_{OA}), and the air–water partition coefficient (K_{AW}). We focus on K_{OW} and K_{AW} and note that to a first approximation, K_{OA} equals K_{OW}/K_{AW}. Dispersion/diffusion processes are driven by concentration gradients in gas and liquid phases of soil. Advection rates are determined by the processes that drive liquids and gases through soil. The link between partitioning and transport by different mechanisms is illustrated in Figure 8.2. In addition to liquid and gas-phase advection, it should be noted that the solid phases of soil are not necessarily stagnant. Vertical transport of the solid phase transport in surface soils occurs by many processes such as bioturbation, cryoturbation, and erosion into cracks formed by soil drying as well as tilling in agricultural lands (McLachlan et al., 2002).

Most models that handle the transport of organic contaminants within soil are based on the landmark publication of Jury et al. (1983). This model treats soil as a mixture of air, water, and soil particles with the assumption of uniform soil properties, linear sorption isotherms, and equilibrium partitioning between the solid, gas, and liquid phases of soil.

Our approach here is essentially that of McKone and Benett (2003), which is based on the Jury et al. (1983) model. It is incorporated within the latest version of the CalTOX model (McKone, 1993) and the original version of the TRIM (USEPA, 2003) model, but also used in other models such as BETR (MacLeod et al., 2001). This approach uses one or more soil compartment layers while maintaining a structure that links easily to other compartments such as air and vegetation in multimedia models. This approach begins by setting up the differential equations describing the mass balance in soil and accounting for diffusion in air and water phases, advection via water, bioturbation, and chemical transformation. From the steady-state analytical solution to these equations, which can be applied stepwise to different layers, one develops an equivalent compartment model that uses compartment-based inventories and transfer factors.

These transfer factors are obtained by matching the fugacity and flux at the boundary between each pair of soil compartments with those obtained from an analytical solution. We consider the appropriate scaling depth for selecting compartment depth and generalize this approach to layers of different composition.

8.6.3 Mass Balance Equations

In any defined horizontal layer of soil, the rate in mol d^{-1} at which the dilute chemical mass inventory in that layer is changing is defined by balancing three processes. First there is dispersion due to both physical (diffusion) and biological (bioturbation) processes. Second is advection in soil fluids—primarily the downward (or upward) movement of water. Third is removal by physical, biological, or chemical reactions. Based on these three processes, Jury et al. (1983) have defined the governing equation for mass balance concentration within any specified region of the soil column in the following form:

$$\frac{\partial C}{\partial t} = \frac{\partial}{\partial z}\left[D_e \frac{\partial C}{\partial z}\right] - v_e \frac{\partial C}{\partial z} - kC,$$

(8.1)

Concentration change in time = Dispersion − Advection losses − Removal losess

where C represents the bulk chemical concentration in soil, mol(chemical) m^{-3} (soil); t is time, days; z is depth in a soil column measured from the top surface, m; k is a removal rate constant, d^{-1}, which accounts for first-order losses by chemical transformation, root uptake, and so on; D_e is the bulk dispersion coefficient in the soil, m^2 d^{-1}; and v_e is the bulk advection velocity of the chemical in the soil, m d^{-1}.

We can easily convert concentration as the state variable in Equation 8.1 to fugacity-based equations using the relationship $C = fZ$, where f is the chemical fugacity in soil, Pa, and Z is the bulk-soil fugacity capacity, mol (m^3 Pa)$^{-1}$. More details on fugacity capacities is given in Chapter 5.

In the case where the soil is being contaminated by the atmosphere such that the soil surface is maintained at constant concentration, C_0, the solution to Equation 8.1 yields a rather complex expression for C as a function of both depth and time:

$$C(z,t) = C_0 \left\{ \frac{1}{2}\exp\left[\frac{(v_e - u)z}{2D_e}\right]\text{erfc}\left[\frac{(z - ut)}{2\sqrt{D_e t}}\right]\right.$$
$$\left. + \frac{1}{2}\exp\left[\frac{(v_e + u)z}{2D_e}\right]\text{erfc}\left[\frac{(z + ut)}{2\sqrt{D_e t}}\right]\right\},$$

(8.2)

where $u = \sqrt{v_e^2 + 4kD_e}$ and erfc is the complimentary error function, and C_0 is the bulk concentration in soil at the soil surface (assumed to be at the same fugacity as the air), mol m^{-3}.

Equation 8.2 is somewhat intimidating and can be difficult to program into simple programs such as spreadsheets. Fortunately this equation takes on a much simpler form as the soil concentration profile approaches steady state, a situation that often applies and is most often of interest. In this case the soil concentration dependence on depth becomes

$$C(z) = C(0)e^{-\gamma z},$$

(8.3)

where γ is a function of D_e, v_e, and k and is obtained from Equation 8.1 in the limit as t goes to infinity.

$$\gamma = \sqrt{\left[\left(\frac{v_e}{2D_e}\right)^2 + \frac{k}{D_e}\right]} - \frac{v_e}{2D_e} \quad \text{or,} \quad \text{if } v_e = 0, \quad \gamma = \sqrt{\frac{k}{D_e}}.$$

(8.4)

The γ parameter is developed in more detail in the following section. In order to apply Equations 8.1 through 8.4 to soil layers, we must consider how D_e, v_e, and k depend on both soil and chemical properties.

8.6.4 Normalized Gradient of Soil Concentration

Equation 8.2 reduces to a very simple form (Equation 8.3) under steady-state conditions and with a fixed concentration at its upper boundary and uniform soil properties. Indeed, Equation 8.3 reveals that concentration in any soil layer at a depth z describes the vertical gradient of concentration in a soil layer when we know the long-term concentration that applies at its surface, $C(0)$. The γ term has units of m^{-1} and tells us how steep the concentration gradient is in a soil layer. It can be viewed as the reciprocal of a characteristic depth of concentration change as discussed below.

At any depth z of a soil layer, there is competition between reaction processes and dispersion/diffusion and advection processes. The ratio of these competing processes is expressed by the Damkoehler number (N_{DA}), which is the ratio of the rate of loss by chemical transformation to the rate of loss by diffusion/dispersion and advection (Cowan et al., 1995) and is defined as

$$N_{DA} = \frac{\text{Transformation loss rate}}{\text{Advection and diffusion/dispersion loss rate}} = \frac{C \times k \times z}{C(v_e + D_e/z)}$$

$$= \frac{k \times z}{(v_e + D_e/z)}. \tag{8.5}$$

When the Damkoehler number is 1, rates of transformation losses and diffusion/advection transport are equal. This happens when z is the characteristic depth or the average depth of penetration for chemical molecules moving into a soil layer from its surface (Cowan et al., 1995). When N_{DA} is 1, Equation 8.5 can be rearranged to find the z corresponding to the characteristic penetration depth, a parameter we label z^*,

$$k(z^*)^2 + v_e z^* + D_e = 0. \tag{8.6}$$

Using the quadratic formula to solve Equation 8.6 gives

$$\frac{1}{z^*} = \sqrt{\left[\left(\frac{v_e}{2D_e}\right)^2 + \frac{k}{D_e}\right]} - \frac{v_e}{2D_e} \quad \text{or,} \quad \text{if } v_e = 0, \quad \frac{1}{z^*} = \sqrt{\frac{k}{D_e}}. \tag{8.7}$$

Comparing Equation 8.7 with Equation 8.4 leads to the realization that

$$\gamma = \frac{1}{z^*} \quad \text{and} \quad C(z) = C(0)e^{-(z/z^*)}. \tag{8.8}$$

In a vertical soil profile with a fixed surface concentration, a unit value of N_{DA} corresponds to the depth, z, at which soil concentration (or fugacity) decreases by $1/e$ relative to the surface concentration (fugacity). So z^* is a "characteristic" depth for

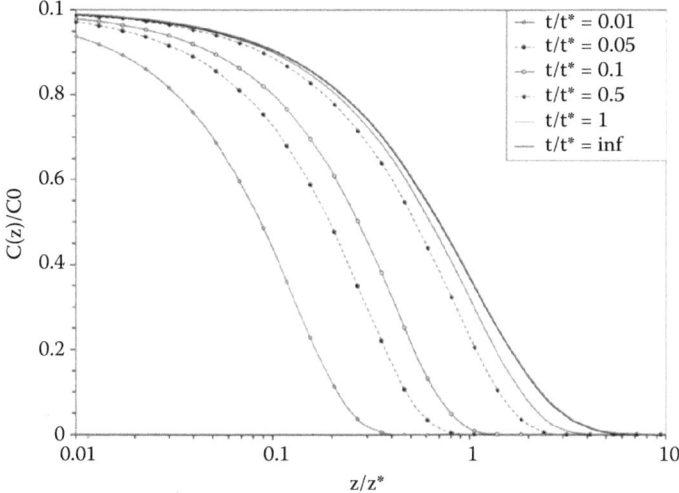

FIGURE 8.3 Concentration curves at various normalized times, t/t^*, plotted against a normalized, z/z^* depth scale.

mass transport and can be interpreted as the average penetration depth of a molecule in vertical cross-section of a soil layer. A large value of z^* indicates a soil with a steep gradient of concentration and reflects a substance that penetrates deeply into soil. A small value of z^* indicates a shallow concentration gradient and a substance that penetrates only to a small depth within soil. We refer to z^* as the "penetration depth" and use it to normalize the scale for mass transfer in soils. This corresponds to the characteristic depth derived from the Jury et al. (1983) steady-state solution to Equation 8.1.

Just as there is a characteristic penetration depth that characterizes the steady-state penetration depth, there is a characteristic time, t^*, that tells how long it takes for the soil compartment to reach steady state. McKone and Bennett (2003) have shown that evaluating Equation 8.2 provides an analytical estimate of t^*:

$$t^* = \frac{4D_e}{(v_e^2 + 4D_e k)}. \tag{8.9}$$

This expression provides us with a parameter to normalize time for any specific chemical and soil system. It is when the normalized time t/t^* approaches infinity that Equation 8.2 simplifies to Equation 8.3. But it is of interest that this simplification effectively applies when $t > 2t^*$. This is illustrated in Figure 8.3, which shows how the concentration profile (concentration versus depth) evolves as t/t^* increases.

8.7 SOIL TRANSPORT PARAMETERS

As demonstrated in the previous section, the key soil transport parameters are the bulk diffusion coefficient, D_e, and the advection velocity, v_e, and their values relative to the reaction rate constant, k, which is a function of the chemical and its reactive environment. We determine the effective bulk diffusion coefficient resulting from vapor and water-phase tortuous diffusion following the approach of Jury et al. (1983). This approach has been widely used by environmental models for more than two decades. We make one adjustment to the Jury et al. (1983) approach by adding a bulk diffusion coefficient, D_{bio}, as a term that represents the bioturbation dispersion, accounting for worms and other detritivores that consume soil as well as burrowing creatures such as ants whose tunneling moves soil and increases the spread of chemicals in soil (Mullerlemans and Vandorp, 1996). Chapter 13 suggests values for D_{bio}. The resulting bulk-soil effective diffusivity, $D_{e,i}$ in $m^2\ d^{-1}$, for any soil layer, i, takes the form

$$D_{e,i} = \frac{Z_{air}}{Z_i}\left(\alpha_i^{10/3}/\phi_i^2\right)D_a + \frac{Z_{water}}{Z_i}\left(\beta_i^{10/3}/\phi_i^2\right)D_w + D_{bio,i}, \qquad (8.10)$$

where the ratios Z_{air}/Z_i and Z_{water}/Z_i are, respectively, the dimensionless air/soil and water/soil bulk partition coefficients; Z_{air}, Z_{water}, and Z_i represent, respectively, the fugacity capacities of pure-phase air, pure-phase water, and a bulk phase soil (i refers to the ith soil layer in multilayered system), mol $(m^3\ Pa)^{-1}$; and D_x is the diffusion coefficient of the chemical in a pure fluid ($x = a$ for air and w for water), $m^2\ d^{-1}$. Other parameters used in Equation 8.3 are described in Table 8.2. Table 8.3 provides definitions of the fugacity capacities expressions used throughout this chapter.

TABLE 8.2
Definitions, Symbols, Units, and Values of Compartment Properties

Property Name	Symbol	Units	Value
Density of air particles	ρ_{ap}	$kg\,m^{-3}$	1000
Concentration of particles in air	C_a	$kg\,m^{-3}$	5.0×10^{-8}
Fraction organic matter associated with particulate matter in air	f_{om}	No units	0.4
Long-term average deposition velocity (wet and dry) of air particles to soil	V_d	$m\,d^{-1}$	400
Long-term average rainfall	rain	$m\,d^{-1}$ (m year^{-1})	0.0027 (1.0)
Flux of water through the soil	v_{water}	$m\,d^{-1}$ (m year^{-1})	0.00082 (0.3)
Thickness of equivalent diffusion boundary layer in air above the surface soil	δ_{ag}	m	0.05
Density of soil particles	ρ_{sp}	$kg\,m^{-3}$	2600
Volume fraction of soil that is gas	α_i	No units	0.2
Volume fraction of soil that is liquid	β_i	No units	0.3
Volume fraction of soil that is void	$\phi_i = \alpha_i + \beta_i$	No units	0.5
Volume fraction of soil that is solid	$(1-\alpha_i-\beta_i)$	No units	0.5
Fraction of organic carbon in soil	f_{oc}	No units	0.02
Equivalent diffusion coefficient for bioturbation	D_{bio}	$m^2\ d^{-1}$	1.7×10^{-5} (2.6×10^{-6} to 8.6×10^{-4})

Source: Adapted from McKone, T. E.; Bennett, D.H. 2003. *Environmental Science & Technology, 33*(14): 2123–2132.

TABLE 8.3
Definitions of Fugacity Capacities and Partition Coefficients Used in the Soil Compartment Model[a]

Name	Symbol	Formula or Value	Units
Fugacity capacity of gas phase	Z_{air}	$= 1/(RT)$	$mol\,(m^3\,Pa)^{-1}$
Fugacity capacity of substances dissolved in water	Z_{water}	$= 1/H$	$mol\,(m^3\,Pa)^{-1}$
Fugacity capacity of air particles[b]	Z_{ap}	$= 0.00123 \times K_{oa}f_{om}Z_{air}\rho_{ap}$	$mol\,(m^3\,Pa)^{-1}$
Fugacity capacity of soil solids	Z_{ss}	$= 0.001\,K_D\rho_{sp}Z_{water}$	$mol\,(m^3\,Pa)^{-1}$
Fugacity capacity of bulk air (gases and particles)	Z_a	$= Z_{air} + (PC/\rho_{ap})Z_{ap}$	$mol\,(m^3\,Pa)^{-1}$
Fugacity capacity of soil compartment i, for the three-compartment example: i = g for ground surface i = s for root zone i = v for vadose and so on.	Z_i	$= \alpha_i Z_{air} + \beta_i Z_{water} + (1-\alpha_i-\beta_i)Z_{sp}$	$mol\,(m^3\,Pa)^{-1}$
Octanol/water partition coefficient	K_{ow}	Literature citations	L(water)/L(octanol)
Octanol/air partition coefficient	K_{OA}	$= RTK_{ow}/H$	m^3(air)/m^3(octanol)
Organic carbon partition ratio	K_{oc}	Measure values or estimated as $0.35K_{ow}$	L(water)/kg(organic carbon)
Soil distribution coefficient (solid/water concentration ratio)	K_D	$= K_{oc}f_{oc}$	L(water)/kg(soil solids)

[a] See Table 8.2 for definitions of parameters other than Z values and partition coefficients.
[b] Harner et al. (1995).

When we consider multiple soil layers, there is an effective bulk advection velocity, $v_{e,i}$ in m d^{-1}, for each soil layer characterizing contaminant transport by the flux of water induced by net rainfall or irrigation infiltration. The value of this bulk property is derived from an assumed equilibrium partitioning of a chemical to the mobile phase relative to the bulk inventory,

$$v_{e,i} = v_{water}\frac{Z_{water}}{Z_i}, \tag{8.11}$$

where v_{water} is the flux of water through the soil, m d^{-1}.

In order to demonstrate the use of chemical and environmental properties to determine MTCs in soil compartments, we use the set of six specimen chemicals that have been used in other chapters. These chemicals and their properties are listed in Table 8.4. Also listed in Table 8.4 are values obtained for the derived estimates of K_D, Z_{air}, k, z^*, and t^*. Notable among these parameters is K_D, the steady-state concentration ratio between soil liquid and soil solids and used to characterize the bulk water/soil partition coefficient. Among the six substances listed in Table 8.4, K_D spans some four orders of magnitude—from 0.93 to 7600 L kg^{-1}. Z_{water} also spans some four orders of magnitude—from 0.0018 to 22 Pa m^3 mol^{-1}. In contrast, the penetration depth z^* varies by only a little over one order of magnitude among the six substances—from 0.09 to 4 m. But the characteristic time (time to steady state) varies over six orders of magnitude. This indicates that, because of the narrow range of z^*, it is not difficult to scale the depth of a soil compartment to capture variability in penetration depth. But determining whether and when steady state is achieved can be more challenging.

TABLE 8.4
Properties Used for the Specimen Chemicals and the Resulting Estimates of K_D, Z values, k_s, v_e, D_e, z^*, and t^*

Chemical	Benzene	Hexachloro-benzene	Lindane	B(a)P	Nickel	Copper
Molecular formula	C_6H_6	C_6Cl_6	$C_6H_6Cl_6$	$C_{20}H_{12}$	Ni	Cu
Molar mass (kg)	78.1	284.8	290.8	252.3	58.71	63.54
Henry's law constant (Pa m^3 mol^{-1})	557	131	0.149	0.0465	0	0
log K_{AW}	−0.65	−1.28	−4.22	−4.73	—	—
log K_{OW}	2.13	5.50	3.70	6.04	—	—
log K_{OA}	2.78	6.78	7.92	10.77	—	—
log K_{OC}	1.67	5.04	3.24	5.58	~2	~5
Transformation half-life (d)	100	1000	300	500	—	—
K_D [L(water)/ kg (soil solids)]	0.93	2190	34.8	7600	2	2000
Z_{air} (Pa m^3 mol^{-1})	0.000425	0.000425	0.000425	0.000425	0.000425	0.000425
Z_{water} (Pa m^3 mol^{-1})	0.0018	0.0076	6.71	21.9	1.0	1.0
Z_{ss} (Pa m^3 mol^{-1})	0.00435	10.2	0.163	35.6	5.20	5,200
Z_{ap} (Pa m^3 mol^{-1})	0.252	2,520	34,800	2.46×10^7	5.20	5,200
k_s, the reaction rate constant d^{-1}	0.00693	0.000693	0.00231	0.00139	0	0
v_e (md^{-1})	5.28×10^{-4}	2.89×10^{-7}	1.80×10^{-5}	8.32×10^{-8}	2.83×10^{-4}	3.16×10^{-7}
D_e (m^{2}d^{-1})	0.00146	1.78×10^{-5}	6.63×10^{-5}	1.73×10^{-5}	2.95×10^{-5}	1.70×10^{-5}
Penetration depth z^* (m)	0.49	0.16	0.09	0.11	4	0.7
Characteristic time t^* (d)	140	1440	423	721	1500	4×10^8

Note: Lindane is gamma hexachlorocyclohexane, B(a)P is benzo(a)pyrene.

8.8   SOIL–AIR MASS TRANSFER RATES

We now exploit the theoretical foundation provided above to calculate mass transfer rates from soils to the overlying atmosphere and vice versa. Parameter values required in these calculations are suggested. Three models are presented that vary in the detail with which vertical concentration profiles in the soil are treated.

Single soil compartment, well mixed. The first and simplest approach assumes that the soil is a single well-mixed "box" with a constant concentration vertically. This situation is most likely to apply when the rate of vertical transport of the chemical in the soil layer is fast relative to the rate of mass transfer through the soil–air interface.

Single soil compartment with a gradient in concentration. The second approach also assumes a single soil layer of constant properties, but there is allowance for concentration gradients within the layer. This situation likely applies when the soil is homogeneous and the rate of vertical transfer of chemical in the soil is equal to, or slower than the rate of mass transfer through the interface.

Multiple soil layers with gradients in concentration. Finally, and most detailed and demanding of parameters, is a soil that varies in properties vertically. In principle any number of layers can be treated, but for illustrative purposes we describe a three-layer soil.

The models are presented in both concentration and fugacity formats. The fugacity format is convenient because at the interface between layers a common fugacity applies whereas there can be differences in concentration, especially at the air–soil interface. This is illustrated in Figure 8.4.

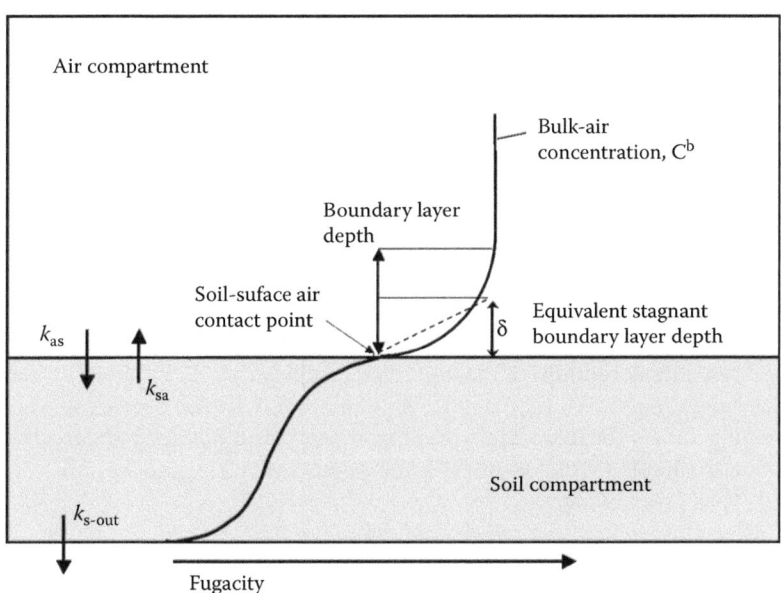

FIGURE 8.4   An illustration of the role of mass transfer coefficients in establishing mass balance in a two-compartment system consisting of soil and air.

In all cases, we assume that the air phase is well mixed vertically, except in the boundary layer immediately above the soil surface in which there is a resistance to mass transfer. The rate of transfer from air to soil influences the chemical levels in the soil and is included in these models, but the reader is referred to Chapter 6 for a more detailed treatment of atmospheric deposition and absorption processes and to Chapter 7 for treatment of absorption to vegetation and subsequent transport to the soil surface.

The equations presented here have been derived in more detail by McKone and Bennett (2003). Readers can refer to this publication for further information. Some changes in nomenclature have been made, especially "T factors" in that paper have been designated here as rate constants with symbol k.

Regardless of the model used there is a single average concentration of chemical in the soil defined as the total chemical quantity in the soil, M mol, divided by the soil volume, that is, M/V mol m^{-3}. The average fugacity is $M/(VZ)$. The models differ in their characterization of vertical spatial differences in concentration or fugacity. An obvious approach is to apply the simplest model first then increase the resolution, and demands for parameter values in the light of experience.

8.8.1 TWO-COMPARTMENT (AIR–SOIL) MODEL ASSUMING A WELL-MIXED SOIL

Mass transfer between soil and air involves diffusion, dry deposition, washout by rain. With one well-mixed soil compartment, the rate of net mass transfer from soil to air is given by 8.12, which contains terms for *net* gaseous diffusion, dry particle deposition, and transfer in precipitation (*rain*) to obtain the *net* flow ($Flow_{sa}$, mol d^{-1}) across the air–soil interface:

$$Flow_{sa} = Area \left[\left(\frac{Z_{air}}{Z_s} C_s - C_a \right) \left(\frac{Z_{air}}{Z_s U_s} + \frac{1}{U_a} \right)^{-1} \right. $$
$$\left. - \left(V_d \frac{PC \times Z_{ap}}{\rho_{ap} Z_a} + rain \frac{Z_{water}}{Z_a} \right) C_a \right]. \tag{8.12}$$

In this expression, *Area* is the horizontal area of contact between two compartments, m^2; C_a is the bulk contaminant concentration in the air compartment, mol m^{-3}; and C_s is the bulk contaminant concentration in the surface soil compartment, mol m^{-3}. Table 8.3 provides definitions of the fugacity capacities Z_{air}, Z_s, Z_a, and Z_{ap}. Table 8.2 provides suggested values for V_d, PC, ρ_{ap}, and rain. Chapter 6 gives more details on obtaining values for these deposition parameters. But the key parameters in this expression are the air-side mass-transfer coefficient (MTC) U_a and the soil-side MTC U_s (m d^{-1}) as given below:

$$U_a = \frac{D_{air}}{\delta_{as}}, \tag{8.13}$$

$$U_s = \frac{D_{e,s}}{\delta_s}, \tag{8.14}$$

where D_{air} as the contaminant diffusion coefficient in pure air, m^2 d^{-1}, δ_{as} is the thickness of equivalent diffusion boundary layer in the air above the soil (illustrated in Figure 8.4), and estimated as 0.005 m; $D_{e,s}$ is the equivalent bulk diffusion coefficient in the ground-surface soil layer, m^2 d^{-1}; and δ_s is the diffusion path length in soil (sometimes estimated as 0.5 times d_s, the thickness of soil layer). Whereas the molecular diffusivities in pure phases (air, water) are well established, values of diffusion path lengths and the equivalent bulk diffusion coefficient in the soil can be in doubt and judgment is required for selection of parameters used to establish MTC values.

If particle and rain deposition are ignored and only diffusion applies, the equation for net soil-to-air transfer becomes

$$Flow_{sa} = Area \left(\frac{C_s}{K_{sa}} - C_a \right) \left(\frac{1}{U_a} + \frac{1}{K_{sa}U_s} \right)^{-1}, \tag{8.15}$$

where K_{sa} is the soil–air partition coefficient equal to Z_s/Z_{air}. Clearly, when C_s equals $K_{sa}C_a$, there is no net transfer and chemical equilibrium applies to the concentration ratio between these two compartments.

We can also present Equations 8.12 and 8.15 by defining the fugacity-based mass-transfer coefficient at the soil–air interface, Y_{sa}, mol (m^2 Pa s)$^{-1}$,

$$Y_{sa} = \left[\frac{1}{Z_s U_s} + \frac{1}{Z_{air} U_a} \right]^{-1} \tag{8.16}$$

and substituting f_a and f_s (fugacities in Pa) for C_a/Z_a and C_s/Z_g to obtain

$$Flow_{sa} = Area \left[Y_{sa}(f_s - f_a) - \left(V_d \frac{PC \times Z_{ap}}{\rho_{ap}} + rain \times Z_{water} \right) f_a \right]. \tag{8.17}$$

Consider the application of the expressions above to mass transfer factors for the situation illustrated in Figure 8.4, which shows two compartments—air and soil—exchanging chemical mass. In this case, the source of contamination, S_a (mol d^{-1}), is in the air so that the fugacity in air is higher than soil, a situation that causes the chemical contaminant to penetrate into the soil layer. In this case, each compartment is represented by a single fugacity or concentration with no spatial variation of concentration. In this system, the concentration of a compartment is simply M_i/V_i and the fugacity is $M_i/(Z_i V_i)$ where M_i is the total mass inventory of the compartment (mol), V_i is the volume of the compartment (m^3), and Z_i is its fugacity capacity [mol (m^3 Pa)$^{-1}$].

In order to illustrate the use of mass transfer coefficients for the two-compartment model, we apply mass balance to the chemical inventory of a single vertical soil compartment designated by the subscript s below an air compartment above soil designated by the subscript a. We consider the case where chemical emissions to the air compartment of volume V_a maintains it at a bulk concentration of C_a (mol m^{-3}) and bulk inventory of $M_a = C_a V_a$ (mol). We assume no emissions to soil, but these could be added as a bulk input of mol d^{-1}. The mass inventory in the soil layer is designated M_s (mol). In setting up the mass balance, we account for losses due to chemical and biological transformation processes with the removal rate constant k_a (d^{-1}) . We use the mass transfer factors, k_{as} and k_{sa} (d^{-1}) to account for the rate of chemical mass transfer at the surface between air and soil compartments. We use the parameter $k_{a\text{-out}}$ (d^{-1}) to account for losses due to leaching and dispersion below the soil layer. Under steady-state conditions the mass balance in the soil compartment is described by the following equation:

$$(k_s + k_{sa} + k_{s\text{-out}})M_s = k_{as}M_a. \tag{8.18}$$

The parameter k_a is obtained from the compartment-specific half-life or residence times that apply to bulk removal by processes such as photolysis, hydrolysis, oxidation/reduction, microbial transformations, and radioactive decay (see Section 8.5). This parameter is estimated as $0.693/T_{1/2}$, where $T_{1/2}$ is the half-time for a given removal process.

By matching the net flow across the air–soil boundary from Equation 8.18, $\text{Flow}_{as} = k_{as}M_a - k_{sa}M_s$, with the flow expressed in Equation 8.17, and substituting $M_i = f_i Z_i V_i$ we obtain the following expressions for the mass transfer factors k_{as} and k_{sa}:

$$k_{as} = \frac{1}{Z_a \times d_a}\left(Y_{as} + V_d\frac{PC}{\rho_p}Z_{ap} + rain \times Z_{water}\right), \tag{8.19}$$

$$k_{sa} = \frac{Y_{as}}{Z_s d_s}. \tag{8.20}$$

The parameter $k_{s\text{-out}}$ must account for diffusion/dispersion and advection losses at the lower boundary of a soil compartment. Advection with water that infiltrates through the soil is typically a unidirectional process, which removes chemicals with the effective velocity obtained in Equation 8.11. However, dispersion and diffusion processes such as molecular diffusion and bioturbation move chemicals both up and down within the soil, making it difficult to define a net loss factor applicable to the bulk soil. However, with a single well mixed compartment receiving chemical input at is surface, we can assume that the net diffusion is in the downward direction and proportional to the concentration gradient in the penetration depth z^*. In this case the parameter $k_{a\text{-out}}$ is obtained from a simple model for mass loss at the lower boundary of the soil compartment:

$$\text{Mass loss (mol } d^{-1}) = k_{s\text{-out}}M_s = Area \times \left(v_{e,s}C_s - D_{e,s}\frac{dC_s}{dz}\right). \tag{8.21}$$

where C_s is the bulk concentration in the soil layer in mol m^{-3}, equal to M_s/V_s, and, as defined above, $v_{e,s}$ is the effective velocity in m d^{-1} of the trace chemical carried through soil by water infiltrating at velocity v_{water}, m d^{-1}, and $D_{e,s}$ is the effective bulk diffusion coefficient is soil, m^2 d^{-1}. We have found that in well mixed soil layers the concentration gradient at the lower boundary of a single soil layer,

dC_s/dz, is approximated reasonably well by C_s/z^*. With these substitutions and the substitution of M_s/V_s for C_s, Equation 8.21 becomes

$$Mass\ loss\ (\mathrm{mol\ d}^{-1}) = k_{\text{s-out}}M_s = \frac{1}{d_s}\left(v_{e,s} + \frac{D_{e,s}}{z^*}\right)M_s. \qquad (8.22)$$

From this we determine that

$$k_{\text{s-out}}\ (\text{no gradient}) = \frac{1}{d_s}\left(v_{e,s} + \frac{D_{e,s}}{z^*}\right). \qquad (8.23)$$

This expression is obtained under the assumption of uniform mixing in a soil layer, with an only an implicit gradient used to estimate diffusion and bioturbation losses. Because we expect a concentration gradient that makes $C_s(z)$ at $z = d_s$ lower than the bulk concentration C_s, we expect this approach can overestimate losses at the lower boundary of the single soil layer. But setting $k_{\text{s-out}}$ to zero can cause our mass balance to ignore a potentially important loss process and underestimate loss from soil. So the absence of an explicit gradient (uniform concentration) simplifies the mass-transfer calculation but introduces uncertainty. In the next section, we provide an approach that is more accurate in capturing both advection and dispersion processes at the lower soil boundary. It should be noted that comparisons of this uniform concentration approach to an approach with an explicit concentration gradient reveals that the advection/dispersion estimates obtained from uniform-concentration approach are typically no more than a factor of 2 larger than the estimates obtained with a concentration gradient. Moreover, advection/dispersion losses are not typically the dominant losses from a single soil layer, such that this simple approach will not result in large errors for estimating surface-soil mass inventories. But this simple approach becomes problematic when it is necessary to estimate the transfer of contaminants to deeper soil layers or to groundwater. Table 8.5 provides a summary of values obtained for the parameters k_{as}, k_{sa}, k_s, and $k_{\text{s-out}}$ for a 15 cm deep soil layer based on chemical properties for the substances listed in Table 8.4 and based on default values for the parameters V_d, PC, ρ_{ap}, and *rain* listed in Table 8.2. This table compares the alternate (gradient/no gradient) approaches. The derivation of $k_{\text{s-out}}$ for a single soil layer with a concentration gradient is provided in the next section.

8.8.2 TWO-COMPARTMENT (AIR–SOIL) MODEL WITH A CONCENTRATION GRADIENT IN THE SOIL

We now illustrate the mass balance derivation for $k_{\text{s-out}}$ using a gradient of concentration in a single soil layer. When there is a concentration gradient in a single soil layer, the concentration $C(d_s)$ at the lower boundary of the soil layer is $C(d_s) = C(0)e^{-\gamma d_s}$. So with a concentration gradient, Equation 8.22 becomes

$$Mass\ loss\ (\mathrm{mol\ d}^{-1}) = k_{\text{s-out}}M_s = Area \times \left(v_{e,s}C(d_s) - D_{e,s}\frac{dC(d_s)}{dz}\right)$$

$$= Area \times C(0)e^{-\gamma d_s}(v_{e,s} + \gamma D_{e,s}). \qquad (8.24)$$

We also note that, with a concentration gradient, the total mass M_s in this compartment must be determined as

$$M_s = Area \times C(0) \int_0^{d_s} e^{-\gamma x} \, dx = Area \times C(0) \frac{1 - e^{-\gamma d_s}}{\gamma} \qquad (8.25)$$

Combining Equations 8.24 and 8.25 results in the following expression for $k_{s\text{-out}}$, which is presented both in terms of the γ parameter and in terms of z^*:

$$k_{s\text{-out}}(\text{with a gradient}) = \frac{v_{e,s}\gamma + D_{e,s}\gamma^2}{(e^{\gamma d_s} - 1)} = \frac{1}{z * (e^{[d_s/z*]} - 1)} \times \left(v_{e,s} + \frac{D_{e,s}}{z*} \right). \qquad (8.26)$$

This expression can be compared to Equation 8.23 in order to see how the $k_{s\text{-out}}$ format changes with the addition of a concentration gradient.

8.8.3 DISCUSSION OF THE TWO APPROACHES TO A SINGLE SOIL COMPARTMENT

An exercise considering MTCs for different specimen chemicals in a two-compartment (air and soil) model provides a number of insights. In order illustrate the application of the expressions in the previous sections, Table 8.5 provides for each of the six specimen chemicals listed in Table 8.4 values for the mass transfer parameters U_a, U_s, Y_{as}, $Y_{s\text{-out}}$, $v_{e,s}$, k_{as}, k_{sa}, and $k_{s\text{-out}}$. In reviewing parameter values in Table 8.5, we note that in this system the only mass-transfer parameter that depends on z^* is $k_{s\text{-out}}$. But $k_{s\text{-out}}$ varies significantly among the specimen chemicals. This variation is

---

TABLE 8.5
Mass-Transfer Parameters for the Six Specimen Chemicals

Chemical: Parameter	Benzene	Hexachloro-benzene	Lindane	B(a)P	Nickel	Copper
U_a (m d^{-1})	100	100	100	100	100	100
U_s (m d^{-1})	0.0195	0.000237	0.000883	0.000230	0.000393	0.000227
Y_{as} [mol (m^2 Pa d)$^{-1}$]	5.40×10^{-5}	0.00118	7.20×10^{-5}	0.00373	0	0
$v_{e,s}$ (m d^{-1})	5.28×10^{-4}	2.89×10^{-7}	1.80×10^{-5}	8.32×10^{-8}	2.83×10^{-4}	3.16×10^{-7}
$D_{e,s}$ (m d^{-1})	0.00146	1.78×10^{-5}	6.63×10^{-5}	1.73×10^{-5}	2.95×10^{-5}	1.70×10^{-5}
k_{as} (d^{-1})	0.000140	0.00285	0.000936	0.233	22,800	23
k_{sa} (d^{-1})	0.130	0.00154	0.00588	0.00140	0	0
k_s (d^{-1})	0.00693	0.000693	0.00231	0.00139	0	0
$k_{s\text{-out}}$ (d^{-1}) [no gradient]	0.0234	0.000743	0.00503	0.00105	0.00194	0.00016
$k_{s\text{-out}}$ (d^{-1}) [with gradient]	0.0200	0.000448	0.00195	0.000489	0.00190	0.00015

important for cases in which this transfer from the lower boundary of the soil compartment is a major loss mechanism or when it is a key process by which deeper soil layers are contaminated. Other issues to consider in looking at these results and at Figure 8.4 are that concentration falls by almost an order of magnitude within a depth of $2z^*$ and that more than 90% of the mass transferred to and retained in the top soil layer is contained above a depth of $2z^*$. From this we see the potential importance of z^* for scaling the system and selecting an appropriate value of soil thickness, d_s, for use in evaluative models.

In order to further explore the impact of MTC values on model performance, we employ a numerical experiment with 300 chemicals reflecting a range of K_{OW}, K_{AW}, and soil half-life values. In this set, K_{OW} varies from 0.01 to 10^8, K_{AW} varies from 10^{-14} to 100, and the chemical half-life in soil varies from 3 h to 10,000 d. The results of this exercise are shown in Figure 8.5, which shows how a key model output, overall persistence T_{ov} (days), varies among these 300 chemicals between a two-compartment model using a fixed depth in the range of 0.1–0.15 m and the same model using a chemical-specific depth, which is $d_s = z^*$. The use of a chemical-specific depth provides a basis for selecting the depth parameter. But this exercise with overall persistence as a measure of model performance reveals that that a single soil compartment of roughly 0.10–0.15 m provides a reliable approximation of the more detailed model for most chemicals. The exception is chemicals with a high mobility in soil and low

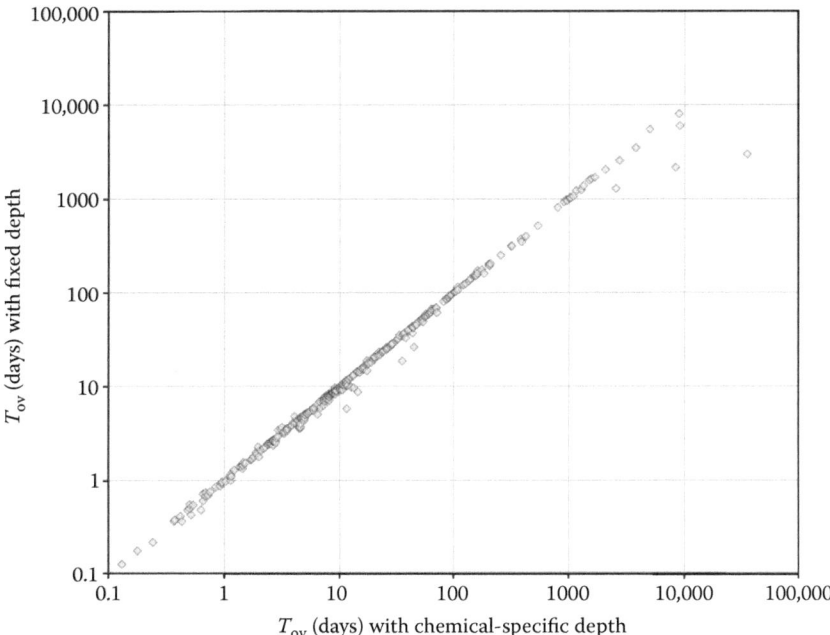

FIGURE 8.5 An illustration of how the model output overall persistence (T_{ov} in days) varies among 300 chemicals between a two-compartment model using a fixed depth in the range of 0.1–0.15 m and the same model using a chemical-specific depth, which is $d_s = z^*$.

reaction rates where the single compartment model can overestimate residence time by a factor of as much as 5. But in considering overall reliability, it is important to recognize that uncertainty in parameters such as the bioturbation factor is more important than the uncertainty introduced by the use of a single soil layer with a fix depth. Nevertheless, this exercise does not negate the need for using the scaling factors z^* and γ in setting chemical-specific MTC values. There also remains the problem of setting up the appropriate soil depth and MTC values in soils having multiple layers with different composition and properties. This is taken up in the next section.

8.8.4 MULTIPLE SOIL LAYERS WITH CONCENTRATION GRADIENTS

As noted above, there are cases where we need more accurate representations of how chemical concentration varies with depth. For example, we may be interested in transfers of chemicals from air to shallow ground water or want to consider how long-term applications of pesticides to the soil surface can impact terrestrial ecosystems—including burrowing creatures. However, we also wish to maintain a simple mathematical mass-balance structure of the multimedia model. To illustrate how we can set up a multilayer model that accurately captures soil mass transport processes, we next derive a vertical compartment structure with an air and three soil compartments, but any number of environmental compartments and soil layers can be employed in this scheme. Figure 8.6 provides a schematic of three soil layers linked to an air compartment and carrying pollutants downward to a saturated zone. We represent the inventory in each vertical compartment i, as M_i (mol), transformation rate constants as k_i, and transfer factors as k_{ij} (d^{-1}). The latter account for the rate of transfer between each i and j compartment pair.

The set of steady-state mass-balance equations describing losses from the air compartment by degradation and net transfer to the ground-surface compartment, the time-dependent mass balance for the air and three soil-layer compartments is described by

$$(k_{ag} + k_a)M_a = S_a + k_{ga}M_g \quad \{\text{Air}\}, \tag{8.27}$$

$$(k_{ga} + k_{gs} + k_g)M_g = k_{ag}M_a + k_{sg}M_s \quad \{\text{Ground-surface soil}\} \tag{8.28}$$

$$(k_{sg} + k_{sv} + k_s)M_s = k_{gs}M_g + k_{vs}M_v \quad \{\text{Root-zone soil}\} \tag{8.29}$$

$$(k_{vs} + k_{v\text{-out}} + k_v)M_v = k_{sv}M_s \quad \{\text{Deeper vadose soil}\} \tag{8.30}$$

where the compartment subscripts used are a for air, g for ground-surface soil, s for root-zone zone soil, and v for deeper vadose-zone soil. S_a represents the source term to air, mol d^{-1}. This set of four equations can be solved to find the inventories M_i, in terms of k's and S_a. The resulting M_i can be converted into either an equivalent fugacity or bulk concentration. But to do this requires that we define the k_{ij} factors in terms of know chemical and compartment properties.

McKone and Bennett (2003) have demonstrated how one can use Equation 8.3 to develop analytical expressions for inventories M_i and MTCs for each soil layer among a set of multiple linked layers and then match these to expressions obtained from Equations 8.27 through 8.30 for the mass flow and fugacity at the boundary between compartments. To define "k" factors in Equations 8.27 through 8.30, McKone and Bennett (2003) show that, when Equation 8.1 is applied to soil layers in which soil properties (distribution coefficient, reaction half-life, etc.) change with depth, Equation 8.1 can be solved stepwise in each layer to obtain the steady-state results:

$$C_i(z) = C_i^* e^{-\gamma_i(z-h_{i-1})}, \tag{8.31}$$

$$C_i^* = \frac{[D_{e,1}\gamma_1 + v_{e,1}]}{[D_{e,i}\gamma_i + v_{e,i}]} C(0) e^{-(\gamma_1 d_1 + \gamma_2 d_2 + \ldots + \gamma_{i-1}d_{i-1})}, \tag{8.32}$$

where i refers to the ith layer in a sequence of different soil layers; h_i is the depth to the lower boundary of the ith soil layer, m; and d_i is the thickness of the ith soil layer, m. Equations 8.31 and 8.32 produce concentrations that can be discontinuous at layer interfaces, but produce fugacities and fluxes that are continuous at layer interfaces.

McKone and Bennett (2003) obtain analytical expressions for the k_{ij} factors between adjacent soil compartments by matching the mass flow obtained from the solution of the compartment mass balance (Equations 8.27 through 8.30) to the mass flow obtained from the exact analytical solution, Equation 8.31. This requires that each soil layer is homogeneous such that Equation 8.31 is continuous with depth in each soil layer, but allows for differences in properties among soil layers i. Table 8.6 provides a summary of the general and specific definitions of k_{ij} factors that are obtained by McKone and Bennett (2003) from the mass-balance matching of the exact and discrete soil layer approach. Figure 8.7 illustrates how the general multilayer approach compares with the mass distribution of the exact analytical model solution for cases with three and four soil compartments.

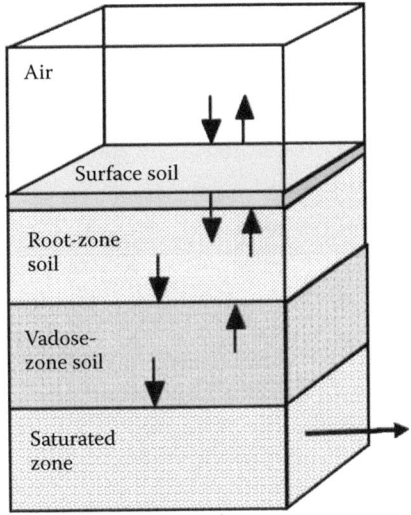

FIGURE 8.6 A schematic of three soil layers linked to an air compartment and carrying pollutants downward to a saturated zone.

TABLE 8.6
Summary of the Mass-Inventory and Fugacity-Based Transfer Factors for the Multilayer Soil Compartment Model

Transfer	Mass-Inventory Based Transfer Factor, d^{-1}	Fugacity-Based Transfer Factors [Mackay "D" values] in mol d^{-1} Pa^{-1}
Air to ground-surface soil	$k_{ag} = \dfrac{1}{Z_a \times d_a}\left(Y_{ag} + V_d \dfrac{PC}{\rho_p} Z_{ap} + rain \times Z_{water}\right)$	$D_{ag} = Z_a V_a k_{ag} = Area \times$ $\left(Y_{ag} + V_d \dfrac{PC}{\rho_p} Z_{ap} + rain \times Z_{water}\right)$
Ground-surface soil to air	$k_{ga} = \dfrac{Y_{ag}}{Z_g d_g},$ $Y_{ag} = \left[\dfrac{1}{Z_{air}U_a} + \dfrac{1}{Z_g U_g}\right]^{-1}$	$D_{ga} = Z_g V_g k_{ga} = Area \times Y_{ag}$
Fugacity-based diffusive-transfer coefficient from g to s		$Y_{sg} \ (\text{mol m}^{-2}\,\text{s}^{-1}) =$ $\left[\dfrac{e^{\gamma_g d_g}-1}{D_{e,g}\gamma_g^2 Z_g d_g} - \dfrac{1-e^{-\gamma_s d_s}}{D_{e,s}\gamma_s^2 Z_s d_s}\right]^{-1}$
Transfer from surface soil to root soil	$k_{gs} = \dfrac{Y_{gs}}{Z_g d_g} + \dfrac{v_{e,g}\gamma_g}{(e^{\gamma_g d_g}-1)}$	$Z_g V_g k_{gs} = Area \times \left[Y_{gs} + \dfrac{v_{e,g}\gamma_g Z_g d_g}{(e^{\gamma_g d_g}-1)}\right]$
Transfer from root soil to surface soil	$k_{sg} = \dfrac{Y_{gs}}{Z_s d_s}$	$Z_s V_s k_{sg} = Area \times Y_{gs}$

Generalized relationship between any two upper (u) and lower (l) soil layers

Transfer	Mass-Inventory Based Transfer Factor, d^{-1}	Fugacity-Based Transfer Factors [Mackay "D" values] in mol d^{-1} Pa^{-1}
Fugacity-based diffusive-transfer coefficient from u to l		$Y_{ul} = \left[\dfrac{e^{\gamma_u d_u}-1}{D_{e,u}\gamma_u^2 Z_u d_u} - \dfrac{1-e^{-\gamma_l d_l}}{D_{e,l}\gamma_l^2 Z_l d_l}\right]^{-1}$ $Y_{u-out} = \left[\dfrac{D_{e,u}\gamma_u^2 Z_u d_u}{e^{\gamma_u d_u}-1}\right]$ (last layer)
Transfer from u to l	$k_{ul} = \dfrac{Y_{ul}}{Z_u d_u} + \dfrac{v_{e,u}\gamma_u}{(e^{\gamma_u d_u}-1)}$	$Z_u V_u k_{ul} = Area \times \left[Y_{ul} + \dfrac{v_{e,u}\gamma_u Z_u d_u}{(e^{\gamma_u d_u}-1)}\right]$
Transfer from l to u	$k_{lu} = \dfrac{Y_{ul}}{Z_l d_l}$	$Z_l V_l k_{lu} = Area \times Y_{ul}$

Source: Adapted from McKone, T. E.; Bennet, D. H. 2003. *Environmental Science Technology*, 33(14): 2123–2132.

8.9　SUMMARY

Quantifying the mass transport of chemicals in soil and between soil and atmosphere is a key step in understanding the role soil plays in controlling fate, transport, and exposure to multimedia pollutants. In existing multimedia mass-balance models, there have been significant variations in both the complexity and structure applied to the soil compartment. Here we have described the conceptual issues involved in the transport and transformation of pollutants in soils. We then describe and evaluate novel approaches for constructing soil transport algorithms and the corresponding MTCs for multimedia fate models. We focus on those algorithms that provide an

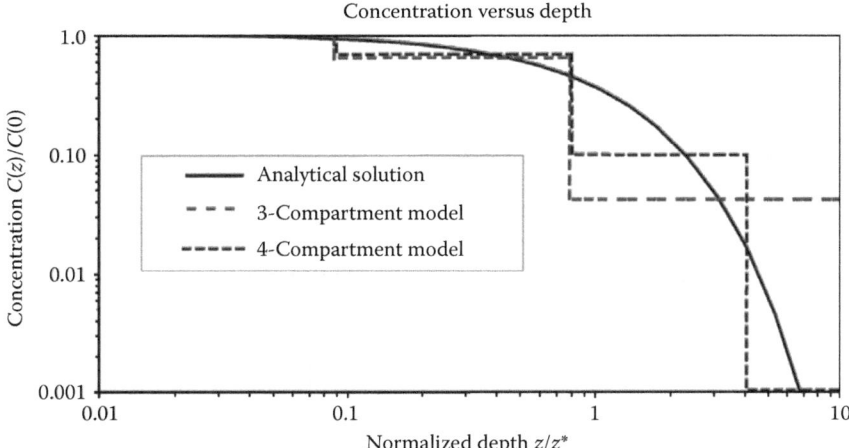

FIGURE 8.7 This graph compares the concentration change (where $C(0)$ is the concentration at the soil surface) versus normalized depth $z/z*$ of the compartment-model solution and analytical solution for the case where soil properties are assumed uniform. (Adapted from McKone, T. E.; Bennet, D. H. 2003. *Environmental Science & Technology*, *33*(14): 2123–2132.)

explicit quantification of the characteristic soil penetration depth in soil layers of different composition. We consider in detail a recently published approach constructed as a compartment model using one or more soil layers. This approach can replicate with high-reliability the flux and mass distribution obtained from the exact analytical solution describing the transient diffusion/dispersion, advection, and transformation of chemicals in soil layers with different properties but a fixed boundary condition at the air–soil surface. We demonstrate and evaluate the performance of the various soil algorithms using specimen chemicals benzene, hexachlorobenzene, lindane (gamma-hexachlorocyclohexane, benzo(a)pyrene, nickel, and copper. We consider the relative advantages of generic versus chemical-specific soil compartment modeling in multimedia models. This example offers important insight into the trade-offs between model complexity and reliability, and provides a case study of quantitative evaluation of model performance. We have learned that this problem can be limited by selecting a shallow (0.1–0.15 m) compartment depth when building regional mass balance models.

ACKNOWLEDGMENTS

The author was supported in part by the U.S. Environmental Protection Agency National Exposure Research Laboratory through Interagency Agreement #DW-89-93058201-1 with Lawrence Berkeley National Laboratory through the U.S. Department of Energy under Contract Grant No. DE-AC02-05CH11231.

LITERATURE CITED

Anderson, T.A.; Beauchamp, J.J.; Walton, B.T. 1991. Fate of volatile and semivolatile organic chemicals in soils: Abiotic versus biotic losses, *Journal of Environmental Quality 20*(2): 420–424.

Brady, N.C.; Weil, R.R. 2004. *Elements of the Nature and Properties of Soils*, 2nd ed. Prentice-Hall, Upper Saddle River, NJ.

Cowan, C.E.; Mackay, D.; Feijtel, T.C.J.; Van De Meent, D.; Di Guardo, A.; Davies, J.; Mackay, N. 1995. *The Multi-Media Fate Model: A Vital Tool for Predicting the Fate of Chemicals*; SETAC Press, Pensacola, FL.

Harner, T.; Mackay, D.; Jones, K. C. 1995. Model of the long-term exchange of PCBS between soil and the atmosphere in the southern U.K., *Environmental Science & Technology 29*: 1200–1209.

Jury, W. A.; Spencer, W. F.; Farmer, W. J. 1983. Behavior assessment model for trace organics in soil: I. Model description, *Journal of Environmental Quality 12*: 558–564.

Jury, W. A.; Russo, D.; Streile, G.; Elabd, H. 1990. Evaluation of volatilization by organic chemicals residing below the soil surface, *Water Resources Research 26*: 13–20.

Mackay, D. 1979. Finding fugacity feasible, *Environmental Science and Technology 13*: 1218–1223.

Mackay, D. 2001. *Multimedia Environmental Models, The Fugacity Approach*, 2nd ed. Lewis Publishers, Chelsea, MI.

Mackay, D.; Paterson, S. 1981. Calculating fugacity, *Environmental Science and Technology 15*: 1006–1014.

Mackay, D.; Paterson, S. 1982. Fugacity revisited: The fugacity approach to environmental transport, *Environmental Science and Technology 16*: A654–A660.

McLachlan, M. S.; Czub, G.; Wania, F. 2002. The influence of vertical sorbed phase transport on the fate of organic chemicals in surface soils, *Environmental Science and Technology 36*: 4860–4867.

MacLeod, M.; Woodfine, D. G.; Mackay, D.; McKone, T.; Bennett, D.; Maddalena, R. 2001. BETR North America: A regionally segmented multimedia contaminant fate model for North America, *Environmental Science and Pollution Research 8*: 156–163.

McKone, T. E. 1993. *CalTOX, A Multimedia Total-Exposure Model for Hazardous-Wastes Sites Part I: Executive Summary*; UCRL-CR-111456PtI, Lawrence Livermore National Laboratory: Livermore, CA.

McKone, T. E. 2005. Pollution, soil, in *Encyclopedia of Toxicology*, 2nd ed. Philip Wexler, ed., Elsevier, Oxford, pp. 489–495.

McKone, T. E.; Bennett, D. H. 2003. Chemical-specific representation of air-soil exchange and soil penetration in regional multimedia models, *Environmental Science & Technology, 33*(14): 2123–2132.

Mullerlemans, H.; Vandorp, F. 1996. Bioturbation as a mechanism for radionuclide transport in soil: Relevance of earthworms, *Journal of Environmental Radioactivity, 31*, 7–20.

National Research Council 2002. *Biosolids Applied to Land: Advancing Standards and Practices, Committee on Toxicants and Pathogens in Biosolids Applied to Land, Board on Environmental Studies and Toxicology*, National Academies Press, Washington, DC.

Roth, K.; Jury, W.A. 1993. Modeling the transport of solutes to groundwater using transfer functions, *Journal of Environmental Quality 22* (3): 487–493.

Sposito, G. 2008. *The Surface Chemistry of Natural Particles*, Oxford University Press, New York.

Thibodeaux, L. J. 1996. *Chemodynamics, Environmental Movement of Chemicals in Air, Water, and Soil*, 2nd ed., Wiley, New York.

USEPA 1999. *TRIM Total Risk Integrated Methodology Status Report*; U.S. Environmental Protection Agency Office of Air Quality Planning and Standards, Research Triangle Park.

Whicker, F. W.; Kirchner, T. B. 1987. Pathway: A dynamic food-chain model to predict radionuclide ingestion after fallout deposition. *Health Physics 52*: 717–737.

PART 2 ADVECTIVE–DIFFUSIVE TRANSPORT PROCESSES FOR CHEMICALS IN SURFACE SOILS

Shannon L. Bartelt-Hunt, Mira S. Olson, and Fred D. Tillman

8.10 INTRODUCTION TO PART 2

Chemicals can be released into the subsurface environment from various sources including leaking landfill liners, improper disposal, accidental spillage, or leaking underground storage tanks (LUSTs). Once in the subsurface, these compounds can be bound to the soil matrix, dissolved in groundwater (or soil water) and/or exist as a separate, residual phase known as nonaqueous phase liquid (NAPL). The physical transport of separate-phase, aqueous, and/or gaseous contaminants within a porous medium occurs by two primary mechanisms: advection and diffusion. Part 1 of this chapter provides a general overview of mass transfer of pollutants within surface soils and includes a description of the nature of soil pollution, the composition of soil, and mass transport processes and transport parameters for contaminants in soil with a primary focus on presenting modeling approaches for incorporating the soil compartment in multimedia fugacity-type fate models. In this section, additional information on the nature of advective and diffusive chemical transport in saturated and unsaturated soils is provided. Chemical transport in the aqueous, gaseous, and NAPL phases is described.

8.11 CONTINUITY EQUATION

8.11.1 LAVOISIER MASS BALANCE AND THE ADVECTIVE–DIFFUSIVE EQUATION

The species continuity equation (CE) is an expression of the Lavoisier general law of conservation of mass. Equation 2.1 presents the CE in vector form and provides the proper context for the various types of chemical mass transport processes needed for chemical modeling and fate analysis. In Section 2.2.2, the mass accumulation portion of the CE is highlighted as the principal term for assessing chemical fate in the media compartments. This term includes reaction, advection, diffusion, and turbulent transport and dispersion processes. Because the magnitude and direction of this term reflect the sum total of all processes, this term uniquely defines chemical fate. In Equation 2.2, the steady-state CE minus the reaction term is commonly referred to as the advective–diffusive (AD) equation. It provides the appropriate starting point for addressing the various transport processes associated with the mobile phases in near-surface soils.

8.11.2 Overview of Closely Related Surface Soil Chemodynamic Processes

A total of 41 individual transport processes are listed in Table 4.1 as being the most significant ones of concern in this handbook. Seven in the list are soil-side transport processes and the main focus of Part 2 of this chapter. Several of these processes including diffusion, advection, and fluid-to-solid mass transfer in porous media are also relevant to many other environmental compartments and are covered in more detail in other chapters of this book. Specifically the chapters are: mass transport fundamentals from an environmental perspective (Chapter 2); molecular diffusion estimation methods (Chapter 5); advective porewater flux and chemical transport in bed-sediment (Chapter 11); diffusive chemical transport across water and sediment boundary layers (Chapter 12); bioturbation and other sorbed-phase transport processes in surface soils and sediments (Chapter 13), and dispersion and mass transfer in groundwater of the near-surface geologic formations (Chapter 15).

8.12 ADVECTIVE CHEMICAL TRANSPORT IN THE MOBILE PHASES OF SOIL

8.12.1 Water-Saturated Soils: Darcy's Law

The subject of advective porewater flux and associated chemical transport is covered in Chapter 11 in the context of aquatic bed-sediment systems, which include surface soil-derived particle layers on the bottom of streams, rivers, lakes, estuaries, and the near-shore marine environment. This "soil" system is a saturated porous medium and therefore the fundamentals of the transport processes and related parameters within this system are identical to that of surface soils. A brief review of advective transport in the subsurface follows.

Advective water flow in soils is driven by hydraulic head gradients within the porous medium (for example, see Figure 11.1a) and is described by Darcy's law (Equation 11.1). To apply Darcy's law, the differential heads are needed as well as the effective hydraulic conductivity of the porous layers. Table 11.1 contains numerical ranges of the hydraulic conductivity of various types of earthen materials including unconsolidated deposits which may best represent surface soils. In addition, Figure 11.3 contains data on measured seepage rates. These are steady-state Darcian velocities through successive layers of the porous materials. The low-end range of the seepage rates in this figure likely reflect the low conductivity values characteristic of the silt and clay fractions that typically dominate the bed-sediment layers. These data are provided to inform the user of the magnitude and range of Darcian velocities that could be expected in various surface soil environments based on published field measurements.

8.12.2 Water-Unsaturated Soils: Richards' Law and Capillary Rise

Water residing in the subsurface above the water table normally occurs in a partially saturated environment, although infiltration events may produce temporary saturated conditions above the water table and the capillary zone may also be mostly saturated

(Chow et al., 1988). In the unsaturated zone, soil pores are partially filled with water and fluid pressure is less than atmospheric pressure. An important implication of partial saturation on fluid flow is that hydraulic conductivity in the unsaturated zone is a function not only of location (x,y,z), but also of pressure (or suction) head (ψ). A reduced form of the continuity equation, assuming that gas-phase pressure remains atmospheric, can be written in terms of pressure head:

$$\frac{\partial}{\partial x}\left[K(\psi)\frac{\partial \psi}{\partial x}\right] + \frac{\partial}{\partial y}\left[K(\psi)\frac{\partial \psi}{\partial y}\right] + \frac{\partial}{\partial z}\left[K(\psi)\left(\frac{\partial \psi}{\partial z}+1\right)\right] = C(\psi)\frac{\partial \psi}{\partial t}, \quad (8.33)$$

where ψ is the pressure head; $K(\psi)$ is the partially saturated hydraulic conductivity that is a function of pressure head; and $C(\psi)$ is specific moisture capacity that is a measure of the change in the amount of moisture stored in the unsaturated zone corresponding to a change in pressure head. This governing equation for unsteady, unsaturated flow in a porous medium is called Richards' equation (see Freeze and Cherry 1979 for derivation). The solution to Richards' equation, $\psi(x,y,z,t)$ describes pressure head at any point in space and time in the unsaturated zone. Applying the nonlinear Richards' equation to simulate moisture movement in porous media may be difficult owing to the necessity of determining $K(\psi)$ and $C(\psi)$. Simplifying assumptions applied to Richards' equation have yielded solutions applied to infiltration including Horton's equation and Philip's equation (see Chow et al., 1988).

At the interface of water and air, there are unbalanced forces owing to the surface tension of the fluid. In the unsaturated zone, this transition zone is known as the capillary fringe or tension-saturated zone and these unbalanced forces result in capillary rise of groundwater and dissolved constituents. Assuming a constant hemispherical fluid meniscus and pore radius, the equation of Young and LaPlace can be combined with an equation for hydrostatic pressure to produce a relation for computing the capillary rise of fluid:

$$\Delta \rho g = (2\gamma \cos \theta)/rh, \quad (8.34)$$

where $\Delta \rho$ is the difference in density between the liquid and gas phase; g is gravitational acceleration; γ is the surface tension of the fluid [force per unit length]; θ is the contact angle of the fluid with the capillary tube; r is the radius of curvature of the meniscus; and h is the capillary rise of the fluid. Ranges of capillary rise for differing porous media were reported in Bear (1972) (see Table 8.7).

TABLE 8.7
Capillary Rise Expected in Typical Porous Media Types

Porous Media Material	Capillary Rise (cm)
Coarse sand	2–5
Sand	12–35
Fine sand	35–70
Silt	70–150
Clay	>200–400

8.12.3   Nonaqueous Phase Liquids

Many contaminants exist as immiscible nonaqueous phase liquids (NAPLs) in soil. These liquids do not fully solubilize in water and exist as a separate phase due to physical and chemical differences from water. NAPLs can be classified as light (less dense than water) nonaqueous phase liquids (LNAPLs) and dense (more dense than water) nonaqueous phase liquids (DNAPLs). A list of typical NAPLs and their important properties is presented in Table 8.8. As described elsewhere in this book (Chapter 15), NAPLs may solubilize, volatilize, and otherwise partition among phases. This section focuses on the advective transport of pure-phase NAPL.

NAPL transport is dependent upon properties of the NAPL (density, viscosity, interfacial tension, and capillary pressure), the porous medium (pore size, pore structure, and saturation), and the groundwater (velocity and chemical water composition). When introduced into the subsurface, NAPL migrates downward through the unsaturated zone via gravity. The released NAPL leaves residual ganglia or blobs in the pore space due to surface tension, and eventually reaches the capillary fringe assuming a sufficiently large volume of NAPL. At the water table, LNAPL will spread atop or slightly depress the water surface while DNAPL will displace the underlying water and continue through the saturated zone. This migration is nonuniform and generally

---

TABLE 8.8
Properties of Typical NAPLs

NAPL Type	Chemical	Density ($g\ cm^{-3}$)	Dynamic Viscosity (cP)	Henry's Law Constant ($atm\text{-}m^3\ mol^{-1}$)
LNAPL	Benzene	0.877	0.647	5.43×10^{-3}
	Ethyl benzene	0.867	0.678	7.9×10^{-3}
	Toluene	0.867	0.58	6.61×10^{-3}
	o-Xylene	0.880	0.802	4.94×10^{-3}
	Methyl ethyl ketone	0.805	0.40	2.74×10^{-5}
	Vinyl chloride	0.910		2.78
DNAPL	Trichloroethylene	1.46	0.570	0.0091
	Tetrachloroethylene	1.62	0.890	0.0153
	Methylene chloride	1.33	0.430	0.010
	Naphthalene	1.16		4.6×10^{-4}
	Phenanthrene	1.20		2.56×10^{-5}
	Pyrene	1.27		1.87×10^{-5}
	Anthracene	1.25		6.51×10^{-5}
Water		1.0	1.14	

Sources: Adapted from Mercer, J.W. and Cohen, R.M. 1990. *Journal of Contaminant Hydrology*, 6: 107–163; Huling, S.G. and Weaver, J.W. 1991. Dense nonaqueous phase liquids, Ground Water Issue, EPA/540/4-91/002, R.S. Kerr Environmental Research Laboratory, USEPA, Ada, OK; Newell, C.J., et al. 1995. Light nonaqueous phase liquids, Ground Water Issue, EPA 540/S-95/500, R.S. Kerr Environmental Research Laboratory, USEPA, Ada, OK; Sharma, H.D. and Reddy, K.R. 2004. *Geoenvironmental Engineering,* Wiley, Hoboken, NJ, 108 pp.

proceeds as preferential fingers (see Section 15.9). The following expressions may be used to estimate the critical NAPL height required to initiate downward NAPL migration in the vadose zone (Equation 8.35) and downward DNAPL or upward LNAPL penetration into a saturated aquifer (Equation 8.36):

$$Z_n(\text{est}) = \frac{2\sigma \cos \phi}{r_t g \rho_n}, \tag{8.35}$$

$$Z_n(\text{est}) = \frac{2\sigma \cos \phi}{r_t g |\rho_n - \rho_w|}, \tag{8.36}$$

where σ is the interfacial tension between NAPL and water; ϕ is the contact angle; r_t is the radius of water-filled pore throat that the NAPL must move through; ρ_n is the NAPL density; ρ_w is the water density; and g is gravity.

In water–NAPL systems, NAPL may exist as a continuous mass (mobile phase or free phase), which can flow under a hydraulic gradient, or as individual ganglia or blobs (residual phase), which are more difficult to mobilize hydraulically. The residual saturation, S_r, defines the NAPL saturation (V_{NAPL}/V_{voids}) below which NAPL distribution is discontinuous. Typical residual saturation values for soils and sands range from 26% to 75% (Thibodeaux 1996), but have been reported as low as 15–25% (Bedient et al., 1997). In separate field studies, free-phase DNAPL (Pankow and Cherry, 1996) and LNAPL (Huntley et al., 1994) were detected at saturation values as low as 15% and 25%, respectively. In unsaturated soil, typical residual saturation values are between 5% and 20% (Bedient et al., 1997), whereas in saturated soil, residual saturation may be 15–50% of the total pore space (Mercer and Cohen, 1990).

Migration of mobile-phase (i.e., continuous) NAPL in saturated soil may be described with Darcy's law (Equation 11.1) by adjusting the hydraulic conductivity parameter to include a relative permeability parameter that accounts for both the reduction in available pore space due to the two fluids and the fluid properties of the NAPL. Depending on the properties of the porous medium and the NAPL, both water-wet and NAPL-wet systems are possible. Typically, water acts as the wetting fluid and NAPL acts as the nonwetting fluid (see Chapter 15 for further explanation of NAPL wettability). The steady-state saturated flow of a wetting fluid (Equation 8.37) and a nonwetting fluid (Equation 8.38) in a two-phase system is given by

$$q_w = \frac{Q_w}{A} = -\frac{k_{rw} k_i \rho_w}{\mu_w} \frac{dh_w}{dl}, \tag{8.37}$$

$$q_{nw} = \frac{Q_n}{A} = -\frac{k_{rnw} k_i \rho_{nw}}{\mu_{nw}} \frac{dh_{nw}}{dl}, \tag{8.38}$$

where q_w and q_{nw} are the Darcy velocities of wetting and nonwetting fluids, respectively; Q_w and Q_{nw} are the volume flow rates of the wetting and nonwetting fluids, respectively; A is the cross-sectional area of flow; k_{rw} and k_{rnw} are the relative permeabilities of the wetting and nonwetting fluids in the two-phase system, respectively;

k_i is the intrinsic permeability; ρ_w and ρ_{nw} are the densities of the wetting and nonwetting fluids, respectively; μ_w and μ_{bw} are the viscosities of the wetting and nonwetting fluids, respectively; and dh_w/dl and dh_{nw}/dl are the gradients of head of the wetting and nonwetting fluids, respectively. An example of relative permeabilities of water and NAPL as a function of water saturation is shown in Figure 8.8.

Mobilization of residual NAPL droplets requires a hydraulic gradient sufficient to overcome the capillary forces holding the NAPL ganglia in place. The potential for NAPL mobilization may be estimated using the capillary number, N_C, a ratio of capillary to viscous forces:

$$N_C = \frac{k_i \rho_w g}{\sigma} \frac{dh}{dl}, \tag{8.39}$$

where k_i is intrinsic permeability; ρ_w is density of water; g is gravity, σ is interfacial tension, and dh/dl is the hydraulic gradient. The critical value of the capillary number, N_C^* indicates the value at which NAPL blob mobilization is initiated. Figure 8.9 shows the hydraulic gradient necessary to initiate NAPL blob mobilization in different media. Generally speaking, a steep gradient is necessary to mobilize NAPL ganglia in coarse gravel, but removal in less permeable media is nearly impossible (Mercer and Cohen, 1990; Wilson and Conrad, 1984).

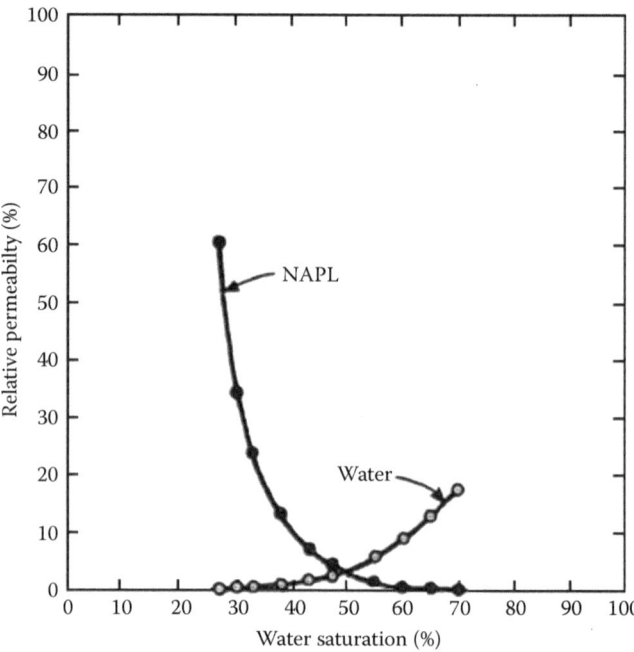

FIGURE 8.8 Relative permeability of water and NAPL as a function of water saturation. (From Mercer, J.W. and R.M. Cohen. 1990. *Journal of Contaminant Hydrology* 6: 107–163. With permission.)

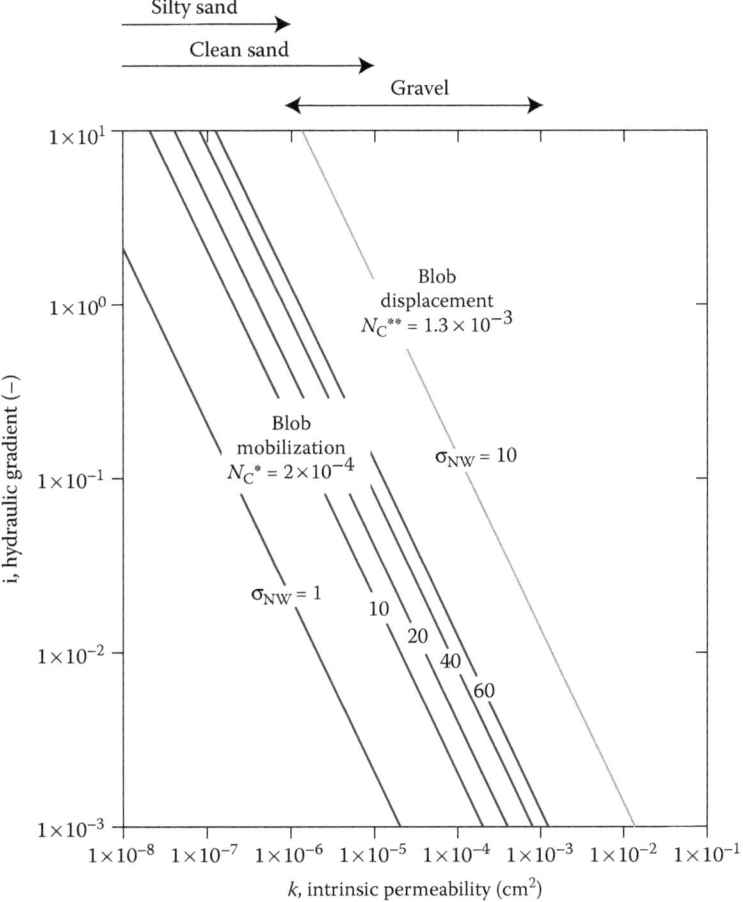

FIGURE 8.9 Hydraulic gradient necessary to initiate blob mobilization at N_C^* for nonwetting NAPLs with various interfacial tensions, σ_{NW} (gr · cm/s²), in different soil types. (From Wilson, J. L. and S. H. Conrad. 1984. In *Proceedings of the National Water Well Association Conference on Petroleum Hydrocarbons and Organic Chemicals in Groundwater*, Houston, TX, Nov. 5–7, 1984, pp. 274–298. With permission.)

8.12.4 SOIL GAS MOVEMENT

Pressure-driven soil-gas advection results from differences in soil-gas pressure, causing soil gas to flow from areas of higher pressure to lower pressure. Subsurface gas pressure gradients may result from several phenomena, including barometric pumping, building underpressurization, vapor density gradients, and landfill gas generation, among others. Diurnal or weather-related atmospheric pressure cycles occurring at land surface may transmit pressure waves into the unsaturated zone and air may flow in response—a process known as barometric pumping. Gas pressure gradients may also form in the subsurface owing to underpressurization of an overlying structure. Soil gas may flow through the structure foundation and into the building in a process known as soil vapor intrusion. Density-driven flow of organic vapors may occur in the vicinity of residual-phase organic compounds whose saturation gas densities are greater than that of air. As organic liquids with high vapor pressures and molecular weights volatilize, the density of the soil gas surrounding the liquid changes. In almost all cases, organic liquids have molecular weights that are greater than air so the resulting density-driven flow will be in a downward direction and will be proportional to soil permeability and density differences between the vapor and air (Falta et al., 1989; Mendoza and Frind, 1990a, 1990b). Organic compounds for which density-driven advection may be significant include methylene chloride, 1,2-dichloroethylene, 1,1,1-trichloroethane, carbon tetrachloride, and 1,1-dichloroethane, among others (Falta et al., 1989; Mendoza and Frind, 1990a, 1990b). Although not covered in further detail in this chapter, soil gas pressure gradients may also be produced by a rapidly rising or falling water table, as in coastal zones (Li et al., 2002), through the buildup of gas pressure from decomposing organic matter inside a landfill (Little et al., 1992) or as a result of natural temperature differences between deep and shallow soil gas (Gustin et al., 1997; Krylov and Ferguson, 1998). This section discusses three advective processes for soil gas transport in the subsurface in greater detail: barometric pumping, soil vapor intrusion, and landfill gas generation.

8.12.4.1 Barometric Pumping

Near land surface, barometric pressure is constantly changing owing both to the passage of high- and-low-pressure weather systems as well as short-term diurnal temperature and gravitational fluctuations. Depending upon soil properties such as air permeability and air-filled porosity, barometric pressure changes at land surface may be transmitted to depths of hundreds of feet in deep unsaturated zones (Weeks, 1978). As subsurface gas pressure reestablishes equilibrium with the changing air pressure at land surface, gas is "breathed" in and out of the unsaturated zone, a process known as barometric pumping. Rossabi and Falta (2002) routinely measured airflow as high as $20 \, L \, min^{-1}$ flowing both into and out of a 1 in. diameter PVC well screened over a permeable layer, which was separated from the surface by an impermeable clay layer. Pressure differences between the screened interval in the confined permeable layer and atmospheric pressure at land surface caused air to flow almost constantly during the entire 25-day measurement period. Periods of falling atmospheric pressure at land surface above a contaminated site or landfill can withdraw soils gas from the unsaturated zone, transporting with it potentially harmful compounds such as organic vapors, mercury vapor, radon, or other radioactive gases. These contaminants would escape from the shallow unsaturated zone faster than they would by pure diffusion alone (Choi et al., 2002). Tillman and Smith (2005) simulated soil-gas flow in response to measured atmospheric pressure changes and found that multilayered unsaturated zone systems experienced significantly more soil-gas flow than homogenous systems.

Periods of increasing atmospheric pressure at contaminated sites may force uncontaminated surface air into the shallow unsaturated zone. This injection of presumably clean surface air into the unsaturated zone may have two consequences. The rate of unsaturated-zone contaminant vapor diffusion will be increased by lowering the surface where the contaminant concentration is zero below true ground level (Auer et al., 1996). Additionally, oxygen from land surface may penetrate into the subsurface during rising atmospheric pressure, providing increased electron acceptors to aerobic bacteria degrading organic compounds. Massmann and Farrier (1992) simulated gas transport in the unsaturated zone, resulting from atmospheric pressure variations using the single-component advection–dispersion equation, and found that "fresh" air can migrate several meters into the subsurface during a typical barometric pressure cycle. These effects may be compounded by the presence of fractures, partially penetrating cracks, or boreholes open to the unsaturated zone (Schery et al., 1988; Holford et al., 1993).

8.12.4.2   Soil Vapor Intrusion

Soil vapor intrusion refers to the transport of vapors from volatile organic compounds (VOCs) or other contaminants of interest from the subsurface into buildings. The potential presence of harmful vapors in buildings is of great importance as the average American spends over 21 hours per day indoors (Olson and Corsi, 2002). VOCs in living spaces can serve as an immediate threat of explosion or acute toxicity, or perhaps more commonly as a long-term source for exposures to potential carcinogenic or toxic compounds. Organic compounds of concern in vapor intrusion are usually divided into two broad categories: chlorinated solvents and petroleum hydrocarbons. Once these contaminants are present near or beneath buildings, they may move as a vapor through soil gas and into the residence.

The effects of overlying buildings play an important role in the subsurface-to-indoor-air pathway. Different building construction techniques may impact the ability of vapors to enter indoor air space. Soil gas flow into buildings can occur through untrapped drains, sumps, perimeter cracks, expansion/settling cracks, or utility conduits. Buildings constructed with basements may have more surface area through which vapors can penetrate, as well as be closer to subsurface sources than slab-on-grade buildings. A single-pour cement foundation may not have the perimeter-crack often associated with foundations whose footers and floor are poured separately, but may still become cracked along stress lines. Foundations and subsurface walls constructed from cement blocks may contain cracks around mortar that can allow subsurface gas to enter the building. Homes built over a crawl space may benefit from the dilution of soil gas by ventilated crawlspace air, but do not have the impedance to vapors that concretes slabs provide.

The underpressurization of buildings relative to subsurface pressure may cause contaminated soil gas to flow into indoor air spaces, increasing exposure over diffusive transport alone. Building underpressurization relative to soil gas pressure can be caused by temperature differences between indoor and outdoor air (known as stack effect), imbalanced air handling systems, wind loading on the building superstructure, or barometric pressure cycles (Nazaroff et al., 1987; Garbesi and Sextro, 1989). Typical values for building underpressurization range from 2 to 10 Pa, but may be as high as 15 Pa during the heating season (Hers et al., 2001). A positive building pressure relative to the subsurface may greatly reduce the intrusion of subsurface vapors by causing air to flow out these same cracks and penetrations of the building envelope.

8.12.4.3 Landfill Gas Generation

Another source of advective gas flows in the subsurface is from gas generated by decomposing solid waste in municipal solid waste landfills. Landfill gas is generated as a result of an anaerobic biological process where microorganisms decompose the organic fraction of waste to produce carbon dioxide, methane, and other gases. Landfill gas generation occurs in several phases with the initial landfill gas composition representing the distribution of gases in the atmosphere: approximately 80% nitrogen and 20% oxygen. A short phase (approximately 3–18 months) of aerobic waste decomposition consumes the oxygen, and produces carbon dioxide with no methane content. Once the landfill enters an anaerobic phase, gas composition is approximately 45–60% methane, 35–45% carbon dioxide with lesser amounts of nitrogen, oxygen, hydrogen, water vapor, and other trace constituents, including nonmethane organic compounds (NMOCs) (Qian et al., 2002). Anaerobic decomposition of solid waste will continue until all of the organic material in the waste is consumed, or until oxygen is reintroduced to the system, which would return the landfill to an aerobic state. Typically, the anaerobic gas production phase of landfills can last for 10–80 or more years. Alternative methods for landfill operation, which enhance gas production, would be expected to have an anaerobic gas production phase on the lower end of the range, while the higher end of the range is likely more indicative of conditions in a conventional landfill.

A number of factors can affect the ability of a landfill to generate gas, including waste composition, age, moisture content, particle size, landfill pH, and temperature. A higher percentage of organic wastes in a landfill will result in a corresponding increase in landfill gas production. Moisture is commonly the limiting factor in the anaerobic decomposition process, therefore, waste with a higher moisture content will typically increase the rate of landfill gas generation. Smaller waste particles have a higher specific surface area, and thus will be more readily degraded. Optimal conditions for landfill gas production are between pH 6 and 8, and at temperatures between 35°C and 65°C (Qian et al., 2002).

Methane yields of 40–120 L CH_4/dry kg of refuse have been reported in laboratory tests, while methane yields reported from field experiments range from approximately 39 to 92 L CH_4/dry kg of refuse. The significant variability reported for methane generation from laboratory studies may be partially due to the wide range of experimental conditions (i.e., pH, temperature, rate of leachate recycle) employed in these tests (Barlaz and Ham, 1993). A theoretical maximum methane yield of 92 L CH_4 per dry kg of refuse has been reported (Eleazer et al., 1997). Commonly, gas production rates of 7.8 L CH_4 kg^{-1} refuse per year (0.1 ft^3 of landfill gas lb waste^{-1} year^{-1}) are used in numerical simulations of landfill gas production.

Landfill gas can migrate from landfills by both diffusion due to concentration gradients and advection due to the pressure differential between the landfill and the surrounding atmosphere. The gas produced by anaerobic decomposition in a landfill will occupy a larger volume than the original waste, which may result in pressure levels at the base of landfills as high as 4 atm (Qian et al., 2002). This pressure may drive landfill gas into a landfill gas collection system, directly into the atmosphere, or potentially into the soil surrounding the landfill. The degree to which landfill gas will migrate vertically or horizontally is dependent on a number of factors including the nature of the landfill cover, the bottom lining system, the gas collection system and the properties of the surrounding soil. A number of numerical models, which include advective gas flow, have been developed to predict potential fate routes for organic contaminants within landfills. These models have previously been used to evaluate the fate of chemical warfare agents (Bartelt-Hunt et al., 2006) and volatile organic compounds (Lowry et al., 2008).

8.13 DIFFUSIVE CHEMICAL TRANSPORT IN MOBILE PHASES

8.13.1 DIFFUSIVE TRANSPORT IN THE AQUEOUS PHASE: SATURATED AND UNSATURATED CONDITIONS

Diffusion in the aqueous phase of porous media is described in several places in this handbook. It first appears in Chapter 5, which focuses on estimating molecular diffusivities in environmental fluids. This chapter contains estimation techniques and data for vapor-phase and aqueous-phase molecular diffusivities. Correction factors and procedures are presented that account for the presence of porous media on the base diffusivities. This section is of particular importance to surface soils as these correction factors account for the fraction of the porosity available for movement and the increasing path length for molecules moving around particles and within open channel (see Equations 5.34 through 5.36). Examples 5.4 and 5.5 illustrate the procedure to calculate diffusion coefficients for naphthalene in saturated sediment and a dry soil, respectively.

As mentioned previously, the bed-sediment of aquatic systems is a saturated porous medium and is similar to soil in many respects. The basic process of chemical transport by diffusion is comparable in these two systems. Section 12.5 contains equations to estimate the effective diffusion coefficients for several chemicals in bed sediment. See specifically Equation 12.18, Archie's law, as well as Figures 12.2 and 12.3. Information on typical porosities in natural sedimentary materials appears in Table 12.7 and Figure 12.4. This section also contains information on pore structure and particle distributions. Tables 12.9 and 12.10 contain compilations of data on this transport parameter in porewaters. Generally speaking, the porous material types, particle sizes, porosities, and structures are comparable to surface soils so that the information and data provided is very relevant to the user.

8.13.2 DIFFUSIVE TRANSPORT BY PARTICLE BIO-DIFFUSION

The solid portion of the porous material primarily consists of unconsolidated sand, silt, and clay particles; either as individual particles or clumps of loosely aggregated grains. Several physical and biological processes occur in the upper surface layers of the soil that mobilize this solid material and are described in Section 13.2. They include the cyclic physical processes of wet–dry and freeze–thaw when liquid water is present. Agricultural plowing or similar periodic surface disturbances may also mix the soil layers. Biomixing by the macrofauna present in the upper layers is acknowledged as likely being the most significant mixing process in near surface soils and the ubiquitous presence of earthworms in many near-surface soil ecosystems is a well-known example of this phenomenon. Section 13.2 is devoted to the theory of chemical transport in soil and sediment due to solid particle mobility and mixing processes, with a primary emphasis on perturbation. Equations 13.2 and 13.3 are the alternative but equivalent forms of the advective–diffusive continuity equation for incorporating particle biodiffusion. Particle biodiffusion is defined analogously to the fluid phase diffusion term but applies to the solid phases. This sorbed-phase parameter is a Fickian-type diffusion coefficient that attempts to capture the random mixing of solid particles; see Equation 13.10. A compilation of numerical values of soil biodiffusion coefficients reflecting measurements with earthworms and other macrofauna appears in Figure 13.2. Section 13.4.2 provides further information for estimating this soil bioturbation transport parameter. The use of an equivalent advective velocity for parameterizing soil turnover by macrofauna has proven to be a less robust approach. The Fickian parameterization approach seems capable of reproducing the observed chemical concentration profile shapes and their evolution with time although there is a lack of theory to support this approach. Section 13.4.3 ends with the presentation of a simple chemical mobility modeling approach that incorporates individual plowing events onto the ongoing soil surface biodiffusion process. Example problem 13.6.1 illustrates the calculation procedure to evaluate chemical transport due to soil bioturbation.

8.13.3   Diffusive Transport in the Vapor Phase

Vapor transport in the unsaturated zone is important in understanding the distribution of chemicals in the subsurface and their exchange between the subsurface environment and the atmosphere. Many studies have shown gaseous diffusion to be the most significant mechanism in the transport of vapors through the subsurface (Currie, 1965; Marrin and Kerfoot, 1988; Fuentes et al., 1991; Shonnard and Bell, 1993; Smith et al., 1996). Although advection of contaminants in the vadose zone periodically occurs due to factors such as vapor extraction systems, water infiltration, barometric pressure changes, and gaseous density differences, diffusional transport is constantly occurring due to localized or regional concentration gradients (Arands et al., 1997).

In addition, many remediation processes require information concerning the diffusion of various gases through the soil. Diffusion may possibly be a rate-limiting step in aerobic biodegradation as these processes require sufficient amounts of oxygen in the soil. Brusseau (1991) noted that diffusion is often the rate-limiting step in contaminant sorption and desorption from soil.

The value of the diffusion coefficient for a particular compound depends on the mass and volume of that compound, the temperature, and the medium through which diffusion is occurring. Empirical relations have been derived to determine the value of diffusion coefficients based on these factors. Diffusion coefficients for nonpolar gases in air at atmospheric pressures and low-to-moderate temperatures may be estimated from the Fuller–Schettler–Giddings (FSG) method (Lyman et al., 1982):

$$D_{air} = \frac{10^{-3} T^{1.75} \sqrt{\dfrac{1}{m_{air}} + \dfrac{1}{m_g}}}{P \left(V_{air}^{1/3} + V_g^{1/3} \right)^2}, \tag{8.40}$$

where V_{air}, V_g, m_{air}, and m_g denote the molar volumes and average molecular masses of air and the diffusing gas, and P is the ambient pressure (atm). Other estimation methods for diffusion coefficients exist, including the Hirschfelder–Bird–Spotz method and the Wilkes–Chang equation (see Treybal, 1980).

In a porous medium, the magnitude of the diffusional flux depends on the mass and volume of the diffusing molecules, the temperature, and the medium through which diffusion takes place. Diffusion is hindered by the tortuous nature of the pores, the diminished cross-sectional area available for movement, and possibly by the pore size (Grathwohl, 1998). The tortuosity of the soil is the impedance to diffusion caused by the porous medium and is defined as the square of the ratio of the effective path length in the pore to the shortest distance through the porous medium:

$$\tau = (l_e / l)^2, \tag{8.41}$$

where l_e is the effective path length and l is the distance through the medium (Grathwohl, 1998). The tortuosity factor is always a number greater than 1, with a more tortuous system having a larger value. The diffusion coefficient through porous media is termed the effective diffusion coefficient (D^*) and is related to the free air diffusion coefficient by the tortuosity:

$$D^* = \tau D_{air}. \tag{8.42}$$

Two factors that can substantially influence the value of diffusion coefficients in porous media are the air-filled porosity and the moisture content of the soil. Since the rate of diffusion through water is nearly 10^4 times slower than the rate of diffusion through air, diffusion through a partially saturated porous media must take place through air-filled pores. The moisture content of the soil can affect the effective diffusion coefficient in a variety of ways, one being the effect of moisture in reducing the air-filled porosity. However, moisture content has less obvious roles in its effect on the diffusion coefficient, including the ability of water to modify the internal pore geometry of a porous medium (Currie, 1961).

Many empirical equations describing the relationship between the effective diffusion coefficient and the air-filled porosity of homogeneous soils have been developed of the form

$$D^* / D_{air} = \tau a, \tag{8.43}$$

where D^* is the gas diffusion coefficient in porous media ($cm^2 \ min^{-1}$); D_{air} is the gas diffusion coefficient in free air ($cm^2 \ min^{-1}$); τ is the tortuosity of the soil, and a is the air-filled porosity.

In studying the diffusion of carbon disulfide and acetone through packed soil cores, Penman (1940) found that for $0.0 < a < 0.7$, τ equals 0.66. Taylor (1949) measured the rate of oxygen diffusion through sand, soil and powdered glass, and found results that coincided with those found by Penman. In soils with a porosity range of 0.10–0.60, Van Bavel (1952) found D^*/D_{air} to be equal to 0.58. Rust et al. (1957) conducted diffusion experiments on mixtures of sand, silica flour, and glass beads. He found τ equal to 0.60 for dry materials and 0.68 for wetted materials. In 1958, Dye and Dallavalle found that for samples of powdered potassium perchlorate in a porosity range of 0.2–0.4, τ ranged from 0.73 to 0.90. A relation was also developed by Currie (1970) that was dependent on both air-filled and total porosity, $D^*/D_{air} = a^4/\varepsilon^{5/2}$, where ε is the total porosity. These relations are summarized in Table 8.9.

Millington (1959) envisioned a porous medium as consisting of solid spheres, which interpenetrate each other, separated by spherical pores that also interpenetrate. Based on his determination of the area available for flow in a given plane of a porous media, and the probability of the continuity of pores within two adjacent planes, Millington (1959) derived the following expression for diffusion in dry isotropic porous media:

$$D^*/D_{air} = \varepsilon^{4/3}, \tag{8.44}$$

where D^* and D_{air} are as previously defined, and ε is porosity.

Millington (1959) derived a separate expression for diffusion if liquid is a component of the porous media system. In this case, he determined that diffusive flow is a function of the number of pore classes drained, assuming spherical pore geometry, such that

$$D^*/D_{air} \propto n^2(\varepsilon_1^{4/3}/m^2), \tag{8.45}$$

where ε, the total porosity, is made up of m equal volume pore classes of which n are drained, and ε_1 is equal to $n\varepsilon/m$, which is equivalent to the air filled porosity.

TABLE 8.9
Literature Relations Predicting Effective Diffusion Coefficients in Porous Media

Relation	Author
$D^*/D_{air} = 0.66a$ for $0.1 < a < 0.7$	Penman (1940)
$D^*/D_{air} = 0.58a$ for $0.1 < a < 0.6$	Van Bavel (1952)
$D^*/D_{air} = 0.60a$ for dry materials	Rust (1957)
$D^*/D_{air} = 0.68a$ for wet materials	
$D^*/D_{air} = 0.73–0.90a$ for $0.2 < a < 0.4$	Dye and Dallavalle (1958)
$D^*/D_{air} = a^4/\varepsilon^{5/2}$	Currie (1970)
$D^*/D_{air} = a^{10/3}/\varepsilon^2$	Millington (1959)

Rearranging and substituting in a for ε_1 yields

$$(m^2/n^2)D^*/D_{\text{air}} = a^{4/3}. \tag{8.46}$$

Multiplying both sides by $1/\varepsilon^2$,

$$(m^2/n^2\varepsilon^2)D^*/D_{\text{air}} = a^{4/3}/\varepsilon^2 = 1/a^2 D^*/D_{\text{air}} = a^{4/3}/\varepsilon^2 \tag{8.47}$$

or

$$D^*/D_{\text{air}} = a^{10/3}/\varepsilon^2. \tag{8.48}$$

An advantage of using a theoretical model to describe diffusion through porous media over empirical models is that they are applicable over a wider range of moisture contents. Theoretical models have been derived (Millington and Shearer, 1971; Troeh et al., 1982; Nielson et al., 1984), but they are generally more mathematically complex, and require data that is not readily available such as pore-size distributions.

Numerous studies have compared experimentally determined effective diffusion coefficients with results from various models. For example, Karimi (1987) measured the vapor-phase diffusion of benzene through a simulated landfill cover at relatively high values of the air-filled porosity. It was determined that diffusion could be accurately predicted using Millington's (1959) equation, although no other models were examined. Sallam et al. (1984) measured freon diffusion through soils at low values of air-filled porosity and compared his results to several models developed in the literature. It was concluded that the expression developed by Millington (1959) was the best fit to the data, and a minor modification was suggested to improve the fit. Millington's equation (1959) is generally considered to be the best predictor of VOC diffusion in heterogeneous soils because it is a theoretical relation applicable over a wide range of moisture contents, and the parameters in the expression are easy to measure (McCarthy and Johnson, 1995).

8.13.4 THE DISPERSION PROCESS IN SATURATED POROUS MEDIA

As presented previously, the steady-state advective–diffusive equation includes a dispersion parameter, D (see Equation 15.6). Chapter 15 provides more information on the subject of dispersion in near-surface geologic media. Parameterized as a mathematical Fickian diffusive process, the dispersion parameter quantifies porewater intermixing and the complex flow pathway lengths and velocities of the fluid moving in the porous medium. It is also referred to as mechanical dispersion and is typically much larger than the aqueous molecular diffusion coefficient (see Section 15.3.1). Values of the coefficients of hydraulic dispersion are linearly related to the average porewater velocity. Sections 15.3.3 and 15.3.4 contain details on the factors that influence the magnitude of these dispersion coefficients. A compiled set of field-scale data for the longitudinal dispersivity appears in Figure 15.2 and these data may be used to estimate dispersivity in surface soils. It is typically appropriate to use horizontal dispersion coefficients when estimating coefficients to describe the downward porewater movement in saturated soils.

8.14   CONTAMINANT MIGRATION BY LEACHING: MASS-TRANSFER BETWEEN MOBILE AND IMMOBILE PHASES

Transfer of contaminants between phases within a porous medium is an important phenomenon to consider when evaluating transport of contaminants in the subsurface. For example, volatile organic compounds (VOCs) are typical contaminants of concern in near-surface soils. Soil, aqueous-phase contamination, and NAPL-phase organics may all be sources of organic vapors in the subsurface. Therefore, organic vapor transport in the unsaturated zone requires understanding of interphase mass-transfer processes as the contaminant can be distributed between soil gas, water, soil, and NAPL phases (Tillman and Weaver, 2005).

Mass transfer from an immobile to a mobile phase can also occur during a precipitation event when water comes in contact with contaminated soil in porous media; during dissolution of NAPL ganglia to a mobile liquid phase; in landfills, when contaminants are transferred from refuse to infiltrating water; and in engineered remediation systems such as soil vapor extraction systems. An important parameter in predicting the mobility of a chemical contaminant in these scenarios is the coefficient describing the mass transfer from the immobile (i.e., solid porous media or immobile NAPL ganglia) to the mobile liquid or gas phase. Mass transfer coefficients are often time-dependent, and influenced by pore water velocity and the length scale of the porous media (Maraqa, 2001).

8.14.1   NAPL-TO-POREWATER TRANSPORT IN POROUS MEDIA

The kinetic chemical mass transfer coefficient for dissolution of immobile packets of nonaqueous phase liquids (NAPLs) in porous media is relevant to the subject of pore water leaching of surface soils. Equation 15.8 defines the mass transfer coefficient for NAPL dissolution, used to describe the transfer of chemicals from the immobile phase due to downward percolating porewaters. Much quantitative information on the subject of NAPL leaching in groundwater has been produced in the last two decades and is the subject of Part 2 of Chapter 15 titled "Mass Transfer Coefficients in Porewater Adjacent to Nonaqueous Liquids and Particles"; the following is a review of the contents of that section pertaining to NAPL dissolution in ground water.

NAPLs in porous media may exist as continuous mass or as discrete ganglia. Section 15.7 includes a description of the distribution and morphology of NAPLs in porous media. The entrapped liquid, both within the unsaturated and saturated zones, produces a fluid–fluid interface through which mass transfer occurs. During its migration, the NAPL behaves as a nonwetting fluid in the presence of water and displaces the nonwetting phase. Once its migration ceases, the NAPL remains suspended as fingers (i.e., ganglia) and/or pools accumulated on lower permeable layers, and may exist at various levels of saturation. Conceptual and theoretical models of chemical flux and mass transfer between NAPL and aqueous phases are presented in Section 15.8. The mass rate equation for flux of a solute from the NAPL phase to the aqueous phase is defined by Equation 15.27, including a lumped mass transfer coefficient, K_c (t^{-1}), defined by Equation 15.28. K_c appears as K (t^{-1}) in the Sherwood number (see Equation 15.31), which is slightly modified to account for the presence of porous media. Section 15.9 provides estimation methods for the mass transfer coefficients between the adjoining fluids. Six Gilland–Sherwood-type correlations from column experiments appear in Table 15.3 for this use. Another correlation representing dissolution in a two-dimensional cell appears as Equation 15.32.

For applications involving chemical leaching from soils to pore water these correlations for the mass transfer coefficients are appropriate to use. In principle, since mass transfer coefficients, such as K_c, are scales, they can be used for chemical adsorption from porewater to the soil solids as well. It is suggested that the length of the soil column being leached, L (m), be substituted for the dissolution length in the correlations.

8.14.2 Solute-to-Solid Phase Transport in Porous Media

Numerous laboratory experiments and a few field studies have been performed on chemical adsorption and absorption mass transfer from porewater moving through soil-like porous materials. Maraqa (2001) compiled several previous studies related to transport of solute through packed columns and developed predictive relationships for the mass-transfer rate coefficients using regression analysis. The review covers 19 previously reported laboratory studies and five field studies. Eight laboratory studies were performed with saturated and unsaturated soils using the nonsorbing tracers Cl, ^3H_2O, ^{36}Cl, and Br with retardation factors ranging from 0.73 to 1.5. The 11 others were performed with sorbing solutes and saturated soils. A total of 20 organic chemicals were used, including hydrocarbons and chlorinated hydrocarbon pesticides with retardation factors ranging from 1.1 to 19.0. Soils investigated included stony soil, sand, sandy loam, and silt loam.

The results reported below are for fine, nonaggregated media only. The studies demonstrated that the mass transfer coefficients were dependent on porewater velocity v_o (m h^{-1}), the length scale, L (m), the retardation coefficient, R ($1 + \rho_b K_D/\theta$) and particle or aggregate size. Relationships were developed through regression analysis that related the mass transfer coefficients to residence time. For the nonadsorbing tracers, the resulting correlation was

$$k = 0.95(LR/v_o)^{-0.8} \qquad (8.49)$$

with k (h^{-1}); the R^2 for the correlation was 0.91. As the residence time, LR/v_o, increased the mass transfer coefficient decreased. For adsorbing nonionic organic chemicals the resulting correlations was

$$k = 1.72(LR/v_o)^{-0.93} \qquad (8.50)$$

with $R^2 = 0.84$. The Maraqa (2001) compilation contains similar correlations for a tracer in aggregated and stony media plus successful simulations demonstrating the potential field applicability of the relationships.

8.14.3 VAPOR-TO-SOLID PHASE TRANSPORT IN POROUS MEDIA

Transfer of contaminants between the vapor and the solid phases is often overlooked in analyses of contaminant movement in the environment. It is commonly assumed that the majority of mass transfer occurs at the air–water or the water–solid interface; however, certain conditions exist in which vapor to solid phase transport of contaminants may be significant, such as near the upper surface of the soil or in arid environments (Schwarzenbach et al., 2002). Vapor to solid phase mass transfer may also dominate in soil vapor extraction systems in the vicinity of air injection wells (Yoon et al., 2003). A number of studies have shown that VOC sorption to dry soils is higher compared with sorption to saturated or partially saturated soils due to competition of the water molecules for adsorption sites on the mineral surface (Smith et al., 1990; Pennell et al., 1992). Although vapor to solid phase mass transfer may be significant in some environments, in a typical partially saturated porous medium, contaminant transfer will occur primarily by partitioning into soil organic matter or dissolution into water films adsorbed to soil mineral surfaces, and thus, coefficients representing chemical transfer between the aqueous and solid phases are appropriate (see Section 8.15.1). Use of liquid–solid mass transfer coefficients for organic chemicals in numerical simulations of contaminant fate in a completely dry porous medium may underpredict contaminant sorption from the vapor phase to soil mineral surfaces.

8.15 CASE STUDY: SOIL GAS MOVEMENT AND CONTAMINANT FLUX UNDER NATURAL CONDITIONS

In a study of natural flux of trichloroethylene (TCE) from the subsurface reported by Tillman et al. (2003), surface flux measurements made by a flux chamber were compared with flux simulations using a transient, one-dimensional gas-flow and transport model incorporating the effects of gas phase diffusion,
equilibrium air–water partitioning of organic vapors, and unsaturated zone airflow caused by atmospheric pressure changes (i.e., barometric pumping). Field data required for flow and transport model calibration and simulation included surface and subsurface air pressure, subsurface moisture content, water table elevation, and subsurface TCE concentrations. These parameters were measured as functions of depth and time at Picatinny Arsenal, NJ during four multiday time periods between July and December 2001. Significant air-pressure gradients between the subsurface and the atmosphere were observed, while little variation in air pressure was measured at depths between 0.5 and 1.7 m. Changing subsurface air pressure in response to varying atmospheric pressure was successfully simulated using a simplified, one-dimensional, finite-difference model. Further model results indicated nonzero subsurface air velocity during most of the simulation period. These results supported the conclusion that airflow was occurring in the unsaturated zone at the site, and that this airflow could be explained by vertical pressure gradients driven by atmospheric pressure variations.

Prior to construction of a new slab-on-grade structure at a USEPA Office of Research and Development facility in Athens, GA, the subsurface beneath the building was outfitted with instruments for measuring real-time soil moisture content and gas pressure at several depths (Tillman and Weaver, 2006, 2007). Data from a weather station installed within 20 m of the building, along with subsurface and surface moisture content, and soil gas pressure data, have been collected at 5-min intervals at the site since 2005. Results indicate that subsurface soils located underneath the edge of the building slab are responsive to rainfall with near-saturation conditions seen during intense rainfall, while center-slab soils show no response; see Figure 8.10. These observations may affect soil vapor intrusion in two ways: floor-wall cracks may have reduced potential as a point of ingress for vapor intrusion during rainy periods due to near-saturation soil conditions, and dry center-slab soils may permit greater vapor intrusion than one-dimensional transport would predict. Some heavy rainfall events also produced positive soil-gas pressure (relative to surface pressure) that were uniform with depth and space, were as high at 60 Pa, and lasted from 30 min to 2 h. Soil-gas movement during these positive pressure events was simulated using a one-dimension gas flow equation and particle tracking code. Results indicate that as much as 17.5 cm of soil gas could be evacuated during one of these brief subsurface pressure spikes. While this amount of advective vapor intrusion into a structure may not significantly contribute to an increase in chronic risk at a VOC site, it may help to explain the reporting of gasoline odors by building occupants at leaking underground storage tanks (LUST) sites during rainfall events.

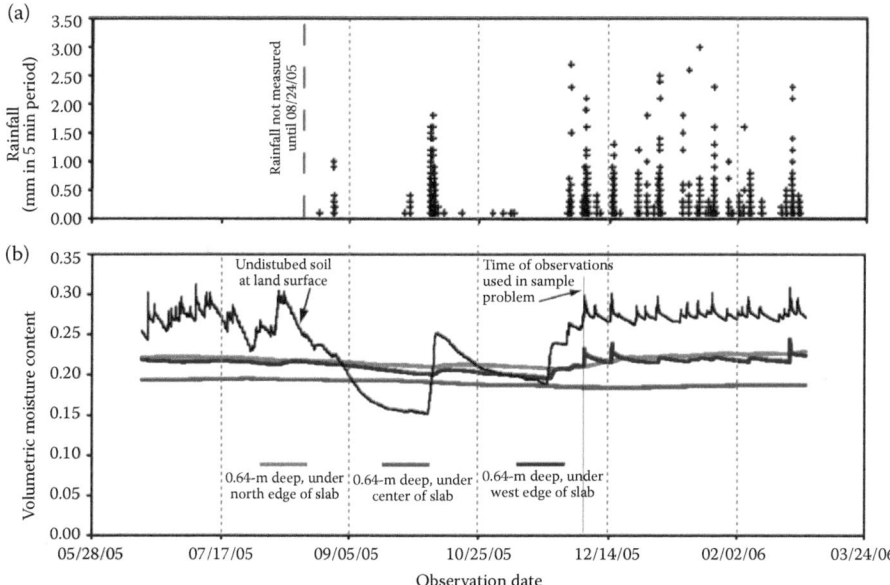

FIGURE 8.10 (a) Time series of subslab and surface moisture content measured using time-domain reflectometry and (b) 5 min rainfall totals at USEPA vapor intrusion research building in Athens, Georgia.

TABLE 8.10
Effect of Moisture Content and Predictive Relationship on Effective Diffusion Coefficients for Trichloroethylene at 22°C

Location	Total Porosity	Volumetric Moisture Content	Air-Filled Porosity	Calculated Effective Diffusion Coefficients ($cm^2\ s^{-1}$)				
				Penman	Van Bavel	Rust	Currie	Millington
Land surface	0.311	0.299	0.012	NA[a]	NA	5.36E−4	2.86E−8	3.04−7
Beneath west edge of slab	0.311	0.211	0.1	4.91E−3	4.32E−3	4.46E−3	1.38E−4	3.57E−4
Beneath center of slab	0.311	0.184	0.127	6.24E−3	5.48E−3	5.67E−3	3.59E−4	7.92E−4
Beneath north edge of slab	0.311	0.21	0.101	4.96E−3	4.36E−3	4.51E−3	1.44E−4	3.69E−4

[a] Not applicable, $a < 0.1$.

8.15.1 EFFECT OF MOISTURE CONTENT AND PREDICTIVE RELATIONSHIP ON ESTIMATES OF EFFECTIVE DIFFUSION COEFFICIENT

In this case study, we can use data generated at the test site to evaluate the influence of variability in soil moisture content and the predictive relationship used on the estimates of effective diffusion coefficients for trichloroethylene. The effective diffusion coefficient is necessary to predict TCE fluxes that could enter a structure due to vapor intrusion. Figure 8.10 presents data showing the observed rainfall measured at the site as well as the moisture contents observed at three locations at a consistent depth (~0.6 m) beneath the concrete slab under the EPA facility in Athens, GA.

To demonstrate the influence of moisture content and the relationship used to predict effective diffusion coefficients, we calculated the effective diffusion coefficient that would be predicted at each of four locations at this test site, using the predictive relationships presented in Table 8.9. In these calculations, the total porosity and the volumetric moisture content were used to calculate the air-filled porosity, and the diffusion coefficient for TCE in air at 22°C was calculated as $0.0744\ cm^2\ s^{-1}$ (Bartelt-Hunt and Smith, 2002). Table 8.10 presents the results of this analysis. For a given location, the predicted diffusion coefficients varied by as much as four orders of magnitude, depending on the relationship used. Results presented in Table 8.10 also demonstrates that using moisture content measured exterior to the building for estimating diffusive transport beneath a building slab may greatly underpredict the actual diffusive flux, and therefore, underestimate the associated risk of VOC vapor transport beneath the structure.

LITERATURE CITED

Arands, R., T. Lam, I. Massry, D. H. Berler, F. J. Muzzio, and D. S. Kosson. 1997. Modeling and experimental validation of volatile organic contaminant diffusion through an unsaturated soil. *Water Resources Research* 33: 599–609.

Auer, L. H., N. D. Rosenberg, K. H. Birdsell, and E. M. Whiteny. 1996. The effects of barometric pumping on contaminant transport. *Journal of Contaminant Hydrology* 24: 145–166.

Barlaz, M. A. and R. K. Ham. 1993. Leachate and gas generation, in *Geotechnical Practice for Waste Disposal*, D. E. Daniel, ed. London: Chapman & Hall, pp. 113–136.

Bartelt-Hunt, S. L., D. R. U. Knappe, P. Kjeldsen, and M. A. Barlaz. 2006. Fate of chemical warfare agents and toxic industrial chemicals in landfills. *Environmental Science and Technology* 40: 4219–4225.

Bartelt-Hunt, S. L. and J. A. Smith. 2002. Measurement of effective air diffusion coefficients for trichloroethene in undisturbed soil cores. *Journal of Contaminant Hydrology* 56: 193–208.

Bear, J. 1972. *Dynamics of Fluids in Porous Media*. New York: Dover Publications, 764pp.

Bedient, P. B., H. S. Rifai, and C. J. Newell. 1997. *Ground Water Contamination*. Upper Saddle River, NJ: Prentice-Hall, 604 pp.

Brusseau, M. L. 1991. Transport of organic chemicals by gas advection in structured or heterogeneous porous media: Development of a model and application to column experiments. *Water Resources Research* 27: 3189–3199.

Choi, J. W., F. D. Tillman, and J. A. Smith. 2002. Relative importance of gas-phase diffusive and advective trichloroethylene (TCE) fluxes in the unsaturated zone under natural conditions. *Environmental Science and Technology* 36: 3157–3164.

Chow, V. T., D. R. Maidment, and L. W. Mays. 1988. *Applied Hydrology*. New York: McGraw-Hill, 572pp.

Currie, J. A. 1961. Gaseous diffusion in porous media. Part 3: Wet granular materials. *British Journal of Applied Physics* 12: 275–281.

Currie, J. A. 1965. Diffusion within soil microstructure; a structural parameter for soils. *Journal of Soil Science* 16: 279–298.

Currie, J. A. 1970. Movement of gases in soil respiration. *SCI Monographs* 37: 152–171.

Dye, R. F. and J. M. Dallavalle. 1958. Diffusion of gases in porous media. *Industrial and Engineering Chemistry* 50: 1195–1200.

Eleazer, W. E., W. S. Odle, Y.-S. Wang, and M. A. Barlaz. 1997. Biodegradability of municipal solid waste components in laboratory-scale landfills. *Environmental Science and Technology* 31: 911–917.

Falta, R. W., I. Javandel, K. Pruess, and P. A. Witherspoon. 1989. Density-driven flow of gas in the unsaturated zone due to the evaporation of volatile organic compounds. *Water Resources Research* 25: 2159–2169.

Freeze, R. A. and J. A. Cherry. 1979. *Groundwater*. Upper Saddle River, NJ: Prentice-Hall, 604pp.

Fuentes, H. R., W. L. Polzer, and J. L. Smith. 1991. Laboratory measurements of diffusion coefficients for trichloroethylene and orthoxylene in undisturbed tuff. *Journal of Environmental Quality* 20: 215–221.

Garbesi, K. and R. G. Sextro. 1989. Modeling and field evidence of pressure-driven entry of soil gas into a house through permeable below-grade walls. *Environmental Science and Technology* 23: 1481–1487.

Grathwohl, P. 1998. *Diffusion in Natural Porous Media: Contaminant Transport, Sorption/Desorption and Dissolution Kinetics*. Norwell, MA: Kluwer Academic.

Gustin, M. S., G. E. Taylor, and R. A. Maxey. 1997. Effect of temperature and air movement on the flux of elemental mercury from substrate to the atmosphere. *Journal of Geophysical Research* 102: 3891–3898.

Hers, I., R. Zapf-Gilje, L. Li, and J. Atwater. 2001. The use of indoor air measurements to evaluate intrusion of subsurface VOC vapors into buildings. *Journal of the Air and Waste Management Association* 51: 1318–1331.

Holford, D. J., S. D. Schery, J. L. Wilson, and F. M. Phillips. 1993. Modeling radon transport in dry, cracked soil. *Journal of Geophysical Research* 98: 567–580.

Huling, S. G. and J. W. Weaver. 1991. Dense nonaqueous phase liquids, Ground Water Issue, EPA/540/4-91/002, R.S. Kerr Environmental Research Laboratory, USEPA, Ada, OK.

Huntley, D. H., R. N. Hawk, and H. P. Corley. 1994. Nonaqueous phase hydrocarbon in a fine-grained sandstone, 1. Comparison between measured and predicted saturations and mobility, *Ground Water*, 32(4): 626–634.

Karimi, A. A., W. J. Farmer, and M. M. Cliath. 1987. Vapor-phase diffusion of benzene in soil. *Journal of Environmental Quality* 16: 38–43.

Krylov, V. V. and C. C. Ferguson. 1998. Contamination of indoor air by toxic soil vapours: The effects of subfloor ventilation and other protective measures. *Building and Environment* 33: 331–347.

Li, L., P. Dong, and D. A. Barry. 2002. Tide-induced water table fluctuations in coastal aquifers bounded by rhythmic shorelines. *Journal of Hydraulic Engineering* 128: 925–933.

Little, J. C., J. M. Daisey, and W. W. Nazaroff. 1992. Transport of subsurface contaminants into buildings: An exposure pathway for volatile organics. *Environmental Science and Technology* 26: 2058–2066.

Lowry, M., Bartelt-Hunt, S. L., Beaulieu, S. M., and M. A. Barlaz. 2008. Development of a coupled reactor model for prediction of organic contaminant fate in landfills. *Environmental Science and Technology* 42: 7444–7451.

Lyman, W. J., W. F. Reehl, and D. H. Rosenblatt, Eds. (1982). *Handbook of Chemical Property Estimation Methods. Environmental Behavior of Organic Compounds.* New York: McGraw-Hill.

Maraqa, M. 2001. Prediction of mass-transfer coefficients for solute transport in porous media. *Journal of Contaminant Hydrology* 53: 153–171.

Marrin, D. L. and H. B. Kerfoot. 1988. Soil–gas surveying techniques. *Environmental Science and Technology* 22: 740–745.

Massmann, J. and D. F. Farrier. 1992. Effects of atmospheric pressures on gas transport in the vadose zone. *Water Resources Research* 28: 777–791.

McCarthy, K. A. and R. L. Johnson. 1995. Measurement of trichloroethylene diffusion as a function of moisture content in sections of gravity-drained soil columns. *Journal of Environmental Quality* 24: 49–55.

Mendoza, C. A. and E. O. Frind. 1990a. Advective–dispersive transport of dense organic vapors in the unsaturated zone 1. Model development. *Water Resources Research* 26: 379–387.

Mendoza, C. A. and E. O. Frind. 1990b. Advective–dispersive transport of dense organic vapors in the unsaturated zone: 2. sensitivity analysis. *Water Resources Research* 26: 387–398.

Mercer, J. W. and R. M. Cohen. 1990. A review of immiscible fluids in the subsurface: Properties, models, characterization, and remediation. *Journal of Contaminant Hydrology* 6: 107–163.

Millington, R. J. 1959. Gas diffusion in porous media. *Science* 130: 100–102.

Millington, R. J. and R. C. Shearer. 1971. Diffusion in aggregated porous media. *Soil Science* 111: 372–378.

Nazaroff, W. W., S. R. Lewis, S. M. Doyle, B. A. Moed, and A. V. Nero. 1987. Experiments on pollutant transport from soil into residential basements by pressure-driven airflow. *Environmental Science and Technology* 21: 459–466.

Newell, C. J., S. D. Acree, R. R. Ross, and S. G. Huling. 1995. Light nonaqueous phase liquids, Ground Water Issue, EPA 540/S-95/500, R.S. Kerr Environmental Research Laboratory, USEPA, Ada, OK.

Nielson, K. K., V. C. Rogers, and G .W. Gee. 1984. Diffusion of radon through soils: A pore distribution model. *Soil Science Society of America Journal* 48: 482–487.

Olson, D. A. and R. L. Corsi. 2002. Fate and transport of contaminants in indoor air. *Journal of Soil and Sediment Contamination* 11: 583–601.

Pankow, J. F. and J. A. Cherry. 1996. *Dense Chlorinated Solvents and Other DNAPLs in Groundwater*, Ontario: Waterloo Press, 522pp.

Penman, H. L. 1940. Gas and vapor movements in the soil I. The diffusion of vapours through porous solids. *Journal of Agricultural Science* 30: 436–462.

Pennell, K. D., R. D. Rhue, S. C. Rao, and C. T. Johnston. 1992. Vapor-phase sorption of *p*-xylene and water on soils and clay minerals. *Environmental Science and Technology* 26: 756–763.

Qian, X., Koerner, R. M., and D. H. Gray. 2002. *Geotechnical Aspects of Landfill Design and Construction.* Upper Saddle River, NJ: Prentice-Hall, 717pp.

Rossabi, J. and R. W. Falta. 2002. Analytical solution for subsurface gas flow to a well induced by surface pressure fluctuations. *Ground Water* 40: 67–75.

Rust, R. H., A. Klute, and G. E. Geiseking. 1957. Diffusion-porosity measurements using a non-steady state system. *Soil Science* 84: 453–465.

Sallam, A., W. A. Jury, and J. Letey. 1984. Measurement of gas diffusion under relatively low air-filled porosity. *Soil Science Society of America Journal* 48: 3–6.

Schery, S. D., Holford, D. J., Wilson, J. L., and Phillips, F. M. 1988. The flow and diffusion of radon isotopes in fractured porous media: Part 2, Semi-infinite media. *Radiation Protection Dosimetry* 24: 191–197.

Schwarzenbach, R. P., P. M. Gschwend, and D. M. Imboden. 2002. *Environmental Organic Chemistry*, Hoboken, NJ: Wiley, 1309pp.

Sharma, H. D. and K. R. Reddy. 2004. *Geoenvironmental Engineering.* Hoboken, NJ: Wiley, 108pp.

Shonnard, D. L. and R. L. Bell. 1993. Benzene emissions from a contaminated air-dry soil with fluctuations of soil temperature or relative humidity. *Environmental Science and Technology* 27: 2909–2913.

Smith, J. A., C. T. Chiou, J. A. Kammer, and D. E. Kile. 1990. Effect of soil moisture on the sorption of trichloroethene vapor to vadose-zone soil at Picatinny Arsenal, New Jersey. *Environmental Science and Technology* 24: 676–683.

Smith, J. A., A. K. Tisdale, and H. J. Cho. 1996. Quantification of natural vapor fluxes of trichloroethene in the unsaturated zone at Picatinny Arsenal, New Jersey. *Environmental Science and Technology* 30: 2243–2250.

Taylor, S. A. 1949. Oxygen diffusion in porous media as a measure of soil aeration. *Soil Science Society of America Proceedings* 14: 55–61.

Thibodeaux, L. J. 1996. *Environmental Chemodynamics.* New York: Wiley, 593pp.

Tillman, F. D. Choi, J. W., and J. A. Smith. 2003. A comparison of direct measurement and model simulation of total flux of volatile organic compounds from the subsurface to the atmosphere under natural field conditions. *Water Resources Research* 39: 1284–1294.

Tillman, F. D. and J. A. Smith. 2005. Site characteristics controlling airflow in the shallow unsaturated zone in response to atmospheric pressure changes. *Environmental Engineering Science* 22: 25–37.

Tillman, F. D. and J. W. Weaver. 2005. Review of Recent Research on Vapor Intrusion. U.S. Environmental Protection Agency, Athens, GA. Publication No. EPA/600/R-05/106, 47pp.

Tillman, F. D. and J. W. Weaver. 2006. Subsurface soil conditions beneath and near buildings and the potential effects on soil vapor intrusion: USEPA Office of Underground Storage Tanks 18th Annual National Tanks Conference, Memphis, Tennessee, March 19–22, 2006.

Tillman, F. D. and J. W. Weaver. 2007. Temporal moisture content variability beneath and external to a building and the potential effects on vapor intrusion risk assessment. *Science of the Total Environment* 379: 1–15.

Treybal, R. E. 1980. *Mass Transfer Operations*. New York: McGraw-Hill.

Troeh, F. R., J. D. Jabro, and D. Kirkham. 1982. Gaseous diffusion equations for porous materials. *Geoderma* 27: 239–253.

Van Bavel, C. H. M. 1952. Gaseous diffusion and porosity in porous media. *Soil Science* 73: 91–105.

Weeks, E. P. 1978. Field determination of vertical permeability to air in the unsaturated Zone. *U.S. Geological Survey Professional Paper*, 1051, pp. 1–41.

Wilson, J. L. and S. H. Conrad. 1984. Is physical displacement of residual hydrocarbons a realistic possibility in aquifer restorations? In *Proceedings of the National Water Well Association Conference on Petroleum Hydrocarbons and Organic Chemicals in Groundwater*, Houston, Texas, Nov. 5–7, 1984, pp. 274–298.

Yoon, H., A. J. Valocchi, and C. J. Werth. 2003. Modeling the influence of water content on soil vapor extraction. *Vadose Zone Journal* 2: 368–381.

ADDITIONAL READING

Chiou, C. T. and T. D. Shoup. 1985. Soil sorption of organic vapors and effects of humidity on sorptive mechanism and capacity. *Environmental Science and Technology* 19, 1196–1200.

Cho, H. J. and P. R. Jaffe. 1990. The volatilization of organic compounds in unsaturated porous media during infiltration. *Journal of Contaminant Hydrology* 6, 387–410.

Cho, H. J. P. R. Jaffé, and J. A. Smith. 1993. Simulating the volatilization of solvents in unsaturated soils during laboratory and field infiltration experiments. *Water Resources Research* 29, 3329–3342.

Cussler, E. L. 1997. *Diffusion: Mass Transfer in Fluid Systems*. New York: Cambridge University Press, 580pp.

Elberling, B., F. Larsen, S. Christensen, and D. Postma. 1998. Gas transport in a confined unsaturated zone during atmospheric pressure cycles. *Water Resources Research* 34, 2855–2862.

Fischer, M. L., A. J. Bentley, K. A. Dunkin, A. T. Hodgson, W. W. Nazaroff, R. G. Sextro, and J. M. Daisey. 1996. Factors affecting indoor air concentrations of volatile organic compounds at a site of subsurface gasoline contamination. *Environmental Science and Technology* 30, 2948–2957.

Fitzpatrick, N. and J. J. Fitzgerald. 2002. An evaluation of vapor intrusion into buildings through a study of field data. *Soil and Sediment Contamination* 11, 603–623.

Hodgson, A. T., K. Garbesi, R. G. Sextro, and J. M. Daisey. 1992. Soil-gas contamination and entry of volatile organic compounds into a house near a landfill. *Journal of the Air and Waste Management Association* 42, 277–283.

Johnson, P. C. and R. A. Ettinger. 1991. Heuristic model for predicting the intrusion rate of contaminant vapors into buildings. *Environmental Science and Technology* 25, 1445–1452.

Jury, W. A., D. Russo, G. Streile, and H. E. Abd. 1990. Evaluation of volatilization by organic chemicals residing below the soil surface. *Water Resources Research* 26, 13–20.

Kim, H., M. D. Annable, and P. S. C. Rao. 1998. Influence of air-water interfacial adsorption and gas-phase partitioning on the transport of organic chemicals in unsaturated porous media. *Environmental Science and Technology* 32, 1253–1259.

Massmann, J. and D. F. Farrier. 1992. Effects of atmospheric pressures on gas transport in the vadose zone. *Water Resources Research* 28, 777–791.

Mercer, J. W. and R. M. Cohen. 1990. A review of immiscible fluids in the subsurface: Properties, models, characterization, and remediation. *Journal of Contaminant Hydrology* 6, 107–163.

Sleep, B. E. and J. F. Sykes. 1989. Modeling the transport of volatile organics in variably saturated media. *Water Resources Research* 25, 81–92.

Thorstenson, D. C. and D. W. Pollock. 1989. Gas transport in unsaturated zones: Multicomponent systems and the adequacy of Fick's Laws. *Water Resources Research* 25, 477–507.

9 Air–Water Mass Transfer Coefficients

John S. Gulliver

CONTENTS

9.1   Introduction 213
    9.1.1   Background 213
    9.1.2   Mass Flux across the Air–Water Interface 214
    9.1.3   Boundary Layer Analogies 217
9.2   Air–Water Mass Transfer Coefficients 218
9.3   Measurement and Prediction of Water-Side Mass Transfer Coefficients 220
    9.3.1   Background 220
    9.3.2   Water-Side Coefficient in Rivers 221
        9.3.2.1   Measurements 221
        9.3.2.2   Prediction 222
    9.3.3   Water-Side Transfer Coefficient in Lakes and Oceans 226
        9.3.3.1   Opportunistic Measurement of Wind Influence 226
        9.3.3.2   Measurement of Wind Influence with Deliberate Tracers ..228
        9.3.3.3   Prediction 233
9.4   Measurement and Prediction of Air-Side Mass Transfer Coefficients 234
    9.4.1   Background 234
    9.4.2   Measurement 238
    9.4.3   Prediction 239
9.5   Summary Guide 239
    9.5.1   User Roadmap 239
    9.5.2   Application Examples 243
Literature Cited 248
Bibliography with Additional References 251

9.1   INTRODUCTION

9.1.1   BACKGROUND

This section is a general introduction to the chemodynamics at the air–water interface, within the context of the natural environment. The air–water interface is important to environmental mass balances, since a large portion of the Earth's surface is covered by water. This includes water bodies of the entire spectrum, from ponds, small streams, rivers, lakes, constructed water treatment systems, lagoons, harbors, wetlands estuaries, inland seas as well as the ocean. Depending upon the locale, the conditions that exist may include chemical species that are moving from the atmosphere to the water or vice versa. Water bodies near urban or industrial centers receive water-borne chemical discharges that cause water concentrations to be greater that their equilibrium with the air. This results in a net volatilization from the water surface. Alternatively, an air mass originating on land is transported to sea by convection, carrying vapor-phase pollutants that result in a net absorption into the sea. The objective of this chapter is to provide the user with background and a methodology for estimating air-side and water-side mass transfer coefficients at air–water interfaces.

This chapter contains the following sections:

1. Introduction.
2. General theory of air–water mass transfer. The reader is also referred to Chapters 2 and 4, which contains fundamental concepts that apply to all interfacial transfer.
3. The approaches and technology needed to make measurements in the field and laboratory.
4. Specific recommended correlations for the various water bodies and atmospheric conditions present in the environment.
5. A list of citations and other texts where additional material on environmental mass transfer coefficients can be found.

This chapter does *not* include the dry and wet deposition of particle/aerosol-bound contaminants. These processes are covered in Chapter 6.

9.1.2 Mass Flux across the Air–Water Interface

Air–water transfer rate of chemicals is dependent upon the rate coefficient and the equilibrium that the concentrations in each phase are moving towards. In environmental air–water mass transfer, the flow is generally turbulent in both phases. However, there is no turbulence across the interface in the diffusive sublayer, and the problem becomes one of rate of diffusion. Temporal mean turbulence quantities, such as eddy diffusion coefficient, provide a semiquantitative description of the flux across the air–water interface, however the unsteady character of turbulence near the diffusive sublayer is crucial to understanding and characterizing interfacial transport processes.

Consider air and water, illustrated in Figure 9.1, where the water contains a given amount of the phenol, C_6H_5OH.

The mass flux is given as

$$J_p = \frac{dm_p}{dt} = V_a \frac{dC_{pa}}{dt} = -V_w \frac{dC_{pw}}{dt}, \qquad (9.1)$$

where m is the mass, V is the volume, J is a temporal mean flux rate from the water to the air, or the mass transfer rate/unit surface area, the subscript p indicates phenol,

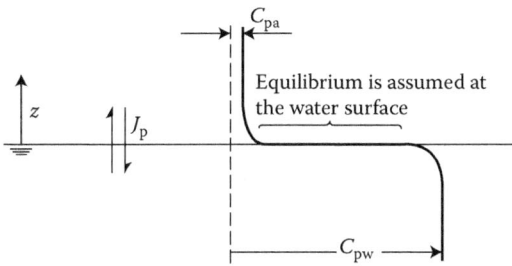

FIGURE 9.1 Illustration of phenol flux rate across a small portion of the water surface.

and subscripts a and w indicate air and water, respectively. Flux of chemical occurs because there is a concentration gradient at the interface, as illustrated in Figure 9.1.

The flux is typically approximated as constant with z, unless the concentrations are very close to equilibrium or there are sources or sinks of phenol in the air or water. At the interface, equilibrium is assumed between the air and water phases. This equilibrium may only exist for a thickness of a few molecules, but is assumed to occur quickly compared to the time scale of interest. The concentrations at this air–water interface are related by the following equation:

$$\frac{C_{pa}^*}{C_{pw}^*} = H_p \text{(dimensionless)} \tag{9.2a}$$

or

$$\frac{P_{pa}^*}{C_{pw}^*} = H_p \left(\frac{\text{atm-m}^3}{\text{mole}}\right), \tag{9.2b}$$

where H_p is the phenol equilibrium partitioning coefficient for air–water equilibrium, also called Henry's law constant, which is either set dimensionless or corresponding to the units of the two concentrations, P_{pa} is the partial pressure of phenol in the air and superscript $*$ refers to the value at equilibrium with the other phase. Now, because J_p is constant with distance from the interface,

$$J_p = \text{Const.} = -(D_{pa} + \varepsilon_{za})\frac{\partial C_{pa}}{\partial z} = -(D_{pw} + \varepsilon_{zw})\frac{\partial C_{pw}}{\partial z}, \tag{9.3}$$

where D is (molecular) diffusivity and ε is turbulent or eddy diffusivity.

Now, examine Equation 9.3 with the knowledge that J_p is constant. Away from the interface, both $\partial C_{pa}/\partial z$ and $\partial C_{pw}/\partial z$ are small. This means that, for J_p to be constant, $D_{pa} + \varepsilon_z$ and $D_{pw} + \varepsilon_z$ must be large. The inverse is also true. When $\partial C_{pa}/\partial z$ and $\partial C_{pw}/\partial z$ are large, $D_{pa} + \varepsilon_z$ and $D_{pw} + \varepsilon_z$ must be small. The only thing that can vary with distance that dramatically in the $D + \varepsilon_z$ terms is ε_z. Thus, the ε_z versus z relationship is seen in the concentration profile.

Consider an adjustment to within a few microns of the interface, and examine the flux. Flux is molecular at the interface between two phases, as illustrated in Figure 9.2.

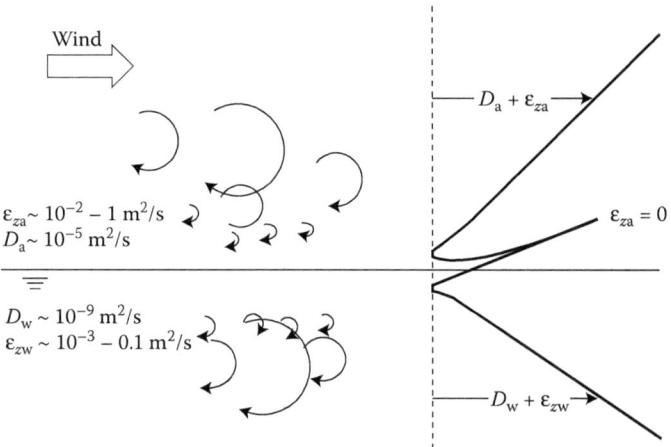

FIGURE 9.2 Illustration of diffusive flux near the air–water interface.

The elimination of turbulence by viscosity means that ε_z is zero and the only active process is diffusion. Note the difference in the order of magnitudes of D and ε given in Figure 9.2. The resistance to mass transfer is much larger near the interface than away from the interface, so the controlling process would tend to be transfer in the diffusive sublayer.

In the diffusive sublayer, flux is due purely to diffusion. Then, the fact that mean flux is constant with distance can be utilized:

$$J_p = -D_{pa}\frac{\partial C_{pa}}{\partial z} = -D_{pw}\frac{\partial C_{pw}}{\partial z}. \tag{9.4}$$

Since J_p is constant and D_{pa} and D_{pw} are also constant, assuming the thermodynamic parameters such as temperature and pressure do not change, the solution to Equation 9.4, applying the boundary conditions at the interface, is

$$C_{pa} = C_{pa}^* + \beta_1 z \tag{9.5}$$

and

$$C_{pw} = C_{pw}^* + \beta_3 z, \tag{9.6}$$

where $z = 0$ at the interface. Both air and water phase concentrations are linear near the interface.

The concentration of phenol in the air and water can be put on the same scale using the Henry's law constant (equilibrium partitioning as defined in Chapters 1 and 5) by dividing the air concentration by H_p, as illustrated in Figure 9.3.

Then, Equation 9.4 may be written as

$$J_p = -D_{pa}H_p\frac{\partial(C_{pa}/H_p)}{\partial z} = -D_{pw}\frac{\partial C_{pw}}{\partial z} \tag{9.7}$$

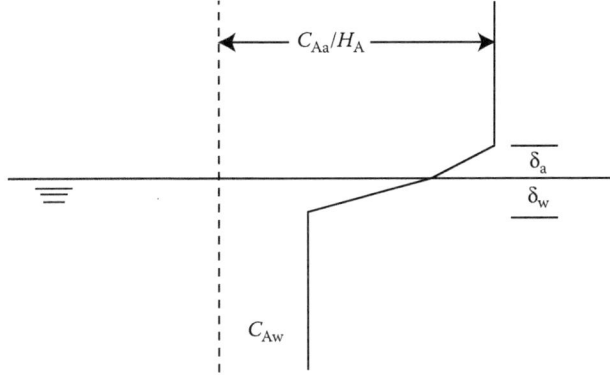

FIGURE 9.3 The result of putting concentrations on the same scale using Henry's law constant.

which does not have a discontinuity in concentration (C_{pa}/H_p or C_{pw}) at the interface. From Equation 9.7, it is seen that compounds with a large value of H will have a small concentration gradient, $\partial(C_{pa}/H_p)/\partial z$, and the equivalent diffusion coefficient, $D_{pa}H_p$, will be large, compared to D_{pw}. These compounds include O_2, N_2, CH_4, carbon tetrachloride, tetrachloroethylene, lower molecular weight PCBs, some aromatic hydrocarbons, C_4H_6, and so on, and are classified as volatile compounds. Compounds with a small value of H' will have a large value of concentration gradient, $\partial(C/H)/\partial z$, and the equivalent diffusion coefficient, D_aH will be small, relative to D_w. These compounds include compounds that have an attraction to dissolution in water, such as acrylic acid, benzidiene, benzoic acid, and water itself. They are classified as nonvolatile compounds. Finally, there are those compounds where D_aH is of the same order of magnitude as D_w, such as phenanthrene, napthalene, heptanone, and the pesticide Aldrin. These are called semivolatile compounds.

9.1.3 BOUNDARY LAYER ANALOGIES

The transport of mass, heat, and momentum are described by analogous transport equations, with the exception of the source and sink terms. There is another significant difference related to the magnitude of the "diffusion coefficient" for mass, heat, and momentum that has an impact on transport in the air and water boundary layers.

The diffusion of momentum is linearly proportional to the kinematic viscosity of the fluid:

$$\frac{\text{Momentum flux rate}}{\text{Unit volume}} = -\nu\frac{\partial(\rho V)}{\partial z}. \tag{9.8}$$

Similarly, in the heat transport equation,

$$\frac{\text{Heat flux rate}}{\text{Unit volume}} = \frac{-k_T}{\rho\,C_p}\frac{\partial T}{\partial z} = \alpha\frac{\partial T}{\partial z} \tag{9.9}$$

TABLE 9.1
Comparison of Momentum, Heat, and Mass Flux per Gradient for Air and Water

	Air (20°C)	Water (20°C)
Momentum flux/Grad. $= \nu$	$1.5 \times 10^{-5} \, \text{m}^2/\text{s}$	$1.0 \times 10^{-6} \, \text{m}^2/\text{s}$
Heat flux/Grad. $= k/(\rho C_p)$	$2.0 \times 10^{-5} \, \text{m}^2/\text{s}$	$1.5 \times 10^{-7} \, \text{m}^2/\text{s}$
Mass flux/Grad. $= D$	$0.8\text{--}7.0 \times 10^{-5} \, \text{m}^2/\text{s}$	$0.6\text{--}6.0 \times 10^{-9} \, \text{m}^2/\text{s}$
Prandtl number, $\text{Pr} = \rho C_p \nu/k$	0.7	7
Schmidt number, $\text{Sc} = \nu/D$	0.2–2	160–1600

and in the mass transport equation,

$$\frac{\text{Mass flux}}{\text{Unit volume}} = -D\frac{\partial C}{\partial z}. \tag{9.10}$$

The boundary conditions can often be made to look similar, through the conversion to dimensionless variables. These equations can be used to compute a momentum boundary layer thickness, a thermal boundary thickness, and a concentration boundary layer thickness. The primary difference within these analogies is shown in Table 9.1, where the momentum flux per momentum gradient, the heat flux per heat gradient, and the mass flux per mass gradient are given. This will provide a good comparison if the boundary conditions are made nondimensional to vary between 0 and 1.

It is seen that we are comparing kinematic viscosity, thermal diffusivity, and diffusivity of the medium for both air and water. In air, these numbers are all of the same order of magnitude, meaning that air provides a similar resistance to the transport of momentum, heat, and mass. In water, there is one order of magnitude or more difference between kinematic viscosity, thermal diffusion coefficient, and mass diffusion coefficient. Also provided in Table 9.1 are the Schmidt and Prandtl numbers for air and water. In water, Schmidt and Prandtl numbers on the order of 1000 and 10, respectively, results in the entire concentration boundary layer being inside of the laminar sublayer of the momentum boundary layer. In air, both the Schmidt and Prandtl numbers are on the order of 1. This means that the analogy between momentum, heat and mass transport are more precise for air than for water, and the techniques applied to determine momentum transport away from an interface may be more applicable to heat and mass transport in air than they are on the liquid side of the interface.

9.2 AIR–WATER MASS TRANSFER COEFFICIENTS

As noted before, most environmental flows are turbulent. The diffusive sublayer, where only diffusion acts to transport mass and the concentration profile is linear, is typically between $10\,\mu\text{m}$ and $1\,\text{mm}$ thick. Measurements within this sublayer are not usually feasible. Thus, the interfacial flux is typically expresses as a bulk transfer coefficient.

$$J_A = K_B \left(C_{Aa}^\infty - C_{AW}^\infty / H_A \right),\tag{9.11}$$

where K_B is the bulk transfer coefficient, C_{Aa}^∞ is the concentration of compound A away from the diffusive sublayer in the air phase, and C_{Aw}^∞ is the concentration of compound A away from the diffusive sublayer in the water phase.

The resistance to interfacial transfer can be visualized as $1/K_B$:

$$\frac{1}{K_B} = \frac{1}{K_L} + \frac{1}{H_A K_G},\tag{9.12}$$

where K_L is the liquid film coefficient and K_G is the gas film coefficient. Equation 9.12 indicates that K_B is dependent upon the chemical being transferred, as it contains H_A. The coefficients, K_L and K_G are less dependent upon the properties of the gas.

Equations 9.4, 9.11, and 9.12 can be combined with the assumption that all resistance to transfer occurs in the diffusive sublayer, as illustrated in Figure 9.2, to show that

$$K_L = \frac{D_{AW}}{\delta_W}\tag{9.13}$$

and

$$K_G = \frac{D_{Aa}}{\delta_a}.\tag{9.14}$$

Note that the unsteady relationships that exist can still be brought into the analysis of Equations 9.13 and 9.14, because $\delta = \delta\ (D, \text{turbulence})$.

If the ratio of K_G to K_L is known, Equation 9.12 can be used to determine whether the compound is volatile, nonvolatile, or semivolatile for a given turbulence characteristics near the interface. First, Equation 9.12 will be written as

$$\frac{1}{K_B} = \frac{1}{K_L} + \frac{1}{H_A K_L \dfrac{K_G}{K_L}} = \frac{1}{K_L} \left(1 + \frac{1}{H_A \dfrac{K_G}{K_L}} \right).\tag{9.15a}$$

If Henry's law constant is provided in Atm-m^3/mol, then Equation 9.12 is written as

$$\frac{1}{K_B} = \frac{1}{K_L} \left(1 + \frac{RT}{H_A \dfrac{K_G}{K_L}} \right),\tag{9.15b}$$

where T is temperature in $^\circ$K and R is the universal gas constant (8.2×10^{-5} atm-m^3/mol/K). Mackay and Yuen (1983) have stated that K_G/K_L values between 50 and 300 are reasonable. Munz and Roberts (1984) found that, for aeration devices generating bubbles to enhance gas transfer, K_G/K_L could be as low as 20. The values given in Appendix A-1 are for percent resistance in the liquid phase, or

$$100\frac{K_B}{K_L} = 100\frac{1}{1 + \dfrac{RT}{H_A K_G/K_L}}. \tag{9.16}$$

In general, water-phase controlled means that $100\,K_B/K_L$ is greater than 90%, and air-phase controlled means that $100\,K_B/K_L$ is less than 10%. Values between 10% and 90% are considered to be controlled by both phases.

9.3 MEASUREMENT AND PREDICTION OF WATER-SIDE MASS TRANSFER COEFFICIENTS

9.3.1 BACKGROUND

The principles of air–water mass transfer are often difficult to apply in field measurements, and thus also in field predictions. The reasons are that the environment is generally large, and the boundary conditions are not well established. In addition, field measurements cannot be controlled as well as laboratory measurements, are much more expensive, and often are not repeatable.

Table 9.2 lists some theoretical relationships for liquid-side transfer coefficient, for example, and the difficulties in applying these relationships to field situations. Eventually, application to the field comes down to a creative use of laboratory and field measurements, with a good understanding of theoretical results, to make sure that the basic principles of the theoretical relationships are not violated.

TABLE 9.2
Theoretical Relationships for Liquid-Side Transfer Coefficient and the Difficulties in Applying Them in the Field

Theoretical Relationship

$J = -D\dfrac{\partial C}{\partial z}\Big\|_{z\to 0}$	Cannot measure $\partial C/\partial z$ within $100\,\mu m$ of the free surface	(9.17)
$J = K_L\left(\dfrac{C_g}{H} - C_w\right)$	Must determine K_L	(9.18)
$J = K_G(C_g - HC_w)$	Must determine K_G	(9.19)
$K_L = D_w/\delta$	Do not know $\delta(t, D_w)$	(9.20)
$K_G = D_G/\delta$	Do not know $\delta(t, D_G)$	(9.21)
$K_L = \sqrt{Dr}$	Do not know r	(9.22)

Difficulties in Field Application

So, how are K_L and K_G determined for field applications? The primary method to measure K_L and K_G is to disturb the equilibrium of a chemical, and measure the concentration as it returns toward either equilibrium or a steady state. Variations on this theme will be the topic of Sections 9.3 and 9.4.

9.3.2 WATER-SIDE COEFFICIENT IN RIVERS

Rivers are generally considered as a plug flow reactor with dispersion (Levenspiel, 1960), where the mixing is provided by the velocity profile and turbulence in the flow field. This concept is applied to the measurement of $K_L/H = K_L a$ in rivers, where H is depth of flow and $K_L a = K_L A/V = K_2$ is generally called the reaeration coefficient, and a is the specific surface area.

9.3.2.1 Measurements

Consider the concentration measurement of two tracers, a volatile tracer that is generally a gas, termed a gas tracer, C, and a conservative tracer of concentration, C_c. The transported quantity will be the ratio of the two tracers, $R = C/C_c$. Then, the convection–dispersion equation becomes (Gulliver, 2007, p. 149)

$$\frac{\partial R}{\partial t} + \frac{Q}{A_{cs}} \frac{\partial R}{\partial x} = D_L \frac{\partial^2 R}{\partial x^2} + K_L a(R_s - R), \tag{9.23}$$

where Q is the river discharge, A_{cs} is the cross-sectional area, D_L is the longitudinal dispersion coefficient, $R_s = C_s/C_c$, and C_s is the concentration of the gas in water at equilibrium. The convection term does not appear when oriented in Lagrangian (moving) coordinates. In addition, the gradient of the ratio R is presumed to be low, because both the gas tracer and the conservative tracer will spread similarly. This means that the longitudinal dispersion term with D_L will be small. Finally, most gas tracers do not have a significant concentration in the atmosphere, so $R_s = 0$. Then, Equation 9.23 becomes

$$\frac{\partial R}{\partial t} = -K_L a R \tag{9.24}$$

with solution,

$$\ln\left(\frac{R_2}{R_1}\right) = -K_L a(t_2 - t_1), \tag{9.25}$$

where subscripts 1 and 2 correspond to times 1 and 2. Most gas transfer measurements are made with a dual tracer pulse, where the pulse requires some distance to mix across the river (Fisher et al., 1979). Thus, a tracer cloud is followed, and the time is generally taken to be the center of mass of the cloud, upon which multiple measurements are made. Then, Equation 9.25 becomes that which is actually used in practice:

$$K_2 = -\frac{1}{(t_2 - t_1)} \ln\left(\frac{\sum\limits_n R_{2i}}{\sum\limits_n R_{1i}}\right), \tag{9.26}$$

where n is the total number of measurements at a given location. There are other considerations with regard to gas transfer measurements in rivers that are detailed elsewhere (Kilpatrick, 1976; Hibbs et al., 1998). Some typical gas tracer pulses are given in Figure 9.4. These measurements are a significant effort, because one needs

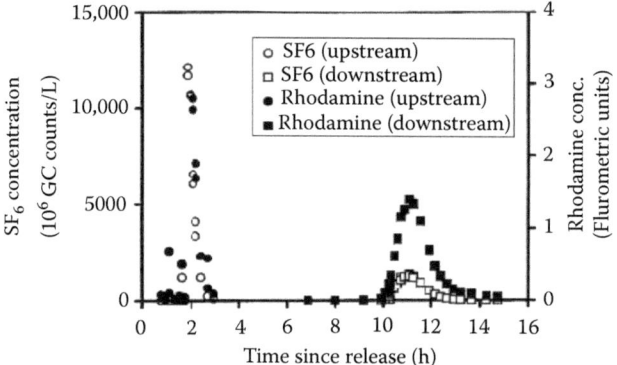

FIGURE 9.4 Gas tracer pulses for the James River, North Dakota, used to measure the reaeration coefficient.

to be out for two or three days and mobilize when the tracer cloud passes, which is typically in the middle of the night.

Of course, the typical objective is the $K_L a$ value for a different gas, such as oxygen, to be used in such things as total daily maximum load (TMDL) calculations, and we just have the $K_L a$ value for the gas tracer. Equation 9.22 can be used to get us from the transfer of one compound to another, because r is similar for all compounds:

$$\frac{K_L a(O_2)}{K_L a(\text{tracer})} = \left(\frac{D_{O_2}}{D_t}\right)^{1/2}, \tag{9.27}$$

where D_t is the diffusion coefficient of the gas tracer.

9.3.2.2 Prediction

Measurements of reaeration coefficients have been made at a number of locations over the years, and it is natural that individuals would try to correlate these measurements with the measured parameters of the river, so that predictions can be made elsewhere. A partial compilation of measurements is given in Figure 9.5. Note that, while there is scatter in flume measurements, this is exceeded by a factor of 10 in field measurements.

Moog and Jirka (1998) investigated the correspondence of a number of equations with the available data, using the mean multiplicative error, MME to determine the fit:

$$\text{MME} = \exp\left[\frac{\sum_{i=1}^{N} \left|\ln \left(K_p/K_m\right)_i\right|}{N}\right], \tag{9.28}$$

where K_p and K_m are the predicted and measures $K_L a$ values, and N is the number of data. The mean multiplicative error compares the ratio of the prediction to the measurement, so that the study would not be biased toward the higher values of $K_L a$. There are orders of magnitude differences in reaeration coefficients, and to simply

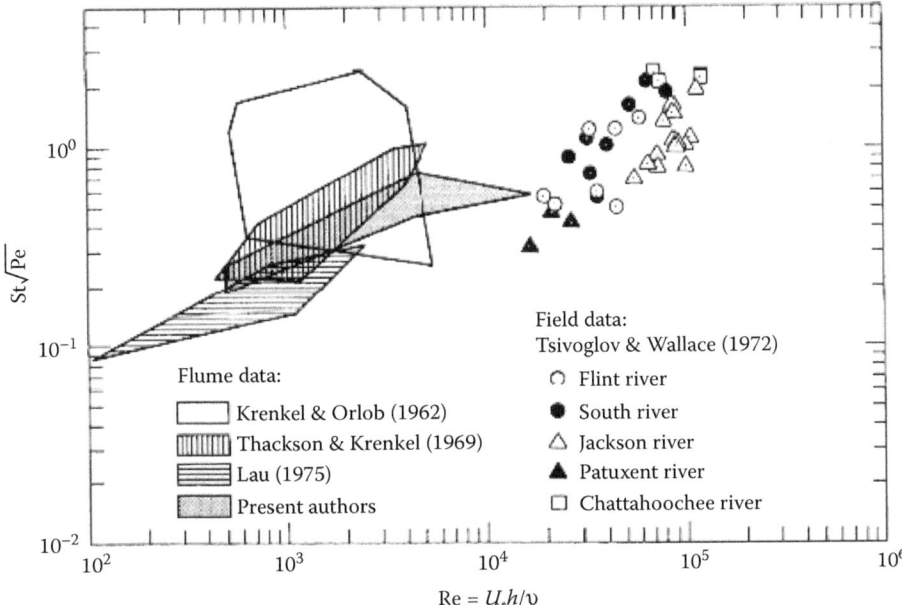

FIGURE 9.5 Measurements of reaeration coefficient in laboratory flumes and in the field. Present authors are Gulliver and Halvorson (1989). St is a Stanton number, K_L/U_*, Pe is a shear Peclet number, U_*h/D, U_* is the bed shear velocity, $(ghS)^{1/2}$, g is the acceleration of gravity, h is average flow depth, S is water surface slope, and v is kinematic viscosity.

take the difference between predictions and measurements as the residuals causes the residuals of the larger reaeration coefficients to dominate the process. The mean multiplicative error thus has the following advantages:

1. The small and large values of K_La are considered equally.
2. The MME is the geometric mean of K_p/K_m. Thus a given equation is on average, in error by a factor equal to the MME.
3. The MME of K_La is equal to the MME of K_L. This occurs because the absolute values of the K_p/K_m ratios were utilized, instead of the square of the ratio.

Moog and Jirka then calibrated the first proportionality coefficient and studied the predictive capability of 10 "calibrated" empirical equations to predict reaeration coefficients that were the result of 331 field studies. The result was surprising, because the best predictive equation was developed by Thackston and Krenkel (1969) from laboratory flume studies, and the comparison was with field equations. In dimensionless form, Thackston and Krenkel's calibrated (multiplying the first coefficient by 0.69) equation can be converted into a dimensionless form utilizing Sherwood, Schmidt, Reynolds, and Froude numbers:

$$\text{Sh} = 4.4 \times 10^{-3}\text{Sc}^{1/2}\text{Re}_*(1 + F^{1/2}), \tag{9.29}$$

where $Sh = K_L h/D = K_L ah^2/D$, $Sc = \nu/D$, $Re_* = u_*h/\nu$, $F = U/\sqrt{gh}$, h is mean stream depth, D is diffusion coefficient of the gas in the water, ν is kinematic viscosity, $u_* = (ghS)^{1/2}$ is shear velocity, S is water surface slope, U is mean stream velocity, and g is the acceleration of gravity. To convert Thackston and Krenkel's equation into Equation 9.29, we have assumed that K_L is proportional to $D^{1/2}$, included the viscosity of water at 20°C, and assumed that $Sc = 476$ for oxygen at 20°C. Equation 9.29 should be functional at all water temperatures where water is a liquid and for all gases. It is probable that the Froude number in Equation 9.29 partially accounts for the additional water surface roughness that occurs above a Froude number of 1.0.

Equation 9.29 results in the following relation for surface renewal rate, r:

$$r = 1.94 \times 10^{-5} \frac{u_*^2}{\nu} \left(1 + F^{1/2}\right). \tag{9.30}$$

Moog and Jirka also found that Equation 9.29, even though it was the best predictor, still had a mean multiplicative error of 1.8. This means that one can expect the predictions of Equation 9.29 to be off the field measurements by either multiplying or dividing by a factor of 1.8. Fifty percent of the predictions will differ by more than this factor, and 50% by less. In addition, they found that below a stream slope of 0.0004, it is just as good to simply use a constant value of $K_L a$ at 20°C of 1.8 day^{-1}, with a 95% confidence interval corresponding to a multiplicative factor of 8. At the low slope values, other factors such as wind velocity and surfactants on the water surface could have an influence on $K_L a$.

Considering that there are 331 "high-quality" measurements of reaeration coefficient in streams, with 54 at $S < 0.0004$, predicting reaeration coefficient is still a challenge. There are four possible reasons for this, indicative of the difficulties that exist when taking detailed results of experiments and analysis from the laboratory to the field.

1. *Field measurements are not as precise as laboratory measurements.* While this is a true statement, some dedicated field experimentalists have improved the field techniques greatly over recent decades (Tsivoglou and Wallace, 1972; Kilpatrick et al., 1979; Clarke et al., 1994; Hibbs et al., 1998). While the implementation of field studies is still a challenge, the accuracy cannot account for a mean multiplicative error of 1.8.
2. *Turbulence is generated at the channel bottom, and reaeration occurs at the top of the channel.* The importance of turbulence to air–water mass transfer was illustrated in Section 9.2. One confounding experimental problem is that the turbulence is generated at the bottom of the channel, and goes through changes in intensity and scale as it moves towards the water surface. This problem exists in both flume and field channel studies, however, and the flume studies are significantly more precise than the field studies.
3. *The independent parameters that are measured are not truly indicative of the important processes for gas transfer.* Tamburrino and Gulliver (2002) found that the process important to surface renewal is two-dimensional divergence

on the free-surface (Hanratty's β):

$$r \approx \beta = -\left(\frac{\partial u}{\partial x} + \frac{\partial v}{\partial y}\right). \tag{9.31}$$

Tamburrino and Gulliver (2002) and Tamburrino et al. (2006) have experimentally visualized β on the free surface of turbulent flows, and indicated that there is a correlation between β and vorticity, ω, on the free surface:

$$\omega = \frac{\partial u}{\partial y} - \frac{\partial v}{\partial x}. \tag{9.32}$$

Now consider a natural river, illustrated in Figure 9.6. There are many sources of vorticity in a natural river that are not related to bottom shear. Free-surface vortices are formed in front of and behind islands, and at channel contractions and expansions. These could have a direct influence upon surface renewal, and the reaeration coefficient, without the dampening effect of stream depth. The measurement of r and surface vorticity in a field stream remains a challenge that has not been adequately addressed.

4. *The mean values that are determined with field measurements are not appropriate.* Most predictive equations for reaeration coefficient use an arithmetic mean velocity, depth, and slope over the entire reach of the measurement. The process of measuring reaeration coefficient dictates that these reaches be long, to insure the accuracy of $K_L a$. Flume measurements, however, have generally shown that $K_L a \sim u_*/h$ or $K_L a \sim (S/h)^{1/2}$ (Thackston and Krenkel, 1969; Gulliver and Halverson, 1989). If this is truly the case, we should be taking the mean of $S^{1/2}$ and the mean of $h^{-1/2}$ in order to use the predictive equations to estimate reaeration coefficient. Example 9.1 will investigate whether this is an important consideration.

The ramifications of the poor $K_L a$ predictive ability are that it is challenging do an adequate job of planning for oxygen concentrations during low flow events or for

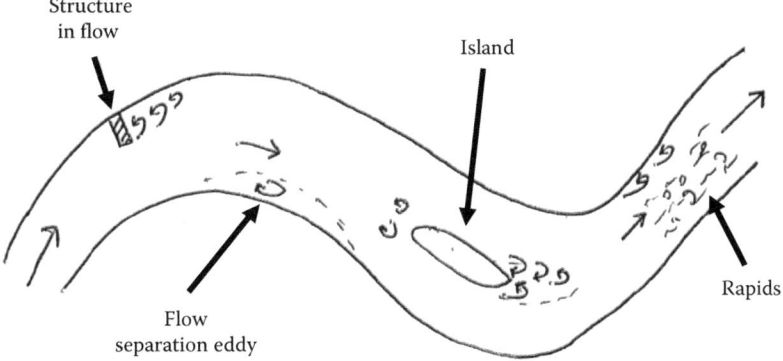

FIGURE 9.6 Illustration of additional surface vortices in a natural river.

spills, unless field measurements of reaeration coefficient have been performed. If there are no measurements, sensitivity analysis on $K_L a$ is needed. The choice of $K_L a$ will typically make a difference.

9.3.3   WATER-SIDE TRANSFER COEFFICIENT IN LAKES AND OCEANS

The influence of wind is predominant in determining the liquid film coefficient for lakes, reservoirs, oceans, and many estuaries. Wind creates a shear on the water surface, and generates turbulence below and on the water surface. Thus, a good independent parameter would be wind shear, which is dependent upon wind speed in the boundary layer of the water surface in question and wave parameters. This is a challenging parameter to predict, although there are individuals who have attempted it, given sufficient data (Hsu, 1974; Mitsuyasu, 1968). Owing to the added complexity of using shear velocity, most predictive equations simply use the wind velocity at a specified height. The user is then required to convert their wind velocity height into the height of the predictions using appropriate boundary layer relations. This section deals with the measurement and prediction of the wind-influence upon liquid film coefficient.

9.3.3.1   Opportunistic Measurement of Wind Influence

The opportunistic measurement techniques generally used are ^{14}C absorption and ^{222}Rn disequilibrium (Asher and Wanninkhof, 1998). First, there is an estimate of a long-term ($\sim$1000 years) global gas transfer coefficient of $K_L = 6 \times 10^{-5}$ m/s, developed by assuming steady state between pre-1950 ^{14}C radioactive decay in the oceans and absorption from the atmosphere (Broecker and Peng, 1982). In addition, nuclear testing since 1950 has increased ^{14}C concentration in the atmosphere. Thanks to the atomic testing "battle" between the United States of America and the Soviet Union a tracer that can be used on an ocean-wide basis currently exists. A box model of an ocean basin is still needed. By using an appropriate oceanic model to estimate the depth of the interactive layer, and taking sufficient measurements of $^{14}CO_2$ at the ocean boundaries and inside the control volume, the fluxes and mean control concentration, respectively, can be determined. Then, the remainder of the flux is assigned to atmospheric fluxes of $^{14}CO_2$, and a liquid film coefficient is determined from a mass conservation equation:

$$V \frac{\partial\, ^{14}C_{basin}}{\partial t} = \text{Ocean Flux Rate}(^{14}C_{In} - \,^{14}C_{Out}) + K_L A(^{14}C_{atm} - \,^{14}C_{basin}),$$

(9.33)

where V is the volume of the interactive layer and A is the surface area of the ocean basin. An important consideration for these estimates is the depth of the interactive layer (Duffy and Caldera, 1995). The response time of Equation 9.33 is on the order of years to decades, so any relationship to wind velocity is a long-term average.

By contrast, the gas transfer estimates utilizing ^{222}Rn measurements assumes steady state between ^{222}Rn production from radioactive decay of nonvolatile ^{226}Rd and gas transfer with the atmosphere. This assumption is possible because ^{222}Rn has a half-life of only 3.8 days, so accumulation and lateral ocean fluxes of ^{222}Rn can be assumed minimal for many basins. Again, a potential problem is the active, versus inactive layer of the ocean, in this case the mixed layer depth which may change during an experiment.

The results of both ^{14}C and ^{222}Rn measurements of liquid film coefficient versus wind velocity are plotted in Figure 9.7, along with two parameterizations that will be discussed in Section 9.3.2.3, prediction of wind influence.

The adjustment of wind velocity to 10 m height is typically performed with the logarithmic boundary layer relation:

$$\frac{W_{10}}{W_z} = \ln\left(\frac{10\,\text{m}}{h_z}\right), \tag{9.34}$$

where W_{10} and W_z are wind velocity at 10 m height and the measured height, respectively, and h_z is the measured height in meters. The measured height must be inside of the boundary layer of the water body, with no influence from the upwind land surface. $h_z = 2$ m is often used for small water bodies.

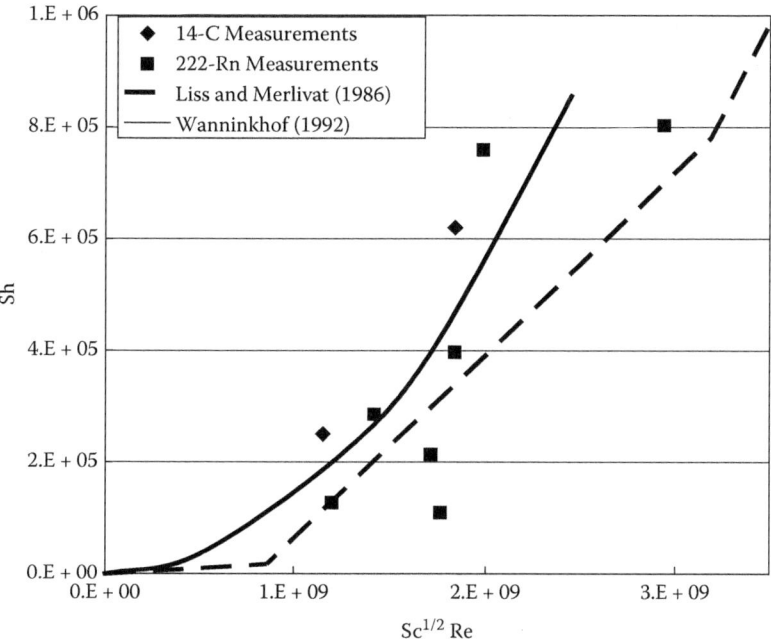

FIGURE 9.7 Sherwood number versus Reynolds number measured with ^{14}C and ^{222}Rn tracers. Solid and dashed lines are empirical relations developed from measurements. Sh $= K_L h/D$, Sc $= \nu/D$, and Re $= Wh/\nu$, where W is wind speed at 10 m height and h is some relevant length scale. (From Broecker, W. S. and Peng, T. H. 1971. *Earth Planet. Sci. Lett.*, 11, 99–108; Peng, T. H., Takahashi, T., and Broecker, W. S. 1974. *J. Geophys. Res.*, 79, 1772–1780; Kromer, B. and Roether, W. 1983. *"Meteor" Forschungsergeb. Reihe A/B*, 24, 55–75; Broecker et al. 1985. *J. Geophys. Res.*, 90, 6953–6970; and Cember, R. 1989. *J. Geophys. Res.*, 94, 2111–2123.)

9.3.3.2   Measurement of Wind Influence with Deliberate Tracers

Batch Technique. As with river reaeration measurements, tracers can also be put into lakes, estuaries, and oceans to measure the influence of wind on liquid film coefficient. If a volatile tracer has been placed in a lake with a well established mixed layer, for example, the batch reactor equation for a well-mixed tank can be applied:

$$V\frac{\partial C}{\partial t} = -K_L A \left(\frac{C_g}{H} - C \right), \tag{9.35}$$

where V is the volume of the mixed layer and A is the interfacial area. Assuming that C_g is a constant value, assigning $C' = C_g/H - C$, separating variables and integrating:

$$\ln\left(\frac{C - C_g/H}{C_0 - C_g/H} \right) = -K_L \frac{A}{V}(t - t_0) = -K_L a(t - t_0), \tag{9.36}$$

where C_0 is the concentration at $t = t_0$ and a is the specific surface area A/V, which has units of length^{-1}. A plot of the log term in Equation 9.36 versus time will result in a straight line of slope $K_L a$. If there are more than 11 points, the standard error of the slope is approximately the precision (random) uncertainty to the 67% confidence interval. That means that 67% of the data, if Gaussian around the mean, will fall within the confidence interval. Bias (systematic) uncertainty, of course, needs to be analyzed separately. There are other means of determining $K_L a$ for a batch reactor, the best known being the ASCE (1992) technique that uses a nonlinear regression to determine C_g/H and $K_L a$.

The batch reactor analysis is a natural one to use for laboratory tanks and wind-wave facilities. In addition, it has been used for lakes, which are either well-mixed vertically or where the surface mixed layer is at close to a constant thickness (Torgersen et al., 1982; Wanninkhof et al., 1987; Upstil-Goddard et al, 1990; Livingstone and Imboden, 1993). Typically, sulfur hexafluoride (SF$_6$) or ^{3}He are used as deliberate gas tracers input to the lakes because they are detectable at low concentrations. The results of selected measurements are plotted versus wind speed Reynolds number, Wh/v where W is wind speed at or adjusted to 10 m. in height, h is some relevant length scale, and v is kinematic viscosity, in Figure 9.8. The primary difference between the two sets of measurements is the fetch length of the water body, where the Upstil-Goddard et al. measurements were taken on a smaller water body. In general, the uncertainty of one given measurement is large, but a number of measurements, taken together while *not* eliminating those that give a negative K_L, results in the summary field measurements seen in Figure 9.8. The problem is that the investigator needs to be involved in deciding which measurements to use in determining an average, so there is a potential bias in the results. Changes in mixed layer depth need to be taken into consideration, as well, if one is not able to assume a well-mixed lake.

Dual Tracer Technique. The dual tracer measurement technique utilizes two gas tracers with diffusion coefficients that are substantially different, such as ^{3}He and SF$_6$. This technique can also be utilized with one volatile (gas) tracer and one nonvolatile tracer. The relevant equations to determine liquid film coefficient from the diffusion equation for both cases, beginning with the two gas tracers, will be derived.

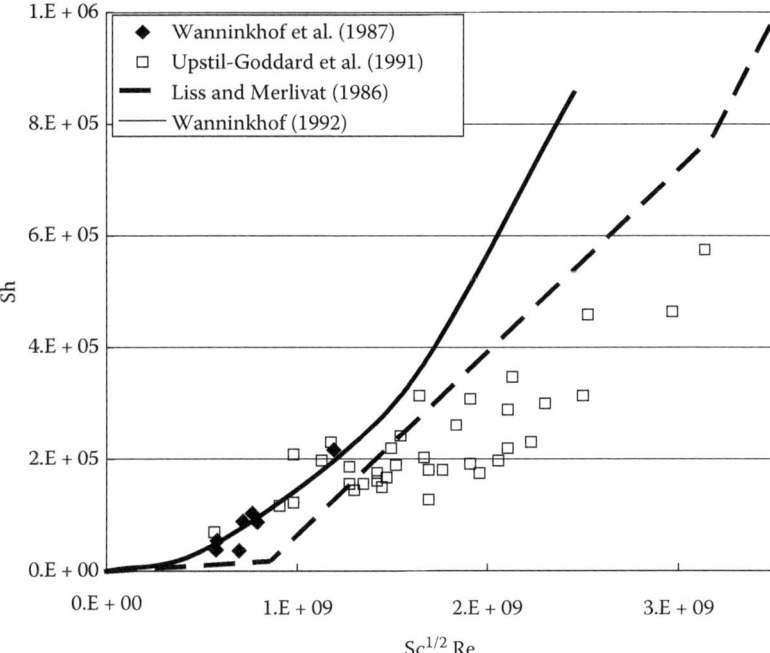

FIGURE 9.8 Sherwood number, $Sh = K_L h/D$ versus $Sc^{1/2} Re = Wh/D$, measured by the batch reactor technique in small lakes. Lines represent equations developed from field measurements.

For a lake or ocean surface, consider a cylindrical control volume of depth h that moves with the mean velocity of the tracer cloud containing two gas tracers, designated A and B, that have different rates of gas transfer. Using the cylinder as our control volume, the transport relation for each of the gas tracers can be written as

$$\frac{\partial \overline{C}_i}{\partial t} = D_x \frac{\partial^2 \overline{C}_i}{\partial x^2} + D_y \frac{\partial^2 \overline{C}_i}{\partial y^2} - \frac{K_L}{h} C_{si}, \tag{9.37}$$

where $\overline{C}_i$ is the mean concentration of compound i over the column depth, D_y and D_x are the dispersion coefficients in the horizontal directions and C_{si} is the concentration of compound i at the water surface. The mass balance will be performed on the ratio, $R = C_A/C_B$. Then, the derivative of R with respect to time is given as

$$\frac{\partial R}{\partial t} = \frac{\partial}{\partial t} \left(\frac{C_A}{C_B} \right) = \frac{1}{C_B} \left(\frac{\partial C_A}{\partial t} - R \frac{\partial C_B}{\partial t} \right). \tag{9.38}$$

Assuming that the dispersion of the ratio of concentrations to be small, the dispersion of R will be ignored, and Equations 9.37 and 9.38 are combined to give

$$\frac{\partial R}{\partial t} = -\frac{1}{h} \left(K_{LA} R_{sA} - K_{LB} R R_{sB} \right), \tag{9.39}$$

where $R_{sA} = C_{sA}/\overline{C_B}$ and $R_{sB} = C_{sB}/\overline{C_B}$. Assuming the $1/2$ power relationship between liquid film coefficient and diffusion coefficient, and applying a surface renewal-type of relationship

$$\frac{K_{LB}}{K_{LA}} = \sqrt{\frac{D_B}{D_A}}. \tag{9.40}$$

Substituting Equation 9.40 into 9.39, and rearranging gives us an equation that can be used to develop K_{LA} from measurements of both tracers:

$$K_{LA} = -\frac{h}{\left(R_{sA} - \sqrt{\frac{D_B}{D_A}} R R_{sB}\right)} \frac{\partial R}{\partial t}. \tag{9.41}$$

R, R_{sA}, and R_{sB} can all be measured at various times. Then, a plot of the various terms in Equation 9.41 will provide us with an estimate of K_L. Typically, gas tracers A and B would have substantially different diffusion coefficients, such that field imprecision in sampling would not mask the ability to determine K_{LA} from Equation 9.41. If tracer B is nonvolatile, $K_{LB} = 0$, and Equation 9.41 becomes

$$K_{LA} = -\frac{h}{R_{sA}} \frac{\partial R}{\partial t}. \tag{9.42}$$

Equation 9.41 applies to two volatile tracers, and Equation 9.42 to one volatile and one nonvolatile tracer.

Gulliver et al. (2002) applied Equation 9.42 to field measurements of SF_6 and Rhodamine-WT, a nonvolatile tracer to Chequamegon Bay of Lake Superior. Their analysis technique was as follows:

1. Regress $\ln(h/R_{sA})$ vs $\ln t$ in a linear regression. Then

$$h/R_{sA} = \alpha_1 t^{\alpha_2}, \tag{9.43}$$

where α_1 and α_2 are fitted constants. A sample for one data set, shown in Figure 9.9, indicates that there is considerable scatter, which is due to sampling uncertainty, that is, one does not know if they are at the peak of the tracer cloud.

2. Regress R vs $t^{1-\alpha_2}$. The α_2 power is used to make the dimensions work out properly in the relation for K_L. Then

$$R = \alpha_3 t^{1-\alpha_2} + \alpha_4. \tag{9.44}$$

A sample of this regression is shown in Figure 9.10, again indicating considerable uncertainty. Equation 9.44 is chosen to result in the proper units for K_L.

Combining Equations 9.42, 9.43, and 9.44 yields

$$K_L = \alpha_1 * t^{\alpha_2} * \left[\frac{\partial}{\partial t} \left(\alpha_3 * t^{(1-\alpha_2)} + \alpha_4 \right) \right] \tag{9.45}$$

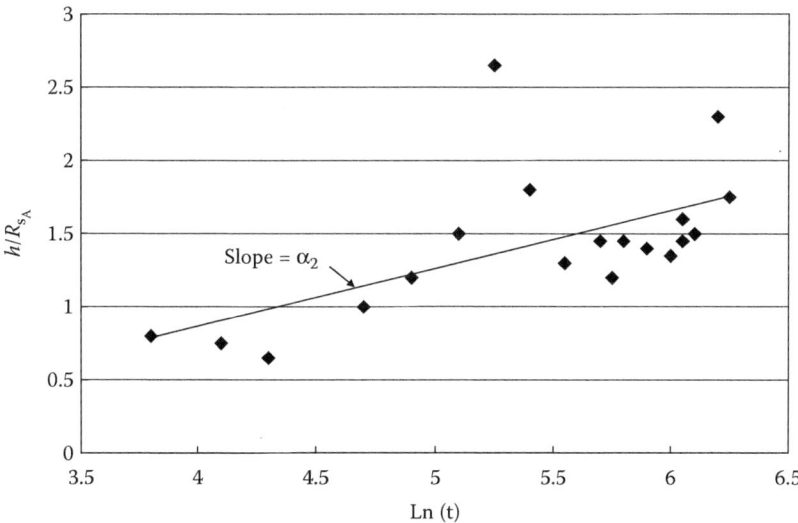

FIGURE 9.9 Measured data from a dual tracer cloud and the curve-fit to the data.

or

$$K_L = -\alpha_1 * \alpha_3 * (1 - \alpha_2). \tag{9.46}$$

With the dimensional formulation of Equation 9.46, α_1 has units of length/time$^{\alpha_2}$ and α_3 has units of time$^{\alpha_2-1}$. Thus K_L has units of length/time.

The precision uncertainty associated with the field sampling is generally much larger than that associated with the analytical technique, which is roughly $\pm 2\%$ to the

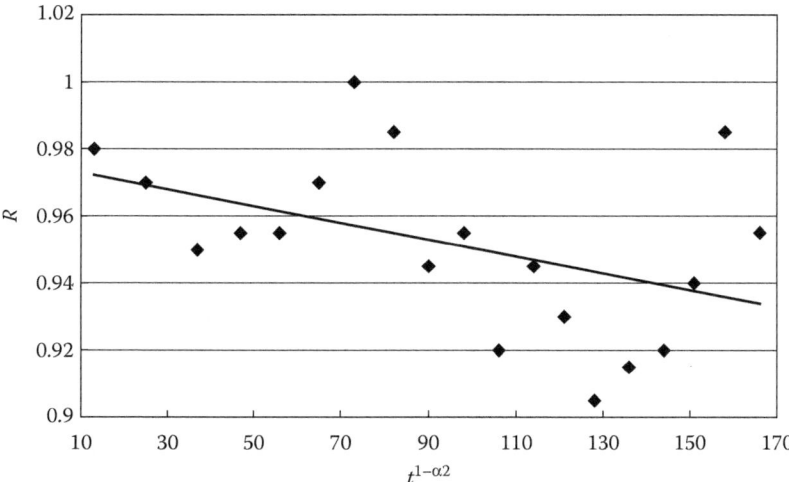

FIGURE 9.10 Measured data from a dual tracer experiment for one day and the curve-fit to the data.

67% confidence interval for the two compounds used as conservative and gas tracers. A technique to determine the precision uncertainty associated with field sampling and incorporated into the mean K_L estimate will therefore be propagated with the first-order-second moment analysis (Abernathy et al., 1985).

$$U_{K_L}^2 = \left(U_{\alpha_1} * \frac{\partial K_L}{\partial \alpha_1} \right)^2 + \left(U_{\alpha_3} * \frac{\partial K_L}{\partial \alpha_3} \right)^2 + \left(U_{\alpha_2} * \frac{\partial K_L}{\partial \alpha_2} \right)^2. \tag{9.47}$$

When the partial derivatives are taken from Equation 9.46, Equation 9.47 becomes

$$U_{K_L}^2 = (\alpha_3 * (1 - \alpha_2) * U_{\alpha_1})^2 + (\alpha_1 * (1 - \alpha_2) * U_{\alpha_3})^2 + (\alpha_1 * \alpha_3 * U_{\alpha_2})^2. \tag{9.48}$$

The variables U_{α_1}, U_{α_2}, and U_{α_3} are the corresponding uncertainty values for each parameter. They are computed to the 67% confidence interval by taking the standard error of each parameter in the regressions, that is, α_1, α_2, and α_3, and multiplying by their Student t-score t_S, that is, $U_{\alpha_1} = t_S * SE_{\alpha_1}$ where t_S is the Student t-score at the confidence level of interest and SE_{α_1} is the corresponding standard error for the parameter α_1. The period can be chosen based upon the maximum r^2 value or another statistical parameter. The results of four experiments taken from larger lakes and the ocean are given in Figure 9.11.

Other investigators have used two measurements of the gas tracers taken over a longer time period (several days), and assumed that the cloud of tracers was well-mixed vertically. In this case, the following conditions must be met: (1) changes in the ratio of tracer concentrations caused by dispersion should be negligible compared to those resulting from gas transfer, (2) the water column should be vertically well-mixed, and (3) the experimental area should be close to constant depth. Criteria number 1 is satisfied by having a tracer cloud that is large, compared to the depth and by staying close to the center of this cloud. With these assumptions, Equation 9.41 becomes

$$\frac{1}{R} \frac{dR}{dt} = \frac{K_{LA}}{h} \left(1 - \sqrt{\frac{D_B}{D_A}} \right) \tag{9.49}$$

which can be integrated to give

$$K_{LA} = \frac{h}{\Delta t} \left(1 - \sqrt{\frac{D_B}{D_A}} \right) (\ln R_{t_2} - \ln R_{t_1}). \tag{9.50}$$

The results of these measurements are also given in Figure 9.11. Unfortunately, the technique utilizing Equation 9.50 cannot provide an estimate of sampling uncertainty.

The data contained in Figures 9.7, 9.8, and 9.11 indicate similarity against wind velocity, even though the fetch length and size of the water bodies are different. This is generally seen as an indication that wind is an important driving factor for lakes, estuaries, and oceans. It has been shown that breaking waves and water surface slicks are important (Asher and Wanninkhof, 1998), and there are other parameters such as

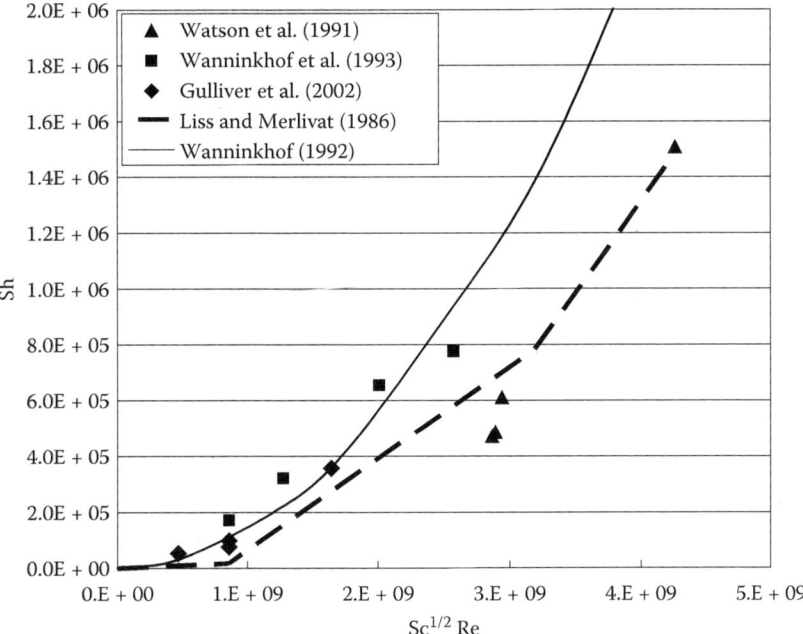

FIGURE 9.11 Sherwood number, $Sh = K_L h / D$ measured by the dual tracer technique versus $Sc^{1/2}Re = Wh/D$, where K_L is gas transfer coefficient, h is some relevant length, D is diffusion coefficient, and W is wind velocity measured at or adjusted to 10 m height.

mean square water surface slope that have been proposed as better indicators (Jahne, 1991). The problem is that our ability to predict these indicators from wind velocity measurements have not been developed and tested for liquid film coefficient.

One other measurement technique that has been used to measure K_L over a shorter time period, and is thus more responsive to changes in wind velocity, is the controlled flux technique (Haußecker et al., 1995). This technique uses radiated energy that is turned into heat within a few microns under the water surface as a proxy tracer. The rate at which this heat diffuses into the water column is related to the liquid film coefficient for heat, and, through the Prandtl-Schmidt number analogy, for mass as well. One problem is that a theory for heat/mass transfer is required, and the Danckwert's surface renewal theory may not apply to the low Prandtl numbers of heat transfer (Atmane et al., 2004). The controlled flux technique is close to being viable for short-period field measurements of the liquid film coefficient.

9.3.3.3 Prediction

There are two predictive relationships based upon wind speed. Liss and Merlivat (1986) used a physical rationale to explain the increase in the K_L versus wind speed slope at higher wind velocities in wind-wave tunnel and lake measurements, resulting in a piecewise linear relationship with two breaks in slope. These breaks are presumed to occur at the transition between a smooth surface and rough surface and between a rough surface and breaking waves. In dimensionless form, this relationship is given as

$$Sh = 3.4 \times 10^{-5} Sc^{1/3} Re, \quad W_{10} \geqslant 3.6 \, m/s, \tag{9.51a}$$

$$Sh = 1.9 \times 10^{-4} Sc^{1/2} (Re - Re_0), \quad 3.6 \, m/s < W_{10} \leqslant 13 \, m/s, \tag{9.51b}$$

$$Sh = 4.1 \times 10^{-4} Sc^{1/2} (Re - Re_b), \quad 13 \, m/s < W_{10}, \tag{9.51c}$$

where W_{10} is the wind speed at 10 m above the mean water surface, $Sh = K_L h/D$, $Re = W_{10} h/\nu$, $Re_o = (3.4 \, m/s) h/\nu$, and $Re_b = (8.3 \, m/s) \, h/\nu$. The length scale, h, could be any relevant length scale because it will drop out of the relationship when determining K_L.

Wanninkhof (1992) developed one relation from ^{14}C data:

$$Sh = 3.4 \times 10^{-5} Sc^{1/2} Re W_{10} (m/s). \tag{9.52}$$

Equation 9.51 does not lend itself to an easy conversion to dimensionless parameters because $K_L \sim W_{10}^2$. It is one equation, however, instead of three, which makes it easier to use in computer programs and spreadsheets. Assuming that $h = 10 \, m$, and $\nu = 1 \times 10^{-6} \, m^2/s$, then Equation 9.52 becomes,

$$Sh = 3.4 \times 10^{-12} Sc^{1/2} Re^2. \tag{9.53}$$

Both Equations 9.51 and 9.53 are compared to the existing field data in Figures 9.7, 9.8, and 9.11. With the scatter in the data, it is difficult to select one equation over the other.

Jähne et al. (1984, 1987) proposed that liquid film coefficient is better related to mean water surface slope. Frew (1997) has found that the K_L relationship using mean square slope can be used to describe gas transfer with and without surface slicks. The problem with mean surface slope is that it cannot be accurately predicted for water bodies, because most investigators have emphasized the larger and longer waves, and the slope is most significant for the small, short waves.

9.4   MEASUREMENT AND PREDICTION OF AIR-SIDE MASS TRANSFER COEFFICIENTS

9.4.1   BACKGROUND

There are two types of convection in an atmospheric boundary layer: (1) forced convection created by air movement across the water body and (2) free convection created by a difference in density between the air in contact with the water surface and the ambient air. If the water body is warmer than the surrounding air, free convection will occur. The combination of these two processes is illustrated in Figure 9.12.

The gas film coefficient due to free convection (Figure 9.12a) is described by the relation of Shulyakovskyi (1969):

$$K_G = 0.14 \left[\frac{g \alpha_a^2 \beta_T \Delta \theta_v}{\nu_a} \right]^{1/3} \quad \text{Positive } \Delta \theta_v, \tag{9.54}$$

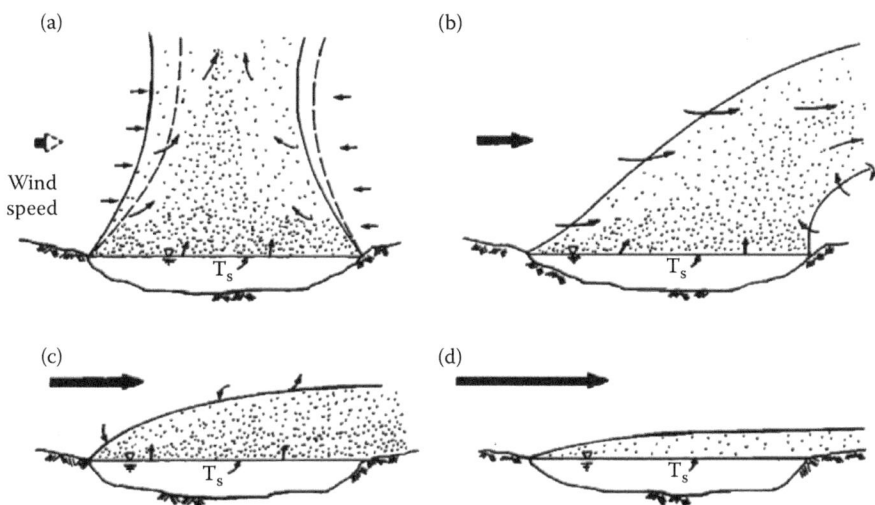

FIGURE 9.12 Free and forced convection regimes: (a) free convective plume, where $T_s > T_{air}$, (b) convective plume deflected by wind, (c) unstable boundary layer, and (d) stable or neutral boundary layer, where $T_s < T_{air}$ (From Adams, E. E., Cosler, D. J., and Helfrich, K. R. 1990. *Water Resources Res.*, 26(3), 425. With permission.)

where g is the acceleration of gravity, α_a is the thermal diffusion coefficient in air, $\beta_T(1/°)$ is the coefficient of thermal expansion for moist air, ν_a is the kinematic viscosity of air, and $\Delta\theta_v$ is the virtual temperature difference between air at the water surface and the ambient air, representing a density difference. Virtual temperature is the density of dry air at otherwise similar conditions. The relation for $\Delta\theta_v$ is

$$\Delta\theta_v = \|T_s(1 + 0.378p_{vs}/P_a) - T_a(1 + 0.378p_{va}/P_a), 0\|, \qquad (9.55)$$

where $\|\ \|$ stands for the greater of the two terms in brackets, T_s and T_a are the temperatures of the water surface and ambient air, respectively, p_{vs} and p_{va} are the vapor pressures of saturated air on the water surface and ambient air, respectively, and P_a is the air pressure. If the air temperature term is greater that the water temperature term, $\Delta\theta_v$ is zero.

Saturation vapor pressure at any air temperature may be computed by the Magnus–Tetons formula:

$$p_{vs}(\text{mb}) = 6.11 \exp\left[\frac{17.3(T(°\text{K}) - 273)}{T(°\text{K}) - 35.9}\right]. \qquad (9.56)$$

The atmospheric vapor pressure can be computed from the saturation vapor pressure and relative humidity, RH:

$$p_{va} = \frac{RH}{100}p_{vs}. \qquad (9.57)$$

Since values of α_a and ν_a can be found from tables such as those in the *Handbook of Chemistry and Physics* (CRC, 2004), all that remains is to determine β_T before we can estimate K_G from free convection. The equation

$$\beta_T = \frac{1}{T_{av}(^\circ C) + 273} \tag{9.58}$$

will provide us with this parameter, where $T_{av} = T_a(1 + 0.378 p_{va}/P_a)$ is the virtual temperature of the air.

Adams et al. (1990) have investigated the application of the free convection relationships for the cooling water pond at the Savannah River thermal power plant, and found that free convection should dominate at wind speeds (at 2 m height) of less than 0.5–1 m/s. At wind speeds greater than 1–1.5 m/s, the evaporation regime is that of an unstable forced convective layer, illustrated in Figure 9.12c. Thus, the deflected plume, Figure 9.12b, is a transition regime that is will be "smoothed over" for evaporation from water bodies.

The unstable boundary layer is a commonly occurring regime in environmental flows, and reverts to the stable boundary layer if the stability terms are zero. Assume a logarithmic velocity profile in the air boundary layer, altered by a stability function (Brutsaert, 1982), and assign $J_m = \rho u_*^2$ to be the flux of momentum at height z. Then, the momentum/unit volume difference between heights z and $z = 0$ is

$$\rho(u(z) - u(z=0)) = \frac{J_m}{\kappa u_*}\left[\ln\left(\frac{z}{z_{0m}}\right) - \psi_m\right] = \frac{u_*}{\kappa}\left[\ln\left(\frac{z}{z_{0m}}\right) - \psi_m\right], \tag{9.59}$$

where $\kappa \approx 0.4$ is von Karmon's constant, z_0 is dynamic roughness, which is a fraction of the true surface roughness, ψ_m is a momentum-flux stability function that is nonzero with stable and unstable boundary layers. For unstable boundary layers, ψ_m is given by (Brutsaert, 1982)

$$\psi_m = \ln\left[\frac{1 + (1 - 16z/L)^{1/2}}{1 + (1 - 16z_{0m}/L)^{1/2}}\right] + 2\ln\left[\frac{1 + (1 - 16z/L)^{1/4}}{1 + (1 - 16z_{0m}/L)^{1/4}}\right]$$

$$- 2\arctan\left((1 - 16z/L)^{1/4}\right) + 2\arctan\left((1 - 16z_{0m}/L)^{1/4}\right) \tag{9.60}$$

and L is Obukhov's stability length,

$$L = -\frac{u_*^3}{\kappa B}, \tag{9.61}$$

where B is the buoyancy flux per unit area, expressed as

$$B = \frac{g\beta_T(\phi_c + 0.07\phi_e)}{\rho C_p}, \tag{9.62}$$

where ρ is the density of air, C_p is the specific heat capacity of dry air, ϕ_c is the heat flux across the water surface by conduction, and ϕ_e is the evaporative heat flux from the water surface. Equation 9.60 results from the integration of a relatively simple equation curve-fit to field data. The integration results in this complicated equation. For a smooth surface, u_*z_{0m}/ν and u_*z_{0A}/D_A are both equal to 9.6. At most wind velocities and wind fetches, however, the water surface is not hydrodynamically smooth, and another relation is needed.

The equation for mass flux of compound A, J_A, and mass of A per unit volume will be written by analogy to Equation 9.11:

$$C_A(z) - C_A(z=0) = \frac{J_A}{\kappa u_*}\left[\ln\left(\frac{z}{z_{0A}}\right) - \psi_A\right], \tag{9.63}$$

where ψ_A is given by (Brutsaert, 1982)

$$\psi_A = 2\ln\left[\frac{1+(1-16z/L)^{1/2}}{1+(1-16z_{0A}/L)^{1/2}}\right]. \tag{9.64}$$

If Equations 9.59 and 9.63 are combined to eliminate shear velocity, an equation can be developed for the flux of compound A:

$$J_A = \frac{\kappa^2(C_A(z) - C_A(z=0))(u(z) - u(z=0))}{\left[\ln\left(\frac{z}{z_{0A}}\right) - \psi_A\right]\left[\ln\left(\frac{z}{z_{0m}}\right) - \psi_m\right]}. \tag{9.65}$$

Equation 9.65 provides us with an expression for the gas film coefficient for an atmospheric boundary layer:

$$K_G = \frac{\kappa^2(u(z) - u(z=0))}{\left[\ln\left(\frac{z}{z_{0A}}\right) - \psi_A\right]\left[\ln\left(\frac{z}{z_{0m}}\right) - \psi_m\right]}. \tag{9.66}$$

The gas film coefficient and gas flux rate can be estimated through iteration on Equations 9.59 through 9.66, if ϕ_c and ϕ_e are known, and a good relationship for z_{0m} and z_{0A} is known. Evaporative heat flux can be determined by applying Equation 9.65 to water vapor, and using the relationship:

$$\phi_e = L_v J_{H_2O}, \tag{9.67}$$

where L_v is the latent heat of vaporization,

$$L_v(\text{cal/g}) = 597 - 0.571T_s(°C), \tag{9.68}$$

where T_s is water surface temperature. Then, Bowen's ratio (Bowen, 1926) may be used to compute ϕ_c:

$$\phi_c = 0.61P_a(\text{bars})L_e K_G(T_s - T(z)), \tag{9.69}$$

where P_a is air pressure. The application of these equations to determine gas film coefficient will be demonstrated in Example 9.2.

9.4.2 MEASUREMENT

The evaporation of water is generally used to determine the gas film coefficient. A loss of heat in the water body can also be related to the gas film coefficient because the process of evaporation requires a significant amount heat, and heat transfer across the water surface is analogous to evaporation. Although the techniques described in Section 9.4.1 can be used to determine the gas film coefficient over water bodies, they are still iterative, location-specific, and dependent upon fetch or wind duration. For that reason, investigators have developed empirical relationships to characterize gas film coefficient from field measurements of evaporation or temperature. Then, the air–water transfer of a nonvolatile compound is given as

$$J_A = K_G(C_A(z) - C_A(z = 0)), \tag{9.70}$$

where $C_A(z)$ is the concentration of compound A at an elevation z, which is typically the elevation of wind and temperature measurements to compute K_G. Conversely, measurements of nonvolatile flux can be used to estimate K_G:

$$K_G = \frac{J_A}{C_A(z) - C_A(z = 0)}. \tag{9.71}$$

The flux J_A is typically measured by pan evaporation, lake evaporation, cooling pond heat transfer or an "eddy correlation" technique that relates turbulent diffusion of water vapor to mass flux. The concentration of water vapor at elevation z, $C_A(z)$, is determined from simultaneous measurements of temperature and relative humidity and an application of Equations 9.56 and 9.57. At the water surface ($z = 0$), it is assumed that the air temperature is equal to water temperature and that relative humidity is 100%, so water temperature measurements and Equation 9.56 are sufficient to determine $C_A(z = 0)$. If the analog of heat transfer is used, then

$$K_G = \frac{\rho C_p J_T}{L_e(T(z) - T(z = 0))}, \tag{9.72}$$

where L_e is the latent heat of vaporization for water, ρ is the density of air, and C_p is the heat capacity of air at constant pressure.

Although pan evaporation data can be used to estimate K_G (Linacre, 1994), the differences in scale between the pan and the water body of interest have resulted in considerable bias with regard to the effect of boundary layer development (wind fetch and sheltering) and water surface roughness (waves) on the concentration boundary layer. There have been detailed, long-term studies on lake evaporation (Kohler, 1954; Harbeck et al., 1958; Turner, 1966; Hughes, 1967; Ficke, 1972) that would incorporate multiple recordings of a few days of evaporation measurements compared to wind velocity.

The use of heat as a proxy on heated water bodies was first reported by Rymska and Dochenko (1958), and followed by a number of investigators. These measurements required less time, and could be more precise in relation to atmospheric parameters.

Because natural convection can be significant with heated water bodies, these measurements were also related to measurements of temperature and density differences between the two elevations.

The eddy correlation technique (Jones and Smith, 1977) measures turbulent vapor fluxes away from the water surface with the rapid response to water vapor and velocity measurements that are possible with current instruments. Then,

$$J_V = \overline{\rho w' C_V'}. \qquad (9.73)$$

9.4.3 PREDICTION

The early measurements were fit to equations that incorporated wind velocity. A revelation occurred in 1969, when Shulyakovskyi brought in buoyancy forces as related to natural convection to explain the heat loss from heated water at low wind velocities. This was picked up by Ryan and Harleman (1973), who realized that natural convection could explain the need for a constant term in front of the relationship for gas film coefficient, as had been found by Kohler (1954), Dochenko (1958), Brady et al. (1969), and Shulyakovskyi (1969). Finally, Adams et al. (1990) rectified the overprediction of Ryan and Harleman's formulation at high wind velocities with the root sum of squares relation and brought in Harbeck's (1962) relation to include a term that represents fetch dependence.

The relationships developed from field measurements have been made dimensionless with the assumptions that $\nu = 1.33 \times 10^{-5}\,\mathrm{m^2/s}$ and $D_{H_2O} = 2.6 \times 10^{-5}\,\mathrm{m^2/s}$, to facilitate comparisons between relations and avoid dimensional problems, and are given in Table 9.3. The equation of Adams et al. (1990) is preferred by the author:

$$\mathrm{Sh} = \left[\left(0.125\mathrm{Ra}_V^{1/3} \right)^2 + \left(0.0061\mathrm{ScRe}^{0.933}\hat{W}^{0.033} \right)^2 \right]^{1/2}, \qquad (9.74)$$

where the parameters in Equation 9.74 are described in Table 9.3. The gas film coefficients from Table 9.3 will be compared in Example 9.3.

9.5 SUMMARY GUIDE

9.5.1 USER ROADMAP

A user roadmap is given in Figure 9.13. For the air–water transfer of a given chemical, one needs to first determine the Henry's law coefficient. Then choose the appropriate ratio of K_L/K_G, and determine whether the compound is water-phase controlled, gas-phase controlled, or controlled by both phases. Then, one simply needs to choose the correct equation(s) for the application.

The data used to develop these equations are summarized in Figures 9.14 through 9.16 for streams, wind-influenced liquid film coefficient, and wind-influenced gas transfer coefficient, respectively. There is a large amount of data with substantial scatter in Figure 9.16, so approximate 67% confidence intervals are given.

TABLE 9.3
Relationships Developed to Characterize Gas Film Coefficient Over Water Surfaces

Investigator	Formula	Water Body
Kohler (1954)	$Sh = 0.011 \left(\dfrac{Ra}{\beta_T \Delta T} \right)^{1/3} + 7.5 \times 10^{-5} Pe$	Lake Hefner, AZ
Marciano and Harbeck (1954)	$Sh = 0.0017 Pe$	Lake Hefner, AZ
Harbeck et al. (1958)	$Sh = 0.002 Pe$	Lake Mead, AZ
Rymsha and Dochenko (1958)	$Sh = 0.042 \left(\dfrac{Ra}{\beta_T \Delta T} \right)^{1/3} + 0.0125 Ra^{1/3} + 0.0016 Pe$	Various rivers, heated in winter
Harbeck (1962)	$Sh = 0.0061 ScRe^{0.933} \hat{W}^{0.033}$	Various reservoirs
Turner (1966)	$Sh = 0.0031 Pe$	Lake Michie, NC
Hughes (1967)	$Sh = 0.0014 Pe$	Salton Sea, CA
Brady et al. (1969)	$Sh = 0.049 \left(\dfrac{Ra}{\beta_T \Delta T} \right)^{1/3} + 9.6 \times 10^{-6} ScRe^{4/3} \hat{W}^{1/3}$	Power plant cooling ponds
Shulyakovskyi (1969)	$Sh = 0.031 \left(\dfrac{Ra}{\beta_T \Delta T} \right)^{1/3} + 0.125 Ra_V^{1/3} + 0.0017 Pe$	Various inland water bodies
Fike (1972)	$Sh = 0.0019 Pe$	Pretty Lakes, IN
Ryan and Harleman (1973)	$Sh = 0.125 Ra_V^{1/3} + 0.0016 Pe$	Cooling ponds
Anderson and Smith (1981)	$Sh = 0.0010 Pe$	Atlantic Ocean
Gulliver and Stefan (1986)	$Sh = 0.125 Ra_V^{1/3} + 0.0013 Pe$	Heated streams
Adams et al. (1990)	$Sh = \left[\left(0.125 Ra_V^{1/3} \right)^2 + \left(0.0061 ScRe^{0.933} \hat{W}^{0.033} \right)^2 \right]^{1/2}$	East Mesa Geothermal Facility and cooling ponds

$Sh = \dfrac{K_G A^{1/2}}{D}$

K_G = gas film coefficient

A = surface area of water body

D = diffusion coefficient of compound in air

W^1 = Wind at 10 m height converted to 2 m with 1/7 power law

$Ra = \dfrac{g \beta_T \Delta T A^{3/2}}{D v}$

W = wind velocity at 2 m above mean water surface

$Pe = \dfrac{W A^{1/2}}{D}$

v = kinematic viscosity of air

α = thermal diffusion coefficient of air

g = acceleration of gravity

$Ra_V = \dfrac{g \beta_T \Delta T_V A^{3/2}}{D v}$

β_T = thermal expansion coefficient of moist air

ΔT = temperature difference between water surface and 2 m height and

$\hat{W} = \dfrac{W^2}{g A^{1/2}}$

ΔT_V = virtual temperature difference between water surface and 2 m height

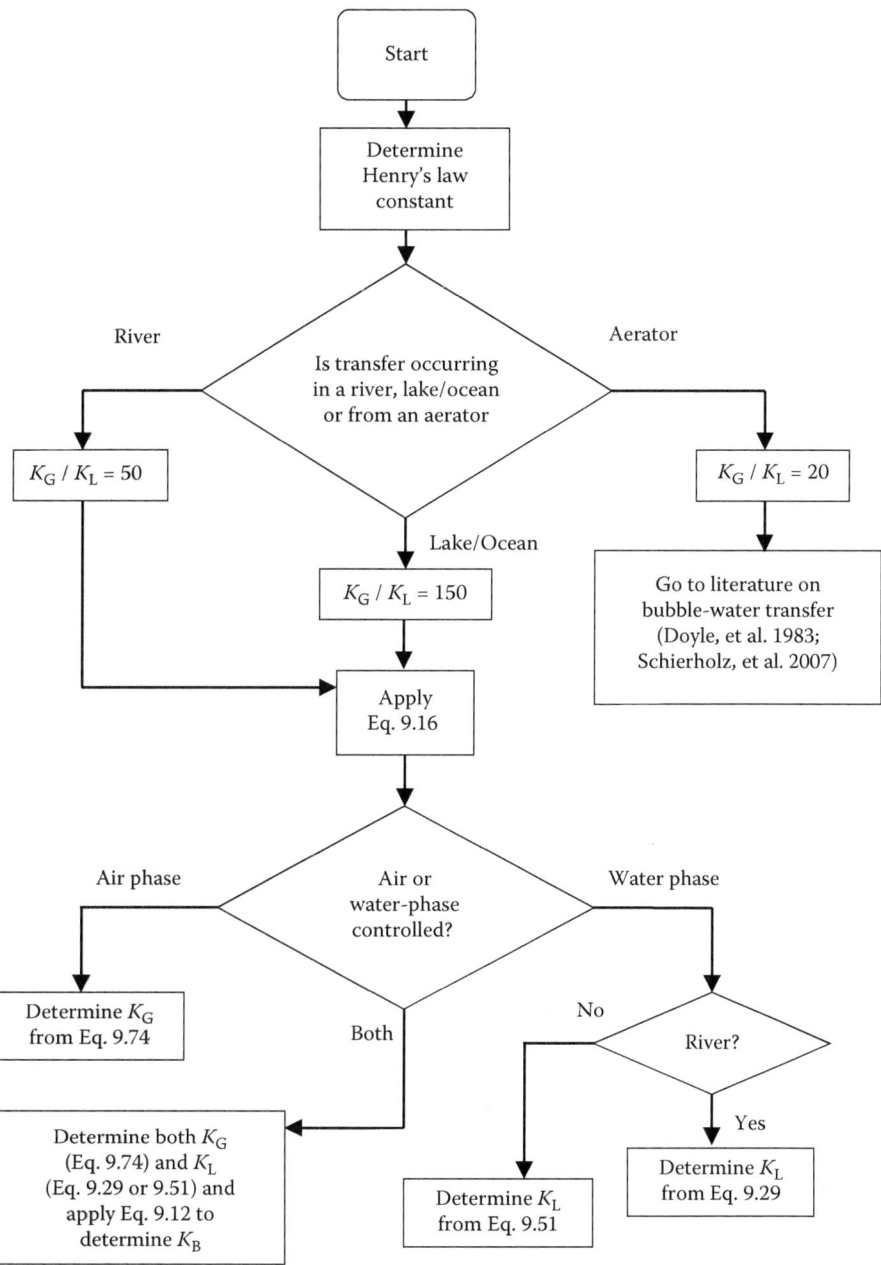

FIGURE 9.13 User roadmap for application of the equations provided in Chapter 9.

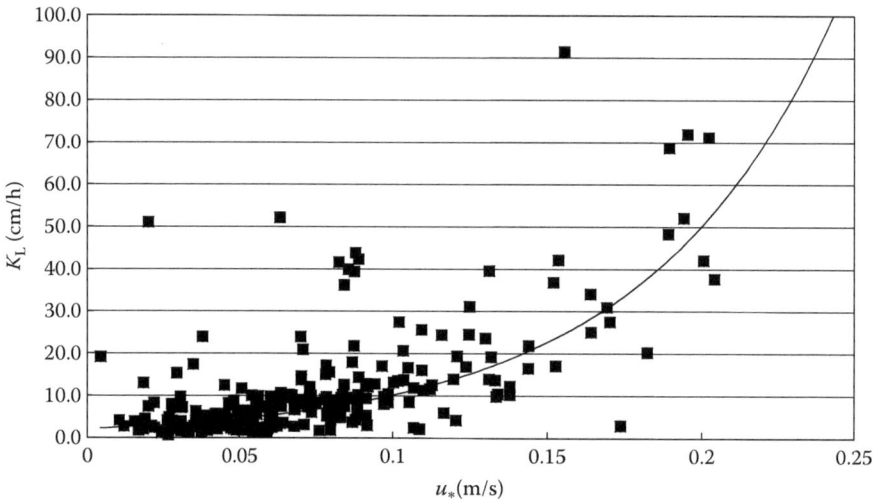

FIGURE 9.14 Measurements of liquid film coefficients for field streams versus bottom shear velocity, $u_* = \sqrt{gHS}$, where g is acceleration of gravity, H is mean stream depth, and S is stream slope. An exponential curve is fit through the data. (Data from Moog, D. B. and Jirka, G. H. 1998. *J. Environ. Eng.*, 124(2), 104.)

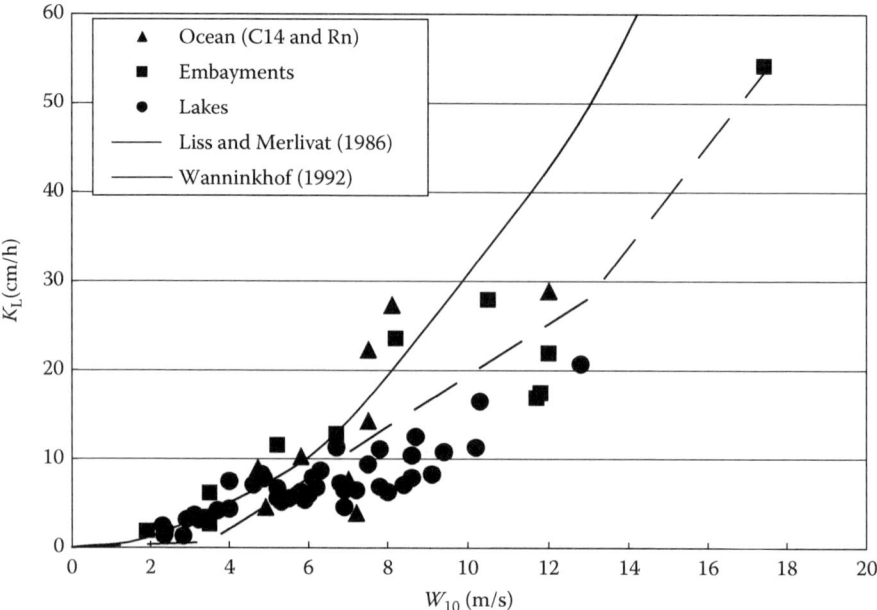

FIGURE 9.15 Measurements of liquid film coefficients influenced by wind. W_{10} is wind velocity at 10 m height and Oceans, Embayments, and Lakes are from Figures 9.7, 9.8, and 9.11, respectively. Schmidt number of 600 and temperature of 20°C are assumed.

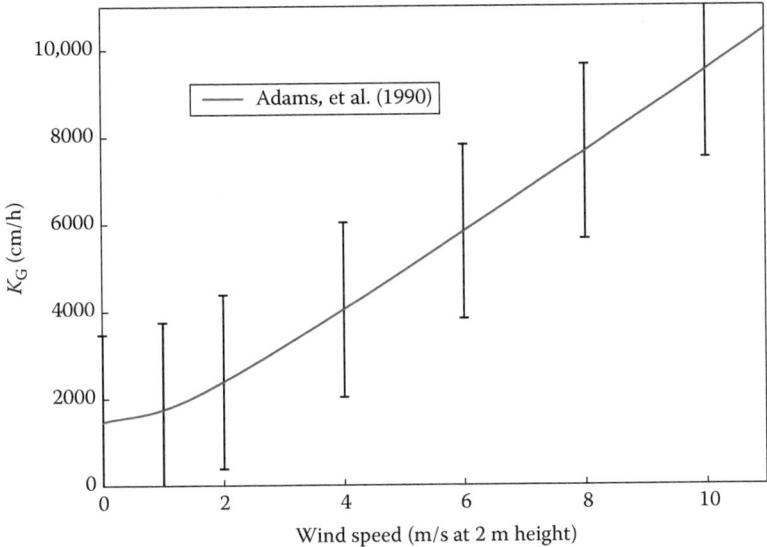

FIGURE 9.16 Prediction equation of Adams et al. (1990) with a rough 67% confidence interval of the measured data from Harbeck (1962), Brady, et al. (1969), Ryan and Harleman (1973), Gulliver and Stefan (1986), and Adams, et al. (1990).

9.5.2 Application Examples

Example 9.1: Estimation of K_2 Using Measured Independent Parameters and Two Types of Mean Values

Most natural rivers and streams are a series of pools and riffles. Calculate the K_2 for one pool riffle pair of equal lengths in a river carrying 12 m³/s at 20°C, using arithmetic means and the mean values weighted according to Equation 9.29. The pool has a width of 60 m, a mean depth of 2 m, and a bottom roughness of 2 mm. The riffle has a width of 10 m, mean depth of 0.5 m and is a gravel bottom with 20 mm roughness. No stream slope can be measured over such a short reach.

Equation 9.29 gives us:

$$K_L a = 4.4 \times 10^{-3} Sc^{-1/2} (1 + F^{1/2}) \frac{u_*}{h}. \tag{E9.1.1}$$

The $Sc^{-1/2}$ relationship will convert for the various gases. For conversion to K_L, use the relationship:

$$K_L = K_L a R_h, \tag{E9.1.2}$$

where $R_h = A/P$ is the hydraulic radius of the stream. In this case, the following equation that results from the definition of the Darcy–Weisbach friction factor will be used to determine u_*:

$$u_* = \sqrt{g R_h S} = \sqrt{\frac{f}{8}} U, \tag{E9.1.3}$$

where U is cross-sectional mean velocity and f is the Darcy–Weisbach friction factor. Moody's diagram can be used to determine f. The Reynolds number, $Re = 4UR_h/\nu$ and the relative roughness, $\varepsilon/4R_h$, are needed, where A is the cross-sectional area, and P is the wetted perimeter. In general, the term $4R_h$ takes the place of the diameter in these calculations for noncircular cross-sections. The mean stream slope, $S = f\,U^2/(8gR_h)$, rather than the mean shear velocity, will be computed, because that is what is measured in practice. Then, Equation E9.1.1 becomes

$$K_La = 4.4 \times 10^{-3}Sc^{-1/2}(1 + Fr^{1/2})\frac{(gR_hS)^{1/2}}{h} \qquad \text{(E9.1.4)}$$

or, assuming that R_h is close to h for a wide stream,

$$K_La \cong 4.4 \times 10^{-3}Sc^{-1/2}(1 + Fr^{1/2})\left(\frac{gS}{h}\right)^{1/2} \qquad \text{(E9.1.5)}$$

and the following parameters can be calculated for the pool and riffle:

Pool

$U = Q/A = 0.1\,\text{m/s}$

$P = 60 + 4 = 64\,\text{m}$, assuming a rectangular cross-section
$R_h = A/P = 1.88\,\text{m}$
$Re = 7.5 \times 10^5$
$\varepsilon/(4R_h) = 2.7 \times 10^{-4}$
$f \sim 0.0155$
$S = 0.0155 * 0.1^2/(8 * 9.8 * 1.88) = 1.1 \times 10^{-6}$
$Fr = 0.1/(9.8 * 1.88)^{1/2} = 0.023$
$Sc = 1 \times 10^{-6}/2.1 \times 10^{-9} = 476$

Riffle

$U = 2.4\,\text{m/s}$
$P = 11\,\text{m}$
$R_h = 0.46\,\text{m}$
$Re = 4.4 \times 10^6$
$\varepsilon/(4R_h) = 0.010$
$f \sim 0.038$
$S = 0.038 * 2.4^2/(8 * 9.8 * 0.46) = 6 \times 10^{-3}$
$Fr = 2.4/(9.8 * 0.46)^{1/2} = 1.13$
$Sc = 476$

These values can now be used to calculate the means and K_La:

Arithmetic mean:

$S = (1.1 \times 10^{-6} + 6 \times 10^{-3})/2 = 3 \times 10^{-3}$
$h = (2 + 0.46)/2 = 1.23\,\text{m}$
$Fr = (0.1 + 2.4)/2/(9.8 * 1.23)^{1/2} = 0.36$

Then, Equation E9.1.3 gives

$$K_L a = 9.3 \times 10^{-5}\,s^{-1}$$

Means weighted according to Equation E9.1.4

$S^{1/2} = ((1.1 \times 10^{-6})^{1/2} + (6 \times 10^{-3})^{1/2})/2 = 3.9 \times 10^{-2}$
$S = 1.5 \times 10^{-3}$
$1/h^{1/2} = (1/2^{1/2} + 1/0.5^{1/2})/2 = 1.06\,m^{-1/2}$
$h = 0.89\,m$
$Fr^{1/2} = (0.023^{1/2} + 1.13^{1/2})/2 = 0.61$
$Fr = 0.37$

and Equation E9.1.3 gives

$$K_L a = 4.2 \times 10^{-5}\,s^{-1}$$

The difference between the two means is a factor of 2.2. This value is larger than the expected error of Equation 9.29. Thus, a channel with a variation in slope and cross-section along its length will have a higher $K_L a$ value computed from arithmetic means than an otherwise equivalent channel, which does not have variation in slope and cross-section. It may not be a coincidence that Moog and Jirka's "calibration" of Thackston and Krenkel's equation for flumes is an adjustment by a factor of 0.69 to represent field measurements.

Example 9.2: Computation of Gas Film Coefficient Over a Water Body

A 10 km long lake is exposed to various wind speeds. On a cold day in fall, the water temperature at 10°C has not cooled yet, but the air temperature is at −5°C. The relative humidity is 100%. Compute the gas film coefficient for water vapor at various wind fetch lengths for wind speeds, at 2 m height, between 2 and 20 m/s.

For these conditions, we have the following water vapor concentrations:

$$C_{H_2O}(z = 0) = p_{vs}\rho/P_a = 13\,g/m^3, \text{ assuming that } P_a = 1.013\,bars, \text{ and}$$

$$C_{H_2O}(z) = 4\,g/m^3$$

In addition, we need to find a parametric relationship or z_0, which will be equal to z_{0H_2O} at velocities above 2 m/s, because the boundary of the water surface will be rough. Gulliver and Song (1986) combined relationships of Hsu (1974) and Mitsuyatsu (1969) for gravity waves to get:

$$\frac{z_0 g}{u_*^2} = 0.00942\pi \left(\frac{gF}{u_*^2}\right), \tag{E.9.2.1}$$

where F is wind fetch, and the relations of Hsu and Long and Hwang (1976) for capillary waves to get

$$\frac{z_0 g}{u_*^2} = 0.0219\pi \left(\frac{gF}{u_*^2}\right)^{1/2} \left(\frac{\sigma}{\rho g F}\right)^{1/3}, \qquad \text{(E.9.2.2)}$$

where σ is the surface tension of clean water exposed to air. The transition is presumed to occur when z_0 from Equation E9.2.2 becomes greater than that of Equation E9.2.1. Finally, Wu (1975) gives

$$u(z = 0) = 0.55u_* \sqrt{\frac{\rho}{\rho_w}}, \qquad \text{(E.9.2.3)}$$

where ρ_w is the water density.

We will assume that $z_{0m} = z_{0A} = z_0$, which is generally true except for a smooth water surface. When we use Equations E9.2.1 through E9.2.3, along with the equations provided above (9.53 through 9.65) to compute K_G for water vapor versus fetch for different wind velocities, the result is that provided in Figure E9.2.1. Iteration was required on the entire set of equations. Also provided in Figure E9.2.1 are similar results for neutral conditions, with $\psi_m = \psi_{H_2O} = 0$. The gas film coefficient decreases with fetch length as z_0 decreases due to an increase in wave velocity. Even though the water surface is not moving quickly, as Equation E9.2.3 can be used to demonstrate, the waves, which provide the roughness, are moving faster as the wind fetch increases. Note that the order of importance in determining K_L is wind speed, wind fetch and finally boundary layer neutrality.

Example 9.3: Application of Characteristic Relations for Gas Film Coefficient

A consulting company has a project that requires determining the evaporation from the 10 ha cooling pond at a thermal power facility, and they realized that they do not know how to determine the gas film coefficient. A table similar to Table 9.3 was found, and they decided to compare the resulting predictions to see if it made a significant difference. Duplicate their results for one such comparison under the following conditions:

Water temperature, T_s, $= 30°C$.
Air temperature, $T = 10°C$
Relative humidity at 2 m height $= 40\%$
Air pressure $= 1$ atm
Various wind velocities up to 15 m/s

The *Handbook of Chemistry and Physics* can be used to determine the following fluid properties:

$\nu = 1.33 \times 10^{-5}$ m²/s
$D = 2.4 \times 10^{-5}$ for water vapor (used in the determination of z_{0A})
$\alpha = 2.0 \times 10^{-5}$ m/s

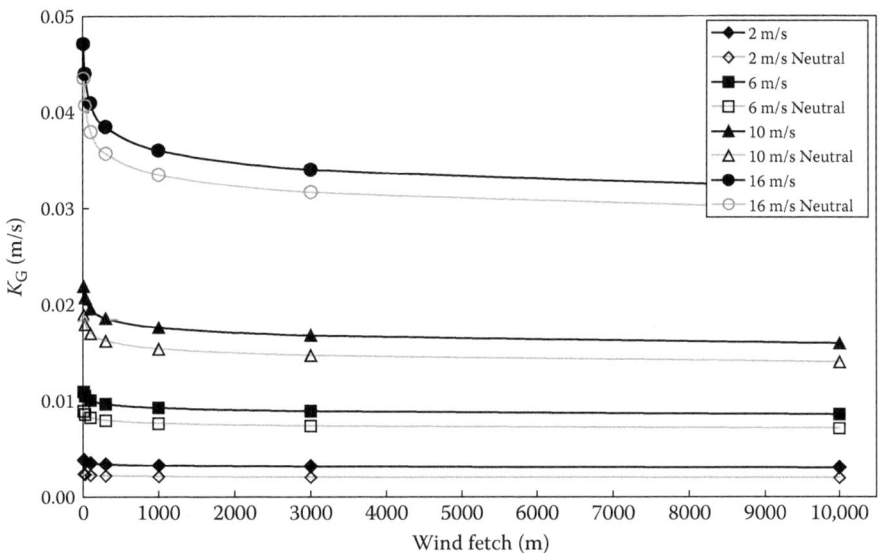

FIGURE E9.2.1 Computations of gas film coefficient for the unstable boundary layer of Example 9.2 and for a neutral boundary layer that is otherwise similar.

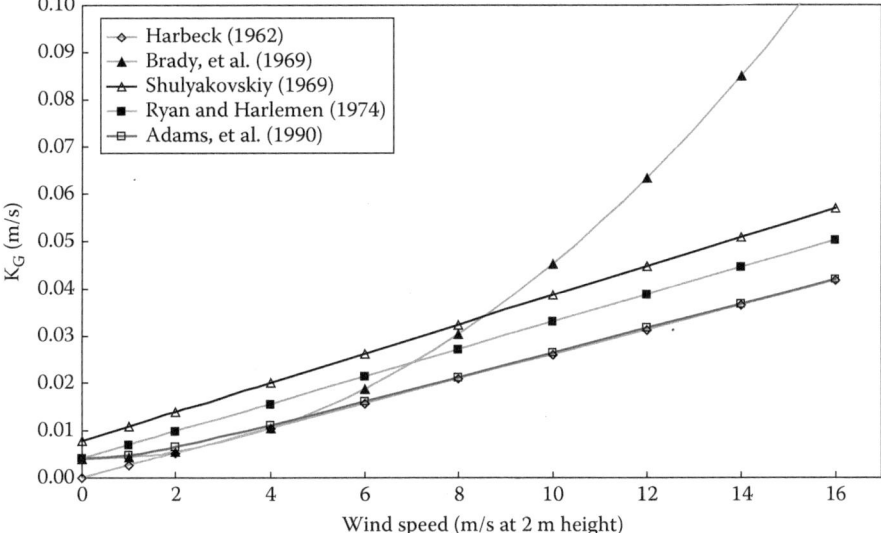

FIGURE E9.3.1 Comparison of five formulas for gas transfer coefficient versus wind speed for Example 9.3.

Now, virtual temperature is given by the equation:

$$\Delta\theta_v = \left\| T_s \left(1 + 0.378 p_{vs}/P_a\right) - T_a \left(1 + 0.378 p_{va}/P_a\right), 0 \right\| \tag{9.54}$$

and saturation vapor pressure will be required to compute virtual temperature:

$$p_{vs}(\text{mb}) = 6.11 \exp\left[\frac{17.3(T(^\circ K) - 273)}{T(^\circ K) - 35.9}\right]. \tag{9.55}$$

Then, $p_{vs}/P_a = 0.042$ at 30°C and $= 0.012$ at 10°C, and

$$p_{va} = \frac{\text{RH}}{100} p_{vs}. \tag{9.56}$$

So that $p_{va}/P_a = 0.005$. Then $\Delta\theta_V = 20.5°C$. Finally,

$$\beta_T = \frac{1}{T_{av}}(^\circ C) + 273. \tag{9.57}$$

So that $\beta_T = 3.29 \times 10^{-3}{}^\circ K^{-1}$. The results for five formulas are given in Figure E9.3.1. The major differences are the slope at high wind speeds and the intercept at a wind velocity of 0. Probably the best documented is the relation of Adams et al. (1990), which used much of the previous field data and tested the results on two additional water bodies at a variety of buoyancy parameters. This relation transitions between the Ryan and Harleman (1974) equation at low wind velocity and that of Harbeck (1962) at higher wind velocity.

LITERATURE CITED

Abernathy, R. B., Benedict, R. P., and Dowdell, R. B. 1985. ASME measurement uncertainty, *J. Fluids Eng.*, 107, 161.

Adams, E. E., Cosler, D. J., and Helfrich, K. R. 1990. Evaporation from heated water bodies: Predicting combined forced plus free convection, *Water Res. Res.*, 26(3), 425.

Anderson, R. J. and Smith, S. D. 1981. Evaporation coefficient for the sea surface from eddy flux measurements, *J. Geophys. Res.*, 86 (C1), 449.

Asher, W. E. and Wanninkhof, R. 1998. The effect of bubble-mediated gas transfer on purposeful dual-gaseous tracer experiments, *J. Geophys. Res.*, 103, 15993.

Atmane, M. A., Asher, W. E., and Jessup, A. T. 2004. On the use of the active infrared technique to infer heat and gas transfer velocities at the air-water free surface, *J. Geophys. Res.*, 109, C08S14.

Bowen, I. S. 1926. The ratio of heat losses by conduction and by evaporation from any water surface, *Phys. Rev.*, 27, 1.

Brady, K. D., Graves, W. L., and Geyer, J. C. 1969. *Surface Heat Exchange at Power Plant Cooling Lakes*. Report RP-49, Johns Hopkins University, Baltimore, MD.

Broecker, W. S. and Peng, T. H. 1971. The vertical distribution of radon in the Bomex area, *Earth Planet. Sci. Lett.*, 11, 99–108.

Broecker, W. S., Peng, T. H., Östlund, G., and Stuiver, M. 1985. The distribution of bomb radiocarbon in the ocean, *J. Geophys. Res.*, 90, 6953–6970.

Cember, R. 1989. Bomb radiocarbon in the red sea: A medium-scale gas exchange experiment, *J. Geophys. Res.,* 94, 2111–2123.

Clarke, J. F., Wanninkhof, R., Schlosser, P., and Simpson, H. J. 1994. Gas exchange rates in the tidal Hudson River using a dual tracer technique, *Tellus, Ser. B*, 46, 274.

Danckwerts, P. V. 1951. Significance of liquid-film coefficients in gas absorption, *Ind. Eng. Chem.*, 23(6), 1460.

Doyle, M. L., Boyle, W. C., Rooney, T., and Huibregtse, G. L. 1983. Pilot plant determination of oxygen transfer in fine bubble aeration, *J. Water Pollution Control Fed.*, 55(12), 1435.

Duffy, P. B. and Caldeira, K. 1995. Three-dimensional model calculation of ocean uptake of bomb 14C and implications for the global budget of bomb 14C, *Global Biogeochem. Cycles*, 9, 373.

Ficke, J. F. 1972. Comparison of evaporation computation methods, Pretty Lakes, lagrange City, Northwestern Indiana, *U.S. Geol. Surv. Prof. Pap.*, 65.

Frew, N. M. 1997. In *The Sea Surface and Global Change*. R. A. Duce and P. S. Liss (Eds). Cambridge University Press, New York, 121.

Friedlander, S. K. 1961. Theoretical considerations for the particle size spectrum of stratospheric aerosol, *AIChE J.*, 7, 317.

Gulliver, J. S. 2007. *An Introduction to Chemical Transport in the Environment*, Cambridge University Press, Cambridge, UK.

Gulliver, J. S. and Halverson, M. J. 1989. Air-water gas transfer in open channels, *Water Res. Res.*, 25, 1783.

Gulliver, J. S. and Song, C. S. S. 1986. Dynamic roughness and the transition between windwave regimes, *J. Geophys. Res.*, 91(C4), 5145.

Gulliver, J. S. and Stefan, H. G. 1986. Wind function for a sheltered stream, *J. Environ. Eng. Div., Am. Soc. Civ. Eng.*, 112(2), 387.

Gulliver, J. S., Erickson, B., Zaske, A. J., and Shimon, K. S. 2002. Measurement uncertainty in gas exchange coefficients, In *Gas Transfer at Water Surfaces*. M. Donelan, W. Drennan, E. Monehan, and R. Wanninkhof (Eds). Geophysical Monograph 127, American Geophysical Union, Washington, DC.

Harbeck, G. E., Jr. 1962. A practical field technique for measuring reservoir evaporation utilizing mass-transfer theory, *U.S. Geol. Surv., Prof Pap.* 272E.

Harbeck G. E., Jr., Kohler, M. A., and Koberg, G. E. 1958. Water loss investigations: Lake Mead studies, *U.S. Geol. Surv., Prof Pap.* 298.

Haußecker, H., Reinelt, S., and Jähne, B. 1995. Heat as a proxy tracer for gas exchange measurements in the field: Principle and technical realization, In *Air-Water Gas Transfer.* B. Jähne and E. C. Monahan (Eds). Aeon Verlag, Hanau, Germany, 405.

Hibbs, D. E., Parkhill, K. L., and Gulliver, J. S. 1998. Sulfer hexafluoride gas tracer studies in streams, *J. Envir. Eng.*, 124(8), 752.

Hsu, S. A. 1974. A dynamic roughness equation and its application to wind stress determination at the air-sea interface, *J. Phys. Oceanogr.*, 4, 116.

Hughes, G. H. 1967. Analysis of methods used to measure evaporation from Salton Sea, California, *U.S. Geol. Surv., Prof Pap.* 272H.

Jähne, B. 1991. Heat as a proxy tracer for gas exchange measurements in the field: Principles and technical realization, In *Air-Water Mass Transfer.* S. C. Wilhelms and J. S. Gulliver (Eds). American Society of Civil Engineers, Reston, VA, 582.

Jähne, B., Fischer, K. H., Imberger, J., Libner, P., Weiss, W., Imboden, D., Lemnin, U., and Jaquet, J. M. 1984. Parameterization of air/lake gas exchange, In *Gas Transfer at Water Surfaces*. W. Brutsaert and G. H. Jirka (Eds). D. Reidel, Norwell, MA, 303.

Jähne, B., Munich, K. O., Bosinger, R., Dutzi, A, Huber, W., and Libner, P. 1987. On parameters influencing air-water gas exchange, *J. Geophys. Res.,* 91, 1937.

Jones, E. P. and Smith, S. D. 1977. A first measurement of sea-air CO_2 flux by eddy correlation, *J. Geophys. Res.*, 82, 5990.

Kohler, M. A. 1954. Lake and pan evaporation, Lake Hefner studies, *U.S. Geol. Surv., Prof. Pap.* 169.

Krenkel, P. A. and Orlob, G. T. 1962. Turbulent diffusion and the reaeration coefficient, *J. Sanit. Eng. Div. Am. Soc. Civ. Eng.*, 88(SA2), 53.

Kromer, B. and Roether, W. 1983. Field measurements of air-sea gas exchange by the radon deficit method during JASIN(1978) and FGGE(1979), *"Meteor" Forschungsergeb. Reihe A/B*, 24, 55–75.

Linacre, E. T. 1994. Estimating U.S. Class-A pan evaporation from few climate data. *Water Int.* 19, 5–14.

Liss, P. S. and Merlivat, L. 1986. Air–sea gas exchange rates: Introduction and synthesis, In *The Role of Air-Sea Exchange in Geochemical Cycling*. P. Buat-Menard (Ed.). D. Reidel, Norwell, MA, 113.

Livingstone, D. M. and Imboden, D. M. 1993. The non-linear influence of wind variability on gas transfer in lakes, *Tellus, Ser. B*, 45, 275.

Mackay, D. and Yuen, A. T. K. 1983. Mass transfer coefficient correlations for volatilization of organic solutes from water, *Environ. Sci. Tech.*, 17, 211.

Marciano, J. J. and Harbeck, G. E. 1954. Mass transfer studies, Lake Hefner studies, *U.S. Geol. Surv. Prof Pap.* 267.

Mitsuyatsu, H. 1968. On the growth of the spectrum of wind generated waves, *Rep. Res. Inst. Appl. Mech., Kyushu Univ.*, 16(55), 459.

Moog, D. B. and Jirka, G. H. 1998. Analysis on reaeration equations using mean multiplicitative error, *J. Environ. Eng.*, 124(2), 104.

Munz, C. and Roberts, P. V. 1984. Analysis of reaeration equations using mean multiplicative error, In *Gas Transfer at Water Surfaces*. W. Brutzert and G. H. Jirka (Eds). D. Reidel, 35, Dordrecht.

Peng, T. H., Takahashi, T., and Broecker, W. S. 1974. Surface radon measurements in the North Pacific Ocean station PAPA, *J. Geophys. Res.*, 79, 1772–1780.

Ryan, P. J. and Harleman, D. R. F. 1973. An analytical and experimental study of transient cooling pond behavior, T. R. 161, R. M. Parsons Laboratory, Massachusetts Institute of Technology, Cambridge, MA.

Rymsha, V. A. and Dochenko, R. V. 1958. The investigations of heat loss from water surfaces in winter time, *Turdy GGI*, 65, l.

Schierholz, E. L., Gulliver, J. S., Wilhelms, S. C., and Henneman, H. E. 2006. Gas transfer from air diffusers, *Water Res.*, 40(5), 1018.

Shulyakovskyi, L. G. 1969. Formula for computing evaporation with allowance for temperature of free water surface, *Sov. Hydrol. Selec. Pap.*, 6, 566.

Tamburrino, A., Aravena, C., and Gulliver, J. S. 2007. Visualization of 2-D divergence on the free surface and its relation to gas transfer, In *Transport at the Air Sea Interface—Measurements, Models and Parameterizations*, C. S. Garbe, R. A. Handler, B. Jähne, (Eds). Springer-Verlag, New York.

Tamburrino, A. and Gulliver, J. S. 2002. Free surface turbulence and mass transfer in a channel flow, *AIChE J.*, 48(12), 2732.

Thackston, E. L. and Krenkel, P. A. 1969. Reaeration prediction in natural streams, *J. Sanit. Eng. Div. Am. Soc. Civ. Eng.*, 95(SAI), 65.

Tsivoglou, E. C. and Wallace, J. R. 1972. Characterization of Stream Reaeration Capacity, Report EPA-R3-72-012, U.S. Environmental Protection Agency, Washington, DC.

Torgersen, T. G., Mathieu, G., Hesslein, W., and Broecker, W. S. 1982. Gas exchange dependency on diffusion coefficient: Direct 222Rn and 3He comparisons in a small lake, *J. Geophys. Res.*, 87, 546.

Turner, J. F. 1966. Evaporation study in a humid region, Lake Michie, North Carolina, *U.S. Geol. Surv. Prof Pap.* 272-G.

Upstill-Goddard, R. C., Watson, A. J., Liss, P. S., and Liddicoat, M. I. 1990. Gas transfer velocities measured in lakes with SF_6, *Tellus, Ser. B*, 42, 364.

Wanninkhof, R. 1992. Relationship between wind speed and gas exchange over the ocean, *J. Geophys. Res.*, 97, 7373.

Wanninkhof, R., Ledwell, J. R., Broecker, W. S., and Hamilton, M. 1987. Gas exchange on Mono Lake and Crowely Lake, California, *J. Geophys. Res.*, 92, 14,567.

Wu, J. 1975. Wind-induced drift currents, *J. Fluid Mech.*, 68, 49.

BIBLIOGRAPHY WITH ADDITIONAL REFERENCES

Brutsaert, W. H. 1982. *Evaporation into the Atmosphere—Theory, History and Application.* D. Reidel, Hingham, MA.

Brutzert, W. H. and Jirka, G. H. Eds. 1984. *Gas Transfer at Water Surfaces.* D. Reidel, Dordrecht.

CRC. 2005. *Handbook of Chemistry and Physics.* CRC Press, Boca Raton, FL.

Donelan, M., Drennan, W., Monehan, E., and Wanninkhof, R., Eds. 2002. *Gas Transfer at Water Surfaces.* Geophysical Monograph 127, American Geophysical Union, Washington, DC.

Garbe, C. S., Handler, R. A., and Jähne, B., Eds. 2007. *Tramport at the Air Sea Interface - Measurements, Models and Parameterizations*, Springer-Verlag, New York.

Gulliver, J. S. 2007. *An Introduction to Chemical Transport in the Environment*, Cambridge University Press, Cambridge, UK.

Jähne, B. and Monahan, E. C., Eds. 1996. *Air-Water Gas Transfer.* Aeon Verlag, Hanau, Germany.

Nezu, I. and Nakagawa, H. 1993. *Turbulence in Open Channel Flow.* Balkema, Rotterdam, The Netherlands.

Wilhelms, S. C. and Gulliver, J. S., Eds. 1991. *Air-Water Mass Transfer.* American Society of Civil Engineers, Reston, VA.

10 Deposition and Resuspension of Particles and the Associated Chemical Transport across the Sediment–Water Interface

Joseph V. DePinto, Richard D. McCulloch, Todd M. Redder, John R. Wolfe, and Timothy J. Dekker

CONTENTS

10.1 Introduction 254
     10.1.1 Scope of the Chapter 254
     10.1.2 Overview 254
10.2 Detailed Discussion of Deposition and Resuspension
     Processes 256
     10.2.1 Water Column Settling and Deposition to
           Bottom Sediments 257
           10.2.1.1 Settling of Individual Particles 258
           10.2.1.2 Formation of Aggregates 259
           10.2.1.3 Settling Speeds of Aggregates 260
           10.2.1.4 Deposition onto the Sediment Bed 261
     10.2.2 Resuspension/Erosion 263
           10.2.2.1 Critical Shear Stress 264
           10.2.2.2 Erosion Rates 265
           10.2.2.3 Bedload and Suspended Load 268
           10.2.2.4 Devices for Measurement of Erosion
                     Properties 270
     10.2.3 Net Sedimentation 271

10.3　Application of Processes..275
　　　10.3.1　Representation in Mathematical Models..........................276
　　　　　　　10.3.1.1　Overview of Sediment Transport Models.............276
　　　　　　　10.3.1.2　Sediment Transport Simulation in WASP.............277
　　　　　　　10.3.1.3　Sediment Transport Simulation in EFDC.............278
　　　　　　　10.3.1.4　Sediment Transport Simulation in SEDZLJ...........284
　　　　　　　10.3.1.5　Comparison of EFDC and SEDZLJ Sediment
　　　　　　　　　　　　Transport..287
　　　　　　　10.3.1.6　Model Calibration Considerations.....................288
10.4　Lower Fox River and Green Bay Case Study...............................289
10.5　Summary and Additional Information Needs..............................292
Literature Cited..295

10.1 INTRODUCTION

10.1.1 Scope of the Chapter

Sediment settling, deposition, and resuspension are sediment–water exchange processes that play an important role in determining contaminant exposures at contaminated sites, under both uncontrolled conditions and following remediation. Because of its importance in the long-term fate of contaminants, considerable research has been conducted in this area. The goal of this chapter is to provide a review of the knowledge base and approaches for quantifying the processes governing the transport of particles and chemicals adsorbed to those particles across the sediment–water interface. The review is not exhaustive, but rather represents the opinions and experiences of the authors aimed at providing the reader with both theoretical and quantitative information on this topic.

10.1.2 Overview

Deposition of particulate matter from the water column to bottom sediments occurs continuously and represents an important process for the removal of particulate matter and associated contaminants from the water column. It is also responsible for the accumulation of bottom sediments in aquatic systems and the associated modification of surface contaminant levels in those sediments. Much of the theory of particle settling in aquatic systems has been developed using noncohesive sediments, which are typically larger sand size particles that do not readily aggregate. Noncohesive sediments are generally of less relevance than cohesive sediments (sediments with a high fraction of silt and clay size particles that have a higher organic carbon content and are much more likely to aggregate) for understanding contaminant fate and transport. According to Stokes' law, particle settling is dictated by particle diameter and density, but additional important factors causing nonideal settling include particle shape and concentration, flow velocity and turbulence, and flocculation. Flocs formed by fluid shear and differential settling differ in formation time, character, including density, and settling rates. Flocs from differential settling are slower to form, but settle faster due to their larger size (Lick, 2009). Deposition and attachment of particles to the sediment bed have been described as probabilistic processes that are affected by turbulence at the sediment–water interface, the roughness of the sediment bed, and the cohesiveness of the depositing particles (Hoyal et al., 1997).

Resuspension of bottom sediments into the water column of aquatic systems represents an important source of particles and particle-associated contaminants into the water column. Unlike deposition, the resuspension process is very sporadic and short-lived, but when it does occur, the flux is generally quite large. Sediment resuspension occurs when hydraulic shear stress at the sediment–water interface rises above a critical level, sufficient to dislodge particles. Shear stress (τ, dyn/cm^2) is calculated as a function of shear velocity (u_*, cm/s) and water density (ρ, g/cm^3):

$$\tau = \rho(u_*)^2. \tag{10.1}$$

Resuspension and scour of noncohesive sediments are well understood as functions of particle diameter, and reasonable estimates of resuspension rates can be determined for noncohesive sediments with information about the system hydraulics and physical properties of the sediment. However, widely applicable relationships predicting cohesive sediment erosion have not yet been developed. Quantifying resuspension of cohesive sediments usually requires development of site-specific data and experimentation. Cohesive sediment resuspension has been observed to depend on sediment bulk density (or porosity), particle size, surface and porewater chemistry, algal colonization, bioturbation and gas formation within the sediments, in addition to bottom shear velocity.

Resuspension in rivers is driven by high bottom shear velocity associated with hydrological events that lead to high flow rates. Resuspension in estuaries is most often driven by tidally induced velocities, so that periodic resuspension and deposition cycles occur as a function of tidal cycles. Resuspension in coastal and nearshore systems is most often driven by wind-generated wave action.

The extent to which *net deposition* (i.e., sedimentation) or *net erosion* (i.e., scour) occurs in aquatic systems depends on the pattern of flow conditions (i.e., typical annual hydrograph), the solids loading magnitude and properties, the geometry of the river, and the time scale over which the net accumulation or erosion is being determined. From a geomorphological perspective, rivers are expected to evolve by scouring down to a benthic layer that has characteristics resistant to further erosion; however, when river systems have been modified by navigational dredging or excessive watershed sediment loading, net sedimentation may occur over long time frames such as may occur in a ponded system. Most often the net sedimentation in estuaries is not significant, but is greatest in the region of the maximum salinity gradient, which leads to what is known as a turbidity maximum.

The effect of sediment deposition and resuspension on net sedimentation/scour of bottom sediments and on associated contaminants is depicted in Figure 10.1. The relative rates of deposition and resuspension at a given location in the system over a specified time period govern whether net sedimentation or net scour occurs. Direct point sources, watershed erosion/washoff, and internal biological

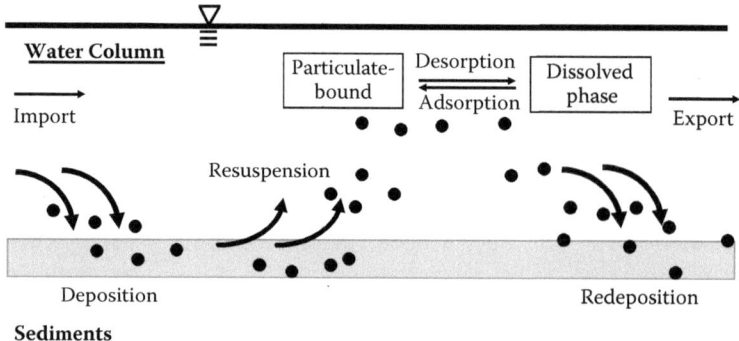

FIGURE 10.1 Illustration of the impact of deposition and resuspension processes on water column and sediment bed chemical concentrations.

production contribute solids for possible deposition. Settling and deposition are preferentially higher for the largest particles and for flow/geometry conditions that produce the lowest overflow rates (i.e., mean depth/hydraulic retention time). At higher velocities and associated shear stresses, resuspension can be initiated, resulting in sediment scour and downstream transport of solids, sorbed contaminants, and porewater.

Contaminant burial in bottom sediments is generally the result of particles depositing at rates exceeding resuspension, typically attenuated by mixing of surficial sediments. However, depending on the external source situation relative to previous contaminant loading to the sediments, it is possible for net contaminant flux from sediments to water to occur even though net solids sedimentation is occurring. If the upstream solids being transported to a specified location are "clean" relative to the existing surface sediments at that location (see Figure 10.1), the result of a net sedimentation will be a decrease in the concentration of surface sediments. While redeposition of resuspended solids may occur at downstream locations, sorbed contaminants are subject to phase partitioning during the time that the particles are suspended in the water column, including transfer to dissolved and vapor phases, as well as binding to dissolved and colloidal carbon.

10.2 DETAILED DISCUSSION OF DEPOSITION AND RESUSPENSION PROCESSES

This section includes a presentation of the theory and mathematical formulation for the quantification of settling, deposition, and resuspension processes for cohesive and noncohesive sediment types. It also includes a discussion of rate governing coefficients/parameters that must be specified for each of these processes. The section concludes with a discussion of net sedimentation and tools available to evaluate the evolution of the sediment bed in depositional environments.

10.2.1 Water Column Settling and Deposition to Bottom Sediments

Solids settling and deposition to bottom sediments is a complex process by which particulate materials, including both individual and aggregate solids, settle from the water column and adhere to the sediment bed. According to Stokes' law, particle settling is dictated by particle diameter and density, but important factors causing nonideal settling include particle shape and concentration, flow velocity and turbulence, and flocculation. Deposition onto and attachment to the sediment bed are usually described as probabilistic processes, affected by turbulence at the sediment–water interface and by the cohesiveness of the solid material.

The mechanisms that describe settling and deposition are controlled by physical and chemical properties of the water column–suspended solids, as well as the hydraulic conditions over the depth of the water column. Significant research has been conducted on these influencing factors and mechanics of settling. Mathematical models describing sediment transport in receiving waters have incorporated aspects of these mechanisms to varying degrees of complexity in order to describe the deposition of solids.

Physical properties which factor into the process of solids settling and deposition include

- particle diameter, or fractal dimension for aggregate (flocculated) material
- particle density
- fluid (water) density and viscosity
- particle shape
- particle concentration
- water velocity, and turbulence
- sediment bed roughness

Of these factors, the size, density, and shape of a particle are the most important determinants of settling velocity. Solids concentration and turbulence indirectly affect settling velocity by influencing formation of flocs, while sediment bed roughness is a factor in deposition. Floc formation is also strongly influenced by particle and surface chemistry. Chemical properties which factor into the process of floc formation include

- particle surface chemistry
- particle mineralogy
- water chemistry (e.g., marine versus fresh water environments, potential for formation of precipitates, etc.)

In the simplest case, uniformly spherical particles of known diameter and density settle at predictable rates (or settling speeds) in accordance with Stokes' law, which balances gravitational forces and drag:

$$w_s = \frac{d^2 g (\rho_s - \rho_w)}{18\eta}, \tag{10.2}$$

where w_s = settling velocity (m s^{-1}); d = particle diameter (m); g = acceleration of gravity (m s^{-2}); ρ_s = particle density (kg m^{-3}); ρ_w = fluid density (kg m^{-3}); and η = kinematic fluid viscosity (kg m^{-1} s^{-1}).

The above equation is based on SI units, but other consistent unit systems can also be used. Experiments have shown Stokes' law to be valid for particles up to about 100 μm in diameter in a laminar flow region, where Reynolds number ($\rho_w V_s d/\eta$) is less than about 0.5. Lower settling velocities than predicted by Stokes' law are observed for larger particles with settling speeds that result in a larger Reynolds number. Additionally, suspended solids in natural systems are generally neither uniform nor spherical, so their drag characteristics and therefore their settling behavior may deviate significantly from this ideal. Therefore, settling speeds for most particles are lower than those given by Stokes' law (Lick, 2009).

Generally, measurable (directly or indirectly) physical properties of the solids and the water in which these solids are suspended effectively determine the rate at which particles settle and whether or not hydraulic shear forces are sufficient to keep the particles suspended in the water column. The chemical properties of the solids and the water can also influence the deposition process through particle aggregation (flocculation) and the effect this has on the effective size, density, and shape of the suspended material.

The mechanics of settling are complex and are influenced by many factors. The studies that have investigated these factors reveal that flocculated or aggregate solids and individual particles have distinct behaviors with regard to settling behavior. Therefore, the research regarding the mechanics of settling is examined within each of these two categories.

10.2.1.1 Settling of Individual Particles

Although much of the suspended solids mass transported in natural aquatic systems likely exists in a flocculated state (Lick, 2009), there are still particles that are not flocculated. Examination of the settling of these particles is a necessary aspect for understanding overall solids settling behavior in aquatic systems. Some research has been performed on the settling of these individual particles. In addition, many existing models of sediment transport are predicated on an assumption of no flocculation in order to simplify the computational requirements.

Based on experimental data obtained with real particles, Cheng (1997a) developed an empirical formula for settling velocities over a range of diameters, temperatures, and Reynolds numbers. The same calculations were also performed using five previously proposed equations (Sha, 1956; van Rijn, 1989; Zhang, 1989; Ibad-zade, 1992; Zhu and Cheng, 1993). Cheng's proposed equation had the highest degree of predictive accuracy when each of these equations was compared to measured data. It is also applicable to a wide range of Reynolds numbers, and it predicts slightly lower settling velocities than Stokes' law, as shown in Figure 10.2 (Cheng's equation is presented in Section 10.3). Using the simplified equation presented in Cheng (1997a), Cheng (1997b) also developed a method to estimate "the effect of sediment concentration on the settling velocity of uniform, cohesionless particles." This hindered settling formulation allows for the lowering of settling velocity for

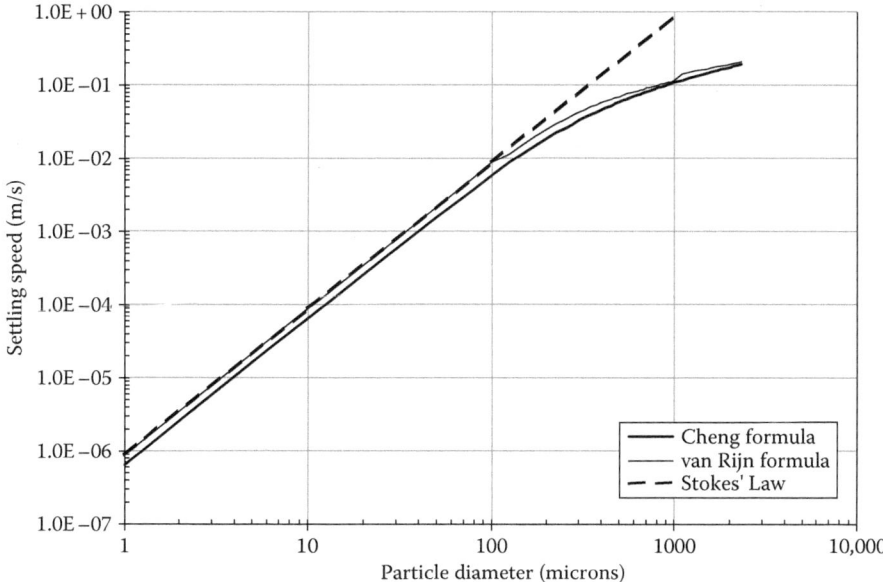

FIGURE 10.2 Settling velocity as a function of particle diameter based on Stokes' law and formulas proposed by Cheng (1997a) and van Rijn (1989).

closely spaced particles in a fluid as compared to an identical isolated particle in a clear fluid.

10.2.1.2 Formation of Aggregates

While actual settling velocities of discrete particles in natural systems differ from the predictions of Stokes' law, the most important deviation is due to particle aggregation, also referred to as flocculation. Most of the suspended material found in rivers, for example, has been found to be in the form of floc (Droppo and Ongley, 1994). Much of the research on mechanisms of settling examines flocculated particles and the process of flocculation, or aggregation, of particles due to the relatively greater abundance of these types of solids in natural systems, and the important effect of flocculation on settling velocity. In contrast to individual particles, flocs have much lower effective densities, larger specific surface areas, and fluids may flow through as well as around the aggregated particles. This section provides a brief summary of the flocculation process and reviews the research related to how the conditions under which flocs form impact floc characteristics such as size and rate of formation.

In order for flocs to form, particles must come into contact and there must be a sufficient attraction to hold them together. Clay and other very fine particles in suspension can be subject to mutual attraction, due to van der Waals forces, and repulsion, due to exposure of surface cations. The presence of anions in solution near the particle surface tends to lessen the repulsive force, promoting particle aggregation.

The greater the likelihood of collisions due to mixing of the fluid, and the higher ionic strength of the fluid, the greater is the tendency for particles to combine into flocs.

Three processes can be distinguished as causes of particle collisions: fluid shear, differential settling, and Brownian motion (Ives, 1978). In theory, fluid shear is the dominant process for forming flocs of particles having similar sizes, and differential settling dominates as a process by which smaller particles aggregate with larger particles (more than about $50\,\mu$m in diameter). Brownian motion is dominant only for the very smallest particles and is therefore of less importance. Lick and Ziegler (1992) developed a numerical model of floc formation that encompasses flocculation due to both fluid shear and differential settling, which provides a time-dependent expression for median floc size as a function of solids concentration and velocity gradient, or shear.

Flocs formed by fluid shear and by differential settling differ in formation time and in character. Differential settling is a slower process, due to a lower rate of particle collisions, and tends to form larger and more fragile flocs. For fine-grained sediments, Lick et al. (1993) showed that hydraulic shear effects dominate in high-turbulence areas, with higher solids concentration and ionic strength leading to more rapid floc formation and smaller steady-state floc size. They also showed that differential settling dominates in open waters, away from shore, resulting in much greater median particle diameters and time of formation than fluid shear effects. Differential settling has also been shown to be important in marine environments (Stolzenbach, 1993). Because of the importance of ionic strength in promoting flocculation, net deposition in estuaries is greatest in the region of the maximum salinity gradient, which leads to what is known as a turbidity maximum.

10.2.1.3   Settling Speeds of Aggregates

The formation of flocs in natural systems impacts the rate at which particles settle to the sediment bed. Researchers have also found that how a floc is formed (e.g., differential settling or fluid shear) influences the settling speed of the aggregates. Lick et al. (1993) found that settling speeds of flocs are much larger when produced by differential settling than when fluid shear is dominant, primarily due to larger floc size. They also found that the settling speed increases more rapidly as a function of floc diameter when produced by differential settling. In this study, settling speeds of flocs produced by differential settling in freshwater increased with floc diameter and were on the order of 10^{-4} to 10^{-2} m/s for floc diameters of approximately $100\,\mu$m to 1 mm. For flocs created by fluid shears, Burban et al. (1990) showed that for a given diameter, flocs formed at higher shears and sediment concentrations have higher settling speeds, and that the effect of salinity is to slightly increase settling speed. In this study, settling speeds of flocs produced by fluid shear in freshwater were on the order of 10^{-5} to 10^{-4} m/s, with floc diameters ranging from 10 to $200\,\mu$m. Comparing results of both studies to the settling rates for individual particles shown in Figure 10.2 demonstrates that flocs settle much more slowly than individual particles of the same diameter largely because of a flocs lower effective density. However, by aggregating very small particles (i.e., less than $10\,\mu$m) into larger diameter flocs, the process of flocculation can significantly increase the rate at which small particles settle.

Drag forces on a floc differ from those exerted on an individual spherical particle, and particle porosity and structure can profoundly affect settling rates. Floc porosity delays boundary layer separation, thus reducing drag, and Wu and Lee (1998) have shown that drag force for a highly porous sphere may be only 20% of that for a nonporous sphere, resulting in higher settling velocities. Settling velocities of fractal aggregates can also be many times higher than predicted by Stokes' law for spherical particles (Johnson et al., 1996).

10.2.1.4   Deposition onto the Sediment Bed

The actual deposition of solids onto the sediment bed is somewhat more complex than is described by the settling speed of the suspended material alone. Turbulence at the sediment–water interface may act as a barrier to the attachment of settling materials. Suspended solids in natural aquatic systems can exhibit a wide range of properties (e.g., size, density, shape, porosity, etc.), so a portion of the material which settles through the water column may remain in suspension or become associated with bed load transport instead of depositing onto the surface of the sediment bed. Because of high fluid shear and solids concentration gradients near the sediment water interface, flocculation may be an important process affecting cohesive sediment deposition.

Krone (1962) developed the empirical concept of "probability of deposition" in order to describe the observed depositional behavior of fine-grained (cohesive) solids. It is assumed that all solids settling near the bed remain on the bed under quiescent conditions, while at some threshold flow-induced bed surface stress, none of the solids settling near the bed will reach and remain on the bed. Between these two conditions the probability of deposition is a decreasing function of flow-induced bed-surface stress. Ariathuri and Krone (1976) used this concept to account for various factors (e.g., hydraulic shear and particle variability) that determine what fraction of a particulate class of solids is truly depositional. Partheniades (1992) developed a nonlinear empirical formula, which is compared graphically to the linear function proposed by Krone (1962) in Figure 10.3 for cohesive sediments. For larger noncohesive particles (i.e., with diameters greater than 200 μm), the function developed by Gessler (1967) based on an assumed Gaussian distribution is often used. Figure 10.4 shows the resulting probability of deposition curves for five noncohesive particle diameters based on Gessler's approach.

The concept of probability of deposition has been applied in transport models for fine-grained sediments that may exist in suspension as both aggregate and individual particles. Notable models include STUDH by Ariathurai and Krone (1976) and SEDZL by Ziegler and Lick (1986), which has also been extended to coarser-grained (noncohesive) sediments by Ziegler and Nisbet (1994), Gailani et al. (1991), and Jones and Lick (2001).

It is an unresolved question, as posed by Lau and Krishnappan (1994), whether deposition and erosion and reentrainment can occur simultaneously for a given cohesive particle size. Lick (1982) has postulated that larger particles settle out while fines stay in suspension, and intermediates are both deposited and resuspended at some equilibrium rate depending on shear stress. In opposition to this point of view, Partheniades et al. (1968) and Mehta and Partheniades (1975) have suggested that

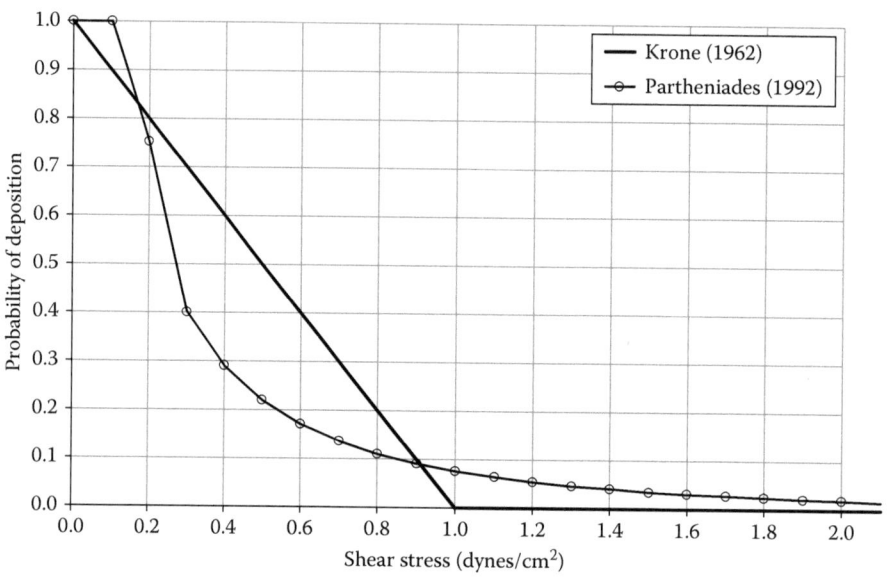

FIGURE 10.3   Probability of deposition for cohesive sediments as a function of shear stress.

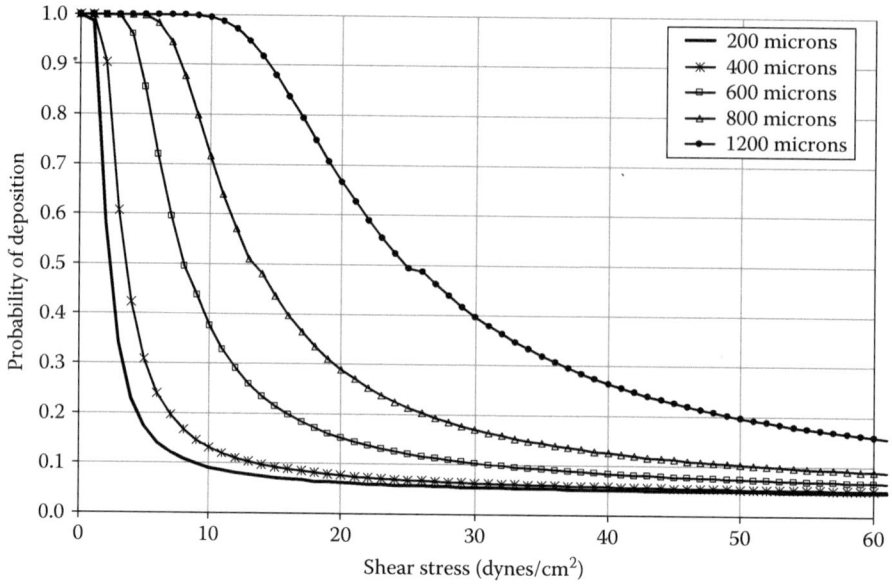

FIGURE 10.4   Probability of deposition for noncohesive sediment as a function of shear stress.

there is no simultaneous erosion and deposition. Rather, particles settle as flocs and bond to the bed if they are strong enough, or are broken up and returned to flow without depositing. In either case, there is a range of shear stresses for any given particle size within which a fraction of settling particles will deposit and the remaining fraction will be reentrained, consistent with the concept of probability of deposition.

10.2.2 RESUSPENSION/EROSION

Sediment erosion is a process by which hydraulic shear forces at the sediment–water interface become sufficient to dislodge particles from the bed. Once this material is scoured, it may either become fully resuspended into the water column or be transported along the bottom as bedload (material that is not fully suspended, but can move along the bed surface due to hydraulic forces). With regard to the transport of sorbed pollutants, many contaminated sediment sites are dominated by fine-grained particles, which may allow for bedload transport to be neglected in contaminant sediment transport model applications. However, in systems with significant coarse-grained particles, both bedload and suspended load can influence the transport and fate of sorbed contaminants.

As with settling and deposition, sediment scour may be influenced by a variety of physical and chemical properties within the bed and the overlying water column. Physical properties that impact the degree to which the sediment bed may be subject to scour include

- bulk properties of the sediment bed (soil strength, plastic limit, bulk density, gas content, etc.)
- particle sizes and their distribution within the bed
- water velocity
- bed roughness
- the presence of submersed aquatic vegetation and macrofauna

Bulk sediment properties and particle sizes affect the resistance to shear, which depends on velocity and bed roughness. Chemical properties that influence the likelihood of scour include those that determine interparticle attraction, so the list is similar to those that affect deposition. These include

- sediment mineralogy
- surface chemistry
- pore water chemistry

There are two distinct erodibility attributes for a given sediment. These are critical shear stress (i.e., the critical level of hydraulic shear stress that is sufficient to dislodge the solid particles), and the erosion rate (i.e., the flux rate of sediment into the water column) as a function of shear stress. Perhaps the most important determinant of erodibility is sediment grain size. Due to larger grain size and lesser interparticle forces, the erosion of coarse sediments is qualitatively quite different than erosion of cohesive sediments. It is also much better understood and empirically characterized.

The following sections provide a more detailed discussion of critical shear stress and erosion rates.

10.2.2.1 Critical Shear Stress

The work of Shields (1936) is often used to relate particle size to the critical shear stress for initiation of motion, referred to as the "Shields curve." The Shields curve relates a dimensionless critical shear stress parameter to another dimensionless number, the grain Reynolds number. The Shields curve is general in nature, allowing for application to a wide range of materials and to liquids other than water. For purposes of estimating the critical shear stress for sediments in an aquatic system, it is useful to develop a more specific curve applicable to a certain particle density, water density, and water viscosity. Using the analytical form of the Shields curve presented by van Rijn (1984a) and assumed values of 2.6 g/cm^3 for particle density, 1 g/cm^3 for water density, and 0.01 cm^2/s for water viscosity, Figure 10.5 plots critical shear stress in dyn/cm^2 against particle size in μm. Many researchers have gathered data on the conditions for initiation of sediment motion since the initial work of Shields. Miller et al. (1977) provide a review of several Shields-type threshold diagrams resulting from these data. Many of the common features of these diagrams are captured by Figure 10.5.

For larger grain sizes (>1000 μm), the critical shear stress is often considered to increase linearly with increasing grain size, although both the grain size where the curve becomes linear and the relationship between grain size and critical shear stress vary amongst the many Shields-type diagrams available in the literature. Two linear

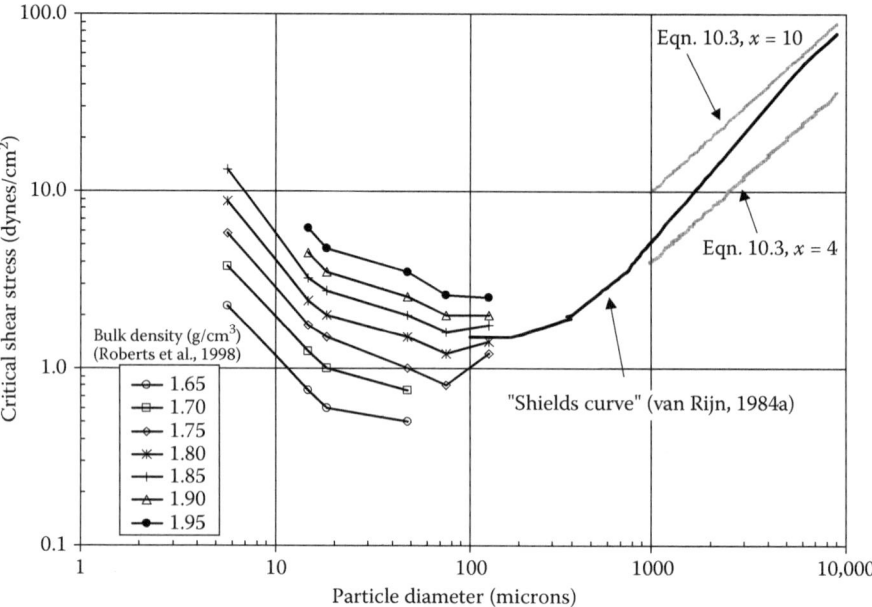

FIGURE 10.5   Critical shear stress for initiation of motion.

relationships for grain size and critical shear are plotted in Figure 10.5 based on the following equation:

$$\tau_{cr} = xd, \tag{10.3}$$

where τ_{cr} is the critical shear stress (dyn/cm^2), d is the particle diameter (mm), and x is the slope relating particle diameter to critical shear stress. Values for x in Equation 10.3 typically range from 4 to 10. Therefore, the lines shown in Figure 10.5 provide reasonable bounds on the critical shear stress for grain sizes larger than 1000 μm.

As the grain size decreases from approximately 1000 to 400 μm, there is a transition to hydraulically smooth conditions where grains do not extend into the turbulent flow. In this region, critical shear stress continues to decline with decreases in grain size, but not as rapidly as for larger grain sizes.

As the grain size decreases from approximately 400 to 200 μm, the critical shear stress levels off to a roughly constant value in the range of 1 to 2 dyn/cm^2. In this region, conditions are hydraulically smooth and cohesive forces are just beginning to become important.

For grain sizes less than 200 μm, cohesive forces and sediment consolidation become important. In this region, critical shear stress begins to increase as grain size decreases, and critical shear stress becomes strongly dependent on bulk density. Data from Roberts et al. (1998) showing the variation in critical shear stress with bulk density for quartz particles less than 200 μm are reproduced in Figure 10.5. For sediments with bulk densities ranging from 1.65 to 1.95 g/cm^3, less consolidated sediments began to erode at shear stresses as low as 0.5 dyn/cm^2, while more consolidated sediments required up to an order of magnitude more shear stress to initiate erosion. This dependence on bulk density translates into a tendency for stability to increase with sediment depth (McNeil et al., 1996; Ravens and Gschwend, 1999) because compression by sediment overburden tends to reduce water content, increasing bulk density.

In addition to bulk density and particle size, other bulk properties have been correlated with critical shear stresses in experimental flume studies, including ionic strength, mineralogy, gas content, and benthic and bacterial colonization. Sediment stability has been found to increase with increasing clay content and porewater ionic strength (Lee and Mehta, 1994). Young (1975) found that bioturbating organisms tend to reduce critical shear stress, while Tsai and Lick (1986) observed that benthic clams can mitigate against sediment compaction, reducing to a few days the transient time during which stability of new sediment increases due to consolidation. Ravens and Gschwend (1999) found that algal mats can significantly increase sediment stability, and Jepsen et al. (2000) found very substantial reductions in stability when gas was generated in sediment pores by increasing temperature. These bulk sediment properties also influence the rate of erosion, which is discussed in the next section.

10.2.2.2 Erosion Rates

The critical shear stress defines the point at which the erosion of sediments is initiated. While there are limitations on measurement of very small amounts of sediment transport, the rate of erosion at the critical shear stress is near zero. As the shear stress increases beyond the critical value, the rate of erosion increases.

Various devices have been used to measure erosion rates. These devices are discussed in Section 10.2.2.4. Erosion rates measured using the Sedflume device are discussed in this section to demonstrate how erosion rates vary with shear stress, particle size, and bulk density. Erosion rate data presented by Roberts et al. (1998) are reproduced as Figure 10.6. This figure shows erosion rate (depth of sediment eroded per time) as a function of particle size and applied shear stress. Particle sizes ranged from 5 to 1350 μm, while applied shear stresses ranged from 2 to 32 dyn/cm^2. Figure 10.6a shows data for sediments with a bulk density of 1.85 g/cm^3. Figure 10.6b shows data for a bulk density of 1.95 g/cm^3.

The erosion rates in Figure 10.6 range from 10^{-5} to 10^{-1} cm/s. The lower end of this range essentially represents no erosion. In fact, 10^{-4} cm/s (roughly 1 mm of erosion in 15 min) is often used operationally to define the critical shear stress using Sedflume data. For each bulk density and shear stress, the maximum erosion rate occurs in the region of 200–400 μm, similar to the region of the minimum critical shear stress shown in Figure 10.5 for all but the most loosely consolidated sediments. The erosion rate declines for larger particles due to the increasing weight of the particles. The erosion rate declines for smaller particles due to the increased significance of cohesive forces. Comparison of Figure 10.6a and b shows that increasing bulk density decreases the erosion rate of particles smaller than 200 μm, but has little impact on larger particles.

Roberts et al. (1998) proposed the following equation to approximate the results of their erosion rate study:

$$E = A \cdot \tau^n \cdot \rho^m, \tag{10.4}$$

where E is the erosion rate (cm/s); τ is the applied shear stress (N/m^2); ρ is the sediment bulk density (g/cm^3); and A, n, and m are constants that depend on the type of sediment. The values reported for this study are reproduced in Table 10.1. The value for m is zero for larger particle sizes for which the erosion rate is not a function of bulk density. The values for A also demonstrate distinct ranges for larger noncohesive particles (on the order of 10^{-3} to 10^{-2}) versus smaller cohesive particles (on the order of 10^3 to 10^4). For this study, n ranged from 1.9 to 3.3.

Equation 10.4 can also be used to describe the critical shear stress (τ_{cr}, N/m^2) for small particles shown in Figure 10.6 by rearranging and defining $E = 10^{-4}$ cm/s at the critical shear stress as follows:

$$\tau_{cr} = \left(\frac{10^{-4}}{A} \right)^{1/n} \rho^{-m/n}. \tag{10.5}$$

The erosion rate data discussed above were measured using quartz particles grouped into size classes with a fairly narrow size distribution, which allowed for examination of the effects of particle size and bulk density on erosion rates. As discussed for the critical shear stress, several other sediment properties also influence the erosion rate, including particle size distribution, ionic strength, mineralogy, gas content, and benthic and bacterial colonization. As a result of the many bulk properties that influence the erodibility of a given sediment, the parameters A, n, and m for Equation

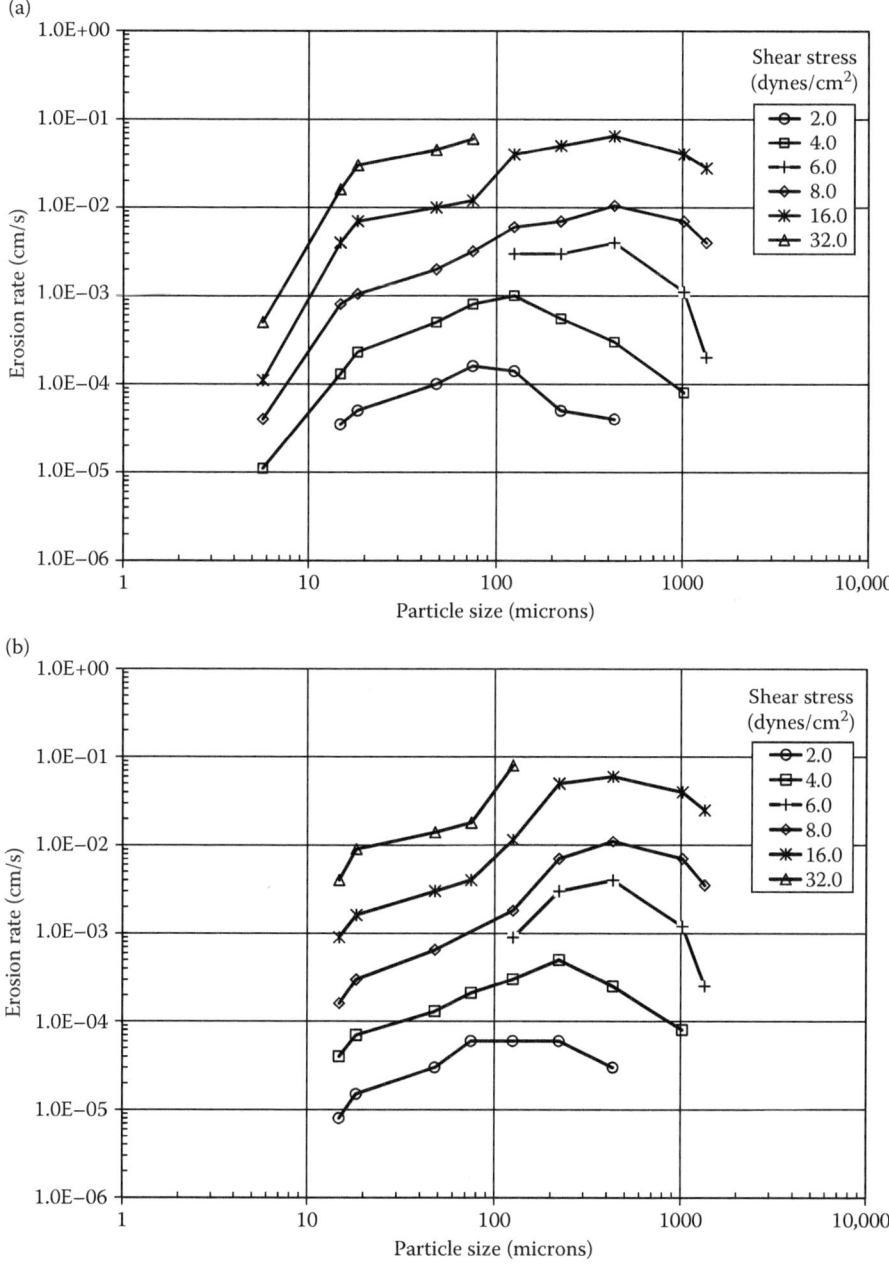

FIGURE 10.6 Erosion rates as a function of particle size for bulk densities of (a) 1.85 g/cm^3 and (b) 1.95 g/cm^3. (Reproduced from Roberts, J.D., R.A. Jepsen, and W. Lick. 1998. *J. Hydraul. Eng.*, 124(12): 1261–1267. With permission from ASCE.)

TABLE 10.1
Parameters Appearing in Equation 10.4 as Function of Sediment Particle Size

Mean Diameter of Sediment Particles (μm)	n	m	A
5.7	1.90	−29.0	3.28×10^3
14.8	2.27	−27.4	2.68×10^4
18.3	2.31	−25.6	1.49×10^4
48	2.23	−23.8	8.27×10^3
75	2.10	−22.3	4.70×10^3
125	2.82	−20.6	4.23×10^3
222	3.32	0	1.25×10^{-2}
432	2.56	0	2.25×10^{-2}
1020	2.51	0	1.14×10^{-2}
1350	2.92	0	6.74×10^{-3}

Source: Roberts, J.D., R.A. Jepsen, and W. Lick. 1998. *J. Hydraul. Eng.*, 124(12): 1261–1267. With permission from ASCE.

10.4 can vary significantly for natural systems. Lick (2009) provides results for sediments from three locations (Fox River, Detroit River, and Santa Barbara Slough). These are all relatively fine-grained sediments, with mean particle diameters less than 35 μm. For these natural sediments, n ranged from 1.89 to 2.23, consistent with the results for quartz particles in Table 10.1. However, results for m (−45 to −95) and A (3.65×10^3 to 2.69×10^6) are outside the range in Table 10.1, highlighting the need for site-specific data to characterize erosion properties, particularly for cohesive sediments or any sediments with unusual bulk properties.

10.2.2.3 Bedload and Suspended Load

After eroding from the sediment bed, particles will settle back to the bed or will be transported with the flow. The transport is generally apportioned into two categories: bedload and suspended load. Bedload refers to the transport of particles that roll, bounce, or slide in a thin layer near the bed and remain in frequent contact with the bed. The bedload layer is on the order of a few particle diameters in thickness. Suspended load refers to the transport of particles within the water column without frequent contact with the bed. Particles transported as suspended load move at the same speed as the surrounding water, while particles transported as bedload move more slowly than the surrounding water.

The mode of transport (bedload, suspended load, or mixed) depends on the relative strengths of turbulent mixing to keep a particle in suspension and the gravitational force causing a particle to settle. Higher flow velocities and smaller particle sizes result in a greater proportion of suspended load. Silt and clay particles are transported entirely as suspended load after disaggregation of any eroded chunks, while coarser sands and gravels are often transported entirely as bedload.

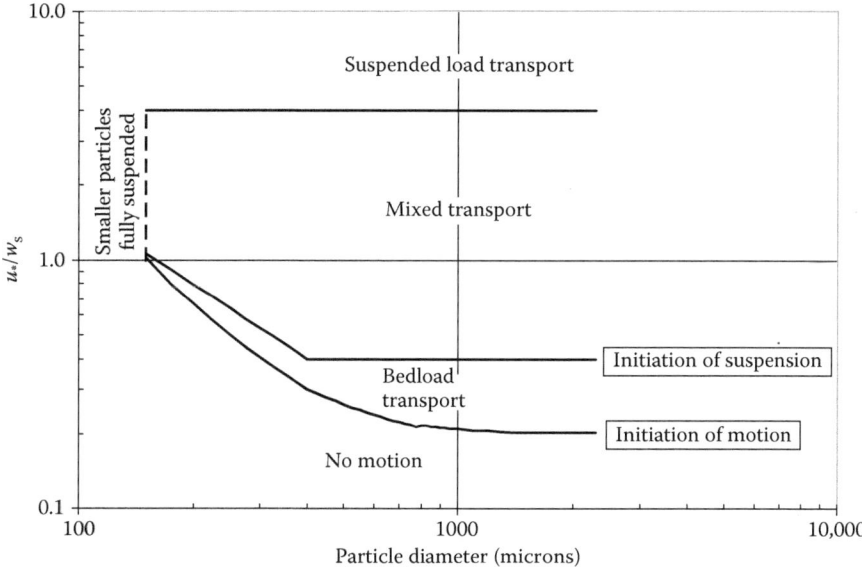

FIGURE 10.7 Sediment transport modes based on shear velocity and settling velocities.

As shown in Figure 10.7, the mode of transport for a given particle size can be estimated from the ratio of shear velocity to settling velocity. The threshold for initiation of motion shown in Figure 10.7 is computed based on Figures 10.2 and 10.5, with critical shear stress (τ_{cr}, dyn/cm^2) converted to critical shear velocity (u_{*cr}, cm/s), defined as

$$u_{*cr} = \sqrt{\frac{\tau_{cr}}{\rho}}. \tag{10.6}$$

For grain sizes larger than 1000 μm, consistent with the region of rough turbulent flow, motion is initiated at $u_*/w_s = 0.2$. A larger u_*/w_s ratio is required to initiate motion for smaller grain sizes where conditions shift to smooth turbulent flow. The threshold for initiation of suspension occurs at $u_*/w_s = 0.4$ for grain sizes greater than 400 μm (van Rijn, 1984b; Julien, 1998) and increases for smaller grain sizes. Between the thresholds for motion and for suspension, transport occurs predominantly as bedload. When u_*/w_s exceeds the threshold for suspension, the transport mode becomes a mix of bedload and suspended load. At $u_*/w_s = 4$, transport occurs predominantly as suspended load. Figure 10.7 is applicable to particle sizes greater than 150 to 200 μm. Smaller particles are transported almost entirely as suspended load. Section 10.3.1.4 presents detailed equations used within the SEDZLJ sediment transport model for computing the relative proportions of bedload and suspended load where the mode of transport is mixed.

10.2.2.4 Devices for Measurement of Erosion Properties

For settling and deposition, mechanistic theories based on discrete particle inter-actions have been developed and fit to experimental data. For erosion, however, mechanistic modeling is less well developed, and reliance has been placed instead on site-specific erosion studies. Site-specific studies are especially important when dealing with erosion of fine-grained sediments, which involves the disintegration of a cohesive sediment matrix into discrete suspended particles. The quantitative effects of bulk properties of sediments on erosion have been inferred from these experiments, rather than from well-established general principles.

Multiple techniques exist to measure sediment erodibility, each with advantages, disadvantages, and potential artifacts. The devices that have been used include annular flumes, Sedflume, and straight flumes. Annualar flames, which can be employed either *in situ* or *ex situ*, apply rotation of an overlying water column to a bed of in-place or reconstituted sediments in a closed circular system. Because they are closed systems, potential entrance and entry effects are avoided. Prior to the development of Sedflume, the annular flume was the leading method of erodibility and critical shear stress measurement for sediment transport studies (Lick et al., 1995). At each velocity and associated shear stress, an experiment was run to establish a steady-state suspended solids concentration in the water column from which net erosion could be inferred. With the annular flume, there is a finite amount of net erosion at any given shear stress.

The reason for the finite amount of net erosion observed with the annular flume depends on the type of sediments being studied. Erosion may cease due to formation of an armored layer at the surface of the sediment bed. An armoring layer can form by selectively eroding smaller particles, leaving behind larger particles that do not erode at the applied shear stress. This type of armoring can occur when sediments include a range of particle sizes. An armoring layer can also form by erosion of less-dense surface sediments until a layer of more compact and difficult to erode sediments is reached. This form of armoring applies to fine-grained cohesive sediments, which typically demonstrate increasing bulk density with depth. Noncohesive sediments of uniform size do not armor by either mechanism. For these sediments, the finite amount of net erosion observed in annular flume experiments is the result of reaching a dynamic equilibrium between a nonzero erosion rate and a nonzero deposition rate (Lick, 2009).

With the annular flume, the net erosion at any given shear stress is interpreted as an event indicative of steady state. The annular flume provides little or no direct information on the rate of erosion. The annular flume provides information on critical shear stress, but use of the annular flume is limited to shear stresses less than about $10\,\text{dyn/cm}^2$. Higher shear stresses result in preferential erosion near the outer wall and a buildup of sediments near the inner wall, thereby tilting the bed surface and further affecting erosion within the device (Lick, 2009). Additional limitations of the device are that it measures only suspended load and can only be used to measure near-surface erosion.

Sedflume estimates a time rate of erosion by subjecting a sediment core to a controlled and adjustable flow in a straight channel. The channel has a false bottom at the center, where a sediment core sample is inserted into the flume. As the sediments erode, the core is moved upward by an operator such that the sediment surface (i.e., the sediment–water interface) remains approximately level with the bottom of the flume channel. This may be complicated by erosion of the core into an irregular surface. For each experimental flow rate and associated shear stress, sediment erosion is measured by the upward rate of movement of the sediments in the coring tube. Critical shear stress is typically defined as the shear stress at which an erosion rate of 10^{-4} cm/s occurs. The rate, which roughly corresponds to 1 mm of erosion in 15 min, was chosen to provide a small, but accurately measurable, rate of erosion to define the initiation of motion.

Sedflume measures gross erosion into both bedload and suspended load rather than net resuspension as measured by the annular flume. Advantages of Sedflume are that it can estimate changes in critical shear stress and sediment erosion rate with depth, limited only by the length of the sample core (McNeil et al., 1996), and it can measure erosion properties at high shear stresses up to about 100 dyn/cm^2.

Straight flumes subject a rectangular patch of sediment to a straight flow at a known velocity, and estimate the resulting erosion rate by measuring the mass of sediment suspended and exiting the flume. These flumes have been employed in *ex situ* (Butman and Chapman, 1989; Lee and Mehta, 1994) and *in situ* (Young, 1975; Gust and Morris, 1989) settings. To avoid entry effects and provide a fully developed bottom boundary layer, a floor of sufficient length should be provided between the flume entrance and the exposed sediment (Ravens and Gschwend, 1999). Possible advantages of the straight flume over Sedflume include the ability to deploy the device *in situ* and the longer test section over which erosion occurs, which could minimize the impact of localized scour pits on estimated erosion rates. A limitation of the straight flume is that the shape of the experimental bed necessarily changes as material is eroded away, creating a depression, so that this technique best measures erodibility at the sediment surface at the beginning of the experiment.

Measured erosion rates using these various techniques range as widely as 10^{-5} to 10^{-1} cm/s. Much of the variation is due to the properties of the sediments, although attempts to reconcile measurements of the same sediments with various devices have also shown considerable measurement variability. For one site where measurements of surficial erosion rates were made with both Sedflume and an *in situ* straight flume, Sedflume produced estimates that were higher than the straight flume by a factor of five (Ravens, 2004, 2007). It is important to understand the advantages and disadvantages of each device, and to know what each is measuring, so that results can be appropriately interpreted and applied. Site-specific estimates of erosion properties can be validated to some degree by incorporating them into sediment mass balances and validating against water column data and net burial rates inferred from the full range of available site evidence. However, the challenge of measuring and verifying erodibility measurements remains a major source of uncertainty in contaminated sediment assessments.

10.2.3 Net Sedimentation

The previous sections describe the component processes relevant to sediment bed evolution, including settling, deposition, burial, and erosion. These processes interact and take on different degrees of significance over the broad range of environmental conditions seen in an aquatic system: high stresses and erosion due to hydrodynamic behavior during high-flow events, deposition due to spikes in watershed solids delivery, or perhaps slow, progressive deposition of fines under low-flow conditions. The sediment bed observed at any point in time is the net product of years of varying flow and sediment load conditions that influence the net accumulation of materials with time, or net sedimentation.

Often, the only data available for assessing that history of influences are the physical and chemical properties of the present-day sediment bed itself, which is an integrated product of the entire history of influences. The net sedimentation rate is the rate at which material accumulates in the sediment bed in response to the historical record of settling, deposition, burial, and erosion processes. Numerous forensic and investigative methods are available to provide insight into the historical record contained in bed sediments. These include physical measurements of the bed such as bed elevation measurements and sediment physical properties, and chemical measurements such as radioisotope analyses and vertical contaminant distributions.

Perhaps the simplest tool for assessing changes in the sedimentation rate is by direct observation of the elevation of the sediment bed, usually by means of an ultrasonic profiler for short-term investigations, or by using comparative bathymetric surveys for longer-term measures of sediment dynamics. Comparative bathymetric profiling must be subjected to very careful scrutiny in order to be effective, as the magnitude of changes in bottom bathymetry is often small relative to the accuracy of bathymetric surveys. However, careful repeat surveys can provide valuable insights into areas of significant gross deposition and erosion, particularly when timed to account for the effects of major flow events. At much smaller scales, a fixed ultrasonic profiler ("downlooker") can be used to monitor very small changes in bottom elevation with high accuracy, making it possible to track the propagation of sand waves and other bed forms, as well as larger changes due to erosion or deposition events.

Profiles of chemical parameters such as radioisotopes and contaminants often provide valuable insight into historical deposition and erosion processes. Radioisotope data have been used extensively to characterize sediments, using both natural Lead-210 (^{210}Pb) and anthropogenic Cesium-137 (^{137}Cs) tracers to provide a historical record of sediment deposition (e.g., Robbins and Edgington, 1975; LimnoTech, 2002a). ^{210}Pb is naturally generated in the atmosphere as a decay product of Radon-222 and is continuously deposited directly into water bodies and indirectly through deposition on to watershed soils and subsequent washoff. ^{137}Cs was generated historically as a by-product of atomic weapons testing, starting in the 1950s and largely ceasing after the mid-1960s. Global atmospheric deposition of ^{137}Cs peaked in 1963, with a smaller peak occurring between 1958 and 1959. Consequently, ^{137}Cs is often used as a benchmark for deposition of sediments deposited during this period and after, as watershed solids slowly eroded and washed off into receiving waters. Inspection of ^{210}Pb and ^{137}Cs profiles can often provide significant insight into the age of sediments, the consistency of deposition, the degree and depth of surficial mixing, and the prevalence of major disruptive events like high flows or wind-wave events.

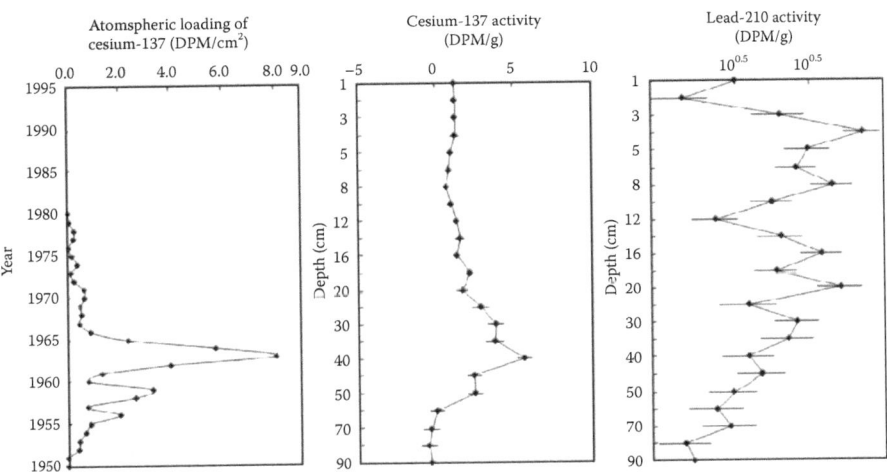

FIGURE 10.8 Historical atmospheric loading of Cesium-137 versus time rates of accumulation of Cesium-137 and Lead-210 in Fox River sediments. (Adapted from LimnoTech. 2002a. Measurement of Burial Rates and Mixing Depths Using High Resolution Radioisotope Cores in the Lower Fox River, for the Fox River Group, Report prepared for the Fox River Group. Ann Arbor, MI, January.)

Figure 10.8 shows the annual atmospheric loading of ^{137}Cs to the Lake Michigan drainage basin, and the corresponding record of ^{137}Cs and ^{210}Pb deposition in a sample of sediment collected from a depositional segment of the Lower Fox River (Robbins, 1985; LimnoTech, 2002a). The profile shows a clear peak at depth and a decrease in concentration with time in the upper part of the sediment core, accompanied by some smearing of the profile that suggests surficial mixing with time. The ^{210}Pb profile shows exponentially decreasing levels with depth, consistent with a decay half-life of 22.3 years and a history of consistent deposition. Other profiles may show discontinuities indicative of an erosional event, a random distribution suggesting a high degree of mixing, or an absence of ^{137}Cs suggesting a high rate of relatively recent sediment deposition containing low concentration of the historical tracer.

Given the relatively well-known history of production and release of many persistent environmental contaminants, the contaminants themselves often can serve as a useful historical marker. Figure 10.9 shows a record of U.S. polychlorinated biphenyl (PCB) sales and a corresponding dated profile of PCB in Lake Ontario sediments (Schwarzenbach et al., 1993). Similar to the radioisotope methods described above, the similarity of the PCB manufacturing history and depositional profiles provides evidence of a long-term, stable deposition of solids-associated PCB with minimal mixing of surficial sediments.

Estimates of net sedimentation rates made using the methods described above can range from negative (erosional) or essentially zero (dynamic equilibrium), to rates in the range of 1 to 4 cm/year, and upward to very high rates of deposition, usually in very localized areas or in response to imposed changes to the sedimentary

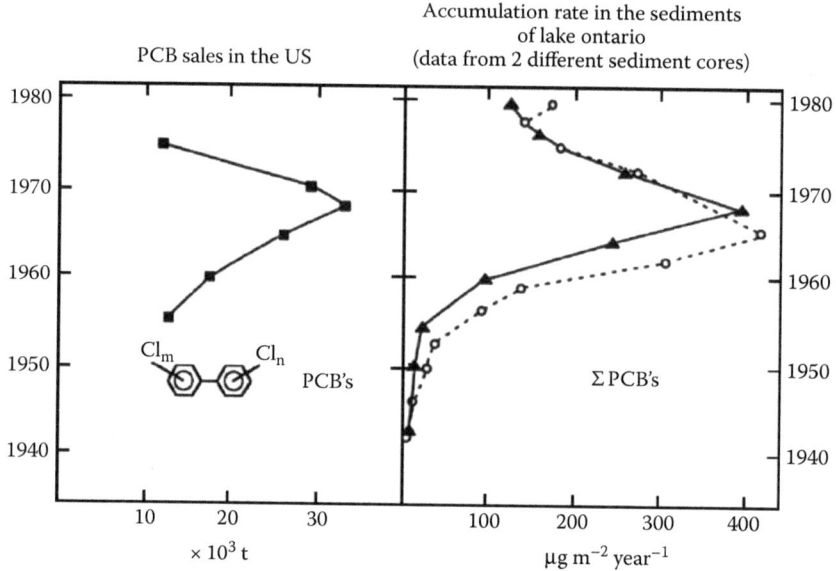

FIGURE 10.9 Historical records of sales/production of PCBs versus time rates of accumulation in Lake Ontario sediments (From Eisenreich, S.J., P.D. Capel, J.A. Robbins, and R. Bourbonniere. 1989. *Environ. Sci. Technol.*, 23: 1116–1126; Adapted from Schwarzenbach, R.P., P.M. Gschwend, and D.M. Imboden. 1993. *Environmental Organic Chemistry*, 2nd ed. John Wiley and Sons, Hoboken, NJ. With permission).

environment (e.g., dredging activities, changes in dam operations, or shoreline alterations). Figure 10.10 presents a summary of erosional and depositional processes operating on a typical sediment core for an estuarine system. Processes operating on the sediment bed in the short term (seconds to days) include gross deposition of watershed sediments and short-term erosion of sediments due to tidal stresses (in an estuary) and freshwater flows (± 0.4 cm/day). The site is also impacted by stress and erosion due to infrequently occurring events like storm surges, wind waves, and boat propeller impacts (<2 cm per event). These different processes are aggregated over time to create a sediment profile that is generally depositional at 0.5 to 2 cm/year, but is also impacted by the many short-term erosional events that mix the sediment profile. The profile is further mixed by the ongoing process such as bioturbation, which primarily affects the upper 10 cm of bed. The present-day result is a mixed contaminant and radionuclide profile in which deeply buried radionuclides and contaminants provide evidence of long-term deposition, but some smearing and irregularity of the profiles suggests the importance of on-going mixing processes like bioturbation and shallow erosion.

    It is important to note that, even in the case of consistently depositional systems, progressive sedimentation does not necessarily prevent a net flux of contaminants to the water column from the sediment bed. As mentioned above, net sedimentation often is accompanied by ongoing surficial mixing process that affects the top 10 to

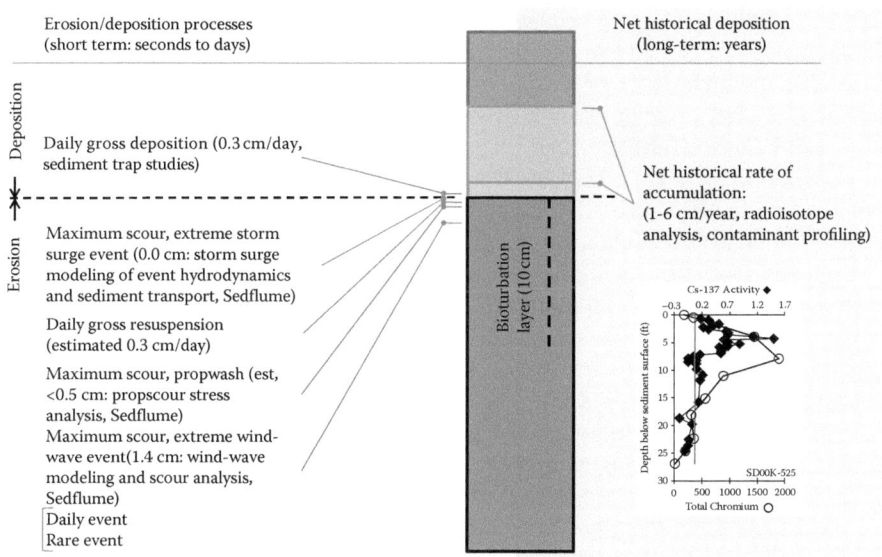

FIGURE 10.10 (See color insert following page 300.) Short-term erosional/depositional processes and long-term bed deposition at an estuarine sediment site.

20 cm of the sediment bed, due to bioturbation and short-term near-surface erosion and deposition events. If the concentration of the contaminant associated with depositing sediments from upstream is less than the concentration on surface sediment particles, then those surficial mixing processes will transfer contaminant mass into the water column and transport it downstream by advection and dispersion processes. This will occur at a given location until the deposition flux of the contaminant is in equilibrium with the erosional flux. In the long term, however, consistently depositional areas will see reduced contaminant flux, as the combined processes of deposition of clean materials and mixing of the upper portion of the sediment bed create a downward trend in surficial contaminant concentrations with time, and burial of elevated concentrations at depth.

10.3 APPLICATION OF PROCESSES

This section contains a discussion of how the processes of deposition and resuspension are applied in practice to estimate fluxes of solids and associated contaminants across the sediment–water interface as a result of particle deposition and bed erosion.

At contaminated sediment sites, contaminant concentrations are often the most thoroughly measured parameters because of their importance in identifying the site as contaminated and for evaluating the prevailing exposures and risks. Suspended solids loads and in-stream concentrations, and their covariation with flow and season, are often less thoroughly measured, but are necessary to constrain estimates of net deposition. The two components of this net quantity (namely deposition and erosion) cannot be individually constrained by ambient water column data because of their simultaneous nature, and can only be estimated by controlled experimentation.

Despite the importance of particle size distributions in determining settling rates, these may not be available, especially for a range of flow rates. Similarly, site-specific information on floc formation and floc settling speeds is usually lacking.

Most important rivers in the United States are gaged by the U.S. Geological Survey, so that continuous flow records are available for use in estimating magnitudes and frequencies of high-flow events. Similar flow records exist in other countries. However, as discussed above, the lack of a general theory of cohesive sediment resuspension often makes it difficult or impossible to predict scour depths on the basis of relatively inexpensive site-specific measurements of sediment bulk properties. Instead, local measurements of critical shear stress and erodibility are obtained by experimental measurements using *ex situ* flumes, or *in situ* flume measurements at multiple locations. Because sediment sites can be spatially heterogeneous, the adequacy of these experimental results for modeling of scour at the desired spatial scale is limited by the resources available for the flume studies. In addition, the magnitudes of the estimated scour can depend profoundly on the measurement technique employed, as discussed in the previous section.

10.3.1 Representation in Mathematical Models

With recent advancements in numerical models of fluid mechanics and sediment transport, modeling has become a key tool for evaluating sediment transport behavior and associated contaminant transport and fate. While state of the art, research-oriented models may contain many of the detailed mechanisms by which solids deposition and scour may occur, the current widely distributed (i.e., public domain) engineering-oriented models for these processes necessarily employ simplifications of some mechanisms due to constraints in computational time, model development time, and site data availability. These models aim to represent the bulk behavior of solids or classes of solids, rather than modeling the forces acting on each individual particle.

10.3.1.1 Overview of Sediment Transport Models

A wide range of numerical models are available for evaluating transport processes for cohesive and noncohesive sediment types. The collection of models represents varying levels of process representation and spatial complexity. One-dimensional (1D) models are generally the simplest to develop and apply and require the least amount of computational time and resources. Models belonging to this class typically simulate steady or unsteady hydraulic behavior, including average cross-sectional velocity and bottom shear stress, total sediment load, and particle size distribution (Papanicolaou et al., 2008). Examples of 1D models include MOBED, HEC-6, and GSTARS. The Water Quality Simulation Program (WASP) model, which is developed and supported by U.S. EPA, provides a generic sediment transport submodel to support simulation of contaminant transport and fate. WASP is commonly applied as a one-dimensional model, but may be applied in two or three dimensions if linked to an independent hydrodynamic model.

Two-dimensional (2D) models represent a second tier of spatial complexity with respect to sediment transport models. Models in this class have become more prevalent during the last couple of decades due to advancements in computer hardware and software capabilities. Two-dimensional models typically solve the depth-averaged flow continuity and Navier–Stokes equations with respect to hydrodynamic behavior and mass balance equations with respect to sediment transport. Computational methods employed in 2D models include finite difference, finite element, and finite volume. Examples of 2D models include Environmental Fluid Dynamics Code (EFDC), SEDZLJ, SEDZL, USTARS, MIKE21, and Delft 2D.

Three-dimensional (3D) models represent the highest tier of spatial and process complexity and solve the flow continuity equation and the Navier–Stokes equations for conservation of mass and momentum in three-dimensional space. Three-dimensional models are favored over depth-averaged models for water body systems where density stratification occurs or hydraulic structures significantly impact hydrodynamic behavior. Examples of 3D coupled hydrodynamic/sediment transport models include EFDC, ECOMSED, MIKE-3, RMA-10, and Delft 3D.

Ultimately, the choice of sediment transport model must be based on (1) the study questions being asked and (2) the availability of data and other resources to support model development and application. One-dimensional models are most appropriate for evaluating study questions that are not fine-scale in nature, while 2D and 3D models are most appropriate when addressing questions that relate to more spatial detail with respect to sediment and contaminant transport. A more complete discussion of available sediment transport models can be found in Papanicolaou et al. (2008). The sediment transport capabilities of three models will be presented in the following sections: water quality simulation program (WASP), environmental fluid dynamics code (EFDC), and SEDZLJ. These models represent a range with regard to spatial and process complexity, and the EFDC and SEDZLJ models incorporate state-of-the-science algorithms for simulating sediment transport. A brief discussion of data needs and considerations for model calibration concludes this section.

10.3.1.2   Sediment Transport Simulation in WASP

The U.S. EPA currently supports and maintains the WASP model, including various versions from 4 through 7 (Ambrose et al., 1988; 1993; Wool et al., 2001). The first EPA public domain WASP applications to examine toxic chemicals in receiving waters and sediments date back to late 1980s with evaluations of volatile organics in the Delaware Estuary (Ambrose, 1987) and heavy metals in the Deep River, North Carolina (JRB, 1984). As a public domain model, various enhancements of the WASP model have been made over the years in order to address site-specific needs and improve on the standard transport and kinetic formulations to simulate a variety of toxics, especially hydrophobic organic chemicals (HOCs) and metals. Several modified versions of the WASP model incorporate settling and resuspension functions that are not available in the EPA-supported model. The WASP model discussion below focuses on the representation of settling and resuspension in the standard EPA-supported model.

WASP is a widely used and adaptable model for simulating chemical constituents in the water column and sediments, but it has a simplified representation with regard to simulating solids settling and deposition to the sediment bed. In WASP, settling is a completely user-specified value, so there is no inherent mechanistic aspect to it. Settling rates are input as a gross settling speed and the segment-to-segment interfacial area over which it applies. Settling rates may be specified as time- and/or space-variable functions.

The WASP model incorporates the ability to input settling rates (and subsequent deposition if settling is from a water column segment to a sediment segment) of up to three types of solids. Additionally, up to three chemical constituent state variables may be associated with any of the solids state variables through user-specified equilibrium partitioning methods, and chemical mass may settle through the water column along with the solids to which they are adsorbed. The fact that WASP can simulate up to three types of solids does allow the model to mimic the mechanistic aspects of particle settling to a limited degree because gross settling rates for each type of solids may be specified based on their representative physical properties (e.g., grain size differences or potential for floc formation). This capability by itself is still insufficient to fully represent settling and deposition behavior for the range of particles that may be represented within each size class. However, an independent sediment transport model can be used to generate the solids settling and deposition rates required by WASP.

The WASP model incorporates erosion in an analogous manner to how it handles settling, relying on user-specified rates of erosion for each solids type in the model. Similar to settling rates, input erosion rates can vary in time and space. As discussed for WASP settling, this is a flexible approach that lends itself well to either very simple representation of sediment dynamics or to coupling of the WASP model to an independent sediment transport model for a more sophisticated representation.

Because erosion and deposition rates are represented in a simplified fashion in WASP, the model is well-suited to simulate sediment and contaminant transport in strongly net depositional environments, such as lake and estuarine systems, where resuspension is a secondary or even negligible process (refer to Section 10.2.3).

10.3.1.3 Sediment Transport Simulation in EFDC

The environmental fluid dynamics code (EFDC) model can be used to simulate hydrodynamics and sediment transport in one-, two-, or three-dimensional space (TetraTech, 2007a, 2007b, 2007c). EFDC can simulate wetting and drying of flood plains, mud flats, and tidal marshes, in addition to surface waters and wetlands. Since 1996, Tetra Tech, Inc. has maintained EFDC with primary support from the U.S. EPA. EFDC provides a range of options for simulating erosion, deposition, and bed armoring and handling for cohesive and noncohesive sediment types. In particular, a comprehensive set of options are available for simulating noncohesive suspended load and bed load transport. Multiple cohesive and/or noncohesive sediment size classes may be represented in a single model simulation. The key model algorithms for cohesive and noncohesive sediments are described in the following sections.

10.3.1.3.1   Settling and Deposition of Cohesive Sediments in EFDC

For cohesive sediment particle types, the current EFDC model allows the user to specify a constant settling velocity or select from a variety of semiempirical expressions that relate particle settling velocity to concentration and/or bed shear stress. The semiempirical expressions are intended to approximate the effects of floc formation and disaggregation on cohesive settling, while avoiding the computational intensity of first principles mathematical modeling of these processes. Once the settling velocity is determined, the deposition flux to the bed (J_d, g/m^2/s) is computed based on the settling velocity (w_s, m/s), the near-bed solids concentration (S_d, g/m^3), a probability of deposition function (P_k, unitless), the applied cohesive grain shear stress (τ_b, dyn/cm^2) predicted by the hydrodynamic model, and the critical shear stress for deposition (τ_{cd}, dyn/cm^2):

$$J_d = \begin{cases} -w_s P_k S_d : & \tau_b \leq \tau_{cd} \\ 0 : & \tau_b > \tau_{cd} \end{cases} \tag{10.7}$$

where the linear probability of deposition function developed by Krone (1962) is often used:

$$P_k = \begin{cases} \left(1 - \dfrac{\tau_b}{\tau_{cd}}\right) : & \text{for } \tau_b \leq \tau_{cd} \\ 0 : & \text{for } \tau_b > \tau_{cd} \end{cases} \tag{10.8}$$

The nonlinear probability function developed by Partheniades (1992) is also an available option in EFDC. This probability of deposition function is based on the concept that under quiescent conditions all solids settling near the bed will reach and adhere to the bed, while at some threshold flow-induced bed surface stress (τ_{cd}) none of the solids settling near the bed will reach and adhere to the bed. In the absence of site-specific data, this critical deposition stress is generally treated as a calibration parameter with a wide range of reported values from laboratory and field observations of 0.6 to 11 dyn/cm^2 (TetraTech, 2007c). The near-bed solids concentration (S_d) is also an estimated value, computed in EFDC based on either the bottom layer concentration (in 3D model applications) or depth-averaged water column concentration (in 1D and 2D models).

10.3.1.3.2   Resuspension of Cohesive Sediments in EFDC

The EFDC model provides several options for representing erosion of cohesive sediments, ranging from a simple constant erosion rate to calculating erosion as a function of shear stress and bed properties, which can vary in time and space. Mass erosion of the bed is computed as the rate of resuspension (w_r, m/s) multiplied by the sediment concentration or dry density of the sediment bed (S_r, g/m^3). This resuspension flux (J_o^r, g/m^2/s) is generally represented in EFDC in the following form:

$$J_o^r = w_r S_r = \frac{dm_e}{dt} \left(\frac{\tau_b - \tau_{ce}}{\tau_{ce}} \right)^a : \tau_b \geq \tau_{ce} \tag{10.9}$$

where dm_e/dt is the "base erosion rate" (g/m^2/s) and τ_{ce} is the critical shear stress for surface erosion (dyn/cm^2). The base erosion rate, critical stress, and the exponent α are generally determined from laboratory or *in situ* field studies. Base erosion rates ranging from 0.005 to 0.1 g/m^2/s have been reported in the literature (TetraTech, 2007c), and the typical range of the exponent α is 2.0 to 3.0 (Ziegler and Nesbit, 1994).

EFDC allows the base erosion rate and critical shear stress for erosion to be user-defined constants or predicted values based on sediment properties (i.e., bulk density or void ratio). Selection of the sediment-dependent formulations requires use of the EFDC bed consolidation simulation to predict time and depth variation in these bed properties. The sediment-dependent formulations in EFDC compute a decreasing base erosion rate and increasing critical shear stress with increases in bulk density and associated decreases in the void ratio.

10.3.1.3.3 Settling and Deposition of Noncohesive Sediments in EFDC

For noncohesive sediment particle types, sediment, deposition, and erosion are tightly coupled processes, with rates determined by the transport capacity of the water column and the availability of a particular particle type in the water column and the surface layer of the sediment bed. The particle settling velocity for each particle size class k ($w_{s,k}$ m/s) can be specified by the user or computed internally based on equations developed by van Rijn (1984a):

$$\frac{w_{s,k}}{\sqrt{g' \cdot d_k}} = \begin{cases} \dfrac{R_{d,k}}{18} & \text{for} : d_j \leqslant 100\,\mu m \\[2mm] \left(\dfrac{10}{R_{d,k}} \right) \cdot \left(\sqrt{1 + 0.01 \cdot (R_{d,k})^2} - 1 \right) & \text{for} : 100\,\mu m < d_j \leqslant 1000\,\mu m \\[2mm] 1.1 & \text{for} : d_j > 1000\,\mu m \end{cases} \tag{10.10}$$

where d_k is the particle diameter for size class k(m), g' is the modified gravitational acceleration based on the particle specific gravity (SSG) and the gravitational acceleration ($g = 9.81$ m/s^2):

$$g' = g * (SSG - 1) \tag{10.11}$$

and $R_{d,k}$ is the dimensionless sediment grain Reynolds number for size class k:

$$R_{d,k} = \frac{d_k \sqrt{g' \cdot d_k}}{\nu}, \tag{10.12}$$

where ν is the kinematic viscosity of water (m²/s). At higher suspended sediment concentrations, which may occur near the sediment bed, the model corrects the particle settling velocity to account for hindered settling. The net deposition flux (J_o^d, g/m²/s) for suspended sediment is calculated as

$$J_o^d = -w_{sc,k} \cdot S_{ne} \cdot \left(1 - \frac{S_{eq}}{S_{ne}}\right), \tag{10.13}$$

where $w_{sc,k}$ is the settling velocity for particle class k corrected for hindering (m/s), S_{eq} is the near-bed equilibrium suspended sediment concentration (g/m³), and S_{ne} is the actual near-bed suspended concentration (g/m³). A negative flux calculated by Equation 10.13 indicates net deposition from the water column to the sediment bed, while a positive flux would indicate net erosion from the bed.

EFDC provides the user with several options for computing the near-bed equilibrium concentration, including algorithms developed by van Rijn (1984b), Smith and McLean (1977), and Garcia and Parker (1991). The method of van Rijn is commonly employed in sediment transport applications and computes the near-bed equilibrium concentration by

$$S_{eq} = 0.015 \cdot \rho_s \cdot \left(\frac{d_k}{z_{eq}^*}\right) \cdot \left(\frac{T^{1.5}}{R_{d,k}^{0.2}}\right), \tag{10.14}$$

where ρ_s is the dry sediment density (g/m³), d is the particle diameter (meter), z_{eq}^* is the near-bed reference level (meter), and T is the dimensionless transport stage parameter. The transport stage parameter is calculated per Equation 10.15 based on the critical shear stress for erosion (τ_{cr}) or the critical shear velocity for erosion (u_{*cr}, m/s), and the grain stress (τ_b) or grain shear velocity (u_{*b}, m/s):

$$T = \left(\frac{\tau_b - \tau_{cr}}{\tau_{cr}}\right) = \left(\frac{u_{*b}^2 - u_{*cr}^2}{u_{*cr}^2}\right). \tag{10.15}$$

10.3.1.3.4   Resuspension of Noncohesive Sediments in EFDC

Erosion of noncohesive sediments depends on the equilibrium suspended load transport capacity and bed load flux relative to local conditions. The mode of transport for a noncohesive particle class k depends on the magnitude of the local shear velocity relative to the particle settling velocity ($w_{s,j}$) and Shields' critical shear velocity for initiation of motion ($u_{*cr,k}$, m/s):

$$u_{*cr,k} = \sqrt{\tau_{cr,j}} = \sqrt{g' \cdot d_k \cdot \theta_{cr,k}}, \tag{10.16}$$

where d_k is the particle diameter (meter) and $\theta_{cr,k}$ is the Shields' dimensionless parameter based on the sediment grain Reynolds number (van Rijn, 1984a). The

Shields' parameter can be estimated as

$$\theta_{cr,k} = \begin{cases} 0.24 \cdot (D_{*,k})^{-1.0} & \text{for :} \quad D_{*,k} \leqslant 4 \\ 0.14 \cdot (D_{*,k})^{-0.64} & \text{for :} \quad 4 < D_{*,k} \leqslant 10 \\ 0.04 \cdot (D_{*,k})^{-0.10} & \text{for :} \quad 10 < D_{*,k} \leqslant 20 \\ 0.013 \cdot (D_{*,k})^{0.29} & \text{for :} \quad 20 < D_{*,k} \leqslant 150 \\ 0.055 & \text{for :} \quad D_{*,k} > 150 \end{cases} \tag{10.17}$$

where $D_{*,k}$ is the dimensionless particle parameter calculated by

$$D_{*,k} = (R_{d,k})^{2/3} = d_k * \left(\frac{g'}{v^2}\right)^{1/3}. \tag{10.18}$$

If the shear velocity exceeds the critical shear velocity and the particle settling velocity, eroded sediment will be at least partially transported as suspended load. For suspended load, the direction and magnitude of the water-bed sediment exchange flux (J_o, g/m^2/s) ultimately depends on the difference between the near-bed actual concentration (S_{ne}, g/m^3) and the near-bed equilibrium concentration (S_{eq}, g/m^3), as described by Equation 10.13.

If the applied grain shear velocity exceeds the particle critical shear velocity but is less than the settling velocity, sediment will be eroded from the bed and transported predominantly as bed load. EFDC provides multiple options for the calculation of bed load transport rates, including algorithms developed by van Rijn (1984a); Engelund and Hansen (1967); Bagnold (1956); and Meyer-Peter and Muller (1948). In general, these bed load formulations are of a similar form and produce roughly similar results. The bed load formulation developed by van Rijn (1984a), which is commonly employed for sediment transport evaluations, calculates the equilibrium horizontal bed load mass transport rate per unit width (q_B, g/m/s) as:

$$q_B = \left(\rho_s \cdot d^{1.5} \cdot \sqrt{g'}\right) \cdot \left(\frac{0.053 \cdot T^{2.1}}{R_d^{0.2}}\right), \tag{10.19}$$

where ρ_s is the dry sediment density (g/m^3), d is the particle diameter (meter), and T is the transport stage parameter given by Equation 10.15. When the transport stage parameter is greater than zero, the grain shear stress exceeds the critical shear stress for bed load transport, and bed load transport will occur at the equilibrium rate calculated by Equation 10.19. The EFDC model calculates the bed load transport rate at the center of each grid cell and then converts the cell-centered rate to fluxes at the cell faces using an upwind difference scheme. For grid cells where the local bed load transport capacity exceeds the incoming bed load transport rate, local net erosion of the bed will occur to meet the bed load demand. Conversely, if the local bed load transport capacity is less than the rate of bed load supplied from upstream cells, a fraction of the bed load will be deposited to the sediment bed and no net erosion of the bed will occur.

10.3.1.3.5   Sediment Bed Representation and Armoring in EFDC

The sediment bed is represented in the EFDC model as a series of discrete vertical layers, which may be described using constant or varying thicknesses. The thickness and bulk properties (e.g., particle size composition) of the surface layer(s) is subject to change over the course of the simulation as erosion and deposition processes remove and/or add net sediment mass to or from the bed. The user specifies the percentage of the original thickness at which an erosion or deposition "event" will occur for the surface layer. For example, the surface layer might be "eroded" and subsequently combined with the next lowest layer when the layer thickness reaches 20% of its original volume, or the surface layer may be subdivided into two new layers if the layer thickness doubles relative to its original value.

Sediment bed armoring, which represents a decrease in the erodibility of the sediment bed over time, is represented within EFDC in a variety of ways, including

1. Coarsening of the bed due to preferential erosion of lighter particles or the deposition of heavier particles.
2. Specification of an empirical "hiding factor" based on the cohesive content of the bed.
3. Consolidation of the bed (increasing bulk density and decreasing the void ratio).

Simulation of bed coarsening in EFDC results from: (1) use of multiple particle classes with differing settling and resuspension properties, and (2) a bed-handling routine that allows time-variation in the mix of particles present in the surface layer of the sediment bed. Under elevated flow conditions, only the more easy-to-erode particles will typically be resuspended and coarse particles will preferentially settle, which will result in a greater proportion of coarser and more difficult-to-erode particles in the surface layer of the sediment bed. This bed coarsening results in a decrease in the overall erosion rate with time. EFDC also includes a specific armoring option that speeds the armoring process by representing a very thin active layer of constant thickness at the top of the sediment bed. This thin layer is typically only a few particle diameters in thickness, and therefore can coarsen very quickly.

Another means of bed armoring available in EFDC is the use of a "hiding factor" based on the cohesive content of the sediment bed. This is a user-specified empirical value that reduces the amount of cohesive sediment resuspension computed in Equation 10.9 by a dimensionless factor (F_{CR}), which is calculated as

$$F_{CR} = (f_{coh})^\beta, \tag{10.20}$$

where f_{coh} is the mass fraction of cohesive sediment in the surficial layer of the sediment bed, and the exponent β is a user-specified "hiding factor." A hiding factor (β) equals to zero results in no adjustment to the cohesive resuspension. For a given nonzero hiding factor, decreases in the cohesive fraction of sediment result in decreases in the resuspension factor (F_{CR}), which simulates armoring of the bed due to larger noncohesive particles protecting ("hiding") smaller cohesive particles from near-bed shear forces.

Consolidation of the sediment bed is the final means by which armoring is represented in EFDC. As discussed above, the base erosion rate and critical shear stress may be predicted based on sediment properties (i.e., bulk density or void ratio). The methodology for representing sediment bed consolidation is described in TetraTech, Chapter 5 (2007c) and varies depending on whether the bed is composed of cohesive sediment, noncohesive sediment or a combination of the two. As the bed consolidates (increasing bulk density and decreasing void ratio), the base erosion rate decreases and the critical shear stress increases. This decreases the erodibility of a consolidating sediment bed layer.

10.3.1.4   Sediment Transport Simulation in SEDZLJ

The SEDZLJ model was developed at the University of California at Santa Barbara and published in 2001. Similar to the EFDC model, SEDZLJ is capable of simulating sediment transport processes for multiple particle size classes in one-, two-, and three-dimensional space. The model is implemented using an independent hydrodynamic model, such as ECOM, to drive the simulation. The model is unique among sediment transport models in that it directly utilizes data obtained using the Sedflume device, which measures changes in erosion rate with depth (Jones and Lick, 2001).

10.3.1.4.1   Erosion of Sediment in SEDZLJ

Erosion in SEDZLJ is parameterized based on data collected using Sedflume, which directly measures erosion rates at various depths within a sediment core as a function of applied shear stress. Rates of erosion, typically expressed in units of cm/s, obtained via Sedflume experiments are used in combination with estimates of sediment dry bulk density, particle size distribution, and critical shear stress to specify the properties of discrete sediment layers in the model simulation. Because erosion rates are measured for discrete shear stresses and sediment depths, interpolation is required to develop continuous functions. Linear interpolation is used between discrete shear stress levels, while logarithmic interpolation is used to specify the erosion as a function of depth.

Similar to EFDC, the SEDZLJ model simulates erosion of sediments into suspended load or bed load depending on the applied bed shear stress and the sediment particle size distribution at the surface of the bed. Fine-grained (i.e., cohesive) sediments are assumed to be transported entirely with the suspended load and therefore do not contribute to the bed load flux. For noncohesive sediments, the ratio of suspended load to total load transport (q_s/q_t) for eroding sediments is estimated using the following relationship, which is based on flume data collected by Guy et al. (1966):

$$
\frac{q_s}{q_t} = \begin{cases} 0 & \text{for } u_* \leq u_{*cs,k} \\[2ex] \dfrac{\ln(u_*/w_{s,k}) - \ln(u_{*cs,k}/w_{s,k})}{\ln(4) - \ln(u_{*cs,k}/w_{s,k})} & \text{for } u_* \leq u_{*cs,k} \text{ and } \dfrac{u_*}{w_{s,k}} < 4 \\[2ex] 1.0 & \text{for } u_{*w_{s,k}} \geq 4 \end{cases} \quad (10.21)
$$

where u_* is the applied bed shear velocity (m/s), $u_{*cs,k}$ is Shields' critical shear velocity for initiation of suspension (m/s) for particle size class k, and $w_{s,k}$ is the discrete particle settling velocity (m/s). Shields' critical shear velocity for initiation of suspension is defined by van Rijn (1984b) as

$$u_{*cs,k} = \begin{cases} \dfrac{4w_{s,k}}{D_{*,k}} & \text{for } d_k \leqslant 400 \, \mu\text{m} \\ 0.4w_{s,k} & \text{for } d_k > 400 \, \mu\text{m} \end{cases}. \qquad (10.22)$$

It should be noted that SEDZLJ assumes that particles less than 200 μm in diameter are transported entirely as suspended load for the purpose of the model calculations. Using Equation 10.21 the suspended load and bed load components of the erosion rate can be calculated as follows:

$$\text{For } \tau_b \geq \tau_{ce,k} : \begin{cases} E_{s,k} = \left(\dfrac{q_s}{q_t} \right) * (f_k * E) \\ E_{b,k} = \left(1 - \dfrac{q_s}{q_t} \right) * (f_k * E) \end{cases} \qquad (10.23)$$

$$\text{For } \tau_b \geq \tau_{ce,k} : \begin{cases} E_{s,k} = 0 \\ E_{b,k} = 0 \end{cases}$$

where E is total erosion rate (cm/s), f_k is the mass fraction of particle size class k in the bed active layer, and $E_{s,k}$ and $E_{b,k}$ are the rates of erosion contributing to the suspended load and the bed load, respectively.

10.3.1.4.2 Settling and Deposition in SEDZLJ

Simulated deposition fluxes of cohesive and noncohesive suspended sediment ($D_{s,k}$, g/m²/s) in SEDZLJ depend on the discrete settling velocity for particle size class $k(w_{s,k}$, m/s), the near-bed suspended sediment concentration ($C_{s,k}$, g/m³), and the probability of deposition function (P_k):

$$D_{s,k} = P_k * w_{s,k} * C_{s,k} \qquad (10.24)$$

The discrete settling velocity (m/s) for particle size class k is calculated based on the formula proposed by Cheng (1997a):

$$w_{s,k} = \left(\frac{\nu}{d_k} \right) * \left(\sqrt{25 + 1.2 \cdot (D_{*,k})^2} - 5 \right)^{1.5}, \qquad (10.25)$$

where ν is the kinematic viscosity of water (m²/s), d_k is the particle diameter (meter), and $D_{*,k}$ is the nondimensional particle parameter.

The calculation for the probability of deposition function (P_k) varies depending on whether a particle class is cohesive or noncohesive in nature. For cohesive particle classes, the relationship developed by Krone (1962) is used (see Equation 10.8). For noncohesive sediment, the probability of deposition is based on a Gaussian distribution developed by Gessler (1967). A simplified derivation of the Gessler relationship that is used in SEDZLJ is described in Jones and Lick (2001). For the bed load transport component, deposition is computed in similar fashion to Equation 10.24:

$$D_{b,k} = P_k * w_{s,k} * C_{b,k} \qquad (10.26)$$

where $D_{b,k}$ is the deposition flux for the bed load component (g/m^2/s) and $C_{b,k}$ is the bed load sediment concentration. The probability of deposition can be calculated as follows based on an assumption of steady-state bed load conditions:

$$P_k = \frac{E_{b,k}}{w_{s,k} * f_k * C_{e,k}}, \qquad (10.27)$$

where $E_{b,k}$ is the erosion flux contributing to bed load for particle size class k(g/m^2/s), f_k is the mass fraction of size class k, $C_{e,k}$ is the equilibrium bed load sediment concentration for size class k(g/m^3), and $w_{s,k}$ is the particle class settling velocity (m/s). Essentially, the gross deposition of bed load to the sediment bed is computed as the difference between the gross rate of erosion contributing to bed load and the local equilibrium bed load transport capacity.

10.3.1.4.3 Sediment Bed Representation and Armoring in SEDZLJ

Similar to the EFDC model, the sediment bed is initialized in SEDZLJ as a series of discrete layers of varying thicknesses. Initial layer properties, including thickness, bulk density, critical shear stress, and erodibility are specified directly based on Sedflume observations. As the sediment transport simulation progresses through time, a net erosion or deposition of the bed will occur relative to the initial bed condition. In the case of net deposition, the model forms a "deposited layer" that exists intermediate to the surficial active layer and the uppermost Sedflume-based layer (see Figure 10.11). In this way, the original Sedflume-based layers remain intact throughout the simulation, although one or more layers may be partially or entirely eroded later in the simulation.

SEDZLJ capabilities for simulating armoring of the sediment bed are generally similar to those for the EFDC model. Armoring of the bed and a resulting decrease in erosion rates can be represented by consolidation of cohesive sediments with depth and time, and local deposition of coarser particle types and/or preferential resuspension of finer particle types during an erosion event.

The use of Sedflume to measure bed erosion rates at depth implicitly accounts for the effects of consolidation on erosion at varying depths within the sediment bed. The armoring of the bed surface due to preferential settling of coarse particles and erosion of relatively finer particles is accounted for by using a thin, completely mixed "active layer." The active layer representation in SEDZLJ is generally similar to that used in EFDC, with gross erosion and deposition occurring at the water-active layer interface and net deposition/erosion applied to the depositional layer immediately below the active layer. One notable difference is that SEDZLJ maintains the original Sedflume-based layer thickness and physical properties, as described above, while

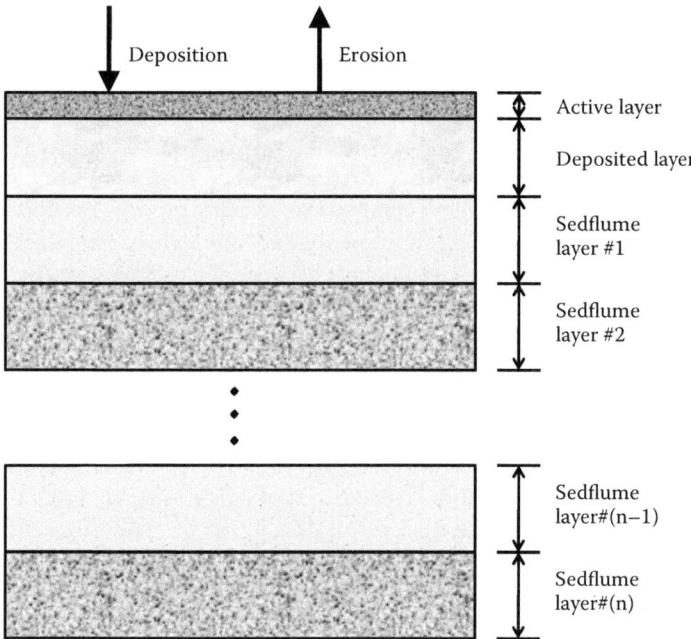

FIGURE 10.11 SEDZLJ sediment bed layer schematic.

EFDC simply adds deposited sediment to the parent bed layer located immediately below and adjacent to the surficial active layer. An additional difference between the models is that EFDC maintains a constant active layer thickness throughout the simulation, while SEDZLJ computes the active layer thickness (T_{AL}, μm) at each timestep based on a formula proposed by van Niekerk et al. (1992):

$$T_{AL} = 2 \cdot d_{50} \left(\frac{\tau_b}{\tau_{ce}} \right), \tag{10.28}$$

where d_{50} is the median particle diameter of the surficial bed layer (μm), τ_b is the applied near-bed shear stress (dyn/cm^2), and τ_{ce} is the critical shear stress for erosion (dyn/cm^2).

10.3.1.5 Comparison of EFDC and SEDZLJ Sediment Transport

The EFDC and SEDZLJ models are based on similar theory for cohesive and noncohesive sediment transport and representation of the sediment bed. Both models have the capability to simulate multiple cohesive and noncohesive particle size classes, as well as bed load transport and suspended load transport. There are some key differences between how these models implement sediment transport, however. EFDC calculates *net* erosion or deposition for a given particle size class at a given simulation timestep. In the case of noncohesive sediments, the net rate of water-bed sediment exchange is defined by the suspended and bed load transport capacities of the water column. The SEDZLJ model, on the other hand, computes gross erosion and gross deposition at each timestep. The specified gross erosion rate for the surface sediment layer is generally determined by Sedflume measurements, and the total rate is partitioned by the model into the separate components that contribute to the suspended load and the bed load. Deposition fluxes for the suspended and bedload components are calculated by SEDZLJ based on the relevant near-bed concentration and the probability of deposition for each of these components. The use of gross erosion/deposition fluxes or net fluxes potentially has important implications for contaminant transport and fate because the cycling of contaminants between the sediment bed and the water column may be understated when relying on net sediment fluxes.

Bed armoring can be implemented in similar fashion in the two models through the use of an surficial "active" layer. This layer is generally a few particle diameters thick and operates as a transitional zone between the water column and the parent bed. In the case of net deposition to the active layer, SEDZLJ handles the parent bed (i.e., bed layer adjacent to the active layer) differently than EFDC. Sediments deposited to the bed are added to a new deposition layer instead of being integrated into the original surficial bed layer, which is the case with EFDC. This has the effect of preserving the original Sedflume-based layers used to initialize the sediment bed in the SEDZLJ model.

Sandia National Laboratories (SNL) has recently led a collaborative effort to incorporate the SEDZLJ algorithms into the EFDC modeling framework. The result of this effort has produced an enhanced version of the EFDC code called "SNL-EFDC" (James et al., 2005; Thanh et al., 2008). SNL-EFDC should provide users with the flexibility to use either the original EFDC sediment transport dynamics or the SEDZLJ-based algorithms coupled to the EFDC hydrodynamic submodel. The selection of sediment transport submodel would depend on the type and quantity of site-specific data available to support model development, calibration, and application.

10.3.1.6 Model Calibration Considerations

As discussed previously, the formulations in sediment transport models depend on empirical relationships between shear stress and transport rates based on experimental data. Although empirical formulations for erosion and deposition are consistent with sediment transport theory, they are not directly based on theoretical considerations. Therefore, regardless of the level of complexity afforded by a particular model framework, model calibration is an essential component of any sediment transport model application. Data need required to support model calibration are intense and ideally include the following:

- Basic sediment properties, such as dry bulk density, porosity, and grain size distribution.
- Total suspended sediment concentrations and particle size distributions based on discrete or continuous vertical measurements through the water column.
- Bed load flux, including total flux and particle size distributions.
- Direct measurement of erosion rates via flume devices such as Sedflume.

- Long-term changes in sediment bed elevations and bed surface particle size distributions, indicating time trends of net deposition, net erosion, or dynamic equilibrium.

Suspended load and bed load data should be collected near model domain boundaries to inform an overall sediment mass budget for the study area. A mass budget provides important context, as well as calibration targets, for the sediment transport simulation model. Measurements of long-term bed elevations provide an important indication of system trends and should be used as a target for longer-term sediment transport model calibration on the scale of weeks, months, or longer depending on the system.

The results of the data collection activities described above ideally should be combined with a robust understanding of hydrodynamics, geomorphology, and an overall conceptual site model to develop an understanding of how transport of cohesive and/or noncohesive sediments affects a riverine, estuarine, or lacustrine study area over various space and time scales.

10.4 LOWER FOX RIVER AND GREEN BAY CASE STUDY

The case study presented below represents an integration of the process and system-level modeling of sediment and associated contaminant fate and transport discussed in this chapter. It represents one of a handful of contaminated sediment sites for which these process were measured and modeled in order to develop a quantitative understanding of how a system would respond to a full range of remedial options, including monitored natural recovery (i.e., no further active *in situ* remedial actions), under a range of potential hydrometeorological forcing conditions.

The Lower Fox River, WI flows from Lake Winnebago to Green Bay, for a distance of about 39 miles. Sediments in the Lower Fox are contaminated with PCBs that were released predominantly through manufacture and recycling of carbonless copy paper, primarily prior to 1971. Risks to anglers and fish-eating predators, due to food-chain bioaccumulation of PCBs, are the most important potential human health and ecological risks at the site. These risks arise through exposure of biota to PCBs that are adsorbed to surficial and resuspended sediments, or dissolved in the water column. Thus, the process depicted in Figure 10.1 (settling and resuspension of sediments, and adsorption and desorption of PCBs) are fundamental determinants of exposure and risk at this site.

The contaminated portion of the river includes steep and gentle bed slopes, numerous dams, and a seiche-impacted embayment of Lake Michigan that comprises the reach between the lowest dam (DePere Dam) and Green Bay. Flows range from 1950 cfs, which is the lowest daily average flow of the year, on September 10(USGS, 2009) to peak annual flows of 12,600 to 19,900 cfs for 2- to 100-year recurrence intervals, respectively (USGS, 2006). The highest flows typically occur in the spring and are associated with releases from Lake Winnebago, a very large lake that is the upstream source of the river.

As a result of these variations in flow energetics, there is considerable spatial and temporal variability in transport of sediment. Operable Unit (OU) 2, which has the steepest bed slope and numerous dams, contains the smallest volume of contaminated sediment, much of it being located immediately upstream of dams. OU 4, which is the seiche-impacted reach connected to Green Bay, and Green Bay contain much of the contaminated sediment volume. Even these more depositional zones are heterogeneous, due in part to navigational dredging, which fosters sedimentation and periodically removes deposited material. Within OU 4, radionuclide dating of sediment cores has indicated that some areas have undergone net sedimentation, at rates of about 0.5–2.0 cm/year, while others are more disturbed (LimnoTech, 2002a). In more depositional areas, peak sediment PCB concentrations occur at depth, consistent with peak PCB loads having occurred in the 1960s, with shallower sediments contaminated by subsequent mixing processes. Figure 10.12 shows an example of Cs-137 dating, where the peak concentration is associated with loads from approximately 1963, and a PCB profile from a collocated core.

Concentrations of total suspended sediment, and of PCBs adsorbed to suspended sediment, are sensitive to daily flow rates, indicating that deposition and scour of contaminated material and water column fluxes of PCBs respond to variations in flow. Hydrodynamic and sediment transport modeling played an important role in the investigation and remedy determination for the Fox River/Green Bay site. Hydrodynamic modeling included one-dimensional steady-state models, one-dimensional dynamic

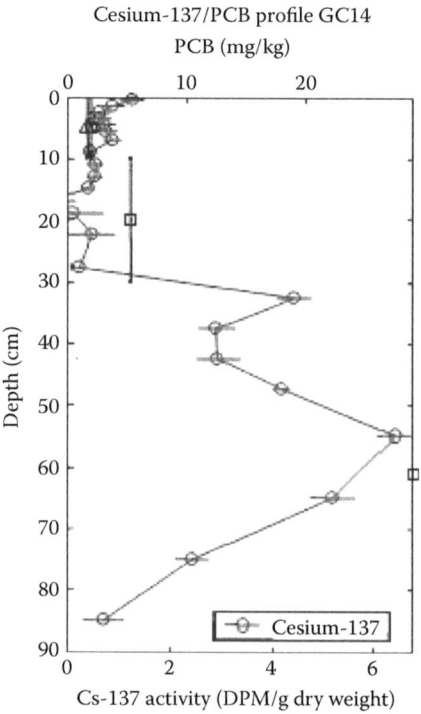

FIGURE 10.12 Example Cesium-137 dated sediment core, with PCB profile from collocated core.

models, and more detailed two- and three-dimensional models of areas of interest within the Fox River-Green Bay system (e.g., Bierman et al., 1992; DePinto et al., 1993; 1994; LimnoTech, 2002b; WDNR, 2001a). External solids loads to the river were estimated using statistical approaches, spreadsheet-based primary production models, and simplified watershed runoff models (e.g., WDNR, 2001b).

Sediment and contaminant transport models were developed by potentially responsible parties (PRPs) and the regulating agencies. Daily measurements of water column solids and PCBs were key calibration targets, augmented by longer-term measurements of net sediment erosion and accretion from bathymetric soundings and radionuclide dating. The model developed by the agencies for remedy selection used three particle sizes, with settling velocities ranging from 0.1 to 470 m/day, and critical shear stresses for deposition ranging from 0.1 to 0.8 dyn/cm^2. Event resuspension was simulated as an exponential function of shear stress as proposed by Lick et al. (1995), with a critical shear stress of 1 dyn/cm^2 and an exponent of 2.3 for shear stress in excess of the critical value. The same event resuspension function was used for all three particle sizes represented. Later model development work by Jones and Lick (2001), with an application to Fox River OU4 sediment transport, includes both settling and resuspension as processes that vary with particle size.

The assumed thickness of the actively mixed upper layer of sediment was an important difference between agency (30 cm) and PRP (10 cm) models (Redder et al., 2005). Sediment porewater diffusion at rates exceeding molecular diffusion rates was also extended to 30 cm in the agency model, and limited to the top 10 cm in the PRP model. The model developed by the PRPs also included contaminant exchange between the water column and the top layer of porewater to help account for water-column PCBs under low-flow conditions, a process that was not included in the agency model (LimnoTech, 2002c).

Figure 10.13 provides an illustrative example of sediment transport-mediated exchange rates for total solids and total PCBs for a single water column segment

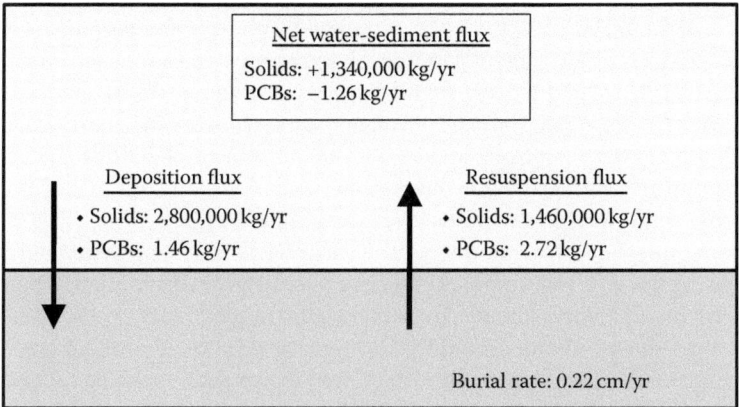

FIGURE 10.13 Example model-predicted sediment-water exchange rates for solids and PCBs in the Lower Fox River.

represented in the PRP model for the Lower Fox River below DePere Dam. The fluxes shown in the figure are averages based on a model simulation extending from January 1, 1989 through July 31, 2001, a period of approximately 12.6 years. Because the depositional flux of solids is predicted to be greater than the resuspension flux during this period, the segment is net depositional, with a net burial rate of 0.22 cm/year. However, the resuspension flux for total PCBs (2.72 kg/year) is higher than the corresponding deposition flux (1.46 kg/year), indicating that sediment transport processes are resulting in a net removal of PCB mass from the upper layer of the sediment bed at a rate of 1.26 kg/year. (Although not shown here, the model also predicts a net loss of PCB mass from the sediment bed to the water column at a rate of 4.06 kg/year due to porewater diffusion.) The mass balance results shown in Figure 10.13 exemplify the scenarios discussed in Sections 10.1.2 and 10.2.3 where net export of contaminant from the sediment bed to the water column occurs despite a long-term net burial of sediment.

The models developed for the Lower Fox River were used to simulate long-term exposures and risks, as determined by daily and seasonal fluctuations in erosion, deposition, and contaminant transport, at high and low flows, in high- and low-energy reaches. These simulations supported comparisons of long-term risks under dredging, monitored natural recovery, and other potential remedial scenarios. The remedy ultimately selected for the site is a combination of dredging, capping, and monitored natural recovery. Hydrodynamic and sediment transport modeling have been used to support design of caps, ensuring stability under conditions that include 100-year flows, wind waves, and extreme seiche activity (Sea Engineering, 2004).

10.5   SUMMARY AND ADDITIONAL INFORMATION NEEDS

Of all the processes that affect the exposure of aquatic systems to contaminated sediments, the deposition and resuspension processes reviewed in this chapter are probably most important in assessing the long-term fate of contaminants in the surface sediments of a system. As a result, these processes are important in the determination of a remediation plan for a contaminated site, because of the need to understand the stability of bottom sediments and the associated stability of contaminants in those sediments. A quantitative understanding of sediment deposition and resuspension processes is needed to address such management questions as: "What are the expected rate and extent of risk reduction at a site under natural recovery (no additional actions), with additional source controls, and/or with *in situ* remedial actions (i.e., dredging and/or capping)?" and "Are risk reductions permanent in response to extreme events (i.e., floods, wind storms, low water level conditions)?"

The above questions may be addressed at various levels of complexity and certainty depending on the size and complexity of the site under consideration. The material presented in this chapter is intended to serve the full range of user needs. This range of users includes those seeking a single value numerical estimate of the deposition and resuspension kinetic parameters, often used in trying to understand the relative importance of particle and associated contaminant transport compared to other contaminant fate and transport processes active at a specific site. In other words, they are trying to build a conceptual site model but are limited as to a complete and coherent data set for their site. The material presented in Section 10.2, which discusses the theory of and provides typical values for particle settling/deposition and resuspension can serve the needs of this user.

On the other extreme, Section 10.2 also accommodates the high-end users requiring detailed procedures and estimating algorithms that will yield particle flux estimates that reflect site-specific conditions that can be used in quantitative models. Specifically these are numerical values that reflect the variations in hydrodynamics, geometry, sediment bed properties and other environmental factors. It is the variations in these factors and other forcing functions at the locales of interest that produce the wide numerical range of particle transport parameters and site-specific fluxes. The formulas and procedures presented can be readily used with these inputs to produce numerical values of the particle fluxes and effective transport parameters, and so on and the ranges so that question of uncertainties in predictions can be addressed and put in a statistical context if needed.

The particle deposition and resuspension processes across the sediment–water interface depicted in Figure 10.1 can be summarized as follows. The well-known Stokes' law settling velocity appears as Equation 10.2. It has been observed that the settling speeds of most particles are slower than predicted by this law (see Figure 10.2 for a comparison). However, most particulate matter exists in the water column in an aggregated state (many individual particles bound together by short range forces). Settling speeds for these aggregated flocs in fresh water undergoing a differential settling process are on the order 0.01 to 1.0 cm/s for diameters 100 μm to 1 mm. Settling speeds of flocs formed by fluid shear range from 0.001 to 0.01 cm/s for diameters 10 to 200 μm diameter.

Particle and aggregate settling speeds are typically larger than the actual deposition rates of these particles onto bottom sediments. The total particle deposition flux to the bed J (g/m^2/s) (see Equation 10.7) is equivalent to the particulate matter settling flux modified by a probability of deposition that is almost always less than one. Turbulence at the interface may act as a barrier to the attachment of settling material. As shown in Figures 10.3 and 10.4, the probability of deposition is a function of shear stress. There is a range of bottom shear stresses for which the probability of deposition is less than one; those particles that do not deposit will remain entrained or suspended in the flow.

Sediment erosion occurs when the applied hydraulic shear stress becomes sufficiently large to dislodge particles from the bed. A critical value shear stress to achieve this condition is required (see Equation 10.5). Typical erosion rates for sediments of density 1.85 and 1.95 g/cm^3 are shown in Figure 10.6. These represent the depth of the bed eroded per unit time. The values range from 10^{-5} to 0.1 cm/s; clearly the erosion rate is a strong function of the applied shear stress, particle density and size. After being dislodged or eroded from the bed, larger particles may settle back to the bed to roll, bounce or slide along the bed. This particle transport process is termed bedload and the speed is slower than the surrounding water. The remaining eroded matter becomes suspended in the water column and moves with the water. Figure 10.7 illustrates the sediment transport mode regimes that are possible. Note that u_* is the so-called friction velocity and w_s is the settling velocity. See Section 10.2.2.3 for details. As a crude approximation, the silt and clay-sized fractions can be assumed to be suspended and the entire erosion rate E used as the numerical estimate.

The models described in Section 10.3 are whole system models that include all processes involved in the transport and fate of contaminants in aquatic systems. Application of these models therefore demonstrates how particle deposition and resuspension affect the transport and fate of contaminants associated with those particles. The case study presented in Section 10.4 for the Lower Fox River is an example of how research knowledge and appropriate field data collection on sediment deposition and resuspension processes (reviewed in Section 10.2) have been incorporated into conceptual and numeric models (reviewed in Section 10.3) that support assessment and remediation of contaminated sediment sites.

The Lower Fox River example also illustrates a very important observation at many contaminated sediment sites that were subjected to considerably higher external loading of the contaminants of concern historically than during the period of remediation assessment and design. That is, that the long-term (e.g., annual average) net flux of particulate matter across the sediment–water interface is often *into* the sediments while the net flux of the contaminant is *out of* the sediments. This occurs when there is net sedimentation of particulate matter, but the contaminant adsorbed to bed sediment is at a much higher concentration than the contaminant adsorbed to particles from upstream. And this condition will persist until such time as the surface bed sediment contaminant concentration is equivalent to the particulate contaminant concentration being delivered from upstream.

Research, data collection, and modeling at large, complex sites such as the Lower Fox River have contributed to the development of a rich knowledge base on the processes discussed in this chapter. However, these same studies have highlighted areas for additional research and data collection necessary to better quantify how these processes respond to both natural and anthropogenic perturbations. Among the areas of research and data needs that have been identified during these investigations are as follows:

- The factors that govern the process of particle aggregation and its impact on settling rates.
- The effect of bottom shear stress and bottom sediment properties and structure on the probability of deposition of particulate matter.
- Further development, implementation, and interpretation of devices for *in situ* measurement of sediment erosion, especially in systems with heterogeneous (cohesive and noncohesive) bottom sediment properties (e.g., grain size distribution, dry bulk density; porosity, organic carbon content); and development of submodels for predicting sediment resuspension as a function of more easily measured sediment properties and modeled bottom shear stress.
- Further development, implementation, and interpretation of bedload transport of sediments and associated contaminants, and comparison of the relative importance of this transport process with resuspension and water column advection.

- More measurement of geochronological profiles in contaminated sediment systems, and analysis of those for determination of net sediment and contaminant burial rates and upper mixed layer depth.
- Research on methods for estimation of spatial variability of deposition and resuspension in large, complex sites, and how to use that information for planning sediment remediation.

This list is neither detailed nor comprehensive, but rather comes from experiences that we have had in working on a variety of site assessments and remedial investigations.

LITERATURE CITED

Ambrose, R.B., Jr. 1987. Modeling volatile organics in the Delaware Estuary. *J. Environ. Engr.*, 113(4): 703–721.

Ambrose, R.B., Jr., T.A. Wool, J.P. Connolly, J.P. Shantz, and R.W. Shantz. 1988. WASP4, A hydrodynamic and water quality model—Model theory, User's Manual, and Programmer's Guide. U.S. EPA, Athens, GA. EPA/600/3-97-039.

Ambrose, R.B., Jr., T.A. Wool, and J.L. Martin. 1993. The Water Quality Analysis Simulation Program, WASP5. Part A: Model Documentation. U.S. EPA Environmental Research Laboratory, Athens, GA, September 20.

Ariathuri, R. and R.B. Krone. 1976. Finite element model for cohesive sediment transport. *J. Hydraul. Div.*, 102(3): 323–388.

Bagnold, R.A. 1956. The flow of cohesionless grains in fluids. *Philos. Trans. R. Soc. Lond.*, Series A, 249(964): 235–297.

Bierman, V.J. Jr., J.V. DePinto, T.C. Young, P.W. Rodgers, S.C. Martin, and R. Raghunathan. 1992. Development and validation of an integrated exposure model for toxic chemicals in Green Bay, Lake Michigan. Final Report for EPA Cooperative Agreement CR-814885, ERL-Duluth, Large Lakes and Rivers Research Branch, Grosse Ile, MI, 350 pp. (September, 1992).

Burban, P.Y., Y. Xu, J. McNeil, and W. Lick. 1990. Settling speeds of flocs in fresh water and seawater. *J. Geophys. Res.*, 95(C10): 18213–18220.

Butman, C.A. and R.J. Chapman. 1989. *The 17-Meter Flume at the Coastal Research Laboratory, Part 1: Description and User's Manual*. WHOI Tech. Rep. 89-10, 31pp.

Cheng, Nian-Sheng. 1997a. Simplified settling velocity formula for sediment particles. *J. Hydraul. Eng.*, 123(2): 149–152.

Cheng, Nian-Sheng. 1997b. Effect of concentration on settling velocity of sediment particles. *J. Hydraul. Eng.*, 123: 728–731.

DePinto, J.V., R. Raghunathan, V.J. Bierman, Jr., P.W. Rodgers, T.C. Young, and S.C. Martin. 1993. Analysis of organic carbon sediment–water exchange in Green Bay, Lake Michigan. *Water Sci. Tech.*, 28(8-9): 149–159.

DePinto, J.V., R. Raghunathan, P. Sierzenga, X. Zhang, V.J. Bierman, Jr., P.W. Rodgers, and T.C. Young. 1994. Recalibration of GBTOX: An Integrated Exposure Model for Toxic Chemicals in Green Bay, Lake Michigan. Final Technical Report to U.S. EPA, Large Lakes and Rivers Research Branch, Grosse Ile, MI, March 1.

Droppo, I.G. and E.D. Ongley. 1994. Flocculation of suspended sediment in rivers of southeastern Canada. *Water Res.*, 28: 1799–1809.

Eisenreich, S.J., P.D. Capel, J.A. Robbins, and R. Bourbonniere. 1989. Accumulation and diagenesis of chlorinated hydrocarbons in lacustrine sediments. *Environ. Sci. Technol.*, 23: 1116–1126.

Engelund, F. and E. Hansen. 1967. *A Monograph on Sediment Transport in Alluvial Streams.* Technical University of Denmark, Copenhagen, Denmark.

Gailani, J., C.K. Ziegler, and W.J. Lick. 1991. Transport of suspended solids in the lower Fox River. *J. Great Lakes Res.,* 17(4): 479–494.

Garcia, M.H. and G. Parker. 1991. Entrainment of bed sediment into suspension. *J. Hydraul. Eng.,* 117(4): 414–435.

Gessler, J. 1967. The Beginning of Bedload Movement of Mixtures Investigated as Natural Armoring in Channels. W.M. Keck Laboratory of Hydraulics and Water Resources, California Institute of Technology, Translation T-5.

Gust, G. and M.J. Morris. 1989. Erosion thresholds and entrainment rates of undisturbed *in situ* sediments. *Journal of Coastal Research*, 5: 87–99.

Guy, H.P., D.B. Simons, and E.V. Richardson. 1966. Summary of Alluvial Channel Data from Flume Experiments, 1956–1961. Geological Survey Professional Paper 462-I. Washington, D.C.

Hoyal, D.C.J.D., M.I. Bursik, J.F. Atkinson, and J.V. DePinto. 1997. Filtration enhances suspended sediment deposition from surface water to granular permeable beds. *Water Air Soil Pollution*, 99: 157–171.

Ibad-zade, Y.A. 1992. *Movement of Sediment in Open Channels.* S. P. Ghosh, translator, Russian Translations Series, Vol. 49, A. A. Balkema, Rotterdam, The Netherlands.

Ives, K.J. 1978. Rate theories, in *The Scientific Basis of Flocculation*, ed. K. J. Ives, pp. 37–61. Alpen aan den Rijn, The Netherlands: Sijthoff and Noordhoff International Publishers, B. V.

James, S.C., C. Jones, and J.D. Roberts. 2005. Consequence of Management, Recovery & Restoration after a Contamination Event. Sandia National Laboratories Report SAND2005-6797. Albuquerque, NM. October.

Jepsen, R., J. McNeil, and W. Lick. 2000. Effects of gas generation on the density and erosion of sediments from the Grand River. *J. Great Lakes Res.,* 26(2): 209–219.

Johnson, C.P., X. Li, and B.E. Logan. 1996. Settling velocities of fractal aggregates. *Environ. Sci. Technol.*, 30: 1911–1918.

Jones, C. and W. Lick. 2001. SEDZLJ: A Sediment Transport Model. Dept. of Mech. and Env. Eng., University of California, Santa Barbara, May 29.

JRB, Inc. 1984. Development of Heavy Metal Waste Load Allocations for the Deep River, NC. JRB Associates, McLean, VA, for USEPA Office of Water Enforcement and Permits, Washington, DC.

Julien, P.Y. 1998. *Erosion and Sedimentation*. Cambridge University Press. New York.

Krone, R.B. 1962. Flume Studies of the Transport of Sediment in Estuarial Processes. Final Report, Hydraulic Engineering Laboratory and Sanitary Engineering Research Laboratory, University of California, Berkeley.

Lau, Y.L. and B.G. Krishnappan. 1994. Does reentrainment occur during cohesive sediment settling? *J. Hydraul. Eng.,* 120(2): 236–244.

Lee, S.-C. and A.J. Mehta. 1994. Cohesive Sediment Erosion. Dredging Research Program Report DRP-94-6, U.S. Army Corps of Engineers.

Lick, W.J. 1982. The entrainment deposition and transport of fine-grained sediments in lakes, in *Interactions between Sediments and Fresh Water*, ed. P. G. Sly, *Hydrobiologia,* 91: 31–40, Dr. W. Junk Publ., The Hague, Netherlands.

Lick, W.J. and C.K. Ziegler. 1992. Flocculation and its effects on the vertical transport of fine-grained sediments. *Hydrobiologia,* 235: 1–16.

Lick, W.J., H. Huang, and R. Jepsen. 1993. Flocculation of fine-grained sediments due to differential settling. *J. Geophys. Res.,* 98(C6): 10279–10288.

Lick, W.J., Y.J. Xu, and J. McNeil. 1995. Resuspension properties of sediments from the Fox, Saginaw, and Buffalo Rivers. *J. Great Lakes Res.,* 21(2): 257–274.

Lick, W.J. 2009. *Sediment and Contaminant Transport in Surface Waters.* CRC Press, Boca Raton, FL.

LimnoTech. 2002a. Measurement of Burial Rates and Mixing Depths Using High Resolution Radioisotope Cores in the Lower Fox River, for the Fox River Group, Report prepared for the Fox River Group. Ann Arbor, MI, January.

LimnoTech. 2002b. Modeling Analysis of PCB Transport for the Lower Fox River, for the Fox River Group. Report prepared for the Fox River Group. Ann Arbor, MI, January.

LimnoTech. 2002c. Evaluation of WDNR Fate and Transport and Food Web Models for the Fox River/Green Bay System, for the Fox River Group. Report prepared for the Fox River Group. Ann Arbor, MI, January.

McNeil, J., C. Taylor, and W. Lick. 1996. Measurement of the erosion of undisturbed bottom sediments with depth. *J. Hydraul. Eng.,* 122(6): 316–324.

Mehta, A.J. and E. Partheniades. 1975. An investigation of the depositional properties of flocculated fine sediments. *J. Hydraul. Eng.,* 12(4): 361–368.

Meyer-Peter, E. and R. Muller. 1948. Formulas for bed-load transport. Proc. Int. Assoc. Hydr. Struct. Res., Report of 2nd Meeting, Stockholm, Sweden, pp. 39–64.

Miller, M.C., N. McCave, and P.D. Komar. 1977. Threshold of sediment motion under unidirectional currents. *Sedimentology,* 24(4): 507–527.

Papanicolaou, A.N., M. Elhakeem, G. Krallis, S. Prakash, and J. Edinger. 2008. Sediment transport modeling review—Current and future developments. *J. Hydraul. Eng.,* 134(1): 1–14.

Partheniades, E. 1992. Estuarine sediment dynamics and shoaling processes. In *Handbook of Coastal and Ocean Engineering*: Vol. 3. J. Herbick, ed., pp. 985–1071.

Partheniades, E.R., H. Cross, and A. Ayora. 1968. Further results on the deposition of cohesive sediments. Proceedings, 11th Conference on Coastal Engineering, London, England, Vol. 2: 723–742.

Ravens, T. 2004. Cohesive sediment stability. Background paper for Contaminated Sediments Workshop, SERDP-ESTCP, Charlottesville, VA, August 10.

Ravens, T.M. 2007. Comparison of two techniques to measure sediment erodibility in the Fox River, Wisconsin. *J. Hydraul. Eng.,* 133(1): 111–115.

Ravens, T. and P. Gschwend. 1999. Flume measurements of sediment erodibility in Boston Harbor. *J. Hydraul. Eng.,* 125(10): 998–1005.

Redder, T.M., J.V. DePinto, H.P. Holmberg, and J.R. Wolfe. 2005. Role of upper mixed layer depth in forecasting contaminated sediment recovery rates. *Aquat. Ecosys. Health Manage.,* 8(1): 53–62.

Roberts, J.D., R.A. Jepsen, and W. Lick. 1998. Effects of particle size and bulk density on erosion of quartz particles. *J. Hydraul. Eng.,* 124(12): 1261–1267.

Robbins, J.A. and D.N. Edgington. 1975. Determination of recent sedimentation rates using lead-210 and cesium-137. *Geoch. Cosmochim. Acta,* 39: 285–304.

Robbins, J.A. 1985. *Great Lakes Regional Fallout Source Functions.* NOAA Technical Memorandum ERL GLERL-56. Great Lakes Environmental Research Laboratory, Ann Arbor, MI.

Schwarzenbach, R.P., P.M. Gschwend, and D.M. Imboden. 1993. *Environmental Organic Chemistry,* 2nd ed. John Wiley and Sons, Hoboken, NJ.

Sea Engineering. 2004. Hydrodynamic Field Measurement and Modeling Analysis of the Lower Fox River, Reach 4. Santa Cruz, CA. September.

Sha, Y.Q. 1956. Basic principles of sediment transport. *J. Sediment Res.* (Beijing, China) 1(2): 1–54 (in Chinese).

Shields, A.F. 1936. Application of similarity principles and turbulence research to bed-load movement. *Mitteilunger der Preussischen Versuchsanstalt f ur Wasserbau und Schiffbau* 26: 5–24.

Smith, J.D. and S.R. McLean. 1977. Spatially averaged flow over a wavy surface. *J. Geophys. Res.*, 82: 1735–1746.

Stolzenbach, K.D. 1993. Scavenging of small particles by fast-sinking porous aggregates. *Deep-Sea Res.*, 40: 359–369.

TetraTech. 2007a. *The Environmental Fluid Dynamics Code User Manual—U.S. EPA Version 1.01.* June. URL: http://www.epa.gov/ceampubl/swater/efdc/EFDC-dl.htm.

TetraTech. 2007b. *The Environmental Fluid Dynamics Code Theory and Computation. Vol. 1: Hydrodynamics and Mass Transport.* June. URL: http://www.epa.gov/ceampubl/swater/efdc/EFDC-dl.htm.

TetraTech. 2007c. *The Environmental Fluid Dynamics Code Theory and Computation. Vol. 2: Sediment and Contaminant Transport and Fate.* June. URL: http://www.epa.gov/ceampubl/swater/efdc/EFDC-dl.htm.

Thanh, P.H.X., M.D. Grace, and S.C. James. 2008. Sandia National Laboratories Environmental Fluid Dynamics Code: Sediment Transport User Manual. Sandia National Laboratories Report SAND2008-5621. September. Albuquerque, NM.

Tsai, C. H. and W. Lick. 1986. A portable device for measuring sediment resuspension. *J. Great Lakes Res.*, 12(4): 314–321.

USGS. 2006. Estimates of Shear Stress and Measurements of Water Levels in the Lower Fox River near Green Bay, Wisconsin. Scientific Investigations Report 2006-5226. In cooperation with the University of Wisconsin-Milwaukee.

USGS. 2009. Surface-Water Daily Statistics of the Nation. Fox River at Oil Tank Depot at Green Bay Wisconsin. URL: http://waterdata.usgs.gov/nwis/dvstat/?format=sites_selection_links&search_site_no=040851385&referred_module=sw.

van Niekerk, A., K.R. Vogel, R.L. Slingerland, and J.S. Bridge. 1992. Routing of heterogeneous sediments over movable bed: Model development. *J. Hydraul. Eng.*, 118(2): 246–262.

van Rijn, L.C. 1984a. Sediment transport, Part I: bed load transport. *J. Hydraul. Eng.*, 110(10): 1431–1455.

van Rijn, L.C. 1984b. Sediment transport, Part II: Suspended load transport. *J. Hydraul. Eng.*, 110(11): 1613–1641.

van Rijn, L. C. 1989. *Handbook: Sediment Transport by Currents and Waves.* Rep. H461, Delft Hydraulics, Delft, The Netherlands.

Wisconsin Department of Natural Resources (WDNR). 2001a. Lower Fox River/Green Bay Remedial Investigation and Feasibility Study: Development and Application of a PCB Transport Model for the Lower Fox River. June 15.

Wisconsin Department of Natural Resources (WDNR). 2001b. Evaluation of Flows, Loads, Initial Conditions, and Boundary Conditions. Model Evaluation Workgroup Technical Memorandum 3a. February 20.

Wool, T.A., R.B. Ambrose, J.L. Martin, and E.A. Comer. 2001. *Water Quality Analysis Simulation Program (WASP), Version 6.0—Draft Users Manual.* U.S. EPA. – Region 4. Atlanta, GA.

Wu, R.M. and D.J. Lee. 1998. Hydrodynamic drag force exerted on a moving floc and its implication to free-settling tests. *Water Res.*, 32(3): 760–768.

Young, R.A. 1975. Flow and sediment properties influencing erosion of fine-grained marine sediments: Sea floor and laboratory experiments. PhD dissertation, Woods Hole Oceanographic Institution, Woods Hole, MA, 202pp.

Zhang, R.J. 1989. *Sediment Dynamics in Rivers*. Water Resources Press (in Chinese).

Zhu, L.J. and N.S. Cheng. 1993. Settlement of sediment particles. Resp. Rep., Dept of River and Harbor Eng., Nanjing Hydr. Res. Inst., Nanjing, China (in Chinese).

Ziegler, C.K. and W. Lick. 1986. A Numerical Model of the Resuspension, Deposition, and Transport of Fine-Grained Sediments in Shallow Water. University of California, Santa Barbara Report ME-86-3.

Ziegler, C. K. and B.S. Nisbet. 1994. Fine grained sediment transport in Pawtuxet River, Rhode Island. *J. Hydraul. Eng.,* 120(5): 561–576.

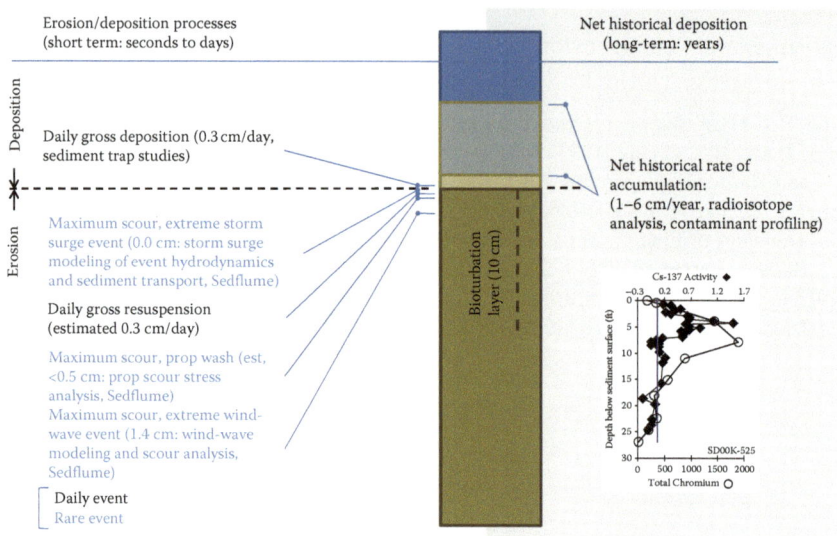

FIGURE 10.10 Short-term erosional/depositional processes and long-term bed deposition at an estuarine sediment site.

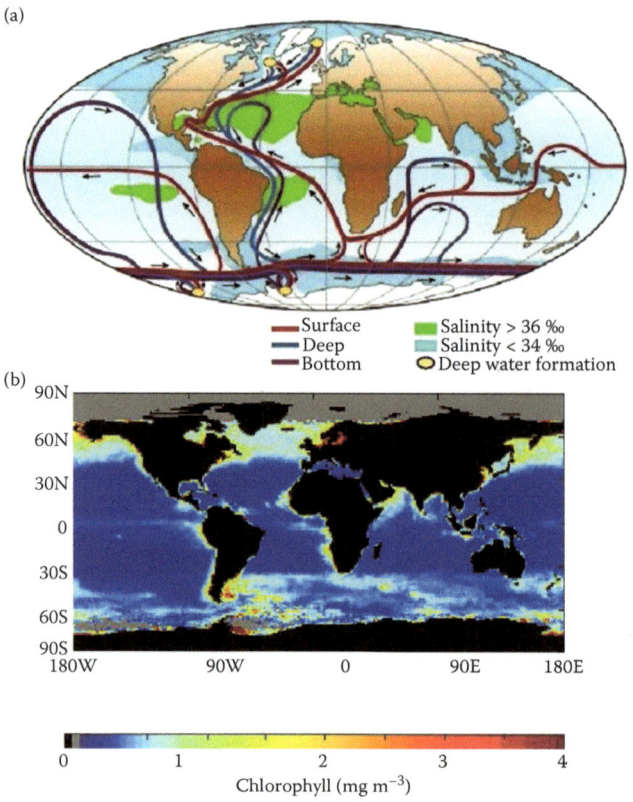

FIGURE 17.1 (a) Principal deep water formation areas; (b) Global chart of chlorophyll *a*, as a proxy for phytoplankton biomass. (From Rahmstorf, S. 2002. *Nature* **419**(6903): 207–214. With permission.)

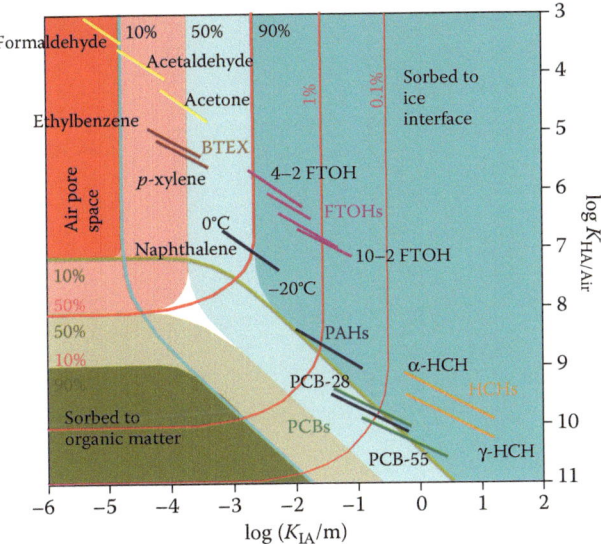

FIGURE 18.2 Chemical space plot of the phase distribution of different chemicals in a dry snow pack as a function of the snow surface/air sorption coefficient (log K_{IA}/m) and the humic acid/air partition coefficient (log $K_{HA/A}$). K_{IA} and $K_{HA/A}$ values were determined based on Roth et al. (2004) and Niederer et al. (2006).

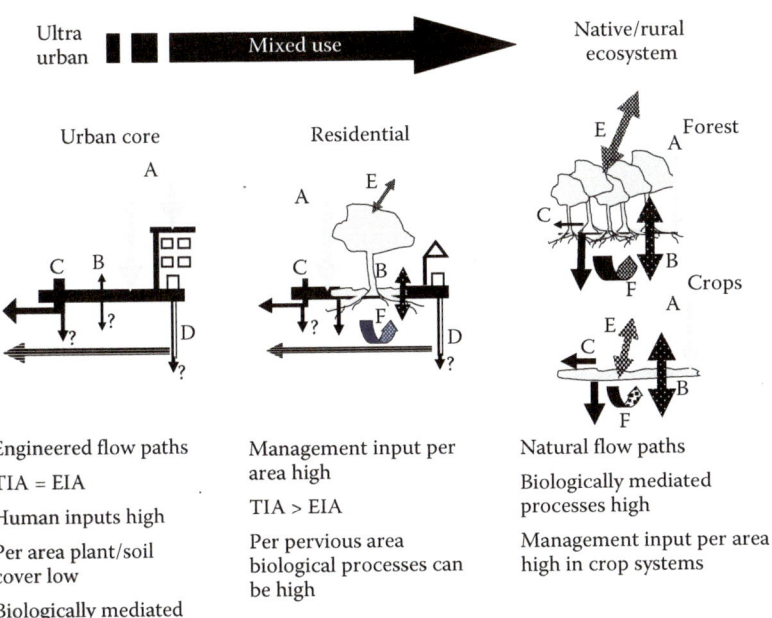

FIGURE 19.1 Conceptual model of ecosystems along an urban to native or rural continuum. Material inputs and flows (A) to an urban system are disconnected from natural processing by impervious surfaces and other features of the built environment, resulting in low connectivity (B) between the processes taking place in the built and natural systems. In residential areas, connectivity between the built and natural system processes can be relatively high resulting in rapid cycling rates (F) for processing and storing often high material and natural inputs and flows (A and E). (Reprinted with permission from Pouyat, R.V., et al. 2007. In: *Terrestrial Ecosystems in a Changing World*. Canadell, J.G., D.E. Pataki, and L.F. Pitelka, eds. Berlin: Springer, pp. 45–58. With permission.)

11 Advective Porewater Flux and Chemical Transport in Bed-Sediment

*Louis J. Thibodeaux, John R. Wolfe,
and Timothy J. Dekker*

CONTENTS

11.1 Introduction 301
11.2 Transport Processes: Hyporheic Exchange 302
11.3 Transport Theory 305
    11.3.1 Darcy's Law Theory 305
    11.3.2 Advection by Darcian Flow in Sediment Beds 306
    11.3.3 Bed-Sediment MTCs with Darcian Advection 306
    11.3.4 Measurement of Transport Parameters 307
11.4 Transport Parameters and Correlations 308
    11.4.1 Darcian Velocity Estimation 308
    11.4.2 Mass-Transfer Coefficient Estimation 313
11.5 Guidelines for Users 316
    11.5.1 Summary of MTC-Recommended Estimation Methods 316
    11.5.2 Example Calculations 317
        11.5.2.1 Grasse River Bed Exchange MTC 317
        11.5.2.2 Florida Straits Bottom Roughness MTC 317
Acknowledgment 318
Literature Cited 319
Further Reading 319

11.1 INTRODUCTION

This chapter focuses on a parameter that regulates the magnitude of the mass-transport coefficient for determining chemical flux across sediment deposits in aquatic systems. Owing to the interactions between the groundwater and the surface water, advective porewater flux moves soluble and colloid-bound chemicals across this environmental interface. Deposited sediments form a dynamic layer of porous solid material on the bottom of streams, lakes, reservoirs, harbors, wetlands, and marine coastal areas. Hyporheic flow is the name given to the process that results in water-driven chemical movement in both directions across a sediment bed. It is termed an advective transport process because the kinetics of chemical movements involves the bulk flow of water. The transport parameter that quantifies the advective process is the average porewater velocity, which is closely related to the well-known Darcian velocity. Three types of hyporheic flows will be described in this chapter, followed by simplistic theoretical models that quantify the key variables controlling the rates. The focus of the chemical transport processes is restricted to the bed sediment layers that separate the surface water body above and the groundwater in the geologic formation materials below.

An outline of the material to be presented is as follows. First, the three hyporheic flow situations relevant to bed-sediment chemodynamics will be reviewed. Next, a brief theory of advection-regulated chemical flux will be presented that provides an analytical connection between the MTC and the Darcian velocity. This will be followed by graphical presentations of field data on measured Darcian velocities in rivers, lakes, coastal estuaries, and so on, which in turn will be followed by a presentation of general correlating algorithms for estimating the Darcian velocities needed at specific sites. The final two sections will contain the needed predictive MTC relationships that accommodate the Darcian velocities and example calculations of the sediment-side MTC.

11.2 TRANSPORT PROCESSES: HYPORHEIC EXCHANGE

Stream beds and banks are unique environments because they are the subsurface of the landscape where groundwater discharges or accumulates and interacts with surface water as shown in Figure 11.1. Although much research related to hyporheic exchange processes has been done on streams, similar processes also occur in the beds of lakes (Winter, 2002) and coastal areas (DePinto et al., 2006). Depending on the type of porous material in the streambed and banks, their variability in slope or hydraulic gradients, and the permeability, the longitudinal size of the hyporheic zone can be up to 10 m or more in length. In these different hyporheic environments the mechanism of groundwater flow and exchange with surface water may differ significantly, depending strongly on the local biochemistry and geochemistry.

In mountainous areas, groundwater is typically tightly connected with surface waters, with relatively highly conductive surface soils and frequently, highly variable rainfall. The groundwater (GW) in these areas is typically recharged by a combination of rainfall and snowmelt. In lower-gradient riverine systems, the GW flow in vicinity of surface waters is often a mix of larger-scale regional flows and local ones that follow smaller scale topography. In particular, relatively flat rivers such as the lower Mississippi can have very dynamic interactions with the adjacent floodplains, including GW-mediated floodwater recession and as a consequence have highly heterogeneous seepage behavior. In coastal areas, GW discharge rates are typically strongly correlated with tidal activity, with periods of highest discharge corresponding with periods

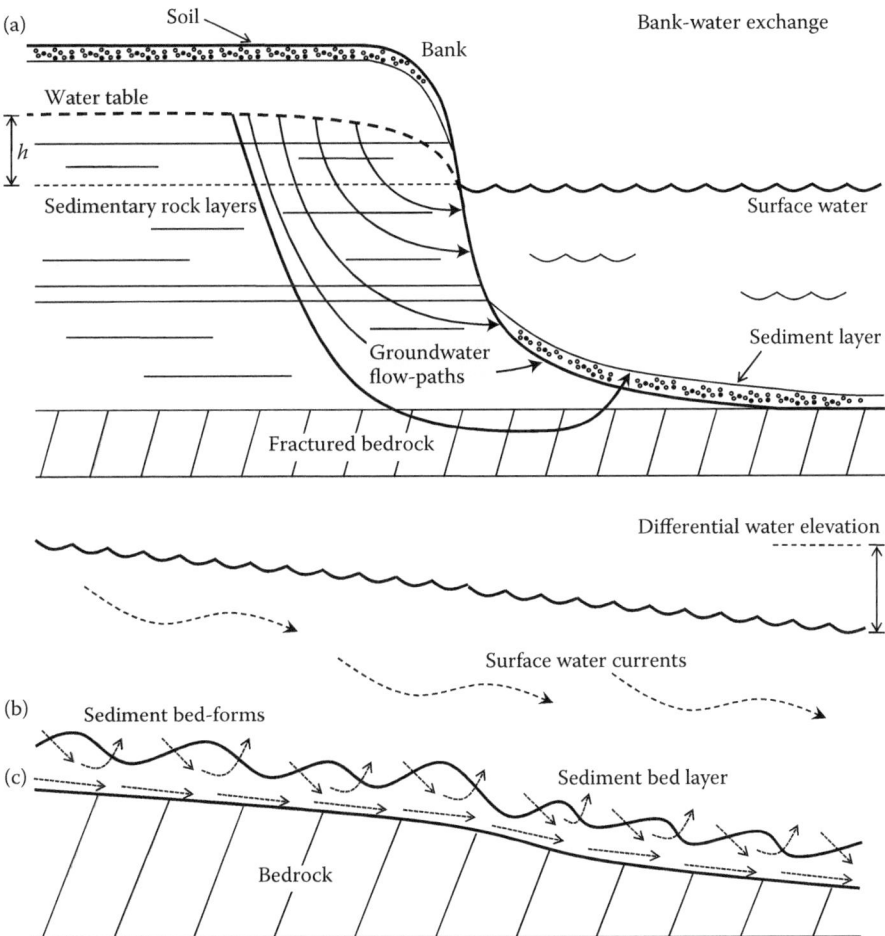

FIGURE 11.1 (a) Hyporheic flow pathway through upland soil and geologic strata into bed sediment layer (several pathway are shown); (b) bottom-form induced exchange flows (arrows in and out of bedforms represent porewater flow paths); and (c) differential surface water elevation induced flow (arrows represent general porewater flow paths).

of low tide. Owing to the periodic nature of tides in these coastal zones "tidal pumping" of marine sediments occurs. A subterranian estuary-type zone (Martin et al., 2007) exists in the seabed at a distance from the beach where the fresh and salt waters contact and mix. Similar to freshwater lakes and impoundments, coastal systems also exhibit high GW discharge near shore and lower rates further from shore, typically decreasing approximately exponentially with distance away from the shoreline. Generally the flux or Darcian, V_D, velocity of water discharging into the water column is the volumetric rate, Q (m^3/s), per unit area of bottom surface, A (m^2).

A single water flow pathway is shown in Figure 11.1a; actually many similar curved pathway lines connect the water table to the sediment layer. Discharge is the movement of groundwater along the flow lines that enter the water column through the sediment layer. Substantial inflow commonly occurs as "springs" and is the result of flows at locations where highly permeable geologic material exists and intersects the beds of the water bodies. As an example of conditions that produce springs, reported fluxes for sand averaged 54 cm/day while those for the silty sands surrounding it averaged 3.6 cm/day. The length of the flow lines can be a few meters to 50 m in length or more and have corresponding travel times of days to years. The longer and deeper flow lines have correspondingly longer travel times. If the elevation of water in the stream is higher than the water table, the groundwater flow direction is reversed; the water originates at the stream bottom and flows down gradient recharging the mass of water in the bank. This situation is not depicted in Figure 11.1a. It is important to note that the direction of GW–surface water interaction can vary significantly across different scales: a river may show net gains of ground water across some reaches and losses along others, while point seepage measurements may show significant viability, including upwelling and downwelling areas, at scales of as little as meters apart.

The hydraulic conductivity is a most important transport parameter in evaluating the rate of surface-water charge or discharge with the subterranean ground water. Geologic heterogeneity within the sediment beds also affects seepage patterns. As noted above (de Marsily, 1986), highly conductive sand beds within finer-grained porous media result in springs. Model studies indicate that relatively thin, either high or low, hydraulic conductivity layers can have a substantial effect on the distribution of seepage flux. Because of the horizontal setting orientation of flat-shaped particles, it is well known that the horizontal (i.e., stream flow direction) hydraulic conductivity of bed material varied five orders of magnitude and the vertical (hydraulic conductivity across the bed) varied by two orders of magnitude.

Three types of hyporheic flows are important with regard to advective chemical transport across bed sediment. (1) Bank water exchange. This first one is depicted in Figure 11.1a and has been described above. (2) Bottom-form roughness induced water exchange is illustrated in Figure 11.1b. These hyporheic flows occur in flowing streams with sand and silt sediment beds containing ripples and dunes or other roughness features on the surface. These features induce pressure differentials, which cause stream water to enter the permeable bed, move within and return to the water column thus continuously irrigating the sediment surface layers. (3) Descending surface water elevation exchange is illustrated in Figure 11.1c. Upstream to downstream elevation differences result in a hydraulic head differential across the permeable sediment bed, which induces flow within and in the same direction as the surface stream flow. Due to sand bars, shoals, pool and riffle features, meanders, etc., an in-and-out flow across these bottom structures occurs as well but on a much larger length scale than the bottom-form roughness elements described in Part 2.

These hyporheic flows can have a significant effect on the chemodynamic transport process at the water sediment interface. Unfortunately, advective water velocities are often incorrectly assumed to be equivalent to be mass-transfer coefficients. The simplicity and convenience of that interpretation propagates the tradition of misuse. Details will be presented in a subsequent section on the correct theoretical approach to apply the MTC interpretation to hyporheic flow-driven processes.

11.3 TRANSPORT THEORY

11.3.1 Darcy's Law Theory

This section outlines briefly the basic hydrodynamic theory of fluid flow through permeable and porous media. The result will provide the basic concept from which to obtain the algorithms for estimating the Darcian velocity for the three most common hyporheic flow situations encountered in nature. These are illustrated in Figure 11.1 and were discussed in a previous section. In each of the three cases, pressure gradients exist across sections of the porous media, which results in water movement according to Darcy's law. The gradients result from differences in water level elevations in the case of situations shown in Figure 11.1a and c. For situation b in Figure 11.1, the pressure gradients across the bedforms, such as sand waves, are due to the Bernoulli effect. Bottom water velocity differences between the upstream and downstream faces of the bottom roughness features along the bed surface induce pressure differences and this in turn induces pore water flow. This flow enters the bed on the upstream face and exits the downstream face, resulting in the semicircular flow pathway illustrated in the figure. Darcy's law applies to all three situations; however, it will be presented in the context of the bank water exchange case only.

As shown in Figure 11.1a, the water table level within the bank aquifer material is elevated with respect to the surface water level. The original Darcy's law contains a specific or intrinsic permeability, which is a function of the porous medium and fluid properties. However, when applied to water a simpler form is used. This form relates the Darcian velocity defined as $V_D = (Q/A)$ in m/s:

$$V_D \equiv K_c \frac{dh}{d\ell} \qquad (11.1)$$

Here $dh/d\ell$ is the water elevation head gradient (m/m) and K_c is the effective hydraulic conductivity (m/s) of the geologic material that water encounters along the flow pathway. The hydraulic conductivity is typically a function of the type of geologic material and the direction of water movement within the geologic strata. Several flow streamlines are shown in part a of Figure 11.1. The flow field between the groundwater in the bank and the sediment layer consists of many such flow lines representing various Darcian velocities, flow residence times, and pathway lengths within the landscape. Where the flow lines terminate at the bed sediment layer, the Darcian velocity is largely directed upward into the surface water column, with V_D assigned as positive (i.e., +).

The reverse hydraulic gradient situation frequently occurs for streams during the seasonal high flow rate time periods. The elevation of the surface water is above the water table elevation in the bank geological material so the directions of the Darcian velocities are reversed. At the sediment bed interface where the flow lines originate they are largely directed downward from the stream water column into the groundwater beneath; the V_D's in this case are given a negative sign (i.e., $-V_D$). The direction of the velocity is important in regulating the magnitude of the chemical MTC in the bed sediment layer and adjoining benthic boundary layer as will be demonstrated in a later section.

11.3.2 Advection by Darcian Flow in Sediment Beds

The advective porewater flux. The preceding discussion was concerned with water flow in both directions across the bed sediment/water interface of aquatic environments. It is necessary at this point to put this water flow into the context of the chemodynamic process, which is occurring across the interface. The flux of a chemical species between different concentrations in water is traditionally expressed by

$$N_y = V_D C_w - D_{eff}\frac{dC_w}{dy},\tag{11.2}$$

where N_y is the chemical flux (g/m^2 s) in the y-direction, C_w is the concentration in porewater (g/m^3), D_{eff} is chemical specific diffusion coefficient in the porous media (m^2/s), V_D is the Darcian velocity in the y-direction (m/s), and y is a length (m). The flux and velocity are defined positive for movement from the bed to the water column. Porewater concentration is a function of position, y, in the bed. Negative chemical fluxes can result from concentration gradients directed into the bed.

The two parts of the above flux expression account for the advection and diffusion processes, respectively. The product of the porewater velocity and concentration is the advective flux component. The diffusive flux component is the other group of terms and is a subject in Chapter 2. The porewater velocity is related to the hyporheic volumetric water flux. This flow is typically determined by measuring the volumetric water flow rate Q, in m^3/s, using a device such as a seepage meter or a chemical tracer. It is then converted into a water flux by using the appropriate bed surface area, A, in m^2, perpendicular to the flow direction. In effect this flux, Q/A (m^3/m^2 s), is consistent with the Darcian velocity V_D, used to describe flow through porous media. The Q/A term has velocity dimensions (m/s) however, it is not the *in situ* porewater velocity. Depending on the average effective flow volumetric porosity of the sediment bed, ε, the two are related by

$$V_y = Q/A\varepsilon.\tag{11.3}$$

The V_y was defined above and is always greater than the Darcian velocity, V_D. For soluble chemicals or those affixed to colloids in the porewater, the $V_D C_w$ product in Equation 11.2 reflects the advective movement rate within the bed. It may be rapid compared to molecular diffusion even for moderate Q/A water fluxes.

The user should realize that V_D or the Darcian velocity is not a MTC. They are vector quantities with both a magnitude and a direction, while MTCs are scalar quantities having only positive numerical values and no implied direction. The noncircular hyporheic flows illustrated in Figures 11.1a and c are vector quantities that drive water through the interface and either enhance or attenuate the active concentration gradient and the diffusive MTC. The next section provides means for incorporating these noncircular flows in modifying the MTC.

11.3.3 Bed-Sediment MTCs with Darcian Advection

The diffusive MTC. The mass-transfer flux expression common throughout this book is used here to describe the chemodynamics transport across a bed-sediment layer of thickness h(m):

$$N = \frac{D_{\text{eff}}}{h} \Delta C_{\text{pw}} \tag{11.4}$$

where N is the chemical flux (g/m² s) into and perpendicular to the plane of the bed, ΔC_{pw} is the porewater concentration difference (g/m³) between the interface and depth h (m), and D_{eff} is the effective diffusion coefficient (m²/s). For information on the sediment-side diffusive MTC and estimating procedures see Section 12.5. Equation 11.4 results from integrating Equation 11.2 over the thickness h and excluding the advective term. In effect D_{eff}/h is the diffusive MTC and the general symbol k_s(diff) in m/s can be used for it as well.

The advection contribution. Porewater advection can affect the flux by altering the effective MTC. In the case of Equation 11.4, for a Darcian velocity directed into the bed, $+V_D$(m/s), in the same direction as the concentration gradient will enhance the MTC. And, a Darcian velocity directed from the bed, $-V_D$(m/s), will attenuate it. A reintegration of Equation 11.2 that includes the advective term results in

$$N = \frac{D_{\text{eff}}}{h} \left[\frac{\text{Pe}\exp(\text{Pe})}{\exp(\text{Pe}) - 1} \right] \Delta C_{\text{Pw}} \tag{11.5}$$

In arriving at this result, the high-end or source concentration in the bed porewater is assumed to be much greater than the low-end or sink concentration. The dimensionless function in the brackets, defined as $f(\text{Pe})$ for simplicity, modifies D_{eff}/h or the sediment-side MTC k_s(diff).

The Pe parameter is the Peclet number; it is defined as: $\text{Pe} = V_D h / \varepsilon D_{\text{eff}}$, or $V_D / k_s(\text{diff})\varepsilon$ and is a dimensionless number, however it is positive or negative depending on the direction of V_D. It is the key independent parameter controlling the magnitude of [ ], which is always positive. The Pe reflects the magnitude ratio of the advective process through V_D to diffusive-type processes through D_{eff}/h or k_s(diff). For a $\text{Pe} = +1$ the processes are equal in magnitude and act in the same direction; in this case [ ] = 1.58. For a $\text{Pe} = -1$ the processes are equal but act in opposite directions; and [ ] = 0.58. For $\text{Pe} = 0$ there is no advection and [ ] = 1.0 and Equation 11.5 becomes Equation 11.4. Figure 11.2 displays the above described behavior of [ ] as a function of Pe. This graph shows that for the positive Pe values, the effective MTC is enhanced. Above about $\text{Pe} = 6.5$, [ ] increases linear with increasing Pe. When advection works against the concentration gradient transport process, as it does for $-\text{Pe}$ values, the effective MTC is modified downward by a factor of $1/100$ at $\text{Pe} = -6.5$. At this condition the advective process effectively stops the diffusive process and hence the flux as well.

11.3.4 MEASUREMENT OF TRANSPORT PARAMETERS

The movement of groundwater to and from surface water can range from slow, diffusive "seepage" across large areas of sediment beds to rapid, concentrated flows at specific locales. Four methods are commonly used to determine these seepage rates, they are: (1) water balance, (2) hydrographic analysis, (3) hydraulic-conductivity and

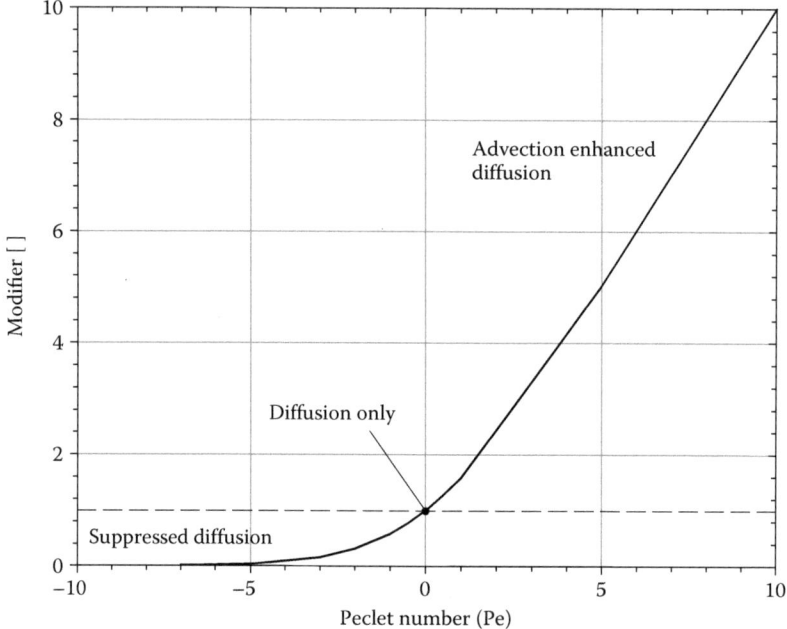

FIGURE 11.2   Advective flow modifier [ ] as a function of Peclet number.

hydraulic-head model method, and (4) direct measures using seepage meters or chemical tracers (Winter 2002). Of the three types of hyporheic flows identified, sufficient data exist only for the bank-water exchange process illustrated in Figure 11.1a. Due to the nature of the in-bed flows in the other two processes illustrated in Figure 11.1b and c, direct measurement is more difficult. However, some limited data are available based on laboratory flume studies, primarily. In the case of the bedform roughness exchange process, laboratory chemical trace studies have been made (Packman et al., 2004; Elliott and Books, 1997; Thibodeaux and Boyle, 1987). And in the case of the differential surface water elevation induced exchange process, field chemical tracer studies have been performed (Bencala as cited in Worman and Packman, 2002). However, too few sets are available to capture the ranges of these advective flows. A recent study by the National Research Council highlighted a need for broad and coherent observations of groundwater recharge and discharge across various hydrologic scales (NRC, 2004).

11.4   TRANSPORT PARAMETERS AND CORRELATIONS

11.4.1   Darcian Velocity Estimation

Bank exchange water fluxes. Data have been collected in groundwater/surface water interaction studies aimed at quantifying advective transport of a wide range of environmentally and geochemically relevant solutes including salt, acidity, pesticides, DOC, and the persistent organic pollutants. In other studies, water fluxes were determined for lake water levels, lake water budgets, riverbank filtration, bank water storage, tidal exchange evaluations, etc. Considerable data have been accumulated for these purposes primarily over the past three decades, and some key sets are presented here. A graphical display of Darcian water velocities, V_D (mm/day), moving in and out of water bodies through the sediment bed appear in Figure 11.3. A cumulative probability distribution (CPD) of log V_D usually conforms to the normal distribution and captures the variability of the data. Recharge refers to water seepage from the bed into the water column and discharge is the reverse. Generally the trend of the data sets suggests that seepages into the coastal waters from the shoreline are largest followed by lakes and then rivers. Numerical mean values of seepage rates are given in Figure 11.3 for each waterbody type. Most of the data are individual seepage meter measurements or other techniques used for obtaining point measurements representing bottom areas of one square meter or less. However, average coastal velocities representing several hectares or more in area are shown in the figure as well. This data represents groundwater modeling, tracers measurements supplemented with mass-balance modeling, remote sensing techniques and so on. These tend to be smaller than the point measurements. Although not a comprehensive review of the data, this figure represents two rivers, 13 lakes including one port, and 12 coastal areas. The average coastal measurements represent seven coastal areas. The data set contains 115 numerical values

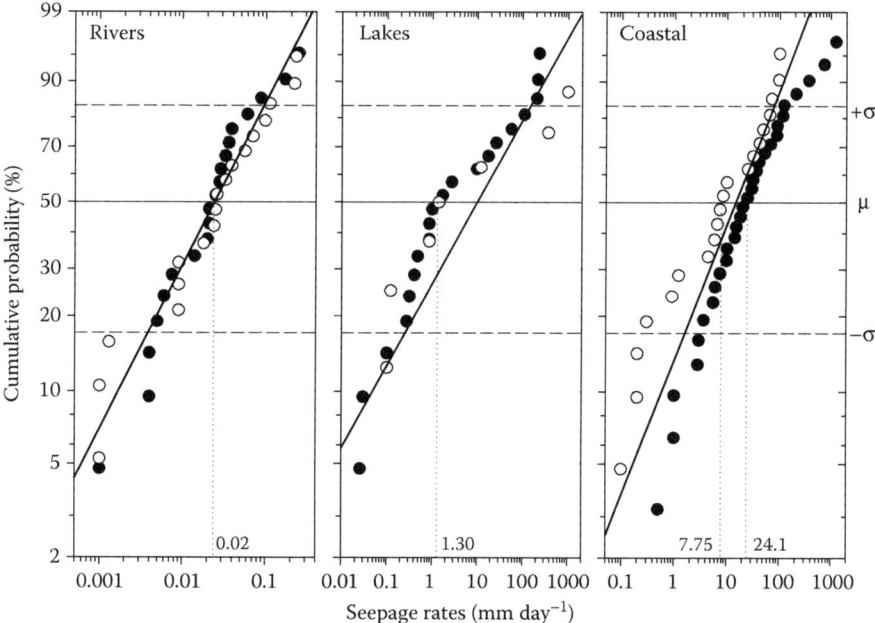

FIGURE 11.3 Distribution of seepage rates. For lakes and rivers, closed circles represent bed recharge and open circles represent bed discharge; for coastal seepage data all points represent bed discharge-closed circles represent point measurements and open circles represent regional averages.

with 25 for lakes, 38 for rivers and 50 for the marine coast. The data was compiled from Winter (2002), BBL (1998), and DePinto et al. (2006).

The purpose of the above paragraph was to present some numerical data on the advective flux of water across the sediment–water interface. The following contains algorithms, correlations, and general conceptual approaches that are needed for obtaining numerical estimates of the Darcian velocity at a particular site or for a particular hypothetical situation. Some sites are data rich with sufficient geologic, hydrographic, and landscape information so as to allow reasonable order-of-magnitude V_D estimates to be made. Other situations are only crudely characterized, having little or no data, only rough and approximate knowledge of the geology and hydrography thus necessitating making reasonable assumptions for a range of key parameters. In this case, the V_D estimating procedure requires a calculation protocol, which gives due attention to the landscape and parameter uncertainties, thus producing individual values that will vary by many orders-of-magnitude. In the final analysis, the V_{DS} that result may be the product of several seemingly reasonable scenarios.

The bank-water exchange process-estimating V_D. This Darcian flow situation is very complex as shown in Figure 11.1a. Arriving at numerical, site-specific values of K_C, Δh, and Δl that realistically characterize any particular situation is challenging. However, rewriting the head gradient in Equation 11.1 in finite difference form allows parceling its parameters, which results in

$$V_D \cong -K_C \frac{\Delta h}{\Delta l} \qquad (11.6)$$

Here Δl denotes an approximate or "characteristic" length of lateral water travel with regard to the groundwater to surface-water landscape. As written Δh (m) is the elevation change over the length of travel; it decreases with an increase in distance Δl (m) to produce a negative gradient and therefore a positive V_D for the water table source illustrated in Figure 11.1a. At the terminus of the flow lines, the porewater moves upward through the sediment bed. Choosing appropriate Δh and corresponding Δl values that reasonably characterize such sites is nonetheless difficult. The flow line pathway characteristic length, Δl, typically consists of n geologic layers including the sediment layer, all with different thicknesses, Δl_i, and hydraulic conductivities, K_{ci}. If these details are available, the overall effective hydraulic conductivity can be estimated by

$$K_C = \Delta \ell \left/ \sum_{i=i}^{n} (\Delta \ell_i / K_{ci}) \right. \qquad (11.7)$$

This term is the so-called resistance-in-series relationship.

Many data exist on measured permeability and hydraulic conductivity of both consolidated and unconsolidated porous material. These data represent the geological material consisting of sand, silt, and clay particles that form the sediment bed plus the sandstone, limestone, and other aquifer material along the flow pathways. Selected data from various sources that represent bank and stream-bed materials appear in Table 11.1. A study of these data indicates that the range of K_{ci} values for any particular material (e.g., sand) is large and may cover three orders of magnitude. Obviously, the use of Equation 11.7 makes the estimate of the effective K_c for the flow-line through the geologic layers even more problematic. The more appropriate use of Equation 11.7 is to focus on the one or two specific layers, which local geological evidence suggests may provide the greatest flow resistance.

Because Equations 11.6 and 11.7 constitute a simple algorithm of a complex hydrodynamic flow process, given the best possible estimate for a number of parameters (i.e., Δh, Δl, and K_c), it is unlikely that the V_D estimate will be accurate to within about one order of magnitude. The situation is improved as the physical scale of the site is diminished, the landscape is made of one or two geologic materials, and some direct measurements can be obtained. Often times the sediment layer itself provides a significant resistance for the movement of bank-water into the stream or the reverse. For example, a layer high in silt content has a significantly lower hydraulic conductivity than the sand within the aquifer material. Even within the categories of sand and silt, it is evident from Table 11.1 that wide variations in K_C values occur for these media. In the case of a specific sediment layer, *in situ* or laboratory measurements on

---

TABLE 11.1
Hydraulic Conductivity, Ranges for Earthen Materials K_c (m/s)

A. Unconsolidated deposits

a. Unweathered marine clay	$2E - 12$ to $9E - 8$
b. Glacial till	$1E - 12$ to $8E - 5$
c. Silt, loess	$9E - 8$ to $8E - 4$
d. Silty sand	$1E - 7$ to $1E - 3$
e. Clean sand	$7E - 5$ to $1E - 2$
f. Course gravel	$1E - 3$ to $1E - 1$

B. Hard rocks

a. Shale	$<1E - 13$ to $1E - 9$
b. Unfractured M&I[a]	$<1E - 13$ to $1E - 10$
c. Sandstones	$2E - 10$ to $9E - 5$
d. Limestone and dolomite	$3E - 9$ to $9E - 5$
e. Unweathered chalk	$1E - 9$ to $1E - 6$
f. Fractured M & I[a]	$7E - 8$ to $1E - 4$
g. Dolomite limestone	$1E - 5$ to $1E - 3$
h. Weathered chalk	$1E - 5$ to $1E - 3$
i. Karst limestone	$1E - 6$ to $1E - 2$
j. Permeable basalt	$1E - 7$ to $1E - 2$

Source: Data compiled from Freeze, R.A. and J.A. Cherry. 1979. *Groundwater.* Prentice-Hall, Englewood Cliffs, NJ, Chapter 2, Section 2.4.; de Marsily, G. 1986. *Quantitative Hydrogeology-Groundwater Hydrology for Engineers.* Academic Press, New York, Chapter 5.

[a] M & I denotes metamorphic and igneous rocks.

samples taken at several locations using piezometers and falling-head permeometers can yield K_C values with a much higher degree of certainty. For example, in a study conducted on Sava Brook, the *in situ* K_C values of the bed-sediment at six locations spaced along the 30 km stretch had an average of 8.9E−4 m/s with standard deviation of 1.0E−4 m/s (Worman and Packman, 2002). With Δl being the measured bed depth and Δh obtained from micropiezometer head differentials, the estimation of V_D is much more direct, with confidence levels within the percent range, rather than the order-of-magnitude range.

    Bottom-form roughness-induced water exchange-estimating V_d. This porous media flow process is illustrated in part b of Figure 11.1 and was described previously. Worman and Packman (2002) have developed a comprehensive, Darcian flow, fluid dynamic theory for quantifying solute exchange in the hyporheic stream bed zone of streams and rivers. The model framework was developed based on the fundamental physics of the process. The subsurface porous media flow follows the continuity equation and Darcy's law. A special periodic variation of hydraulic head is specified along the bed surface to mimic that produced by sand wave roughness. Roughness is due to irregularities on the bed surface. Some roughness features are formed naturally due to stream flow over a loose sediment bed. These sand wave shapes can be reproduced in the laboratory and are used in experimental studies that result in useful correlations for estimating V_d. Additional irregularities can be formed by biotic processes (i.e., fish nest mounds) or by obstacles like stones, tree branches, subaquatic vegetation, and so on. Any such obstructions may produce high-pressure and low-pressure regions that force water and solutes to enter and exit the bed. This phenomenon has been termed "pumping" because of the in then out water exchange across the bed surface. The symbol V_d, with lowercase subscript, will be used for the bedform roughness induced Darcian velocity because it has no consistent direction of movement, being generally semicircular in pathway as shown in Figure 11.1b.

    The bottom-form roughness water exchange process is very amenable to laboratory flume studies and these provide a very realistic simulation of field stream-bed conditions. Several research groups have used this approach (Packman et al., 2004; Elliot and Brooks, 1997; Savant et al., 1987; Thibodeaux and Boyle, 1987) and the process is well described and quantified. A key result of this work has been the development of Darcian theory-based algorithms for estimating V_d. Based on separate studies, two are available for making V_d estimates. Using an analogy theory for flow conditions around solid objects and a published correlation for friction on a nonpermeable wave form, Thibodeaux and Boyle (1987) proposed the following for the superficial porewater Darcian velocity, V_d(m/s):

$$V_d = 0.50 \left(\frac{\Delta}{d}\right)^{3/8} K_c V^2 / g \lambda \tag{11.8}$$

where Δ is the height of the sand wave bed forms (m), d is stream water depth (m), K_c is the bed-sediment hydraulic conductivity (m/s), V is the stream average velocity (m/s), g is the gravitational constant 9.81 m/s^2, and λ (m) is the length of the sand waves from crest-to-crest. Extensive and detailed laboratory measurements by Elliot and Brooks (1997) have resulted in the following algorithm for estimating the characteristic Darcy velocity V_d:

$$V_d = 0.88 \left(\frac{\Delta}{0.34d} \right)^r K_c V^2 / g\lambda \qquad (11.9)$$

with $r = 3/8$ for $\Delta/d \leqslant 0.34$ and $r = 3/2$ for $\Delta/d > 0.34$. Equation 11.9 gives numerical values two to three times larger than Equation 11.8. The confidence interval on the V_d estimates is typically in the low percentage range as one would expect from controlled laboratory experiments.

The differential surface water elevation-induced flow water exchange process. This process is illustrated in Figure 11.1c as flow pathway lines in the lower region of the sediment layer. The Darcian velocity representative of this hyporheic flow process can be estimated by

$$V_D = K_c s, \qquad (11.10)$$

where s (m/m) is the slope of the surface water interface plane. This V_D is in effect the longitudinal velocity of the groundwater in the bed and deeper geologic strata underlying the water column. It is the subterranean portion of the stream flow and remains largely submerged through much of its pathway downstream, emerging occasionally at locations on the bottom where abrupt changes in bottom elevation or permeability occur. Within a short distance, it becomes submerged again. The role played by this submerged flow in the overall chemical exchange process is unknown at this time; however, due to the flow connectivity with the bottom-form induced flows, some exchange interaction is expected.

11.4.2 Mass-Transfer Coefficient Estimation

The preceding section was concerned with data and algorithms for use in estimating the appropriate Darcian velocities across the sediment–water interface. This advective process plays an important role in regulating the magnitude of the chemical MTC for the two significant situations of bank water exchange and bottom roughness induced water exchange. The following paragraphs give the key algorithms and procedures needed for obtaining the appropriate sediment-side MTCs in the presence of these hyporheic flow situations. For information on the water-side MTC and estimating procedures see Section 12.2.

Estimating the MTC for the bank water exchange process. Details concerning the theory were presented above in Sections 11.3.2 and 11.3.3. The chemical flux concept in equation form, such as Equation 11.2, requires that advection be connected to other on-going in-bed transport processes such as diffusion. Advection in a chemodynamic context cannot be considered a stand-alone process. The molecular diffusion transport process is used in Equation 11.5; however, it can be generalized to accommodate any diffusive-type process and the appropriate MTC. As such the appropriate sediment-side MTC is

$$k_s(\text{Dar}) = k_s(\text{diff}) \left[\frac{\text{Pe} \exp(\text{Pe})}{\exp(\text{Pe}) - 1} \right], \qquad (11.11)$$

where Dar signifies the Darcian velocity contribution to k_s(diff). The latter is any diffusive-type or concentration gradient defined MTC in the bed-sediment surface layer. It includes D/h, which is specific to solute molecular diffusion in porewater but also accommodates the Brownian diffusion of aquasols, bioturbation, and so on. The Peclet number, Pe, is defined for Equation 11.5 above. Rather than using the Pe function [ ] in Equation 11.11 numerical estimates for it can be obtained from Figure 11.2. Because the Pe can be both positive and negative the user must have some sense of the direction and magnitude of the advection water flux or Darcian velocity vis-à-vis the operative concentration gradient direction at the sediment–water interface of concern so as to properly account for the numerical effect of any hyporheic porewater movement direction on the MTC. Example calculation 11.5.1 contains some information in this regard.

Estimating the MTC for the bottom roughness induced water exchange process. In 1983, Bencala and coworkers proposed a mass-transfer process for simulating the transport of solutes between the water column and the stream-bed porewater in a mountain pool and riffle stream. Worman and Packman (2002) reviewed the subject of mass transfer in hyporheic zones and noted that first-order mass-transfer relationships were used early on to describe hyporheic solute exchange. They note that it is a convenient parameterization of all mechanisms governing solute mixing in the hyporheic zone and it has also been represented as a diffusive process however, both are a considerable abstraction of reality as the diffusive exchange coefficient has been found to be orders of magnitude larger than molecular diffusivity of solutes in porous beds. Nevertheless, this early work highlighted the chemodynamics of the hyporheic process in streams, which led to further research.

As discussed and documented in Section 11.4.1, experimental evidence from laboratory flume studies indicates the process is Darcian and that Equation 11.1 applies to the individual flow lines that trace a semicircular pathway in and out of the bed and this flow pattern has been documented experimentally (Thibodeaux and Boyle, 1987; Packman et al., 2004). One quantitative result is the numerical flow line pathways of the advective pumping, which show infiltration into and exfiltration from the interface. Velocities are known as well, which allows the computation of the net average flux of a dissolved chemical across the interface for an area encompassing the length of the sand wave. A factor of $1/2$ is introduced into the flux expression because half the surface is subject to infiltration and half is exfiltration. A distribution of porewater path length and residence times characterizes the subsurface. These and other computations (Savant et al., 1987) show the flow lines are indeed semicircular and the porewater velocities are highest in the porous bed nearest the sediment–water interface. This theoretical approach provides additional evidence that the water pumping exchange model is responsible for the solute exchange across the bed-sediment interface. The Worman–Packman model connects the physics of the process to the solute chemodynamics across the bed-sediment interface and its result can be used as the conceptual basis for formulating the following general mass-transfer approach.

If the bed material is contaminated, having a higher effective pore water concentration than the water column, the in-and-out water flow shown in Figure 11.1b results in a net downstream direction chemical flux from the bed. From a chemodynamic perspective it can be argued that the semicircular flows result in V_d having the scalar properties of a MTC. In other words, the V_d associated with the bottom-form roughness has no implied direction when averaged across a large area of stream bottom, since the water exchange occurs continuously in the up-down directions as well as lateral. Because of this random flow situation, the direction of the chemical flux is determined by the chemical concentration gradient or concentration difference between porewater and surface water. If the concentration is higher in the surface water, V_d regulates the magnitude of the flux into the bed. However, if the concentration gradient is reversed, the same V_d applies for the flux from the bed to the water column. This being the case there is no need for the advective flow modifier, [ ], or Figure 11.2 or Equation 11.11. The sediment-site MTC for bottom roughness, $k_s(bro)$, can therefore be defined as follows:

$$k_s(bro) \equiv V_d/\varepsilon \cdot 2. \tag{11.12}$$

Here ε is the bed porosity and either Equation 11.8 or 11.9 can be used to estimate V_d. The MTC takes the units of V_d.

Although measured MTCs have not been located, Elliot and Brooks (1997a) showed that the initial rate of bed form-induced advective chemical pumping exchange does mimic diffusive behavior. Packman and Salehim (2003) presented experimental data of the effective diffusion coefficients $D_{eff}(cm^2/s)$ measured using sodium chloride, lithium ion, and a fluorescent dye as tracers "early" in the experiments. The data appear in Figure 11.4 where it is clear that the numerical values of D_{eff} are much greater that the molecular diffusion coefficients, D_m, of these solutes in this type of porous material. Typically for these solutes, the D_m values range between $3.3E-6$ and $1.4E-5$ cm^2/s in a bed with 35% porosity. These values are $10\times$ to $100\times$ smaller than the smallest measured D_{eff} values shown in Figure 11.4. It is possible to

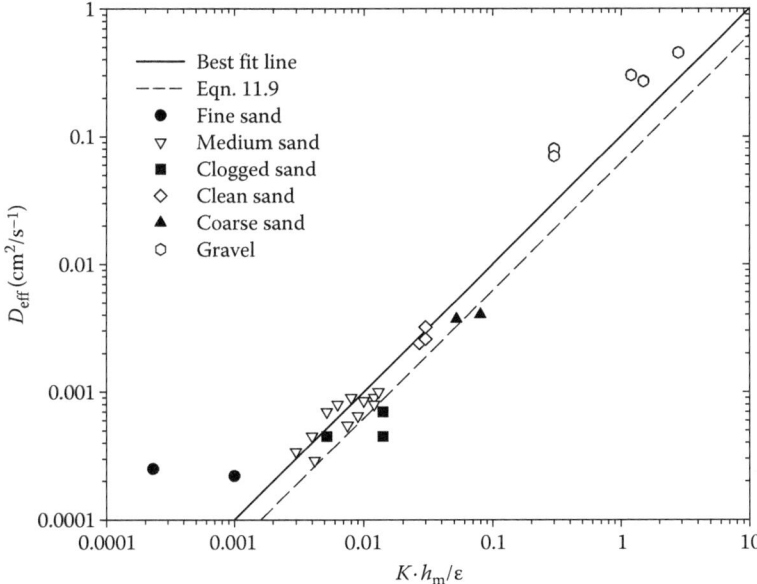

FIGURE 11.4 Effective diffusivities for experimental bedforms compared to pumping theory. (Modified from Packman, A.I. and M. Salehin. 2003. *Hydrobiologia*, 494, 291–297.)

transform the MTC Equation 11.12 to its equivalent D_{eff} form so as to compare model predictions with the laboratory data in Figure 11.4. The transformation is as follows.

The equations for estimating the Darcian velocity, Equation 11.8 and 11.9, can be put in the simple form of $V_d = h_m K_c/\lambda$ where h_m is the pressure head distribution on the bed surface. For most sandwave bed forms the wave length-to-height ratio, λ/Δ, ranges from 5 to 10; a value of 8 will be assumed. Using these relationships, Equation 11.12 is then transformed to: $D_{eff} = k_s(\text{bro})\Delta = h_m K_c/16$. The result appears graphed as a single line in Figure 11.4 along with the data. The estimates under predict the measured data by a factor of approximately 2.0. In using D_{eff} as a surrogate for $k_s(\text{bro})$ it appears that the correlations proposed for V_d, along with Equation 11.12, give a very reasonable representation of the experimental data. Interestingly, 2 is the infiltration-to-exfiltration factor introduced earlier.

The fact that the Darcian-based pumping model algorithms represented by Equations 11.8 and 11.9 can reproduce such a wide variety of observed exchange behavior shows that the underlying mechanism of hyporheic exchange is well understood and that the quantitative description for flows induced by common dune-shaped bedforms applies for a wide range of stream and sedimentary conditions. The model cannot necessarily be expected to apply directly to natural streams, which can have a range of different bed features. Even so, the underlying process will drive exchange with other stream bottom form features and the MTC correlations presented above should give reasonable estimates in these situations.

Many chemodynamic processes occur in parallel in the sediment bed. For example, molecular diffusion is always present and occurs simultaneously with the bottom roughness induced Darcian process. So, the combined processes result in a MTC which is the sum of $k_s(\text{diff})$ and $k_s(\text{bro})$. See Section 2.2 for more information on parallel and series processes. Also see Example Calculation 11.5.2 below.

Estimating the MTC for the differential surface water elevation exchange process. The significance of mass-transfer for this submerged flow process illustrated in Figure 11.1c is unknown at this time; however, due to the flow connectivity with the bottom-form-induced flows, some exchange interaction must occur. Based on a theoretical study, it was found that for typical stream conditions the longitudinal groundwater flow induced by the mean stream slope does *not* have a significant impact of the porewater residence times and it was concluded that the advective pumping flows due to the bottom-forms dominate the hyporheic exchange in Sava Brook (Worman et al., 2002). In effect, Equation 11.10 is the rate of porewater movement directed $90°$ from and not across the sediment-water interface so as to likely be a minor contributor to the overall chemodynamics of solute exchange across the sediment bed. No algorithms or procedures are available to estimate its magnitude.

11.5 GUIDELINES FOR USERS

11.5.1 SUMMARY OF MTC-RECOMMENDED ESTIMATION METHODS

Hyporheic flow gives rise to two known and significant advective transport processes that influence mass transfer at and across the bed-sediment layer in aquatic bodies. One is termed bank-water exchange; it induces a noncircular Darcian flow through the bed. Measured data, on the magnitudes and ranges of V_D for various water bodies,

TABLE 11.2
Estimated PCB MTCs for the Grasse River, New York

Bed Sediment Seepage	V_D (mm/d)	Pe	[]*	k_s(Dar) (mm/d)
Advective water flow through bed	+0.10	1.05	1.61	0.306
into water column	+0.03	0.316	1.16	0.220
	+0.01	0.105	1.05	0.200
Diffusion only; no advective flow	0.0	0.00	1.000	0.190
	−0.01	−0.105	0.948	0.180
Advective water flow from water	−0.03	−0.316	0.850	0.162
column into sediment bed	−0.10	−1.05	0.565	0.107

* $[] = P_e \exp (P_e)/(\exp(P_e) - 1)$.

appear in Figure 11.3 as a source of approximate values. Alternatively, more definitive and site-specific values can be obtained by use of Equations 11.6 and 11.7. In either case, use Equation 11.11 to obtain the MTC, k_s(Dar). The Peclet number is needed in the final calculation. The other is termed the bottom roughness water exchange, which induces a semicircular flow path in the bed. Use Equations 11.8 or 11.9 to estimate V_d and Equation 11.12 to obtain the MTC k_s(bro). Add coefficients to the other parallel MTCs occurring to obtain the combined sediment-side MTC.

11.5.2 EXAMPLE CALCULATIONS

11.5.2.1 Grasse River Bed Exchange MTC

Using the average river data for V_D and ±1 SD shown in Figure 11.3, obtain estimates of the MTCs for the PCB contaminated bed-sediment of the Grasse River, New York. Based on the computed result, comment on significance of advection on soluble transport from bed to water column. Assume tetrachlorobiphenyl with $D_{eff} = 2.20E - 6 \, cm^2/s$ in bed with porosity $\varepsilon = 0.5$, and $h = 10 \, cm$ active transport layer thickness.

Solution: The diffusion-based MTC, k_s(diff)$= D_{eft}/h = 2.20E - 7 \, cm/s = 0.19 \, mm/d$. V_D values selected appear in column 2 of Table 11.2. See Equations 11.5 and 11.11 for the Pe, [ ] and k_s(Dar) formulations. $P_e = V_D h/D_{eft}\varepsilon = V_D/(.190)(0.5) = 10.5 V_D$. The Pe, [] and k_s(Dar)$= k_s$(diff) [] computed results appear in columns 3, 4, and 5, respectively.

The k_s(Dar) values vary from 0.11 to 0.31 mm/d. Note that all the MTCs are positive while V_D varies from a positive high of 0.1 mm/d to a negative high of −0.1 mm/d. In reality, the $\sim 3\times$ variation in the k_s(Dar) values is of borderline significance. Advection at this site on the Grasse River appears not to be a dominant mechanism transporting solubilized PCBs from the bed sediment to the water column. Its effect is to slightly enhance or attenuate the diffusive transport process which has a MTC of 0.190 mm/d. A review of recent measurements indicates that the Grasse River MTC ranges from 10 to 70 mm/d (Thibodeaux, 2005).

11.5.2.2 Florida Straits Bottom Roughness MTC

A current meter and camera were placed at 710 m depth on the bottom of the eastern Florida Straits to record interaction of sea-floor sediment with overlying current flow

TABLE 11.3
Example Calculation of Florida Straits
Mn^{2+} MTC (cm/d)

Current speed (cm/s)	5	5	15	15
Equation used for V_d	11.8	11.9	11.8	11.9
k_s(bro)		0.0182	0.0480	0.164	0.432
k_s		0.0389	0.0686	0.185	0.453

Note: k_s (diff) $= 0.0207$ cm/d.

(Wimbush, et al., 1982). An illuminated 1 sq. m patch of the seabed showed "...
sediment ripples of 15 cm wave length were always present." Similar ripples were
photographed at 815 m depth in the Straits of Gibraltar. The sediments were 98%
(wt) carbonate with grain density 2.85 g/cm^3. The average grain size was 250 μm
but a sizeable silt-clay size-fraction (<62 μm) was present as well. Current speeds at
0.82 m above the bottom ranged 5–15 cm/s with occasional "storms" up to 40 cm/s.

Assuming a 2 cm sand wave height and 50% bed porosity, estimate the sediment-
side MTC for the manganese cation for a bed depth $h = 5$ cm. The molecular
diffusivity of Mn^{2+} is $3E − 6$ cm^2/s.

Solution: The effective diffusivity in the saturated porous media, D_{eff} is
$D_m \varepsilon^{4/3} = 3E − 6(0.5)^{4/3} = 1.2E − 6$ cm^2/s. The diffusion MTC is: $k_s(diff) = D_{eff}/h =$
$2.4E − 7$ cm/s (0.0207 cm/d). Without a measurement of the hydraulic conductiv-
ity, a value $K_c = 1E − 5$ m/s is assumed. Using Equation 11.8 with $d = 0.82$ m, the
estimated hydraulic boundary layer depth, and $V = 5$ cm/s:

$$V_d = 0.5 \left(\frac{2}{82} \right)^{3/8} (1E − 5)(0.05)^2/(9.81)(0.15) = 2.11E − 9 \text{ m/s } (0.0182 \text{ cm/d}).$$

For $V = 15$ cm/s the $V_d = 0.164$ cm/d. A recalculation using Equation 11.9 yields
$V_d = 0.0480$ and 0.432 cm/d for the two currents speeds. The bottom roughness MTC,
k_s(bro), is obtained from Equation 11.12. The calculated results appear in Table 11.3.
Since k_s(bro) and k_s(diff) are for parallel processes, they are added to obtain the
combined MTC, k_s, which also appears in the table.

In this example, the magnitude of k_s(bro) is dependent on the numerical K_c chosen.
Given that, the k_s(bro) values are approximately equal to k_s (diff) indicating the
semicircular hyporheic flow in the surficial sediments is a significant contributor to
the sediment side MTC, even though the bottom current speeds are relatively low.
The results also indicate that the MTC is very sensitive to the bottom current speed.

ACKNOWLEDGMENT

The authors kindly acknowledge Danny Fontenot for the word processing of this
chapter and the review, comments, and graphic illustration provided by Justin
Birdwell.

LITERATURE CITED

BBL. Blasland, Bouck, and Lee, Inc. 1998. ALCOA Remediation Project Grasse River, Massena, NY.

Committee on Hydrologic Science. 2004. *Groundwater Fluxes across Interfaces.* National Academy Press, Washington, DC, 85pp.

de Marsily, G. 1986. *Quantitative Hydrogeology-Groundwater Hydrology for Engineers.* Academic Press, New York, Chapters 5 and 6.

DePinto, J.V., D. Rucinski, N. Barabas, S. Hinz, R. McConllough, T.J. Dekker, B. Valente, H. Holmberg, J. Kaur and J. Wolfe. 2006. Review and evaluation of significance and uncertainty of sediment-water exchange processes at contaminated sediment sites. Appendix D. Limno-Tech, Inc., Ann Arbor, MI.

Elliott, A.H. and N.H. Brooks. 1997. Transfer of non-adsorbing solutes to a stream bed with bed forms: Theory. *Water Resources Research*, 33: 137–151.

Freeze, R.A. and J.A. Cherry. 1979. *Groundwater.* Prentice-Hall, Englewood Cliffs, NJ, Chapter 2, Section 2.4.

Martin, J.B., J.E. Cable, C. Smith, M. Roy and J. Cherrier. 2007. Magnitudes of submarine groundwater discharge from marine and terrestrial sources: Indian River Lagoon, Florida. *Water Resources Research*, 43: 1–15.

Packman, A.I. and M. Salehin. 2003. Relative roles of streamflow and sedimentary conditions in controlling hyporheic exchange. *Hydrobiologia*, 494: 291–297.

Packman, A.J., M. Salehin and M. Zaramella. 2004. Hyporheic exchange with gravel beds: Basic hydrodynamic interactions and bedform-induced advective flows. *Journal of Hydrology and Engineering*, 130(7): 647–656.

Savant, S.A., D.D. Reible, and L.J. Thibodeaux. 1987. Convective transport within stable river sediment. *Water Resources Research*, 23(9): 1763–1768.

Thibodeaux, L.J. 2005. Recent advances in our understanding of sediment-to-water contaminant fluxes. The soluble release fraction. *Aquatic Ecosystem Health & Management*, 8(1): 1–9.

Thibodeaux, L.J. and J.D. Boyle. 1987. Bedform-generated convective transport in bottom sediment. *Nature*, 325(6102): 341–343.

Wimbush, M., L. Nemeth and B. Birdsall. 1982. *The Dynamic Environment of the Ocean Floor*, Eds. K.A. Fanning and F.T. Manheim. Lexington Books, Lexington, MA, Chapter 4.

Winter, T.C. 2002. Subaqueous capping and natural recovery: Understanding the hydrogeologic settings at contaminated sites. DOER Technical Notes Collection (TN DOER-C26), U.S. Army Engineer Research and Development Center, Vicksburg, MS.

Worman, A. and A.I. Packman. 2002. Effect of flow-induced exchange in hyporheic zones on longitudinal transport of solutes in streams and rivers. *Water Resources Research*, 38(1), 2-1–2-15.

FURTHER READING

de Marsily, G. 1986. *Quantitative Hydrogeology–Groundwater Hydrology for Engineers.* Academic Press, New York, Chapter 5.

Freeze, R.A. and J.A. Cherry. 1979. *Groundwater.* Prentice-Hall, Englewood Cliffs, NJ, Chapter 2, Section 2.4.

12 Diffusive Chemical Transport across Water and Sediment Boundary Layers

*Louis J. Thibodeaux, Justin E. Birdwell,
and Danny D. Reible*

CONTENTS

12.1   Introduction 322
12.2   Forced Convective Transport on the Water Side............................323
      12.2.1   A Qualitative Description of the Water-Side Boundary Layer 323
      12.2.2   MTC Definition, Measurements, and Data Correlations 324
            12.2.2.1   Data Correlations......................................326
      12.2.3   Estimating the Fluid-to-Bed Surface Friction Velocity u_* 328
      12.2.4   Conclusion...331
12.3   Wind-Generated Convective Mass Transport on Water-Side of Sediment
      Interface: Lakes and Similar "Quiescent" Water Bodies....................332
      12.3.1   The Origin and Structure of In-Lake Water Currents 332
            12.3.1.1   Surface Water Movements 332
            12.3.1.2   Internal Water Movements 333
      12.3.2   The Bottom-Water MTC for Nonstratified Water Bodies........334
            12.3.2.1   Laboratory Simulation of Wind-Enhanced Bottom
                    Water MTC for a Nonstratified Waterbody 335
      12.3.3   Model of Stratified Lake Bottom-Water MTC 336
            12.3.3.1   Wind-Generated Bottom Water Currents...............336
            12.3.3.2   Natural Convection Density Driven Bottom Water
                    MTCs 337
12.4   Theoretical Correction Factors for *In Situ*, Solid Plate MTC
      Measurements..338
12.5   Diffusive Chemical Transport on the Sediment Side of the Interface 339
      12.5.1   A Qualitative Description of Upper Sediment Layers 339
      12.5.2   The Molecular Diffusion Process in Porous Media 339
            12.5.2.1   Solute Molecular Diffusive Transport 340
            12.5.2.2   Porosities of Sediment Beds 342
            12.5.2.3   Natural Colloids and Brownian Diffusive Transport...345

                12.5.2.4 Sediment Bed Pore Sizes.............................346
                12.5.2.5 Conclusion...347
        12.5.3 The MTC for Bed Sediment Diffusive Transport.................348
                12.5.3.1 Solute Flux MTC.......................................348
                12.5.3.2 The Colloidal Particle Flux MTC......................348
12.6 A Guide for Users...349
12.7 Example Problems..350
        12.7.1 Upscaling Tracer Measured Water-Side MTCs
                To Field Conditions...350
        12.7.2 Smooth and Rough Surface MTCs...................................351
        12.7.3 Water-Side MTC for the Hudson River-I...........................352
        12.7.4 Water-Side MTC for the Hudson River-II..........................354
        12.7.5 Water-Side MTC Estimate for Stratified Lock Earn..............354
        12.7.6 Natural Convection MTC for Gypsum...............................355
        12.7.7 In-Bed Naphthalene MTC...356
Literature Cited..356
Further Reading...358

12.1 INTRODUCTION

Chemical transport across the water–sediment interface takes place through numerous chemical, biological, and physical mechanisms (Reible et al., 1991; Thibodeaux and Mackay, 2007). These reflect in large part the characteristics of the particular aquatic system, which include flowing freshwater streams, lakes, estuaries, and the marine, both near shore and beyond. This chapter is focused on so-called "diffusive type" processes on either side of the sediment–water interface. Specifically, it covers the water-side mass transfer coefficient in the fluid boundary layer above the bed and diffusion within the interparticle pore spaces in the near-surface bed sediment layers.

Information on several other transport processes and MTCs at the sediment–water interface appear elsewhere in this handbook. Due to special nature of these subjects and the complexity of the processes, individual chapters are devoted to three of them. Chapter 10 is concerned with particle resuspension and deposition as it affects chemical transport in flowing water streams. Chapter 11 is concerned with water advection processes that contribute to enhanced chemical transport in the various aquatic-sediment bed systems. Chapter 13 is concerned with chemical biodiffusion processes in the sediment bed as a consequence of the presence of macrofauna in the surface layers (i.e., bioturbation). These three processes do not fit nicely into a chapter devoted to the more conventional diffusive process; the contents of this chapter are as follows.

Sections 12.2 and 12.3 discuss the mass transport on the water-side of the sediment–water interface and address MTCs influenced by aquatic stream flow and wind-generated convective transport. After a brief description of some key characteristics of this fluid boundary layer, one direct technique of measuring the water-side MTCs are presented. This helps focus the user as to precisely what this subject is about.

A brief review of the limited number of in situ measurements is then presented. This is followed by a review of the available correlations proposed for estimating these kinds of MTCs in Section 12.4. Section 12.5 is concerned with the subject of classical molecular diffusion in porous media at steady state. The presentation includes a brief description of the upper sediment layers, measurement techniques, laboratory measurement data of effective diffusion coefficients, and models for prediction and extrapolation. A guide appears in Section 12.6 to steer users to suggested procedures for estimating these two types of MTCs. The chapter ends with some example problems and their solutions in Section 12.7.

12.2   FORCED CONVECTIVE TRANSPORT ON THE WATER SIDE

12.2.1   A QUALITATIVE DESCRIPTION OF THE WATER-SIDE BOUNDARY LAYER

A brief description of the benthic boundary layer is necessary to address the complexity of the fluid dynamics and solid surface interactions. The typical hydrodynamic situation involves the flow of water parallel to the bed surface. In the case of freshwater streams and estuaries, the water depth varies from around 1 to tens of meters, and longitudinal water elevation gradients drive the flow magnitude. In the case of lakes, the bottom water movement is often directly related to the wind direction, magnitude and duration, as well as the presence or absence of thermal stratification in the water column. The bottom waters in estuaries respond to these same forces as well as the presence or absence of salinity stratification and tidal forces. In off-shore, deeper marine environments, a combination of wind factors, riverine input flows, tidal fluctuations, and other long distance forces that create pressure gradients along the bed surface all influence water motion at the sediment–water interface. Depending on the particular conditions, the above-bed water velocities can be dominated by individual processes or a combination thereof. In addition, all water bodies are subject to transient forces that vary on daily to annual cycles and results in similar time variations of the bottom-water velocities.

The physical state of the surface of the bed also plays a significant role in regulating the magnitude of the water-side MTC. These include the sediment particle size distribution, bottom surface geometry (bedforms), and biological factors. This simplest bed form is one made of sand or finer particles with a flat surface. However, this is expected to be the exception rather than the general condition. The particles on the surface move as a consequence of the shear stress applied by the flowing water, leading to the development of a wave-form patterns or dune structures. This rippled and wavy surface occurs where unconsolidated material exists, even when low water velocities are typical. Macrofauna living both on the bed (i.e., epifauna) and within the bed (i.e., infauna) produce structures that increase surface roughness. Examples include fish nesting, which produces depressions in the surface, and worm burrowing, which produces mounds composed of subsurface source material moved to the interface. Sunlight reaching the bottom in shallow waters stimulates the growth of algae in mats that can blanket the bottom surface. Sunlight also stimulates the growth of rooted subaquatic vegetation (SAV) with grass-like forms that extend some distance upward into the water column. Some of the most extraordinary surface perturbations result from the presence of coral reefs, whose biological structures result in extremely convoluted surface shapes and roughness.

Considering the preceding qualitative description of sediment-water environment, the reader should have an appreciation for the potential complexities of flow and bed structures that can exist at the benthic boundary layer. The levels of fluid turbulence and the existence of laminar flow regions in this layer are quite variable and together control the magnitudes of the MTCs. The present state of the science and engineering knowledge is limited with regard to our understanding of chemical transport in these types of boundary layers. Limitations exist in the theory as well as the interpretation of the empirical data, both laboratory and field environments, making it difficult to account for the many factors that affect the magnitudes of MTCs. Many factors have not been studied in a systematic way and therefore their quantitative influence on MTCs is unknown. Some related theoretical approaches exist but are of limited utility. The available laboratory and field measurements are presented in the following section.

12.2.2   MTC Definition, Measurements, and Data Correlations

MTC Definition. The appropriate rate equation to quantity chemical flux across the water-side boundary layer is Equation 2.11 in Section 2.4.8. It contains the convective MTC, k_c (m s^{-1}), and is widely accepted and used for fluid–solid interfaces. Section 2.5 contains a review of the conceptual models used to provide some theoretical foundation. It also contains some correlations for obtaining first-order, numerical estimates for k_c under ideal flow and geometric conditions. The user is urged to review these sections in Chapter 2.

MTC Measurements. *In situ* estimates of bottom water mass-transfer coefficients can be obtained by measurements of aragonite (i.e., $CaCO_3$) and alabaster (i.e., $CaSO_4$), also called gypsum, dissolution rates (Santschi et al., 1983, 1990, 1991). Flat plates made of these solids sealed on the bottom with the upper surface in contact with water have been found to provide reasonable representations of the bed surface for mass transport measurements. In one field experiment, the slightly soluble alabaster panels 1 cm thick occupied the central 20% of the 30 cm × 30 cm plate area. After four days at a site in the northeastern Pacific Ocean, the panels were retrieved and flux determined from weight loss, which averaged about 10%. If the rate-limiting step is assumed to be molecular diffusion through the water boundary layer, the film-theory flux equation applies (i.e., Equation 2.12) and the ratio of D/δ_c is the MTC k_c. Using this model for the boundary layer resistance, the measured fluxes were converted to sublayer thickness, δ_c, in microns (μm). An average value of 475 $\pm50\,\mu$m for the equivalent stagnant boundary layer film thickness was reported for this site. Using the molecular diffusivity of $CaSO_4$ in the sea water yields a MTC of 0.31 cm h^{-1}.

In situ marine measurements of this MTC are very limited in number. Table 12.1 contains a list of a few of the available datasets. The top half of the table contains numerical values representative of marine sites and the bottom half for freshwater sites. As noted above, marine geochemists report the results of in situ measurements

TABLE 12.1
***In Situ* Measure MTCs Using Gypsum (k_c in cm h^{-1})**

Aquatic System	n	Depth (m)	$\bar{k}_c - \sigma$	$\bar{k}_c$	$\bar{k}_c + \sigma$
Pacific Ocean, Coast of Central California[a]	8	3330	0.60	0.80	1.00
	9	2010	0.46	0.84	1.22
	11	1185	1.06	1.22	1.39
	11	795	0.86	1.31	1.75
Atlantic Ocean, Outer Shelf Mid-Atlantic Bight[a]	4	133	0.66	0.77	0.89
	9	1095	0.28	0.38	0.48
Atlantic Ocean, Inner Slope Mid-Atlantic Bight[a]	1	125	0.33	0.90	0.43
	5	925		0.38	
Mississippi River Bank, Baton Rouge, LA	13	0.67	3.98	4.74	5.49
University Lake, Baton Rouge, LA	14	0.4–1.2	1.76	2.28	2.79
Natural convection Tests (bucket at field conditions)	42	0.15	0.74	1.23	1.71

[a] Santschi, P.H. et al. 1991. *Journal of Geophysical Research*, 96(10), 641–657.

as a film thickness. These have been converted to their equivalent MTCs for Table 12.1. The data presented in Table 12.1 are point values of k_c and are strongly influenced by bottom conditions on and around the solid tracer chemical source material. The length of this solid material, which is typically 5–20 cm, is a key independent variable as is the local water velocity. These MTCs may overestimate k_c by as much as an order of magnitude.

There are experimental and theoretical issues associated with using semisoluble substances, like gypsum and aragonite, for making in situ measurements. These concerns have to do with (1) changes in bottom roughness due to the presence of the device (i.e., plate) and (2) the accumulation of gypsum in the boundary layer from such a finite size source. Therefore, the representativeness of these localized measurements to overall sediment–water interface is questionable. In the first case, the uneven positioning of the solid plate may create artificial edge-effects and localized enhanced turbulence. Santschi et al. (1983) investigated this factor experimentally in a flume apparatus and found differences in flux of less than 10%. There is also the question of the plate smooth surface. Typically, these surfaces are ∼30 cm in length. Whether or not turbulence levels in the boundary layer are reduced compared to the rough surrounding/upstream sediment surface so as to change the transport characteristics is an open question. In the second case, the chemical concentration boundary layer above the gypsum plate is thin and unlike that of the longer sediment surface sources typically encountered in nature. Santschi et al. (1990) have addressed this issue and proposed a procedure for theoretical correction factors. It seems clear based on the experimental data and boundary layer theory that the short sections of gypsum give artifically high MTCs. As illustrated in Figure 12.1, the k_c values for finite plates of length L are larger than those for semi-infinite lengths ($L \rightarrow \infty$) by nearly an order of magnitude. This subject will be revisited in a later section.

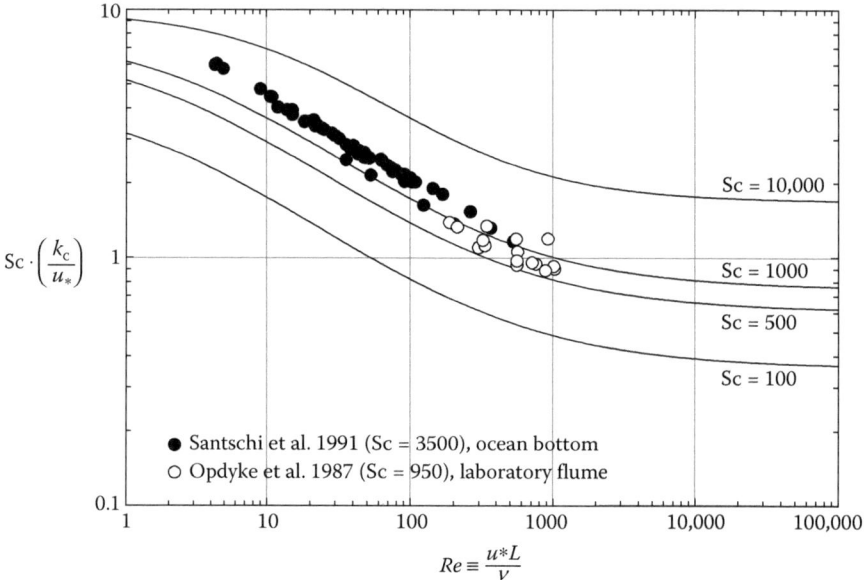

FIGURE 12.1   Product of the mass transfer Stanton number ($St = k/u_*$) and Schmidt number ($Sc = v/D$) vs plate Reynolds number ($Re = u_* L/v$) with Schmidt number as parameter (Modified from the original Higashino, M. and M.G. Stefan. 2004. *Water Environmental Research* 76, 292–300.); k_c is the water-side mass transfer coefficient at the sediment-water interface ($cm\,h^{-1}$), v is the kinematic viscosity of water ($cm^2\,h^{-1}$), D is the diffusivity of gypsum in water ($cm^2\,h^{-1}$), u_* is the friction velocity at the sediment–water interface ($cm\,h^{-1}$), L is the gypsum plate length (cm).

12.2.2.1   Data Correlations

Solute transport in flowing fluids adjacent to a solid surface is a mature subject. Understanding the process and having reliable MTC correlations is key for many chemical and environmental engineering applications. The subject was reviewed and corresponding correlations were compiled with regard to applications at the sediment–water interface (Boudreaux and Jorgensen, 2001). The correlation used by all investigators was of the general form:

$$k_c = \frac{au_*}{Sc^b} \qquad (12.1)$$

where u_* is the friction velocity ($m\,s^{-1}$) and Sc the Schmidt number (dimensionless). However, many of the solid surfaces were generally not representative of bed sediment materials or porous media. Values for the empirical parameters a and b needed for Equation 12.1 appear in Table 12.2. In all cases, the MTC is directly proportional to the fluid friction velocity with the adjoining surface. Also appearing in Table 12.2 are numerical values for k_c computed from each of the seven correlations. The average value is $2.68\,cm\,h^{-1}$ for $u_* = 1.0\,cm\,s^{-1}$ and $S_c = 1000$. All correlations represented

TABLE 12.2
Equations for Estimating the Water-Side MTCs[a]

Investigator(s)	Surface Characteristics	a	b	k_c^b (cm h^{-1})
1. Shaw and Hanratty (1977)	Flat plate	0.0889	0.704	2.47
2. Pinczewski and Sideman (1974)	Flat plate	0.0671	2/3	2.42
3. Wood and Petty (1983)	Flat plate	0.0967	0.7	2.76
4. Deissler (1954)	Smooth tube	0.11	3/4	2.23
5. Opdyke et al. (1982)	Smooth alabaster	0.078	2/3	2.81
6. Hondzo (1968) with Steinberg (1999)	Oxygen, sediment	0.0558	2/3	2.0
7. Thibodeaux, Chang, and Lewis (1980)	Organic liquids, sediments	0.114	2/3	4.10

[a] Form of equation: $k_c = au_*/Sc^B$, $u_* =$ friction velocity and Sc $=$ Schmidt number.
[b] Calculated results for $u_* = 1.0$ cm s^{-1} and Sc $= 1000$.

by Equation 12.1 and Table 12.2 are for $L \rightarrow \infty$ except in the organic liquids case, where $L = 10$ cm. Excluding the liquid value, the average is 2.45 cm h^{-1} ($\sigma = 0.31$). Opdyke et al. (1982) notes that in oceanic applications u_* for the BBL is seldom much greater than 1.0 cm s^{-1}.

With the exception of two correlations in Table 12.2, smooth surfaces characterized were the solid source materials and not rough sediment surfaces. Boudreaux and Jorgensen (2001) also considered this aspect. Mass transfer roughness correction factors, r_c (dimensionless) were included in their review; the correlating equations were of the form:

$$r_c = \frac{\alpha Sc^b}{Re_*^a} \tag{12.2}$$

where Re$_* \equiv u_* \Delta/\nu$. The term Δ (m) is the average height of the sediment roughness elements and ν is the fluid kinematic viscosity (m^2 s^{-1}). The reviewers note that these formulas predict substantial increases in the MTC with increasing roughness. Table 12.3 contains the three constants for use in Equation 12.2. It also contains the computed MTC correction factors for specific flow conditions. The average r_c is 1.83; this means the rough surface MTC is 83% greater than for the smooth surface.

TABLE 12.3
Bottom Roughness Correction for Smooth Surface MTCs (Equation 12.2)

Investigators	Surface characteristics[a]	α	a	b	r_c^b
1. Dawson and Trass (1972)	?	1.94	0.10	0.09	2.25
2. Grimanis and Abedian (1985)	Rough tubes	2.00	0.20	0.10	1.55
3. Postlethwaite and Lotz (1988)	Erosion-corrosion	1.64	0.14	0.10	1.69
4. Zhao and Trass (1997)	Rough pipe	1.79	0.17	0.12	1.84

[a] $25 \leqslant$ Re$_* \leqslant 120$ generally.
[b] Re$_* = 112$, $u_* = 1$ cm s^{-1}, $\Delta = 10$ cm, and Sc $= 1000$ used in calculations of r_c.

The surface roughness represented by Equation 12.2 may not be realistic for natural sediment beds. The values of Δ are representative of engineered solid surface and are very small in magnitude compared to sediment bed-forms. Thibodeaux et al. (1980) and Christy and Thibodeaux (1982) found that organic liquids used as tracers positioned in the sand wave valley have MTCs lower than the flat sediment surfaces. The in-valley MTC correlation for $\Delta = 5$ cm and 15 cm sand waves were described by the following equation:

$$k_c = \frac{0.114 u_* \text{Sc}^{2/3}}{1 + 9.6\Delta^{1/2}} \tag{12.3}$$

$16 \leq \text{Re} \leq 1600$ with Δ in meters. In this case, $r_c = [1 + 9.6\Delta^{1/2}]^{-1}$. Based on the parameter values used in Table 12.3, the wave valley $r_c = 0.25$. This correction factor is less than unity and is significantly lower than those presented in Table 12.3. Recent studies by the authors support the finding that local gypsum-measured MTCs in valleys are lower than those on the crests of sand waves. This point warrants further study.

12.2.3 ESTIMATING THE FLUID-TO-BED SURFACE FRICTION VELOCITY u_*

The correlations for estimating the water-side MTC presented above are dependent on knowing the friction velocity, $u_* (\text{m s}^{-1})$. It is defined $u_* \equiv \sqrt{\tau_0/\rho} \, (\text{m s}^{-1})$ where $\tau_0 \, (\text{N m}^{-2})$ is the shear stress applied to the bed surface and $\rho \, (\text{kg m}^{-3})$ is the density of the fluid flowing parallel to the bed surface. The quadratic drag law is

$$\tau_0 = \rho C_D u^2 \tag{12.4}$$

It is used to relate the flow speed $u \, (\text{m s}^{-1})$ in the BBL to the shear stress. The dimensionless coefficient of proportionality, C_D, is called the quadratic drag coefficient. It is referenced to a particular elevation y (m) above the bed at which the flow speed, u (m s^{-1}) is measured. Typically, $C_D(1 \text{ m})$ applies at elevation $y = 1$ m where the flow speed is $u(y)$ in m s^{-1}. The speed of the fluid or its average velocity is the necessary independent variable required to estimate the MTC. Typical mid-latitude BBL flows measured at 1 m above the bed are 3 cm s^{-1} for the deep-sea and 30 cm s^{-1} for the shelf (Wimbush, 1976). Those in freshwater streams can be as high as 500 cm s^{-1}.

Estimating the quadratic drag coefficient, C_D. Unfortunately this parameter is not constant and depends upon the BBL flow interaction with the roughness of the bed surface. However, there are two convenient means of estimating drag effects. The first method requires velocity profile measurements. On a vertical transect from the bed surface several measurements of fluid velocity u with elevation y within the turbulent portion of the BBL are obtained and fitted to:

$$\ln y = \frac{\kappa_*}{u_*} u + \ln y_0 \tag{12.5}$$

where the parameter κ is von Karman's constant. Its numerical value is 0.4 as long as suspended solids are not present at significant concentrations in the BBL fluid. With such velocity profile data, numerical values of u_* and y_0 can be obtained. Obviously the average and range of the friction velocity observations can be used directly from these profile measurements to estimate u_* values.

The second method for estimating u_* requires an approximation for C_D. Equations 12.4 and 12.5 can be transformed to

$$C_D = [\kappa / \ln(y/y_0)]^2 \tag{12.6}$$

Numerical values of the C_D can be obtained; clearly C_D depends on y_0. The parameter y_0 is known as a roughness length and its magnitude is related to the dimensions of the bed roughness, or the elevation where $u = 0$ according to Equation 12.5. The present knowledge of typical values of C_D and y_0 in the marine environment is based largely on profile measurements. The values generally indicate considerable scatter even for measurements made at a particular location. For flow ranging from hydrodynamically transitional to rough, C_D values range from 1×10^{-4} to 0.01 and y_0 values ranged from 1×10^{-6} to 0.1 cm (Heathershaw, 1976). Unfortunately, no compilation of numerical values for y_0 is available for aquatic flows.

The above does not apply for flow conditions in a shelf setting, estuarine, or freshwater stream flows where the BBL may become depth limited and subject to turbulent mixing from both upper and lower boundaries. Under such conditions, alternative means for estimating C_D are needed. In Chapter 2, a general equation for the skin friction factor, C_f, for open channel flow was presented (see Equation 2.25). The equivalent formulation for the quadratic drag coefficient is

$$C_D = \frac{8gn^2}{R^{1/3}} \tag{12.7}$$

where g is the gravitational constant (9.81 m s^{-2}), n is Manning's roughness coefficient (s m$^{-1/3}$), and R is the hydraulic radius of the channel (m). Using the defining equation for the friction velocity and Equation 12.4 for τ_0 yields

$$u_* = \sqrt{C_D}\, u \tag{12.8}$$

This final result allows for conversion of stream velocity, u, to the equivalent friction velocity.

The bed surfaces in the deep sea, shelf areas, estuaries, and terrestrial freshwater streams exhibit a range of roughness conditions. In the cases of the seabed and estuaries, roughness is related to the size of the sand, silt, or clay particles, the sizes of sand ripples and other bed forms, biogenic pellets, tracks, trails, tubes and mounds, seabed nodules, and so on. Table 2.2 contains the roughness characteristics representative of freshwater streams and channels and the corresponding numerical values of the Manning roughness coefficient. Several of these channels have roughness characteristics that may translate to equivalent ones on the seabed and estuary beds. In the case of seabeds, the R in Equation 12.7 may be approximated by the thickness of the turbulent boundary layer; typical values range from 1 to 4 m. Using the higher value results in

$$C_D = 5gn^2 \tag{12.9a}$$

Use SI Units in this equation for estimating the quadratic drag coefficient in cases where bed form roughness characteristics can be approximated by n appearing in Table 12.4. The values selected for inclusion in the table represent surfaces that are somewhat smoother than those in Table 2.2. These are channel characteristics that may approximate the roughness of many seabed and estuary bed surfaces. Santschi et al. (1990) suggest using the approximation of Hickey et al. (1986) for estimating C_D; it is

$$C_D = 0.001 \cdot (2.33 - 0.00526u + 0.00365u^2) \qquad (12.9b)$$

where u (cm s^{-1}) is the current velocity 5.0 m off the bottom. This velocity is approximately 1.33 times that measured at 1.0 m off the bottom.

It may be of interest to the user to know that the use of Equations 12.7 and 12.9 are consistent with the general observations that measured water-side MTCs increase with increasing surface roughness; see Equations 12.1 and 12.2. The use of either formula presented above for estimating drag effects (C_D) reflects roughness through the n values. Since Equation 12.8 applies and the MTC is directly proportional to u_*

TABLE 12.4
Manning's Roughness Coefficient n (s m$^{-1/3}$) for Stream Channel Surfaces

Type of Channel and Description	n
Cement with neat surface	0.011
Cement with mortar	0.013
Concrete trowel finish	0.013
Concrete float finish	0.015
Concrete finished with gravel on bottom	0.017
Concrete unfinished	0.017
Concrete on good excavated rock	0.020
Concrete on irregular excavated rock	0.027
Metal, smooth steel surface, unpainted	0.012
Metal, smooth steel surface, painted	0.013
Metal corrugated	0.025
Gravel bottom with sides of formed concrete	0.020
Gravel bottom with sides of stone in mortar	0.023
Gravel bottom with sides of dry rubble or riprap	0.033
Brick, glazed	0.013
Brick, in cement mortar	0.015
Masonry, cemented rubble	0.025
Masonry dry rubble	0.032
Asphalt, smooth	0.013
Asphalt, rough	0.016
Vegetal lining	0.030

Source: Adapted from tables presented by Yang, C.T. 2003. *Sediment Transport Theory and Practice.* Kreiger Publishing Company, Malabar, FL.

(see Equation 12.1) it is therefore linearly related to Manning's roughness coefficient. In other words $k_c \sim n$.

12.2.4 Conclusion

Section 12.2 was focused on outlining the theory and providing methods for estimating forced convection MTCs on the water-side of aquatic sediment beds. The methods depend on having some knowledge of the water velocities in the BBL or the speed of flow in streams. As indicated by theory, this independent variable is key as it directly affects the magnitude of the MTC. It is obviously a highly variable parameter. Ranges of values may be large for freshwater streams, estuaries, shelf and deep sea environments and it is a challenge to arrive at representative values.

To a large degree, the choice and range of velocities or current/speeds will be obtained from reported or acquired measurements. Such measurements were performed at 710 m in the eastern Florida Straits (Wimbush, 1976). The current near the bottom was typically $5–15\,\mathrm{cm\,s^{-1}}$ with occasional "storm" speeds of up to $40\,\mathrm{cm\,s^{-1}}$. It was also observed that sporadic sediment ripple migrations occurred with intermittent quiescent periods during which there was a gradual rounding and smoothing of the ripple crests. Further analysis indicated that the mean critical speeds for sediment ripple migration and suspension were 19 and $22\,\mathrm{cm\,s^{-1}}$, respectively. A nearby sample station found a bimodal distribution of particle sizes with peaks at 250 and $62\,\mu\mathrm{m}$. Clearly, in a general way the composition of the bed sediment reflects a time integrated bottom current and bed interaction process. The small size particles entrain in the water column and are therefore removed, leaving those that are part of the bedload; it then slowly migrates down current. Wetzel (2001) presents data reflecting the current velocity and sediment composition; it is reproduced in Table 12.5. This table provides an alternative approach to estimating ranges of boundary layer current speeds where the bed particle diameter range can be used as the proxy.

Table 12.6 presents calculated MTCs based on two estimation techniques to provide the reader with estimates of water-side MTC magnitudes. The half dozen numerical values presented in top down order reflect general conditions in marine seabeds, deep freshwater streams, and shallow, rapid flowing streams respectively and the resulting MTCs. The two large values for the latter have not been observed in nature.

TABLE 12.5
Relationship of Current Velocity to Sediment Composition (Wetzel, 2001)

Velocity Range ($\mathrm{cm\,s^{-1}}$)	General Bottom Composition	Approximate Diameter (mm)
3–20	Silt, mud small organic debris	<0.02
20–40	Fine sand	0.1–0.3
40–60	Course sand to fine gravel	0.5–8
60–120	Small, medium, to large gravel	8–64
120–200	Large cobbles to boulders	>128

TABLE 12.6
Water-Side MTCs k_C (cm h^{-1}) Estimates

Velocity (cm s^{-1})	Bottom Roughness Character, n (s m$^{-1/3}$)	Depth R (m)	S[a]	H&S[a]
2	Smooth, 0.011	4[b]	0.433	0.310
10	Sand ripples, 0.017	4[b]	3.35	2.49
20	Gravel, 0.020	10	6.76	4.83
50	Sand ripples, 0.017	10	14.4	10.3
100	Stones, 0.025	2	166	119
300	Rocky, 0.027	2	178	128

[a] Equations from Table 12.2. S = Santschi; H & S = Handzo and Steinberg; Sc = 1000. Equation 12.7 for C_D estimate.

[b] Assumed BBL thickness.

12.3 WIND-GENERATED CONVECTIVE MASS TRANSPORT ON WATER-SIDE OF SEDIMENT INTERFACE: LAKES AND SIMILAR "QUIESCENT" WATER BODIES

12.3.1 THE ORIGIN AND STRUCTURE OF IN-LAKE WATER CURRENTS

12.3.1.1 Surface Water Movements

Currents are nonperiodic water movements generated by forces external to the water bodies in which they occur (Wetzel, 2001). The responsible forces include the wind, changes in atmospheric pressure, horizontal density gradients caused by differential heating or by diffusion of dissolved minerals from sediments and the influx of water. Wind-generated currents are typically dominant and are highlighted in this introduction.

In general, the velocity of wind-driven currents is about 2–3% of wind speed and is largely independent of the height of the surface waves. Once in motion, the surface currents are subject to geostrophic re-directing, due to the Earth's rotation (i.e., Coriolis forces). Water currents in moderate to large lakes in the Northern Hemisphere are deflected to the right relative to the direction of the wind. The surface currents of very large lakes such as Lake Michigan in the United States tend to circulate in large swirls or gyrals. These are the combined results of geostrophic forces, waves, and especially by the shifts in the duration of strong prevailing winds, as the moving water is shepard ed and deflected by the enclosing land mass.

Turbulence generated by surface water movement and wave action is often sporadic. It occurs, for example, in the dispersion of wave energy in the mixing of the epilimnion when a lake is thermally stratified or during the mixing of the total water column mass. The epilimnion is the warm layer of water that lies on the surface of the lake during the stratified period. Langmuir circulations in the upper layers of lakes occur under some wind and wave circumstances moving water in vertical helical currents. Although surface waves and Langmuir circulations cause turbulence, the thermal density gradients in the metalimnion (i.e., thermocline) suppresses the transmission of kinetic energy into the deeper parts, namely to the metalimnion and the hypolimnion. The hypolimnion is the layer of cold bottom waters of a stratified water body, while the metalimnion is the transitional thermal layer separating the warm and cold layers. The metalimnion is a very stable water mass that resists most turbulence and currents moving between the other layers. The thermodine is the plane running through the middle of the metalimnion roughly parallel to the air–water interface that generally separates the warm and cold water layers. A stratified lake contains these layers. A non-stratified lake would, ideally, have a single layer of approximately uniform temperature.

Wind movements in the atmospheric BL directly impact water turbulence and currents in the surface layers. However, the role of wind in generating turbulence and currents in the bottom-water (i.e., hypolimnion) above the sediment bed is the issue of interest here. Wind stress on the surface moves water masses in bulk generating internal water movements that are transmitted deep into the water column resulting in bottom currents. However, these are also sporadic in direction, magnitude, and duration.

12.3.1.2   Internal Water Movements

As the wind blows over the surface for a sufficient period of time, wind drift causes water to pile up resulting in an elevated surface level at the lee end (i.e., downwind end) of the lake. During one instance during the first decade of the twentieth century, a persistent, brisk westward wind blew over Lake Erie for several days. The Niagara River nearly ran dry and the water over Niagara Falls became a virtual trickle as the water piled-up against the western shoreline. Such standing waves of water on the surface are referred to as seiches, a term originally referring to the periodic "drying" of the exposed sediment bed. This pile-up generates internal water movements in stratified and nonstratified lakes.

The idealized, nonstratified lake system is the simplest case and will be presented first. The accumulated mass of water in the pile-up is pulled down by gravity due to its elevation and produces a downward current. On contacting the bottom of the lake, the current is deflected and flows parallel to the bottom surface in the opposite direction of the wind. If the wind condition remains for a sufficient length of time, a large gyre current of internal water movement is produced that continuously circulates from one end of the lake to the other.

A benthic boundary layer forms on the water side of the sediment–water interface with the same fluid dynamic and transport characteristics described in Section 12.2 for the marine BBL.

In the case of a persistent wind blowing over a stratified lake containing a warm layer above a cold layer or other quiescent waterbody, the water pile-up occurs in the same way. However, the internal circulating water current develops only in the warm waters of the epilimnion. The metalimnion has a slippery surface of low frictional shear and little mixing occurs across this interface. In addition, the pile-up produces a deeper epilimnion. The downward flowing water is deflected by the presence of the metalimnion. In the thermocline, the plane becomes tilted downward toward the pile-up end. The titling causes the thickness of hypolimnion to be reduced. The net effect is a large mass of water must move under the thermocline in the up-wind direction within the hypolimnion, transferring large amounts of water from one end of the lake to the other. This migration of water produces a benthic boundary layer on the water-side of the sediment interface.

The surface layer seiches are small in magnitude. In other words, the water pile-up with air above typically is limited to elevations of a few centimeters to a meter. Even so, vast areas of bottom surface may be exposed. The internal seiches occur along the thermocline. As the warm and cold layers have only slightly different densities, the depression of the hypolimnion thickness, Δh (m), and the amplitude of the internal standing wave is much greater than those on the surface. The lee end of the thermocline may tilt downward by several meters because of the large quantities of water being moved laterally in the hypolimnion parallel to sediment bed. These movements are responsible for the internal seiches.

When the wind stops blowing, both the surface layer and the internal layer rush toward level (i.e., equilibrium) positions. The internal flows quickly cease and followed by all flows rapidly reversing direction in both layers. Overshooting equilibrium of a horizontal thermocline plane position produces bottom water currents that rhythmically move back and forth in opposing directions and giving rise to the major deep water currents in lakes. These currents lead to vertical and horizontal transport of heat as well as dissolved substances and particulate matter. Once set in motion, frictional and gravity damping of the fluid oscillations begins and the water mass gradually settles into static equilibrium. In deep lakes, weakly damped oscillations may persist, slowly diminishing long after the storm that set them in motion has passed. Internal current speeds have been reported. Small lakes, a few kilometers in length, have speeds of $1 \, \mathrm{cm \, s^{-1}}$ while larger lakes or long basins have horizontal currents greater than $10 \, \mathrm{cm \, s^{-1}}$.

In summary, water movements in lakes occur in response to various natural forces, particularly wind, that transfer energy to the water. Both steady winds and rhythmic internal wave motions (i.e., oscillations) produce water currents in the surface layer as well as within the deeper internal layers of the water column, when they exist. These motions and their attendant currents may be in phase or in opposition, creating a complex mosaic and ever changing pattern of current magnitudes and directions over the sediment–water interface. The ultimate fate of these movements is to degrade into arrhythmic turbulent motions, which disperse the water and the chemicals within the water body (Wetzel, 2001).

12.3.2 The Bottom-Water MTC for Nonstratified Water Bodies

Unfortunately, there is limited information concerning the bottom-water mass transport parameter. There have been a few direct measurements and mass-balance studies in which its magnitude has been quantified. Apparently, only one theoretical study aimed specifically at lake-bottom water mass transport chemodynamics has been reported. This section contains a brief review of the available data and presents an equation for estimating the MTC.

12.3.2.1 Laboratory Simulation of Wind-Enhanced Bottom Water MTC for a Nonstratified Waterbody

Compared to a stratified lake, the nonstratified lake wind–water interaction current generating process is simpler to simulate in the laboratory and describe theoretically. Simply viewed, the process begins with the wind blowing over the surface, transferring energy to the water. This moves surface water in the same direction toward the lee shore where the surface water current is deflected downward. It is deflected again when encountering the bottom; the bottom current is then directed upwind. Two more deflections find the current once again near the surface and being energized again by the wind. The net effect is an idealized circulating mass of moving water constrained on five sides by solid sediment-like boundaries and one exposed to the atmosphere, a process akin to surface-driven gravity flow. Although this represents a too idealized rendition of complex, real-world lakes for most limnologists, the model has key features in common with natural systems.

The above described wind–water interaction model can be used to obtain a theoretical relationship between wind speed and the average current speed. The quadratic drag law is used to equate the wind-imposed stress on to the surface water. For an assumed steady-state process, the input energy is dissipated by the circulating water current at the solid surfaces by skin friction and can be described using a mechanical energy balance for an incompressible fluid (i.e., water). The result is a fluid dynamic–based MTC correlating equation containing one adjustable parameter.

Benzoic acid wafers as solid dissolution tracers placed on the bottom of a laboratory-scale wind-water tank were used to obtain mass-transfer coefficient, k_c (m s^{-1}), measurements (Thibodeaux and Becker, 1982). The MTCs ranged from 0.42 to 1.8 cm h^{-1}. The resulting MTC (m s^{-1}) correlation for benzoic acid is

$$k_c = 1.71 C_D \left(\frac{\rho_a}{\rho_w} v_a^2 \frac{h^{1.25}}{l} \right) \tag{12.10}$$

where C_D is the quadratic drag coefficient at the air-water interface, ρ_a and ρ_w are the air and water densities, respectively, v_a (m s^{-1}) is air speed, h (m) is water depth, and l (m) is wind–water fetch length. The drag coefficient may be estimated as follows: for wind velocities between 1 and 7 m s^{-1}, $C_D = 1.66 \times 10^{-3}$ and for 4–12 m s^{-1}, $C_D = 2.37 \times 10^{-3}$. A measured wind speed at a height of 10 m is appropriate, however if a land-wind velocity is used, it must be multiplied by 1.5 to get the equivalent over-water velocity. The turbulent boundary layer theory result (Equation 2.18) has a diffusivity functionality that is recommended for use to adjust the benzoic acid correlation to other substances; it is $k_c \sim D^{2/3}$.

A few field measurements for this coefficient have been reported (Thibodeaux and Becker, 1982). The transient build-up of phosphorus in an on-bottom microcosm placed in Lake Warner yielded a MTC of 0.38 cm h^{-1}. Using local wind and lake conditions, Equation 12.10 yielded a value of 0.80 cm h^{-1}. Continuous observations of phosphorous in Lake Washington over a nearly 20-year time-period coupled with a process mass balance resulted in an average value of 0.41 cm h^{-1}. In this case, the calculated value is approximately twice that of the measured MTC for the site because Equation 12.10 is based on the dissolution of solid benzoic acid plates of finite size. As most in situ field and laboratory measurements of k_c are made using solid tracer sources of some finite length, L (m), the measured values are larger than k_c for the semi-infinite sediment-water interface ($L \to \infty$). Figure 12.1 displays this theoretical behavior and shows that as L increases, k_c decreases. Since k_c ($L \to \infty$) is needed in most environmental chemodynamic situations, Figure 12.1 can be used to correct measured values. The procedure is demonstrated in Example Problem 12.7.1 (see Section 12.7).

12.3.3 Model of Stratified Lake Bottom-Water MTC

12.3.3.1 Wind-Generated Bottom Water Currents

The wind–water interactions that lead to water movement in a stratified lake were discussed in the later part of Section 12.3.1. In summary, as the thermocline becomes tilted due to wind stress, a large volume of water ΔV (m^3) must move laterally along the length of the hypolimnion from one end of the lake to the other. This volume is quantified by $(L/2) \times (\Delta H/2) \times W$ where L (m) length, W (m) is width, and ΔH (m) is the vertical downward displacement of the thermocline. Average or characteristic values of the lake dimensions will suffice for this model. Wetzel (2001) presents the time-period, τ (s), of the first internal seiche mode for a simple rectangular lake after which the wind stress ceases:

$$\tau = 2L\sqrt{\frac{\rho_h/H_h + \rho_e/H_e}{g(\rho_h - \rho_e)}} \qquad (12.11)$$

where ρ_h and ρ_e (kg m^{-3}) are the hypolimnion and epilimnion water densities, H_h and H_e (m) are the respective thicknesses, and g (9.81 m s^{-2}) in the gravitational constant. The volume ΔV is moved twice during this time period. Using Wetzel's formula for τ and the appropriate rectangle volumes and areas yields the maximum average current velocity, v (m s^{-1}), in the hypolimnion:

$$v = 0.25 \left(\frac{\Delta H}{H_h}\right) \sqrt{g(\rho_h - \rho_e)H/\rho_e} \qquad (12.12)$$

where h (m) is the total average lake depth.

If the time-period, τ (s) of the metalimnic seiches reflect hypolimnic water movements Equation 12.12 simplifies to

$$v = \left(\frac{\Delta H}{H_h}\right) \frac{L}{2\tau} \qquad (12.13)$$

Wetzel (2001) reports some numerical values for seiche generated horizontal currents. In Lunzer Untersee, Austria, with a fetch of 1.6 km, an internal seiche with period of 4 h had a $\Delta H = 1$ m, with antinodal amplitude and generated horizontal currents of $1\,\mathrm{cm\,s^{-1}}$ at the node. Similarly, currents of $1\,\mathrm{cm\,s^{-1}}$ were reported for Lake Mendota, United States. In larger lakes, for example at the end of the long basin of Loch Ness, Scotland, and in Lake Michigan, amplitudes of $\Delta H > 10\,\mathrm{m}$ are common, with horizontal currents of $>10\,\mathrm{cm\,s^{-1}}$ near the nodes. Calculated values using Equation 12.12 yielded similar results. For data given by Wetzel (2001) on Lock Earn, Scotland: $L = 9.6\,\mathrm{km}$, $\Delta H = 10\,\mathrm{m}$, $H = 80\,\mathrm{m}$, $H_\mathrm{h} = 55\,\mathrm{m}$, $T_\mathrm{e} = 11°\mathrm{C}$, and $T_\mathrm{h} = 9°\mathrm{C}$, yielded $v = 1.7\,\mathrm{cm\,s^{-1}}$. Such field data is necessary in order to use Equations 12.12 and 12.13 for estimating hypolimnic velocities. Based on current estimates given by Equations 12.12 and 12.13, the friction velocity, $u_*\,(\mathrm{m\,s^{-1}})$ can be estimated (see Equation 12.8). The data correlations in Section 12.2.2 are appropriate for use estimating the MTC from the estimated u_* values; see Equation 12.1.

The above velocities represent currents at their maximum estimated values following periods of high wind speeds or storm events. During quiescent periods after such events, the currents slowly subside and layers maintain fairly stable positions with much quieter bottom water conditions and much lower currents. The next section is devoted to estimating MTCs during the quiet bottom-water time periods.

12.3.3.2 Natural Convection Density Driven Bottom Water MTCs

The so-called free or natural convection processes will continue to drive water movement and therefore the magnitude of MTCs when external or forced convection influences, such as wind, are absent. Some fundamentals of both the thermal gradient and the concentration gradient driven free convection process are presented in Section 2.5.7.

A limited number of natural convection experiments have been performed with chemicals in flat bottom orientations within aquatic microsms simulating solid sediment surfaces. Solid sodium chloride blocks positioned in upward-facing orientations of $60°$ from vertical yielded data for the following correlation:

$$\mathrm{Sh_A} = a(\mathrm{Gr_A\,Sc_A})^{1/3} \tag{12.14}$$

where a ranged from 0.05 to 0.11. The slabs were $1\,\mathrm{m}$ in length. Because of the nonhorizontal orientation this correlation will yield high values. Furfural, a dense liquid, dissolving into water from shallow dishes facing up resulted in

$$\mathrm{Sh_A} = 9.22(\mathrm{Gr_A\,Sc_A})^{1/6} \tag{12.15}$$

The dishes were 5, 7, and $10\,\mathrm{cm}$ in diameter (Thibodeaux et al., 1980). Using heat transfer data for a cold plate facing upward, Equation 2.34 in Table 2.3 can be applied to assess natural convection as well. In the above cases, the correlations relate chemical dissolution at the sediment–water interface, which forms a boundary layer with fluid density slightly greater than that of pure water. This particular mass transfer process is very slow since a high-density fluid accumulates on the bottom surface and forms a stable layer, which resist the generation of BL turbulence. The resulting estimated MTCs should be the lowest for the water-side bed sediment surfaces, and appropriate for waterbodies in the absence of bottom of currents.

12.4　THEORETICAL CORRECTION FACTORS FOR *IN SITU,* SOLID PLATE MTC MEASUREMENTS

The technique of and results from using pure solid and liquid chemical tracers for measuring water-side MTCs was presented in sections on forced convection. It was presented in Sections 12.2 and 12.3 where the use of gypsum and aragonite was described for both marine and lake environments. The issue was raised as to whether or not such small devices (tens of cm long) could be used to obtain MTCs representative of the much larger source areas found in aquatic systems. Boundary layer theory suggests they cannot without some adjustment because it has been clearly shown, experimentally and theoretically, that as the source length increases the numerical MTC values decreases. See Section 2.5.5 on boundary layer theory for details. Theoretically, MTCs decrease with plate length to negative one-half power under laminar flow conditions and to the negative one-fifth power for turbulent flow. Opdyke et al. (1982) and Santschi et al. (1991) provide evidence from field and laboratory measurements that the length dependency is $L^{-1/3}$. The experimentally determined power law can be used to adjust other measured MTCs from tracer plates of finite size to the scale of sediment bed sources. Since MTCs are scalar quantities, the same applies whether the sediment surface is a source or sink for a dissolved material in the water column.

Opdyke et al. (1982) argue that for turbulent flow, advection dominates diffusion at some height above the surface so that the thickness of the concentration boundary layer approaches a finite limit downstream and the MTC becomes approximately independent of source length; as seen in Equations 2.19 and 2.20 in Section 2.5.6 of Chapter 2. Santschi et al. (1991) propose a methodology for extrapolating a MTC measured with a chemical source length L, $k_c(L)$ (m s^{-1}), to one that is representative of a great length, $L \to \infty$, or $k_c(L \to \infty)$. It is based on knowing the bottom friction velocity, u_* (m s^{-1}), for the $L \to \infty$ condition:

$$k_c(L \to \infty) = k_c(L)0.096(u_*L/\upsilon)^{1/3} \tag{12.16}$$

where L (m) is the source length and υ (m^2 s^{-1}) the kinematic viscosity of the water.

A theoretical study of diffusive boundary layer development above the sediment–water interface, such as those found in lakes, reservoirs, rivers and estuaries, was performed by Higashino and Stefan (2004). The development was based on a steady-state advective–diffusive transport model using molecular and turbulent diffusivity in the turbulent boundary layer. The study produced a numerical solution to determine when the diffusive boundary layer is fully developed. The concentration boundary layer above small chemical transport areas, such as tracer plates used for *in situ* MTC measurements, does not represent the steady-state, fully developed boundary layers representative of the large chemical transport surface areas typically encountered in aquatic sediment beds. One particularly useful result is a graphical technique for extrapolating a MTC measured with finite length tracer sources to the MTC representative of a semi-infinite length sediment bed source. The graph, modified from the original, appears in Figure 12.1. It shows that a for Reynolds number between 1×10^4 to 1×10^5 the Stanton number, k_c/u (dimensionless), approaches a constant or steady state value. See Example 12.7.1 for an illustration of the procedure.

12.5   DIFFUSIVE CHEMICAL TRANSPORT ON THE SEDIMENT SIDE OF THE INTERFACE

12.5.1   A QUALITATIVE DESCRIPTION OF UPPER SEDIMENT LAYERS

The bed surface, in the context of this section, is considered to be nonmobile porous material made of various size solid particles. The particles typically range in diameter from a few millimeters to less than a micron. The various bed types generally reflect the fluid dynamic nature of the water column above the bed. Table 12.7 contains typical porosity values for sediment. Section 12.2.1 contains an overview description of the types of aquatic streams and currents above the beds. These aquatic systems include rivers, lakes, estuaries, shelf, and marine environments. Unlike the air–water interface, the sediment-water interface has a single fluid (i.e., aqueous phase) on either side. Water, the continuous phase, exists from within the column above, through the imaginary interface plane and into the porous bed where it is termed porewater. The interface plane is not a sharp one. It can be considered a thin mixed layer of finite thickness in the context of mass-transfer modeling (DiToro, 2001). Visual and physical examination of thin-sliced (0.1 mm) layers of a frozen core sample from a lake sediment bed microcosm showed the presence of a finite flocculent layer positioned between the water side and the particles on the bed surface (Formica et al., 1988). Little is known about this layer; from a mass-transfer perspective, it will not be considered further. Mass transport in those bed surface layers at and below the first layer of solid particles will be the subject of this section.

12.5.2   THE MOLECULAR DIFFUSION PROCESS IN POROUS MEDIA

The appropriate parameter for quantifying chemical mobility in pore water channels of the upper sediment bed layers is the flux N_z (kg s^{-1} m^{-2}). Its use commences with Equation 2.3:

$$n'_z = D^{(\ell)} \frac{dC}{dz'} + v_z C + j^{(t)} \tag{2.3}$$

where C (kg m^{-3}) is the interparticle porewater concentration in the bed, z' (m) is distance into the bed, $D^{(\ell)}$ (m^2 s^{-1}) is the molecular diffusivity, v_z (m s^{-1}) is the

---

TABLE 12.7
Typical Porosity Values of Natural Sedimentary Materials[a]

Material	ε (%)	Material	ε (%)
Peat soil	60–80	Medium to course mixed sand	35–40
Soils	50–60	Uniform sand	30–40
Clay	45–55	Fine-to-medium mixed sand	30–35
Silt	40–50	Gravel	30–40
		Gravel and sand	30–35

[a] Modified from Bear, J. 1972. *Dynamics of Fluids in Porous Media*. Dover, New York. pp. 45–47.

Darcian pore water velocity, and $j^{(t)}$ (kg m^{-2} s^{-1}) is the turbulent flux contribution. The advective flux term containing the Darcian velocity is the subject of Chapter 11. The turbulent flux contribution in the bed can be relevant in interparticle porespaces very near the sediment–water interface; see Chapter 14. It will be neglected in this discussion. Only the diffusive flux term will be addressed.

It is not convenient to use the flux within individual porewater channels but to base it upon the plane area, A (m^2), of the bed surface oriented perpendicular to the direction of the flux vector. In other words, the flux equation involves coordinates for a total cross-section (solid plus voids) and not just the void fraction. The conventional approach is the following porous media diffusive flux expression:

$$n_z = -D_e \frac{dC}{dz} \tag{12.17}$$

where D_e (m^2 s^{-1}) is the effective diffusion coefficient, which reflects the key physical characteristics of the porous media as well as the molecular and physical properties of the mobile chemical species. Diffusive transport theory for porous media is a very mature subject (Bear, 1972; Dullien, 1979; Cunningham and Williams, 1980) and has been extensively applied to geochemical processes (Lerman, 1979) and chemical contaminant transport (Thibodeaux, 1996; Grathwohl, 1998).

12.5.2.1   Solute Molecular Diffusive Transport

Under water-saturated conditions, solute diffusion in natural porous media is hindered by the tortuous nature of the pores and reduced cross-sectional area available for diffusion; both factors are influenced by the pore size distribution within the medium. If the pores are larger than the mean free path of the molecules in solution ($\sim$1 nm for organic chemical contaminants in water), the molecular diffusivity applies and Knudsen diffusion is negligible (Grathwohl, 1998). Lerman (1979) also consider these factors along with laboratory measurements for water-filled sediments, natural clays, silty-clay soil and sedimentary rocks using sodium chloride, trichloroethylene and iodine. Generally, tortuosity and pore size distribution are difficult to assess, so overall volumetric porosity ε_1 (m^3 m^{-3}), which can be readily determined from relatively simple measurements, is used to assess the porous media effects on the diffusive process. Therefore, the effective diffusion coefficient is related to the aqueous diffusivity, D_{aq} (m^2 s^{-1}) and an empirical function of ε alone:

$$D_e = D_{aq}\varepsilon^m \tag{12.18}$$

where m is an empirical exponent. Such empirical correlations are analogous to Archie's law, which describes electrical conductivity in porous rocks. Figures 12.2 and 12.3 contain some diffusion coefficient data and the Archie's Law functions with various values of m.

Lerman (1979) proposes $m = 2$ for the porewaters of sediments, for lack of a better model. Grathwohl (1998) found that the experimental data show that solute transport through porous samples follows the diffusion models reasonably well and that the porosity-based Archie's law estimates of effective diffusion coefficients are in good

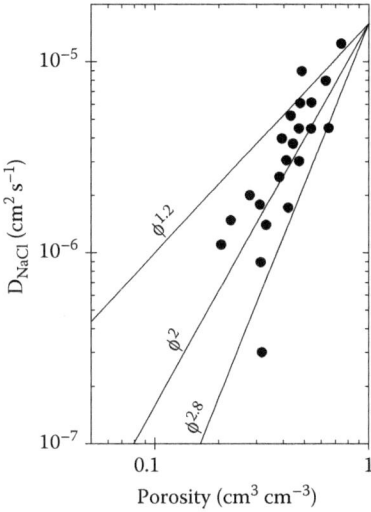

FIGURE 12.2 Diffusion Coefficients for NaCl in water-filled sediments of different porosity. (Reproduced from Lerman, A. 1979. *Geochemical Processes—Water and Sediment Environments*. John Wiley & Sons, NY, pp. 92. With permission.)

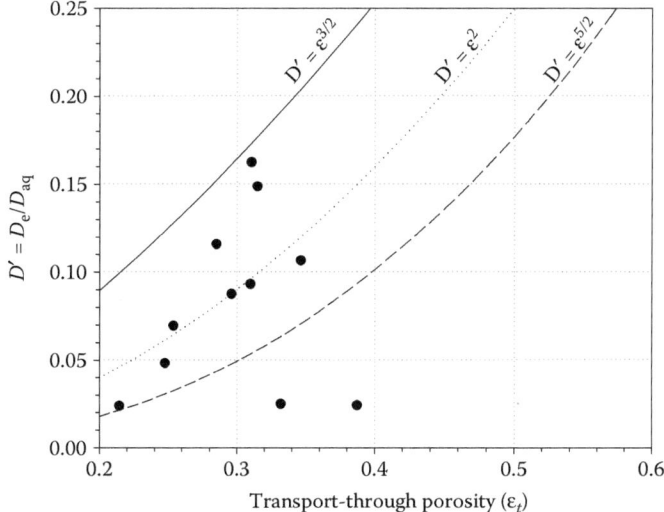

FIGURE 12.3 Normalized diffusion coefficients (D') versus ε_t (ε corrected based on water adsorbed onto dry samples at a relative humidity of 98.8%). The lines represent Archie's law. (Reproduced from Grathwohl, P. 1998. *Diffusion in Natural Porous Media*. Kluwer Academic Publisher, MA, p. 92. With permission).

agreement with the experimental data when $m = 2.2$. However, in the case of high surface area clays, the overall water saturated pore volume (ε) does not represent the transport-through porosity (ε_τ). The water absorbed onto the mineral surfaces (water coatings) and water present in between clay layers is not accessible for pollutant diffusion. Some data suggest that ε_τ, the transport-through porosity, may be the preferred parameter (see Figure 12.3).

Values of m, as reported in the literature, can vary significantly (Grathwohl, 1998). Taken together, data from diverse sources support an Archie's law correlation. The exponent m varies from 1.8 to 2.0 for unconsolidated material. For unconsolidated sands, m values between 1.3 and 1.64 have been reported for Fontainbleu sandstone. In silver catalyst pellets and sedimentary rocks (sandstones and limestones), values close to 2 where observed. Loose pocking of catalyst particles yielded $m = 1.43$. Similar m values have been reported for the diffusive transport of gases in soils ($m = 4/3$) and compacted sands ($m = 1.5$). Generally, for materials of lower porosities (e.g., $\varepsilon < 0.2$) larger m values have been observed ($m \geq 2$); this has led to its designation as the "cementation factor" in sedimentary rocks.

The porosity or pore water volume fraction of total bed volume ε ($m^3\ m^{-3}$) is obviously a key independent variable for assessing diffusive transport in porous media. The water that is contained in the bed is called the porewater or interstitial water because it fills the pores or interparticle spaces. It is the key phase for describing chemical mass transport interactions with the overlying water. Hence, all in-bed fluxes of dissolved constituents are transported in this fluid phase.

12.5.2.2   Porosities of Sediment Beds

Packed beds are a special case of consolidated beds. Dullien (1979) notes that prior knowledge of bed pore structure is usually impossible, though packed beds of uniform spheres in ordered (regular) arrangements are amenable to theoretical analysis. Most notable are the regular packs of uniform spheres: cubic, orthorhombic, tetragonal-sphenoidal, and rhombohedral packing. Based on theoretical calculations the cubic arrangement with six contact points per sphere has $\varepsilon = 0.476$. The rhombohedral, with twelve contact points, has $\varepsilon = 0.260$. Random packing arrangements of identical spheres have ε values that depend on the method of formation. Measurements of the porosity on the sedimentation of identical spheres produce a bed of about $\varepsilon \simeq 0.44$, while minimum ε values of 0.36–0.38 are obtained when the bed is vibrated or shaken down vigorously.

Natural porous media, such as bed sediments, are usually composed of a heterogeneous collection of particle sizes and mineralogies. The mass or volume distribution of particular particle size (diameter) classes is generally the preferred way to describe sediment texture and knowledge of the size distribution can be used to describe porosity ranges. The most common standard size classification system includes three categories: clay-sized particles of any solid material are those with diameters $<2\ \mu m$; silt-sized particles have diameters that range from 2 to 60 μm; and sand-sized particles range from 60 to 2000 μm. All larger discrete particles ($>2,000\ \mu m$) are considered gravel. The particle size distribution can vary markedly, depending on the the depositional environment; see Table 12.5 for examples.

TABLE 12.8
Colloidal Matter in Natural Waters

Type	Approximate Size (μm)
Organic macro molecules Humic substances, polysaccharides 10,000–100,000 Da	0.001–0.01
Viruses	0.01–0.1
Oxides of Fe and Al; clay minerials (e.g., illite, kaolinite, and montmonillonite)	0.01–1.0
Sewage particles	0.10–100
Bateria (living or dead); biologically ppt. MnO_2	1.0–10
Detrital SiO_2(quartz)	1.0–100
Algae (living or dead); biologically ppt. SiO_2	10–100
$CaCO_3$ chemically and biologically ppt.	10–1000
Fecal pellets, zoo plankton	100–1000

Source: Adapted from Thibodeaux, L.J. 1996. *Environmental Chemodynamics*. Wiley, New York.

Natural sedimentary materials are made of a mixture of discrete large and small grains. The quantity of finer particles has a marked effect on the porosity. The porosity of unconsolidated materials depends on the packing of the grains, their shape arrangement and size distribution. The latter may appreciably affect the resulting porosity, as small particles may occupy pores formed between those that are larger, thus reducing porosity. Table 12.8 gives typical porosity values of various natural sedimentary materials.

Berner (1980) notes that the initial porosity of a sediment at the time of sedimentation is primarily a function of grain size. Clays, due in part to their micron-to-submicron particle size, have surface properties that lead to electrostatic repulsion between platelets leading to an open structure that is exhibited by the sediment as a whole. This results in very high initial porosities in the range of $\varepsilon = 0.7$ to 0.9. Sand-sized particles are much less affected by surface chemistry. Simple geometric packing takes place, as observed for uniform spheres leading to porosities of 26–48%, depending on the closeness of the packing. Real, well sorted sands have measured porosities ranging from $\varepsilon = 0.36$ to 0.46. Sorting in the case of sediment containing clay, silt, and sand particles is the most important factor affecting the initial porosity. Poor sorting enables interstices between larger grains to be filled by smaller grains, thereby lowering the porosity. The combined effect of these factors on the initial porosity of natural particle beds is shown in Figure 12.4. The standard classification system for particle sizes is superimposed on the grain size axis.

At many locales, other factors operate to influence and change the initial porosity in the top 1 m zone of the bed, which is of primary interest with regard to contaminant chemodynamics. These factors include bioturbation, particle resuspension/deposition, and compaction. The "initial porosities" term used in the previous paragraph generally denotes the sediment layers in the top 1–20 cm of the bed. Bioturbation, when present, is primarily a particle and porewater mixing process caused by macrofauna species living on the bed, within it or both, which continually "perturbs" the physical and chemical structure of the bed. (See Chapter 13 for specific

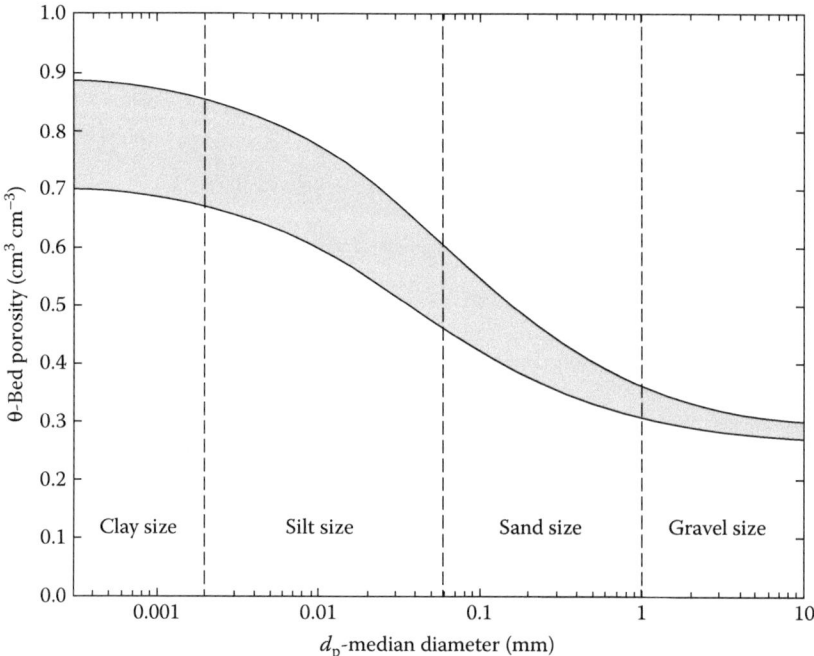

FIGURE 12.4 Initial porosity (θ_0) as a function of grain size for terrigenous surficial sediments. (Redrawn from Berner, R.A. 1980. *Early Diagenesis*. Princeton University Press, Princeton, NJ, pp. 28–31.)

information about this transport process.) The construction of burrows and particle relocation by macrofauna results in higher sediment water content than would otherwise be observed due to compaction. Particle resuspension followed by redeposition induced by high bottom currents during "storms" also contributes to maintaining high porewater contents at numerical valves not unlike the initial porosities reported above.

In the absence of surface layer mixing processes, particle compaction results in decreasing porosity with depth. The adjustment of porosity at increasing distance from the sediment–water interface is due to an applied overburden mass that forces the particles into tighter packing arrangements, displacing interparticle pore water. This process affects clay-rich beds more so than sandy beds. Sediment core samples that display typical values of $\varepsilon = 0.7$–0.9 in the top few centimeters decrease with depth; sediment porosities deeper in the bed range from 0.4 to 0.7 for clay-rich beds. Over depths as little as 20 cm, significant differences in porosities may occur. For Archie's law (Equation 12.18, $m = 2$), the diffusion coefficient is reduced by a factor of five as porosity changes from 0.9 to 0.4. For sandy sediment, the reduction is less; as porosity decreases from 0.5 to 0.3, the diffusivity is reduced by a factor of three. These variations underscore the importance of having *in situ* or core sample measurements of porosity for use in Equation 12.18.

12.5.2.3 Natural Colloids and Brownian Diffusive Transport

Natural colloids are particles found in essentially all aquatic systems. Table 12.8 lists some common types. High concentrations are found in bed sediment pore waters and lower ones found in the water column. Though their size ranges considerably, typically colloids are much smaller than particles dominated by mineral content. However, some large colloidal structures, such as fecal pellets and soil aggregates, approach the size range of sand (60–2000 μm).

A significant mass fraction of bed sediment is represented by organic carbon. Excluding black carbon derived from combustion sources, the material is mostly of plant and microbial origin. Most sediments contain only a few percent organic matter but some peat soils can contain up to 30% or 40% by weight. Diagenetic processes are continuously active in the bed producing dissolved organic carbon (DOC) in porewater that is transported out and into the adjoining water column (Aguilar and Thibodeaux, 2005; Thibodeaux and Aguilar, 2005). Table 12.9 contains some DOC size fraction diffusion data. Typical porewater concentrations are 10–$200\,mg$-$C\,L^{-1}$. Because of its mobility, DOC is a major colloidal constituent in bed sediment porewater and surface waters. Due to the ability of DOC to interact with organic and metal species, thereby enhancing their apparent "solubility" of these materials in aqueous solutions, DOC transport chemodynamics need to be quantified.

DOC production is the result of microbial breakdown of large organic molecules and aggregated organic matter into smaller components. The lower molecular weight compounds can then undergo a series of condensation reactions to reform higher molecular weight compounds such as fulvic acids and other humic substances. The final mixture is composed of organic compounds, which vary in size from 1000 to 10,000 Da. Operationally, DOC is defined to be nonsettlable organic matter in the 0.001–$0.45\,\mu$m size range. However, a significant portion of this operationally defined

TABLE 12.9
Brownian Diffusion Coefficients in Water

(a) Diffusion Coefficients in Water (Shaw, 1978)		(b) DOC Diffusion Coefficients[a] in Water (Thibodeaux, 1996)	
Colloid Radius or Type	D_B ($T = 20^\circ$C) (cm^2 s^{-1})	DOC Size Traction	D_B ($T = 25^\circ$C) (cm^2 s^{-1})
10^{-6} m (1 μm)	2.1×10^{-9}	0.05–0.4 μm	2.48×10^{-8}
10^{-7} m (100 nm)	2.1×10^{-8}	15–50 nm	1.97×10^{-7}
Tobacco mosaic virus	3×10^{-8}	3–15 nm	7.11×10^{-7}
Bushy stunt virus	1.15×10^{-7}	1.3–3 nm	2.98×10^{-6}
10^{-8} m (10 nm)	2.1×10^{-7}	0–1.3 nm	9.8×10^{-6}
Haemoglobin (horse)	6.3×10^{-7}		
10^{-9} m (1 nm)	2.1×10^{-6}		
Sucrose	3.6×10^{-6}		
Urea	1.29×10^{-5}		

[a] Based on Stokes–Einstein equation using the average radius of the size fraction.

TABLE 12.10
Bed Sediment DOC Diffusivities: Model-Derived from Microcosm Data

Sediment and Lab Measurements	D^a (cm^2 s^{-1})	Source
EPA5 (1.6%OC) bed porewater profile	5.7×10^{-6}	Thoma et al. (1991)
EPA5 (1.6%OC) water column	1.78×10^{-5}	Thoma et al. (1991)
University Lake (2.9%OC) water column	2.43×10^{-6}	Thoma et al. (1991)
Peat soil (0.65%C) water column	7.20×10^{-7}	Thibodeaux and Aguilar (2005)
Peat soil (10%C) water column	1.24×10^{-6}	Thibodeaux and Aguilar (2005)
Peat soil (19%C) water column	1.85×10^{-6}	Thibodeaux and Aguilar (2005)
University Lake (3%OC)-1 flux to water	1.6×10^{-5b}	Valsaraj et al. (1996)
University Lake (3%OC)-2 flux to water	2.6×10^{-5b}	Valsaraj et al. (1996)
Bayou Manchac (2.1%OC) flux to water	1.4×10^{-5}	Valsaraj et al. (1993)
University Lake (4.3%OC) flux to water	1.04×10^{-6}	Valsaraj et al. (1993)

^{a} $D = D_p \varepsilon^{4/3}$.

^{b} Denotes D_p values. $\varepsilon =$ bed porosity.

"dissolved" organic carbon is generally present in the form of small colloids (Thoma et al., 1991; Reible et al., 1991; Valsaraj et al., 1993; Sajitra et al., 1995; Sajitra et al., 1996; Valsaraj et al., 1996; Aguilar and Thibodeaux, 2005; Thibodeaux and Aguilar, 2005). Table 12.10 contains Brownian diffusion coefficients obtained from DOC measurements fitted with an appropriate Fickian-type transport model.

The continuous and erratic motion of individual particles (i.e., pollen grains) as the result of random collisions with the adjoining molecules of fluid (water) was observed by the botanist Robert Brown in 1827. The term Brownian diffusion is used for colloidal particles to distinguish it from solute molecular diffusion. However, both are end members of a continuum of particle sizes and a fundamental consequence of kinetic theory is that all particles have the same average translational kinetic energy. The average particle velocity increases with decreasing mass (Shaw, 1978). The Stokes–Einstein's equation for particle diffusivity is based on this concept. It is

$$D_p = \frac{k_B T}{6\pi\mu r} \tag{12.19}$$

where k_B is Boltzman's constant (1.3805×10^{-6} g cm^2 s^{-2} K^{-1}); T (K) is temperature, μ (g cm^{-1} s^{-1}) is water viscosity and r (cm) is particle radius. Using these units yields the Brownian diffusivity, D_P in cm^2 s^{-1}. In a similar fashion to the solute molecular diffusivity, D_{aq}, this particle diffusivity must be modified for the presence of the porous media in the bed using an Archie-type function; see Equation 12.18.

12.5.2.4　Sediment Bed Pore Sizes

In addition to the reduced cross-sectional area available for diffusion and pore space tortuosity, Grathwohl (1998) describes a factor that accounts for the constrictiveness of pores relative to solute molecular size; the same idea applies to diffusing particles. The porous medium may contain small pores which are not accessible due to size exclusion and restricted diffusion. Unfortunately, the study of bed sediment pore structure has received little attention. However, some information is available on selected porous media as packing and geological materials such as sedimentary rocks (Dullen, 1997). Measurements on sandstones indicated pores diameters between ~ 2 and $110\,\mu m$ with an average of about $50\,\mu m$. Such broad ranges are to be expected given the small openings surrounding the contact points of spherical particles. Theoretical studies of the void shape and void size in regular packing of identical spheres provides some insight into the size of open passageways for particles. For example, the interconnected channels in a pack have been characterized by the size of a sphere that can pass through the windows in the channel. For rhombohedral packing this maximum sphere size is $[1/\sqrt{3} - 1/2]d_p$, where d_p is the diameter of the identical sphere packing pieces. This packing arrangement is the densest regular packing possible, with each sphere having 12 contact points with neighboring particles and a mean bulk porosity of 0.26, which is independent of sphere diameter.

The final point illustrates the problem of using the Archie-type porosity correction (Equation 12.18) and the porewater Brownian diffusivity obtained using Equation 12.19. Although the porosity may be sufficient for diffusion, the ideal maximum sphere size relation given above ($\sim 0.0773\,d_p$) places size restrictions based on packing considerations. For identical sand-size spheres $50\,\mu m$ in diameter, only colloids $4.6\,\mu m$ or less in diameter can pass through. For clay-size identical spheres with $2\,\mu m$ diameter only $0.15\,\mu m$ particles or smaller are not excluded. By this idealized theoretical analysis, only the smaller types of colloids listed in Tables 12.8 and 12.9 will find sufficiently large pore openings through which to diffuse.

Most natural and even closely graded industrial materials will involve a particle size variation of between 2 and $100\,\mu m$, the range being smaller at the large sizes. The smaller particles tend to fill the voids, reducing porosity and the size of interparticle pore openings. In-filling of voids in a rhombohedral packing by five successive specific sizes gave a minimum porosity of 0.15. Further addition of finer particles reduced it to 0.038 (Dullien, 1979). However, the natural sorting or grading process that occurs when particles settle from a water column apparently precludes such low sediment bed porosities. Nevertheless, it is likely that some of the smaller colloids represented by the particle size ranges in Tables 12.8 and 12.9 may be denied diffusive transport passage through some porous media.

There is currently no viable alternative to Equation 12.18 for correction of porewater Brownian diffusivities. Hopefully, advances in colloid transport assessment in bed sediment will result in more reliable procedures and algorithms.

12.5.2.5   Conclusion

In summary, Equation 12.17 is appropriate for estimating colloidal flux based on the colloid concentration gradient (i.e., dC_p/dz). The proceeding material addresses ways of estimating the required colloid or particle diffusivity, D_p. However, modifications are needed for estimating the chemical flux due to the presence of colloids. Since chemicals, organic pollutants and metals in particular, can be highly concentrated in the colloidal phase, a common approach is to assume local chemical equilibrium exists between the colloidal and adjoining aqueous phases. Typically, a colloid-to-water chemical partition coefficient defined as $C_p = K_{cw}C_w$ is used where the p and w subscripts denote the particle/colloid phase and the aqueous solute concentrations, respectively. See Example Problem 12.7.7 for the MTC computation procedure for colloids in porewater.

12.5.3 THE MTC FOR BED SEDIMENT DIFFUSIVE TRANSPORT

Due to the dynamic nature (i.e., transient) of chemical transport within sediment beds the concentration gradient usually changes with time. In this case, the flux equation presented above (Equation 12.17) is the appropriate since dc/dz is a function of time. It also contains the appropriate solute or colloid diffusion coefficient. However, there are situations where the use of a steady-state model is appropriate and the gradient is assumed to be constant. In this case, the so called quasisteady-state approach developed for this application is appropriate (Valsaraj et al., 1997). The concept is based on the idea that the chemical mass capacity of the sediment is typically much larger, due to the adsorptive/sequestering character of the solid material, compared to that of the porewater column. This is generally true, so the bed has a relatively long chemical response time; in other words changes in bed sediment concentration occur slowly whether it is either a sink or a source. Therefore, a quasisteady-state (QSS) flux condition is assumed to apply as a reasonable approximation in the case of hydrophobic chemicals bed sediments.

12.5.3.1 Solute Flux MTC

The integrated form of the diffusive components of the flux, commences with the following equation:

$$n_z = \frac{D_e}{h} \Delta C \tag{12.20}$$

where ΔC (kg m^{-1}) is the chemical concentration difference in the bed porewater at the interface plane $z = 0$ to depth $z = h$ (m). Over the time-period of the QSS assumption ΔC remains constant. It is appropriate to define the sediment-side MTC, k_c (m s^{-1}) as

$$k_c = \frac{D_e}{h} \tag{12.21}$$

The effective diffusion coefficient, D_e (m s^{-1}), represents the solute molecules transport in the porous media.

12.5.3.2 The Colloidal Particle Flux MTC

The steady-state diffusive flux function for organic colloidal particles in bed porewaters is similar to Equation 12.20. The concentration difference in units of mass organic carbon per volume of water is the driving force, ΔC, described using the particle concentration at depth $(z = h)$, C_p, g-OC m^{-3}, subtracted from that at the interface $(z = 0)$, C_{pi}. The chemical flux from the bed is the product of the particle flux and the loading of chemical onto the particles, W_{pi} (mg g^{-1}), at the interface plane. Assuming the solute-to-particle equilibrium exists at the plane, $W_{pi} = K_{oc}C_{si}$, where C_{si} (g m^{-3}), is the solute concentration and $K_{oc,}$ m^3 g-OC^{-1} is the chemical partition coefficient for the colloidal particles. Combining these ideas yields the chemical flux from the bed due to the flux of colloidal particles:

$$n = \frac{DK_{oc}}{h}(C_p - C_{pi})C_{si} \qquad (12.22)$$

D is the Brownian diffusion coefficient of the particles. Within the porous media, an Archie's law type expression (Equation 12.18) is appropriate for estimation purposes.

As presented, Equation 12.22 does not allow for a simple expression of an overall water-side MTC that combines both the soluble and particle fractions. However, an approximate value can be obtained by the following procedure. Replacing C_{si} in Equation 12.22 with C_s and assuming $C_{pi} = 0$ maximizes the chemical colloidal flux. Assuming $C_s \gg C_{si}$ in Equation 12.20 maximizes the chemical solute flux. These assumptions result in the flux expression $n = k_{cx}(C_s - C_{si})$ for the overall

$$k_{cx} \equiv \frac{D_e}{h} + \frac{D K_{oc} C_P}{h} \qquad (12.23)$$

This final result provides a conservative MTC estimate. In addition it allows for the assessment of the relative contributions of the solute and particle fractions to the chemical flux from the bed. See Example Problem 12.7.7.

The most appropriate means of estimating the appropriate value of h is to use field data. Based on concentration profiles, a linear function is usually discernable just below the interface in the top most centimeters of sediment. Because of bed mixing, typical values are in the 1–2 cm range with 5 cm being a high-end value of h. It is unlikely that chemical mass originating deeper within the bed will become mobilized to the water column by molecular or Brownian diffusion processes over the assumed QSS time-period. So, for those steady-state model applications used to describe chemical transport across the upper bed layers, Equation 12.23 may be used to estimate the appropriate bed-side MTCs.

12.6   A GUIDE FOR USERS

This chapter is concerned with molecular diffusion-related chemical transport parameter estimation procedures at the water–sediment interface. Transport coefficients for particle resuspension and settling are in Chapter 10. Chapter 11 is concerned with the bed porewater advective process and Chapter 13 covers macrofauna driven biodiffusion (i.e., bioturbation).

The water-side estimating procedures fall into three broad categories: hydraulic forced convection, wind-generated forced convection and natural or free convection. For terrestrial freshwater streams, Equation 12.1 is recommended. The MTC requires the bottom friction velocity, a highly variable and uncertain parameter. Two methods of estimating it are presented. See Equations 12.5 and 12.8 as well as the bottom roughness Equation 12.2 and Table 12.4. Estimates of the MTC appear in Table 12.6. Marine bottom waters and estuaries have currents as well and Equation 12.1 applies.

In the absence of measurements bed sediment composition and/or particle sizes may serve as current velocity proxies, see Table 12.5.

Bottom water currents in sluggish streams (i.e., bayous), lakes, estuaries, and other near-shore marine waters are moved by the wind at the surface. Both thermal and salinity stratification in these waters is a factor influencing the magnitudes of the bottom-water transport coefficients. Although this subject of MTCs has received limited study, some estimation methods are proposed. For unstratified water bodies, Equation 12.10 is useful; wind speed is a key independent variable. For stratified lakes surface winds cause seiches that generate bottom water currents. Equations 12.11 through 12.13 can be used with seiche water displacement heights. To estimate bottom currents, these values are converted to bottom friction velocities with Equation 12.8, Equation 12.1 is then used for the MTC estimate. Bed characteristics can be used as proxies for bottom currents; see Table 12.5.

In the absence of hydraulic or wind forces, the water becomes "quiescent" but natural or free convection processes remains operative. Driven by bottom residing thermal or concentration gradients, Equations 12.14 and 12.15 may be used for estimating these low-end MTCs. The chemical diffusion coefficient in the porewaters of the upper sediment layer is the key to quantifying the sediment-side MTC. Use Archie's law, Equation 12.18, to correct the aqueous chemical molecular diffusivity for the presence of the bed material. Bed porosity is the key independent variable that determines the magnitude of the correction factor. See Table 12.7 for typical porosity values in sedimentary materials. For colloids in porewaters, Equation 12.18 applies as well. The Stokes-Einstein equation (Equation 12.19) is recommended and some reported particle Brownian diffusion coefficients appear in Tables 12.9 and 12.10. Under quasisteady-state conditions, Equation 12.23 is appropriate for estimating the bed-side MTCs.

12.7 EXAMPLE PROBLEMS

12.7.1 Upscaling Tracer Measured Water-Side MTCs To Field Conditions

Various solid and liquid tracers placed on sediment beds yield in situ local MTC measurements, some are given in Section 12.2 (see Table 12.1).

(a) Alabaster plates deployed on the outer shelf and upper slope of the Middle Atlantic bight in the Delmarva Peninsula at 1095 m depth (Santschi, et al., 1991) yielded MTC equal $3.02 \, \text{cm} \, \text{h}^{-1}$. The plate length was 10 cm length and bottom friction velocity $0.12 \, \text{cm} \, \text{s}^{-1}$.

Estimate $k_c(L \rightarrow \infty)$; the Schmidt number is 3700.

Solution:

Figure 12.1 provides a convenient graphical means for extrapolating k_c to $L \rightarrow \infty$. The $\text{Re}_* = 0.12(10)/0.0166 = 72.3$. From Figure 12.1, $Sc k_c/u_* = 2.2$. At $\text{Re}_* = 1 \times 10^5$ it is 1.2. So, the $k_c(L \rightarrow \infty) = 3.02(1.2/2.2)$. The up-scaled value is $1.6 \, \text{cm} \, \text{s}^{-1}$ and 55% of the measured one.

(b) Furfural pools 10.1 cm in length placed on the sand in re-circulating flume yielded correlation 7 in Table 12.2. For $Sc = 642$ and $u_* = 3$ cm s^{-1} correct $a = 0.114$ in the table for $L \to \infty$.

Solution:
Figure 12.1 is appropriate again. $Re_x = 3(10.0)/0.00893 = 3390$. The correction factor is the k_c ($Re_x = 1 \times 10^5$)/k_c($Re = 3,390$), which is approximately $0.70/0.80 = 0.875$. The corrected factor, a, is $0.875(0.114) = 0.10$ and is in general agreement with the others in Table 12.2.

12.7.2 SMOOTH AND ROUGH SURFACE MTCs

(a) Table 12.2 contains seven data correlations for estimating $k_c(L \to \infty)$ for smooth surfaces. Verify the MTC values in the table and that the average is 2.68 cm h^{-1} ($\sigma = 0.31$) excluding correlation 7, which is for a liquid tracer.

Solution:
See Table 12.2.

(b) Figure 12.1 modified from Higashino and Stefan (2004) can be used to estimate $k_c(L \to \infty)$ as well. For $Sc = 1000$ at $Re_x = 1 \times 10^5$, $k_c Sc/u_* = 0.8$. Therefore $k_c = 0.8u_*/Sc = 0.8(1.0)3600/1000 = 2.89$ cm h^{-1}. This result is in agreement with those in Table 12.2.

(c) Four data correlations appear in Table 12.3 that may be used to correct k_c values for bottom roughness. Verify the r_c values in the final column.

Solution:
Note $Re_x = u_* \Delta/\nu$ where $\Delta = 10$ cm, the height of the bottom roughness elements. The range of correction factors is 1.6–2.3 with 1.8 the average. All values of r_c indicate bed surface roughness enhances the MTCs.

(d) Manning's equation for estimating the average stream flow velocity contains a roughness coefficient which reflects the general characteristics of the bed; see Equation 2.24. Values of the roughness coefficient, n (s m$^{-1/3}$) appear in Table 2.2, Chapter 2 and Table 12.4. The Chilton–Colburn analogy result, Equation 2.20 in Chapter 2, offers a correlation of key parameters similar to Equation 12.1. With this information derive a relationship for a rough surface MTC estimating equation.

Solution:
Step 1: Start with Equation 2.20.
Step 2: Convert u_∞ to its equivalent friction velocity using Equation 12.8 and assume C_f and C_D, the quadratic drag coefficient, are equivalent friction parameters.
Step 3: Use Equation 12.7 to represent C_D and
Step 4: Compare the final result to Equation 12.1 and discuss common and different qualities of the two. Executing each step:

$$k_c = \frac{C_f}{2}u_\infty/Sc^{2/3} = \frac{\sqrt{C_D}}{2}u_*/Sc^{2/3} = \sqrt{\frac{2g}{R^{1/3}}}\,nu_*/Sc^{2/3} \qquad (12.24)$$

TABLE 12.11
MTCs vs Manning's Roughness Coefficient, k_c in cm h^{-1}

Bottom Characteristics	n (s m$^{-1/3}$)	$R = 1$ m[a]	$R = 10$ m[a]
1. Cement with mortar	0.013	2.07	1.41
2. Unfinished concrete	0.017	2.71	1.85
3. Earth, straight and uniform	0.018	2.87	1.96
4. Gravel bottom	0.023	3.67	2.50
5. Earth, winding w/o vegetation	0.025	3.99	2.72
6. Vegetal lining	0.030	4.78	3.26
7. Main stream, irregular and rough	0.035	5.58	3.80

[a] R denotes water or benthic boundary layer hydraulic depth. Also $u_* = 1.0$ cm s^{-1} and Sc $= 1000$.

This modified Chilton–Colburn result is very similar to Equation 12.1 but b is a constant. The "a factor" equivalent in Equation 12.24 contains the bed surface roughness, whereas, Equation 12.2 is needed to correct Equation 12.1. Otherwise, Equation 12.24 seems to be a theoretically sound alternative that more directly reflects natural bed surface roughness characteristics.

(e) Select various n values that represent commonly encountered bottom/channel lining characteristics and obtain estimates of MTCs using Equation 12.24 above.

Compare them with those in Table 12.2 modified by $r_c = 1.8$, the average from Table 12.3, so as to reflect rough sediment surfaces.

Solution:
The selected bottom characteristics appear in Table 12.11 column 1 and the corresponding n values appear in column 2. Two values of the hydraulic radius or stream depth, R, were used; the MTCs appear in columns 3 and 4.

The r_c-modified k_c values in Table 12.2 excluding correlation 7 for liquids was 4.41 cm h^{-1} with a range of 3.6–5.1 cm h^{-1}. In comparing these with the tabulated values it appears that the Chilton–Colburn result incorporating Manning's roughness factors Equation 12.24, yields MTC estimates comparable to those obtained from engineering apparatus derived correlations; see citations in Tables 12.2 and 12.3.

12.7.3 WATER-SIDE MTC FOR THE HUDSON RIVER-I

Flux measurements for 12 polychlorinated biphenyl congeners from a mass balance across the Thompson Island Pool (TIP) yielded the average MTCs shown in Table 12.12 (Erickson et al., 1991). The data represented 512 observations on MTCs over a 3-½ year study period, 1996–1999. The average MTC was 1.3 cm h^{-1} and no discernable correlation of MTCs to stream flow or water temperature was found. The TIP is 9.8 km length, average width 190 m and volume 429 hectare-meter (ha m) formed by the TI Dam. During the flux observations the hydraulic residence times varied from 4 to 14 h.

TABLE 12.12
Average PCB Water-Side MTCs
Hudson River

Seasons and Julian Days[a]	MTC $(cm\,h^{-1})$
Early spring, 105–130	0.78
Spring, 131–190	1.4
Summer, 191–265	2.2
Fall, 266–320	0.44
Winter, 321–365	1.5

[a] Ice conditions; no measurements $J = 1$–104.

Eight aquatic vascular plant species were observed with *V. americana* being the most common throughout the main channel. Open sediment areas were covered with a layer of periphyton characterized as algal mats. High-density vegetation occurred at depths of 2.7 m or less and fine sediment with particle sizes ≤ 0.5 mm was usually present. Vegetation observed at depths greater than 4.0 m were low density. Algal mats and vegetation areas extended from the shoreline inward occupying half the channel width. The center of the channel had an average depth of 4.6 m $\pm$ 1 m standard deviation. The in-bed mass of PCB contributing to the flux mostly resides in the fine grained sediments occupied by vegetation along the shallow shoreline areas of the TIP. The deeper part of the channel contained gravel, cobble and rock.

Use the information given above to corroborate these reported MTCs with appropriate estimation methods available in this chapter and reasonable assumptions about process variables.

Solution:
As a place to begin, Equation 12.24 will be used in the following approximate form for the average stream velocity, u_∞ $(m\,s^{-1})$

$$k_c = \left(\frac{4n^2 g}{R^{1/3}}\right) u_\infty / Sc^{2/3} \tag{12.25}$$

for estimating the MTC. Based on the TIP dimensions, hydraulic residence times, and volume the average pool velocities range from 0.19 to 0.68 m s^{-1} and average depth 2.3 m. These values are likely not representative of the shallow near shore areas containing SAV and algal mats. Low water depths and surface roughness contribute to increased flow resistance and low current speeds. Most of the volumetric flow that contributes to the average current speeds will occur in the deeper parts of the center of the channel where the maximum depths range from 4 to 6 m.

Currents in the shallows will be assumed one-fifth of the average, 0.038–0.14 m s^{-1}, and water depth 1.5 m. Grassy stream bottoms have $n = 0.030$ values as seen in Tables 12.4 and 2.2. The average Schmidt number is 1800 for the PCB mixture in the stream. Using these numbers, Equation 12.25 yields MTCs of 2.9 and

$10.5\,\mathrm{cm\,h^{-1}}$. Due to the location of fine sediment, half the bottom materially contributes to the PCB flux so the MTCs in Table 12.12 can be doubled. When this is done, the range is 0.88–$4.3\,\mathrm{cm\,h^{-1}}$ and not inconsistent with the model estimates but on the low end of that range.

12.7.4 WATER-SIDE MTC FOR THE HUDSON RIVER-II

Due to its structure the TIP is a run-of-the-river reservoir and displays lake-like conditions. The width of the TIP is 190 m on average. Some sections of the pool have open-water "fetches" up to approximately 1.5 km over which the wind has an unobstructed path. Assume the pool is an unstratified lake, estimate the bottom water MTC. Use a wind speed of $3.9\,\mathrm{m\,s^{-1}}$ and appropriate information from Problem 12.7.3 above.

Solution:
Equation 12.10 for nonstratified waterbodies will be used. With a 3.9 wind speed $C_D = 1.66 \times 10^{-3}$. The water depth average is 2.3 m. Substituting the required parameters into Equation 12.10 with $L = 190$ m is

$$k_c = 1.71(1.66\mathrm{E} - 3)\left(\frac{1.183}{997}\right)(3.9 \cdot 1.5)^2 (2.3)^{5/4}/190$$

$$k_c = 1.72\mathrm{E} - 6\,\mathrm{ms^{-1}}(0.62\,\mathrm{cm\,h^{-1}})$$

For $L = 1500$ m, $k_c = 0.078\,\mathrm{cm\,h^{-1}}$.

These results are for benzoic acid with molecular weight of $122\,\mathrm{g\,mol^{-1}}$. The estimated PCBs have an average molecular weight of $269\,\mathrm{g\,mol^{-1}}$. Assume for solutes in water, $D \sim 1/M_A^{6/10}$ to yield $k_c \sim 1/M_A^{4/10}$ where M_A is the molecular weight of the compound of interest. The estimated PCB MTCs are therefore $k_c = 0.62(122/269)^{4/10} = 0.45\,\mathrm{cm\,h^{-1}}$ and $0.057\,\mathrm{cm\,h^{-1}}$ respectively. These values are approximately an order of magnitude lower than those observed in the TIP. Multiply the table values by 2 for a direct comparison.

The results of this calculation suggest that wind driven water current contribute little to the TIP water-side MTCs. The average observed for Table 12.12 is $1.26 \times 2 = 2.5\,\mathrm{cm\,h^{-1}}$, so the contribution is 2–14%.

12.7.5 WATER-SIDE MTC ESTIMATE FOR STRATIFIED LOCK EARN

Key information concerning the Lock appears in Section 12.3.3.

(a) Using that information and data, estimate the maximum MTC that may occur at the peak time-period during a seiche. First verify the average water current speed in the hypolimnion is $0.017\,\mathrm{m\,s^{-1}}$.

Solution:
Equation 12.12 is approximate for this stratified lake. Substituting the

appropriate parameter yields

$$\bar{u} = 0.25 \left(\frac{10}{56}\right) \sqrt{\frac{(9.81 \text{ m s}^{-1})(999.781 - 999.605)80 \text{ m}}{999.605}}$$

$$\bar{u} = 0.0166 \text{ ms}^{-1}$$

(b) It will be assumed that the lake bed sediment material is consolidated so that the surface roughness is equivalent to "concrete float finish" in Table 12.4 with $n = 0.015$. The effective hydraulic radius is assumed to be 27.5 m, half the hypolimnion depth. The chemical present has a $Sc = 1000$. Equation 12.25 applies with u_∞ twice the average current value 0.017 m s^{-1}. Substituting into Equation 12.25 yields

$$k_c = \left[\frac{(0.015)^2 4(9.81)}{(27.5)^{1/3}}\right]^{1/2} 2(0.017)/1000^{2/3} = 9.95\text{E} - 07 \text{ m s}^{-1}$$

This is equivalent to 0.35 cm h^{-1} and likely represents a maximum bottom water MTC for this lake under these conditions. As the event that generated the seiche subsides quieter bottom waters will prevail with much lower MTC.

12.7.6 NATURAL CONVECTION MTC FOR GYPSUM

Alabaster plates are used to make in situ measurements of bottom water MTCs in various water bodies. However, these measured values contain MTC contributions due to the natural or free convection process created by the presence of gypsum. It has a solubility of 0.0011 kg L^{-1}, therefore increasing the local fluid density by a factor of approximately 0.11%. Its Schmidt number is 2180.

(a) Estimate the MTC for natural convection produced by the presence of a 10 cm plate on the surface of bottom water.

Solution:
Although other equations may be appropriate as well Equation 12.15 will be used. The water-gypsum diffusivity is $4.1 \times 10^{-6} \text{ cm}^2 \text{ s}^{-1}$ and water viscosity $8.93 \times 10^{-3} \text{cm}^2 \text{ s}^{-1}$. From Chapter 2, Equation 2.30, the Grashoff number is

$$\text{Gr}_A \equiv \frac{L^3 g}{\upsilon^2} \frac{|\Delta\rho|}{\rho} = (.1)^3 \, 9.81 \left(\frac{0.0011}{1.0}\right) /(8.93\text{E} - 7)^2 = 1.35\text{E}7$$

Substituting into Equation 12.15

$$k_c = \frac{4.1\text{E} - 10 \text{ m}^2\text{s}^{-1}}{0.1 \text{ m}} \times 9.22(1.35\text{E}7 \times 2180)^{1/6} = 2.10\text{E} - 6 \text{ m s}^{-1}$$

The k_c in cm h^{-1} is 0.75 and represents a 10 cm plate.

(b) Project what the natural convection MTC may be for a surface $L \to \infty$.

Solution:

Figure 12.1 will be used to make the projection. Natural convection produces advective flow in the vicinity of the 10 cm plate, see Section 2.5.7. It will be assumed the Re_x for $L = 0.1$ m is 1.0 and that for $L \to \infty$ the $Re_x = 1E5$. From Figure 12.1 for $Sc = 1000$ $k_c(L \to \infty)/k_c(0.1 \text{ m}) = 0.75/6.1 = 0.12$. Therefore $k_c(L \to \infty) = 0.75(0.12) = 0.090 \text{ cm h}^{-1}$. This MTC is a low value which it should be. Compare it to the field observed values in Table 12.1.

12.7.7   IN-BED NAPHTHALENE MTC

The presence of 54 mg-C L^{-1} as "dissolved organic carbon" (i.e., DOC) concentration in bed-sediment porewater is presumed to be colloids with Brownian diffusivity 1.86×10^{-6} cm^2 s^{-1} (Thoma et al., 1991). Naphthalene in aqueous solution has a diffusivity of 7.3×10^{-6} cm^2 s^{-1} and an organic carbon base partition coefficient, $K_{OC} = 3000$ L kg^{-1}. For an assumed 2 cm diffusion path length estimate the bed-side MTC for one containing silt-size particles with porosity of 75%. Separate the MTC into its soluble and colloidal contributing fractions.

Solution:
Archie's law, Equation 12.18, with $m = 2$ will be used. For estimating the overall MTC on the bed-side Equation 12.23 will be used. Substituting the appropriate parameters yields:

$$k_c = \frac{7.3E - 6(.75)^2}{2} + \frac{1.86E - 6(.75)^2}{2}(3000)(54.3E - 6)$$
$$k_c = 2.05E - 6\text{cm s}^{-1} + 8.52E - 8\text{cm s}^{-1} = .00739 + .000304 \text{ cm h}^{-1}$$
$$k_c = 0.0077 \text{ cm h}^{-1}$$

In this case the colloid particle diffusion contribution is small and only 4% of the total.

LITERATURE CITED

Aguilar, L. and L.J. Thibodeaux. 2005. Kinetics of peat soil dissolved organic carbon released from bed sediment to water. Part 1. Laboratory simulation. *Chemosphere* 58, 1309–1318.

Bear, J. 1972. *Dynamics of Fluids in Porous Media*. Dover, New York. pp. 45–47.

Berner, R.A. 1980 *Early Diagenesis*. Princeton University Press, Princeton, NJ, pp. 28–31.

Boudreau, B.P. and B. Barker Jorgensen. 2001. *The Benthic Boundary Layer-Transport Processes and Biogeochemistry*. Oxford University Press, Oxford, UK, Chapter 5.

Christy, P.S. and L.J. Thibodeaux. 1982. Spill of soluble, high-density immiscible chemicals on water. *Environ. Progress* 1(2), 126–129.

DiToro, D.M. 2001. *Sediment Flux Modeling*. Wiley Interscience, NY, Chapter 2.

Erickson, M.J., C.L. Turner, and L.J. Thibodeaux. 2005. Field observation and modeling of dissolved fraction sediment-water exchange coefficients for PCBs in the Hudson River. *Environ. Sci. Technol.* 39, 549–556.

Formica, S.J., J.A. Baron, L.J. Thibodeaux, and K.T. Valsaraj. 1988. PCB transport into lake sediments, conceptual model and laboratory simulation. *Environmental Science & Technology*. 22, 1435–1440.

Heatershaw, A.D. 1976. Measurements of turbulence in the Irish Sea benthic boundary layer. Chapter 2 in *The Benthic Boundary Layer*, N. McCave, Editor. Plenum Press, New York.

Hickey, B., E. Baker, and N. Kachel. 1986. Suspended particle movement in and around Quinault submarine Canyon. *Marine Geology* 71, 35–83.

Higashino, M. and M.G. Stefan. 2004. Diffusive boundary layer development above a sediment-water interface. *Water Environmental Research* 76, 292–300.

Lerman, A. 1978. Chemical exchange across sediment-water interface. *Annual Review of Earth and Planetary Sciences* 6, 281–303.

Opdyke, B.N., G. Gust, and J.R. Ledwell. 1982. Mass Transfer from smooth alabaster surfaces in turbulent flows. *Geophysical Research Letters* 14, 1131–1134.

Reible, D.D., K.T. Valsaraj, and L.J. Thibodeaux. 1991. Chemodynamic models for transport of contaminants from sediment beds. *The Handbook of Environmental Chemistry*, Editor O. Hutzinger. Vol. 2, Part F. Springer-Verlag, Berlin, pp. 185–228.

Santschi, P., P. Bower, and U.P. Nyffeler. 1983. Estimates of the resistance to chemical transport posed by deep-sea boundary layer. *Limnol. Oceanography* 28(5), 899–912.

Santschi, P., P. Hohener, G. Benoit, and M.B. Brink. 1990. Chemical processes at the sediment–water interface. *Marine Chemistry* 30, 269–315.

Santschi, P.H., R.F. Anderson, M.Q. Fleisher, and W. Bowles. 1991. Measurements of diffusivity sublayer thicknesses in the ocean by alabaster dissolution and their implications for the measurement of benthic fluxes. *Journal of Geophysical Research*, 96(10), 641–657.

Shaw, D.J. 1978. *Introduction to Colloid and Surface Chemistry*, Chapter 2. Butterworths, London.

Sojitra, I., K.T. Valsaraj, D.D. Reible, and L.J. Thibodeaux. 1995. Transport of hydrophobic organics by colloids through porous media. 1. Experimental results. *Colloids and Surfaces A. Physicochemical and Engineering Aspects* 94, 197–211.

Sojitra, I., K.T. Valsaraj, D.D. Reible, and L.J. Thibodeaux. 1996. Transport of hydrophobic organics by colloids through porous media. 2. Commercial humic acid macromolecular and polyaromatic hydrocarbons. *Colloids and Surfaces A: Physicochemical and Engineering Aspects* 110, 141–157.

Thibodeaux, L.J., L.-K. Chang, and D.J. Lewis. 1980. Dissolution rates of organic contaminants located at the sediment interface of rivers, streams and tidal zones, Chapter 16 in R.A. Baker, Editor. *Contaminants and Sediments*, Vol. 1, Ann Arbor Science, MI.

Thibodeaux, L.J. and B. Becker 1982. Chemical transport rates near the sediment in wastewater impoundments. *Environmental Progress* 1, 296–300.

Thibodeaux, L.J. and L. Aguilar. 2005. Kinetics of peat soil dissolved organic carbon release to surface water. Part 2. A chemodynamic process model. *Chemosphere* 60, 1190–1196.

Thibodeaux, L.J. and D. Mackay, 2007. The importance of chemical mass-transfer coefficients in environmental and geochemical models. *Society of Environmental Toxicology and Chemistry—Globe*, July–August issue, pp. 29–31.

Thoma, G.J., A.C. Koulermos, K.T. Valsaraj, and L.J. Thibodeaux. 1991. The effect of porewater colloids on the transport of hydrophobic organic compounds from bed sediment. In. R.A. Baker, Editor. *Organic Substances and Sediments in Water*, Volume 1. Humics and Soils. Lewis Publishers, Chelsea, MI.

Valsaraj, K.T., G.J. Thoma, C.L. Porter, D.D. Reible, and L.J. Thibodeaux. 1993. Transport of dissolved organic carbon-derived natural colloids from bed sediments to overlying water: Laboratory simulations. *Water Science Technology* 28, 139–147.

Valsaraj, K.T., S. Verma, I. Sojitra, D.D. Reible, and L.J. Thibodeaux. 1966. Diffusive transport of organic colloids from sediment beds. *Journal Environmental Engineering.* 122, 722–728.

Valsaraj, K.T., L.J. Thibodeaux, and D.D. Reible. 1997. A quasi-steady-state pollutant flux methodology for determining sediment quality criteria. *Environmental Toxicology and Chemistry* 16, 391–396.

Wetzel, R.G. 2001. *Limnology—Lake and River Ecosystems*, 3rd ed. Academic Press, San Diego, CA. p. 95.

Winbush, M. 1976. The physics of the benthic boundary layer, Chapter 1 in *The Benthic Boundary Layer*, N. McCave, Editor. Plenum Press, New York.

Yang, C.T. 2003. *Sediment Transport Theory and Practice.* Kreiger Publishing Company, Malabar, FL, Table 3.3.

FURTHER READING

Bear, J. 1972. *Dynamics of Fluids in Porous Media*, Dover, New York.

Boudreau, B.P. and B.B. Jorgensen. 2001. *The Benthic Boundary Layer.* Oxford University Press. New York. See specifically Chapter 5.

Cunningham, R.E. and R.J.J. Williams. 1980. *Diffusion in Gases and Porous Media.* Plenum Press, New York.

DiToro, D.M. 2001. *Sediment Flux Modeling.* John Wiley, New York.

Dullien, F.A.L. 1979. *Porous Media-Fluid Transport and Pore Structure.* Academic Press, New York.

Grathwohl, P. 1998. *Diffusion in Natural Porous Media: Contaminant Transport Sorption/Desorption and Dissolution Kinetics.* Kluwer Academic, Boston, MA.

Kullenberg, G., Editor. 1982. *Pollutant Transfer and Transport in the Sea*, Volume 11. See Chapter 3 by E.K. Duurasma and M. Smies, CRC Press, Boca Raton, Florida.

Lerman, A. 1979. *Geochemical Processes—Water and Sediment Environments.* Wiley, New York.

Reible, D.D., K.T. Valsaraj, and L.J. Thibodeaux. 1991. Chemodynamic models for transport of contaminants from sediment beds. *The Handbook of Environmental Chemistry*, Volume 2, Part F. Editor O. Hutzinger, Springer-Verlag, Berlin, Heidelberg, pp. 185–228.

Thibodeaux, L.J. 1996. *Environmental Chemodynamics.* Wiley, New York. See specifically Chapter 5.

13 Bioturbation and Other Sorbed-Phase Transport Processes in Surface Soils and Sediments

Louis J. Thibodeaux, Gerald Matisoff, and Danny D. Reible

CONTENTS

13.1 Introduction 360
13.2 Bioturbation and Sorbed-Phase Transport Processes........................361
    13.2.1 Aquatic Sediments 361
    13.2.2 Soil Sorbed-Phase Transport Processes 362
    13.2.3 Soil Bioturbation 363
    13.2.4 Agriculture and Tillage Mixing...................................365
13.3 Transport Theory...366
    13.3.1 Instantaneous Uniform Mixing Model 366
    13.3.2 Diffusive Mixing Models 367
    13.3.3 Nonlocal Mixing by Conveyor-Belt Species........................369
    13.3.4 Summary and Conclusions 369
13.4 Estimating Bioturbation Transport Parameters 370
    13.4.1 Sediment Bioturbation Transport Parameters......................370
        13.4.1.1 The Particle Biodiffusion Coefficient 371
        13.4.1.2 Particle Mixing Depth.................................372
        13.4.1.3 Bed Surface Properties 373
        13.4.1.4 The Sedimentation Velocity or Burial Velocity 373
        13.4.1.5 The Porewater Biodiffusion Coefficient................374
        13.4.1.6 The Mass-Transfer Coefficient 375
    13.4.2 Soil Bioturbation Transport Parameters..........................377
        13.4.2.1 The Surface Soil Particle Biodiffusion Coefficient 378
        13.4.2.2 The Surface Soil Particle "Mixing" Depth 379
        13.4.2.3 The Soil-Side MTC 380
    13.4.3 Plowing Enhancement to Soil Bioturbation Transport 380
    13.4.4 Simultaneous Parallel and Series Transport Processes 382
13.5 A Summary Guide for Users 382
13.6 Example Calculations..383

13.6.1   Baltic Sea Sediment Bioturbation MTC 383
13.6.2   Chemical Role and the Soil Bioturbation MTC 384
Acknowledgment..386
Literature Cited 386
Further Reading 388

13.1   INTRODUCTION

Chemicals become associated with surface soils as a result of various human activities and other natural processes. These activities include the direct placement of waste materials onto the soil for storage or treatment and the application of pesticides for the control of agricultural pests. In addition, some chemicals arrive at the soil surface by the process of deposition from the air phase. These substances may originate from nearby or remote sources including power plant stack emissions and volcanoes in either vapor or particulate form and are transported via the atmospheric pathway prior to deposition. Once in place on the soil surface mixing and transport processes occur within the upper layers to move this material deeper into the soil column. Once contaminated with chemical substances, it is not uncommon for the soil surface layers to be the source of chemical transport back to the atmosphere. Evaporation and particle resuspension contribute to atmospheric transport but mixing in the upper soil column actively moves contaminants upwards and maintains an elevated concentration level at the surface. Consequently soil mixing in the upper layers is important to quantifying the transport chemodynamics in either direction. This chapter examines soil mixing processes in more detail.

In a somewhat analogous fashion, the upper layers of aquatic sediments become contaminated by human activities and natural processes. Direct discharges from municipal, industrial, and agricultural sources deliver chemical contaminants to nearby water bodies where they deposit from the water column onto the sediment surface and eventually become mixed into the upper sediment layers. Once these contaminant sources are reduced or stopped the continued mixing of the sediments reverses the transport chemodynamics and reintroduces the contaminants into the water column. This chapter examines mixing processes on the sediment side of the interface that are capable of mobilizing chemicals in both directions across the sediment–water interface.

The movement and mixing of solid particles (i.e., clay, silt, sand, and composite soil lumps) and adjoining fluids by macrofauna is one of the most significant processes that affects contaminant chemodynamics. It is particularly significant for those chemicals that are strongly sorbed to the particles ("sorbed-phase" transport). Perturbation of the sediment column by macrofauna and plants is termed "bioturbation." "Pedturbation" is used by soil scientists for the analogous action on surface soils. Here the term bioturbation is used for both. By extension, particle movement by drying/wetting and freezing/thawing processes is termed "dryoturbation" and "cryoturbation," respectively, and mixing by agricultural plowing is termed "tillage mixing" regardless of the agricultural implement used. All these "turbation" processes influence both material transport and chemical reaction rates. These are the primary sorbed-phase transport processes that will be described in this chapter and numerical values of the transport coefficients given. A section on the qualitative descriptions of the solid-phase turbation processes will be given first; it will be followed by review of the models available for quantifying the relevant transport parameters. The section will include data, correlations, and modifying factors for estimating these parameters in a variety of realistic applications. The chapter ends with a simple guide and example calculations.

13.2 BIOTURBATION AND SORBED-PHASE TRANSPORT PROCESSES

13.2.1 AQUATIC SEDIMENTS

Sediment bioturbation is not a single process but the net effect of locomotion, feeding and tube construction activities of the host of benthic organisms that inhabit the floor and bottom sediments of rivers, lakes, estuaries, and oceans. Organisms that influence sediments can be divided into two basic groups: infauna and epifauna. Those that live within sediments are termed infauna and epifauna is the name given to organisms that live either temporarily or permanently on the sediment surface. Epifauna are generally mobile and can include insects, crabs, lobsters, bottom-living fish, and birds. Both types of organisms mix sediments by random or highly ordered movement of particles.

Collectively, sediment mixing processes by benthic invertebrates have the effect of enhancing the rate of "diffusion" of a chemical in the sediment porewater and those sorbed to solid particles. Some organisms, such as crabs and snails, mix surface sediment by crawling or plowing through it. Other organisms, especially polycheate and oligochaete worms and bivalves, burrow into and ingest sediment particles. Although smaller in size polycheates and oligochaetes are the aquatic sediment versions of earthworms. Such burrowing and ingestion can extend to several tens of centimeters and produce structures on the bed surface such as fecal pellet mounds and funnel-type depressions. Burrow constructing organisms remain in their constructed tubes and chambers and flush or "bioirrigate" these permanent habitats with overlying water. The literature on bioturbation including the effects on the sediment bed typology, stability, porosity, porewater chemistry, composition, and so on is exceedingly rich. Detailed descriptions of sediment bioturbation are contained in reviews performed by Thoms et al. (1995), DePinto et al. (2006), McCall and Tevesz (1982), Matisoff (1982), and Boudreau and Jorgensen (2001).

The first type of mixing—(bio) diffusive mixing—is not the result of a particular feeding strategy, but is the net effect of burrowing, feeding, and locomotion of suspension feeders, omnivores, and some deposit feeders. Organism that mix sediments by random movement of particles are often termed eddy diffusion mixers, using the analogy from fluid dynamics. For example, random burrowing enables sediment mixing to occur at all depths within the bioturbation zone. Such local mixing is analogous to Fickian diffusion; over many mixing events the particle motions appear to be random and the distances particles are displaced are smaller than the typical distance of change of the property of interest such as chemical concentrations in the bed. It has been hypothesized to be the more common mode of mixing for biogenic organisms.

Demersal fish, cephalopods, squids, cuttlefish and octopi, insects, and bivalves are examples of bioturbators that mix sediments in a net biodiffusive manner.

Advective transport of sediment by ingestion at depth and defecation on the surface is called "conveyor-belt feeding" and the transport process is modeled as "nonlocal mixing" (Boudreau, 1997). In conveyer-belt (nonlocal) feeding organisms mix sediments in a highly ordered mode. These may be subsurface (head-down) deposit feeders that orient themselves vertically, ingest sediment at depth, and egest fecal material at or above the sediment surface. Since particle column downward setting and burrow infilling eventually return the particulate matter to subsurface layers the resulting semiclosed particle loop is termed a "conveyer belt," a name applied to these organisms. Polychaetes and tubificids are typical head-down feeders and are important in both marine and freshwater environments. The so-called reverse conveyer belt organisms feed at the surface and defecate at depth. In contrast there is no return of particles to the surface, however the interface rises to make room.

Bioirrigation enhances solute exchange between sediments and the overlying water. It is more rapid than molecular diffusion-driven processes with bioirrigating fauna increasing the flux of solutes into overlying waters by as much as an order of magnitude. Freshwater bioirrigating organisms include mayflies, chironomids, and unionid clams. Ghost shrimp and heart urchins are recognized as bioirrigating benthos in coastal environments. Bioresuspension occurs when organisms eject fluidized pellets directly into the water column. Marine deposit feeders such as bivalves, polychaetes and mud shrimp have all been observed to inject watery sediment directly into the water column. Demersal fish can also be important agents for bioresuspension. Benthivorous bream and carp suspend particles in their feeding activities; suspended sediment concentrations were found to increase linearly with the biomass of the benthivorous fish. The chemodynamics of bioresuspension is less well developed as a process and not considered farther in this chapter.

13.2.2 Soil Sorbed-Phase Transport Processes

The soil column mixing process, pedbioturbation, functions similarly to sediment bioturbation. The upper soil layers contain a gas phase with a composition much like the atmosphere, as well as water and solid particles. The mixing that occurs involves all three phases. In addition, two additional processes occur in soils that are absent in sediments. The presentation that follows is largely extracted from Rodriguez (2006).

The objective in this section is to focus on the solid phase, and mechanisms that enhance the movement of soil particles, as well as substances retained in them. The main three transport/mixing processes that influence the vertical distribution of chemicals sorbed to soil particles are bioturbation, cryoturbation, and movement into cracks formed due to soil drying. Bioturbation, the most significant and most studied of these three processes, refers to the disturbance of soil or sediment layers by biological activity. Some species disturb the soil by burrowing and feeding, enhancing the transport of chemicals in this compartment. This process is thoroughly explained in the next section.

Cryoturbation is the process of stirring, heaving, and thrusting of soil material resulting from frost action, characteristic of areas at high latitudes with cold artic or alpine climate. It encompasses frost heave, thaw settlement, and differential mass movements, which are responsible for downslope soil movements in these areas. The extent of cryoturbation features in high-altitude areas depends on the amount of available moisture, the rate of freezing, and the types of rocks and soils present in a given area. Freezing and thawing influences profoundly the stability, hydrology, chemistry, biology, and ecology of soils. Chemicals within the soil profile are redistributed due to the presence of temperature gradients and nonuniform freezing. Freeze–thaw cycling can also cause cracks in a soil system, which may open further during subsequent cycles. Freeze–thaw rates vary with locations. For example in northwest Greenland, soils thaw completely in the summer and freeze completely in the winter. Therefore, the active part of the year occurs in fall and spring when soils may freeze and thaw on a diurnal basis, probably forming soil cracks. This is different than for example alpine areas, where freezing and thawing may occur on a daily basis for longer periods of time. Due to the presence of water, drying cracks are also formed in soils, such as clayey soils or vertisols, with high shrink–swell potential. When a body of clay sorbs water and then dries, shrinking and swelling occur forming numerous cracks. The ability of a soil to crack during shrinking or drying influences many of the transport processes that occur in the soil profile. Typical cracks in vertisols are at least 1 cm wide and reach depths of 50 cm or more. Shrinkage cracks expose considerable hidden subsurface soil. This shrink–swell behavior is also termed dryoturbation. Cryoturbation and dryoturbation, similarly to bioturbation, contribute to macropore flow and to the transport of sorbed chemicals downward and upward through the soil. However, due to the limited amount of information on their importance to chemodynamics both cryoturbation and dryoturbation will not be considered further in this chapter.

13.2.3 Soil Bioturbation

Mixing of soils due to agricultural practices, such as plowing, harrowing, disking, among others is an additional important soil turnover process. This process has been termed "tillage mixing" although it has been primarily studied to understand soil erosion and downslope transport by tillage (tillage translocation) and not vertical mixing within a soil profile (Govers et al., 1996; Lobb and Kachanoski, 1999a, 1999b). Soil bioturbation is the biologically driven mixing of material in the soil layer between the underlying geological formations and the overlying atmosphere. Effects of bioturbation are evident in the upper soil horizons; in soil science these are termed O and A. Often equivalent to the A-horizon or topsoil is the biomantle, it is also called the "biologically active" of "biologically mixed zone." It constitutes the upper part of soil produced by biota, essentially by bioturbation, and may include deeper levels in some soils. Soils accommodate an extensive variety of biota. Although not generally visible to the naked eye, it is one of the most diverse habits on the earth and contains one of the most diverse assemblies of living organisms. It provides a range of habitats for a multitude of fauna and flora ranging from macro- to micro-levels depending on climate, vegetation, and physical and chemical characteristics of the given soil. Species number, composition, and diversity depend on many factors including aeration, temperature, acidity, moisture, nutrient content, organic matter, and toxic substance concentrations of pollutants.

Bioturbation by trees, the process of tree throw of soil from toppling and the process of root growth and decay, is a factor in geologic landscape evolution (Johnson, 1990; Gabet et al., 2003). Although "floralturbulation," as it is called, is an emerging field of study, important to environmental chemodynamics, it is not fully developed and not considered further at this time.

Macrofauna are generally >2 mm in diameter, visible to the naked eye, and large enough to disrupt the structure of mineral and organic soil horizons through their feeding and burrowing activities. It is the most mobile fauna group, moving through macro- and micro-pores in the soil. They include vertebrates (snakes, lizards, mice, squirrels, badgers, armadillos, prairie dogs, and others) that primarily dig within the soil for food or shelter, and invertebrates that live in or upon the soil, the surface litter and their components (ants, termites, millipedes, centipedes, earthworms, snails, spiders, scorpions, crickets, cockroaches, and crawfish).

Invertebrates including earthworms, ants, and termites are the most common and the numbers of individuals are staggering, ranging from tens to tens of millions per square meter. Because earthworms are large, abundant and active their effects on soil have been widely studied. Charles Darwin was one of the first who recognized their important role in affecting soil structure, organic matter processing, and nutrient cycling. He observed how they caused solid objects such as stones to migrate downward by ingesting soil at depth and regularly depositing it upon the surface. Three major ecological groups have been defined. Epigeic species ~3–9 cm in length, live in or near the surface litter and feed on it and other organically enriched soil layers. Endogeic species, 2–10 cm in length, live within the 0–40 cm soil mineral zone. They feed on soil associated matter, form persistent lateral branching burrows with vertical components and openings to the surface. Anecic species are large bodied earthworms >10 cm in length that burrow up to 1–2 m deep. They live in permanent vertical burrows that extend several meters and feed primarily on fresh surface litter.

In many situations earthworms are the most important animals that live in soils. Worldwide, over 3500 species have been described and divided into 23 families and several genera. Although not numerically dominant, the large size of earthworms makes them one of the main contributors to invertebrate biomass in soils and therefore to bioturbation. They are found in most recent regions of the world except those with extreme (dry, cold) climates. In arid and semiarid areas where vegetation cover is often low, ants and termites play an important role in bioturbation. In a range of land management types the proportion of ant biomass is minor compared with other soil macrofauna. Their abundance and effects depend on climate, soil type, moisture, land cover and management, among others. They make complicated burrows deep into the soil and construct nests by pulling soil particles and carrying loosened particles to another place where they deposit it. Termites also excavate large galleries below ground and construct surface mounds, altering the soil profile. They are considered the tropical analogs of earthworms since they reach large abundances in the tropics and process large amounts of litter. A principle difference between earthworms and termites is that earthworms ingest and transforms much of the soil whereas termites primarily move soil and organic matter into nests and mounds. The feeding and burrowing activities of these invertebrates influence the physical structure of soils, water flux pathways and transport materials. These activities affect the basic patterns of water and chemical movement in soils.

Many vertebrate species of mammals burrow into and through the soil. Most excavate dens that serve as protection when they are not active. These dens can be extensive, and usually stable through time. Rodents are the most important vertebrates with substantial impact on soils, including pocket gophers, kangaroo rats, ground squirrels, and prairie dogs. Gophers occur over much of the North America continent west of the Mississippi River and in the Southeast United States. Their impacts on soil can be profound. Rodents excavate long burrows and place the loose soil on the surface as mounds or deposit it into abandoned burrows. The net effect of gophers is to mix soil vertically and generate irregular soil conditions horizontally.

In the Western United States, American badgers are a major predator of ground squirrels and other burrowing fauna. They produce extensive soil disturbances that initiate notable changes in landscape structure. While preying, badgers enlarge the small squirrel holes producing a large mound at the entrance. These holes become sinks for soils, seeds, and litter. Excavations by animals such as wombats, kangaroo rats, and pocket gophers are important in biomantle evolution and redistribution of materials in soil profiles. In deserts, bioturbation by small mammals may be the most important mechanism for depositing soluble nutrients from deep soil layers onto the surface and may be the only mechanism that brings insoluble materials to the surface. Prairie dogs are an additional example of these mammals; they burrow extensively and develop substantial mounds 0.6 cm wide $\times$ 10 cm deep around their burrow entrances. These abandoned mounds contribute to soil mixing and their burrows serve as litter and seed traps. According to the volume of soil removed by these rodents, they should be recognized as significant controllers of ecosystem processes. A major effect of prairie dogs is to destroy soil structure and therefore aggregate stability through digging and scratching.

13.2.4 AGRICULTURE AND TILLAGE MIXING

Apart from actions and effects of organisms living in soils, human activities also cause mixing in soils. Even though humans do not live in soils, they make use of land for a variety of purposes, especially agriculture. The agricultural preparation of land to receive seeds, known as tillage, involves mechanical actions exerted on soil that modify soil conditions using various combinations of equipment, such as conventional moldboard plough, disk plough, harrow, hoe, and tillers, among others. Species can be altered or destroyed after mixing or complete turn over of the soil when plowing or harrowing. Abundance and prevailing species could also be affected; for example, a field with 17 years of no-till cropping was compared with a conventionally tilled field and found to have from 3.5 to 6.3 times more earthworms (House and Parmalee, 1985). During tillage, earthworms are often brought to the surface, increasing their exposure to predation by birds, desiccation, and mechanical damage. However, the degree of physical damage appears not to be significant for most earthworm species, compared to the effects of tillage on the incorporation of surface crop residue, which may otherwise provide food resources and protective cover. The abundance and diversity of other invertebrates and vertebrates are similarly affected.

13.3 TRANSPORT THEORY

In the previous section some relevant transport features of the bioturbation process were described for macrofauna that occupy the upper layers of both soils and sediments. Different modes of bioturbation are possible and model formulations should reflect the various sediment reworking mechanisms. Models developed for benthic organisms in sediment are quite advanced and described in a recent literature review (Thoms et al., 1995). An overview of those sediment mixing models will be presented in the following section. Although comparable and extensive developments do not exist for soil bioturbation, the similarities in the bioturbation transport mechanisms permit the application of sediment models to describe soil bioturbation.

Models have been developed for the two dominant types of sediment reworking mechanisms observed in nature. They tend to treat chemical transport in sediments as either predominantly diffusive mixing or advection. The random burrowing of organisms over time scales much shorter than that of the observation time leads to rapid exchange of neighboring parcels of sediment and interstitial water within the mixing zone which mimics a diffusive transport process. On the other hand, transport over distances larger than the scale of observation by specific directional activities such as that of head-down conveyor-belt feeders, is primarily advective; that is, the organisms move sediments in preferential directions over measurably large distances. Therefore, a biogenic advection term must be incorporated into the model. Separate models are developed for the diffusive and conveyor-belt species. In addition to particle mixing these models incorporate diffusive mixing of porewaters, degradation of chemicals or radioactive decay of radionuclides, sediment compaction, sedimentation, and solute sediment–water exchange.

The models presented in this section are mathematical descriptions that mimic particle tracer concentration profiles in the upper layers and reflect very little of the underlying biological processes that produced the profiles. However, from the model fitting of the data useful numerical transport parameters are obtained. Bernard Boudreau (1997) notes that; "there is still no fundamental theory that would enable one to predict values of the biodiffusion coefficient, D_B, in a prior manner, and it's likely that there never will be, considering the complexity of the biological processes that are involved." The same can be said of several other model parameters included in the following descriptive models. Thus it is presently impossible to estimate the numerical magnitude of the key transport parameters based on local observables such as organism types and numbers, sediment food and nutrient quality, particle size distribution, water chemistry, water depth, predators, etc. In the absence of a comprehensive theory for bioturbation, it is necessary for the user to understand the key aspects of the biology so as to appreciate its connection to the chemodynamics. What follows is a review of the mathematical models that have been proposed to describe these key aspects of sediment bioturbation as they relate to chemical transport.

13.3.1 Instantaneous Uniform Mixing Model

This box model approach assumes the macrofauna mix uniformly and instantaneously a layer of sediment of thickness, h, from the sediment water interface downward to the depth limit of the bioturbation. A depositional flux input occurs at the sediment–water interface and depositional burial occurs at the bottom of the box. Early modeling work employed this concept (Berger and Heath, 1968) and showed how box modeling could be used to describe the exponential time decay of the concentration of volcanic ash suddenly deposited as a thin layer on to the sediment surface (Berner, 1980). Most mixing rates are not infinitely fast or even fast relative to the frequency of observation and the timescales of chemical decay and other chemical and physical processes often do not lend themselves to a box model description. The governing equation for the completely mixed layer model is a special case of the diffusive–advective models presented in the following paragraphs. The magnitude of the sedimentation rates vs. the biodiffusion mixing rates is examined through the Peclet number. The special case occurs for small Peclet numbers, which reflect relatively high mixing rates to low particle settling rates.

13.3.2 DIFFUSIVE MIXING MODELS

By far the most popular models of sediment mixing are the diffusive mixing models. In many situations, bioturbation in the upper sediment layers appears like a random particle mixing process where complete bed homogenization does not occur. Due to the apparent random movement of the particles by the macrofauna, the process can be considered predominantly dispersive and over long time scales appears as a Fickian diffusion process. In these models, the sediment column is represented as a one-dimensional system undergoing sedimentation and surface sediment mixing. The mixing process is parameterized using an eddy diffusivity or dispersion coefficient (e.g., particle biodiffusion coefficient), which quantifies the magnitude of the particle mixing. The average distance a particle will travel in a given time interval is incorporated into eddy diffusion equations through a diffusion coefficient for either sediment grains or solute molecules in porewater solution. All diffusive mixing models are formulated as an advective–diffusive process in which the diffusion term, D (m^2/s) represents biological mixing of particles (D_B) or solutes (D_S) and the advective term can take on different meanings depending on whether it is sedimentation, bed compaction, or porewater flow.

An equation for the mixing of a tracer associated with the solid phase in a sediment profile is derived from the general diagenic equation (Berner, 1980). For a radioactive tracer with steady-state conditions and where bed porosity changes with depth can be considered negligible, the defining equation is the familiar equation of advective–diffusive transport:

$$D_B \frac{\partial^2 C}{\partial z^2} - V \frac{\partial^z C}{\partial z} - \lambda C = 0, \tag{13.1}$$

where C is the concentration of the tracer per unit volume (g/m^3), λ the decay constant ($1/s$), V (m/s) is advective velocity (bed sedimentation rate), and z (m) is distance into the bed. It was first used by Goldberg and Koide (1962) for a two-layer, diffusive mixing model with constant particle biodiffusion coefficient, D_B (m^2/s), for the surface layer of depth, h (m), and no particle mixing in the lower layer consisting of the remaining sediment profile. Steady-state solutions are useful for estimating D_B

if the depth of penetration of the radionuclide is less than the depth of the mixed layer, which is estimated from the depth of the penetration of ^{137}Cs or ^{210}Pb$_{xs}$, for example. Values of the Peclet number, Pe $(= Vh/D_B) > 3$ allow the tracer pulse to remain within the biologically active layer because mixing levels are low relative to sedimentation. Under these conditions, the key parameters D_B and h can be estimated. This model and similar ones have been widely used to determine biological mixing rates from fitting data on vertical distribution of radiotracers in sediment cores to the appropriate solution of Equation 13.1. Numerical estimates of D_B and h for various tracer concentration profiles will be presented in a later section.

Mathematical models that simulate the exchange of solutes in the presence of macrofauna typically have the same form as Equation 13.1 except that the particle mixing parameter is replaced by an apparent chemical biodiffusion coefficient that represents the influence of the biota. This replacement must accommodate active water exchange through irrigated tube structures in addition to molecular diffusion in porewater and the exchange of solutes between sediment and the overlying water. These and similar diffusive–advective models were first applied to quantifying the fate of particulate opaline silica that settle out of the water column and become the source of enriched concentrations of dissolved silica in sediment porewater surface layers (Schink and Guinasso, 1978). Similar studies involved calcium carbonate dissolution on the ocean floor and the biogeochemical cycling of carbon in bioturbated sediments (Schink and Guinasso, 1977). Schink and Guinasso (1978) also developed a model to simulate the chemodynamics of chemical species that exchange between the solid and porewater phases through adsorption and desorption.

Simplifying the general diagenetic equation (Berner, 1980) for the mixing of a tracer at equilibrium within the solid and aqueous phase in a one-dimensional sediment profile of nonuniform density results in

$$(\varphi + K_D^0)\frac{\partial C_w}{\partial t} = \frac{\partial}{\partial z}[j] + R. \tag{13.2}$$

The flux of the solute within the bed, j (g/m^{2}s), incorporates six effective diffusion and advection terms:

$$j = -(\Phi D_M + D_{bs}K_D^0 + D_{bl})\frac{\partial C_w}{\partial z} + \left(V_w + vK_D^0 - D_{bl}\frac{\partial K_D^0}{\partial Z}\right)C_w, \tag{13.3}$$

where D_M (m^2/s) is the molecular diffusion coefficient of the solute, D_{bs} (m^2/s) is the particle biodiffusion and D_{bl} (m^2/s) is the biological mixing of porewaters, $K_D^0 = K_D$ $(1 - \Phi)$ ρ_S where ρ_S is particle density (kg/L), V_w (m/s) is the advective velocity of porewater movement, v (m/s) is the bed sediment velocity, K_D is the linear equilibrium partition coefficient between the particles adsorbed and the aqueous dissolved phase concentrations, and Φ is bed porosity. The last group in Equation 13.3 is a curious term that is neglected. The R in Equation 13.2 accommodates production or degradation by reactions of the solute in porewater (g/m^3/s).

Application of this model, using realistic values for marine sediments and a range of hypothetical adsorption coefficients, revealed substantial large sediment–water flux of adsorbed compounds in the presence of biological mixing. The authors (Thibodeaux and Bierman, 2003; Erickson et al., 2005) attributed this effect to the tendency of bioturbation to move particles with adsorbed material from depth at high concentration to low concentration at the interface, where some material will desorb, thus increasing the sediment–water flux.

13.3.3   NONLOCAL MIXING BY CONVEYOR-BELT SPECIES

Advective or nonlocal mixing models apply where sediment occupied by macroinvertebrates exchange particles over distances equal to or greater than the scale over which concentrations change substantially or when specific mechanisms of mixing are the focus. Most of the models of advective mixing describe head-down, conveyor-belt deposit feeders. The first models that considered transport analogies other than diffusion for these organisms started with the advective–diffusion formalism except that a vertical velocity term, v (m/s) was included as a function of sediment layer depth and a rate of sediment mass loss from a layer, R (g/m^3/s), due to feeding was included as well (Rice, 1986). The vertical velocity at various depths was assumed to be nonuniform and dependent on the feeding rate and the depth location and distribution of feeding. Data on these parameters is limited, however. Later model developments included a complex feeding function and feeding on the tracer was assumed to be dependent on the fraction of ingestible fines in the sediment (Robbins, 1986). Using this approach, the number of model parameters increased but the model reflected the biological realism that existed in the sediment column. The parameters included the standing crop of deposit feeding organisms and the biomass-specific particle ingestion rate plus the depth of feeding. An additional level of realism was introduced in the modeling of conveyor belt species when the nonlocal biological mixing allowed for the cycling of material from the sediment–water interface both upward and downward (Boudreau, 1986). This movement of material was quantified as a stochastic process in which biodiffusivity was made up of a fluctuating component and a local advective velocity to arrive at integro-differential equations. Parameters were needed for: Ingested/removal, advective and reinjected nearby, for removal and advected to the sediment–water interface, and another for the advection of material from the interface to the sediment, plus specified depths for delivery and removal of material. The resulting model results were tracer profile concentrations reflecting various removal rates and feeding function depths, etc. that were compared to field and laboratory data. The matching of predicted versus data profiles allowed hypothesis testing of the relative importance of various model parameters and provided important insight into how organisms process particles and porewaters in the upper layer of sediment connected to the water column (Boudreau, 2004).

13.3.4   SUMMARY AND CONCLUSIONS

Most of the models used for bioturbation are based on a diffusion analogy, which best describes mixing of materials over short distances or local mixing. These models treat mixing as diffusive and divide sediments into a surface layer, which is mixed at a uniform rate and a lower unmixed layer so as to incorporate a biodiffusion coefficient and mixing depths. These models also accommodate fluid mixing, adsorption, and desorption of solutes and allow chemicals on particle surfaces to exchange with the water column which makes these models useful for many chemodynamic applications. Biodiffusion coefficients for particles and some mixing depths have been compiled for various locations and for various organisms.

Models that simulate nonlocal mixing by conveyor-belt species are available and are used primarily for process hypothesis testing. Those that use advection of particles are the most realistic and many retain a diffusive term as well. Few data are available to establish the nonlocal reworking rates and related feeding depth parameters needed for these models.

It remains true that bioturbation has no overarching theory coupling the biology to the chemistry and physics, resulting in algorithms that allow predictive quantitative estimates of key model parameters. These parameters, which include the density and types of organisms present, biodiffusion coefficients, feeding rates, feeding locations, organism selection of particle size, effective depth of the bioturbed zone, etc. must be measured and/or observed at the specific sites of interest. The next generation of mathematical models needs to better mimic the physiology of the epifauna and infauna that occupy the sediment; this is a daunting task as demonstrated by the models proposed for the conveyor-belt species. However, some progress has occurred. For example, a limited theoretical base exist that relates the functionality for the observed standing crop of biomass or organism size, in number of individuals to per unit sediment area, to the relative magnitude of the biodiffusion coefficient (Reible and Mohanty, 2002; McCall et al., 2010).

13.4 ESTIMATING BIOTURBATION TRANSPORT PARAMETERS

The previous two sections contained a narrative description of mixing and other sorbed-phase transport processes for sediment bioturbation. The purpose of this section is to apply that general information and give specific data and guidance for estimating the chemodynamic mass-transport parameters for bioturbation. A brief introduction and overview of the general flux equation will be presented and includes surficial aquatic sediments bioturbation, surface soil bioturbation, and soil surface plowing. The final subsection will contain guidance for incorporating these processes simultaneously with bioturbation.

13.4.1 Sediment Bioturbation Transport Parameters

Flux equation 13.3 reflects the most significant processes that allow the quantification of chemical mobility in surficial bed sediments. Since some of these processes are covered in other chapters only a subset related to sorbed-phase and bulk phase mixing will be covered here. Equation 13.3 relates the chemodynamics to the chemical concentration in the porewater of the sediment. The aqueous phase is continuous from the porewater surrounding the surfaces of the solid particles, through the sediment–water interface plane and into the overlying water column. Therefore the concentration in the water, C_W (mol/m^3), is the most appropriate variable for quantifying the driving potential for the flux and the related mass transfer coefficients (MTCs). The appropriate form of Equation 13.3 to describe this process is

$$n = -D_{bs}K_D^o \frac{\partial C_w}{\partial z} + vK_D^o C_w - D_{bl} \frac{\partial C_w}{\partial z}, \tag{13.4}$$

where n (mol/m^2 s) is the flux and all other terms were previously defined following Equation 13.3. Absent in the above flux expression is molecular diffusion and advection within the porewater; this is the subject of Chapter 11. It is assumed that K_D^o is not a function of distance, z, into the bed so the last term in Equation 13.3 vanishes. In its final form Equation 13.4 reflects the diffusive and advective transport of chemical species sorbed onto particles and the diffusive transport of bioturbed porewater in the sediment.

13.4.1.1 The Particle Biodiffusion Coefficient

Particle-reactive radioisotopes are commonly used to estimate the values of the bioturbation model parameters in both coastal and freshwaters. Profiles of ^{210}Pb and ^{137}Cs are among the most frequently used; concentration within sediment cores can be used to quantify particle biodiffusion coefficients as well as porewater biodiffision coefficients, sediment accumulation rates, and biological mixing depths. Multiple tracers are necessary to provide evidence of mixing on different time scales. The short lived isotopes ^{234}Th and ^{7}Be reflect very recent changes in benthic systems on time scale of seasons of the year whereas ^{210}Pb and ^{137}Cs are used to determine events over the past 50–100 years. Other materials including chlorophyll-a, lumino-phores (painted fluorescent particles), samarium oxide, and magnetite (Libbers, 2002) have been added to sediment systems to measure bioturbation-related parameters.

It is most common to use mathematical models in extracting the various bioturbation parameters from measured sediment concentration profiles. It is important to note that the values of the bioturbation parameters are model specific, since different models account for mixing in different ways (Wang and Matisoff, 1997; Matisoff and Wang, 1998). Both numerical and analytical solutions to diffusive–advective models such as Equation 13.1 have been used. The comprehensive review performed by Thoms et al. (1995) should be consulted for additional details. In addition the document contains numerous tabularized data on particle biodiffusion coefficients and other bioturbation parameters. These data have been sorted by water body and are displayed graphically in Figure 13.1 as log-normal cumulative probability distributions (CPD) and can be used to estimate the average D_{bs} (CP = 50%). The lower and upper numerical values capture the expected numerical variability in D_{bs}. Hundreds of measured values have been published (Thoms et al., 1995) and tabulated; users may be able to locate site-specific values of D_b in that tabulation.

It is difficult to make generalizations about D_{bs}; it does reflect the intensity of bioturbation in sediments and its value generally decreases with increasing depth. This trend is thought to correlate with increasing "energy costs" of burrowing at depth due to the decreasing sediment water content. It may also reflect the lack of oxygen penetration that will certainly keep some organisms near the surface. Regardless, some exceptions do occur; increasing mixing rates have been observed with depth. D_{bs} varies widely between sites and over time due primarily to the heterogeneous and

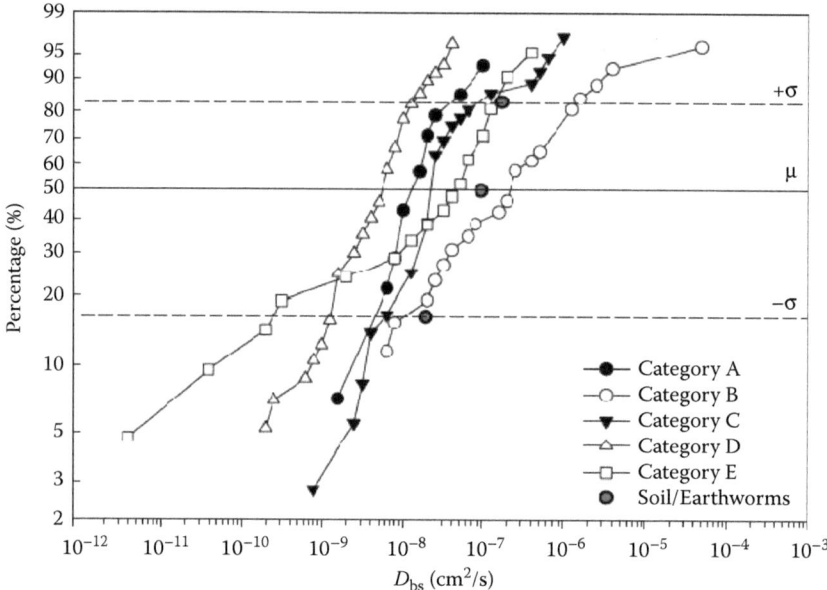

FIGURE 13.1 Sediment particle biodiffusion coefficients. Data for: A = rivers, lakes, tarns, and ponds. B = harbors, bays, havens, sounds, intertidal areas, and beaches. C = bights, shelfs, trenches, slopes, troughs, basins, canyons, and rises. D = oceans, abyssal plains, and plateaus. E = Gulf of Mexico, Mediterranean Sea, and the Great Lakes. Ordinate plotted using Weibull scaling.

dynamic nature of macobenthic communities. Differences in biodiffusion rates may reflect subtle changes in the distribution of biota types. No general trend appears evident when freshwater and marine organisms of similar size and feeding mechanisms are compared. Some differences in reported biodiffusion coefficients may be related to size with larger organisms expected to dominate particle movement (Reible and Mohanty, 2002; McCall et al., in preparation).

13.4.1.2   Particle Mixing Depth

While D_{bs} varies over many orders of magnitude the biological mixing depth, h, is in comparison fairly well constrained. Boudreau (1994) defines the biological mixing depth as the thickness of surficial sediment that is most frequently mixed. The use of the adjective "mixing" for describing this parameter is somewhat unfortunate. It should be interpreted as a slow diffusive-type mixing rather than the rapid and complete-mixed term commonly used in mathematical modeling. The depth of mixing is typically delineated by a perceived break in the slope of the vertical tracer profile and therefore statistical methods are rarely used. However, he found that the worldwide average for deposit-feeding organisms is 9.8 ± 4.5 cm (1-σ error). Biological depths vary considerably and variations may be great between sites in the same water body.

The above results are for mixing depths in marine settings. In freshwater, values of D_{bs} and h are limited; some data exist for lakes but studies in rivers are virtually nonexistent. In general, bioturbation appears to mix freshwater sediments to depths of 10 cm or less. Mixing depth studies do not account for temporal changes or quantitatively characterize the benthic population numbers at the study sites. Such inadequacies make it difficult to extrapolate bioturbation parameters to other conditions (DePinto et al., 2006).

13.4.1.3 Bed Surface Properties

The bed surface properties in Equation 13.4 are very site-specific. Bed surface porosities range from 0.8 to 0.9 at the sediment water interface for clays, 0.36 to 0.46 for well-sorted sands and those for silts are in between (see Section 12.5 in Chapter 12). Particle densities range from 2.0 to 2.5 kg/L. The latter appears in the relationship $K_{DI}^{o} = K_D \rho_s (1 - \varphi)$, which contains the solid-to-water chemical partition coefficient, $K_D(m^3/kg)$, which relates the chemical partitioning between porewater and the particle solids.

13.4.1.4 The Sedimentation Velocity or Burial Velocity

The sedimentation rate (v in Equation 13.4) is the particle advection contribution to the chemical flux. Conversion between the rate of sediment mass flux onto the bed, $n_s(kg/m^2/s)$ and v is simple if we view the latter as the volume of sediment added to each unit of sediment surface per unit time. Thus the flux is $n_s = \rho_s v(1 - \varphi)$ and constant for steady state at each depth, z, whether or not there is compaction. Berner (1980) notes that this advective flow of solids and porewater relative to the sediment–water interface, is primarily due to depositional burial, which ranges from 0.1 to 10 cm per thousand years in deep sea sediments to 0.1–10 cm per year in near-shore fine-grained muds. Lerman (1979) distinguishes between terrigenous and biogenic forms of sedimentation. Terrigenous sedimentation is dominated by the land derived, solid products of coastal weathering deposited as oceanic and lake sediments. Biogenic fractions enter as solution and are formed in the water column through biological production and chemical reaction prior to settling onto the bottom. Boudreau (1997) presents nearly 200 measured values of marine sediment burial velocities, which range from 1.0E–4 to 10 cm/year. Rates in oceans vary from 0.03 to 6.0 cm per 1000 years. In lakes the rate of sedimentation ranges from 0.05 to 10 mm per year with rates in millimeters per year most common.

The mass accumulation rates of minerals, organic debris, nutrients, trace metals, and anthropognic organic chemicals require accurate and meaningful sedimentation rate determinations. Sedimentation plays a key role burying water borne contaminants that arrive at the sediment–water interface. It also retards the release of a sediment source of contaminants by providing a cover of cleaner material (i.e., natural capping). Consequently, measured site-specific sedimentation rates are obviously preferred. However, without measured values it is possible to assume a sedimentation rate. Some suggested default values are: Lakes and reservoirs 1.0 cm/year, coastal environments 0.5 cm/year, continental slopes 0.1 cm/year, and abyssal settings 0.01 cm/year. The marine region estimates are from Reed et al. (2006).

13.4.1.5 The Porewater Biodiffusion Coefficient

The porewater biodiffusion coefficient is defined as D_{bl} to denote by subscript that liquid is being mixed and moved by the benthic macrofauna. This is the last component of the chemical flux in Equation 13.4. Porewater biodiffusion plays an important role in the flux of solutes such as nutrients, carbonate and opaline silica, organic matter, oxygen, total carbon dioxide, and so on. It therefore is expected to be an important factor in transporting the soluble fraction of organic chemicals and heavy metals that partition loosely onto solid particles. The process of translocating material between reaction zones within the bed is termed accelerated diffusion or biologically induced fluid motion caused by bioturbation. The macrobenthos increase sediment porosity by burrow formation, repackaging particles and creating surface roughness. Measured fluxes are significantly greater than those calculated assuming only molecular diffusion (Matisoff et al., 1985) their other reported findings as follows. Porewater concentration profiles of N and P were observed to be nearly uniform due to rapid mixing. The ventilation of burrows by organisms has been observed to increase the flux of dissolved constituents 2.5–3.5 times those in defaunated cores. Burrows in sediments were found to enhance the flux of arsenic species by a factor of 5 relative to control sediments without worms in part due to the irrigation of burrows.

Schink and Guinasso (1978) introduced a term for the biological mixing of porewaters, D_{bl}, as part of the overall effective diffusion coefficient that simulates the exchange of chemicals between the solid and porewater phases. Typically the porewater biodiffusion coefficient is approximated by comparing it to the water molecular diffusion value or to the D_{bs} value. Reported values of D_{bl} are few in comparison to D_{bs} values. Solute biodiffusion has been modeled as vertical and radial diffusion in burrows and burrow walls as well as an excretion and ventilation process. Typically ^{234}Th and ^{7}Be have been used as tracers for D_{bl} measurements although Br^{-} and ^{222}Rn also are often used (Martin and Banta, 1992). The influence of biota has been incorporated by increasing the value of the apparent diffusion coefficient in the bioturbated zone over its calculated molecular diffusion value in the bulk sediments. However, this may be insufficient to account for active exchange of water through irrigated tube structures (Aller, 1980; Boudreau, 1984; Matisoff and Wang, 1998) and alternative models using an advective term and in-bed MTCs have been proposed (Marinelli, 1992).

At this time, the state of the model development for bioturbation driven or influenced soluble transport in porewaters is less developed than the particle phase-sorbed chemical transport counterpart. On the one hand, it can be argued that there should be a direct connection and a quantitative relationship between D_{bs} and D_{bl}. In a theoretical sense, the fluid mixing is more likely a random process such as diffusion rather than an advective process. Being an undirected process, it appears that quantifying it using a scalar parameter such as a diffusion coefficient rather than a vector such as an advective velocity is a better theoretical approach. Only if there is a known Darcian velocity through the bed should the fluid advection term be used as well; see Chapter 11 for details. As the macrobenthos move the solid particles mixing also occurs in the fluid surrounding the particles. Since the fluid is indeed more "fluid" (i.e., less viscous) and the fluid movement transports colloidal-size particles, it seems reasonable that

D_{bl} should be greater than D_{bs}. As a first approximation, it may be assumed that D_{bl} is proportional to D_{bs}. This can be evaluated using previously published data (Table 13.1), which lists values for D_{bl} and D_{bs} for freshwater bioirrigators (*Chironomus* sp., *Hexagenia* spp., and *Coelotanypus* sp.) and a freshwater conveyor-belt deposit feeder (*Branchiura sowerbyi*). In all cases D_{bl} is much greater than D_{bs}. The ratio D_{bl}/D_{bs} ranges from about 150 to almost 9000. While these data are limited and only for a few freshwater organisms, they indicate that as a conservative approximation D_{bl} is at least $100 \times D_{bs}$.

It is clear that the bioirrigators move fluids because of their relatively high values of α the nonlocal exchange coefficient, and their relatively low values of γ, the fraction of sediment that undergoes feeding, which results in nonlocal transport (Table 13.1). Conversely, the conveyor-belt deposit feeder has a relatively low value of α and a relatively high value of γ. This is well illustrated by the α/γ ratio, which is ~ 2 for the conveyor belt deposit feeder but ranges from 4 to 86 for the bioirrigators.

13.4.1.6 The Mass-Transfer Coefficient

Four transport-related parameters are contained in Equation 13.4 and all are involved in the sediment-side bioturbation MTC. Assuming a steady-state chemical flux directed into a bed of thickness h (m) with concentration in water C_{wi} (kg/m^3) and C_{Wh} (kg/m^3) at $z = h$ yields upon integration:

$$n = \frac{D}{h} \left[\frac{Pe \exp(Pe)}{\exp(Pe) - 1} \right] (C_{wi} - C_{wh}), \tag{13.5}$$

TABLE 13.1
Laboratory Determined Values of D_{bl} and D_{bs} for Freshwater Bioirrigators (*Chironomus* sp., *Hexagenia* spp., and *Coelotanypus* sp.) and a Freshwater Conveyor-Belt Deposit Feeder (*Branchiura sowerbyi*)

Model Parameter	*Chironomus* sp.	*Hexagenia* spp.	*Coelotanypus* sp.	*B. sowerbyi*
D_{mol} cm^2/year	36.4[a]	42.6[a]	49.2[a]	49.2[b]
D_{bl} cm^2/year/indiv	—	42.2–89.6[a]	19.9–43.1[a]	45.9–58.7[b]
$\alpha \times 10^{-7}$/s/indiv	9.5–19.0[a]	1.59–7.93[a]	1.59–4.76[a]	2.5[b]
L cm	5–9[a]	5[a]	3[a]	11.7–13.6[b]
D_{bs} cm^2/year/indiv	0.18[c]	0.01[c]	0.04[c]	0.26–0.39[d]
Γ/year/indiv	—	0.29[c]	1.16[c]	3.47–4.55[d]
D_{bl}/D_{bs}	$\sim$202	4220–8960	498–1078	151–158
α/D_{bs} cm^{-2}	166–333	501–2498	125–374	20.2–30.2
α/γ	—	17.3–86.1	4.32–12.9	1.73–2.27

[a] Matisoff and Wang (1998).
[b] Wang and Matisoff (1997).
[c] Matisoff and Wang (2000).
[d] Matisoff et al. (1999).

where n is the chemical flux (kg/m^2 s), $D = D_{bs}K_D(1 - \varphi)\rho_s + D_{bl}$ and Pe $= vK_Dh/D$. The first definition sums the effective particle and porewater biodiffusion coefficients since both processes operate simultaneously. The second definition is the Peclet number denoted as Pe; it compares particle sedimentation and bioturbation rates. In developing Equation 13.5 it was assumed that $C_{wi} >> C_{wh}$, that is, that there is a substantial difference in concentration across the particle mixed depth, h. In the case where the flux is outward from the bed Equation 13.5 applies as well. In this case, v and Pe are negative numbers. The term in brackets in Equation 13.5 is a part of the MTC, which is termed the bioturbation MTC, k_b (m/s). It is defined as

$$k_b \equiv \frac{D}{h}\left[\frac{Pe\,\exp(Pe)}{\exp(Pe) - 1}\right], \qquad (13.6)$$

and used with the above porewater concentration difference. As suggested by Equation 13.4 diffusion and advection are not stand-alone mechanisms, however some simplifications of Equation 13.6 are possible. In the absence of sedimentation $v = 0$ and Pe $= 0$; Equation 13.6 reduces to the purely biodiffusion form of the MTC:

$$k_b(P_e = 0) = \frac{D_{bs}K_D(1 - \phi)\rho_s}{h} + \frac{D_{bl}}{h}. \qquad (13.7)$$

For Pe $= 1$, the sedimentation transport process is equal in direction and magnitude to the diffusion driven transport process and the term in brackets in Equation 13.6 $= 1.58$ so the chemical flux into the bed MTC is 58% larger that represented by Equation 13.7. At Pe $= 1.6$, the term in brackets $= 2.00$ and the advective sedimentation rate is $2 \times$ the purely diffusive rate. At higher values of Pe, sedimentation increasingly dominates the numerical value of the MTC; at Pe $= 9$, it is 90% of the MT process in the bed. Under this limiting condition

$$k_b(Pe > 9) \cong vK_D(1 - \varphi)\rho_s. \qquad (13.8)$$

For chemical transport from the bed into the water column the sedimentation rate is counter to diffusion and Pe is negative. At Pe $= -1.25$, the term in brackets $= 0.50$ and the purely diffusive transport process represented by Equation 13.7 is reduced by half. With larger negative values of Pe, the diffusive transport process is increasingly diminished; at Pe $= -3.45$ the term in brackets $= 0.113$ and the purely diffusive MTC is reduced to 11.3% of its original value. A graphical presentation of [Pe exp(Pe)/(exp(Pe) $-$ 1)] vs Pe appears in Figure 11.2 on page 306. The graph illustrates the effect of sedimentation on the diffusion transport process. The right side of the figure shows how sedimentation enhances chemical diffusion into the bed and the left side of the figure shows how it retards diffusion occurring from the bed to the water column.

From the above it is clear that four basic transport parameters need to be specified in order to estimate the sediment bed MTC: D_{bs}, D_{bl}, h, and v, and information and data were presented above in this section to aid users in selecting appropriate values of each. However, without a theory as a guide and the limited data base available from which to obtain site-specific values, plus the considerable variability inherent in biological processes some users must nevertheless select numerical values from what's available. In this case, only suggested generic default values are available to users. The following is recommended as a crude approximation: $h = 10$ cm; select D_{bs} and v based on waterbody type (i.e., lake, coastal, etc.) and assume $D_{bl} = 100 D_{bs}$.

A complete understanding of the factors that regulate the magnitude of activity of bed sediment macrofauna is not available. However, some theory and evidence exist that allows the extrapolation of site-specific D_{bs} data to other locales or conditions. In the case of temperature, it is safe to assume the "standard" correction $f_T = 1E[(T - 20°)/33]$ where T is °C. This predicts a $2 \times$ increase in D_{bs} for each 10°C rise in temperature. In the case of carbon food source in the bed the D_{bs} vs $v^{0.6}$ burial velocity empirical correlation (Boudreau, 1997) may be used to adjust between site conditions. This relationship assumes the energy available to sustain the macrofauna community is derived from material settling from the water column onto the bed surface. In the case of corrections for population density, the Levy flight-random walk theoretical model (Reible and Mohanty, 2002) result of $D_{bs} \sim \sqrt{n}$, where n is the number of organism per unit area, may be used. Alternatively, D_b and γ may be estimated from the size of the benthos, $D_{bs} = 5.57E - 7 * (\text{length}^{4.74})$ where length is in cm and for the feeding rate, $\gamma = 0.000287 * (\text{length}^{2.95})$ (McCall et al., 2010). In addition, the theory has a degree of empirical data support (McCall and Tevesz, 1982). Many other and possibly more important factors such as species type, freshwater vs marine water (i.e., salinity), prey–predator, etc. need to be developed into corrections to estimate or adjust D_{bs} and D_{bl} values between sites.

13.4.2 SOIL BIOTURBATION TRANSPORT PARAMETERS

The following flux expression characterizes the dominant processes of chemical transport in surface soils of particle density, ρ_s (kg/L); it is similar in structure to the sediment bioturbation formulation, Equation 13.3:

$$n = -\left[D_{ma} \frac{\partial C_a}{\partial Z} + D_{mw} \frac{\partial C_w}{\partial Z} + (1 - \varphi_a - \varphi_w) D_{bs} \frac{\partial (W_A \rho_s)}{\partial Z} \right] + v_a C_a + v_w C_w, \tag{13.9}$$

where C_a and C_w are the air, and water concentrations (mg/m^3), and W_A the loading ratio on soil (mg/kg) dry soil, respectively. The three terms in brackets account for molecular diffusion, D_{ma}, within the gas (i.e., air) phase, molecular diffusion, D_{mw}, within the aqueous phase and the solid sorbed-phase biodiffusion coefficient D_{bs}. The ϕ_a and ϕ_w are volume fractions air and water in the soil and lastly, the advective flux components account for gas-phase and liquid water Darcian velocity (i.e., v_a and v_w) driven fluxes, respectively. This is a convenient choice since the gas phase is continuous from the atmosphere downward through the air–soil interface and into the upper porous layers of the surface soils. Because of the interphase chemodynamic transport considerations all chemical concentrations need to be expressed as equivalent gas phase concentrations, C_a (mg/m^3). Assuming equilibrium-phase partitioning

Equation 13.9 can be transformed to

$$n = -\left[D_{ma} + \frac{1}{H}D_{mw} + (1 - \varphi_a - \varphi_w)\frac{\rho_s K_d}{H}D_{bs} \right]\frac{dC_a}{dz} + \left(v_a + \frac{v_w}{H} \right)C_a,$$

(13.10)

where H is the Henry's law constant. The above result contains several terms that are of minor significance in comparison to others. This typically includes the terms containing the aqueous molecular diffusivity and gas Darcian velocity.

Traditionally the upper layer soil-phase chemodynamic processes have been the concern of pesticide chemists and chemical engineers involved with the land application and disposal of hazardous waste, respectively. The vapor-phase diffusive transport is the dominant pathway of chemical movement from the soil to the air for these volatile species. Due to its low numerical value the aqueous diffusivity contribution is usually negligible. Air movement through surface soils without the presence of a huge subterranean void space to be alternative pressurized and de-pressurized by the atmosphere is a very limited process. However, water infiltration can result in significant downward leaching and the capillary rise of liquid water in fine grained media (rising damp) can transport chemical quantities at depth upward into the surface layers. Estimates of positive and negative advective porewater velocities will be needed to quantify the significance of this process at specific sites, and they may be weather and season dependent. Nevertheless, the focus here is on the bioturbation driven sorbed phase flux and its key transport parameters.

13.4.2.1 The Surface Soil Particle Biodiffusion Coefficient

Charles Darwin in 1881 reported that by ingesting soil at depth and depositing it upon the surface earthworms caused surface objects to migrate downward. Soil turnover rates, v' in cm/year, and depths of activity h, in cm, are the measurements typically taken on cast production by earthworms. A random particle displacement concept applied over time as the conventional definition of a diffusion coefficient allows using the two measurements to yield a biodiffusion coefficient, $D_{bs} = v'h$ (cm^2/year). This approach was first used by Mclachlan et al. (2002); the results that appear here are those of Rodriguez (2006), who extended the earlier work. Unlike sediment bioturbation where chemical tracers and Fickian diffusion-type mathematical models, such as Equation 13.1, are used with chemical profile data to yield D_{bs} and h values directly, no such approaches have been used for estimating surface soil bioturbation parameters.

Rodriguez (2006) was able to locate numerous sets of data on both the soil turnover rates and depth of soil activity/disturbance for several types of macrofauna. These were converted to particle biodiffusion coefficients and summaries of the data appear graphically in Figure 13.2. Shown in this figure are cumulative probability distributions of this coefficient for earthworms, ants and termites and vertebrates. Included within the earthworm particle data set are four sorbed-phase chemical data points. These D_{bs} data are for PBCs, which were extracted from concentration profiles using a Fickian diffusion model. The PCB profiles were obtained in soils with abundant earthworm populations. The reader should note that the range of D_{bs} values for soils is within those for sediments; see Figure 13.1 in comparison.

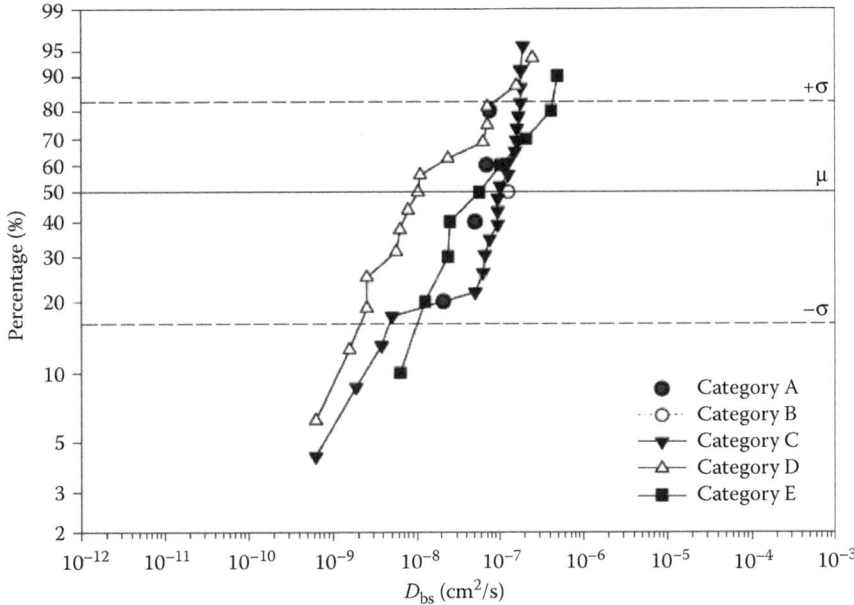

FIGURE 13.2 Soil biodiffusion coefficients. Data for: A = earthworms—PCB-contaminated soils. B = earthworms—forests soils. C = earthworms—grasslands and pastures. D = ants and termites. E = vertebrates. Ordinate plotted using Weibull scaling.

13.4.2.2 The Surface Soil Particle "Mixing" Depth

Published averages and ranges of soil macrofauna occupation depths were obtained for three types of earthworms, ants, termites, and four types of vertebrates (Rodriguez, 2006). The depths reported for each were characterized as "definite" or "likely." The former designation refers to the penetration depth below the surface observed in all the reports studied. The average definite depths for nine species appear in Table 13.2a. It is assumed that these data also represent the mixing length for biodiffusion.

The definite mixing depths for all species were combined and a cumulative probability vs depth graph produced. The data displayed a normal distribution and the result appears in Table 13.2b. Based on this limited set of data the average mixing depth is 20 cm; this is to be compared with 9.8 cm for sediment macrofauna. However, the depths in Table 13.2 reflect organism occupancy and may overestimate the effective biodiffusion "mixing" depths. In addition, it is important to note that the entire surface isn't bioturbated at the same time, that is, the burrows are localized and distributed over the landscape and that only over long periods of time is the entire surface bioturbated. This is particularly true for the larger, less populous animals. Technically, this is also true in the aqueous environment, but those benthos tend to be much smaller and with much higher population densities, so that the distances between perturbed sediment is small with respect to the sample size. However, that may not be so in all cases, especially for organisms that have lower population densities. Clearly more

TABLE 13.2
Surface Mixing Depths for Nine Types of Soil Macrofauna

(a) Species	Average Depth (cm)	(b) CP^a (%)	Definite Depth (cm)
Earthworms		90	1
Epigeic	2	83	5
Endogeic	8	74	10
Anecic	20	50	20
Ants	25	23	30
Termites	25	15	35
Vertebrates			
Kangaroo	10		
Squirrels	12		
Badgers	20		
Gophers	22		

^{a} CP = cumulative probability all species.

research and data is needed so as to better characterize both soil surface bioturbation transport parameters D_{bs} and h.

13.4.2.3 The Soil-Side MTC

Selecting only the third term from Equation 13.10 and integrating it across z yields the bioturbation sorbed-phase flux n_{bs} (mg/s/m^2):

$$n_{bs} = - \left\{ \left[(1 - \varphi_a - \varphi_w) \frac{\rho_s K_d}{H} \right] \frac{D_{bs}}{h} \right\} (C_{ai} - C_{ah}) \qquad (13.11)$$

The effective MTC is the group of terms within the braces in Equation 13.11. The concentration differences across the mixed surface layer of thickness h (m), is the interface vapor-phase concentration, C_{ai} (mg/m^3), minus the vapor-phase concentration at $z = h$, C_{ah} (mg/m^3). In addition to the kinetic transport parameter D_{bs}/h, the MTC contains the thermodynamic parameter ratio K_d/H, which imparts the sorbed-phase chemical loading fraction characteristic of the solid particles. When mobilized by the macrofauna this fraction significantly enhances the magnitude of the MTC. For certain strongly sorbed chemicals the coefficient can be very large so that this mechanism dominates the rate of chemical movement from within the soil layers to the soil-air interface. See Example 13.6.1 below for numerical verification of this behavior.

13.4.3 PLOWING ENHANCEMENT TO SOIL BIOTURBATION TRANSPORT

A narrative description of plowing was presented above in Section 13.4.3. Unlike bioturbation, which is virtually continuous due to the large number of events and the rapid frequencies of event occurrence, plowing is a single annual or biannual event. Between the plowing events the soil undergoes the normal bioturbation process although the intensities may be lower because tilling appears to reduce the earthworm populations slightly (House and Parmake, 1985). However, the incorporation of crop residues within the soil column may provide food resources and protective cover resulting in slight increases in populations and an increase in D_{bs} (El Titi, 2003).

Plowing events are best viewed as a chemodynamic process that instantly mixes the soil column over the bioturbated (plowing) depth h and establishes a uniform chemical concentration throughout. A transient diffusion period is initiated that is well modeled by the semi-infinite slab solution to Fick's second law (see Section 2.7.1). The outward chemical flux within the soil column can be closely modeled by

$$n = \sqrt{D/\pi t}(C_{ah} - C_a), \tag{13.12}$$

where $C_a(mg/m^3)$ is the concentration in air within the constant flux atmospheric layer, t (seconds) is the lapsed time since the plowing event occurred, and D represents the three terms appearing in the brackets in Equation 13.10. Although all occur simultaneously only biodiffusion is considered here.

The MTC portion of Equation 13.12 takes on large numerical values but very quickly decreases in magnitude with increasing time t after the plowing event. For a time period, the effect of plowing is to enhance the soil bioturbation-driven process above the quasi-steady-state value represented by the MTC in the braces in Equation 13.11. The time-period, t_e (seconds), of this enhancement is obtained by equating the coefficients and solving for time:

$$t_e = Hh^2/\pi(1 - \varphi_a - \varphi_w)\rho_s K_d D_{bs}. \tag{13.13}$$

Because of the large numerical values of the group of terms $\rho_s K_d/H$ in Equation 13.13 for the sorbed-phase class of chemicals, t_e is of short duration and typically a fraction of a year. For volatile chemicals, t_e is much larger. Nevertheless, t_e serves to apportion the time lapsed between plowing events into two MTCs. For the plowing enhanced time-period, which is $t = 0$ to t_e, the average MTC is obtained from integrating Equation 13.12 over the period to obtain

$$k_{bs} = 2\sqrt{(1 - \varphi_a - \varphi_w)\rho_s K_d D_{bs}/H\pi t_e}. \tag{13.14}$$

For the remainder of time until the next plowing event the MTC is

$$k_{bs} = (1 - \varphi_a - \varphi_w)\rho_s K_d D_{bs}/Hh. \tag{13.15}$$

With these two relationships for estimating MTCs available the effective value for the period between events is the time fraction ratio summation. If t_e from Equation 13.11 is greater than the time period between plowing events, t_p (seconds), the latter is used in Equation 13.14 to estimate the plowing enhanced MTC; Equation 13.15 is omitted.

The above development is not restricted in application to plowing *per se* nor to sorbed-phased chemicals. Numerous chemodynamic situations arise in which the soil surface layers undergo a mixing or surface renewal process similar to plowing. These include the agricultural practice of applying pesticides onto and into the upper soil column. Later plowing remixes the column. Solid waste is periodically applied as new layers onto the surface of the ground in the case of land filling or land farming operations and similar transport models have been developed (Thibodeaux, 1996). The disposal of material dredged from the sediment beds of rivers, lakes, harbors, and so on. is also deposited upon the land and the chemicals undergo similar transport process. From perspectives of both historical and geological timescales and geographic scales human activities continuously bioturbate the landscape so that the above model is applicable in these contexts as well. Equations 13.14 and 13.15 are appropriate to use but the two additional diffusion terms appearing in Equation 13.10 need to be included for completeness.

13.4.4 SIMULTANEOUS PARALLEL AND SERIES TRANSPORT PROCESSES

In addition to particle bioturbation chemical transport other possibly significant simultaneous transport processes should be explored. Equation 13.4 contains the three most appropriate to the subject matter of this chapter. For sediments Equation 13.3 contains two additional bed-side processes that do become significant for highly soluble chemicals and colloidal matter. These are molecular or Brownian diffusion and advection in the porewater. They occur simultaneously with sedimentation processes and should be included using the "in-parallel" procedures for combining transport coefficients; see Chapter 4. Also see Chapter 13 on the subject of advective transport in bed sediment. Equation 13.10 contains similar terms for processes in surface soils, where it is well known that the gas-phase molecular diffusion is a significant process for volatile and many semi-volatile chemicals including pesticides. The water movement within the soil column, characterized by v_w, can be a significant process for soluble chemicals transport. The "in-parallel" procedures of combining the transport coefficients should be used here as well.

It is a common misconception to assume that the fluid-side of these interfaces provide little or no resistance to the overall chemical transport between sediments and water column or soils and the atmospheric boundary layers. In fact, the significance of the fluid side resistance has been found to be the most significant factor in the solubilization of high molecular weight PCB congeners from bed-sediment undergoing bioturbation in the Hudson River, New York (Erickson et al., 2005). The bed-side particle biodiffusion is very rapid, this makes the sorbed-phase chemicals is readily available at the interface where they desorb rapidly so that the rate of their transport through the water-side boundary layer becomes the slowest step in the overall chemodynamic process. The "in-series" procedures presented in Chapter 4 provide the algorithms for combining the sediment-side and fluid-side transport coefficients to account for both resistances. The air-side resistance across the air–surface soil boundary can become the controlling one as demonstrated in Example 13.6.2. This example also illustrates how multiple processes need to be combined in order to yield an overall MTC for chemicals moving across the air–soil interface.

13.5 A SUMMARY GUIDE FOR USERS

Bed-sediment particle biodiffusion coefficients, D_{bs}, for several aquatic bodies appear graphically in Figure 13.1. From these the user may choose the average values using the 50% probability and a range of values based on ± 1 σ. These values of D_{bs} are selected from the 83.3% and 16.7% probability points on the y-axis. Values of the porewater biodiffusion coefficient, D_{bl}, are based on the estimate derived above in Section 13.4.1.5 as $100 \times$ the D_{bs} values. These two transport parameters may be converted to MTC equivalents by dividing each by a mixing length of $h = 10$ cm. Bed sedimentation rates, v, appropriate as first estimates for aquatic bodies are: Lakes and reservoirs 1.0, coastal 0.5, continental slope 0.1, and abyssal 0.01 cm/year. Literature review compilations of both D_{bs} and h are available; see Thoms et al. (1995). With luck site specific values of D_{bs} and h, such as for the Great Lakes, for example, may be obtained from the literature.

The MTC representing the combined diffusive and advective (i.e., sedimentation rate) process k_b must be estimated using Equations 13.6 and 13.7 by accounting for the Peclet number.

The soil surface layers undergo a similar bioturbation process although the macro-fauna responsible are drastically different. Values of D_{bs} are graphed in Figure 13.2. Averages and ranges (i.e., ± 1 σ) may be obtained for three general classes of macro fauna. The particle MTCs are estimated using the D_{bs} to h ratio. Appropriate esti-mates of h values for earthworms, ants and termites, and vertebrates are: 15, 25, and 16 cm, respectively. Equation 13.11 may be used to estimate the chemical specific MTC. The soil bioturbation transport parameters are derived from very few studies, they represent soil mass turnover rates and observed macrofauna occupation depths. Chemical-specific data are woefully lacking.

The user must be aware that the key mass-transport parameters in this chapter depend directly on the presence of living biological forms, which is a requirement very unlike those in all the other chapters. In this regard, the user needs to verify that the sites for MTC application do contain the necessary conditions of food, nutrients, temperature, habitat, etc. to adequately sustain appropriate numbers of organisms and the chemical levels in the soil or sediment are below toxic or debilitating effects.

13.6 EXAMPLE CALCULATIONS

13.6.1 Baltic Sea Sediment Bioturbation MTC

The Baltic, much like the Mediterranean and Black seas, is a large nearly land-locked waterbody with little power to flush itself quickly and thus prone to accumulate pol-lutants entering via European interior rivers, streams, and canals. Pollutants including PCBs entering the Stockholm Archipelago before the 1970 ban are presently lying in "laminated" sediments under hypoxic bottom waters. Reoxygenization of the waters is enabling the bioturbating polychaeta *Marenzelleria viridis* to recolonize the bottom at a rate of 30–480 km/year (Granberg et al., 2008). Estimate (1) the bioturbation MTC (cm/d) and its enhancement over a presumed molecular diffusion MTC and (2) the attenuation effect that land-derived particles may have on the bioturbation MTC for 2,2′,4,4′,5,5′-hexachlorobenzene with sediment-to-water partition 35,300 L/kg and diffusivity in H_2O of 4.31E–6 cm^2/s.

1. Without average basin-wide data the following approximate values will be used: Bed porosity 85% particle density 2.5 kg/L and bioturbated depth

$10\,\mathrm{cm}\,D_m\varepsilon^{4/3}/h = 0.030\,\mathrm{cm/d}$. The maximum and average D_{bs} reported for the Mediterranean (Thoms et al., 1995), the nearest equivalent waterbody, yield particle biodiffusion coefficients of $5.7E{-}3\,\mathrm{cm^2/d}$ (max.) and $3.15E{-}4\,\mathrm{cm^2/d}$ (ave). From Equation 13.7 this yields chemical bioturbation MTCs of

$$k_b = \frac{D_{bs}}{h}\rho_3(1-\varphi)K_d = (5.70E-4)2.5(.15)35,300 = 7.5\,\mathrm{cm/d},$$

(maximum) and 0.42 cm/d (average). These are factors of 14–250 times larger than molecular diffusion in porewater. By comparison, reported laboratory measured rates with *M. viridis* range from 6.3 to 148 cm/d (Granberg et al., 2007).

2. Assuming the coastal Baltic can be categorized as continental slope, its sedimentation rate is 0.1 cm/year and the Pe = $(-0.1\,\mathrm{cm/year})$ $(10\,\mathrm{cm})/$ $(5.70E{-}4)365\,\mathrm{cm/year} = -4.81$ and term in brackets in Equation 13.6 equals 0.0396, which reduces the max MTC to 0.30 cm/d. For comparison, an "abyssal" sedimentation rate 0.01 cm/year reduces the MTC to 5.9 cm/d. For the average MTC the values are 6.1E–4 and 0.26 cm/d when using the slope and abyssal sedimentation rates. Clearly, burial of the sediment by arriving clean soil particles from the land will help attenuate the solubilization rate of PCBs from the bioturbed bed.

13.6.2 Chemical Role and the Soil Bioturbation MTC

The three phases present in surface soils (i.e., air, water, and solids) accentuate the role of the sorbed-phase chemistry of the individual substances. The group of parameters $(\rho_B K_d/H)$ regulates the magnitude of the MTCs when bioturbation is one of the active processes as seen next in Equation 13.16. (1) Describe the relationship and discuss what

$$K_a' = \frac{1}{\{1/k_a + 1/[D_{ma}/h + (\rho_B K_d/H)D_{bs}/h]\}}, \tag{13.16}$$

it represents. (2) Using the chemical data in Table 13.3 estimate the K_a's (cm/d) for surface soils bioturbated by earthworms. In doing so, fill in the columns beginning with 3, 4, and 5 in the table. Use 4800 cm/d for the air-side MTC. Even though there are slight differences in the air diffusivities for chemicals, use $D_{ma} = 1300\,\mathrm{cm^2/d}$ for all chemicals in the table. (c) When done with the calculations inspect the numerical values of K_a's and identify the likely processes controlling the magnitudes of these MTCs.

Solution:

1. Equation 13.16 above is a relationship for the overall MTC, K_a', in terms of the three most active transport processes in the interface region. The k_a is the air-side MTC; D_{ma}/h is that resulting from molecular diffusion in air-filled soil pore spaces and D_{bs}/h is the particle biodiffusion MTC due to earthworm

TABLE 13.3
Soil Bioturbation MTCs (cm/d)

Chemical	$(\rho_B K_d/H)$	$(\rho_B K_d/H)D_{bs}/h$	[·]	K_a
EPS[a]	1.3	6.2E–6	65.0	64.1
diBrClPropane	50	0.024	65.0	64.2
Toxaphene	292	0.14	65.1	64.3
Lindane	3.3E5	158	232	213
Mirex	1E7	4800	4870	2420
2,3,7,8-TCDD	4.7E10	2.3E7	2.3E7	4800

[a] A hypothetical equilibrium-phase partitioning substance.

activities. The latter two processes occur simultaneously and are summed in Equation 13.16.

2. From Figure 13.2 $D_{bs} = 3.5\,\text{cm}^2/\text{year}$ at the 50% probability and $h = 20\,\text{cm}$ from Table 13.2. This yields a particle biodiffusion MTC $= 4.8\text{E}{-}4\,\text{cm/d}$. For the air porespaces $D_{ma}/h = 1300/20 = 65\,\text{cm/d}$. The various MTCs and the group $\rho_B K_d/H$ is entered into Equation 13.16 to give

$$K_a = 1/\{1/4800 + 1/[65 + (\rho_B K_d/H)4.8\text{E} - 4]\}.$$

Column 2 of Table 13.3 contains numerical values for the group $\rho_B K_d/H$, which represents the equilibrium, sorbed-phase-to-air chemical partition coefficients. It converts D_{bs}/h to the individual chemical bioturbation MTCs, which are listed in Column 3. The sum of the air pore space diffusion and chemical biodiffusion is the bracketed terms in Equation 13.16; it appears in column 4. Column 5 is the overall MTC.

3. Using Equation 13.16a the following controlling processes are identified. The molecular diffusion in the air-filled pores of the soil has an MTC $= 65\,\text{cm/d}$ and controls the soil-side transport and provides a larger resistance than $1/k_a$. So, it determines the magnitude of the k_a for the first three on the list. For lindane and mirex the chemical biodiffusion overpowers the soil-side transport MTC with the value for mirex approaching, numerically half the air-side MTC, k_a. For TCDD the soil-side chemical bioturbation MTC dominates this coefficient. Its resistance is therefore much smaller than that in the air-side boundary layer so that the magnitude of k_a regulates the overall MTC to 4800 cm/d. Besides illustrating the calculation procedure to obtain the bioturbation MTC, this example demonstrates the role that chemical partitioning between the air, water, and soil microphase has on regulating the magnitude of the overall MTC. Furthermore it illustrates that for highly sorbed (i.e., hydrophobic) chemicals such as TCDD the air-side MTC controls the kinetics of the vaporization process. Although the MTC is very large numerically the effective partial pressure is reduced due to its hydrophobicity and therefore the net result is a low flux to air.

ACKNOWLEDGMENT

The authors kindly acknowledge the word processing of this chapter by Danny Fontenot and Darla Dao. The figures were prepared by Justin Birdwell.

LITERATURE CITED

Aller, R.C. 1980. Quantifying solute distributions in the bioturbated zone of marine sediments by defining an average microenvironment. *Geochem. Cosmochem. Acta* 44: 1955–1965.

Berger, W.H. and G.R. Heath. 1968. Vertical mixing in pelagic sediments. *J. Mar. Res.* 26: 134–143.

Berner, R.A. 1980. *Early Diagenesis*. Princeton University Press, Princeton, NJ, pp. 43–47.

Boudreau, B.P. 1984. On the equivalence of nonlocal and radial-diffusion models for porewater irrigation. *J. Mar. Res.* 42: 731–735.

Boudreau, B.P. 1986. Mathematics of traces mixing in sediments. II. Nonlocal mixing and biological conveyor-belt phenomena. *Amer. Jour. Sci.* 286: 199–238.

Boudreau, B.P. 1994. Is burial velocity a master parameter for bioturbation? *Geochim. Cosmochim. Acta* 58(4): 1243–1249.

Boudreau, B.P. 1997. *Diagenetic Models and Their Implementation*. Springer-Verlag, Berlin, pp. 138, 141.

Boudreau, B.P. and B.B. Jorgensen. 2001. *The Benthic Boundary Layer-Transport Processes*. Oxford University Press, Oxford, UK.

Boudreau, B.P. 2004. What controls the mixed-layer depth in deep-sea sediments? The importance of POC flux. COMMENT. *Limnology Oceanogr.* 49, 620–622.

DePinto, J.V., D. Rucinski, N. Barabas, S. Hinz, R. McColloch, T.J. Dekker, B. Velente, H. Holmberg, J. Kaur, and J. Wolfe. 2006. Review and evaluation of significant and uncertainty of sediment-water exchange processes at contaminated sediment sites. Appendix E. Limno-Tech., Inc., Ann Arbor, MI.

Erickson, M.J., C.L. Turner, and L.J. Thibodeaux. 2005. Field observation and modeling of dissolved fraction sediment-water exchange coefficients for PCBs in the Hudson River. *Environ. Sci. Technol.* 39: 544–556.

El Titi, A. 2003. *Soil Tillage in Agroecosystems*. CRC Press, Boca Raton, FL, pp. 2–14, 148–151, 230–245.

Gabet, E.J., J. Reichman, and E.W. Seabloom. 2003. The effects of bioturbation on soil processes and sediment transport. *Ann. Rev. Earth Planet Sci.* 31: 249–73.

Goldberg, E.D. and M. Koide. 1962. Geochronological studies of deep sea sediments by the ionium/thorium method. *Geochem. Cosmochem. Acta* 26: 417–450.

Govers, G., T.A. Quine, P.J.J. Desmet, and D.E. Walling. 1996. The relative contribution of soil tillage and overland flow erosion to soil redistribution on agricultural land. *Earth Surf. Process. Landforms* 21: 929–946.

Granberg, M.E., J.S. Gunnarson, J.E. Hedman, R. Rosenberg, and P. Jonsson. 2008. Bioturbation-driven release of organic contaminants from Baltic Sea sediments by the invading polychaete *M. varidis*. *Environ. Sci. Technol.* 42, 1058–1065.

House, G.J. and R.W. Parmake. 1985. Comparison of soil anthropods and earthworms from conventional and no-tillage agroecosystems. *Soil Tillage Res.* 5: 351–360.

Johnson, D.L. 1990. Biomantle evolution and the redistribution of earth materials and artifacts. *Soil Sci.* 149(2): 84–102.

Lerman, A. 1979. *Geochemical Processes—Water and Sediment Environments*. Wiley, New York.

Libbers, P.D. 2002. Development of a magnetite tracer protocol for seasonal measurements of bed sediment biodiffusion coefficients. M.S. Thesis, Louisiana State University, Baton Rouge, LA, 84pp.

Lobb, D.A. and R. G. Kachanoski. 1999a. Modeling tillage erosion in the topographically complex landscapes of southwestern Ontario, Canada. *Soil Tillage Res.* 51: 261–277.

Lobb, D.A. and R. G. Kachanoski. 1999b. Modeling tillage translocation using step, linear-plateau and exponential functions. *Soil Tillage Res.* 51: 317–330.

Marinelli, R.L. 1992. Effects of polychaetes on silica dynamics and fluxes in sediments: Importance of species, animal activity and polychaete effects on benthic diatoms. *J. Mar. Res.* 50: 745.

Martin, W.R. and G.T. Banta. 1992. The measurement of sediment irrigation rates: A comparison of the Br- tracer and ^{222}Rn/^{226}Ra disequilibrium techniques. *J. Mar. Res.* 50: 125–154.

Matisoff, G. 1982. Mathematical models of bioturbation. In McCall, P.L. and M.J.S. Tevesz, eds. *Animal–Sediment Relations, The Biogenic Alteration of Sediment.* Plenum Press, New York.

Matisoff, G. and X. Wang. 1998. Solute transport in sediments by freshwater infaunal bioirrigators. *Limnol. Oceanogr.* 43: 1487–1499.

Matisoff, G. and X. Wang. 2000. Particle mixing by freshwater bioirrigators—Midges (Chironomidae: Diptera) and Mayflies (Ephemeridae: Ephemeroptera). *J. Great Lakes Res.* 26(2): 174–182.

Matisoff, G., J. B. Fisher, and S. Matis. 1985. Effects of benthic macro-invertebrates on the exchange of solutes between sediments and fresh-water. *Hydrobiologia* 122: 19–33.

Matisoff, G., X. Wang, and P.L. McCall. 1999. Biological redistribution of lake sediments by tubificid oligochaetes: *Branchiura sowerbyi* and *Limnodrilus hoffmeisteri/Tubifex tubifex. J. Great Lakes Res.* 25(1): 205–219.

McCall, P.L. and M.J.S. Tevesz, eds. 1982. *Animal–Sediment Relations, The Biogenic Alteration of Sediment.* Plenum Press, New York.

McCall, P.L., G. Matisoff, X. Wang, and J.A. Robbins. 2010. Particle bioturbation by the marine bivalve *Yoldia limatula* (Say), in press.

Mclachlan, M.S., G. Czub, and F. Wania. 2002. The influence of vertical sorbed phase transport of the fate of organic chemicals in surface soils. *Environ. Sci. Technol.*, 36: 4860–4867.

Reed D.C., K. Huang, B.P. Boudreaux, and F.J.R. Meysman. 2006. Steady-state traced dynamics in a lattice-automation model of bioturbation. *Geochim. Cosmochim. Acta* 70: 5855–5867.

Reible, D. and S. Mohanty. 2002. A Levy flight-random walk model for bioturbation. *Environ. Toxicol. Chem.* 21: 875–881.

Rice, D.L. 1986. Early diagenesis in bioadvective sediments: Relationships between the diagenesis of BET, sediment reworlsing rates, and the abundance of conveyor-belt deposit feeders. *J. Mar. Res.* 44: 149.

Robbins, J.A. 1986. A model for particle-selective transport of tracers in sediments with conveyor belt deposit feeders. *Jour. Geophys. Res.* 91: 8542–8558.

Rodriguez, M. 2006. The Bioturbation Transport of Chemicals in Surface Soils. M.S. Thesis in Chemical Engineering, Middleton Library, LSU, Baton Rouge, LA. 81pp.

Schink, D.R. and Guinasso, N.L., Jr. 1977. Modeling the influence of bioturbation and other processes on carbonate dissolution at the sea floor. In N.R. Anderson and A. Malahoff, eds, *The Fate of Fossil Fuel CO$_2$ in the Oceans*, pp. 375–399.

Schink, D.R. and Guinasso, N.L. 1978. Redistribution of dissolved and adsorbed matarials in abyssal marine sediments undergoing biological stirring, *Am. J. Sci.* 278: 687.

Thibodeaux, L.J. 1996. *Environmental Chemodynamics*, John Wiley, New York, Chapter 6, pp. 408–416.

Thibodeaux, L.J. and V.J. Bierman. 2003. The bioturbation-driven chemical release process. *Environ. Sci. Technol.*, July 1, pp. 253A–258A.

Thoms, S.R., G. Matisoff, P.l. McCall, and X. Wang. 1995. Models for alteration of sediments by benthic organisms. Project 92-NPS-2. Water Environmental Research Foundation, Alexandria, VA.

Wang, X. and G. Matisoff. 1997. Solute transport in sediments by a large freshwater oligochaete, *Branchiura sowerbyi. Environ. Sci. Technol.* 31: 1926–1933.

FURTHER READING

Berner, R.A. 1980. *Early Diagenesis*. Princeton University Press, Princetion, NJ, 241pp.

Boudreau, B.P. 1997. *Diagenetic Models and their Interpretation*. Springer-Verlag, New York, 414pp.

Boudreau, B.P. and D.B. Jorgensen, Editors. 2001. *The Benthic Boundary Layer*. Oxford University Press, New York, 404pp.

McCave, I.N. 1976. *The Benthic Boundary Layer*. Plenum Press, New York, 323pp.

Paton, T.R., G.S. Humphreys, and P.B. Mitchell. 1995. *Soils—A New Global View*. Yale University Press, New Haven and London. See Chapter 3 on Bioturbation.

Thibodeaux, L.J. 1996. *Environmental Chemodynamics*. John Wiley, New York, Chapter 5, 501pp.

Thoms, S., G. Matisoft, P.L. McCall, X. Wang, A. Stoddard, M. Martin, and V.J. Banks. 1995. Models for alteration of sediments by benthic organisms. *Water Environ. Res.* Foundation, Alexandria, VA.

14 Mass Transfer from Soil to Plants

William Doucette, Erik M. Dettenmaier,
Bruce Bugbee, and Donald Mackay

CONTENTS

14.1   Introduction 389
14.2   Root Uptake and Translocation 390
14.3   Plant Bioconcentration Factors 392
      14.3.1   Root Concentration Factors 394
      14.3.2   Plant Tissue Water Sorption Coefficients.......................395
      14.3.3   Transpiration Stream Concentration Factors 396
14.4   Phytovolatilization...398
14.5   Plant Biotransformation or Metabolism 399
14.6   Soil Pore Water Concentrations 400
14.7   Transpiration Rates 401
14.8   Quantifying Mass Transport into Plants 403
14.9   Case Study 1: Sulfolane Removal from Wetland by Cattails 405
14.10 Case Study 2: Uptake into Apple Trees and Fruit........................405
14.11 Conclusion 406
Literature Cited...406

14.1   INTRODUCTION

Quantifying and predicting the mass transfer of organic chemical contaminants into plants is critical for assessing the potential human, wildlife, and ecological health risks associated with the consumption of plants growing in contaminated environments. Understanding the extent of plant uptake is also important for evaluating the use of plants as biomonitors of environmental contamination (Larsen et al., 2008) and predicting the effectiveness of phytoremediation efforts (McCutcheon and Schnoor, 2003).

Organic chemicals can directly reach above ground plant tissues by vapor and particle deposition as discussed in Chapter 7, or they many reach below ground tissues by way of the roots, the emphasis of this chapter. Atmospheric deposition pathways to above ground plant components are likely to be more important for hydrophobic and less volatile compounds, while the root sorption and uptake pathway is generally more important for more water-soluble or less hydrophobic compounds. The relative importance of these two routes depends on the chemical's partitioning properties, on plant physiology and on the relative contamination levels of the soil and the atmosphere. If the edible portion of the plant is below ground (e.g., carrots, potatoes, sugar beets) then diffusion of the contaminant from the soil vapor or solution is likely the dominant mechanism for uptake of, and exposure to, these compounds. However, in this case, very hydrophobic compounds may be of less concern due to their strong sorption to soils, low soil pore water concentrations, and slow diffusion into the plant in comparison to the rate of plant growth.

Plant uptake of chemicals by roots and the subsequent translocation or transport into above ground tissues is generally divided into *passive* and *active* mechanisms. Passive transport requires no energy expenditure by the plant and is driven by diffusive and advective processes, while active transport uses proton pumps and redox gradients to move charged substrates (primarily N, P, and K) across membranes (Marschner, 1995). For passive transport, the greater the amount of water transpired, the greater amount of organic contaminants that moves into the plant. This may not, however, increase contaminant concentrations in the above ground plant tissues over time because of losses due to metabolism and volatilization. Factors such as growth dilution and the uneven distribution of chemicals within specific plant tissue types (i.e., leaves, stems, fruits) must also be considered when predicting final tissue concentrations for risk assessments.

Unlike organic contaminants, nutrient and metal ions are generally actively transported into roots since the rate of passive transport of charged compounds across hydrophobic root membranes is limited. Plants take up most nonnutrient metals incidentally while acquiring nutrients for growth (Pulford and Watson, 2003; Saison et al., 2004). We do not address uptake of metals and nutrients in this chapter. The reader is referred to the above references and to phytoremediation reviews by Salt et al. (1995), Raskin and Ensley (2000), and Weis and Weis (2004).

The aim of this chapter is to provide the reader with an entry to the literature addressing root uptake of organic chemicals from soil and transport and fate within plants. Starting from a simple conceptual model, we describe the various processes and parameters important in understanding and quantifying the rate of root uptake and subsequent fate of organic chemicals in plants including various plant bioconcentration factors, transpiration rates, phytovolatilization, and metabolism. A brief description of plant uptake models is followed by illustrative examples of plant uptake. It should be appreciated that the state of the science regarding mass transport of organic contaminants into plants from soil is much less developed than other transport processes described in this handbook. This reflects both a relative lack of comparable empirical data and the number and complexity of the plant physiological and environmental variables associated with the uptake processes as discussed by Legind and Trapp (2009) and McKone and Maddalena (2007).

14.2   ROOT UPTAKE AND TRANSLOCATION

Transpiration is defined as the root uptake of water, the subsequent transfer to leaves, and loss due to evaporation, and it drives the passive root uptake of organic compounds.

Water loss during transpiration occurs mainly through the leaves as the stomates open to allow the exchange of CO_2 and O_2 during photosynthesis. However, water loss can also occur through stems, flowers, and fruits. Dissolved organic solutes are drawn toward the roots along with water and nutrients (e.g., Bowling, 1979; Briggs et al., 1982; Bromilow and Chamberlain, 1995; McFarlane, 1995). For organic compounds with relatively high Henry's law constants, movement to the roots through the soil vapor phase may also be important. Uptake is facilitated by root hairs that penetrate into the soil pores and influence the volume of soil that is available for chemical uptake.

The two main pathways that water and organic chemicals follow into the roots are referred to as *symplastic* and *apoplastic*. Symplastic transport is characterized by water and solutes moving through the plasma membranes and interior of individual root cells, while apoplastic transport occurs when water and solutes pass between the roots cells until reaching the Casparian strip, a waxy band associated with the endodermis as illustrated in Figure 14.1. The Casparian strip consists of lipophilic compounds called suberin as well as assorted carbohydrates and cell wall proteins (Schreiber et al., 1999).

In addition to the *symplastic* and *apoplastic pathways*, water and a few organic solutes such as glycerol, can move from cell to cell through water channels within cellular membranes referred to as aquaporins (Steudle and Peterson, 1998; Javot and Maurel, 2002; Maurel et al., 2002). The ability to control water transport by cell-to-cell mechanisms, allows plants to control solute concentrations within the plant more effectively by forcing water and solutes to cross multiple membranes. The aquaporin pathway is more important under low water transport conditions (Henzler et al., 1999) while the apoplastic pathway begins to play a much more important role in water transport under high transpiration conditions (Steudle and Peterson, 1998).

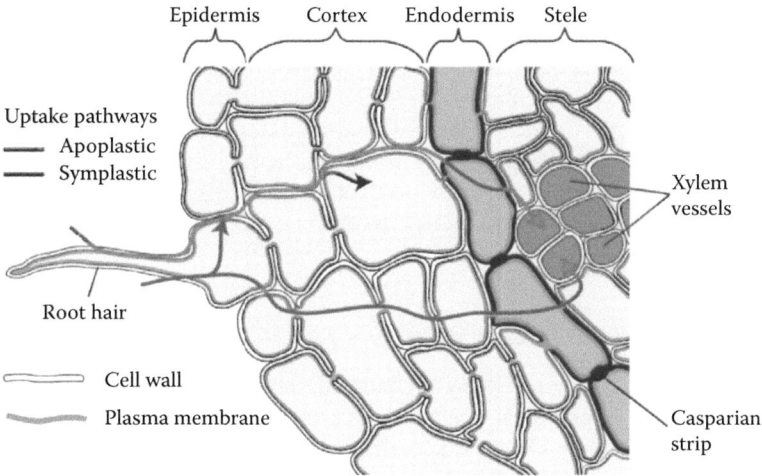

FIGURE 14.1 Diagram of root illustrating symplastic and apoplastic transport pathways. (Adapted from Campbell, N. A. et al. 1999. *Biology*. Harlow, UK: Pearson Benjamin Cummings.)

The impact of these pathways on organic compound uptake has not yet been fully investigated.

Once the chemical passes through the root membrane, and depending on its properties, it can be transported to other parts of the plant by way of the flow of sap in the xylem and phloem channels. Xylem channels conduct the unidirectional flow of water and nutrients from roots to the photosynthetic sections of the plant while phloem is the bidirectional flow that distributes sugars and other photosynthetic products throughout the plant. However, within the xylem, lateral movement to adjacent cells occurs and may provide a pathway for contaminants to partition into the phloem (Hendrix, 1995; Marschner, 1995). Xylem transport rates are on the order of 10 cm/min while phloem transport rates are much slower, $\sim$1 cm/min (Lang, 1990). Xylem transport rates are directly related to transpiration rates while the rate of phloem transport is governed by differences in solute concentrations between sites of solute synthesis (source) and consumption (sink) (Marschner, 1995).

Xylem and phloem both contribute water and solutes to fruits, but their relative contributions can fluctuate seasonally. For example, while xylem and phloem contribute roughly equally to apple growth early in the growing season, from mid-to-late-growing season the phloem dominates (Lang and Thorpe, 1989; Lang, 1990).

When modeling the transport of organic contaminants within plants, the potential impact of the xylem and phloem solution composition (Yeo and Flower, 2007) on the solubility of these contaminants is not generally considered. There are, however, organic and inorganic solutes in solution that can affect the capacity of the water to solubilize organic substances. Inorganic anions and cations generally "salt out" organics to an extent expressed using the Setschenow equation. Sugars and other polar organics can increase the solubility, but the effect is usually relatively small (Copolovici and Niinemets, 2007). Translocation into the above ground plant tissues via the xylem is believed to be less important for hydrophobic compounds (log K_{OW} > 3.5) while phloem transport is primarily significant for polar and neutral organics; ionized weak acids; and/or glycolylated organic metabolites.

During transport within the plant, organic chemicals can be metabolized and sequestered within various plant tissues and volatilized from the plant surfaces, especially the foliage. A conceptual model summarizing the main organic chemical transport pathways and fate processes within a plant is illustrated in Figure 14.2. A brief description follows of each process and the distribution or partitioning descriptors used to quantify mass transport.

14.3 PLANT BIOCONCENTRATION FACTORS

Empirical data describing the extent of chemical uptake by plants roots are generally expressed as ratios of chemical concentrations in the plant compartment of interest (e.g., shoots, roots, xylem sap) to that in the exposure medium (soil, soil pore water, hydroponic solution) measured at the time the samples are collected. These ratios are generally referred to as bioconcentration factors (BCFs) but they may or may not reflect equilibrium conditions. Plant BCF values are widely used to provide direct and approximate estimates of plant tissue concentrations from measured exposure

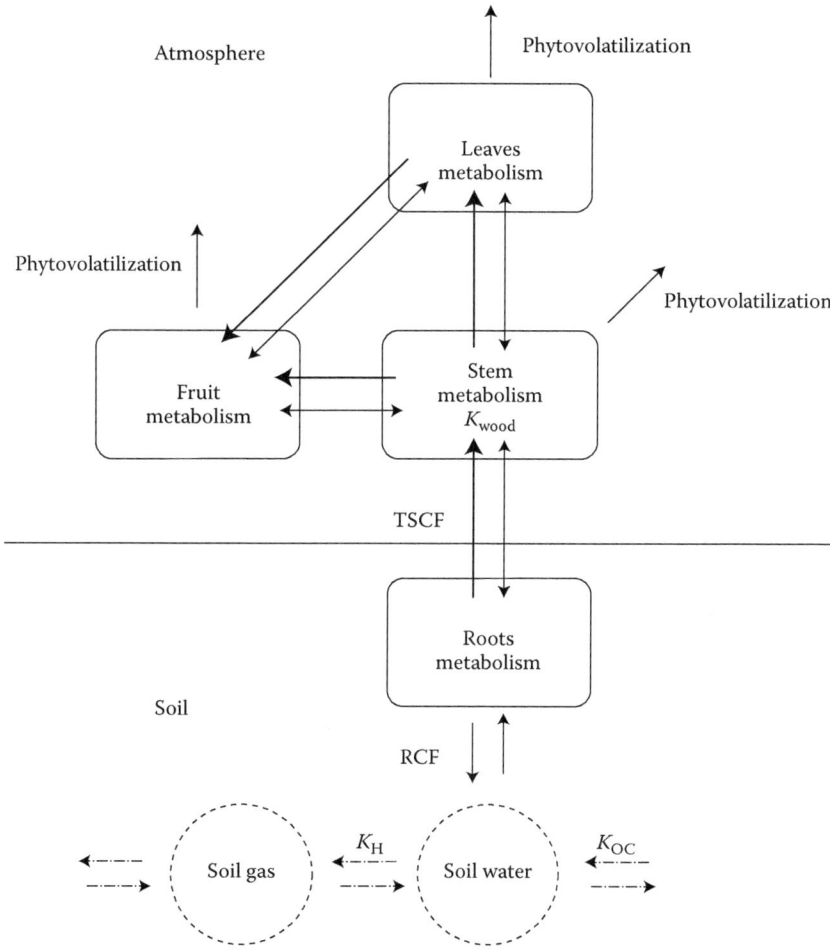

FIGURE 14.2 Conceptual model of organic chemical root uptake and fate within plant.

concentrations and have also been used as input into mechanistic models of plant uptake.

Relatively few experimental BCF values exist, due in part to the cost associated with the determinations and the lack of regulatory requirements for such data. When data exist for chemicals having more than one literature BCF, the variability is often relatively large due to the lack of standardized or generally accepted methods. This lack of consistent methods hampers the development of quantitative methods for predicting plant uptake by roots.

As briefly discussed in the following sections, the most commonly reported plant BCFs are for roots, stem or wood, shoots (above ground tissues), and xylem sap. The chemical concentration ratio between xylem sap and the external exposure solution (usually a hydroponic solution) is specifically referred to as a transpiration stream concentration factor (TSCF). Care must be exercised when using plant BCF values because both dry and wet weights of plant tissue and soil have been used to express the concentrations (McKone and Maddalena, 2007). Correlations between measured BCFs and chemical contaminant properties such as the octanol–water partition coefficient (K_{OW}) have also been reported.

14.3.1   ROOT CONCENTRATION FACTORS

Partitioning of nonionized organic chemicals into the lipophilic root membranes and the root sap has been described as a root concentration factor (RCF), the ratio between the chemical concentration in the roots and that in the exposure media (water or soil) contacting the roots (Briggs et al., 1982, 1983; Topp et al., 1986).

Briggs et al. (1982) measured RCF values by hydroponically growing 10-day-old barley plants in solutions containing 18 individual ^{14}C-labeled chemicals selected from two series of nonionized O-methylcarbamoyloximes and substituted phenylureas. The plants were exposed for 48 h, the concentrations in roots and solution measured directly, and the RCF values expressed as a ratio of the chemical concentration in the roots on a fresh weight basis divided by the chemical concentration in solution. The RCF values were believed to reach equilibrium within 24 h of treatment and generally increased as a function of increasing chemical lipophilicity expressed as the octanol–water partition coefficient (K_{OW}). However, the RCF values were relatively constant for the more polar compounds, ranging from 0.6 to 1.0.

Briggs also compared the RCF values to root water sorption coefficients obtained using macerated roots. The sorption coefficients were found to be similar to the RCF values for the hydrophobic compounds but continued to decline for the more polar compounds unlike the RCF values. Based on these results, Briggs et al. (1982) suggested that two components determine the total amount of chemical in the roots, the chemical associated with the water in the roots and the chemical associated with the lipophilic root solids, leading to the correlation (Briggs et al., 1982):

$$\log(\text{RCF} - 0.82) = 0.77 \log K_{OW} - 1.52. \tag{14.1}$$

Topp et al. (1986) conducted similar uptake experiments with barley plants grown from seed for 7 days in soil containing individual ^{14}C-labeled organic chemicals at concentrations of 2 mg/kg. Root concentration factors (CFsoil), defined similarly as the concentration of chemical in the root (fresh weight) divided by the concentration of the chemical in the soil pore water, were determined and found to be correlated with log K_{OW} as shown:

$$\log \text{RCF} = 0.959 + 0.630 \log K_{OW}. \tag{14.2}$$

A correlation between root/soil partition coefficients with log K_{OC} was also obtained and it was reported that in long-term studies, a constant RCF was approached only after 100 days or more, presumably reflecting the slow transport of sparingly soluble more hydrophobic substances.

The relationships presented above suggest that the more hydrophobic the compound and the greater the lipid content of the roots, the greater the concentration of organic chemical contaminant in the roots. In principle, it should be possible to estimate chemical root concentrations from a mass transport rate of water, a root volume and appropriate partitioning information. It should, however, be appreciated that soil–root equilibrium is unlikely to be reached because of the rate of root growth, but a steady-state concentration may be achieved in which the uptake rate is balanced by the rates of growth and transport from the root. Chemical concentrations in the roots may not be uniform and higher concentrations are often observed near the soil–root interface. Relatively little information on the lipid content of roots is available but ' typical values range from 0.001 to 0.025 kg/kg (Trapp, 2002; Trapp et al., 2007). Partitioning into other lipophilic materials such as waxes, suberin and lignins is likely. Uptake from air can also be significant (Mikes et al., 2008).

Ionized compounds also enter the roots, but do not readily cross the root membranes and enter the xylem. For organic acids and bases, the pH of the system and the pK_a of the compound can be used to determine the fraction of the neutral form of the chemical most available for uptake (Briggs et al., 1987; Trapp, 2000, 2004). These ionic substances undergo a variety of additional processes such as dissociation, ion trapping, and electrical interactions with soils and plant cells (Trapp, 2004).

14.3.2 PLANT TISSUE WATER SORPTION COEFFICIENTS

Stem, wood, or wood component water sorption coefficients have been reported for several organic xenobiotic compounds (e.g., Garbarini and Lion, 1986; Severtson and Banerjee, 1996; Mackay and Gschwend, 2000; Trapp et al., 2001; Ma and Burken, 2003a) along with several predictive expressions relating these sorption coefficients to $\log K_{OW}$.

Using macerated stems from 11-day-old barley plants and two series of nonionized chemicals (O-methylcarbamoyloximes and substituted phenylureas), Briggs et al. (1983) reported the following relationship between stem concentration factors (SCF), defined as the ratio of chemical concentrations in stem tissue on fresh weight basis and the exposure solution, and $\log K_{OW}$:

$$\log \text{SCF (macerated stems)} = 0.95 \log K_{OW} - 2.05. \qquad (14.3)$$

Tam et al. (1996) measured the uptake equilibria and kinetics of chlorobenzenes by soybean plants and similarly correlated the results against K_{OW}.

The sorption of 2,4-dichlorophenol and 2,4,5-trichlorophenol by pulped wood fibers over a pH range of 2–12 was investigated by Severtson and Banerjee (1996), who found that the only significant interaction was between the neutral (protonated) form of these compounds and lignin. Combining these data with the lignin normalized wood water sorption coefficients reported for trichloroethylene and toluene (Garbarini and Lion, 1986) the following relationship between lignin water sorption coefficients (K_{lignin}) and octanol–water partition coefficients ($\log K_{OW}$) for neutral organic chemicals was presented:

$$\log K_{lignin} = 0.95 \log K_{OW} - 0.48. \qquad (14.4)$$

A similar expression was later reported by Mackay and Gschwend (2000), who investigated the sorption of benzene, toluene, and o-xylene to Douglas fir and Ponderosa pine chips.

$$\log K_{\mathrm{lignin}} = 0.74 \log K_{\mathrm{OW}} - 0.04. \tag{14.5}$$

Ma and Burken (2002) also found that wood–water partitioning depended mainly on the fraction of lignin when evaluating the sorption of trichloroethylene (TCE), 1,1,2,2-tetrachloroethane, and carbon tetrachloride to poplar tree cores.

Instead of normalizing experimental wood water sorption coefficients to lignin, Trapp et al. (2001) reported two separate expressions for oak and willow that related wood water sorption coefficients (K_{wood}) to $\log K_{\mathrm{OW}}$ for 10 organic chemicals ranging in $\log K_{\mathrm{OW}}$ from 1.48 to 6.20:

$$\log K_{\mathrm{Wood}} = -0.27(\pm 0.25) + 0.632(\pm 0.063) \log K_{\mathrm{OW}} \quad \text{for oak} \tag{14.6}$$

and

$$\log K_{\mathrm{Wood}} = -0.28(0.40) + 0.668(0.103) \log K_{\mathrm{OW}} \quad \text{for willow.} \tag{14.7}$$

Overall, these results indicate that the lignin fraction of the stem is the most sorptive for nonionized organic chemicals and that the sorption is approximately proportional to $\log K_{\mathrm{ow}}$.

14.3.3 TRANSPIRATION STREAM CONCENTRATION FACTORS

The relative tendency of a chemical to move from root to shoot (above ground tissue) tissue is often described by the transpiration stream concentration factor (TSCF), the ratio of the chemical's concentration in the xylem sap to that in the solution to which the plant root is exposed.

Values of TSCF are usually measured by one of two general approaches. Firstly, living plants are exposed to a constant root-zone concentration of the chemical of interest (e.g., Briggs et al., 1982; Burken and Schnoor, 1998; Doucette et al., 2005a,b). A hydroponic system is often used because exposure concentrations are more easily measured and controlled than in soil environments. Since the direct collection of a sufficient amount of xylem sap for analysis is problematic for most living plants, TSCFs are generally indirectly determined from measured shoot concentrations normalized to the amount of water transpired during the exposure period [i.e., concentration in xylem sap is deduced as the total mass (mg) of the compound in shoots divided by the volume of water transpired (L)]. Any loss of the chemical due to metabolism or volatilization once it reaches the foliar tissue should be corrected for but this is often analytically difficult and is not always done.

Secondly, the roots of a de-topped plant (i.e., the above-ground tissues are removed just above the lowest leaves) are sealed in a pressurized chamber containing a solution of a known concentration of the chemical of interest (e.g., Hsu et al., 1990; Sicbaldi et al., 1997; Ciucani et al., 2002; Dettenmaier et al., 2009). The xylem fluid is forced through the roots as the chamber is pressurized and the xylem is collected and analyzed as it exits the cut stem. Ideally, the pressure is adjusted to provide a constant flow rate corresponding to the transpiration rate of an intact plant of similar size without damaging the roots. A TSCF is calculated from the ratio of the steady-state xylem concentration to the root exposure concentration.

Like the RCF, the TSCF has also between related to log K_{OW}. Two different types of general relationships have been reported. Bell-shaped curves relating TSCF to log K_{OW} (Briggs et al., 1982; Hsu et al., 1990; Burken and Schnoor, 1998) suggest that compounds that are either highly polar or are highly lipophilic will not be significantly taken up by plants. However, more recently, Dettenmaier et al. (2009) presented the following empirical relationship between TSCF and log K_{OW} (shown in Figure 14.3), that indicates that nonionizable, polar, highly water soluble organic compounds are most likely to be taken up by plant roots and translocated to shoot tissue.

$$\text{TSCF} = \frac{11}{(11 + 2.6^{(\log K_{OW})})}. \tag{14.8}$$

The reason for the difference between the two types of TSCF and log K_{OW} relationships is not clear, but could be due to a variety of factors including differences in methods used to generate TSCF (intact vs. pressure chamber), plant growth conditions (hydroponic vs soil), plant species, plant age, test duration, transpiration rates, losses due to metabolism and volatilization, and perhaps even nonpassive, compound-specific uptake mechanisms. However, the trend observed in Figure 14.3 is consistent with the increasing number of observations in the literature reporting high root uptake for low log K_{ow} compounds such as sulfolane (Doucette et al., 2005a), MTBE (Hong et al., 2001; Rubin and Ramaswami, 2001; Yu and Gu, 2006), *tert*-butyl alcohol

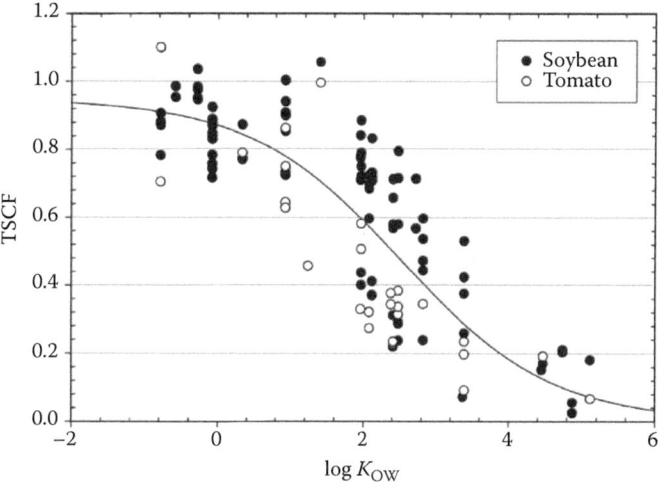

FIGURE 14.3 Plot of measured TSCF in relation to log K_{OW} using a modified pressure chamber apparatus with empirical nonlinear least squared fit. (Adapted from Dettenmaier, E. et al. 2009. *Environ. Sci Technol.* 43(2), 324–332.)

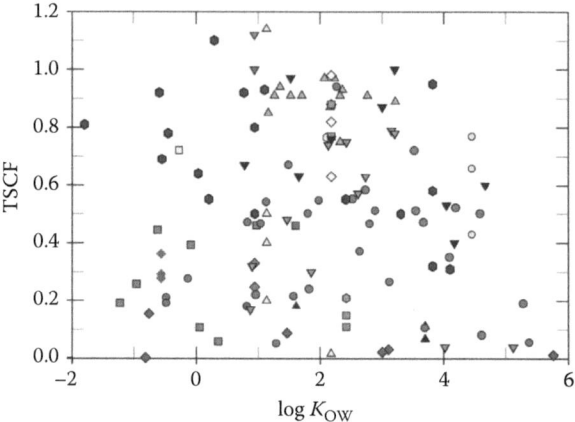

FIGURE 14.4 One hundred thirty-two TSCF values compiled from 26 referred publications for 93 compounds. (Adapted from Dettenmaier, E., W.J. Doucette, and B. Bugbee. 2009. *Environ. Sci. Technol.* 43(2): 324–332.)

(TBA), and 1,4-dioxane (Aitchison et al. 2000) in addition to the results of a recently published mechanistic plant uptake model (Trapp, 2007).

14.4 PHYTOVOLATILIZATION

For chemicals having high Henry's law constants or air–water partition coefficients, volatilization from leaves and stems, often referred to as phytovolatilization, can be a significant loss mechanism for chemicals that have been translocated to the above-ground tissues (Ma and Burken, 2003b). However, direct measurements of phytovolatilization from leaves (e.g., Doucette et al., 2003) and stems (Ma and Burken, 2003b) reported in the literature are relatively limited, especially in field settings. Differences between the age and species of plants tested and in the methods used to quantify phytovolatilization can make quantitative comparisons between the various studies difficult to interpret.

Initial reports of phytovolatilization were often based on indirect evidence such as a decline in chemical concentration within the trunks of trees as a function of height during phytoremediation or groundwater monitoring applications (e.g., Vroblesky et al., 1999; Ma and Burken, 2003b). However, in other situations decreasing concentrations with height have not been observed (Doucette et al., 2003, 2007; Vroblesky et al., 2004). Baduru et al. (2008) speculated that volatilization through the trunks of tall, large diameter trees is probably much slower than losses from small, thinner trees and may account for the observational differences.

Phytovolatilization from leaves has been measured using static chambers or bags (e.g., Arnold et al., 2007), flow through chambers (e.g., Doucette et al., 2003), and open path FTIR (Davis et al., 1994). Static chambers are placed over a section of leaves until enough transpired water is collected to perform analysis for the compounds of interest. Such data provide evidence that the chemical has reached the leaves but the information is difficult to interpret or scale to the whole plant since the leaves slow or stop transpiring when the humidity in the enclosure is high and the lack of evaporative cooling artificially increases the leaf temperature. Flow-through chambers in which the air exiting the chamber is analyzed for the compound of interest and water can avoid this problem if the flow rates are high enough to maintain ambient humidity. Dividing the mass of the compound collected by the volume of water collected yields a transpiration stream concentration that can be multiplied by the total amount of water transpired by the plant to provide an estimate of the amount of chemical phytovolatilized (Doucette et al., 2003). Open path FTIR has been used to measure the flux of volatile chemicals from plants and since no enclosure is needed, it eliminates potential artifacts associated with artificial environments. However, this approach is generally not sensitive enough for most applications.

In addition to the direct measurement of phytovolatilization from leaves, recent focus has been on volatilization through the stems prior to the chemical reaching leaves. Using diffusion traps to quantify the volatilization of TCE from poplar stems, Ma and Burken (2003b) observed that the TCE diffusion flux decreased with height and concluded that TCE volatilization from stems via diffusion to the atmosphere was the major fate process for TCE after uptake by plants, although volatilization can also occur from leaves. The two concurrent processes thought to influence the eventual fate of volatile organic compounds taken up by trees are advective transport within the xylem sap stream and diffusive transport from stem to atmosphere.

To generate diffusion coefficients necessary to model the volatilization from stems, Baduru et al. (2008), measured the diffusion of several VOCs commonly found as groundwater contaminants through freshly excised hybrid poplar tree tissue. Diffusivities in xylem tissue, determined using a one-dimensional diffusive flux model developed to mimic the experimental arrangement, were inversely related to molecular weight and ranged from 0.32 to 0.05×10^{-7} cm^2/s for tetrachloroethylene to 2.98 to 1.1×10^{-7} cm^2/s for benzene. A model was presented that can be used for the prediction of diffusion coefficients for other compounds on the basis of chemical molecular weight and plant-specific properties such as water, lignin, and gas contents.

In summary, phytovolatilization from both leaves and stems has been reported. However, the importance of phytovolatilization depends on both the properties of the chemical contaminant and the plant of interest. While phytovolatilization has been incorporated into several mechanistic plant uptake models such as that of Trapp (2007), additional experimental data collected under a variety of field settings are needed before the general significance of this process can be adequately assessed.

14.5 PLANT BIOTRANSFORMATION OR METABOLISM

Plants have been shown to biotransform xenobiotic organic compounds (Komossa and Sandermann, 1995; Pflugmacher and Sandermann, 1998). Historically, much of the information on plant biotransformation of organic compounds has been obtained from studies examining intentionally applied chemicals such as herbicides and pesticides. More recently, the biotransformation of nonherbicide organics has been investigated in support of phytoremediation studies. Experimental evidence regarding the bio-transformation capabilities of plants comes from cell culture and intact plant studies. These studies have shown that plants have the ability to degrade a wide range of compounds ranging from the highly polar herbicide glyphosate to very hydropobic chemicals such as DDT and hexachlorobenzene (Komossa and Sandermann, 1995).

Plant biotransformation parallels liver biotransformation and is conceptually divided into three phases. Phase I typically consist of oxidative transformations in which polar functional groups such as OH, NH_2, or SH are introduced. However, reductive reactions have been observed for certain nitroaromatic compounds. Phase II involves conjugation reactions that result in the formation of water soluble compounds such as glucosides, glutathiones, amino acids, and malonyl conjugates or water-insoluble compounds that are later incorporated or "bound" into cell wall biopolymers. In animals, these water-soluble Phase II metabolites would typically be excreted. In Phase III, these substances are compartmentalized in the plant vacuoles or cell walls. For additional details, the reader is referred to reviews on the subject by Komossa and Sandermann (1995), Pflugmacher and Sandermann (1998), and Burken (2003). Enzymatic conversion rates typically follow Michaelis–Menten kinetics and are temperature-dependent (Larsen et al., 2005; Yu et al., 2004, 2005, 2007).

Recent studies suggest that endophytes (bacterium or fungus that live symbiot-ically within plants) are also partially responsible for the degradation of organic contaminants taken up by plants (e.g., Taghavi et al., 2005; Newman et al., 1997). Endophytic fungi and bacteria live within most plant species without any apparent negative impacts and colonize the plant vascular systems with the densities decreasing from roots to stem to leaves.

Unfortunately, few measured rates of transformations are available for plants making it difficult to quantitatively incorporate metabolism into plant fate models.

14.6 SOIL PORE WATER CONCENTRATIONS

A critical variable when estimating the amount of chemical passively taken up by the root is the concentration of the chemical in the water used by the plant. This must be measured or estimated. In laboratory uptake experiments, typically performed in hydrophonic systems, concentrations are directly measured. In field situations, soil concentrations are often determined and the concentration of chemical in the soil solution used by the plants is estimated soil–water partition coefficients as discussed in Chapter 8. For a given organic chemical, sorption coefficients vary considerably depending on the properties of the soil. However, for many organic chemicals, sorption has been observed to be directly proportional to the quantity of organic matter associated with the solid (Hamaker and Thompson, 1972; Chiou et al., 1979; Rao and Davidson, 1980; Briggs, 1981; Karickhoff, 1981). Normalizing soil-specific sorption coefficients (K_d) to the fraction of organic carbon (f_{oc}) of the sorbent yields the organic carbon normalized sorption coefficient (K_{oc}) that is generally considered constant for the sorbed organic chemical.

$$K_{oc} = \frac{K_d}{f_{oc}}. \tag{14.9}$$

If the K_{oc} approach is used to estimate site-specific sorption coefficients, the reader should be aware that considerable variation in K_{oc} values has been observed in the literature. For example, Mackay et al. (1992) provided 24 values of K_{oc} for benzene ranging from 0.11 to 2.08 L/kg. Goss and Schwarzenbach (2001) used literature data for atrazine (log K_{oc} ranged from 1.3 to 3.5) to highlight the variability associated with K_{oc} values. Delle Site (2001) provides a comprehensive summary of literature K_{oc} values for a variety of compounds that further illustrates the fact that K_{oc} is not an unvarying constant.

Variation of K_{oc} values in the literature is likely the result of differences in the sorption characteristics of the soil organic matter (SOM), sorption kinetics, variation in the methods used to determine K_{oc} (separation of phases, mass balance, single point, or isotherm) impact of other soil properties, and the properties of chemicals being sorbed (Seth et al., 1999). The use of K_{oc} to estimate site-specific soil water sorption coefficients is most appropriate for predicting the sorption of nonpolar organic compounds to soils having greater than 0.1% organic matter content. Additional information pertaining to the soil sorption of organic chemicals, the K_{oc} approach, and the estimation of K_{oc} values can be found in the review by Doucette (2003).

For organic acids and bases, the K_{oc} approach can still be applied if the neutral form of the compound dominates at the pH of the soil solution. If both neutral and ionized forms of the chemical are present in significant quantities, then the extent of sorption will depend on the fraction of each form present (Bintein and Devillers, 1994; Franco and Trapp, 2008; Franco et al., 2009).

14.7 TRANSPIRATION RATES

Transpiration is the evaporation of water from the interior airspaces of leaves and the subsequent diffusion through the stomata (pores) into the atmosphere. The flux of water moving through plants controls the transport of soil contaminants to the root surfaces and to the whole plant. More than 90% of transpiration occurs through the leaves as the stomates open to allow the inward diffusion of CO_2 for photosynthesis. Some transpiration can also occur through stems, flowers, and fruits. For every gram of organic matter made by the plant in photosynthesis, 200–800 g of water is taken up and transpired (Taiz and Zeiger, 1998).

The most important factors controlling transpiration are climatic including soil water availability, solar radiation, vapor pressure deficit, wind speed, and air temperature (Allen, 1998). However, plant characteristics can also be important. For example, the quantity of leaves, expressed as leaf area index (the ratio of leaf area to ground area), greatly impacts transpiration with the greater the leaf area index the greater the rate of transpiration. Even with the same leaf area index, fast growing annual plants use more water per unit leaf area than slower growing perennial plants. Broad-leaved plants and trees (dicots) can have higher transpiration rates than grasses or conifer species under equal precipitation conditions (Lambers et al., 1998). Species with tap roots that can use deeper groundwater sources (phreatophytic species) can maintain higher transpiration rates when precipitation and surface water is minimal (Lambers et al., 1998).

Direct measurement of transpiration is difficult for all but relatively small plants that can be grown in weighing lysimeters or enclosed in chambers in which the flux of water can be calculated from the humidity increase in the enclosure. Common approaches used to quantify transpiration in the field include precipitation minus runoff on gauged watersheds, energy balance equations (Allen et al., 1998), eddy covariance (water vapor gradients above and below canopy) (Baldocchi, 2003), hydrologic models, and sap flow measurements (Vose et al., 2003).

Transpiration rates in the field vary from 1 to $10 \, \mathrm{L \, m^{-2} \, d^{-1}}$ (corresponding to 1–$10 \, \mathrm{mm \, d^{-1}}$) depending on the soil water availability and evaporative demand. Potential transpiration rates, calculated from weather data or pan evaporation rates, are widely used to schedule irrigation of crop plants. A high potential transpiration rate of $10 \, \mathrm{L \, m^{-2} \, d^{-1}}$ can occur on hot days in well-watered soils in dry climates, but the cumulative annual transpiration rate is more useful in risk assessment or phytoremediation calculations. The potential annual transpiration rate can be as high as $1800 \, \mathrm{L \, m^{-2} year^{-1}}$ in hot desert climates such as Arizona, and as low as $200 \, \mathrm{L \, m^{-2} \, year^{-1}}$ in cool, moist environments such as Alaska (Camp et al., 1996). However, even well-watered crops can fail to attain the potential transpiration rate in the summer because of partial stomatal closure during periods of high evaporative demand. The difference between the actual and potential transpiration rates is usually greater in hot, low-humidity environments. During winter months, deciduous trees drop leaves and evergreen trees have low transpiration rates as the result of shorter days, lower light levels, and colder temperatures. When forced to use groundwater, phreatophytic plants typically do not achieve the high transpiration rates that occur with vegetation that uses surface water (Camp et al., 1996). Thus, the actual annual transpiration rate is usually less than the potential rate. Depending on the climate, 200–$1400 \, \mathrm{L \, m^{-2} \, year^{-1}}$ probably represents a reasonable range of values for annual transpiration.

It can also be important to estimate the transpiration of isolated trees in various phytoremediation and fruit tree risk assessment scenarios. In a review of 52 water use studies since 1970, Wullschleger et al. (1998) found that 90% of the observations for maximum rates of daily water use were between 10 and $200 \, \mathrm{L \, d^{-1}}$ for individual trees that averaged 21 m in height. The fraction of groundwater used by plants is often critical to determining phytoremediation potential. Unfortunately, the groundwater use fraction is difficult to measure and is poorly characterized. As expected, groundwater use tends to decrease as the availability of surface water increases (Nilsen and Orcutt, 1996). Additional studies using stable isotope techniques (Nilsen and Orcutt, 1996) are necessary to determine a reasonable range of values for this parameter. Until such data are available, a range of groundwater use fractions from 0.1 to 0.5 is probably realistic for climates with more than 40 cm of precipitation per year.

Some 20 empirical methods have been developed over the past 60 years to estimate transpiration and soil surface evaporation as a function of meteorological variables. Most of these are adequate for the region in which they were developed, but they often have significant errors when applied to other climatic regions. In 1990, the American Society of Civil Engineers (ASCE) undertook a major study to assess the accuracy of 20 methods for estimating transpiration from measured environmental variables (Allen et al., 1998). The models were compared across 11 unique locations that had detailed measurements from lysimeters. The goal was to determine a standard method for calculating potential evapotranspiration (ETo, the maximum transpiration rate for well watered plants) from a vegetated surface. The Penman–Monteith (PM) equation emerged as the most accurate across a wide range of climates. The PM equation was adopted as the standard method and is now widely used to predict potential transpiration from uniform plant communities, which are typical of agricultural crops. The PM equation is biophysically based and requires input measurements of radiation, air temperature, air humidity, and wind speed. The derivation of the equation and the associated equations to calculate each of the component terms can be found at: http://www.fao.org/docrep/x0490e/x0490e06.htm.

The PM equation provides an estimate of the potential rate of transpiration of a closed plant canopy for specific environmental conditions. The ETo provides an estimate of the volume of water transpired per unit ground area. After canceling units, this is typically expressed as a depth of water. The peak ET rate, which occurs during the hottest days of the year in well-watered plants, is from 5 to 10 mm per day (5–$10 \, L \, m^{-2} d^{-1}$). The highest ET rates occur only on windy days in low humidity climates. The daily potential ET rate for a region is usually available from publicly maintained weather stations, and is often published in local newspapers to encourage seasonally appropriate irrigation practices.

Because the PM equation was developed for a grass reference crop it is necessary to use a crop coefficient (K_c) to better predict the potential evapotranspiration rate for a wide variety of plant communities. Because of the value of water in agriculture, extensive tables of crop coefficients have been developed. The coefficient for healthy turf grass is 0.8. Coefficients tend to increase with plant height. Values for alfalfa and corn at full ground cover are about 1.05 in humid, low-wind conditions and up to 1.4 in dry, windy conditions. Tree crops like apples, peaches, and pears are considerably taller than corn, but they tend to have more closed stomates and their crop coefficients are similar to corn (Allen et al., 1998). More detail on crop coefficients can be found at: http://en.wikipedia.org/wiki/Crop_coefficient.

14.8 QUANTIFYING MASS TRANSPORT INTO PLANTS

Since the passive root uptake of organic chemicals is driven by the amount of water transpired, the simple lipid–water equilibrium partitioning approach that is applied so successfully to animals and fish is less applicable to plants. Transpiration rates vary diurnally and seasonally as plants respond to changes in climatic conditions (e.g., humidity, temperature, sunlight, and wind speed). The simplest and obvious response to these complexities has been to measure chemical concentrations in specific plant fluids or tissues at specific times or over specific intervals and relate them to external exposure concentrations in hydroponic solution, soil or soil pore water. The BCFs and RCFs described earlier are examples of these ratios and have been correlated to properties of the chemical such as K_{OW}. An example of one of earliest and the most widely used plant BCF—chemical properties relationship was derived by Travis and Arms (1988):

$$\log B_v = 1.588 - 0.578 \log K_{ow} \qquad (14.10)$$

where B_v is the ratio of concentration in above ground plant parts (mg of compound/kg dry plant) to the concentration in the soil (mg compound/kg dry soil). This relationship is consistent with the general trend reported more recently by Dettenmaier et al. (2009).

The second general approach to predicting tissue concentrations resulting from chemical exposure are mechanistic mass balance models in which one or more compartments are defined and rates of input, output and accumulation are expressed using equations containing parameters describing partitioning, degradation, and flow or diffusion rates (e.g., Topp et al., 1986; Ryan et al., 1988; Trapp et al., 1994; 2003; Trapp and Matthies, 1995; Hung and Mackay, 1997; Paterson et al., 1991, 1994; Trapp, 2000, 2002, 2007; Chiou et al., 2001; Ouyang, 2002; Ouyang et al., 2005; Trapp and Legind, 2009).

A recent evaluation of six plant uptake models with five different sets of experimental data was conducted by the Environment Agency of the UK (Collins et al., 2006a, Science Report SC050021, http://www.environment-agency.gov.uk/static/documents/Research/sc050021_2029764.pdf). The report concluded that the majority of models overpredicted root concentrations but predictions of shoot concentrations were varied. Performance was not related to model complexity and simple empirical models were as effective as multiple compartment models. The lack of high-quality data for a broad range of organic chemicals may have contributed to this conclusion.

The parameters required for estimating the rate of mass transfer into plants from soils fall into three categories.

First are chemical properties including K_{ow}, K_{aw}, K_{oa}, K_{oc}, pK_a, solubility and vapor pressure. Also required, but rarely available, are biotransformation half-lives or rate constants for plant compartments.

Second are environmental properties including soil composition (volume fractions of mineral and organic matter, water and air), soil–water K_d, soil pH, temperature, and relative humidity in air. Third are plant properties including the dimensions, masses and volumes of all compartments and their growth rates, transpiration rates and flow rates in xylem and phloem, lipid-equivalent or wax contents and volume fractions of water and air. Time to harvest, leaf area and root depth may also be required.

Trapp and Legind (2009) have suggested values for many of these parameters, but have pointed out that because about 20,000 plant species are used by humans and some 600 are cultivated, the data requirements are potentially enormous. One option for advancing the systematic acquisition of required data is to identify a number of key plant guilds, for example, grasses and cereals such as wheat and corn, root vegetables such as carrots, tubers such as potatoes, foliage vegetables such as lettuce and cabbage, plants exploited for phytoremediation purposes such as poplar and finally plants bearing edible fruits and nuts such as apples and walnuts. Generic plant properties and indications of variation between species could be compiled, thus providing a basis for development and parameterization of a variety of mass balance models. It is likely that models of organic chemical uptake in plants will improve as more empirical data become available and ultimately the most successful models of plant uptake will likely be mechanistic in nature.

The following case studies give examples of approaches used to estimate plant uptake for phytoremediation and risk assessment purposes. The first uses a minimal number of inputs to estimate the amount of a highly water soluble compound removed by cattails growing in a contaminated wetland. The second example is a brief summary of a mechanistic model for estimating the concentration of neutral organic chemicals in apples growing in contaminated soil.

14.9 CASE STUDY 1: SULFOLANE REMOVAL FROM WETLAND BY CATTAILS (DOUCETTE ET AL., 2005a)

Sulfolane (tetrahydrothiophene 1,1-dioxide, $C_4H_8O_2S$) is a nonionizable, organic compound used in the Sulfinol^{TM} process to remove hydrogen sulfide from natural gas. It is highly water-soluble (290 g/L) with a log K_{OW} of -0.77 and a low Henry's Law constant (4.85×10^{-6} atm m^{-3}mol^{-1}). It been identified in wetland vegetation near a sour gas processing facility in Alberta, Canada (Headley et al., 2002).

To estimate the annual amount of sulfolane, N mg m^{-2}year^{-1}, which can potentially be removed by cattails growing in a contaminated wetland in Alberta Canada, Doucette et al. (2005a) used the following expression:

$$N = C_s(\text{TSCF})(V)(f), \tag{14.11}$$

where C_s is the average sulfolane water concentration in the wetlands (40 mg/L), TSCF is a laboratory study derived value of 0.68, V is the estimated volume of water transpired per unit area per year (up to 10 L/m^2 year, Pauliukonis and Schneider, 2001), and f is the fraction of the plant water needs met by contaminated water (assumed to be 1.0 for wetland plants). This assumes that C_s is constant but a more realistic calculation would incorporate the reduction in concentration that may occur over time. Assuming a 120-day growing season, the estimated sulfolane removal by cattails via root uptake was as high as to 32,800 mg m^{-2} year^{-1}.

Simple calculations of this type can be used to determine the potential impact of plant uptake at a particular site. In this example, the estimated uptake values for sulfolane suggest that wetland plants can play a significant role in its natural attenuation, except during periods of winter dormancy. It is noteworthy that since sulfolane is relatively nonvolatile, chemical transpired to the foliage may accumulate there and reach high concentrations.

14.10 CASE STUDY 2: UPTAKE INTO APPLE TREES AND FRUIT (TRAPP, 2007)

Building on earlier related efforts, Trapp (2007) developed a mechanistic model to predict uptake of neutral organic chemicals from soil and air into fruits. A more recent version has been described by Legind and Trapp (2009) and Trapp and Legind (2009). The 2007 model includes eight compartments (two soil compartments, fine roots, thick roots, stem, leaves, fruits, and air) with defined chemical equilibrium expressions, advective transport rates in xylem and phloem, diffusive exchange to soil and air, and growth dilution as the main processes considered. An example data set for apple orchard (1 m^2) was used to illustrate model output and the sensitivity to the input parameters. A listing of data sources is beyond the scope of this chapter, but the above papers suggest values for many of the parameters required.

In general, the model predicted that polar, nonvolatile compounds will be most effectively transported from soil to fruits, while lipophilic compounds will preferably accumulate from air into fruits. The model results also showed reasonable agreement to experimental data generated from a greenhouse study using ^{14}C-labeled TCE and sulfolane (Chard et al., 2006).

14.11 CONCLUSION

Quantitatively estimating the extent of organic contaminant uptake by the roots of plants and the subsequent fate within the plant is critical for assessing the potential of phytoremediation and for assessing rates of vegetation consumption by humans and wildlife for risk evaluations. Unfortunately, the lack of consistent reproducible empirical data has greatly limited model development and validation. Currently, approaches to estimate root uptake are based on simple bioconcentration factors and measured exposure concentrations. More promising are mechanistic models such as those of the Trapp group that require chemical, plant, and environmental properties as input. It is hoped that as standard protocols for the measurement of plant uptake data become available and used to generate consistent, high-quality data and models based on these data are tested, will there emerge more accurate models and methods for estimating mass transport of chemicals from soils to plant tissues.

LITERATURE CITED

Aitchison, E. W., S. L. Kelley, P. J. J. Alvarez, and J. L. Schnoor. 2000. Phytoremediation of 1,4-dioxane by hybrid poplar trees. *Water Environ. Res.* 72(3): 313–321.

Allen, R., L. Pereira, D. Raes, and M. Smith. 1998. Crop evapotranspiration: Guidelines for computing crop water requirements. FAO Irrigation and Drainage Paper number 56. Food and Agriculture Organization of the United Nations, Rome.

Arnold, C. W., D. G. Parfitt, and M. Kaltreider. 2007. Phytovolatilization of oxygenated gasoline-impacted groundwater at an underground storage tank site via conifers. *Int. J. Phytoremediation* 9(1–3): 53–69.

Baduru, K. K., S. Trapp, and J. G. Burken. 2008. Direct measurement of VOC diffusivities in tree tissues: Impacts on tree-based phytoremediation and plant contamination. *Environ. Sci. Technol.* 42(4): 1268–1275.

Baldocchi, D. 2003. Assessing the eddy covariance technique for evaluating carbon dioxide exchange rates of ecosystems: Past, present and future. *Global Change Biology* 9: 479–492.

Bintein, S. and J. Devillers. 1994. QSAR for organic chemical sorption in soils and sediments. *Chemosphere* 28(6): 1171–1188.

Bowling, D. J. F. 1980. Uptake and transport of nutrients. *Mineral Nutrition of Fruit Trees*. D. Atkinson, J. E. Jackson, R. O. Sharples, and W. M. Waller, eds. Butterworths, London.

Briggs, G. 1981. Theoretical and experimental relationships between soil adsorption, octanol–water partition coefficients, water solubilities, bioconcentration factors and the parachor. *J. Agric. Food Chem.* 29: 1050–1059.

Briggs, G. G., R. H. Bromilow, and A. A. Evans. 1982. Relationships between lipophilicity and root uptake and translocation of non-ionized chemicals by barley. *Pesticide Sci.* 13(5): 495–504.

Briggs, G. G., R. H. Bromilow, A. A. Evans, and M. Williams. 1983) Relationships between lipophilicity and root uptake and translocation of non-ionized chemicals in barley shoots following uptake by the roots. *Pesticide Sci.* 14: 492–500.

Briggs, G. G., R. L. O. Rigitano, and R. H. Bromilow. 1987. Physicochemical factors affecting uptake by roots and translocation to shoots of weak acids in barley. *Pesticide Sci.* 19(2): 101–112.

Bromilow, R. H. and K. Chamberlain. 1995. Principles governing uptake and transport. *Plant Contamination: Modeling and Simulation of Organic Chemical Processes.* S. Trapp, and J.C. McFarlane, eds. Boca Raton, FL: Lewis Publishers, pp. 37–68.

Burken, J. G. 2003. Uptake and Metabolism of Organic Compounds. Green-Liver Model. *Phytoremediation: Transformation and Control of Contaminants.* S.C. McCutcheon and J.L. Schnoor, eds. Indianapolis, IN: John Wiley & Sons, pp. 59–84.

Burken, J. G. and J. L. Schnoor. 1998. Predictive relationships for uptake of organic contaminants by hybrid poplar trees. *Environ. Sci Technol.* 32(21): 3379–3385.

Camp, C. R., E. J. Sadler, and R. E. Yoder. 1996. Evapotranspiration and Irrigation Scheduling American Society of Agricultural Engineers International Conference. San Antonio, TX, p. 1166.

Campbell, N. A., J. B. Reece, and L. G. Mitchell. 1999. *Biology.* Harlow, UK: Pearson Benjamin Cummings.

Chard, B. K., W. J. Doucette, J. K. Chard, B. Bugbee, and K. Gorder. 2006. Trichloroethylene uptake by apple and peach trees and transfer to fruit. *Environ. Sci. Technol.* 40(15): 4788–4793.

Chiou, C., L. Peters, and V. Freed. 1979. A physical concept of soil–water equilibria for nonionic organic compounds. *Science* 206: 831–832.

Chiou, C. T., G.Y. Sheng, and M. Manes. 2001) A partition-limited model for the plant uptake of organic contaminants from soil and water. *Environ. Sci. Technol.* 35: 1437–1444.

Ciucani, G., M. Trevisan, G. A. Sacchi, and S. A. J. Trapp. 2002. Measurement of xylem translocation of weak electrolytes with the pressure chamber technique. *Pest Manage. Sci.* 58(5): 467–473.

Collins, C., M. Fryer, and A. Grosso. 2006a. Evaluation of models for predicting plant uptake of chemicals from soil. Science Report—SC050021/SR. Environment Agency, Rio House, Waterside Drive, Aztec West, Almondsbury, Bristol BS32 4UD.

Collins, C., M. Fryer, and A. Grosso. 2006b. Plant uptake of non-ionic organic chemicals. *Environ. Sci. Technol.* 40(1): 45–52.

Copolovici, L. and U. Niinemets. 2007. Salting-in and salting-out effects of ionic and neutral osmotica on limonene and linalool Henry's law constants and octanol–water partition coefficients. *Chemosphere* 69(4): 621–629.

Davis, L. C., N. Muralidharan, V. P. Visser, C. Chaffin, W. G. Fateley, L. E. Erickson, and R. M. Hammaker. 1994. Alfalfa plants and associated microorganisms promote biodegradation rather than volatilization of organic-substances from ground-water. *Bioremediation Rhizosphere Technol.* 563: 112–122.

Delle Site, A. 2001. Factors affecting sorption of organic compounds in natural sorbent/water systems and sorption coefficients for selected pollutants. A review. *J. Phys. Chem. Ref. Data* 30(1): 187–439.

Dettenmaier, E., W. J. Doucette, and B. Bugbee. 2009. Chemical hydrophobicity and plant root uptake. *Environ. Sci .Technol.* 43(2): 324–332.

Doucette, W. J. 2003. Quantitative structure–activity relationships for predicting soil-sediment sorption coefficients for organic chemicals. *Environ. Toxicol. Chem.* 22(8): 1771–1788.

Doucette, W. J., B. G. Bugbee, S. C. Hayhurst, C. J. Pajak, and J. S. Ginn 2003. Uptake, metabolism, and phytovolatilization of trichloroethylene by indigenous vegetation: Impact of precipitation. *Phytoremediation: Transformation and Control of Contaminants.* S. C. McCutcheon and J. L. Schnoor, eds. Indianapolis, IN: Wiley, p. 968.

Doucette, W. J., J. K. Chard, H. Fabrizius, C. Crouch, M. R. Petersen, T. E. Carlsen, B. K. Chard, and K. Gorder. 2007. Trichloroethylene uptake into fruits and vegetables: Three-year field monitoring study. *Environ. Sci. Technol.* 41(7): 2505–2509.

Doucette, W. J., J. K. Chard, B. J. Moore, W. J. Staudt, and J. V. Headley (2005a. Uptake of sulfolane and diisopropanolamine (DIPA) by cattails (*Typha latifolia*). *Microchem. J.* 81(1): 41–49.

Doucette, W. J., B. R. Wheeler, J. K. Chard, B. Bugbee, C. G. Naylor, J. P. Carbone, and R. C. Sims. 2005b. Uptake of nonylphenol and nonylphenol ethoxylates by crested wheatgrass. *Environ. Toxicol. Chem.* 24(11): 2965–2972.

Edwards, N. T., B. M. Ross-Todd, and E. G. Graver. 1982. Uptake and metabolism of 14C anthracene by soybean (glycine max). *Environ. Expt. Bot.* 22(3): 349–357.

Franco, A. and S. Trapp. 2008) Estimation of the soil–water partition coefficient normalised to organic carbon for ionizable organic chemicals. *Environ. Toxicol. Chem.* 27 (10): 1995–2004.

Franco A., W. Fu, and S. Trapp. 2009. Soil pH and sorption of ionisable chemicals: Effects and modelling advance. *Environ. Toxicol. Chem.* 28: 458–464.

Garbarini, D. R. and L. W. Lion. 1986. Influence of the nature of soil organics on the sorption of toluene and trichloroethylene. *Environ. Sci. Technol.* 20(12): 1263–1269.

Goss, K. U. and R. P. Schwarzenbach. 2001. Linear free energy relationships used to evaluate equilibrium partitioning of organic compounds. *Environ. Sci. Technol.* 35(1): 1–9.

Hamaker, J. W. and J. Thompson. 1972. Adsorption. *Organic Chemicals in the Soil Environment.* C. Goring and J. W. Hamaker, eds. New York: Marcel Dekker, pp. 49–143.

Headley, J. V., L. C. Dickson, and K. M. Peru. 2002. Comparison of levels of sulfolane and diisopropanolamine in natural wetland vegetation exposed to gas-condensate contaminated ground water. *Commun. Soil Sci. Plant Anal.* 33(15–18): 3531–3544.

Hendrix, J. E. 1995. Assimilate transport and partitioning. *Handbook of Plant and Crop Physiology.* M. Pessarakli, ed. New York: Marcel Dekker.

Henzler, T., R. N. Waterhouse, A. J. Smyth, M. Carvajal, D. T. Cooke, A. R. Schaffner, E. Steudle, and D. T. Clarkson. 1999. Diurnal variations in hydraulic conductivity and root pressure can be correlated with the expression of putative aquaporins in the roots of *Lotus japonicus. Planta* 210(1): 50–60.

Hong, M. S., W. F. Farmayan, I. J. Dortch, C. Y. Chiang, S. K. McMillan, and J. L. Schnoor. 2001. Phytoremediation of MTBE from a groundwater plume. *Environ. Sci. Technol.* 35(6): 1231–1239.

Hsu, F. C., R. L. Marxmiller, and A. Y. S. Yang. 1990. Study of root uptake and xylem translocation of cinmethylin and related-compounds in detopped soybean roots using a pressure chamber technique. *Plant Physiol.* 93(4): 1573–1578.

Hung, H. and D. Mackay. 1997) A novel and simple model of the uptake of organic chemicals by vegetation from air and soil. *Chemosphere* 35: 959–977.

Javot, H. and C. Maurel. 2002. The role of aquaporins in root water uptake. *Ann. Bot.* 90(3): 301–313.

Karickhoff, S. 1981. Semi-empirical estimation of sorption of hydrophobic pollutants on natural sediments and soils. *Chemosphere* 10: 833–846.

Komossa, D. and H. Sandermann. 1995. Plant metabolic studies of the growth-regulator maleic hydrazide. *J. Agric. Food Chem.* 43(10): 2713–2715.

Lambers, H., F. Chapin III, and T. Pons. 1998. *Plant Physiological Ecology.* New York: Springer-Verlag.

Lang, A. 1990. Xylem, phloem and transpiration flows in developing apple fruits. *J. Exp. Bot.* 41(227): 645–651.

Lang, A. and M. R. Thorpe. 1989. Xylem, phloem and transpiration flows in a grape— application of a technique for measuring the volume of attached fruits to high-resolution using Archimedes' principle. *J. Exp. Bot.* 40(219): 1069–1078.

Larsen M., Burken J., Macháčková J., Karlson U.G., and Trapp S. 2008. Using tree core samples to monitor natural attenuation and plume distribution after a PCE spill. *Environ. Sci. Technol.* 42: 1711–1717.

Legind, C. N. and S. Trapp. 2009. Modeling the exposure of children and adults via diet to chemicals in the environment with crop-specific models. *Environ. Pollution.* 157(3): 778–785.

Ma, X. M. and J. G. Burken. 2002. VOCs fate and partitioning in vegetation: Use of tree cores in groundwater analysis. *Environ. Sci. Technol.* 36(21): 4663–4668.

Ma, X. M. and J. G. Burken. 2003a. TCE diffusion to the atmosphere in phytoremediation applications. *Environ. Sci. Technol.* 37(11): 2534–2539.

Ma, X. M. and J. G. Burken. 2003b. VOCs fate and partitioning in vegetation: Potential use of tree cores in groundwater analysis. *Environ. Sci. Technol.* 37(9): 2024–2024.

Mackay, A. A. and P. M. Gschwend. 2000. Sorption of monoaromatic hydrocarbons to wood. *Environ. Sci. Technol.* 34(5): 839–845.

Mackay, D., W. Shiu, and K. Ma. 1992. *Illustrated Handbook of Physical–Chemical Properties and Environmental Fate for Organic Chemicals.* Boca Raton, FL: Lewis Publishers.

Marschner, H. 1995. *Mineral Nutrition of Higher Plants,* 2nd ed. New York: Academic Press.

Maurel, C., H. Javot, V. Lauvergeat, P. Gerbeau, C. Tournaire, V. Santoni, and J. Heyes. 2002. Molecular physiology of aquaporins in plants. *Int. Rev. Cytol.,* 25(215): 105–148.

McCutcheon, S. C. and J. L. Schnoor, eds. 2003) *Phytoremediation: Transformation and Control of Contaminants.* Hoboken, NJ: Wiley. .

McFarlane, J. C. 1995. Anatomy and physiology of plant conductive systems. *Plant Contamination: Modeling and Simulation of Organic Chemical Processes.* S. Trapp and J. C. McFarlane, eds. Boca Raton, FL: Lewis Publishers, pp. 13–36.

McKone, T. E. and R. L. Maddalena. 2007. Plant uptake of organic pollutants from soil: Bioconcentration estimates based on models and experiments. *Environ. Toxicol. Chem.* 26(12): 2494–2504.

Mikes O., P. Cupr, S. Trapp, and J. Klanova. 2008. Uptake of polychlorinated biphenyls and organochlorine pesticides from soil and air into radishes (*Raphanus sativus*). *Environ. Pollut.,* doi:10.1016/j.envpol.2008.09.007.

Newman, L. A., S. E. Strand, N. Choe, J. Duffy, G. Ekuan, M. Ruszaj, B. B. Shurtleff, J. Wilmoth, P. Heilman, and M. P. Gordon. 1997. Uptake and biotransformation of trichloroethylene by hybrid poplars. *Environ. Sci. Technol.* 31(4): 1062–1067.

Nilsen, E. T. and D. M. Orcutt. 1996. *Physiology of Plants Under Stress.* New York: Wiley.

Orchard, B. J., W. J. Doucette, J. K. Chard, and B. Bugbee. 2000. Uptake of trichloroethylene by hybrid poplar trees grown hydroponically in flow-through plant growth chambers. *Environ. Toxicol. Chem.* 19(4): 895–903.

Ouyang, Y. 2002. Phytoremediation: Modeling plant uptake and contaminant transport in the soil–plant–atmosphere continuum. *J. Hydrol.* 266(1–2): 66–82.

Ouyang, Y., D. Shinde, and L.Q. Ma. 2005. Simulation of phytoremediation of a TNT-contaminated soil using the CTSPAC model. *J. Environ. Quality* 34: 1490–1496.

Paterson, S., D. Mackay, and A. Gladman. 1991. A fugacity model of chemical uptake by plants from soil and air. *Chemosphere* 24: 539–565.

Paterson, S., D. Mackay, and C. McFarlane. 1994. A model of organic chemical uptake by plants from soil and the atmosphere. *Environ. Sci. Technol.* 28 (13): 2259–2266.

Pauliukonis, N. and R. Schneider. 2001. Temporal patterns in evapotranspiration from lysimeters with three common wetland plant species in the eastern United States. *Aquatic Bot.* 71(1): 35–46.

Pflugmacher, S. and H. Sandermann. 1998. Taxonomic distribution of plant glucosyltransferases acting on xenobiotics. *Phytochemistry* 49(2): 507–511.

Pulford, I. D. and C. Watson. 2003. Phytoremediation of heavy metal-contaminated land by trees—a review. *Environ. Int.* 29(4): 529–540.

Rao, P. and J. Davidson. 1980. Estimation of pesticide retention and transformation parameters required in non-point source pollution models. In *Environmental Impact of Nonpoint Source Pollution*. M. Overcashand and J. Davidson, eds. Ann Arbor, MI: Ann Arbor Science, pp. 23–67.

Raskin, I. and B. D. Ensley, eds. 2000. *Phytoremediation of Toxic Metals: Using Plants to Clean Up the Environment*. New York: Wiley.

Rubin, E. and A. Ramaswami. 2001. The potential for phytoremediation of MTBE. *Water Res.* 35(5): 1348–1353.

Ryan, J. A., R. M. Bell, J. M. Davidson, and G. A. O'Conner. 1988. Plant uptake of non-ionic organic chemicals from soils. *Chemosphere* 17: 2299–2323.

Saison, C., C. Schwartz, and J. L. Morel. 2004. Hyperaccumulation of metals by *Thlaspi caerulescens* as affected by root development and Cd–Zn/Ca–Mg interactions. *Int. J. Phytoremediation* 6(1): 49–61.

Salt, D. E., M. Blaylock, N. P. Kumar, V. Dushenkov, B. D. Ensley, I. Chet, and I. Raskin. 1995. Phytoremediation: A novel strategy for the removal of toxic metals from the environment using plants. *Biotechnology* 13(5): 468–474.

Schreiber, L., K. Hartmann, M. Skrabs, and J. Zeier. 1999. Apoplastic barriers in roots: Chemical composition of endodermal and hypodermal cell walls. *J. Exp. Botany* 50(337): 1267–1280.

Seth, R., D. Mackay, and J. Muncke. 1999. Estimating the organic carbon partition coefficient and its variability for hydrophobic chemicals. *Environ. Sci. Technol.* 33(14): 2390–2394.

Severtson, S. J. and S. Banerjee. 1996. Sorption of chlorophenols to wood pulp. *Environ. Sci. Technol.* 30(6): 1961–1969.

Shone, M. G. T. and A. V. Wood. 1974. A comparison of the uptake and translocation of some organic herbicides and a systemic fungicide by barley. *J. Exp. Bot.* 25(85): 390–400.

Sicbaldi, F., G. A. Sacchi, M. Trevisan, and A. A. M. Del Re. 1997. Root uptake and xylem translocations of pesticides from different chemical classes. *Pestic. Sci.* 50: 111–119.

Steudle, E. and C. A. Peterson. 1998. How does water get through roots? *J. Exp. Botany* 49(322): 775–788.

Taghavi, S., T. Barac, B. Greenberg, B. Borremans, J. Vangronsveld, and D. Van Der Lelie. 2005. Horizontal gene transfer to endogenous endophytic bacteria from poplar improves phytoremediation of toluene. *Appl. Environ. Microbiol.* 71: 8500.

Taiz, L. and E. Zeiger. 1998. *Plant Physiology*. Sunderland, MA, Sinauer Associates.

Tam, D. D., W. Y. Shiu, and D. Mackay. 1996. Uptake of chlorobenzenes by tissue of the soybean plant: Equilibria and kinetics. *Environ. Toxicol. Chem.* 15(5): 489.

Topp, E., I. Scheunert, A. Attar, and F. Korte. 1986) Factors affecting the uptake of C-14-labeled organic-chemicals by plants from soil. *Ecotoxicol. Environ. Safety* 11: 219–228.

Trapp, S. 2000. Modelling uptake into roots and subsequent translocation of neutral and ionisable organic compounds. *Pest Manage. Sci.* 56(9): 767–778.

Trapp, S. 2002. Dynamic root uptake model for neutral lipophilic organics. *Environ. Toxicol. Chem.* 1(1): 203–206.

Trapp, S. 2004. Plant uptake and transport models for neutral and ionic chemicals. *Environ. Sci. Pollution Res.* 11(1): 33–39.

Trapp, S. 2007. Fruit Tree model for uptake of organic compounds from soil and air. *SAR QSAR Environ. Res.* 18: 367–387.

Trapp, S., A. Cammarano, E. Capri, F. Reichenberg, and P. Mayer. 2007. Diffusion of PAH in potato and carrot slices and application for a potato model. *Environ. Sci. Technol.* 41(9): 3103–3108.

Trapp, S. and C.N. Legind. 2009. Uptake of organic contaminants from soil into vegetation. In: *Dealing with Contaminated Sites. From Theory towards Practical Applications.* F. Swartjes, ed. Dordrecht: Springer.

Trapp, S. and M. Matthies. 1995. Generic one-compartment model for uptake of organic-chemicals by foliar vegetation. *Environ. Sci. Technol.* 29: 2333–2338.

Trapp, S., C. McFarlane, and M. Matthies. 1994. Model for uptake of xenobiotics into plants—validation with bromacil experiments. *Environ. Toxicol. Chem.* 13(3): 413–422.

Trapp, S., K. S. B. Miglioranza, and H. Mosbaek. 2001. Sorption of lipophilic organic compounds to wood and implications for their environmental fate. *Environ. Sci. Technol.* 35(8): 1561–1566.

Trapp, S., D. Rasmussen, and L. Samsoe-Petersen. 2003. Fruit tree model for uptake of organic compounds from soil. *SAR and QSAR Environ. Res.,* 14 (1): 17–26.

Travis, C. C. and A.D. Arms. 1988. Bioconcentration of organics in beef, milk, and vegetation. *Environ. Sci. Technol.* 22 (3): 271–274.

Vose, J., G. Harvey, K. Elliott, and B. Clinton. 2003. Measuring and modeling tree and stand level transpiration. In: *Phytoremediation: Transformation and Control of Contaminants.* S. McCutcheon and J. Schnoor, eds. New York: Wiley, Chapter 8.

Vroblesky, D. A., B. D. Clinton, J. M. Vose, C. C. Casey, G. Harvey, and P. M. Bradley. 2004. Ground water chlorinated ethenes in tree trunks: Case studies, influence of recharge, and potential degradation mechanism. *Groundwater Monitoring Remediation* 24(3): 124–138.

Vroblesky, D. A., C. T. Nietch, and J. T. Morris. 1999. Chlorinated ethenes from groundwater in tree trunks. *Environ. Sci. Technol.* 33(3): 510–515.

Weis and Weis. 2004) Metal uptake, transport and release by wetland plants: Implications for phytoremediation and restoration. *Environ. Int.* 30(5): 685–700.

Wullschleger, S. D., F. C. Meinzer, and R. A. Vertessy. 1998. A review of whole-plant water use studies in trees. *Tree Physiol.* 18(8–9): 499–512.

Yeo, A. and T. J. Flowers, 2007. *Plant Solute Transport.* Oxford, UK: Blackwell.

Yu, X. Z. and J. D. Gu. 2006. Uptake, metabolism, and toxicity of methyl tert-butyl ether (MTBE) in weeping willows. *J. Hazardous Mater.* 137(3): 1417–1423.

Yu X., S. Trapp, Z. Puhua, W. Chang, and Z. Xishi. 2004. Metabolism of cyanide by Chinese vegetation. Chemosphere 56: 121–126.

Yu, X., S. Trapp, P. Zhou, and H. Hu. 2005. The effect of temperature on the rate of cyanide metabolism of two woody plants. *Chemosphere* 59: 1099–1104.

Yu X. Z., S. Trapp, P. Zhou, and L. Chen. 2007. The effect of temperature on uptake and metabolism of cyanide by weeping willows. *Int. J. Phytorem.* 9: 243–255.

Zhang, Q. Z., L. C. Davis, and L. E. Erickson. 2001. Transport of methyl tert-butyl ether through alfalfa plants. *Environ. Sci. Technol.* 35(4): 725–731.

15 Dispersion and Mass Transfer Coefficients in Groundwater of Near-Surface Geologic Formations

Tissa H. Illangasekare, Christophe C. Frippiat, and Radek Fučík

CONTENTS

Part 1   Dispersivity in Soil and the Up-Most Groundwater of
         Geologic Formations
15.1   Introduction 414
15.2   Transport Process: Differential Advection 415
15.3   Transport Theory...416
       15.3.1   Hydrodynamic Dispersion at the Microscopic Pore Scale 416
       15.3.2   Governing Equations..418
       15.3.3   Hydrodynamic Dispersion at the Macroscale.....................419
       15.3.4   Upscaling Models for Dispersion Coefficients 420
15.4   Estimation of Transport Parameters...421
       15.4.1   Regression Laws to Estimate Dispersivity Coefficients..........422
       15.4.2   Laboratory Methods for the Determination of Dispersivity.......425
             15.4.2.1   Methods for Column Tests 425
             15.4.2.2   Device for Transverse Dispersivity 426
       15.4.3   Field Methods for the Determination of
             Dispersivity Coefficients...427
             15.4.3.1   Natural Gradient Tests 427
             15.4.3.2   Forced Gradient Test 428
15.5   Example Calculations...429
       15.5.1   Laboratory Estimation of α_L for an Unsaturated
             Medium Sand..429
       15.5.2   Analysis of Single-Well Withdrawal Tests in a
             2D Heterogeneous Medium 431
Literature Cited 433

Additional Reading 437
Part 2 Mass Transfer Coefficients in Pore-Water Adjacent to Nonaqueous
        Liquids and Particles..438
15.6 Introduction 438
15.7 Distribution and Morphology of NAPLs in Porous Media 438
15.8 Conceptual Models of Mass Transfer......................................439
      15.8.1 One-Dimensional Vertical Dispersion Model 439
      15.8.2 Linear Driving Model for Interphase Mass Transfer 440
      15.8.3 Stagnant Film Model...441
15.9 Empirical Mass Transfer Rate Coefficients................................442
15.10 Example Problems 445
      15.10.1 Dissolution using Spherical Blobs Model 447
      15.10.2 Dissolution using Tubular Model 447
      15.10.3 Numerical Experiments...449
Literature Cited 450
Additional Reading 451

PART 1 DISPERSIVITY IN SOIL AND THE UP-MOST GROUNDWATER OF GEOLOGIC FORMATIONS

15.1 INTRODUCTION

This section focuses on dispersion, a primary process that contributes to the transport of dissolved chemicals (solutes) in porous media. The specific porous medium that is of focus is the upper water-bearing zones of subsurface geologic formations. This part of the subsurface is bounded by the ground surface as the upper boundary, where the intergranular spaces of the soil are only partially filled with water with the rest of the pore spaces occupied by air. This zone is referred to as the unsaturated, partially saturated or vadose zone of the aquifer. When water, the wetting fluid, and air, the nonwetting fluid, occupy the same pore space, the surface tension at the water/air interfaces results in the water pressure to be less than the air pressure (negative gauge pressure). The bottom boundary of the unsaturated zone below which the pores are filled with water is the water table. In the absence of fluid interfaces, the water pressure is higher than atmospheric pressure (positive gauge pressure). This aquifer zone is referred to as the saturated zone. When water-soluble chemicals enter unsaturated or saturated zones of aquifers, they are transported through two primary mechanisms, namely advection and dispersion. The process of advection that is a result of water flow was discussed in Chapter 11. A second process that contributes to the transport of dissolved chemicals both in the unsaturated and saturated zones of aquifers is associated with hydrodynamic mixing, resulting from the velocity variations that occur at the microscopic pore scale. This process is parameterized through a relationship that contains the pore-water velocity and a parameter that is referred to as dispersivity. In most practical field situations, the dispersivity cannot be measured at the pore scale. The dispersivity values that are estimated from field-scale observations or tracer tests depend on the spatial variability of soil characteristics in space and the scale of the measurement. Hence, the dispersivity is considered to be scale-dependent.

The outline of the material to be presented is as follows. The physical process that contributes to hydrodynamic dispersion and how the process is parameterized at the macroscopic scale is reviewed. A summary of existing knowledge on the scale dependence of dispersivity is presented. This will be followed by a discussion on how the parameter is estimated in the field using various field testing methods. Finally, two example applications will be presented to demonstrate how this process is modeled.

15.2 TRANSPORT PROCESS: DIFFERENTIAL ADVECTION

Consider two fluids of equal viscosity and equal density. One of the fluids is displacing the other one from a porous medium. Initially, also assume that the flow is one-dimensional. The mean position of the front of the second fluid will evolve according to the mean advective velocity. However, as the displacement progresses, both fluids will mix due to diffusion and mechanical dispersion.

Mechanical dispersion is the tendency for fluids to spread out from the flow lines that they would be expected to follow according to the advective hydraulics of the flow system. This spreading process results from microscopic velocity variations, causing fluid particles to move at various velocities through the tortuous paths of the medium. There are three basic mechanisms producing these pore-scale velocity variations: (1) the variability in pore lengths, which causes fluid elements starting at a given distance from each other and proceeding at the same velocity not to remain the same distance apart, (2) friction along soil grains and viscous shear forces, yielding a smaller velocity at the border of a pore, and a maximum velocity at its center, and (3) the variability in pore sizes, which results in a variability of pore-scale velocity. Mechanical dispersion is a nonsteady and irreversible process, as initial fluid distributions cannot be recovered by reversing the flow direction.

Figure 15.1 describes the classical laboratory column experiment used to determine mechanical dispersion. Steady-state flow is established in a column packed with a homogeneous granular medium. A nonreactive tracer at concentration C_0 [ML^{-3}] is continuously introduced at the upstream end of the column from time t_0 [T]. If the column is initially solute-free, the tracer input can be represented as a step-function (Figure 15.1b). The relative concentration C/C_0 [−] of the column outflow is plotted as a function of time (Figure 15.1c). This type of curve is called a *breakthrough curve*. If there is no mixing of any sort, the plot of C/C_0 is a step change from 0 to 1 at $t = t_m$, where t_m corresponds to advective transport through the column. If the only mixing process taking place is molecular diffusion, sharp concentration gradients will be smoothened out and the plot of C/C_0 will slightly spread. In real situations, mechanical mixing will cause a significantly larger spreading of concentration distributions. An early breakthrough will be observed for $t \ll t_m$ as a result of microscopic velocities larger than the mean velocity. Reciprocally, the concentration distribution will also exhibit a long tail for $t \gg t_m$ due to fluid particles moving along slow-velocity flow lines. When diffusion can be neglected, the plot of C/C_0 is therefore a representation of the pore-scale velocity distribution.

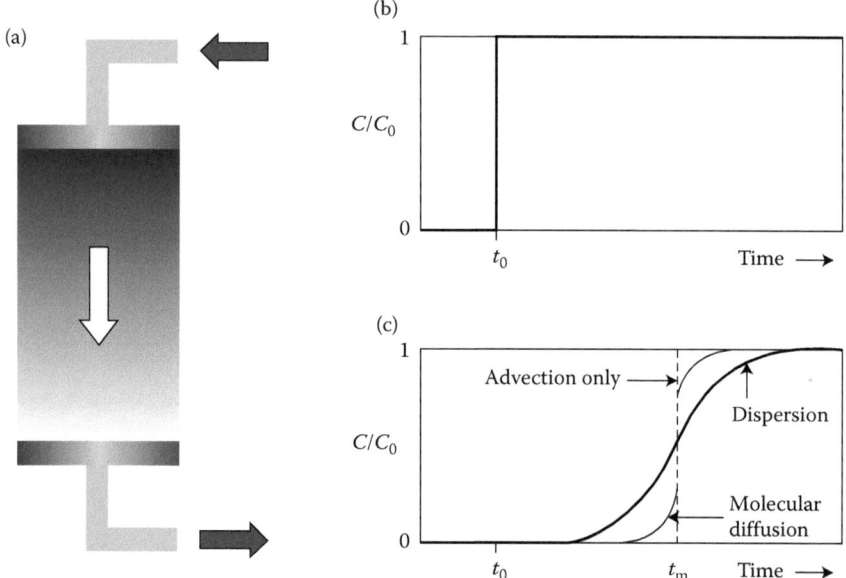

FIGURE 15.1 One-dimensional column experiment: (a) Sketch of the column device; (b) step-function input of tracer; and (c) Relative tracer concentration at column outlet and the effect of advection, diffusion, and dispersion. (After Freeze, A.R. and Cherry, J.A. 1979. *Groundwater*, Upper Saddle River, Prentice Hall, NJ.)

When the transport problem is multidimensional, even if the flow system remains one-dimensional, a solute plume originating from a point source will disperse both longitudinally and transversely to mean flow direction. Transverse dispersion is caused by the fact that the flow paths can split and branch out to the side to bypass soil grains as a fluid flows through a porous medium. This will occur even in the laminar flow conditions that are prevalent in groundwater flow.

15.3 TRANSPORT THEORY

15.3.1 Hydrodynamic Dispersion at the Microscopic Pore Scale

As the effect of dispersion is similar to that of diffusion, the dispersive solute flux is classically represented using a diffusion-like or Fickian law:

$$J_m = -\theta D_m \nabla C, \tag{15.1}$$

where J_m [$ML^{-2}T^{-1}$] is the dispersive solute mass flux in direction, θ [$-$] is the volumetric water content, and D_m [L^2T^{-1}] is a fictitious diffusion coefficient called mechanical dispersion. As mechanical dispersion is mathematically analogous to diffusion at the microscopic scale and as both processes cannot be separated from each other in flowing groundwater, they are usually combined into a single parameter called hydrodynamic dispersion coefficient:

$$D = D_m + D_d, \qquad (15.2)$$

where D_d $[L^2 T^{-1}]$ an effective diffusion coefficient. In a three-dimensional system, the hydrodynamic dispersion coefficient is a second-order tensor that takes the form

$$D = \begin{bmatrix} D_{xx} & D_{yx} & D_{zx} \\ D_{xy} & D_{yy} & D_{zy} \\ D_{xz} & D_{yz} & D_{zz} \end{bmatrix}, \qquad (15.3)$$

In a uniform flow field, if the principal directions of the dispersion tensor are aligned with the principal directions of the velocity flow field, the dispersion coefficient tensor can be reduced to

$$D = \begin{bmatrix} D_L & 0 & 0 \\ 0 & D_{TH} & 0 \\ 0 & 0 & D_{TV} \end{bmatrix}. \qquad (15.4)$$

where D_L $[L^2 T^{-1}]$ is a longitudinal hydrodynamic dispersion coefficient, and D_{TH} and D_{TV} $[L^2 T^{-1}]$ are horizontal and vertical transverse hydrodynamic dispersion coefficients, respectively. When horizontal and vertical transverse dispersion coefficients are equal, one defines $D_T = D_{TH} = D_{TV}$.

The relative contribution of mechanical dispersion and diffusion to solute transport is evaluated using Peclet numbers. A Peclet number is a dimensionless number that relates the effectiveness of mass transport by advection to the effectiveness of mass transport by diffusion or dispersion. Peclet numbers have the general form

$$\text{Pe} = \frac{vd}{D_d} \quad \text{or} \quad \text{Pe} = \frac{vL}{D_L}.$$

$v = q/\theta [LT^{-1}]$ is the average pore-water velocity and q $[LT^{-1}]$ is the specific discharge of water through the porous medium, or the Darcy velocity. d [L] is a characteristic grain size and L [L] is a characteristic transport distance. For low Peclet numbers, D_L and D_T are both equal to the effective diffusion coefficient. At larger Peclet numbers, longitudinal and transverse coefficients of hydrodynamic dispersion are found to depend strongly on the average pore-scale water velocity. The exact relationship between pore-scale dispersion and velocity can obtained from theoretical considerations for simple or hypothetical pore systems (Saffman, 1959). Except in the case of very simple conceptual models, one can generally find that the coefficients of hydrodynamic dispersion are linearly related to velocity

$$D_L = D_d + \alpha_L v, \qquad (15.5a)$$

$$D_{TH} = D_d + \alpha_{TH} v, \qquad (15.5b)$$

$$D_{TV} = D_d + \alpha_{TV} v, \qquad (15.5c)$$

where α_L, α_{TH}, and α_{TV} [L] are characteristic lengths called longitudinal dispersivity, horizontal transverse dispersivity, and vertical transverse dispersivity, respectively. Since dispersivities quantify mechanical dispersion resulting from pore-scale velocity variations, they are characteristic properties of a medium. Field-studies have shown that Equations 15.5a through 15.5c are also valid at large scale, for typical groundwater flow conditions. For example, Klotz et al. (1980) investigated a more general relation $D_L = Av^B + D_d$ and found that exponent B should be close to 1. They also showed the dependence of dispersivity to soil sedimentological properties.

Equations 15.5a through 15.5c have been shown to accurately model dispersion in saturated porous media and for a stationary flow in unsaturated media. In transient conditions, however, the relationship between hydrodynamic dispersion coefficients and velocity becomes more complicated. In unsaturated media, the water content of the soil changes with the water flux. Hence, the structure of the water-filled pore space also changes with the water flux. The flow field, and therefore the distribution of pore velocities, depends on the saturation of the medium (Flury et al., 1994). As a consequence, dispersivity coefficients are strongly impacted by the volumetric water content. Usually, dispersivity is found to increase when the water content decreases as a result of the larger tortuosity of solute trajectories and a disconnection of continuous flow paths (Vanclooster et al., 2006). In some cases, especially when the activation of macropores significantly enhances pore-water variability, dispersivity is found to increase with volumetric water content. Currently, there is no unique validated theoretical model available for dispersivity in transient unsaturated flow.

15.3.2 GOVERNING EQUATIONS

Combining advective flux and Fickian hydrodynamic dispersive flux, and applying the principle of mass conservation over a representative elementary volume of soil yields

$$\frac{\partial \theta C}{\partial t} = \nabla \cdot (\theta D \cdot \nabla C - C \cdot q(\theta)), \tag{15.6}$$

where the specific discharge of water through the porous medium depends on the volumetric water content of the medium. Equation 15.6 is the governing equation for solute transport in unsaturated porous media. It is usually referred to as the *advection–dispersion equation* (ADE) or the *convection–dispersion equation*. The initial and boundary value problem obtained by combining the above second-order PDE with the initial concentration distribution in the medium and appropriate boundary conditions is solved to obtain space–time distributions of solute concentrations. It must be noted that specific discharge and volumetric water content to be used in Equation 15.6 must be obtained by separately solving the unsaturated flow equation.

In unsaturated medium, especially at low water content, the liquid phase is not fully connected, and therefore not fully participating to the flow. In such a situation, Equation 15.6 must be augmented by a sink term that accounts for mass exchange by diffusion toward stagnant zones. This type of model is usually referred to as a *mobile–immobile model*, or a *two-region model* (Coats and Smith, 1964; van Genuchten and

Wierenga, 1978).

$$\frac{\partial \theta_m C_m}{\partial t} + \frac{\partial \theta_{im} C_{im}}{\partial t} = \nabla \cdot (\theta_m D \cdot \nabla C_m - C_m \cdot q(\theta_m)), \tag{15.7}$$

where θ_m [$-$] and θ_{im} [$-$] are the volumetric fraction of mobile and immobile water, respectively. C_m [ML^{-3}] and C_{im} [ML^{-3}] are solute concentrations in the mobile and immobile zone respectively. In this case, D refers to hydrodynamic dispersion in the mobile zone. As an additional unknown appears in Equation 15.7, an additional relationship is required to solve the problem. Usually, it comes from the assumption of linear nonequilibrium or rate-limited mass transfer (Coats and Smith, 1964)

$$\frac{\partial C_{im}}{\partial t} = \omega(C_m - C_{im}), \tag{15.8}$$

where ω [T^{-1}] is a mass transfer rate coefficient (see Part 2). A one-dimensional diffusion model can also be used (Rao et al., 1980). Breakthrough curves computed from Equation 15.7 are characterized by a significant tailing and longer times to reach a unit relative concentration as a result of slow diffusion exchange of solutes between the mobile and the immobile zone.

15.3.3 HYDRODYNAMIC DISPERSION AT THE MACROSCALE

The traditional approach to modeling transport in natural formations is to assume that the advection–dispersion equation also holds at large scale. However, field investigations show in a consistent manner that the values of dispersion coefficients derived under laboratory conditions do not apply to large scale transport. Whereas typical values of dispersivity from column experiments range between 0.01 and 0.1 m, values of macroscopic dispersivity (or macrodispersivity) are in general three to four orders of magnitude larger (Gelhar et al., 1992; Lallemand-Barres and Peaudecerf, 1978). It has also been widely observed that field-scale dispersion coefficients increase with distance and with time (Sauty, 1980).

The main key to understanding this scale effects is heterogeneity. Dispersion is an advective process, as it is caused by variations in fluid velocity. However, variations in fluid velocity do not only take place at the pore scale, but also occurs at larger scales, ranging from macroscopic to megascopic. At the field scale, commonly encountered geological structures influence contaminant transport drastically, leading to velocity variations over several orders of magnitude. This includes the effects of stratification and the presence of lenses with higher or lower permeability. At the megascopic scale, differences between geologic formations also cause nonideality in solute transport. As the flow path increases in length, a solute plume can encounter greater and greater variations in the aquifer, causing the variability of the velocity field to increase. Because dispersivity is related to the variability of the velocity, neglecting or ignoring the true velocity distribution (i.e., by replacing the heterogeneous medium by an equivalent homogeneous one) must be compensated for by a corresponding higher apparent (or effective) dispersivity, leading to what is commonly called the scale effect of dispersion.

15.3.4 Upscaling Models for Dispersion Coefficients

During the past three decades, a number of theoretical studies have been carried out to describe field-scale dispersive mixing as a function of soil heterogeneity and develop upscaling methods for the estimation of macrodispersivities. These upscaling methods can usually be categorized into deterministic or stochastic methods.

Deterministic upscaling methods require the spatial variability of the hydraulic conductivity of the soil to be fully characterized. Flow and transport are solved for a given set of initial and boundary conditions, either using analytical or numerical methods. Macroscopic mixing properties of the heterogeneous medium are then obtained by assuming that the solute plume is migrating in an equivalent homogeneous medium. Historically, deterministic upscaling models turned out to be mostly applied to compute macrodispersion coefficients of perfectly stratified aquifers (Berentsen, 2005, 2007; Guven et al., 1984; Marle et al., 1967; Mercado, 1967).

The idea behind stochastic models is that soil properties cannot be practically fully characterized. To a certain extent, the hydraulic conductivity exhibits random patterns, which result in a statistical uncertainty of concentration distributions. Stochastic analysis enables the variability in flow and transport to be related to the variability and the spatial structure associated to hydraulic properties of the heterogeneous medium considered. Let us define Y as the natural logarithm of the hydraulic conductivity K, and assume that Y is normally distributed. This assumption accommodates the large hydraulic variations that can be found in the field and excludes negative values, which is consistent with the physical requirement that permeability is positive. The distribution of Y is fully characterized by its mean and its covariance function. The covariance function describes the variability of Y, based on two parameters: The variance σ_Y^2 [−] and the correlation length λ [m]. The variance is a measure of the degree of variability of Y, whereas λ quantifies its spatial variability. A large value of λ indicates that Y values are correlated over large distances. On the contrary, a small value of λ indicates that there is no particular spatial structure for Y. λ can therefore be understood as a characteristic length of heterogeneity.

Stochastic upscaling theories are found to be attractive since they allow the estimation of macrodispersion coefficients based on a statistical description of soil heterogeneity. They also allow the demonstration of the scale-dependence of macrodispersion coefficients. For large scale, stochastic theories usually predict the convergence of macrodispersion coefficients toward constant asymptotic values. For example, the asymptotic value of longitudinal macrodispersivity α_L^* for a saturated isotropic medium with $\alpha_L \ll \lambda$ is given by (Dagan, 1984; Gelhar and Axness, 1983)

$$\alpha_L^*(\infty) = \sigma_Y^2 \frac{\lambda}{\gamma^2}, \tag{15.9}$$

where γ [−] is a flow factor accounting for the dependency of effective permeability on dimensionality. In two-dimensional situations, $\gamma = 1$, whereas in three-dimensional situations, $\gamma = \exp(\sigma_Y^2/6)$. It must be noted that Dagan (1984) states that γ should be kept equal to 1 in all situations. Equation 15.9 shows that macrodispersivity is directly linked to the structure of the log-hydraulic conductivity field, and increases when the variability of Y increases. Similar analytical expressions can be obtained for transverse macrodispersivity α_T^*

$$\alpha_T^*(\infty) = \sigma_Y^2 \frac{\alpha_L + 3\alpha_T}{8} \, (2D), \tag{15.10a}$$

$$\alpha_T^*(\infty) = \sigma_Y^2 \frac{\alpha_L + 4\alpha_T}{15\gamma^2} \, (3D), \tag{15.10b}$$

where Equations 15.10a and 15.10b are applicable to two-dimensional (2D) and three-dimensional (3D) plumes, respectively. As for longitudinal macrodispersivity, transverse macrodispersivity is thus found to depend on the dimensionality of the problem considered. Equations 15.10a and 15.10b also show that when local mixing can be neglected, heterogeneity does not produce any macroscale transverse spreading. Gelhar and Axness (1983) have computed exact expressions for α_L^* and α_T^* under various conditions. Time-dependent analytical solutions of α_L^* in two- and three-dimensional isotropic media are given by Dagan (1988). Other authors have derived analytical expressions in other specific cases (see the reviews by Dagan, 1989; Gelhar, 1993; Rubin, 2003). Stochastic theories are typically limited to $\sigma_Y^2 \ll 1$ and to situations where λ is much smaller than the scale of the problem.

The authors have also applied stochastic methods to situations where hydraulic conductivity is not log-normally distributed. Rubin (1995) and Stauffer and Rauber (1998) propose analytical expressions for macrodispersion coefficients in aquifers made of two materials of different hydraulic conductivity. Stochastic methods have also been applied to situations where heterogeneity cannot be characterized using a single finite correlation scale (Di Federico and Neuman, 1998; Rajaram and Gelhar, 1995; Zhan and Wheatcraft, 1996).

Most of the results presented above are related to solute transport in saturated heterogeneous media. In the vadose zone, the variability in water saturation usually contributes to enhance the variability in water velocity, and therefore solute spreading (Russo, 1998). However, it has been shown that macrodispersion coefficients for solute transport in unsaturated soils characterized by strong stratification are usually smaller than saturated values, especially at low water content (Harter and Zhang, 1999).

15.4 ESTIMATION OF TRANSPORT PARAMETERS

Although the theoretical studies reported in previous section have generated some important answers to key questions regarding scale effects, the estimation of dispersivities from the practitioner's point of view still faces a lot of difficulties. Stochastic upscaling methods for dispersivities require a significant amount of data to determine the statistical characteristics of hydraulic conductivity variations for a given site. Considering the costs of field investigation, it is generally rare to find a site that has enough data points for this kind of statistical evaluation.

Currently, the only practically viable method to obtain a priori estimates of dispersivities is by means of empirical approaches, which are based on regression curves fitted on dispersivity data. In this section, major compilations of existing data on dispersivity are reported, and regression laws are provided as rule-of-thumb estimations of core- and field-scale dispersivity coefficients. Then, laboratory methods to determine longitudinal and transverse dispersivity are described. Finally, field-scale tracer testing methodologies are presented. Indeed, sound field-scale modeling of solute transport cannot rely only on bulk a priori values or laboratory-scale estimates of solute transport parameters. *In situ* tracer tests must be conducted in order to understand site-specific advection and dispersion processes.

The methods reported below all assume that the medium under investigation is homogeneous. Hence, the methods allow the estimation of effective dispersivity coefficients at the scale of interest. However, for the sake of simplicity, the notation α_L will be used throughout this section, instead of α_L^*.

15.4.1 REGRESSION LAWS TO ESTIMATE DISPERSIVITY COEFFICIENTS

There are currently large controversial views regarding the interpretation of compiled field data to obtain universal scaling laws for dispersivity coefficients. Whereas some authors say that a single universal regression line would ignore the fact that different aquifers may have different degrees of heterogeneity at a given scale (Gelhar et al., 1992), others state that, on average, all aquifers have a similar behavior at a give scale and individual departures from the universal scaling rule must be viewed as local fluctuations around the mean behavior (Neuman, 1990). Moreover, uncertainty is often attached to field dispersivity values. Numerous factors, such as actual injection conditions, solute density effects, or even temporal variations of the advective flow regime or biased interpretation techniques, are likely to be interpreted as dispersion. Even at the laboratory scale, Bromly et al. (2007) showed that dispersivity values were highly dependent on the type and on the size of experimental device. The empirical laws presented in this section should therefore be used with extreme caution.

Table 15.1 reports several empirical laws to estimate core-scale dispersivities based on other physical properties of the soil, such as porosity n [−], median grain size d_{50}

TABLE 15.1
Regression Laws for Core-Scale Dispersivity Coefficients (mm)

Regression Law	Applicability	Source
$\alpha_L = 1.75\,d_{50}$		Perkins and Johnston, 1963
$\alpha_L = 3.49\,C_u - 1.41$		Xu and Eckstein, 1995
$\alpha_L = -3.51 + 4.41\,C_u$	Glass beads	Xu and Eckstein, 1997
$\alpha_L = -25.47 + 12.40/n$	Glass beads	Xu and Eckstein, 1997
$\alpha_L = 0.46 + 0.85\,d_{50}$	Glass beads, $C_u = 1$	Xu and Eckstein, 1997
$\alpha_L = -3.15 + 0.85\,d_{50} + 3.55\,C_u$	Glass beads, $C_u < 2$	Xu and Eckstein, 1997
$\alpha_L = -2.17 + 0.81\,d_{50} + 2.73\,C_u$	Glass beads, $C_u < 3$	Xu and Eckstein, 1997
$\alpha_L = -2.75 + 4.08\,C_u$	Glass beads, $C_u < 4$	Xu and Eckstein, 1997
$\alpha_L = 1.25\,d_{50}S_w^{-1.2}$	Glass beads, $S_w > 0.8$	Haga et al., 1999
$\alpha_L = 1.11\,d_{50}S_w^{-3.1}$	Glass beads, $S_w < 0.8$	Haga et al., 1999
$\alpha_L = d_{50}a^{-1}S_{we}^{-2}$	$a = 6a^*/n^* + 0.015$	Sato et al., 2003
$\alpha_T = 0.055\,d_{50}$		Perkins and Johnston, 1963

[mm] or coefficient of uniformity $C_u = d_{60}/d_{10}$ [−]. Mean grain size and uniformity of grain size are usually considered as the two most important factors affecting grain size. For relatively uniform materials, dispersivity is directly proportional to median grain size. For less uniform materials, the shape of the particle size distribution is the dominant factor for dispersivity and α_L is directly proportional to the coefficient of uniformity. Dispersivity is also found to be inversely proportional to porosity.

In unsaturated media, the estimation of pore-scale dispersivity is complicated by its additional dependence on the saturation degree S_w [−]. The saturation degree is the volume of water per unit pore volume of the medium. It is usually related to the capillary pressure. In an isolated unsaturated soil pore, a curved interface appears between air and water phases and a pressure difference exists across the interface. This pressure difference depends on the interfacial forces between air and water and on the radius of the pore. Following standard conventions, the capillary pressure is defined as the difference between the pressure in the air phase and in the water phase. At the continuum scale, there exists a relationship between capillary pressure and the saturation of the porous medium. A porous medium consists of a distribution of pores with different radii. If an increasing macroscopic capillary pressure is applied to a porous media, the air phase would invade the larger pores and the water phase would be present in smaller pores. The larger sized pores could not support the capillary pressure and would release water. Thus, the larger the capillary pressure, the smaller amount of water will be present in the porous medium. The relationship between capillary pressure and the water phase content is referred to as the capillary pressure curve or the retention function, which is an intrinsic property of a porous medium. A well-known model for the retention function of a porous medium is the model of van Genuchten (van Genuchten, 1980, 1991). It is characterized by two parameters a^* [cm^{-1}] and n^* [−]. a^* is related to the threshold capillary pressure required to start draining the porous medium. Hence, it is also related to the smallest pore size of the medium. n^* is related to the distribution of pore size. A small value of n^* reflects a large distribution of pore size, while a small value of n^* would apply to a relatively uniform porous medium (Lu and Likos, 2004). In the model of Sato et al. (2003), the pore-scale dispersivity is expressed as a function of the van Genuchten parameters and as a function of effective saturation degree (see Table 15.1). The effective saturation degree S_{we} [−] links with the saturation degree using $S_{we} = (S_w - S_{wr})/(1 - S_{wr})$, where S_{wr} [−] is the residual saturation degree. The longitudinal dispersivity tends to increase when the saturation of the medium decreases. For example, the equations provided by Haga et al. (1999) reported in Table 15.1 predict that dispersivity is 1.8 times larger at a saturation degree of 80% as compared to full saturation, 4.3 times larger at a saturation degree of 60%, and 15.2 times larger at a saturation degree of 40%.

Figure 15.2 shows one of the most recent compilations of longitudinal dispersivity values in field-scale saturated-flow situations. The trend for α_L to increase with scale L is relatively clear. Field data typically range between 0.01 m and 5500 m at scales of 0.75 m to 100 km. Also, the values for porous (unconsolidated) and fractured (consolidated rock) media tend to scatter over a similar range. At a given scale, the longitudinal dispersivity typically ranges over 2–3 orders of magnitude. This degree of variation can be explained in terms of stochastic macrodispersion theories presented in Section 15.3.3. When the reliability of the data is accounted for, the scale dependence

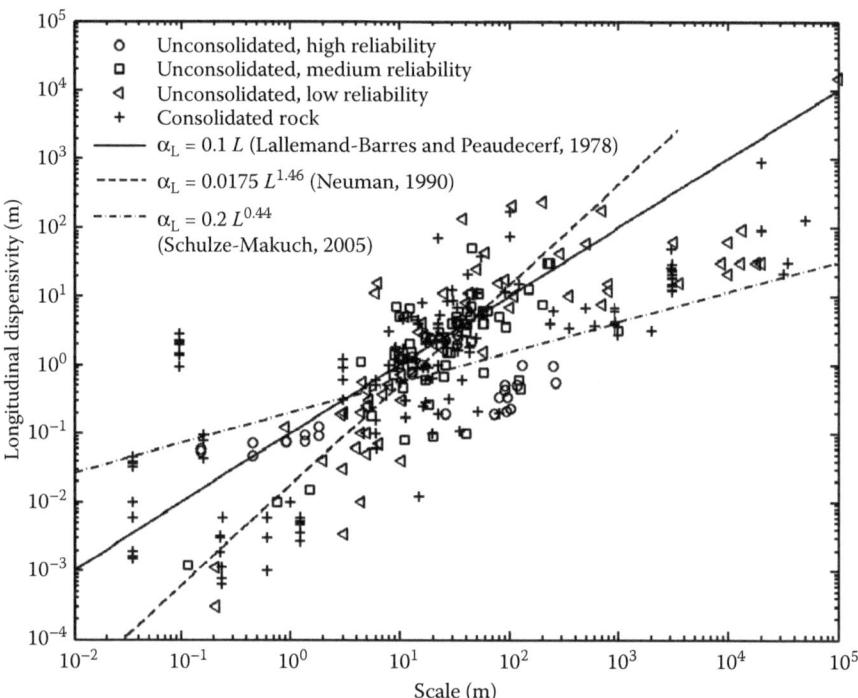

FIGURE 15.2 Field-scale longitudinal dispersivity coefficients as a function of scale. [Adapted from Schulze-Makuch, D. 2005. Longitudinal dispersivity data and implications for scaling behavior. *Ground Water* 43(3): 443–456.]

of longitudinal dispersivity is less obvious: There are no high-reliability points at scales larger than 300 m. This reflects the fact that large-scale α_L values are almost exclusively obtained from contamination plume simulations or environmental tracer studies. As large-scale controlled tracer experiments require a very long period of time, such experiments have not been conducted.

Scaling relationships for longitudinal dispersivities are usually described using power laws of the form

$$\alpha_L = cL^d, \tag{15.11}$$

where c $[L^{1-d}]$ is a characteristic property of the medium and d $[-]$ a scaling exponent. Early attempts to fit a regression law on compiled field data yielded a simple rule $\alpha_L = L/10$ (Lallemand-Barres and Peaudecerf, 1978). Later, Neuman (1990) found $\alpha_L = 0.175 L^{1.46}$, valid for $L < 3500$ m. He also fitted two separate regression lines for $L < 100$ m and $L > 100$ m. He found $\alpha_L = 0.0169 L^{1.53}$ for $L < 100$ m and $\alpha_L = 0.32 L^{1.83}$ for $L > 100$ m. Recently, Schulze-Makuch (2005) performed regressions on field data accounting for their reliability. He found $\alpha_L = 0.2 L^{0.44}$

using high-reliability dispersivity data only for unconsolidated sands. He also established regression laws for consolidated rocks of various types. Other authors have also provided other regression laws (Arya, 1986; Xu and Eckstein, 1995).

Gelhar et al. (1992) caution users routinely adopting α_L values from Figure 15.2 or from a linear representation of the data. Instead, users should favor the use of dispersivity values in the lower half of the range at any given scale. If values in the upper part of the range are adopted, excessively large dilution may be predicted and the environmental consequences misrepresented.

Field data on transverse dispersivity are relatively scarce, and available data are generally of a lower reliability compared to longitudinal dispersivity. Only Gelhar et al. (1992) provide a compilation of α_{TH} and α_{TV} values, but they do not provide regression laws. Typically, values of α_{TH} are found to be about one order of magnitude smaller than α_L, while values of α_{TV} are about two orders of magnitude smaller than α_L. The smaller values of α_{TV} reflect the roughly horizontal stratification of hydraulic conductivity in permeable sedimentary materials. Small α_{TV} values also imply that contaminant plumes will potentially show very limited vertical mixing with high concentrations at given horizons. The trend for transverse dispersivity coefficients to increase with scale is usually less clear due to the low reliability of larger-scale data, generally based on contaminant events, for which sources are ill-defined.

15.4.2 LABORATORY METHODS FOR THE DETERMINATION OF DISPERSIVITY

15.4.2.1 Methods for Column Tests

Pore-scale longitudinal dispersivity can be determined in the laboratory using columns packed with the porous media under investigation. The device is similar to that depicted in Figure 15.1 for saturated flow experiments. Under unsaturated flow conditions, flow boundary conditions must be adapted. Usually, the column is placed vertically, with an irrigation system on top, imposing a constant discharge. One appropriate analytical solution to the advection-equation is (Kreft and Zuber, 1978)

$$\frac{C}{C_0} = \frac{1}{2}\mathrm{erfc}\left(\frac{L - vt}{2\sqrt{D_L t}}\right) + \frac{1}{2}\exp\left(\frac{vL}{D_L}\right)\mathrm{erfc}\left(\frac{L + vt}{2\sqrt{D_L t}}\right), \qquad (15.12)$$

where, in unsaturated conditions, $v = v(\theta)$ and $D_L = D_L(\theta)$. erfc is the complementary error function. It assumes as initial condition $C(x,0) = 0$, and as boundary conditions $C(0,t) = C_0$ and $C(\infty,t) = 0$. Fitting of this solution (e.g., using a least-square criterion) onto observed breakthrough curves allows the simultaneous determination of D_L and v.

For high Peclet numbers, the second term in Equation 15.12 can be neglected. Rewriting this equation for saturated conditions using the number of pore volumes $U = vt/L\ [-]$ as temporal variable yields

$$\frac{C}{C_0} = \frac{1}{2}\mathrm{erfc}\left(\frac{1 - U}{2\sqrt{\frac{UD_L}{vL}}}\right). \qquad (15.13)$$

This equation has several properties that render the estimation of v and D_L easier: The plot of the outlet concentration curve as a function of $J = (U - 1)/\sqrt{U}$ corresponds to a normal probability distribution with a mean $\mu_J = 0$ and a standard deviation $\sigma_J = \sqrt{2D_L/vL}$. The plot C/C_0 versus J on normal probability paper should therefore be linear. The mean pore velocity is estimated using $v = L/t_m$, t_m corresponding to $C/C_0 = 0.5$. The value of D_L is found from

$$D_L = \frac{vL}{8}(J_{0.84} - J_{0.16})^2, \tag{15.14}$$

where $J_{0.16}$ and $J_{0.84}$ are the values of J corresponding to 16% and 84% of relative concentration, respectively.

This method is typically valid when effluent concentrations are measured. When using a measurement device that allows the measurement of pore-water concentrations, other boundary conditions apply to the advection equation and Equation 15.12 is not valid anymore (Kreft and Zuber, 1978; van Genuchten and Parker, 1984). Attention must also be paid to experimental artifacts arising from specific laboratory devices. For example, if injection is performed in a volume of water outside of the column (like a device to maintain the piezometric head), the actual injection condition is not an instantaneous step variation. Due to mixing with the volume of water, the injection is actually exponential. Not accounting for such effects can result in a serious bias in the estimated values of dispersivity (Novakowski, 1992).

15.4.2.2   Device for Transverse Dispersivity

Existing methods to estimate transverse dispersion are usually based either on tracer tests or on dissolution tests. Dissolution tests generally imply groundwater flow along a stagnant zone containing constant concentration gas (Klenk and Grathwohl, 2002; McCarthy and Johnson, 1993), NAPL (Oostrom et al., 1999a; 1999b; Pearce et al., 1994) or solid (Delgado and Guedes de Carvalho, 2001; Guedes de Carvalho and Delgado, 1999; 2000). Transverse dispersivity can then be inferred from the rate of dissolution of the third phase that is obtained through solute breakthrough curve measurements at the laboratory model outlet.

Most of laboratory tracer tests designed to determine transverse dispersion coefficients are performed in a uniform flow at constant mean velocity. Blackwell (1962), Hassinger and von Rosenberg (1968), and recently Frippiat et al. (2008) used the so-called "annulus-and-core" approach, in which the inlet and the outlet cross-sections of a column are divided into two concentric zones. The concentration of the solution flowing in the inner inlet zone (the core) is rapidly increased, while the solution in the outer inlet zone (the annulus) is kept solute-free. Transverse dispersivity is computed by comparing steady-state concentration of effluent solutions in the outlet annulus and core zones. Divided inlets were also adopted in several other column studies involving intrusive local concentration measurements (Bruch, 1970; Grane and Gardner, 1960; Han et al., 1985; Harleman and Rumer, 1963; Perkins and Johnston, 1963; Zhang et al., 2006). Other authors preferred point injection (Pisani and Tosi, 1994; Robbins, 1989). A few specific devices imply nonuniform flow: Cirpka and Kitanidis (2001) and Benekos et al. (2006) investigate flow and transport in a helix and in a cochlea to determine transverse dispersivity. Kim et al. (2004) determined local longitudinal and transverse dispersivities in a laboratory aquifer model with a local recharge zone.

15.4.3 FIELD METHODS FOR THE DETERMINATION OF DISPERSIVITY COEFFICIENTS

In theory, velocity and dispersivities can be estimated from virtually any test where tracer is added in a controlled way to the groundwater. However, a few standard tests are generally preferred because simple procedures are available to interpret the results. The choice of which test configuration to adopt then results from practical or economical constrains, from the duration of test, to the number of observation wells, to the spatial scale to investigate.

Standard tracer tests are typically used at relatively small field scales. Estimates of dispersivity at scales larger than several hundred of meters usually rely on different methods, either using historical contamination data or exploiting natural variations in the chemistry of natural recharge of the aquifer. However, estimates of advection and dispersion based on data from contaminant plumes or environmental tracer measurements are less reliable than field tracer tests, since there is a larger uncertainty in the location and the intensity of source zone. Often, there is also an inadequate number of sampling points.

15.4.3.1 Natural Gradient Tests

The natural gradient test involves monitoring a small volume of tracer as it moves down the flow system. The resulting concentration distributions provide the data necessary to determine advective velocities, dispersivities, but also chemical parameters. This type of test is usually considered to be of a high reliability. When the test is performed in a supposedly homogeneous formation using a fully penetrating well, a two-dimensional analytical solution of the advection–dispersion equation can be used (Domenico and Schwartz, 1997):

$$\frac{C}{C_0} = \frac{V/b}{4\pi t \sqrt{D_L D_T}} \exp\left(-\frac{(x - vt)^2}{4D_L t} - \frac{y^2}{4D_T t}\right), \quad (15.15)$$

in which it is assumed that the injection well is located at the origin of the coordinate system and that velocity is constant and aligned with the x-axis. V [L^3] is the volume of tracer solution injected and b [L] is the thickness of the aquifer. When the injection well is screened on a very small portion of its length, the three-dimensional solution of the advection–dispersion equation must be used (Domenico and Schwartz, 1997):

$$\frac{C}{C_0} = \frac{V}{8(\pi t)^{3/2} \sqrt{D_L D_{TH} D_{TV}}} \exp\left(-\frac{(x - vt)^2}{4D_L t} - \frac{y^2}{4D_{TH} t} - \frac{z^2}{4D_{TV} t}\right). \quad (15.16)$$

Dispersivity coefficients can be estimated by fitting Equation 15.15 or 15.16 on concentration data monitored in observation wells. Other analytical solutions are also available when the lateral extent of the source cannot be neglected or for sorbing or decaying tracer species (see, e.g., Domenico and Schwartz, 1997).

15.4.3.2 Forced Gradient Test

Forced gradient tests are conducted using injection and/or pumping wells, locally increasing hydraulic gradients at levels significantly larger than those naturally occurring in aquifers. The advantage is that test duration is greatly diminished. Compared to a natural gradient test of the same duration, the tested volume of the aquifer is also usually larger. As a result, due to the unavoidable heterogeneity of the soil, dispersivities obtained from forced-gradient tests are also usually larger than the values obtained from natural gradient tests performed in the same aquifer (Fernandez-Garcia et al., 2005; Tiedeman and Hsieh, 2004).

15.4.3.2.1 Single-Well Injection or Withdrawal Test

The single-well injection test involves pumping at a constant rate. In case an observation well is located close to the injection well, observed concentrations during the injection phase can be fitted using (Gelhar and Collins, 1971)

$$\frac{C}{C_0} = \frac{1}{2}\mathrm{erfc}\left(\frac{r^2 - R^{*2}}{\left(\frac{16}{3}\alpha_L\left(R^{*3} - r_w^3\right)\right)^{1/2}}\right), \tag{15.17}$$

R^* [L] being the mean radial position of the tracer front

$$R^* = \sqrt{\frac{Qt}{\pi b\theta}}, \tag{15.18}$$

and r [L] is the radial position from the injection well, r_w [L] is the radius of the well, Q [L^{3}T^{-1}] is the injection rate and b [L] the thickness of the aquifer. Values of longitudinal dispersivity determined using this type of test are usually considered to be relatively reliable (Gelhar et al., 1992).

In the single-well withdrawal test, a radially converging steady-state flow field is established by pumping at a constant rate in a well. A fixed amount of tracer is injected in a second well. A solute plume develops and starts to migrate toward the pumping well. Analytical solutions for concentration in the pumping well under such conditions are not straightforward to evaluate and often imply semianalytical expressions with power series (Chen, 1999; Moench, 1989). Simplified solutions are given by Sauty (1980), and Welty and Gelhar (1994). One of the advantages of the single-well withdrawal test is that it allows the simultaneous estimation of longitudinal and transverse dispersivities. However, since the converging flow field tends to counteract spreading due to longitudinal dispersion, α_L estimates are thought to be of a lower reliability.

15.4.3.2.2 Single-Well Push–Pull Test

This type of test involves two distinct flow phases. First, tracer is injected in a well at a constant flow rate. The tracer is moving radially from the well. After a certain period of injection, flow is reversed and the tracer is pumped out of the soil at the same rate. The tracer is moving radially toward the pumping well. The single-well push–pull test does not require any observation well: Concentrations are monitored at the well during the recovery phase. Measured data can be analyzed using the analytical solution developed by Gelhar and Collins (1971):

$$
\frac{C}{C_0} = \frac{1}{2}\mathrm{erfc}\left(\frac{\dfrac{V_p}{V_i} - 1}{\left(\dfrac{16}{3}\dfrac{\alpha_L}{R^*}\left[2 - \left(1 - \dfrac{V_p}{V_i}\right)^{3/2}\right]\right)^{1/2}}\right). \tag{15.19}
$$

where V_i [L^3] is the total volume of water injected in the aquifer during the first phase, and $V_p = V_p(t)$ [L^3] is the volume of water withdrawn from the aquifer at time t of phase 2.

The single-well push–pull test is a small-scale test and is generally found to have a limited applicability in estimating macrodispersivities. The dispersion process in a single-well push–pull test is significantly different from that of unidirectional flow: Macrodispersion near the injection well results from differential advection caused by vertical variations in hydraulic conductivity. As a result, the tracer travels at different velocities as it radiates outwards. But it will also travel with the same velocity pattern as it goes back to the production well. This means that the mixing process is partially reversible and that the dispersivity might be underestimated compared to that of unidirectional flow.

15.4.3.2.3   Two-Well Tracer Test

In the two-well test, water is pumped from one well and injected into the other at the same rate to create a steady-state flow regime. The tracer is added in the injection well and monitored in the withdrawal well. Also, dispersivity estimates can be improved by adding more observation wells between the pumping-injection doublets. In general, these tests can be performed over several hundreds of meters in sandy formations. Analytical solutions are provided by Grove and Beetem (1970) and Maloof and Protopapas (2001).

Dirac input should be preferred rather than step input. A potential problem with the step input test configuration is that the breakthrough curve is not strongly influenced by dispersion except in the early stages, when concentrations are low. For this reason, tests based on this approach are generally considered to produce low-reliability dispersivity data.

15.5   EXAMPLE CALCULATIONS

15.5.1   Laboratory Estimation of α_L for an Unsaturated Medium Sand

Problem description: Figure 15.3 shows the results of two one-dimensional laboratory test performed by Sato et al. (2003). The porous medium consists of a

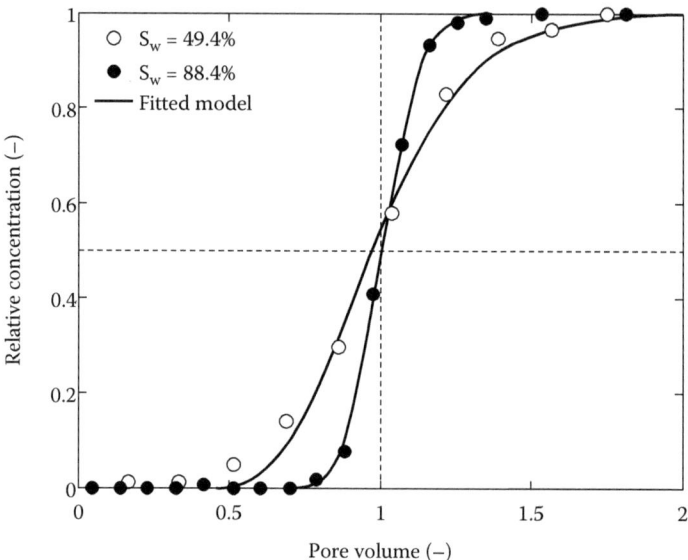

FIGURE 15.3 Experimental breakthrough curves for Toyoura sand and fitted solutions of the advection–dispersion equation. (After Sato, T., Tanahashi, H. and Loaiciga, H.A. 2003. Solute dispersion in a variably saturated sand. *Water Resources Research* **39**: doi: 10.1029/2002WR001649.)

repacked sample of Toyoura sand, a medium sand characterized by a median grain size $d_{50} = 180\,\mu m$ and van Genuchten parameters $a^* = 0.036\,cm^{-1}$ and $n^* = 4.2$. The column has an internal diameter of 5 cm and a length of 12 cm. Steady-state unsaturated flow was established by injecting water at the top of the column at a constant rate and draining water at the bottom of the column. Controlled air suction was applied at the bottom of the column to suppress boundary effects and establish a constant vertical profile of water content through the column. Two tests are reported: Test 1 was performed at a saturation of $S_w = 49.4\%$, and Test 2 was performed at a saturation of $S_w = 88.4\%$. A pore-water velocity of 0.5 cm/min was used for each test.

Question: Determine experimental values of longitudinal dispersivity from the time series of concentration recorded during each column test.

Solution: For a homogeneous medium with a constant saturation degree, Equation 15.12 can be used to analyze the experimental data. Fitting of this equation to the data (e.g., by minimizing the sum of the squared residuals between the equation and the data) yields experimental dispersivity values of $\alpha_L = 0.4$ cm and $\alpha_L = 0.06$ cm for Test 1 and Test 2, respectively. The corresponding column Peclet numbers are Pe = 30 and Pe = 195 for Test 1 and Test 2, respectively. These values are large, so the use of Equation 15.13 instead of Equation 15.12 yields relatively similar results. Using the van Genuchten parameters of the saturation curve, the value of parameter a was used in the empirical law established by Sato et al. (2003) is $a = 0.066$ (see Table 15.1). The empirical values of longitudinal dispersivity are then 0.27 cm and 0.08 cm for Test 1 and Test 2 respectively, which is reasonably close to actual values.

15.5.2 ANALYSIS OF SINGLE-WELL WITHDRAWAL TESTS IN A 2D HETEROGENEOUS MEDIUM

Problem description: Chao et al. (2000) carried out intermediate-scale tracer experiments in a two-dimensional horizontal laboratory tank (244 cm × 122 cm × 6.35 cm). The tank was packed with five different sands, in order to create a heterogeneous medium with well-defined statistical properties. The sands used were crushed silica sands. The heterogeneous packing was designed to simulate a lognormal distribution of saturated hydraulic conductivity (K) using five different sands. The resulting ln K distribution had a mean value of 4.75, a variance of 1.81, and an isotropic correlation length of 10 cm (Figure 15.4). Convergent tracer tests were performed, using potassium bromide as a conservative tracer. A total of 36 tracer tests were carried out, in order to investigated the effect of (1) pumping rate; (2) distance between injection well and pumping well; and (3) direction between injection well and pumping well.

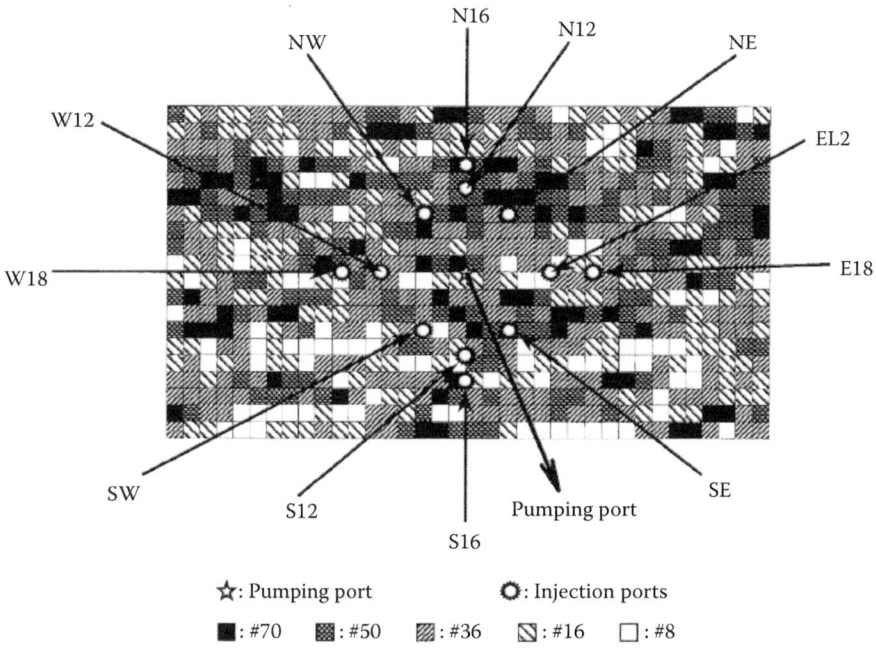

FIGURE 15.4 Experimental representation of the two-dimensional laboratory tank, showing the heterogeneous pattern of the hydraulic conductivity field and the locations of injection and pumping ports. Each block has dimensions of 6.1 × 6.1 cm. The sands are referred to using their respective sieve size. (From Chao, H.-C., Rajaram, H., and Illangasekare, T.H. 2000. *Water Resources Research* 36(10): 2869–2884. With permission.)

Question: Determine experimental values of longitudinal dispersivity from the time series of concentration recorded during each tracer test.

Solution: The breakthrough curves from the tracer tests were analyzed using a two-dimensional analytical solution provided by Welty and Gelhar (1994):

$$\frac{C}{C_0} = \frac{V}{2\pi\theta bR^2 \sqrt{\dfrac{16\pi\alpha_L}{3R}\left(1 - \left|1 - \dfrac{t}{t^*}\right|^{3/2}\right)}} \exp\left(\frac{-\left(1 - \dfrac{t}{t^*}\right)^2}{\dfrac{16\alpha_L}{3R}\left(1 - \left|1 - \dfrac{t}{t^*}\right|^{3/2}\right)}\right)$$

(15.20)

where R [L] is the radial distance between the injection well and the pumping well. The mean transit time t^* [T] can be computed using Equation 15.18. A least-square criterion was adopted to determine α_L by fitting Equation 15.20 onto experimental data. As summarized in Table 15.2, the dispersivity estimates ranged from 0.065 to 0.953 cm. The high variability of α_L reflects the variability of the hydraulic conductivity field. Even at the same scale, different tracer tests in the same heterogeneous medium yield widely different estimates of transport parameters. The estimated dispersivities at the same scale varied by a factor of 2–6.

Chao et al. (2000) also performed a one-dimensional tracer test in their tank. They obtained a longitudinal dispersivity of 12.0 cm. The theoretical value computed using Equation 15.9 is 18.1 cm. Several factors account for this discrepancy:

TABLE 15.2
Dispersivity (cm) Estimates from Radial Flow Tracer Experiments

Injection Port	Radius (cm)	Q = 25 mL/min	Q = 50 mL/min	Q = 75 mL/min
NE	25.4	0.207	0.201	0.140
NW	25.4	0.182	0.245	0.146
SE	25.4	0.116	0.109	0.065
SW	25.4	0.329	0.370	0.399
E12	30.4	0.362	0.363	0.485
W12	30.4	0.439	0.333	0.225
S12	30.4	0.392	0.465	0.369
N12	30.4	a	0.482	0.499
E18	45.7	0.741	0.567	0.395
W18	45.7	0.658	0.777	0.953
S16	40.6	a	a	a
N16	40.6	0.368	0.439	0.303

Source: From Chao, H.-C., Rajaram, H., and T.H. Illangasekare. 2000. *Water Resources Research* 36(10): 2869–2884. With permission.

a Unreliable data.

- Dispersivity is actually scale-dependent, and the theoretical value computed using Equation 15.9 is a large-scale asymptotic value, whereas the experimental value corresponds to a finite displacement of about 22 correlation lengths.
- The presence of lateral no-flow boundaries tends to decrease the overall variability of flow, and therefore the macroscale value of dispersivity.
- The tracer test was carried out in a single realization of the heterogeneous hydraulic conductivity field. There are number of other realizations satisfying the same statistical properties, each potentially yielding different α_L values. Since Equation 15.9 yields a theoretical value that represents the average over all possible realizations of the K field, one could expect the value of a single realization to be different.

The main reason for the discrepancy between the one-dimensional tracer test and the radial tracer test lies in the dimension of the source zone. One-dimensional tracer tests are characterized by a large source zone (i.e., the full cross-section of the medium), whereas convergent tracer tests have a point source. In the latter case, solute plumes do not sample the full variability of aquifer properties, and therefore undergo smaller dispersion processes. This mostly highlights that, even if theories are available to predict macroscale dispersion coefficients, they are bounded to certain limitations which could make them unsuited to given situations.

LITERATURE CITED

Arya, A. 1986. Dispersion and reservoir heterogeneity. PhD dissertation. University of Texas at Austin.

Benekos, I.D., Cirpka, O.A. and P.K. Kitanidis. 2006. Experimental determination of transverse dispersivity in a helix and a cochlea. *Water Resources Research* 42: W07406, doi:10.1029/2005/WR004712.

Berentsen, C.W.J., Verlaan, M.L. and C.P.J.W. van Kruijsdijk. 2005. Upscaling and reversibility of Taylor dispersion in heterogeneous porous media. *Physical Review E* 71: doi: 10.1103/PhysRevE.71.046308.

Berentsen, C.W.J., van Kruijsdijk, C.P.J.W. and M.L. Verlaan. 2007. Upscaling, relaxation and reversibility of dispersive flow in stratified porous media. *Transport in Porous Media* 68: 187–218.

Blackwell, R.J. 1962. Laboratory studies of microscopic dispersion phenomena. *Society of Petroleum Engineers Journal* 2(1): 1–8.

Bromly, M., Hinz, C., and L.A.G. Aylmore. 2007. Relation of dispersivity to properties of homogeneous saturated repacked soil columns. *European Journal of Soil Science* 58: 293–301.

Bruch, J.C. 1970. Two-dimensional dispersion experiments in a porous medium. *Water Resources Research* 6(3): 791–800.

Chen, J.-S., Liu, C.-W. and C.-M. Liao. A novel analytical power series solution for solute transport in a radially convergent flow field. *Journal of Hydrology* 266: 120–138.

Chao, H.-C., Rajaram, H. and T.H. Illangasekare. 2000. Intermediate-scale experiments and numerical simulations of transport under radial flow in a two-dimensional heterogeneous porous medium. *Water Resources Research* 36(10): 2869–2884.

Cirpka, O.A. and P.K. Kitanidis. 2001. Theoretical basis for the measurement of local transverse dispersion in isotropic porous media. *Water Resources Research* 37(2): 243–252.

Coats, K.H. and B.D. Smith. 1964. Dead-end pore volume and dispersion in porous media. *Society of Petroleum Engineers Journal* 4: 73–84.

Dagan, G. 1984. Solute transport in heterogeneous formations. *Journal of Fluid Mechanics* 145: 151–177.

Dagan, G. 1988. Time-dependent macrodispersion for solute transport in anisotropic heterogeneous aquifers. *Water Resources Research* 24(9): 1491–1500.

Dagan, G. 1989. *Flow and Transport in Porous Formations*. Berlin: Springer-Verlag.

Delgado, J.M.P.Q. and J.R.F. Guedes de Carvalho. 2001. Measurement of the coefficient of transverse dispersion in flow through packed beds for a wide range of values of the Schmidt number. *Transport in Porous Media* 44: 165–180.

Di Federico, V. and S.P. Neuman. 1998. Transport in multiscale log conductivity fields with truncated power variograms. *Water Resources Research* 34(5): 963–973.

Domenico, P.A. and F.W. Schwartz. 1997. *Physical and Chemical Hydrogeology*, 2nd ed. New York: Wiley.

Fernandez-Garcia, D., Illangasekare, T.H. and H. Rajaram. 2005. Differences in the scale-dependence of dispersivity and retardation factors estimated from forced-gradient and uniform flow tracer tests in three-dimensional physically and chemically heterogeneous porous media. *Water Resources Research* 41: doi:10.1029/2004WR003125.

Flury, M., Fluhler, H., Jury, W.A. and J. Leueberger. 1994. Susceptibility of soils to preferential flow of water: A field study. *Water Resources Research* 31: 2443–2452.

Freeze, A.R. and J.A. Cherry. 1979. *Groundwater*. Upper Saddle River, NJ: Prentice-Hall.

Frippiat, C.C., Conde Pérez, P. and A.E. Holeyman. 2008. Estimation of laboratory-scale dispersivities using an annulus-and-core device. *Journal of Hydrology* 362: 57–68, doi: 10.1016/j.jhydrol.2008.08.007.

Gelhar, L.W. 1993. *Stochastic Subsurface Hydrology*. Englewood Cliffs, NJ: Prentice-Hall.

Gelhar, L.W. and M.A. Collins. 1971. General analysis of longitudinal dispersion in nonuniform flow. *Water Resources Research* 7(6): 1511–1521.

Gelhar, L.W. and C.L. Axness. 1983. Three-dimensional stochastic analysis of macrodispersion in aquifers. *Water Resources Research* 19(1): 161–180.

Gelhar, L.W., Welty, C. and K.R. Rehfeldt. 1992. A critical review of data on field-scale dispersion in aquifers. *Water Resources Research* 28(7): 1955–1974.

Grane, F.E. and G.H.F. Gardner. 1961. Measurement of transverse dispersion in granular media. *Journal of Chemical and Engineering Data* 6(2): 283–287.

Grove, D.B. and W.A. Beetem. 1971. Porosity and dispersion constant calculations for a fractured carbonated aquifer using the two well tracer test method. *Water Resources Research* 7(1): 128–134.

Guedes de Carvalho, J.R.F. and J.M.P.Q. Delgado. 1999. Mass transfer from a large sphere buried in a packed bed along which liquid flows. *Chemical Engineering Science* 54: 1121–1129.

Guedes de Carvalho, J.R.F. and J.M.P.Q. Delgado. 2000. Lateral dispersion in liquid flow through packed beds at Pem< 1400. *AIChE Journal* 46(5): 1089–1095.

Guven, O., Molz, F.J. and J.G. Melville. 1984. An analysis of dispersion in a stratified aquifer. *Water Resources Research* 20(10): 1337–1354.

Haga, D., Niibori, Y. and T. Chida. 1999. Hydrodynamic dispersion and mass transfer in unsaturated flow. *Water Resources Research* 35(4): 1065–1077.

Han, N.-W., Bhakta, J. and R.G. Carbonell. 1985. Longitudinal and lateral dispersion in packed beds: effect of column length and particle size distribution. *AIChE Journal* 31(2): 277–288.

Harleman, D.R.F. and R.R. Rumer. 1963. Longitudinal and lateral dispersion in an isotropic porous medium. *Journal of Fluid Mechanics* 16: 385–394.

Harter, T. and D. Zhang. 1999. Water flow and solute spreading in heterogeneous soils with spatially variable water content. *Water Resources Research* 35(2): 415–426.

Hassinger, R.C. and von Rosenberg, D.U. 1968. A mathematical and experimental investigation of transverse dispersion coefficients. *Society of Petroleum Engineers Journal* 8(2): 195–204.

Kim, S.-B., Jo, K.-H., Kim, D.-J. and W.A. Jury. 2004. Determination of two-dimensional laboratory-scale dispersivities. *Hydrological Processes* 18: 2475–2483.

Klenk, I.D. and P. Grathwohl. 2002. Transverse vertical dispersion in groundwater and the capillary fringe. *Journal of Contaminant Hydrology* 58: 111–128.

Klotz, D., Seiler, K.-P., Moser, H. and F. Neumaier. 1980. Dispersivity and velocity relationship from laboratory and field experiments. *Journal of Hydrology* 45: 169–184.

Kreft, A. and A. Zuber. 1978. On the physical meaning of the dispersion equation and its solutions for different initial and boundary conditions. *Chemical Engineering Science* 33: 1471–1480.

Lallemand-Barres, A. and P. Peaudecerf. 1978. Recherche des relations entre la valeur de la dispersivity macroscopique d'un milieu aquifere, ses autres caracteristiques, et les conditions de mesure. Etude bibliographique. *Bulletin du BRGM 2e serie*, section III, 4: 277–284.

Lu, N. and W.J. Likos. 2004. *Unsaturated Soil Mechanics*. Hoboken, NJ: John Wiley and Sons.

Maloof, S. and A.L. Protopapas. 2001. New parameters for solution of two-well dispersion problem. *Journal of Hydrologic Engineering* 6(2): 167–171.

Marle, C.M., Simandoux, P., Parcsirszky, J. and C. Gaulier. 1967. Etude du deplacement de fluids miscibles en milieu poreux stratifie. *Revue de l'Institut Francais du Petrole* 22: 272–294.

McCarthy, K. and R.L. Johnson. 1993. Transport of volatile compounds across the capillary fringe. *Water Resources Research* 29(6): 1675–1683.

Mercado, A. 1967. The spreading pattern of injected waters in a permeability stratified aquifer. *Artificial Recharge of Aquifers and Management of Aquifers* 72: 23–36.

Moench, A.F. 1989. Convergent radial dispersion: A Laplace transform solution for aquifer tracer testing. *Water Resources Research* 25(3): 439–447.

Neuman, S.P. 1990. Universal scaling of hydraulic conductivities and dispersivities in geologic media. *Water Resources Research* 26(8): 1749–1758.

Novakowski, K.S. 1992. An evaluation of boundary conditions for one-dimensional solute transport. 2. Column experiments. *Water Resources Research* 28(9): 2411–2423.

Oostrom, M., Hofstee, C., Walker, R.C. and J.H. Dane. 1999a. Movement and remediation of trichloroethylene in a saturated homogeneous porous medium. 1. Spill behavior and initial dissolution. *Journal of Contaminant Hydrology* 37: 159–178.

Oostrom, M., Hofstee, C., Walker, R.C. and J.H. Dane. 1999b. Movement and remediation of trichloroethylene in a saturated homogeneous porous medium. 2. Pump-and-treat and surfactant flushing. *Journal of Contaminant Hydrology* 37: 179–197.

Pearce, A.E., Voudrias, E.A. and M.P. Whelan. 1994. Dissolution of TCE and TCA pools in saturated subsurface systems. *Journal of Environmental Engineering* 120(5): 1191–1206.

Perkins, T.K. and O.C. Johnston. 1963. A review of diffusion and dispersion in porous media. *Society of Petroleum Engineers Journal* 3(1): 70–84.

Pisani, S. and N. Tosi. 1994. Two methods for the laboratory identification of transversal dispersivity. *Ground Water* 32(3): 431–438.

Rajaram, H. and L. W. Gelhar. 1995. Plume-scale dependent dispersion in aquifers with a wide range of scales of heterogeneity. *Water Resources Research* 31(10): 2469–2482.

Rao, P.S., Rolston, D.E., Jessup, R.E. and Davidson, J.M. Solute transport in aggregated porous media: Theoretical and experimental evaluation. *Soil Science Society of America Journal* 44: 1139–1146.

Robbins, G.A. 1989. Methods for determining transverse dispersion coefficients of porous media in laboratory column experiments. *Water Resources Research* 25(6): 1249–1258.

Rubin, Y. 1995. Flow and transport in bimodal heterogeneous formations. *Water Resources Research* 31(10): 2461–2468.

Rubin, Y. 2003. *Applied Stochastic Hydrogeology.* Oxford: Oxford University Press.

Russo, D. 1998. Stochastic analysis of flow and transport in unsaturated heterogeneous porous formation: Effects of variability in water saturation. *Water Resources Research* 34(4): 569–581.

Saffman, P.G. 1959. A theory of dispersion in a porous medium. *Journal of Fluid Mechanics* 6: 312–349.

Sato, T., Tanahashi, H. and H.A. Loaiciga. 2003. Solute dispersion in a variably saturated sand. *Water Resources Research* 39: doi: 10.1029/2002WR001649.

Sauty, J. 1980. An analysis of hydrodispersive transfer in aquifers. *Water Resources Research* 16(1): 145–158.

Schulze-Makuch, D. 2005. Longitudinal dispersivity data and implications for scaling behavior. *Ground Water* 43(3): 443–456.

Stauffer, F. and M. Rauber. 1998. Stochastic macrodispersion model for gravel aquifers. *Journal of Hydraulic Research* 36(6): 885–896.

Tiedeman, C.R. and P.A. Hsieh. 2004. Evaluation of longitudinal dispersivity estimates from simulated forced- and natural-gradient tracer tests in heterogeneous aquifers. *Water Resources Research* 40: doi:10.1029/2003WR002401.

Vanclooster, M., Javaux, M. and J. Vanderborght. 2005. Solute transport in soil at the core and field scale. *Encyclopedia of Hydrological Science, Part 6: Soils*, doi: 10.1002/0470848944.hsa073.

van Genuchten, M.T. and P.J. Wierenga. 1976. Mass transfer studies in sorbing porous media. I. Analytical solutions. *Soil Science Society of America Journal* 40: 473–480.

van Genuchten, M.T. 1980. A closed-form equation for predicting the hydraulic conductivity of unsaturated soils. *Soil Science Society of America Journal* 44: 892–898.

van Genuchten, M.T. and J.C. Parker. 1984. Boundary conditions for displacement experiments through short laboratory soil columns. *Soil Science Society of America Journal* 48: 703–708.

van Genuchten, M.T., Leij, F.J. and S.R. Yates. 1991. *The RETC Code for Quantifying the Hydraulic Functions of Unsaturated Soils*, version 1.0. EPA/600/2-91/065. U.S. Salinity Laboratory, Riverside, CA.

Welty, C. and L.W. Gelhar. 1994. Evaluation of longitudinal dispersivity from nonuniform flow tracer tests. *Journal of Hydrology* 153: 71–102.

Xu, M. and Y. Eckstein. 1995. Use of weighted least-squares method in evaluation of the relationship between dispersivity and field scale. *Ground Water* 33(6): 905–908.

Xu, M. and Y. Eckstein. 1997. Statistical analysis of the relationships between dispersivity and other physical properties of porous media. *Hydrogeology Journal* 5(4): 4–20.

Zhan, H. and S.W. Wheatcraft. 1996. Macrodispersivity tensor for nonreactive solute transport in isotropic and anisotropic fractal porous media: Analytical solutions. *Water Resources Research* 32(12): 3461–3474.

Zhang, X., Qi, X., Zhou, X. and H. Pang. 2006. An *in situ* method to measure the longitudinal and transverse dispersion coefficients of solute transport in soil. *Journal of Hydrology* 328: 614–619.

ADDITIONAL READING

Aris, R. 1956. On the dispersion of a solute in a fluid flowing through a tube. *Proc. Royal Soc. London A* 235: 67–77.

Batu, V. 2006. *Applied Flow and Solute Transport Modeling in Aquifers. Fundamental Principles and Analytical and Numerical Methods*. Boca Raton, FL: CRC Press.

Bear, J. 1972. *Dynamics of Fluids in Porous Media*. New York: American Elsevier Publishing Company.

Bedient, P.B., Rifai, H.S. and C.J. Newell. 1999. *Ground Water Contamination: Transport and Remediation*. Upper Saddle River, NJ: Prentice-Hall.

Berkowitz, B., Cortis, A., Dentz, M. and H. Scher. 2006. Modeling nonFickian transport in geological formations as a continuous time random walk. *Reviews of Geophysics* 44: doi: 10.1029/2005RG000178.

de Marsily, G. 1986. *Quantitative Hydrogeology—Groundwater Hydrology for Engineers*. New York: Academic Press.

Davis, S.N., Campbell, D.J., H.W. Bentley and T.J. Flynn. 1985. *Ground Water Tracers*. Wothington: National Water Well Association.

Elfeki, A.M.M., Uffink, G.J.M. and F.B.J. Barends. 1997. *Groundwater Contaminant Transport: Impact of Heterogeneous Characterization*. Rotterdam: Balkema.

Fetter, C.W. 1999. *Contaminant Hydrogeology*, 2nd ed. Upper Saddle River, NJ: Prentice-Hall.

Fried, J.J. 1975. *Groundwater Pollution*. Amsterdam: Elsevier.

Frippiat, C.C. and A.E. Holeyman. 2008. A comparative review of upscaling methods for solute transport in heterogeneous porous media. *Journal of Hydrology* 362: 150–176, doi: 10.1016/j.jhydrol.2008.08.015

Gaspar, E. 1987. *Modern Trends in Tracer Hydrology*. Vol. I. Boca Raton, FL: CRC Press.

Gaspar, E. 1987. *Modern Trends in Tracer Hydrology*. Vol. II. Boca Raton, FL: CRC Press.

Greenkorn, R.A. 1983. *Flow Phenomena in Porous Media*. New York: Marcel Dekker.

Javaux, M. and M. Vanclooster. 2003. Scale- and rate-dependent solute transport within an unsaturated sandy monolith. *Soil Science Society of America Journal* 67(5): 1334–1343.

Kass, W. 1998. *Tracer Techniques in Geohydrology*. Rotterdam: Balkema.

Matheron, G. and G. De Marsily. 1980. Is transport in porous media always diffusive? A counterexample. *Water Resources Research* 16(5): 901–917.

Ptak, T., Piepenbrink, M. and Martac, E. 2004. Tracer tests for the investigation of heterogeneous porous media and stochastic modeling of flow and transport—a review of some recent developments. *Journal of Hydrology* 294: 122–163.

Scheidegger, A.E. 1960. *The Physics of Flow through Porous Media*. New York: Macmillan.

Taylor, G.I. 1953. Dispersion of soluble matter in solvent flowing slowly through a tube. *Proceedings of the Royal Society of London A* 219: 186–203.

Taylor, G.I. 1954. Conditions under which dispersion of a solute in a stream of solvent can be used to measure molecular diffusion. *Proceedings of the Royal Society of London A* 225: 473–477.

Tindal, J.A. and J.R. Kunkel. 1999. *Unsaturated Zone Hydrology for Scientists and Engineers*. Upper Saddle River, NJ: Prentice-Hall.

Wheatcraft, S.W. and S.W. Tyler. 1988. An explanation of scale-dependent dispersivity using concepts of fractal geometry. *Water Resources Research* 24(4): 566–578.

PART 2   MASS TRANSFER COEFFICIENTS IN PORE-WATER ADJACENT TO NONAQUEOUS LIQUIDS AND PARTICLES

15.6   INTRODUCTION

Organic chemicals and hydrocarbons with very low aqueous solubility remain a separate phases or as nonaqueous phase liquids (NAPLs) for long periods of time in the subsurface contributing to soil and groundwater contamination. At a very fundamental level, the mass transfer occurs at the NAPL–water interfaces within the pores. However, when the NAPLs enter the soil, they produce complex entrapment morphologies and architecture that makes the mass transfer process complex. The morphologies of entrapment at the pore scale and the spatial distribution that defines the architecture are controlled by many factors that include the spill configuration, type, and physical and chemical properties of NAPL and the subsurface heterogeneity. Uncertainty associated with all these controlling factors contributes to the prediction uncertainty of how much dissolved mass flux is generated from source zones of NAPL-contaminated sites. The focus of this section is to discuss and present modeling methods that are used to predict mass transfer from entrapped NAPL sources taking into consideration the various entrapment morphologies and architecture that occur at naturally heterogeneous field sites. To be of practical value, methods have to be developed to up-scale this mass transfer process from smaller measurement scales (laboratory) to the field.

The outline of the presented material is as follows. The mass transfer that occurs at the NAPL–water interfaces at the pore scale is generally approximated using a linear model based on film theory. In extending this formulation to the representative elemental volume (REV) scale in porous media, it is necessary to define an overall mass transfer rate coefficient. The theoretical development of phenomenological models that are used to estimate these overall mass transfer coefficients is presented. Methods to up-scale these REV scale models to field scale are presented. A set of examples based on intermediate-scale laboratory tests is presented to demonstrate the use of these methods.

15.7   DISTRIBUTION AND MORPHOLOGY OF NAPLs IN POROUS MEDIA

Nonaqueous phase liquids are classified into two groups depending on their specific gravity. Hydrocarbons and petroleum products that are less dense than water are referred to as light nonaqueous phase liquids (LNAPLs). Solvents, cold tar, wood preservatives that are heavier than water are referred to as dense nonaqueous phase liquids (DNAPLs). They are classified as nonaqueous phase liquids because of their very low solubility. They stay as a separate fluid phase when in contact with water for a very long period of time. When introduced on to the ground surface because of accidental spills, improper disposal or leaking from storage systems, NAPLs migrate through the unsaturated zone of the subsurface where the wetting water phase partially occupies the pore space. Because of surface tension, the water pressure stays at less than atmospheric (capillary suction). During migration, the NAPL that behaves as a nonwetting fluid in the presence of water displaces the nonwetting air phase. After the

NAPL front has propagated, a fraction of the NAPL remains entrapped within the soil pores. Before reaching the water table (where the water pressure is at atmospheric), the NAPL front will penetrate the zone that is referred to as the capillary fringe where the pore water is close to saturation but is under less than atmospheric pressure. The NAPL front will penetrate the capillary fringe and reaches the water table. After reaching the water table, the behavior of NAPL depends on the relative density compared to water. Lighter than water LNAPLs tend to float on the water table and the dense DNAPLs will penetrate the water table and enters the saturated zone of the aquifer where the water pressures are higher than atmospheric. Depending on the conditions as determined by the spill volume, rate, fluid, and porous media properties, the DNAPL behaves unstably and will migrate preferentially as fingers (Held and Illangasekare, 1995). Laboratory and theoretical studies suggest that, even in the most homogeneous of porous media, the infiltration and dissolution of dense NAPL solvents into the saturated zone will tend to occur as a number of scattered fingers and not along one uniform plug or front. Once a sufficient amount of NAPL accumulates and the NAPL solvent enters the porous medium, downward movement will continue until all of the NAPL solvent is present as suspended fingers, ganglia, and/or as pools of NAPL accumulated on lower-permeable layers. Because fingers tend to have small dimensions in the saturated zone (usually occupying single pore throat), significant fraction of NAPL mass in the saturated zone may be present as NAPL pools (Anderson et al., 1992). However, deep penetration of downward moving fingers and subsequent formation of new pools of NAPL flowing through the finger result in a complex spill morphology and thus the prediction of finger penetration into the porous medium is also important (Illangasekare et al., 1995).

The entrapped NAPL, both in the unsaturated and saturated zones produce fluid–fluid interfaces through which mass transfer occurs. The unsaturated zone NAPLs produce NAPL–water and NAPL–air interfaces. Water infiltrating through the unsaturated zone picks up the dissolved mass and transports the solute to the saturated zone. The mass transfer that occurs through the NAPL–air interface contributes to the vapor migration through the air phase. In the following sections, we will only focus on the mass transfer that occurs at NAPL–water interfaces, thus focusing only on the problem of groundwater contamination by NAPLs.

15.8 CONCEPTUAL MODELS OF MASS TRANSFER

A variety of models exist to describe mass transfer phenomena among phases in a multiphase system. Here, an overview from simple dispersion model (Johnson and Pankow, 1992) to solute mass flux models (Miller et al., 1990; Powers et al., 1992; Geller and Hunt, 1993; Imhoff et al., 1993), including various models for Sherwood transfer rate number is presented in this section.

15.8.1 ONE-DIMENSIONAL VERTICAL DISPERSION MODEL

Johnson and Pankow (1992) presented a simple analytical model for dissolution of pools of a NAPL by treating the mass transfer to be a vertical transport process. The general two-dimensional mass transport equation can be simplified by assuming:

(1) the time required for total pool dissolution is exceedingly longer in comparison with the contact time between the pool and the flowing groundwater, therefore, a steady-state form of the advection–dispersion equation can be used, (2) sorption is not important at steady state, and (3) groundwater flows with the velocity v in the horizontal direction. Thus, a two-dimensional steady-state equation can be used. The governing equation is given by

$$v \frac{\partial C}{\partial x} = D_T \frac{\partial^2 C}{\partial z^2}, \tag{15.21}$$

where $C [ML^{-3}]$ is the concentration of NAPL, v $[LT^{-1}]$ is the groundwater velocity in the horizontal direction, $z[L]$ is the vertical distance above the pool and $x[L]$ is the horizontal distance along the length of the pool with the origin at the beginning of the pool. The vertical dispersion coefficient D_T $[L^2 T^{-1}]$ is given either by Equation 15.5b or Equation 15.5c.

Hunt et al. (1988) presented the analytical solution of Equation 15.21 with the boundary conditions $C(x, +\infty) = 0$ and $C(x, 0) = C_s$ for all $x \in (0, L_p)$ and $C(0, z) = 0$ for all $z \in (0, L_p)$, where L_p is the length of the pool. Based on this solution, the vertical concentration profile at the downgradient edge of the pool $(x = L_p)$ is given by

$$C(L_p, z) = C_s \text{erfc} \left(\frac{z\sqrt{v}}{2\sqrt{D_v L_p}} \right), \tag{15.22}$$

where C_s $[ML^{-3}]$ is the solubility limit of the dissolving NAPL component. Based on the above solution, the time for complete dissolution of the pool can be estimated. If a pool consists of constant thickness with a thickness/length ratio r, then the time to complete dissolution τ_p will be given by

$$\tau_p = \frac{r \rho L_p^{3/2} \sqrt{\pi}}{C_s \sqrt{4 D_v v}}, \tag{15.23}$$

where ρ $[ML^{-3}]$ is the density of the NAPL. The above derivation assumes a surface area averaged mass transfer across the pool length.

15.8.2 Linear Driving Model for Interphase Mass Transfer

A common concept implicit in many mass transfer theories is to describe the mass transfer across two phase boundaries through a mass transfer rate coefficient. The driving force in this case is determined by the difference in the concentration at the phase boundary (e.g., NAPL surface) and the bulk phase (e.g., water or air). This linear relationship is given by

$$J = k_l (C_s - C), \tag{15.24}$$

where $J [ML^{-2} T^{-1}]$ is the mass flux rate from the NAPL, k_l $[LT^{-1}]$ is the mass transfer rate coefficient, C_s $[ML^{-3}]$ is the aqueous phase concentration under conditions

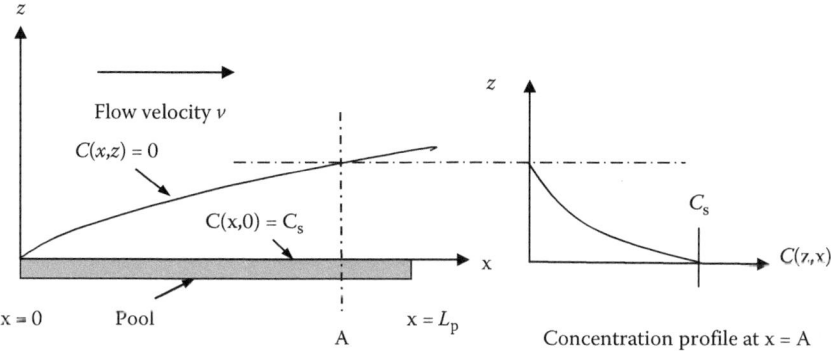

FIGURE 15.5 Dissolution of a NAPL pool.

when the NAPL is in thermodynamic equilibrium with the solute in the aqueous phase (solubility limit of NAPL in water) and C [ML^{-3}] is the aqueous phase solute concentration in the bulk solution. The subscript l denotes that the driving force acts along the longitudinal direction normal to the direction of flux. Note that this model does not assume the presence of a porous medium.

15.8.3 STAGNANT FILM MODEL

A conceptual model that describes mass transfer across two phases is assumed to occur through a stagnant aqueous film adjacent to the interface has been adopted for porous media applications. A schematic illustration of the process that occurs is shown in Figure 15.6.

As there is no mass storage within the film, the concentration gradient has to be linear. Applying Fick's law, an expression for steady mass flux J [ML^{-2}T^{-1}] is obtained as

$$J = -D_l \frac{dC}{dl} = \frac{D_l}{\delta}(C_s - C), \qquad (15.25)$$

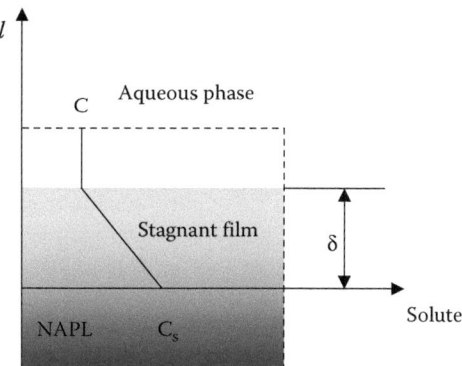

FIGURE 15.6 Stagnant film model.

where D_l $[L^2 T^{-1}]$ is the diffusion coefficient in free liquid and δ [L] is the thickness of the assumed stagnant film. By introducing a mass transfer coefficient $k_l = D_l/\delta$ $[LT^{-1}]$ an equation similar to the linear driving model can be written as

$$J = k_l(C_s - C). \tag{15.26}$$

The macroscopic groundwater flow equations are written at the representative elemental volume (REV) scale. As (Δ) is defined as the pore scale (NAPL–water interface within a pore) and cannot be measured, a lumped mass transfer coefficient defined at the REV scale is used (Pankow and Cherry, 1995). Hence, a linear driving force model similar to Equation 15.24 can be used to describe mass flux from entrapped NAPL sources in porous media. This is accomplished by introducing a lumped mass transfer rate coefficient K_c $[T^{-1}]$ (Miller et al., 1990). The mass rate equation for mass rate J' $[ML^{-3}T^{-1}]$ then takes the form

$$J' = K_c(C_s - C). \tag{15.27}$$

The pore-scale mass transfer coefficient k_1 and the lumped mass transfer coefficient K_c are related by

$$K_c = k_l \frac{A_n}{V}, \tag{15.28}$$

where A_n $[L^2]$ is the total NAPL$^-$ water surface area within the REV of volume V $[L^3]$. As A_n cannot be measured or estimated in practical situations in involving ground water contamination, K_c is treated as an empirically determined parameter.

15.9 EMPIRICAL MASS TRANSFER RATE COEFFICIENTS

As was explained in the previous section, the lumped mass transfer rate coefficient needs to be empirically determined because the basic parameters that describe the mass transfer process cannot be measured for porous media systems. However, an insight to the processes that contribute to mass transfer can be obtained by studying simple settings, where the governing equations can be solved to obtain closed form analytical solutions. Chemical engineering literature provides number of examples of such closed form solutions (e.g., Bird et al., 1961). By identifying the driving forces and mechanisms that contribute to mass transfer, the terms that appear in these closed from solutions can be arranged into dimensionless groups. One example has relevance to understanding dissolution from a pool and entrapped NAPLs, it is the case of the dissolution of the wall when the water flows through a tube of length L and internal radius R (Figure 15.7). The flow through the tube is assumed to be laminar.

The closed-form solution for the advection–dispersion equation expressed in dimensionless group that is referred to as the Graetz–Nusselt problem solution is given as

$$\text{Sh} = \left(\frac{2}{3\Gamma(4/3)}\right) \text{Re}^{1/3} \text{Sc}^{1/3} \left(\frac{L}{R}\right) \tag{15.29}$$

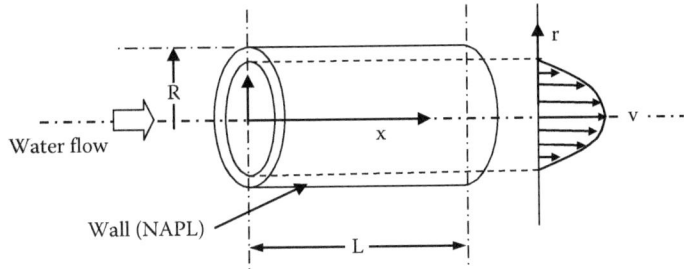

FIGURE 15.7   Dissolution of wall during laminar flow.

where Γ is the gamma function, Re $[-]$ is the Reynolds number, and Sc $[-]$ is the Schmidt number. The dimensionless Sherwood number Sh $[-]$ is related to the mass transfer coefficient k_l as Sh $= k_l d_p / D_l$, where d_p [L] is the geometric mean particle diameter. The relationships such as the one given by Equation 15.29 are referred to as Gilland–Sherwood models.

Saba and Illangasekare (2000) proposed a model for two-dimensional flow conditions. This model introduced a dissolution length along the flow path. Also, the appearance of a tube radius allows for the introduction of the volumetric NAPL content into the phenomenological model. Conceptually, as the NAPL gets depleted, the effective radius of the flow tube changes. The Gilland–Sherwood model that was proposed by Saba and Illangasekare (2000) is of the form

$$\mathrm{Sh}' = a\,\mathrm{Re}^\beta\,\mathrm{Sc}^\alpha \left(\frac{\theta_n d_{50}}{\tau L} \right)^\eta, \qquad (15.30)$$

where Sh$'$ $[-]$ is a modified form of Sherwood number used for porous media applications defined as

$$\mathrm{Sh}' = \frac{K d_p^2}{D_l}. \qquad (15.31)$$

The terms a, β, α, and n are empirical coefficients $[-]$, τ $[-]$ is the tortuosity factor of the flow path, L [L] is the dissolution length, d_{50} [L] is the mean grain size, and θ_n $[-]$ is the volumetric NAPL content. The mass transfer coefficient K [T^{-1}] that appears in Equation 15.31 is the lumped mass transfer coefficient and it contains the NAPL/water interface area as introduced in Equation 15.28.

The correlation based on Equation 15.30 that was fitted to NAPL dissolution data obtained in a two-dimensional dissolution cell by Saba and Illangasekare (2000) is

$$\mathrm{Sh}' = 11.34\,\mathrm{Re}^{0.28}\,\mathrm{Sc}^{0.33} \left(\frac{\theta_n d_{50}}{\tau L} \right)^{1.037}. \qquad (15.32)$$

Gilland–Sherwood correlations were developed for a number of test systems and configurations by a number of investigators. These correlations are listed in Table 15.3.

TABLE 15.3
Gilland–Sherwood Correlations Reported by Different Investigators

Reference	Correlation	Valid Range
Estimated from Geller and Hunt (1993) in Imhoff et al. (1993)	$\mathrm{Sh}' = 70.5\,\mathrm{Re}^{1/3}\theta_n^{4/9}S_{ni}^{1/9}\varphi^{-2/3}\left(\dfrac{d_p}{d_{ni}}\right)^{5/3}$	$\theta_n \in (0, 0.056)$ $\mathrm{Re} \in (0, 0.014)$
Miller et al. (1990)	$\mathrm{Sh}' = 12(\varphi - \theta_n)\mathrm{Re}^{0.75}\theta_n^{0.60}\mathrm{Sc}^{0.5}$	$\theta_n \in (0.016, 0.07)$ $\mathrm{Re} \in (0.00015, 0.1)$
Est. from Parker et al. (1991) in Imhoff et al. (1993)	$\mathrm{Sh}' = 1240(\varphi - \theta_n)\mathrm{Re}^{0.75}\theta_n^{0.60}$	$\theta_n \in (0.02, 0.03)$ $\mathrm{Re} \in (0.1, 0.2)$
Powers et al. (1992)	$\mathrm{Sh}' = 57.7(\varphi - \theta_n)^{0.61}\mathrm{Re}^{0.61}d_{50}^{0.64}U_i^{0.41}$	$\mathrm{Re} \in (0.012, 0.21)$
Imhoff et al. (1993)	$\mathrm{Sh}' = 340\,\mathrm{Re}^{0.71}\theta_n^{0.87}\left(\dfrac{x}{d_p}\right)^{-0.31}$	$\theta_n \in (0, 0.04)$ $(\varphi - \theta_n)\mathrm{Re} \in (0.0012, 0.021)$ $x/d_{50} \in (1.4, 180)$
Powers et al. (1994)	$\mathrm{Sh}' =$ $4.13\,\mathrm{Re}^{0.598}\left(\dfrac{d_{50}}{d_p}\right)^{0.673}U_i^{0.369}\left(\dfrac{\theta_n}{\theta_{ni}}\right)^{\beta}$ $\beta = 0.518 + 0.114\left(\dfrac{d_{50}}{d_p}\right) + 0.1U_i$	$\mathrm{Re} \in (0.015, 0.023)$

Note: θ_n [−] is the NAPL content, φ [−] is the porosity, S_{ni} [−] is the initial NAPL saturation, d_p [L] is the diameter of the porous media mean particle, d_{ni} [L] is the mean value of the initial NAPL ganglia, d_{50} [L] is the particle diameter such that 50% of the porous media are finer by weight (median particle size), U_i [−] is the uniformity index, and x/d_p [−] is the dimensionless distance into the region of residual NAPL.

The correlation by Geller and Hunt (1993) was developed for variable volumetric content, that is, S_n or n are not constant. The phenomenological model for mass transfer was based on the correlation developed by Wilson and Geankoplis reported by Imhoff et al. (1993) with an assumed spherical shape of NAPL ganglia. In that case, it was necessary to choose the initial NAPL saturation S_{ni} and the initial NAPL ganglia diameter d_{ni}, which are not needed in other models, since a shrinking NAPL ganglia was examined in this study. In the development of the correlation by Miller et al. (1990) the residual NAPL within the porous medium was established by mechanical stirring glass beads, water, and NAPL. The laboratory created NAPL ganglia were more spherical and smaller in size than that obtained by the displacement mechanism as in the correlation by Powers et al. (1992), who explain that dissolution is fast in the work of Miller et al. (1990) because these relatively small spherical NAPL ganglia have larger interfacial contact area for an equivalent NAPL volumetric content. A constant volumetric NAPL content and a steady state dissolution experimental data were correlated. In a study by Parker et al. (1991) the residual NAPL distribution was created by mechanical mixing of sand and NAPL. This technique results in a similar residual NAPL morphology as in the previous case, that is, likely small spherical NAPL ganglia with different structure than in the case of natural NAPL displacement mechanism. As in the case of (Miller et al., 1990), a constant volumetric NAPL content and steady-state dissolution were considered. To develop the correlation by Powers et al. (1992), the residual NAPL distribution was achieved by an immiscible displacement process: NAPL first flooded a water saturated medium and then it was followed by water flush to displace the mobile NAPL. This process creates NAPL ganglia in a similar way they are created under natural conditions, that is, the ganglia are nonuniformly displaced and variously shaped. Imhoff et al. (1993) used a different regression techniques based on the Gauss–Newton nonlinear least-squares algorithm and linear least-squares regression. The authors conclude that the simplest model which adequately described TCE dissolution for the porous medium is that obtained by nonlinear regression of power law, where the exponential variation of Sh with x/d_p was suggested. The major difference between this and other models (see Table 15.3) lies in the inclusion of the x/d_p coefficient in the correlation of the mass transfer rate. The correlation by Powers et al. (1994) included the uniformity index U_i. Since the shrinking of the NAPL blobs is considered in this model, the modified Sherwood number correlation includes the initial volumetric NAPL content θ_{ni}.

Saba and Illangasekare (2000) compared some of the models listed in Table 15.3 based on one-dimensional testing systems and demonstrated that the correlation based on two-dimensional data result in significant errors. This finding suggested that the flow dimensionality has to be taken into consideration when upscaling the models based on one-dimensional systems to multidimensional flow systems in the field.

Saenton and Illangasekare (2007) proposed a method to upscale the mass transfer rate coefficient for numerical simulation of mass transfer in heterogeneous source zones where NAPLs are entrapped. The basic approach involves the use of geostatistical parameters of the heterogeneity and the statistics that describes how the NAPL saturation is distributed in the source zone. Through numerical experiments the authors demonstrated that the mass transfer is most sensitive to the vertical smearing of the NAPL that is represented by the second moment of the saturation distribution. The upscaled mass transfer correlation is given by

$$\overline{Sh} = Sh_0(1 + \sigma_Y^2)^{\varphi_1} \left(1 + \frac{\Delta z}{\lambda_z}\right)^{\varphi_3} \left(\frac{\hat{M}_{II,z}}{\hat{M}_{II,z}^*}\right)^{\varphi_5}, \qquad (15.33)$$

where $\overline{Sh}$ is the up-scaled Sherwood number containing the effective mass transfer rate coefficient, σ_Y^2 is the variance of the ln K field, Δz is the vertical dimensions of the simulation grid, λ_z is the vertical correlation length and the last set of terms is the dimensionless second moment of the vertical saturation distribution. This method of upscaling was validated using data from an intermediate scale tank experiment.

15.10 EXAMPLE PROBLEMS

A cleanup of a contaminated porous medium by complete dissolution is one of the many applications of the presented models. Based on different geometrical entrapment

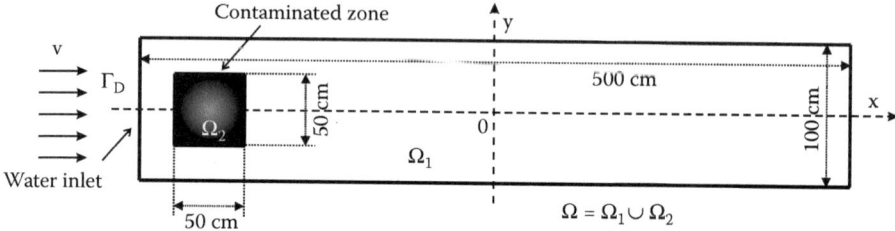

FIGURE 15.8 Illustration of two-dimensional complete dissolution problem.

of the NAPL, the Graetz–Nusselt model given by Equation 15.29 or the Powers et al. (1994) model (see Table 15.3) for the Sherwood number Sh is used.

The contaminated sand is initially assumed to contain NAPL at its residual saturation, that is, $S_{ni} = S_{nr}$, and thus it cannot be cleaned otherwise than by dissolution. Hence, the contaminated sand is put into a horizontally placed tube that contains clean sand of the same properties as it is shown in Figure 15.8. Constant flux of water v_x is introduced to the inlet and only laminar Darcian flow is considered.

The dissolution of the entrapped NAPL system is modeled by the following two-dimensional transport equation:

$$\frac{\partial \theta_w C}{\partial t} + \nabla(D\nabla C) - v\nabla C + J = 0, \qquad (15.34)$$

where C [ML^{-3}] is the solute concentration in water, θ_w [−] is the volumetric water content given as $\theta_w \varphi S_w$, ϕ is the porosity, and S_w [−] is the water saturation related to the NAPL saturation as $S_w + S_n = 1$. D [L^{2}T^{-1}] is the hydrodynamic dispersion tensor given as

$$D = \begin{bmatrix} D_m + v_x \alpha_L & 0 \\ 0 & D_m \end{bmatrix}, \qquad (15.35)$$

where v[LT^{-1}] is the velocity of the horizontal flow of water, that is, $v = (v_x, 0)$, D_m [L^{2}T^{-1}] is the mechanical dispersion introduced in Equation 15.1, α[L] the longitudinal dispersivity introduced in Equation 15.5, and J [ML^{-2}T^{-1}] is the dispersive solute mass flux given by Equation 15.24.

Since the saturation of NAPL changes during the dissolution process, the mass conservation equation is added in the form

$$\rho_n \frac{\partial S_w}{\partial t} = S_w J, \qquad (15.36)$$

where ρ_n[ML^{-3}] is the NAPL density.

Equations 15.34 and 15.36 are supplemented by initial and boundary conditions. Initially, the saturation of water and the concentration of NAPL are given as

$$S_w = \begin{cases} 1 & \text{in } \Omega_1 \\ 1 - S_{ni} & \text{in } \Omega_2 \end{cases}, \quad C = 0 \quad \text{in } \Omega. \qquad (15.37\text{ic})$$

The boundary conditions are given as

$$S_w = 1 \quad \text{on } \partial\Omega \qquad (15.37.\text{bc1})$$

$$\nabla_n C = 0 \quad \text{on } \partial\Omega/\Gamma_D, \qquad (15.37.\text{bc2})$$

$$C = 0 \quad \text{on } \Gamma_D, \qquad (15.37.\text{bc3})$$

where $\nabla_n C$ denotes the derivative of C in the direction of the outer normal of the boundary.

15.10.1 DISSOLUTION USING SPHERICAL BLOBS MODEL

Powers et al. (1994) considered a spherical TCE entrapment in a porous medium and the respective model for the modified Sherwood number is shown in Table 15.3. In order to correctly illustrate the dissolution process, dissolution of TCE in the Ottawa sand used by Powers et al. (1994) is discussed in this section (refer to Table 15.4 for fluid and sand properties).

15.10.2 DISSOLUTION USING TUBULAR MODEL

As introduced in Equation 15.29, the Graetz–Nusselt closed-form solution for the Sherwood number models dissolving walls of a single tubular TCE entrapment. In

---

TABLE 15.4
Ottawa Sand and TCE Fluid Properties

	Symbol	Units (SI)	Value
Property of Ottawa Sand			
Porosity	ϕ	–	0.37
Median particle size	D_{50}	m	7.1×10^{-4}
Medium particle size	d_p	m	2×10^{-4}
Darcian velocity	V	m s^{-1}	9.796×10^{-5}
Initial NAPL saturation	S_{ni}	–	0.01
Uniformity index	U_i	–	1.21
Property of TCE			
Density	ρ_n	kg m^{-3}	1470
Solubility limit in water	C_s	kg m^{-3}	1.27
Molecular diffusivity	D_m	m^2 s^{-1}	8.8×10^{-10}
Longitudinal dispersion	α_L	m	1
Dynamic viscosity	μ_n	kg m^{-1} s^{-1}	5.9×10^{-4}
Property of Water			
Density	ρ_w	kg m^{-3}	1000
Dynamic viscosity	μ_w	kg m^{-1} s^{-1}	0.001

Source: Based on Powers, S.E., Abriola, L.M. and W.J. Weber Jr. 1994. *Water Resources Research* 30: 321–332.

order to extend the use of such a model to the macroscopic scale, mean tube length L[L], outer radius R_{out} [L], and initial radius R_{ini} [L] have to be chosen such that NAPL can assumed to be uniformly redistributed in these tubes. If the number of the tubes, mean length L and mean outer radius R_{out} remain constant during the dissolution, the following upscaled formula for the Graetz–Nusselt model can be used:

$$\text{Sh} = \left(\frac{2}{3\Gamma(4/3)}\right) \text{Re}^{1/3} \text{Sc}^{1/3} L^{1/3} \left(R_{out}^2 - \frac{S_n}{S_{n,ini}}(R_{out}^2 - R_{ini}^2)\right)^{-\frac{1}{6}}. \qquad (15.38)$$

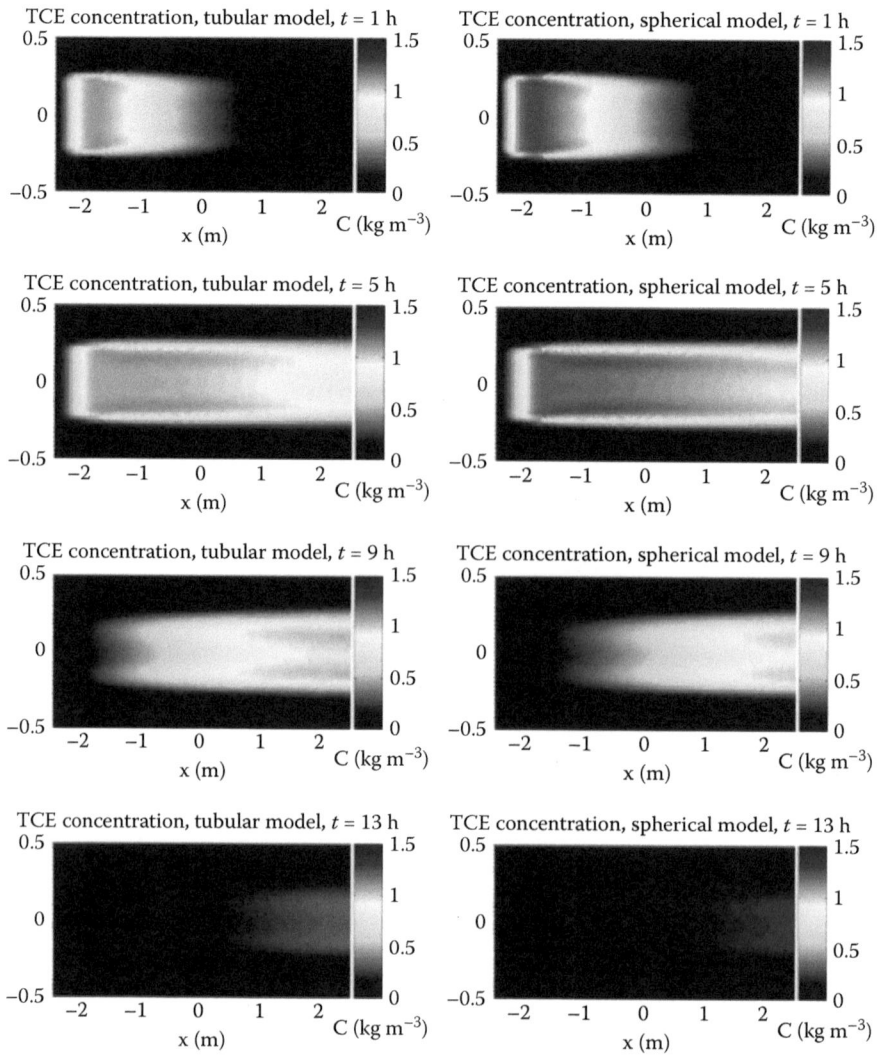

FIGURE 15.9 Concentration distribution in the domain after 1, 5, 9, and 13 h obtained by tubular (left side) and spherical (right side) models for the Sherwood number Sh.

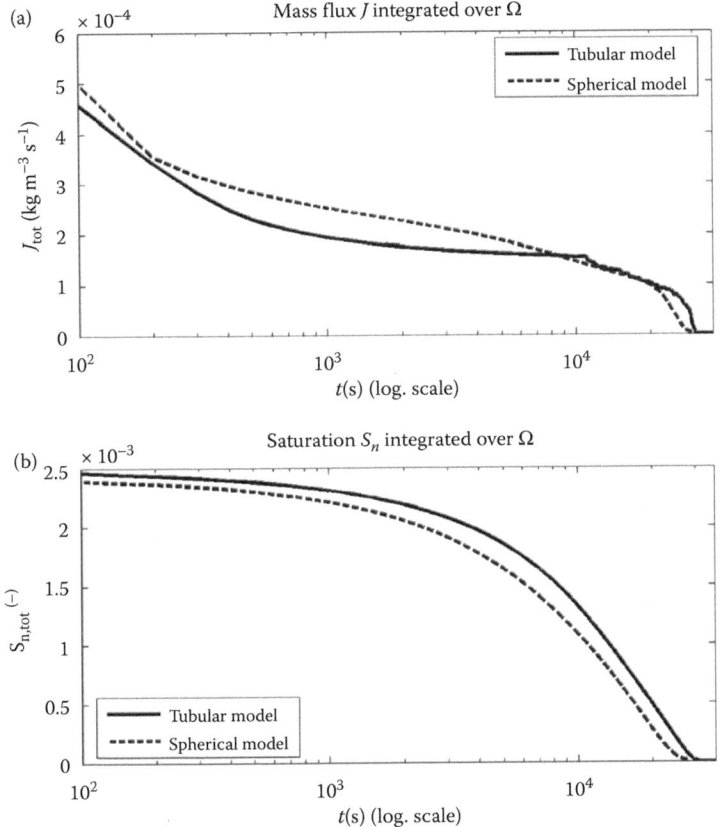

FIGURE 15.10 Time evolution of the total integrated mass flux J (a) and total integrated TCE saturation S_n (b).

Trivial algebraic manipulations in Equation 15.38 reveal that only values of L/R_{out} and R_{ini}/R_{out} have to be known and, consequently, there are only two degrees of freedom to determine. In the numerical simulations, the mean length of a tube is given as $L = d_p/100$ and the last parameter R_{out} is chosen such that the Sherwood number has the same value as in the case of the spherical blobs model by Powers et al. (1994) in the previous section.

15.10.3 NUMERICAL EXPERIMENTS

The problem defined by Equations 15.34 and 15.36 together with the initial and boundary conditions (15.37) is solved by the finite element method.

The concentration distribution for the spherical and tubular model are shown in Figure 15.9. The evolution of the total saturation $S_{n,tot}$ and the total mass flux J_{tot} of the TCE is shown in Figure 15.10. The total saturation or the total mass flux is given by the integration of S_n or J over the domain Ω, respectively.

As a consequence of the calibration of the Graetz–Nusselt problem to give exactly the same value of the initial Sherwood number, both models behave in A similar way. There is a slight difference in the dynamics of the dissolution as the spherical blobs model by Powers et al. (1994) gives stronger TCE flux than the Graetz–Nusselt model. However, the complete dissolution time of the residual TCE content is the same for both models as it is shown in Figure 15.10.

LITERATURE CITED

Anderson, M.R., Johnson, R.J. and J.F. Pankow. 1992. Dissolution of dense chlorinated solvents into groundwater. *Environmental Sciences and Technology* 30: 250–256.

Bird, R.B., Stewart, W.E. and N. Lightfoot. 1961. Transport Phenomena. *Journal of the Electrochemical Society* 108(3): 78C–79C.

Geller, J.T. and J.R. Hunt. 1993. Mass transfer from nonaqueous phase organic liquids in water saturated porous media. *Water Resources Research* 28: 833–846.

Held, R.J. and T.H. Illangasekare. 1995. Fingering of dense nonaqueous phase liquids in porous media. 1. Experimental investigation. *Water Resources Research* 31(5): 1213–1222.

Hunt, J.R., Sitar, N. and K.S. Udell. 1988. Nonaqueous phase liquid transport and cleanup. 1. Analysis and mechanisms. *Water Resources Research* 24: 1247–1258.

Illangasekare, T.H., Ramsey, J.L., Jensen, K.H. and M.B. Butts. 1995. Experimental study of movement and distribution of dense organic contaminants in heterogeneous aquifers. *Journal of Contaminant Hydrology* 20(1–2): 1–25.

Imhoff, P.T., Jaffe, P.R. and G.F. Pinder. 1993. An experimental study of complete dissolution of nonaqueous phase liquids in saturated media. *Water Resources Research* 30(2): 307–320.

Johnson, R.J. and J.F. Pankow. 1992. Dissolution of dense chlorinated solvents into groundwater. 2. Source functions for pools of solvent. *Environmental Sciences and Technology* 26: 896–901.

Miller, T.C., Poirier-McNeill, M.M. and A.S. Mayer. 1990. Dissolution of trapped NAPLs: Mass transfer characteristics. *Water Resources Research* 26(11): 2783–2796.

Pankow, J.F. and J.A. Cherry. 1996. *Dense Chlorinated Solvents and Other DNAPLs in Groundwater*. Waterloo Educational Services, Rockwood, Ontario, Canada.

Parker, J.C., Katyal, A.K., Kaluarachchi, J.J., Lenhard, R.J., Johnson T.J., Jayaraman, K., Unlu, K. and Zhu, J.L. 1991. *Modeling Multiphase Organic Chemical Transport in Soils and Ground Water*. U.S. Environ. Protection Agency, Washington, DC, Rep. EPA/600/2-91/042.

Powers, S.E., Abriola, L.M. and W.J. Weber Jr. 1992. An experimental investigation nonaqueous phase liquid dissolution in saturated systems: Steady state mass transfer rates. *Water Resources Research* 28: 2691–2705.

Powers, S.E., Abriola, L.M. and W.J. Weber Jr. 1994. An experimental investigation of nonaqueous phase liquids dissolution in subsurface systems: Transient mass transfer rates. *Water Resources Research* 30: 321–332.

Saba, T. and T.H. Illangasekare. 2000. Effect of groundwater flow dimensionality on mass transfer from entrapped nonaqueous phase liquids. *Water Resources Research* 36(4): 971–979.

Saenton, S. and T.H. Illangasekare. 2007. Upscaling of mass transfer rate coefficient for the numerical simulation of dense nonaqueous phase liquid dissolution in heterogeneous aquifers. *Water Resources Research* 43: W02428, doi:10.1029/ 2005WR004274.

ADDITIONAL READING

Glover, K., Munakata-Marr, J. and T.H. Illangasekare, 2007. Biologically-enhanced mass transfer of tetrachloroethene from DNAPL in source zones: Experimental evaluation and influence of pool morphology, *Environmental Science and Technology* 41(4): 1384–1389.

Illangasekare, T.H., Munakata-Marr, J.J., Siegrist, R.L., Soga, K., Glover, K.C., Moreno-Barbero, E., Heiderscheidt, J.L. et al., 2007. *Mass Transfer from Entrapped DNAPL Sources Undergoing Remediation: Characterization Methods and Prediction Tools.* SERDP Project CU-1294, Final Report, 435pp.

Kaplan, A.R., Munakata-Marr, J. and T.H. Illangasekare. 2008. Biodegradation of residual dense nonaqueous-phase liquid tetrachloroethene: Effects on mass transfer. *Bioremediation Journal* 12(1): 23–33.

Oostrom, M., Dane, J.H. and T. W. Wietsma. 2006. A review of multidimensional, multifluid intermediate-scale experiments: Nonaqueous phase liquid dissolution and enhanced remediation. *Vadose Zone Journal* 5: 570–598.

16 Dust Resuspension and Chemical Mass Transport from Soil to the Atmosphere

Cheryl L. McKenna-Neuman

CONTENTS

16.1 Introduction 453
16.2 Contaminant Fluxes 457
16.3 Transport Theory and Measurement 458
    16.3.1 Transport Modes: An Overview 458
    16.3.2 The Surface Wind 460
    16.3.3 Measurement of Wind Speed 461
    16.3.4 Particle Entrainment by Wind 464
    16.3.5 Constraints on Particle Entrainment by Wind 467
    16.3.6 Measurement of Dust Concentration 468
16.4 Estimation of Aeolian Mass Transport: Correlations and Parameters 469
    16.4.1 Computation of the Dust Emission Rate: An Analytical
           Approach 469
    16.4.2 Relations from Empirical Studies of Particle Resuspension 472
16.5 Chemical Flux to Air 474
    16.5.1 The Chemical Flux 475
    16.5.2 The Aeolian Process Mass Transport Coefficient...........475
16.6 Short Guide to Users..476
16.7 Conclusions 477
16.8 Case Studies...478
    16.8.1 Field Estimation of Emissions from Slag 478
    16.8.2 Wind Tunnel Estimation of Emissions from
           Tailings as Slurry...483
Literature Cited 487
Additional Resources 493

16.1 INTRODUCTION

Large quantities of dust are transported great distances within the Earth's atmosphere each year, while related environmental and human health concerns have drawn much attention to dust emission and dispersion processes. Estimates of annual dust transport tend to increase with the number of studies carried out and the sophistication of the technology; but at a minimum, quantities well exceed 3 billion tons per year (Tegen and Fung, 1994). About 20% is mineral debris (Junge, 1971). Concentrations of dust range from 10^5 $\mu g\,m^{-3}$ near the source to well under 1 $\mu g\,m^{-3}$ over the ocean. The most common forms of injection are dust storms that occur in arid and semiarid regions (Figure 16.1), volcanic eruptions, and anthropogenic processes.

Dust is a subject of considerable scientific interest for several reasons. Deposits of Quaternary loess or windblown silt currently cover 5–10% of the world's dryland surface (Pye and Sherwin, 1999, p. 214) and form some of the world's most productive soils. The depletion of fine particles from the source regions by wind erosion leads to gradual land degradation, since these fines are usually rich in nutrients and organic matter. Dust is a major source of atmospheric aerosols, impacting climate and weather. Air quality is strongly linked to dust concentration and therefore, human health. Finally, all air-borne particles are potential sorption sites for chemical contaminants. Dust emission, diffusion, and advective transport in the atmosphere all contribute to a suite of processes important to the transfer of chemical contaminants (Figure 16.2). The study of dust and its chemical constituents then is highly interdisciplinary in nature, involving geomorphologists, soil scientists, agricultural engineers, environmental chemists, remote sensing and air pollution specialists, climatologists, and meteorologists.

In the context of the present chapter, dust is treated as particulate matter (PM) of either geologic (oxides of silicon, aluminum, calcium, and iron), organic, or synthetic origin that is or can be *suspended* into the atmosphere as a result of aeolian (wind) processes. We do not include nongeologic particulate matter emitted directly by internal and external combustion processes.

FIGURE 16.1 Dust storm at Silver Peak Nevada, following hail that broke up the protective surface crust. Sources of dust included an upwind playa and local alluvial fans.

FIGURE 16.2   Dust plume emission from a lithium mine located on a desert playa.

Dust may be further subdivided into sources or emission events that are natural, and those that are anthropogenic. They may occur simultaneously in the atmosphere and usually are indistinguishable. For example, a barren desert playa will emit some natural dust during high wind events, but even more when the surface is disturbed by human activities. On a given day, dust concentrations measured in the atmosphere can be part natural and part anthropogenic. Dust that is not emitted in a confined flow stream through a stack, chimney, vent, or other functionally equivalent opening is commonly referred to as *fugitive dust*. That is, such dust typically is emitted from open sources. Many anthropogenic activities are associated with fugitive dust production such as agriculture, mining, construction, manufacturing, and transportation. Dust generated by road traffic (especially from dry, unpaved roads) is a major source of fugitive dust.

Total suspended particulate (TSP) has a relatively coarse size range up to approximately $100\,\mu m$ as detected by a standard high-volume (HiVol) air sampler. Suspended particulate (SP) is often used as a surrogate for TSP, but has an aerodynamic diameter less than $30\,\mu m$ and is usually denoted as PM_{30}. Inhalable particulate (IP) is associated with particles finer than $15\,\mu m$ (PM_{15}) and fine particulate (FP) addresses particles of diameter no greater than $2.5\,\mu m$ ($PM_{2.5}$). $PM_{2.5}$ travels to the deepest parts of the lungs and may cause nose and throat irritation in addition to respiratory illnesses, such as bronchitis, lung damage, and asthma. Health effects vary widely depending on the chemical composition and physical mineralogy of the individual particles. On average, PM_{10} is assumed to represent about 50% of TSP, although this ratio is governed by the source of the PM, and so, large site-to-site variations are reported throughout the literature. Moreover, TSP tends to settle by gravity near to the source according to Stokes law, while PM_{10} tends to be transported over much larger distances. Therefore, the TSP:PM_{10} ratio is expected to decrease with distance away from the source.

Any level of dust generation is considered air pollution. However, PM_{10} is currently the particle size of greatest regulatory interest and is the basis for the National Ambient Air Quality Standards (NAAQS) in the United States. As compared to coarser dust particles, PM_{10} is not overly susceptible to gravitational settling. Emission factors and emission inventories are fundamental tools used in air quality management.

As defined in US EPA AP 42 (*Compilation of Air Pollutant Emission Factors*), Volume I (*Stationary Point and Area Sources*) Fifth Edition:

> An emission factor is a representative value that attempts to relate the quantity of a pollutant released to the atmosphere with an activity associated with the release of that pollutant. These factors are usually expressed as the weight of pollutant divided by a unit weight, volume, distance, or duration of the activity emitting the pollutant (e.g., kilograms of particulate emitted per megagram of coal burned). Such factors facilitate estimation of emissions from various sources of air pollution. In most cases, these factors are simply averages of all available data of acceptable quality, and are generally assumed to be representative of long-term averages for all facilities in the source category (i.e., a population average).

The general equation for emission estimation is given as

$$E = A \cdot \text{EF} \cdot \left(\frac{100 - \text{ER}}{100} \right), \tag{16.1}$$

where E represents the total mass of material emitted over some characteristic length of time, A the activity rate or disturbance frequency, and EF the emission factor. ER (%) represents the removal or capture efficiency of any control systems in place. AP 42 (updated online) documents emission factors (EF) for a very large number of activities associated with the emission of fugitive dust. These factors are determined from published references, although some activities are much better documented than others. In many instances, the EF provided represents an average based on data from numerous sources.

Ultimately, it is widely recognized by air quality specialists, as well as stated in AP 42, that direct data obtained from source-specific emission tests or continuous emission monitoring does provide the best representation of the emissions from a given source. If representative source-specific data cannot be obtained, the use of emission factors is regarded as a last resort. Whenever factors are used, one should be aware of their limitations in accurately representing a particular facility, and the risks of using emission factors in such situations should be evaluated against the costs of further testing or analyses (AP 42, p. 3, Introduction).

This chapter addresses the physical processes that govern the mass of dust particles emitted to the atmosphere from a unit area of surface in a given period of time, otherwise known as the emission rate (E). Much of this work originates from an interdisciplinary field of science known as aeolian research. The instruments, methods, and models required for estimation of the emission rate at a given site are outlined. Two brief case studies are provided concerning fugitive dust emission and related contaminant transport in the mining industry. The dispersion of fugitive dust over long distances is relatively well understood, and is not addressed here. The intent of this chapter is to provide a brief introduction to the topic of dust suspension and modeling with several supporting examples. For additional detail, the reader is referred to more comprehensive review papers regarding grain-scale processes involved in wind erosion (e.g., Nickling and McKenna-Neuman, 2009) and methodologies for estimating fugitive windblown dust (e.g., Countess, 2001).

16.2   CONTAMINANT FLUXES

Dust particles can have both local and global climate effects by modifying the Earth's radiation budget (e.g., Ackerman and Hyosang, 1992), altering cloud properties and providing condensation nuclei, and by affecting the oxidative capacity of the atmosphere due to ozone uptake. Terrestrial dust emissions are an important pathway for iron transport to the world's oceans, affecting ocean biogeochemistry and biological productivity (Jickells et al., 2005). Excessive dust deposition can cause reduced crop yields through contamination by its chemical composition. Industrial pollution is believed to accompany the trans-Pacific transport of mineral dust from China to North America (e.g., Jaffe et al., 2003), as well as Saharan dust that impinges upon regions bordering the Mediterranean Sea (e.g., Israel, Formenti et al., 2001; Spain, Avila et al., 1997, 1998). Furthermore, viable microorganisms, trace metals, and an array of organic contaminants that are carried in these dust-laden air masses and deposited in the oceans are speculated to play important roles in complex changes observed for coral reefs worldwide (Garrison et al., 2003). In short, contaminant flux modeling for dust requires not only estimation and/or measurement of the particle emission rate as outlined in Sections 16.3–16.5 to follow, but also, detailed knowledge of the chemistry and biogeochemistry of the particles themselves.

A great deal of work has been carried out in characterizing the trace element content of aeolian dusts as required for determination of the source and trajectory of the suspended particles (back-trajectory calculations), and in the investigation of seasonal controls on emissions. Mineral dust is a major component of airborne particles in the atmophere globally, which accounts for the prevalence of quartz and aluminosilicates with elemental compositions reflecting the geology/mineralogy and climate of the source region. Numerous studies have shown that the concentration of trace metals, and in particular anthropogenic metal pollutants such as lead, iron, zinc, and cadmium, is inversely related to dust particle size (e.g., Fergusson and Ryan, 1984; Al-Rajhi et al., 1996). Castillo et al. (2008, p. 1042) offer the following explanation for this relation:

> The effect of both physical and chemical processes on the parental materials prior to and during transport will influence particle size ranges and therefore accessory mineral and clay content, which in turn control many aspects of trace element chemistry…. The higher concentrations of most trace elements in the finer dust fractions can be explained in terms of reduced quartz dilution effects and higher phyllosilicate and hydrous oxide contents.

Some dust particles are agglomerates of even smaller particles coated by both inorganic and organic materials, as shown in the work of Falkovich et al. (2004). In-cloud processes and reactions that take place within the atmopshere as a whole can substantially alter the surface composition of mineral particles. Adsorption of organic compounds occurs when dust plumes pass over regions that are forested; for example, or enter areas associated with large amounts of industrial pollution. Until very recently the presence of adsorbed organic compounds in air masses has largely been neglected, and at best, mineral dust and organic aerosols have been studied separately. Falkovich et al. (2004) suggest that the adsorption of low-vapor-pressure organic species on mineral dust may be a key process that affects their long-range transport and atmospheric fate as well as their own impact on the atmosphere. The authors report on a detailed study of adsorbed species identified for Saharan dust particles captured on impactor plates at a site within an urban region on Israel's coastal plain. They identify pesticides and their related decomposition products (e.g., dihydro-5-pentyl-2(3H)-furanone and 2-(1H-tetrazol-5yl)-pyridine) in high-volume dust samples with maximum concentrations found on particles 2–6 µm in diameter. The concentration of pesticides residing in the coarse fraction was noted to increase with either an intensification of dust storm activity or an increase in atmospheric humidity. None of these pesticide compounds were detected on days during which dust storms were not reported for the region, nor were they detected on days when the dust plume came in from over the Mediterranean Sea.

With the advancement of technologies (e.g., GC/MS, ICP-MS, ICP-AES) available for analysis of the biogeochemistry of dust particles captured in cyclones and on impactor plates, one might hope that modeling of the air-borne flux of identified contaminants is a relatively straightforward operation, provided the windspeed field is known or can be estimated. Unfortunately, this is not the case. Empirical relationships between the vertical dust flux (E) and windspeed (u_z) are notorious for order of magnitude variation in E for a given airflow rate.

The following sections in this chapter principally review the physics of particle entrainment by wind and introduce a range of predictive models which often require careful field calibration.

16.3 TRANSPORT THEORY AND MEASUREMENT

16.3.1 Transport Modes: An Overview

The size of particle entrained determines the mode of aeolian transport (Bagnold, 1941), either as suspension, saltation, or creep (Figure 16.3).

In order for suspension of a particle to occur, its terminal velocity (u_f) must be less than the velocity at which the air parcels are dispersed vertically by turbulence, as scaled by the friction velocity (u_*). The ratio of u_f to u_* (both in m s^{-1}) is typically less than 0.1 for dust particles that remain suspended for long periods and are carried great distances (Hunt and Nalpanis, 1985). Long-term suspension events last several days and involve particles with an upper size limit of about 20 µm (Tsoar and Pye, 1987). During this period, these dust particles can travel thousands of kilometers from the source area. In comparison, coarse dust particles (20 µm $< d <$ 70 µm) remain in short-term suspension for only minutes to hours, travelling only a few meters to several kilometers given $0.1 < u_f/u_* < 0.7$.

In saltation, large particles (approximately 70–500 µm) skip across the surface following long, parabolic trajectories (Figure 16.3a). The characteristic height (h) of given particle trajectory may be approximated as $0.81\,u_*^2/g$ and its length (l) as $10.3\,u_*^2/g$ (Owen, 1964), where the friction velocity is assumed to be similar in magnitude to the initial take-off speed. Such particles have a high probability of rebounding from the surface and rising to a height sufficient for acceleration by the wind (Figure 16.3a). Saltation comprises as much as 95% of the total mass transport in an aeolian

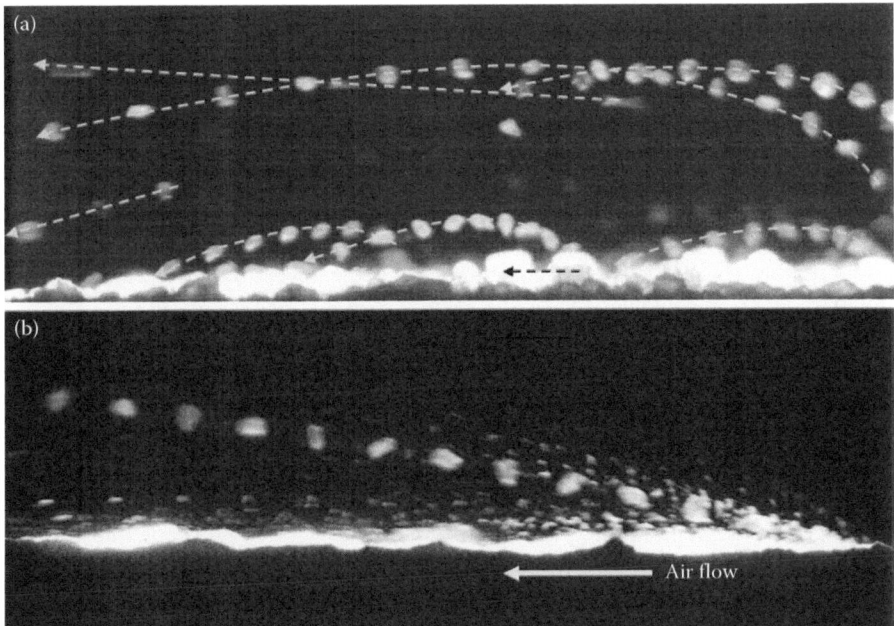

FIGURE 16.3   Particle tracking velocimetry (PTV, 380 Hz) images illustrating varied aeolian transport modes: (a) 300 μm sand particles bounce along the bed in high trajectories known as *saltation*. Low-energy particles traveling very near the bed are incapable of ejecting other particles upon impact with the surface. These move in *reptation*. Particles pushed or rolled within the bed (black arrow) move as *creep*. (b) Sand particle ricocheting from a bed of pulverized silt (see Figure 16.11 for the particle size distribution). Note the fine particles (many in the order of 10–30 μm diameter) that are splashed into the airflow as a result of the impact. Some fraction of these small particles will be dispersed in *suspension* in the turbulent airflow.

system. A sharp distinction does not exist between saltation and true suspension. Instead, Hunt and Nalpanis (1985) define a transitional state called modified saltation that is characterized by semirandom particle trajectories, influenced by both inertia and settling velocity.

The impact of saltating particles upon a bed surface can initiate high-energy ejection of additional saltators, as well as low-energy ejections of adjacent grains in intermittent short-distance hops known as reptation. Reptating particles do not carry sufficient momentum to cause further ejections. Finally, large particles (>500 μm in diameter) are too heavy to be lifted by the wind and therefore are pushed or rolled along as surface creep. Dust particles often are intermingled in the soil matrix around these larger particles (or particle aggregates), so that saltator impacts on the bed surface as shown in Figure 16.3b actually serve as the primary mechanism for release of these fines to the atmosphere.

Contaminant transport occurs with all modes of particle motion under windy conditions. However, suspended particles are the focus of regulatory interest since in remaining aloft and moving at the speed of the airflow, they can be transported over

great distances. With direct inhalation, they pose a threat to human health. With the
eventual deposition of air-borne contaminants upon either agricultural, urban, or water
surfaces located downwind of the source region, entire ecosystems may be affected.

16.3.2 THE SURFACE WIND

The horizontal wind near the surface of the Earth is retarded by viscous frictional
effects imposed on the flow by the rough surface, so that the change in velocity with
height (Figure 16.4) follows a log-linear relation in the lowest 10% of the atmospheric
boundary layer, as characterized by the Prandtl–von Karman equation:

$$u_z = \frac{u_*}{k} \ln\left(\frac{z}{z_0}\right),$$ (16.2)

where u_z (m s^{-1}) is the velocity at height z, z_0 (m) is the aerodynamic roughness length
of the surface, u_* (m s^{-1}) is the friction velocity, and κ is von Karman's constant
(≈ 0.4). On a semilogarithmic plot, the vertical wind velocity profile is transformed
into a straight line with the intercept on the ordinate axis representing z_0 (Figure
16.4c). The friction velocity (u_*) is proportional to the slope of this line and indicates
the fluid drag on the bed surface (τ_0, kg m^{-1} s^{-2}) as follows:

$$\tau_0 = \rho\, u_*^2,$$ (16.3)

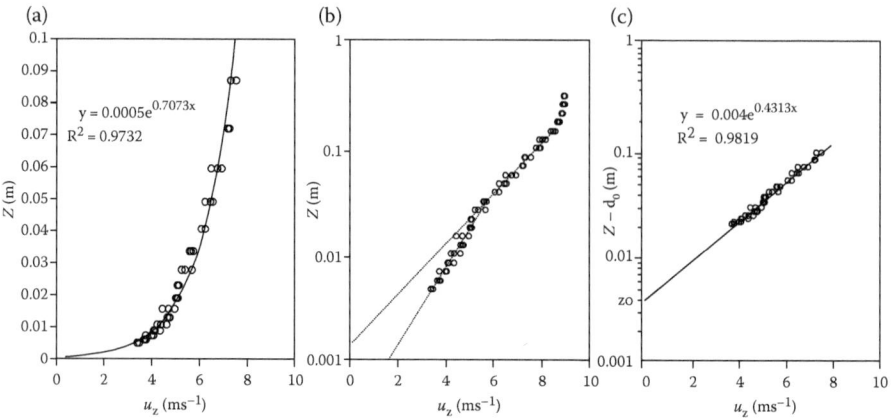

FIGURE 16.4 Vertical profile of wind speed in the lower atmospheric boundary layer over
coarse, crushed gravel. The data were collected in a wind tunnel at a free stream velocity of
8 m s^{-1}. (a) The plot has a linear vertical scale; (b) the height above the surface is shown on a
logarithmic scale; (c) the trend of the data points displays an outward kink, which disappears in
the plot when the vertical height above a given reference point is corrected for the displacement
height (d_0), which in this case is negative (-0.015 m) as determined through iteration to give
the maximum r^2 value. That is, the effective aerodynamic surface lies below the tops of the
gravel elements. The regression equation then can be easily manipulated to determine z_0 (0.004
m) and u_* (0.4213/$\kappa = 1.03$ m s^{-1} where $\kappa = 0.41$).

where ρ (kg m^{-3}) is the air density. No matter how strong the wind blows at a given height above any surface, all associated velocity profiles will extrapolate to converge upon approximately the same intercept. So, the parameter z_0 is an index of the aerodynamic response to the drag of the rough surface on the boundary layer flow. Its value is surface specific. Table 16.1 provides typical values of z_0 for a wide range of surfaces potentially affected by wind erosion.

Where the surface is covered by tall vegetation or high densities of large roughness elements (e.g., large boulders, man-made structures), the vertical profile of wind velocity is displaced (Oke, 1978) to a new reference plane (d_0) that varies with the height, density, porosity, and flexibility of the elements (Figure 16.4c). The wind profile equation then is modified as follows:

$$u_z = \frac{u_*}{k} \ln \left(\frac{z - d_0}{z_0} \right),$$ (16.4)

where d_0 is termed the zero plane displacement height. The exact nature of the relationship among z_0, d_0, and surface roughness remains poorly defined (Lancaster et al., 1991), and as a consequence, usually is determined through direct measurement of the wind velocity profile at a given site (e.g., Figure 16.4c).

16.3.3 MEASUREMENT OF WIND SPEED

Wind speed, and thereby u_*, is typically measured in aeolian transport studies using between 3 and 10 cup anemometers mounted on towers in an exponential height progression, beginning a few centimeters above the surface and ending somewhere between 5 and 10 m. A wide range of commercial anemometer types is available that varies in size, quality, and sensitivity. As a general rule, small anemometers with light cups and high-quality bearings are used for detailed measurements close to the surface, although they can be damaged by the abrasion of saltating sand or dust particles entering the bearings. Data loggers are used to collect and store the continuous data output from the anemometers and wind direction vanes.

Although cup anemometry is a simple, relatively inexpensive and highly reliable method, it only provides the time-averaged horizontal component of the wind flow, as the instruments are slow to respond to gusty/turbulent conditions. Recent research (e.g., Wiggs et al., 1996; Butterfield, 1999; Walker, 2000, 2005; Wiggs, 2001; van Boxel et al., 2004) has placed increasing emphasis upon the role of turbulence in the entrainment and transport of sediment. Hot-wire anemometry is one of the oldest and most frequently used methods for measurement of the instantaneous velocity in clean airflows, most often in wind tunnel experiments (e.g., Wiggs et al., 2002; Walker and Nickling, 2003). Several suppliers now manufacture ruggedized or hot film probes that are suitable for use in sediment laden airflows. These have been used successfully in both field (e.g., Bauer et al., 1998; Rasmussen and Sørensen, 1999; Namikas et al., 2003) and wind tunnel experiments (e.g., Bauer et al., 2004). More recently, sonic anemometers have been employed in aeolian transport studies (e.g., Leenders et al., 2005; Anderson and Walker, 2006). These very accurate instruments are available in several different configurations and sizes, and are quite rugged since they have no moving parts. All three directional components of wind velocity can be measured

TABLE 16.1
Typical Values of Aerodynamic Roughness Length for Surfaces Affected by Wind Erosion

Terrain Type	Mean z_0 (m)	# of profiles	Standard Deviation	Source
General				
Mountainous regions	1.0–100.0	N/A	N/A	Stull (1988)
Urban centers	1.00000	N/A	N/A	Stull (1988)
Dense forest	0.90000	N/A	N/A	Stull (1988)
Moderate tree/scrub cover	0.25000	N/A	N/A	Stull (1988)
Sparse tree cover	0.05500	N/A	N/A	Stull (1988)
Long grass/Crops	0.05000	N/A	N/A	Stull (1988)
Uncut grass	0.02000	N/A	N/A	Stull (1988)
Cut grass	0.00600	N/A	N/A	Stull (1988)
Flat desert	0.00055	N/A	N/A	Stull (1988)
Mud flats	0.00002	N/A	N/A	Stull (1988)
Forests	1.0–6.0	N/A	N/A	Oke (1987)
Orchards	0.5–1.0	N/A	N/A	Oke (1987)
Crops	0.04–0.1	N/A	N/A	Oke (1987)
Grasses	0.003–0.01	N/A	N/A	Oke (1987)
Typical soils	0.001–0.01	N/A	N/A	Oke (1987)
Sand	0.00030	N/A	N/A	Oke (1987)
Desert Soils				
Sand-mantled pahoehoe	0.03014	4	0.030666	Blumberg and Greeley (1993)
Playa	0.00008	1	N/A	Blumberg and Greeley (1993)
Playa/Gravel	0.00013	1	N/A	Blumberg and Greeley (1993)
Pahoehoe	0.00132	1	N/A	Blumberg and Greeley (1993)
Alluvium	0.00175	10	0.000864	Blumberg and Greeley (1993)
Lava flows	0.01480–0.02860	N/A	N/A	Lancaster et al. (1991b)
Alluvial fans	0.00084–0.00167	N/A	N/A	Lancaster et al. (1991b)
Playas	0.00013–0.00015	N/A	N/A	Lancaster et al. (1991b)
Desert pavements	0.00035	N/A	N/A	Lancaster et al. (1991b)
Coastal plain deposits	0.00300	N/A	N/A	Draxler (2001)
Smooth sand sheet	0.00050	N/A	N/A	Draxler (2001)
Deflated sand sheet	0.00040	N/A	N/A	Draxler (2001)
Gravel lag	0.00020	N/A	N/A	Draxler (2001)
Covered desert floor	0.00020	N/A	N/A	Draxler (2001)

TABLE 16.1   (continued)
Typical Values of Aerodynamic Roughness Length for Surfaces Affected by Wind Erosion

Terrain Type	Mean z_0 (m)	# of profiles	Standard Deviation	Source
		Desert Soils		
Active sand sheet	0.00002	N/A	N/A	Draxler (2001)
Playa deposits	0.00002	N/A	N/A	Draxler (2001)
Golden Canyon—active alluvial fan/mudflow	0.00287	6	0.001649	Lancaster et al. (1991)
Trail Canyon—coarse desert pavement	0.00128	3	0.001005	Lancaster et al. (1991)
Kit Fox—coarse alluvial fan	0.00127	4	0.000745	Lancaster et al. (1991)
Confidence Mill—clay/silt playa	0.00041	2	0.000318	Lancaster et al. (1991)
Stovepipe—alluvial fan	0.00041	2	0.000205	Lancaster et al. (1991)
Desert pavement	0.00268	4	0.002068	Gillette et al. (1980)
Aeolian deposit	0.00205	2	0.002051	Gillette et al. (1980)
Salt Crust—playa	0.00180	1	N/A	Gillette et al. (1980)
Crusted playa soils	0.00059	15	0.000991	Gillette et al. (1980)
Desert flat	0.00028	4	0.000184	Gillette et al. (1980)
Dry wash	0.00023	2	0.000240	Gillette et al. (1980)
River bottom	0.00020	1	N/A	Gillette et al. (1980)
Sand dune	0.00010	2	0.000000	Gillette et al. (1980)
Alluvial fan	0.00009	4	0.000080	Gillette et al. (1980)
Pediment	0.00001	1	N/A	Gillette et al. (1980)
Flat prairie	0.00001	2	0.000001	Gillette et al. (1980)
		Agricultural Soils		
Tilled field—chisel ploughing	0.00901	4	0.001401	Lopez (1998b)
Tilled field—mouldboard ploughing	0.00454	10	0.000763	Lopez (1998b)
Untilled agricultural field	0.00043	3	0.000153	Zoebeck and Van Pelt (2006)
Bare, flat agricultural field (Spain)	0.00055	11	0.000053	Gomes et al. (2003)
Bare, flat agricultural field (Niger)	0.00042	31	0.000452	Gomes et al. (2003)
		Industrial Materials		
Ripped slag—graded into furrows	0.00260	1	N/A	Sanderson et al. (2009)
Aging slag—cobble-sized clasts	0.00110	1	N/A	Sanderson et al. (2009)
Aging slag—gravel-sized clasts	0.00050	1	N/A	Sanderson et al. (2009)
Unpaved road	0.00010	1	N/A	Sanderson et al. (2009)
Coal mine overburden	0.00300	N/A	N/A	AP-42
Scoria (roadbed material)	0.00300	N/A	N/A	AP-42
Uncrusted coal pile	0.00300	N/A	N/A	AP-42
Coal dust on concrete pad	0.00200	N/A	N/A	AP-42
Scraper track on coal pile	0.00060	N/A	N/A	AP-42
Ground coal	0.00010	N/A	N/A	AP-42

TABLE 16.1   (continued)
Typical Values of Aerodynamic Roughness Length for Surfaces Affected by Wind Erosion

Terrain Type	Mean z_0 (m)	# of profiles	Standard Deviation	Source
Models and Analogues				
Forest trees, houses (modeled)	2.14000	N/A	N/A	Lettau (1969)
Field crops, tall grasses (modeled)	0.13800	N/A	N/A	Lettau (1969)
Lower grasses, weeds (modeled)	0.00800	N/A	N/A	Lettau (1969)
Bare soil (modeled)	0.00058	N/A	N/A	Lettau (1969)
Sand flats (modeled)	0.00004	N/A	N/A	Lettau (1969)
Artificial roughness elements—λ=0.016	0.00218	1	N/A	Gillies et al. (2006)
Artificial roughness elements—λ=0.030	0.00279	1	N/A	Gillies et al. (2006)
Artificial roughness elements—λ=0.038	0.00705	2	0.002376	Gillies et al. (2006)
Artificial roughness elements—λ=0.050	0.00881	2	0.000955	Gillies et al. (2006)
Artificial roughness elements—λ=0.062	0.01194	2	0.001563	Gillies et al. (2006)
Artificial roughness elements—λ=0.095	0.02006	2	0.002970	Gillies et al. (2006)

at very high frequencies (see van Boxel et al., 2004 and Walker, 2005 for detailed reviews on the operation and use of sonic anemometers).

16.3.4   PARTICLE ENTRAINMENT BY WIND

A particle is moved by the wind when the fluid drag forces ($\sim d^2$) exceed both the particle's weight ($\sim d^3$) and the sum of the cohesive forces ($\sim d^{-1}$) that bind it to adjacent particles within the bed. These resisting forces are related to physical properties of the particle, including its size, density, mineralogy, shape, and packing as well as the presence or absence of bonding agents such as adsorbed water films (or as pendular rings), soluble salts, organic matter, and synthetic surfactants, which occasionally are applied in industrial settings. The friction velocity (m s^{-1}) at the threshold for particle motion (u_{*t}) was first modeled by Bagnold (1941) as

$$u_{*t} = A\sqrt{\left(\frac{\sigma - \rho}{\rho}\right) gd},   \tag{16.5}$$

where A is a dimensionless, empirical coefficient, σ (kg m^{-3}) is the grain density, d (m) is the grain diameter, and g (m s^{-2}) is gravitational acceleration. This relation is ilustrated in Figure 16.5. The value of A is constant at approximately 0.1 for particle friction Reynolds numbers (Re$_P = u_* d/\nu$) exceeding 3.5 (or $d \geq 80\,\mu$m) where ν (m^2 s^{-1}) is the kinematic viscosity. Large particles protrude directly into the airflow and are aerodynamically "rough." Equation 16.5 suggests that u_{*t} essentially increases with the square root of the grain diameter, given that the natural variation in both the mineral and air densities are small in comparison.

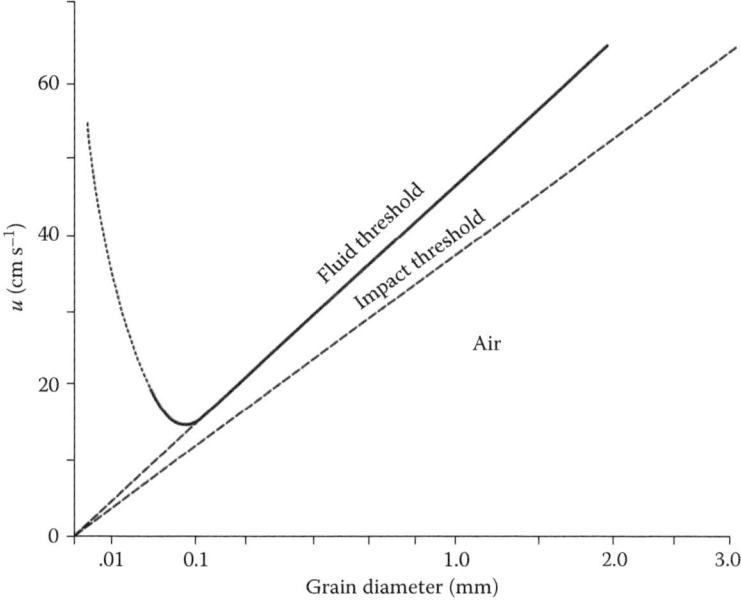

FIGURE 16.5 Variation in threshold friction velocity with particle diameter. The plot distinguishes between fluid and impact threshold.

In contrast, small silt particles lie entirely within the viscous sublayer for $Re_P < 3.5$ and the surface is aerodynamically "smooth." As a result u_{*t} is no longer solely proportional to the square root of grain diameter, but also is dependent on the value of A, which rises rapidly as particle size diminishes. Wind tunnel studies (e.g., Bagnold, 1941; Chepil, 1945a) confirm that very high wind speeds are required to entrain these fine-grained materials. However, Iversen et al. (1976), Greeley et al. (1976), and Iversen and White (1982) question the assumption that the coefficient A in the Bagnold threshold model is a unique function of particle friction Reynolds number. The relative importance of the forces acting on sedimentary particles also changes significantly with size, so that the mechanisms leading to the entrainment of sand and dust particles differ profoundly. As particle diameter ($<80\,\mu$m) and therefore the influence of the gravitational force diminishes ($\sim d^3$), interparticle cohesion ($\sim d^{-1}$) becomes increasingly dominant. Unfortunately, significant uncertainties still exist in the estimation of these binding forces. Particularly for particles $<20\,\mu$m, the cohesive forces attributed to electrostatic charges and moisture films begin to dominate so that such particles are not easily lifted by aerodynamic forces alone (Shao et al., 1993; Shao, 2000, p. 201).

During their downwind flight, saltating particles extract fluid momentum from the wind. Upon their descent, they impact the bed surface and may either ricochet off other grains, or become embedded in the surface. In both cases, momentum is transferred to the bed in the disturbance of one or more stationary grains. As a result of the impact, several particles may be ejected into the airflow at friction velocities that are lower than required to move a stationary grain by direct fluid drag alone.

Bagnold (1941) defined this lower threshold as the dynamic or impact threshold. For large sand particles, it follows the same general relationship as the fluid threshold (Equation 16.5), but with a lower value for the coefficient $A \sim 0.08$ (Figure 16.5). Most important, it continues to decline with decreasing particle diameter, even in the case of dust/clay-sized particles.

Therefore, even though fine particles with diameters $<80\,\mu m$ are inherently resistant to entrainment by fluid drag alone, major dust storms primarily driven by dynamic impact processes are common occurrences in many parts of the world. Under natural conditions, dust either coats sand particles or exists as aggregates in soils with high clay content. These fines generally cannot be entrained under light winds. In strong winds, however, large amounts of dust (silt and clay size) may be emitted as a direct result of the abrasion carried out by the impact of saltating sand particles as shown in Figure 16.3 (Chepil and Woodruff, 1963; Chepil, 1945b; Gillette, 1974; Nickling and Gillies, 1989). The abrasion breaks up surface crusts and aggregates, releasing fine particulates into the air stream (e.g., Chepil, 1945c; Gillette et al., 1974; Nickling, 1978; Hagen, 1984; Shao et al., 1993; Shao and Lu, 2000; Roney and White, 2006; and McKenna-Neuman et al., 2009). It therefore can be argued that the surface texture and degree of cohesion is crucial to the prediction of emission events, while the dependence of dust emission on wind velocity arises through the saltation flux (Q).

As reviewed by Shao (2000, p. 202), a physical dust emission model should depend on three basic components:

1. A model of saltation intensity for the given wind speed, surface, and soil conditions
2. A statistical representation of the saltation bombardment; inclusive of the particle trajectory, impact particle velocity, and angle
3. A description of the binding strength of the dust particles as well as some relationship between the dust emission rate and the intensity of saltation

While items 1 and 2 have been addressed to a great extent over decades of fundamental research (see reviews by Shao, 2000; Nickling and McKenna-Neuman, 2009 in press), much work remains to be carried out regarding item 3. Energy-based dust emission models face two intrinsic challenges in particular (Shao, 2000, p. 208). It is extremely difficult to estimate accurately the dust particle binding energy on either experimental (e.g., Alfaro et al., 1997) or theoretical grounds. The uncertainty alone spans several orders of magnitude. Also, the kinetic energy of the saltating sand particles is not entirely conserved, since an unknown portion is converted into heat. Lu and Shao (1999) propose a volume removal model, which describes the gouging of a surface by the impact of saltating sand particles, and in association with this, the ejection of dust as a plume into the airflow as illustrated in Figure 16.3. This model appears to provide a reasonable representation of empirical data collected by Gillette (1977), and Gordon and McKenna-Neuman (2009, accepted for publication). It also explains the wide divergence observed in the power relationship between the vertical dust flux (F, $kg\,m^{-2}\,s^{-1}$) and u_*, and confirms that the soil surface conditions play a vital role in determining the dependence of the dust emission rate.

16.3.5 Constraints on Particle Entrainment by Wind

The mass of sand particles moved by the wind in unit time is self-regulated by the associated partitioning of a portion of the total fluid stress to that borne by the particles. In the absence of interparticle forces, such systems are treated as transport limited and are amenable to the development of analytical and numerical models of the mass transport rate associated with saltation (Q); that is, the mass of sediment transported through a unit width of surface in unit time ($kg\ m^{-1}\ s^{-1}$). As a general approximation, Q scales with u_*^3 as modified by some function of u_*/u_{*t} representing the degree hy which the wind speed exceeds that required for particle entrainment. The transport limited case represents an upper limit for Q in assuming an unrestricted supply of loose particles for the airflow to carry. Such conditions can be attained in wind tunnel studies, but are rarely met in nature.

Much work has been carried out over the last decade concerning surface constraints on wind erosion associated with interparticle forces, inclusive of soil moisture, silt and clay, organic matter, and precipitated soluble salts. High friction velocities are required for the entrainment and transport of either very damp or aggregated particles. In the extreme case of continuous surface crusts, entrainment usually arises only through rupture from the impact of loose particles. Only a small fraction of this body of work addresses dust emission directly, while the focus has primarily been on the entrainment of sand particles and their transport via saltation. Nonetheless, since particle impact plays a key role in dust emission, such work is relevant.

The physical mechanisms underlying the effect of moisture on interparticle cohesion have received considerable attention in the soil physics literature and now are well understood. The total potential energy of soil water arises from both capillary and adsorptive forces and is represented as the matric potential (ψ_m). The relation of ψ_m (Pa) to the humidity of the air within the soil voids is modeled by the Kelvin equation:

$$\psi_m = \frac{R}{V_w}\, T \ln\left(\frac{e}{e_s}\right),\qquad(16.6)$$

where R is the ideal gas constant ($8.314\ mol^{-1}\ K^{-1}$), T is temperature (K), V_w is the molar volume of water ($1.8 \times 10^{-5}\ m^3\ mol^{-1}$), and e/e_s represents the relative humidity (RH), giving ψ_m in Pa. When a completely dry soil is exposed to an atmosphere containing water vapor at $35\% < RH < 40\%$, a single layer of water is adsorbed onto the charged faces of mineral surfaces by hydrogen bonding. As the RH increases, additional water is adsorbed in an ice-like structure, with the second layer completed at about 60% RH. For any two sedimentary particles that share a thin film of adsorbed water where they contact one another, the interparticle cohesion varies as $\psi_m A_c$. As a first approximation, the contact area (A_c, m^2) is assumed to increase with the film thickness (δ). For larger amounts of pore water that accumulate in the form of pendular lenses at the interparticle contacts, the magnitude of cohesion can be estimated directly from an explicit model for an ideal soil consisting of spheres of equal diameter, as developed by Haines (1925) and modified by Fisher (1926).

An extensive review of the role of soil moisture is provided in Nickling and McKenna-Neuman (2009). As a rule, gravimetric moisture contents of approximately

0.6% can more than double the threshold velocity of medium-sized sands (Belly, 1964), while for those exceeding 5%, sand cannot be entrained by most natural winds (McKenna-Neuman and Nickling, 1989). Empirical studies in a very wide range of settings all demonstrate that for a given wind speed, the sand transport rate (Q) for damp surfaces can vary by several orders of magnitude. Sherman et al. (1998) compare measurements of Q on a beach at Kerry, Ireland with predictions from five transport models adjusted for slope and mean gravimetric water content. None performed satisfactorily. The authors report that "the influence of sediment moisture content appeared to be the critical factor in degrading model viability." Despite a high degree of control in wind tunnel simulation, McKenna-Neuman and Langston (2006) also observed order of magnitude variation in the mass transport rate measured over damp surfaces at constant u_*.

Similarly, crusting is a key factor in reducing soil erosion and dust emission in arid and semiarid regions. The soil constituents that contribute to particle aggregation and crusting are highly variable in nature, found under a wide range of environmental conditions, and can include any or all of the following: Silts and clay minerals (physical), organic materials (biological), and precipitated soluble salts (chemical). Physical crusts can be formed by raindrop impact on bare soils (Chen et al., 1980), so that as the soil dries, the fine clay particles cement the larger particles together (Rajot et al., 2003). In dryland soils, these clay minerals often originate from dust precipitation as compared to rock weathering in more temperate settings.

A growing body of work has addressed the physical characteristics of crusts in relation to the threshold for entrainment, their response to particle impact, and the mechanisms by which they disintegrate. As reported by Gillette et al. (1980, 1982) for surfaces in the Mojave Desert, even weak crusts (modulus of rupture <0.07 mPa) significantly increase the threshold velocity for wind erosion. Crust strength varies with the composition and distribution of the binding media, and therefore, is usually spatially heterogeneous. This heterogeneity is important since wind-borne particles that strike the surface will rupture weaker areas of the crust, provided the impact force exceeds the binding force. This abrasion, in turn, liberates greater numbers of saltating particles and creates further ruptures downwind (McKenna-Neuman et al., 1996, 2005; McKenna-Neuman and Maxwell, 2002). Studies by Cahill et al. (1996) and Houser and Nickling (2001a, 2001b) clearly demonstrate that the emission of dust from crusted playa surfaces is dependent upon the abrasive action of saltating sand grains. Unfortunately, disturbance of crusted surfaces by off-road vehicles, industrial/agricultural machinery, and cattle trampling is a growing problem in many arid regions of the world that contributes widely to dust emission. McKenna-Neuman et al. (2009) demonstrate that mechanical disturbance of crusted mine tailings leads to a significant increase in the emission rate (see case study 2).

16.3.6 Measurement of Dust Concentration

The development of techniques for determining fugitive dust emissions presents numerous challenges and remains the subject of much investigation. Past approaches have included field-based monitoring stations (e.g., Gillette, 1977; Nickling, 1978; Nickling and Gillies, 1989, 1993) and field and lab wind tunnel studies (e.g., Gillette,

1978; Ciccone et al., 1978; Houser and Nickling, 2001a, 2001b; Roney and White, 2006, 2010; McKenna-Neuman et al., 2009), while more recently, highly portable field units (e.g., PI-SWERL, Etyemezian et al., 2007; Sweeney et al., 2008) and vehicle-based approaches to road dust measurement (Etyemezian et al., 2003) have been added to our toolkit of instruments and methodologies.

The need to measure and meet air quality standards has been a major driving force in the development of standardized techniques/technologies. A higher level of rigor generally is applied in particulate air pollution research than has been the case for aeolian geomorphology. The ideal dust sampling instrument for use in field and wind tunnel studies should have high accuracy while allowing for rapid sample collection and continuous recording. It also should be capable of isokinetic and omnidirectional sampling. Finally, such an instrument should be inexpensive (accessible), portable, and require little maintenance, while providing long-term operation with low power consumption. Technologies currently used in the measurement of dust concentration fall into two major groups: Direct mass samplers (gravimetric method) and optical samplers (laser photometry, beta ray attenuation, mass spectrometry). Direct mass samplers generally have good accuracy. Since the dust particles are captured on glass fiber filters, this method is well suited to subsequent analysis of the trace element and contaminant composition via AA and ICP. However, gravimetric sampling provides only course temporal resolution since many hours of operation often are required to collect a sufficient amount of dust to weigh on a microbalance. This method also requires high maintenance in the sense that the filters must be manually inserted, removed, and weighed. Errors can arise in the transfer, drying, and weighing of the filters, and from sampling at a fixed flow rate which is not representative of isokinetic conditions. In comparison, optical samplers provide rapid dust concentration measurements in real-time, and are reasonably accurate provided very careful attention is paid to cleaning and calibration. One important limitation is that the majority of these instruments are designed only to monitor particles sizes corresponding to the inhalable fraction (e.g., $<10 \mu m$, not TSP). They generally do not sample isokinetically, except at some optimal wind velocity that matches the intake flow rate. Table 16.2 reviews a small number of commercially produced instruments. Recent studies comparing the accuracy and performance of these instruments in a range of environments include Baldauf et al. (2001), Chung et al. (2001) and Heal et al. (2000).

16.4 ESTIMATION OF AEOLIAN MASS TRANSPORT: CORRELATIONS AND PARAMETERS

16.4.1 COMPUTATION OF THE DUST EMISSION RATE: AN ANALYTICAL APPROACH

The dust emission rate (E) is defined as the vertical mass flux of dust particles *leaving* a given surface. With reference to a control volume approach (Figure 16.6), E can be estimated from measurements of either the net vertical ($F_{diff} - F_g = \mathrm{d}m_{DIFF}/\mathrm{d}t - \mathrm{d}m_{DEPOS}/\mathrm{d}t$) or the net horizontal mass flux ($\mathrm{d}m_{out}/\mathrm{d}t - \mathrm{d}m_{in(ambient)}/\mathrm{d}t$).

TABLE 16.2
Comparison of Selected Dust Monitoring Instruments

Instrument	TSP	PM$_{10}$	PM$_{2.5}$	Technology	Flow Rate (L min^{-1})	Real-Time Sampling	USEPA-FRA Compliant
MiniVol	Yes	Yes	Yes	Gravimetric	5	No	No
TEOM	Yes	Yes	Yes	Gravimetric Internal oscillating balance weighs filter	0.5–4	Yes	Yes
DustTrak	No	Yes	Yes	Optical Laser photometry	1.4–2.4	Yes	No
HiVol	Yes	Yes	Yes	Gravimetric	750–1600	No	Yes
BAM 1020	Yes	Yes	Yes	Optical—β-ray attenuation Filter tape transport system	Up to 20	Yes	Yes
Met One E-sampler	Yes	Yes	Yes	Dual technology optical—mass spectrometry gravimetric	1–3.5	Yes	No
Grimm Monitor	Yes	Yes	Yes	Optical—mass spectrometry sizes and counts dust particles	1.2	Yes	No
Partisol	Yes	Yes	Yes	Gravimetric optional optical scan	16.7	Yes, with optical scan only	Yes

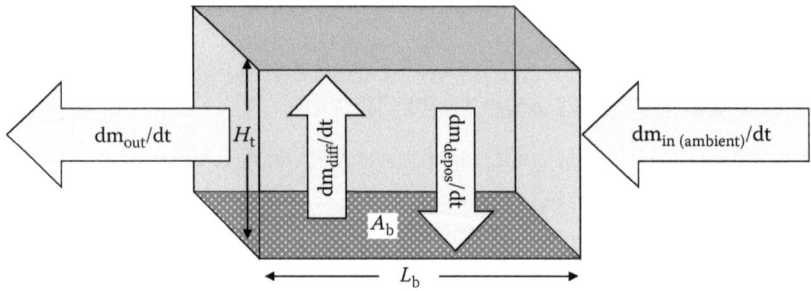

FIGURE 16.6 Horizontal and vertical mass fluxes defined for a hypothetical control volume.

The diffusive component F_{diff} associated with turbulent motions is determined from the vertical profile of dust concentration (c, mg m^{-3}), as

$$F_{\text{diff}} = -K_p \frac{\partial c}{\partial z}, \tag{16.7}$$

where K_p is the eddy diffusivity. Although K_p is affected by atmospheric stability, in most models it is approximated by $\kappa u_* z$ for neutral boundary layer flows, similar to the flux of momentum. Under steady-state conditions, the net vertical dust flux is approximately constant at all levels in the atmosphere with both gravitational settling (F_g) and turbulent diffusion (F_{diff}) increasing as the surface is approached. Shao (2000, p. 197) makes the important distinction that it is F_{diff}, which is measured in experimental studies as an approximation of E, while F_g is disregarded.

Unfortunately, there also is no consensus in the literature at present on the best relation to represent the vertical distribution of suspended particulate concentration. Using theoretical arguments, Shao (2000) suggests that under steady-state conditions, the vertical profile of dust concentration obeys a power law

$$c = c_r \left(\frac{z}{z_r} \right)^{w_t / k u_*}, \tag{16.8}$$

with the concentration of dust increasing toward the source of emission at the surface. Herein, c_r is the dust concentration at reference height z_r, and w_t is the vertical (terminal) velocity of the dust particles, which in vector form is –ve, indicating motion toward the surface. This outcome is not strictly upheld in all empirical work, with some studies suggesting a logarithmic relation between c and z, similar to the vertical profile of the fluid momentum (Equation 16.2).

In order to accommodate individual measurements from a meteorological tower, Equation 16.7 often is written and solved in discrete form as follows:

$$F_{\text{diff}} = -\kappa u_* z_2 \frac{c(z_2 + \Delta z; x) - c(z_2 - \Delta z; x)}{2 \Delta z}. \tag{16.9}$$

This ubiquitous method requires at least two vertical locations (z_1, z_2) for the measurement of dust concentration (c_1, c_2) for a given instrument tower in order to estimate the vertical dust flux. Gillette (1979) and Saxton et al. (2000) make a direct analogy between the mass and momentum fluxes in suggesting

$$F_{\text{diff}} = \frac{-k u_* (c_2 - c_1)}{\ln \left(\frac{z_2}{z_1} \right)}. \tag{16.10}$$

The accuracy of this approach is highly dependent on the turbulent diffusion scaling with u_*, and the assumption that all dust measured is emitted from the surface of interest rather than from an unrelated, upwind source.

Measurements of the net horizontal mass flux have been carried out in both wind tunnel and field studies. As for example in recent wind tunnel simulations, both Roney and White (2006) and McKenna-Neuman et al. (2009) apply a mass balance approach to a defined control volume ($W_b \times L_b \times H_t$) in order to measure E directly. Herein, W_b is defined as the width of the soil bed, L_b its length, A_b its area, and H_t as the height of the working section in the tunnel (Figure 16.6). The difference between the mass flux into (m_{in}) and out of (m_{out}) the designated control volume provides the emission rate,

$$E = \frac{1}{A_b}(m_{out} - m_{in}), \qquad (16.11)$$

where the flux rates can be determined from vertical profiles of the dust concentration (c_z) and velocity (u_z) as

$$m_{OUT} = \int_0^{Ht} c_{OUT}\, u_{OUT}\, W_b\, dz, \qquad (16.12)$$

$$m_{IN} = \int_0^{Ht} c_{IN}\, u_{IN}\, W_b\, dz, \qquad (16.13)$$

giving

$$E = \frac{1}{L_b} \int_0^{Ht} (c_{OUT}\, u_{OUT} - c_{IN}\, u_{IN})\, dz. \qquad (16.14)$$

Similar to the vertical concentration gradient observed by Gillies and Berkofsky (2004), detailed PM_{10} concentration profiles were found by McKenna-Neuman et al. (2009) to follow a log-law with very high concentrations within a few centimeters of the bed surface and little to no differentiation from the background within the free stream. The advantage of this methodology is that it accounts for all dust emitted from the test surface and transported by both horizontal advection and turbulent diffusion.

16.4.2 Relations from Empirical Studies of Particle Resuspension

Data collected in empirical studies show that the dependence of the vertical dust flux upon friction velocity generally follows a power law (e.g., Figure 16.7),

$$F = a\, u_*^b, \qquad (16.15)$$

where a and b are empirical constants. However, all such data contain a very large amount of scatter as governed by the varied binding strength of the dust particles and related constraints on saltation impact. Gillette and Passi (1988) suggest from theory that the dust emission rate should scale approximately with u_*^4, and specifically as

$$F = \alpha\, g\, u_*^4 \left(1 - \frac{u_{*t}}{u_*}\right), \qquad (16.16)$$

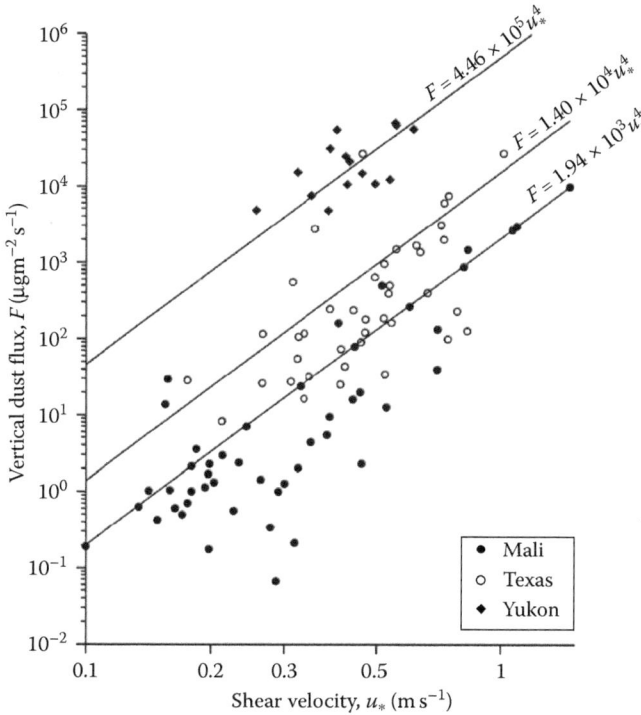

FIGURE 16.7 Variation of the vertical dust flux with friction velocity. [After Nickling, W.G. and Gillies, J.A. 1993. *Sedimentology* 40(5):859–868.]

where $u_* \geq u_{*t}$. In comparison, Shao et al. (1993) argue that since dust emission is primarily a function of saltation bombardment, which varies as the cube of shear velocity, F also should be a function of u_*^3. While sand-sized aggregates of silt and clay can be transported in saltation, these tend to break up or abrade during transport, and so, provide fines that are carried away in suspension (Gillette and Goodwin, 1974; Gillette, 1977; Nickling, 1978; Hagen, 1984). Table 16.3 summarizes, for a wide range of site conditions, a number of empirical relations reported between the dust flux and u_*. In general, undisturbed (natural) desert sites, and mine tailings have emission rates that scale with the cube of the friction velocity, as compared to highly disturbed sites (i.e., construction sites) with a larger exponent ($4 < b < 6$).

Not surprisingly, many of the relations suggested for F (Table 16.3) are identical in structure to well-established mass flux equations for saltation (Q). A wide range of site dependent values are reported for the scaling between F and Q, some indicating a strong correlation (e.g., wind tunnel simulations by Roney and White, 2006; McKenna-Neuman et al., 2009) and others indicating little to no correlation (e.g., Gillette, 1977).

TABLE 16.3
Forms of Equations Used to Predict the Flux of Windblown Dust

Measurement	Relation	Material Characteristics/Comments	Source
PM$_{10}$—wind tunnel	$32,000u_*^{1.3}$	Aluminum powder, no saltation	Fairchild and Tillery (1982)
PM$_{10}$—wind tunnel	$570,000u_*^{1.8}$	Aluminum powder, 100 μm glass bead saltators	Fairchild and Tillery (1982)
PM$_{10}$—wind tunnel	$16,000u_*^{2.1}$	Aluminum powder, 200 μm glass bead saltators	Fairchild and Tillery (1982)
PM$_{10}$—wind tunnel (portable)	$9.836u_*^{2.14}$	Crusted playa, frequently disturbed	Houser and Nickling (2001)
PM$_{10}$—wind tunnel	$0.0036u_*^3$	Arizona Test Dust, no saltation	Loosemore and Hunt (2000)
PM$_{10}$—wind tunnel	$8\times10^{-9}u_*^{5.4}$	Freshly ripped coarse slag, no saltation	Sanderson et al. (2009)
TSP	$0.031u_*$	Coal dust on concrete paddock	AP-42
TSP—tower	$0.591u_*^{5.67}$	Tilled field—mouldboard ploughing	Lopez et al. (1998)
TSP—tower	$0.101u_*^{5.98}$	Tilled field—chisel ploughing	Lopez et al. (1998)
PM$_{10}$—wind tunnel	$1.392u_*^{4.4}$	Dry, pulverized tailings	McKenna-Neuman et al. (2009)
PM$_{10}$—wind tunnel	$0.289u_*^{4.2}$	Dry tailings, crust disturbed	McKenna-Neuman et al. (2009)
PM$_{10}$—wind tunnel	$0.014u_*^{2.7}$	Damp tailings—2–10 %WC	McKenna-Neuman et al. 2009
TSP—tower	$u_*^{3.55}$	Slims River delta	Nickling (1978)
TSP—wind tunnel (portable)	$u_*^{4.27}$	Sites >25% silt/clay	Nickling and Gillies (1989)
TSP—wind tunnel (portable)	$u_*^{3.02}$	Sites <15% silt/clay	Nickling and Gillies (1989)
PM$_{10}$—tower	$u_*^{4.94}$	Australian rangeland	Nickling et al. (1999)
TSP—wind tunnel	$u_*^{3.0}$	Varying loamy soils	Gillette (1978)
TSP—tower	$u_*(u_*^2-u_{*t}^2)$	Owens (dry) lake playa soils	Gillette et al. (1997)
TSP—tower	$(u_*^3-u_{*t}^3)$	Owens (dry) lake playa soils	Gillette et al. (1997)
TSP—tower	$u^2(u-u_t)$	Heavy saltation bombardment	Gillette (1974)
TSP—combined results	$u_*^3(u_*-u_{*t})$	Combined field/wind tunnel study	Saxton et al. (2000)
TSP—combined results	$u_*^2(u_*-u_{*t})$	Combined field/wind tunnel study	Saxton et al. (2000)
PM$_{10}$—wind tunnel	$(u_*-u_{*t})^2$	Owens (dry) lake playa soils	Roney and White (2006)
<PM$_{20}$—combined results	$u_*^4(1-u_{*t}/u_*)$	Deduced from communication with P.R. Owen	Gillette and Passi (1988)
TSP—wind tunnel	$u_*^3(1-u_{*t}^2/u_*^2)$	Kaolin clay (mean d = 2 μm), heavy saltation	Shao et al. (1993)

16.5 CHEMICAL FLUX TO AIR

Several consequences associated with aeolian transport of contaminated soil particles are presented in Section 16.2. As outlined, dust generation is a significant chemical mobilization process. Besides the TSP being the air contaminant, the chemical loadings also contribute to the overall air pollution burden. Once mobilized to the atmosphere, chemical reactions with gases and vapors, as well as moisture, may alter the composition of the particles. They may also catalyze the production of additional air contaminants. Nevertheless, the entrained fine particles that are transported by the local and regional winds are eventually deposited into water bodies and onto agricultural soils, crops, urban surfaces, park areas, and so on. At these locales, selected biota will become exposed to the particles and the sorbed chemicals. The purpose of this section is to make the theoretical and mathematical connection between chemical mobility and dust resuspension for use in multimedia chemodynamic models.

16.5.1 The Chemical Flux

Aeolian mass transport is quantified by the dust emission flux of particle mass, F_p ($\mathrm{kg\,m^{-2}\,s^{-1}}$), defined as the vertical mass rate of particle type-p leaving a given surface; see Section 16.2. The aeolian chemical flux is dependent on particle type; this being its size, density and chemical composition. It has been observed that the fine particle fraction, which includes the clays and silts primarily, contributes a disproportionately large part of the chemical loading. The chemical mass loading on the particle is denoted by w_{Ap} (mg/kg). Therefore, the chemical flux for particle type-p is

$$n_{Ap} = F_p w_{Ap}. \tag{16.17}$$

The total flux is the sum of each type or classification with units of ($\mathrm{mg\,m^{-2}\,s^{-1}}$). This relationship assumes the flux is the advective rather than a diffusive type. This means the chemical "piggy-backs a ride" on the particle flux and it is independent of the TSP or total chemical concentration existing in the background air.

16.5.2 The Aeolian Process Mass Transport Coefficient

The dust resuspension process is one of two very significant processes that mobilize chemicals from soil surfaces to the atmospheric boundary layer. The other is termed chemical vaporization or volatilization. In the case of creating and using multimedia chemodynamic models, the flux expression must be either an advective or diffusive type relationship, containing a chemical fugacity or concentration as the state variable; see Chapters 3 and 4. With the availability of F_p, it is possible to obtain the dust resuspension MTC as follows. It is defined as the ratio of the particle flux to the soil particle density or grain density σ_p ($\mathrm{kg\,m^{-3}}$):

$$k_p = \frac{F_p}{\sigma_p} \tag{16.18}$$

This MTC has units of velocity ($\mathrm{m\,s^{-1}}$). The alternative chemical flux expression to Equation 16.17 is therefore

$$n_{Ap} = k_p \sigma_p w_{Ap}. \tag{16.19}$$

Table 16.4 contains some useful soil particle densities. The product $\sigma_p w_{Ap}$ is the equivalent chemical concentration ($\mathrm{mg\,m^{-3}}$) of the surface soil particles being eroded.

TABLE 16.4
Soil Materials and Specific Gravity

Material	Sp. Gr.	Material	Sp. Gr.	Material	Sp. Gr.
Clay	2.23	Gypsum rock	2.69	Pyrite ore	3.48
Clay, calcined	2.32	Magnitite	3.88	Sandstone	2.68
Coal	1.63	Lead ore	3.44	Shale	2.58
Coke	1.53	Limestone	2.69	Silica	2.71
Copper ore	3.02	Mica	2.89	Silica sand	2.65
Gneiss	2.71	Nickel ore	3.32	Slag	2.93
Gold ore	2.86	Oil shale	1.76	Slate	2.48
Granite	2.68	Phosphate fertilizer	2.65	Uranium ore	2.70
Gravel	2.70	Phosphate rock	2.66	Zinc ore	3.68

Source: Adapted from *Perry's Chemical Engineering Handbook*, 7th ed, 1997. McGraw-Hill, New York, pp. 20-14.

Note: Specific gravity reference is water at $0°C$ 1000 $(kg\,m^{-3})$.

Such alternative concentration-dependent MTC flux equations are convenient in combining the individual processes formulating the overall flux expressions needed in multimedia models; see Chapter 4. As basic rate of transport parameters, the MTCs are convenient to use in evaluating and interpreting the controlling chemical transport process mechanisms in specific situations.

16.6 SHORT GUIDE TO USERS

Estimation of the emissions from a given surface primarily depends on knowledge of

1. *Wind speed (u_z) and the aerodynamic roughness length (z_0) to derive the friction velocity (u_*).* If the surface is spatially heterogeneous, the wind direction must also be known. The aerodynamic roughness length can be estimated from published values for surfaces having similar properties (e.g., Table 16.1), but also, is straightforward to measure in the field from a vertical array of 4–5 anemometers. If the surface material is relatively homogeneous and not sheltered extensively by vegetation, then a wind tunnel simulation can provide a reasonable approximation for z_0 as well. For convenience, model estimates often depend upon wind speed data obtained from anemometers mounted at 10 m on meteorological towers at airports and weather stations. However, turbulence, atmospheric stability, and topographic forcing can contribute to large errors in the estimation of drag from u_{10} at the source of the emissions. Investment in the set up and maintenance of one or more microclimate stations directly over the site in question will provide significantly improved accuracy in the temporal measurement of u_z, u_* and therefore, the model estimation of F from Equation 16.16. This is especially important given that the majority of the time, the near surface wind speed will lie below the threshold (u_{*t}) for the emission of dust.

2. *Particle size distribution of the surface materials.* If the material under consideration is coarse with very few particulates in the silt range having diameters under $70\,\mu m$ (e.g., only a few percent), then it is reasonable to expect that the severely limited supply of particles will principally govern the emission rate, which should be among the lowest values in the published literature (e.g., $\ll 1\,\mu g\ m^{-2}\ s^{-1}$). As well, there should be little dependency upon wind velocity (*b* small, around 1 or 2). If the surface contains a very large proportion of silt (e.g., exceeding about 30%) as well as some sand to create ballistic impacts, and the sediment is dry and loose, then the emission rates will not only be extremely high (e.g., 10^3–$10^5\,\mu g\ m^{-2}\ s^{-1}$), but also show a strong dependence upon wind speed with the exponent *b* in Equation 16.15 having a value of about 4. Empirical relations such as those reported in Table 16.3 perform well under such circumstances.

3. *Surficial factors.* Fortunately, there are numerous surficial factors which can limit or prevent dust emission over time and space. These include moisture content, aggregation, crusting, vegetation cover, and surface armouring. Anthropogenic disturbance is a key cause of dust emission (e.g., road traffic, plowing, excavation, etc.) from surfaces that would otherwise be stable. These represent the most challenging conditions in which to model emission rates. At the present time, the best practice in the estimation of emissions for complex conditions, apart from direct measurement, is to identify a suitable analogue within the published literature. Through the cumulative efforts of academic researchers, environmental consultants, and government researchers working in the agriculture, air quality, and transportation sectors, the database of emission rates is slowly expanding in scope. However, there has not been a coordinated effort to compile all such data, while selected works demonstrate order of magnitude uncertainty even within measurement replicates at the same field site.

16.7 CONCLUSIONS

Aeolian dust systems have few spatial or temporal boundaries. They can operate anywhere at any time, with the sediments transported in any direction over distances between a few meters to several thousand kilometers. Wind speed is only one of a large number of governing factors that determine whether or not dust is liberated from a surface. Binding agents such as electrostatic forces, moisture films, biotic materials, clay minerals and salts, all play a role in determining the energy required to entrain dust particles. In many cases, these particles either form aggregates or coat larger sand-sized particles that saltate and release dust through abrasion. Since the suspension of dust extracts very little momentum from the airflow, as compared to the saltation flux, the amount transported in the atmosphere is largely governed by the textural properties of the surface as compared to other transport modes.

As a result, these aeolian systems are far more difficult to model than the movement of contaminants in channelized (fluvial) flows. Although the mass flux of dust is well demonstrated to vary as a power function of the friction velocity, the value of the exponent appears to be determined by the soil properties. Even so, order of magnitude variation from the expected relation is quite common, even at the same site under similar air flow conditions.

While numerous algorithms have been published for the prediction of dust emissions, inclusive of those listed in AP-42 for a wide range of sites and activities, model calibration via on-site measurement remains the cornerstone of emissions evaluation and mitigation. In the context of contaminant transport modeling, it is similarly important to know the composition of the dust particles which varies with weathering processes as determined by geology/geochemistry and climate, in addition to anthropogenic activities. Such processes also have a direct bearing on particle size and therefore the mode and distance of aeolian transport. Again, it is usually necessary to obtain benchmark measurements of the dust particle composition for a given setting.

16.8   CASE STUDIES

16.8.1   FIELD ESTIMATION OF EMISSIONS FROM SLAG

Setting

Slag disposal site at smelter (example of low PM_{10} emissions from a rough, coarse surface).

Problem

Slagard Corporation is preparing an emissions inventory for the entire smelter operation. The TSP emission rates for the "ripped" (crushed and graded) slag are unknown. The closest analogues are the rates published for coal in USEPA AP-42. Considering the dissimilarity of the materials, and the relatively low specific gravity of coal as compared to slag in particular (Table 16.4), these rates are believed to overestimate vertical fluxes and chemical emissions (Equations 16.7 through 16.19) at the site. Similarly, there are substantial differences in the physical properties of the slag across the mine site. The company believes that direct measurement of the emission rates for selected sites will provide more accurate representation, and ultimately, will contribute to a reduction in their emissions inventory. Such information is expected to be useful in designing cost-effective mitigation strategies where required, for example, watering. Slagard Corporation further intends to use the emissions data, in combination with ICP metals assay data (e.g., Cd 0.40 μg/mg; Pb 0.210 μg/mg), to evaluate air quality on the work site, along with AERMOD dispersion modeling to aid in the prediction of off-site fugitive dust concentrations based on prevailing weather conditions.

Goals

1. Determine the aerodynamic roughness length (z_0) for the slag as required for atmospheric dispersion modeling.
2. Determine emission rates (TSP and PM_{10}) for freshly ripped (dusty) slag, and "weathered" (rain-washed) slag, for varied wind speeds.

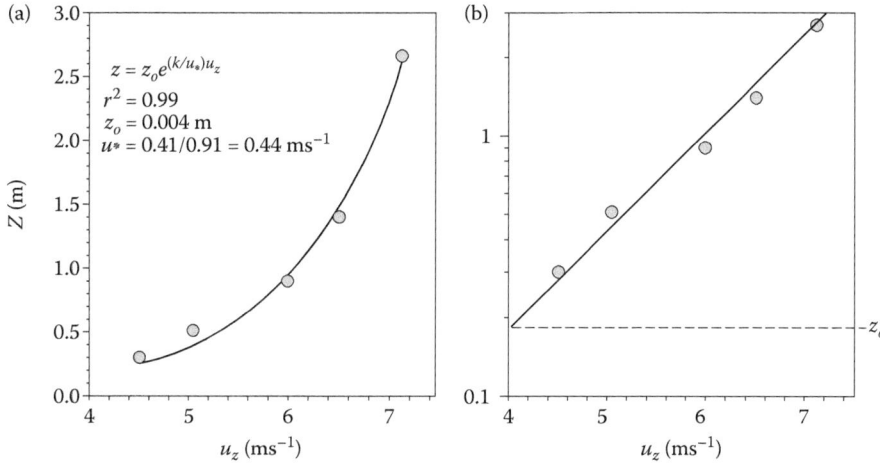

FIGURE 16.8 Sample wind speed profile over ripped slag at a smelter site. (a) Linear z; (b) $\log z$.

Methods

1. Measure z_0 from vertical profiles of wind velocity obtained over the slag.
2. Measure the vertical dust flux (F) from fugitive dust concentrations obtained synchronously at a minimum of two elevations (z_1 and z_2) at each of the field sites in question. In the present case study, vertical arrays of four TSI DustTraks™ measured PM_{10} concentrations continuously during the monitoring program, providing a high degree of temporal and spatial resolution. TSP was not measured due to a lack of instrument capability. In such situations, a scaling factor (approximately $2\times$) can be used to estimate TSP from PM_{10}.
3. Include wind tunnel emissions measurements to cover a broad range of wind speed under constant conditions of humidity and temperature.

Selected results

1. Figure 16.8 illustrates the analysis of a vertical profile of wind velocity obtained over freshly ripped slag using an array of five cup anemometers mounted on a portable tower. As expected for the law of the wall when the vertical axis is transformed from an arithmetic to a logarithmic scale, the data points fall along a straight line with slope proportional to u_*. During the sampling period on this particular day, the wind speed at 2.5 m above the surface averaged 7 m s^{-1}. Least-squares regression of an exponential relation to the data provides a value of 0.004 m for the aerodynamic roughness length (z_0) and 0.45 m s^{-1} for the friction velocity (u_*) with $r^2 = 0.99$. A value of 0.41 is assumed for von Karman's constant (κ).

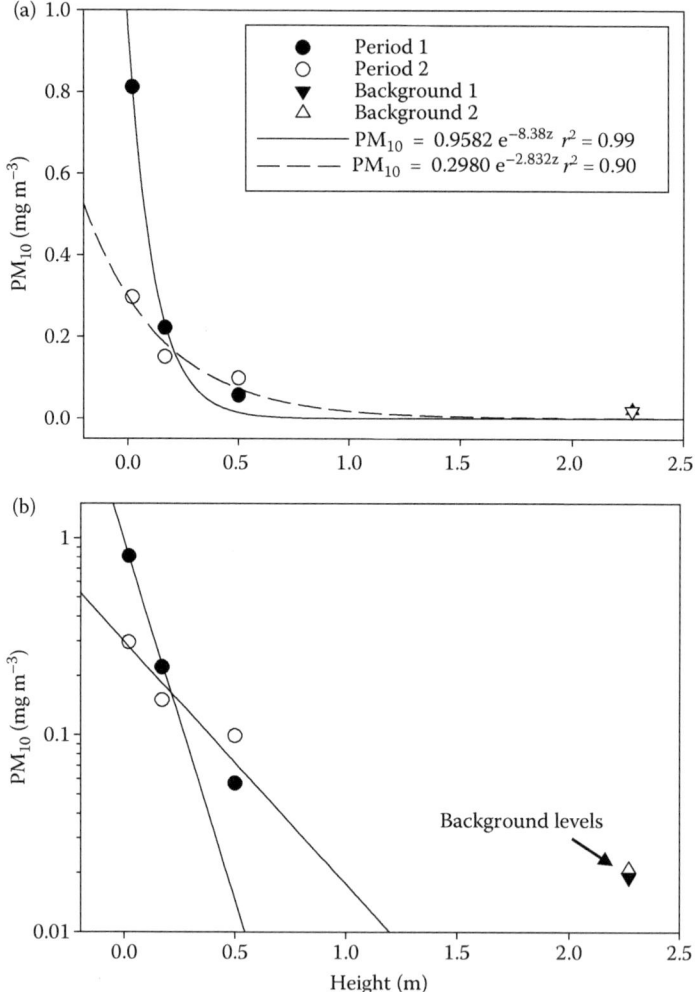

FIGURE 16.9 Selected vertical PM_{10} profiles obtained over freshly rippled slag. Wind speed at 2.5 m averaged $7\,m\,s^{-1}$. (a) The plot is on an arithmetic vertical scale; and (b) the plot is on an logarithmic scale. The profiles do not provide a good estimation of fluxes at $z > 2\,m$, suggesting that the sources of particulates is not the same.

2. Two sample vertical profiles of PM_{10} from field measurement are shown in Figure 16.9. A least-squares regression of the exponential function,

$$c = a\,e^{-bz}, \tag{16.20}$$

is found to fit both profiles well. The diffusive component of the vertical flux of PM_{10} then can be estimated either using a finite difference approximation (Equation 16.9), or the log-law relation provided in Equation 16.10. Table 16.5 summarizes the flux estimates derived from both methodologies, based on $u_* = 0.45\ \mathrm{m\ s^{-1}}$. Notably, the vertical fluxes (F_{diff}) are extremely low as compared to emission rates reported in the second case study below (Section 16.8.2). Equations 16.9 and 16.10 for estimation of F_{diff} provide similar values for $0.17\ m < z < 0.5\ \mathrm{m}$. Above 2 m, the PM_{10} concentration consistently appears to diminish toward background PM_{10} levels at the smelter site, so that the gradient is reduced and the vertical dispersion becomes trivial in magnitude at this particular site.

Initially, a power model (Equation 16.8) was applied to the vertical profiles of dust concentration, $c(z)$, but was found to provide a poor fit to the data for Period 2. However when both data sets are combined, the relation az^{-b} fits reasonably well and so for comparison with the log-law model (Equation 16.10), a vertical flux rate can be computed from

$$F = -\kappa u_* z \partial (a z^{-b}) / \partial z \tag{16.21}$$

$$= \kappa u_* z a b (z^{-b}/z). \tag{16.22}$$

This relation, based on values of 0.423 and 0.674 for the empirical coefficients a and b respectively, suggests a high flux immediately adjacent to the surface where the difference methods do not provide an estimate, and well approximates the log-law values for F above $z = 0.15\ \mathrm{m}$ (Table 16.5). With regard to Equation 16.8, the empirical model fit would suggest that the settling velocity (w_t) for the sampled dust from the nickel slag is around $0.1\ \mathrm{m\ s^{-1}}$.

The question of which model to use is largely academic, and for the practitioner, it may not be of central importance as compared to order of magnitude, temporal and spatial variation that is characteristic of many field data sets. As the data in Table 16.5 further indicate, poor instrument placement in a given setting can introduce substantial error, particularly for sparse arrays under conditions that may well not be *steady state*, and/or have a strong advective

TABLE 16.5
Summary of F_{diff} Estimations ($\mathrm{mg\ m^{-2}\ s^{-1}}$)

			Equation 16.9 Difference Method		Equation 16.10 Log Model		Equation 16.22 Power Model
	Period 1 [PM10]	Period 2 [PM10]	Period 1	Period 2	Period 1	Period 2	Combined
$z(m)$	$(\mathrm{mg\ m^{-3}})$	$(\mathrm{mg\ m^{-3}})$	F_{diff}	F_{diff}	F_{diff}	F_{diff}	F_{diff}
0.02	0.812	0.297	—	—	—	—	0.077
0.17	0.222	0.151	0.018	0.073	0.013	0.054	0.018
0.50	0.057	0.099	0.010	0.033	0.09	0.030	0.009
2.27	0.019	0.021	0.012	0.006	0.010	0.005	0.003

Note: Varied solutions are provided. Positive values of F_{diff} reported here indicate the upward diffusion of dust *away* from the surface.

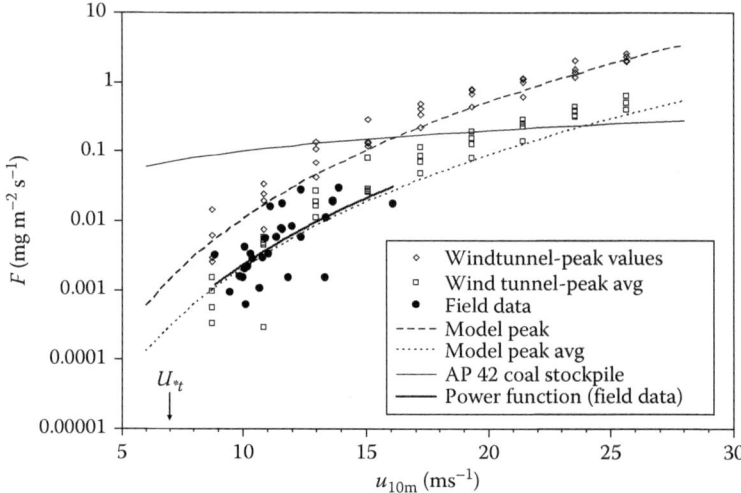

FIGURE 16.10 Vertical PM_{10} flux from ripped slag. A comparison of measurements from laboratory wind tunnel simulation, as well as onsite field data with the model estimate for active coal stockpiles (AP 42).

component. Unfortunately there are no clear guidelines in regard to this issue, so that it is largely up to the experience and judgement of the investigator to design and establish an instrument array that is suitable for the setting.

3. Figure 16.10 summarizes the relation between wind speed at 10 m and the vertical PM_{10} flux from freshly ripped slag. The measurements from the laboratory wind tunnel simulations compare very well with those from the field study, particularly for the temporally averaged data. The least squares power relation representing the field data set is provided below:

$$F = 8 \times 10^{-9} u_{10}^{5.4} \qquad (16.23)$$

as compared to the linear model for active coal stockpiles affected by wind erosion (AP-42),

$$E = 0.031 u_{10}, \qquad (16.24)$$

where $F \sim E$ is given in $mg\,m^{-2}\,s^{-1}$ and u_{10} is wind velocity ($m\,s^{-1}$) at a standard height of 10 m. As expected for the varied particle properties (σ_p), the AP-42 emissions model for coal overestimates the emission rate by one to two orders of magnitude at wind speeds near threshold (u_{*t}) and up to $15\,m\,s^{-1}$ ($54\,km\,h^{-1}$), but performs reasonably well for wind speeds well above threshold, between 20 and $30\,m\,s^{-1}$ (72 and $108\,km\,h^{-1}$). In the case of the "weathered" or rain-washed slag that was found to not emit any measurable quantities of PM_{10} at $u_{10} < 11$–$12\,m\,s^{-1}$ in wind tunnel testing (i.e., $F \cong 0$), the AP-42 model is not applicable. Finally, the emission rate for any given metal can be determined by multiplying the estimation of F by the respective concentration value from the metals assay, as described in Equation 16.17.

16.8.2 WIND TUNNEL ESTIMATION OF EMISSIONS FROM TAILINGS AS SLURRY

Setting

Tailings "beach" deposited as slurry at a gold mine within a desert region. (Example of high PM_{10} emissions from a smooth, silty surface of varied water content.)

Problem

Goldiggers Corporation is in the process of designing and establishing a mining operation within a desert region. The slurry produced from milling of the extracted rock contains a great deal of silt and clay (Figure 16.11). It is expected that these fines will be highly susceptible to entrainment by wind and will contribute to poor air quality in the region. The company is proposing to mitigate this potential problem by wetting the tailings on an ongoing basis. Given the arid setting and the large scale of the operation, there is a high cost associated with this strategy. The company wishes to determine the minimum level of gravimetric water content needed to prevent dust emission.

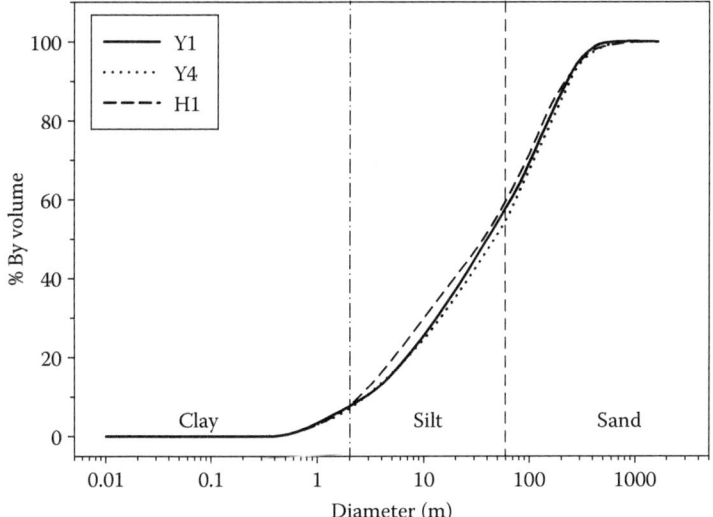

FIGURE 16.11   Particle size distribution of gold mine tailings.

Goals

1. For varied wind speeds, determine the emission rates (PM_{10}) from the tailings for the following surface conditions: (i) dry, (ii) dry pulverized, (iii) wetted to 10% and 2% gravimetric water content, and (iv) disturbed crust.
2. Repeat the experiments (described above for clean air) with particle impact on the surface in order to determine E during sand storms in the desert region.

Methods

1. Measure the dust emission rate for samples of the slurry in wind tunnel simulations using the control volume approach to correct for advection transport. Given the remote location of the site, and the fact that the mine is not in operation as yet, field work is ruled out.

Results

1. Figure 16.12 illustrates the basis of the control volume approach for the calculation of E.

   Figure 16.13 shows a sample plot of the relation between E and u_* for dry, pulverized slurry with particle impact, and in clean air. Each of the data series are well described by a power function with a high degree of correlation (see Table 16.6). The exponent averages around 4, similar to other published values for construction sites, and shows more variability for clean air. In comparison, the variation with u_* of PM_{10} emission rates from pulverized silt wetted to 10% gravimetric water content is illustrated in Figure 16.14, again in clean air and with abrasion from an upwind sand feed. In the case of emissions produced by sand abrasion, the data series are reasonably well described by power functions but with an exponent around 3. However, for a majority of the

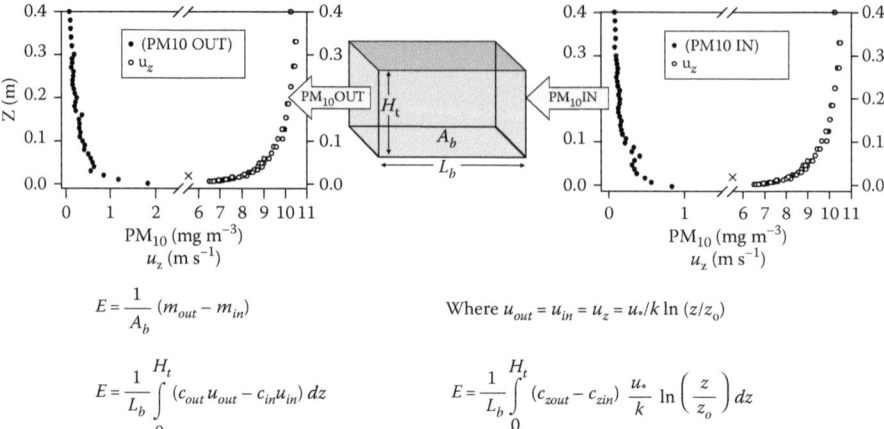

FIGURE 16.12 Control volume framework illustrating basis of flux assessment in wind tunnel simulation.

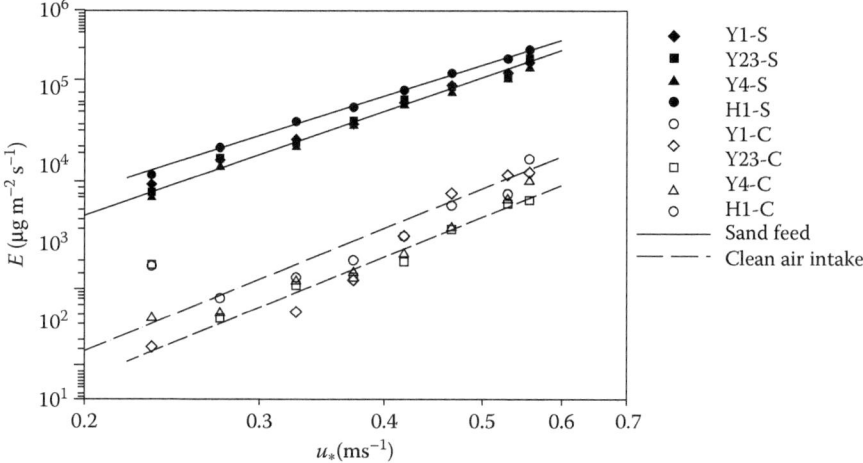

FIGURE 16.13 Variation with wind speed of PM_{10} emission rates from dry pulverized silt in clean air and with abrasion from an upwind sand feed in wind tunnel simulation. The data series are well described by power functions, although the high degree of correlation shown here is unusual for field data, and particularly, particle supply limited systems. [After McKenna-Neuman, C., Boulton, J. W., and Sanderson, S. 2009. *Atmospheric Environment* 43(3):520–529.]

TABLE 16.6
Listing of Flux Equation ($E = au_*^b$) Coefficients

	a	b	r^2
Pulverized Tailings			
H1-S	3511	4.15	1.00
Y23-S	2633	4.27	0.99
H1-C	56	0.79	
Y23-C	15	3.1	0.67
Slurry Dried then Disturbed			
H1-S	472	4.21	0.98
Y23-S	883	5.56	0.90
Y23-C	36	5.21	0.87
Slurry Dried to 2% Gravimetric Water Content			
H1-C	8	2.64	0.71
Slurry Dried to 10% Gravimetric Water Content			
H1-S	8	2.64	0.88
Y23-S	2	3.5	0.67
H1-C	3	1.35	0.85

Source: After McKenna-Neuman, C., Boulton, J. W., and Sanderson, S. 2009. *Atmospheric Environment* 43(3):520–529. S, upwind sand feed; C, clean air intake.

Note: E is given in $\mu g\, m^{-2}\, s^{-1}$ and u_* in $m\, s^{-1}$. Coefficients are not reported for experiments with $r^2 < 0.5$.

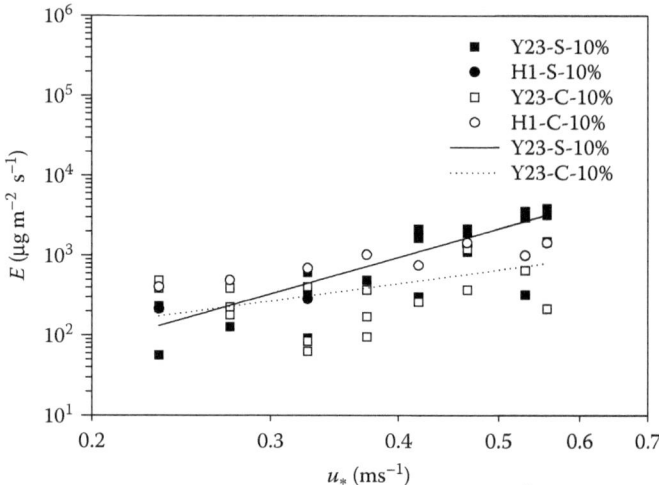

FIGURE 16.14 Variation with wind speed of PM_{10} emission rates from pulverized silt (wetted to 10% gravimetric water content) in clean air and with abrasion from an upwind sand feed. The data series are reasonably well described by power functions ($r^2 \sim 0.65$) in the case of PM_{10} produced by sand abrasion, but in the context of a clean air intake, there is arguably little to no dependency upon the friction velocity. (After McKenna-Neuman, C., Boulton, J. W., and Sanderson, S. 2009. *Atmospheric Environment* 43(3): 520–529.)

TABLE 16.7
Wind Tunnel Emission Results (E, mg m^{-2} s^{-1}) for a Clean Air Intake

U_*	0.19	0.23	0.27	0.33	0.37	0.41	0.47	0.53	0.56
U_{10m}	7.0	7.5	8.8	10.4	11.9	13.3	14.9	17.0	17.8
Pulverized Tailings									
H1	0.030	0.516	0.198	0.367	0.614	1.240	3.090	4.332	12.23
Y23	0.323	0.537	0.110	0.290	0.374	0.587	1.527	3.265	3.607
Slurry Dried then Disturbed									
H1	0.038	0.060	0.164	—	0.081	0.067	0.201	0.584	
Y23	—	—	0.074	—	0.0765	0.739	0.927	1.656	
Slurry Dried to 2% Gravimetric Water Content or Lower									
H1	0.973	1.397	1.648	2.195	2.123	2.147	2.275	2.270	
H1	0.968	1.245	1.358	1.772	1.694	2.086	3.289	2.570	
Y23	0.781	0.944	1.115	1.307	1.114	1.468	1.348	1.622	
Y23	1.087	0.899	0.827	1.065	0.881	1.032	0.969	1.148	
Slurry Dried to 10% Gravimetric Water Content									
H1	0.404	0.488	0.692	1.021	0.752	1.437	0.994	1.438	
Y23	0.443	0.224	0.084	0.170	—	0.372	—	0.216	
Y23	0.488	0.178	0.063	0.095	—	—	—	—	

Source: After McKenna-Neuman, C., Boulton, J.W., and Sanderson, S. 2009. *Atmospheric Environment* 43(3): 520–529.

TABLE 16.8
Wind Tunnel Emission Results (E, $mg\,m^{-2}\,s^{-1}$) with Particle Feed ON

U_*	0.19	0.23	0.27	0.33	0.37	0.41	0.47	0.53	0.56
U_{10m}	7.0	7.5	8.8	10.4	11.9	3.3	14.9	17.0	17.8
Pulverized Tailings									
H1	0.592	7.65	17.17	37.26	56.87	93.98	155.7	238.9	311.1
Y23	0.541	4.57	12.62	20.10	38.41	71.57	105.0	139.8	243.4
Slurry Dried then Disturbed									
H1	0.804	1.787	5.365	6.592	13.36	16.08	31.82	43.66	
Y23	1.180	1.012	0.639	5.030	9.537	12.47	20.94	38.54	
Slurry Dried to 2% Gravimetric Water Content or Lower									
H1	1.532	1.814	2.847	2.901	3.519	2.689	1.098	—	
H1	1.715	2.279	4.272	3.747	3.716	4.547	1.905	—	
Y23	0.628	0.257	0.719	—	0.138	0.264	—	3.714	
Y23	0.745	0.300	0.756	—	0.013	0.934	—	—	
Slurry dried to 10% gravimetric water content									
H1	0.214	—	0.286	—	—	1.318	—	—	
Y23	0.231	0.127	0.330	0.491	2.125	1.103	2.969	3.827	

Source: After McKenna-Neuman, C., Boulton, J. W., and Sanderson, S. 2009. *Atmospheric Environment* 43(3): 520–529.

experiments with a clean air intake, there is arguably little to no dependency upon the friction velocity.

Tables 16.7 and 16.8 present PM_{10} emissions data ($\mu\,g\,m^{-2}\,s^{-1}$) determined from the wind tunnel simulations for the four samples of slurry provided. All data reported in Table 16.7 refer to experiments with a clean air intake, as compared to those in Table 16.8 for which the test surfaces were abraded with particles from an upwind sand feed. As a "rule of thumb," the abrasion increases PM_{10} emissions by about two orders of magnitude, and 10^3 in the case of very damp tailings. The dry, pulverized tailings released the most dust, followed by dry, crusted tailings that had been mechanically disturbed (increased crack density). Surfaces that were damp still released small quantities of dust, but at rates indistinguishable from the undisturbed, crusted tailing surfaces. This would suggest that the formation of a cohesive crust overrides the effects of moisture adsorption.

LITERATURE CITED

Ackerman, S.A. and Hyosang, C. 1992. Radiative effect of airborne dust on general energy budget at the top of the atmosphere. *Journal of Applied Meteorology* 31(2):223–233.

Al-Rajhi, M.A., al-Shayeb, S.M., Seaward, M.R.D., and Edwards, H.G.M. 1996. Particle size effect for metal pollution analysis of atmospherically deposited dust. *Atmospheric Environment* 30(1):145–153.

Anderson, J.L. and Walker, I.J. 2006. Airflow and sand transport variations within a backshore-parabolic dune plain complex: NE Graham Island, British Columbia, Canada. *Geomorphology* 77(1–2):17–34.

Avila, A., Queralt-Mitjans, I., and Alarcon, M. 1997. Mineralogical composition of African dust delivered by red rains over northeastern Spain. *Journal of Geophysical Research* 102(D18):21977–21996.

Avila, A., Alarcón, M., and Queralt, I. 1998. The chemical composition of dust transported in red rains: Its contribution to the biogeochemical cycle of a holm oak forest in Catalonia (Spain). *Atmospheric Environment* 32(2):179–191.

Bagnold, R.A. 1941. *The Physics of Blown Sand and Desert Dunes*. London: Methuen, 265pp. Reprinted 1954; 1960; 2005, by Dover, Mineola, NY.

Baldauf, R.W., Lane, D.D., Marotz, G.A., and Wiener, R.W. 2001. Performance evaluation of the portable MiniVOL particulate matter sampler. *Atmospheric Environment* 35(35):6087–6091.

Bauer, B.O., Yi, J., Namikas, S.L., and Sherman, D.J. 1998. Event detection and conditional averaging in unsteady aeolian systems. *Journal of Arid Environments* 39(3): 345–375.

Bauer, B.O., Houser, C.A., and Nickling, W.G. 2004. Analysis of velocity profile measurements from wind-tunnel experiments with saltation. *Geomorphology* 59(1–4):81–98.

Belly, P.-Y. 1964. Sand movement by wind. United States Army, Corps of Engineers, Coastal Engineering Research Center, Washington DC, Technical Memorandum TM-1, 80pp.

Blumberg, D.G. and Greeley, R. 1993. Field studies of aerodynamic roughness length. *Journal of Arid Environments* 25:39–48.

Butterfield, G.R. 1999. Near-bed mass flux profiles in aeolian sand transport: High-resolution measurements in a wind tunnel. *Earth Surface Processes and Landforms* 24(5):393–412.

Cahill, T.A., Gill, T.E., Reid, J.S., Gearhart, E.A., and Gillette, D.A. 1996. Saltating particles, playa crusts and dust aerosols at Owens (dry) Lake, California. *Earth Surface Processes and Landforms* 21(7):621–640.

Castillo, S., Moreno, T., Querol, X., Alastuey, A., Cuevas, E., Herrmann, L., Mounkaila, M., and Gibbons, W. 2008. Trace element variation in size-fractionated African desert dusts. *Journal of Arid Environments* 72(6):1034–1045.

Chen, Y., Tarchitzky, J., Brouwer, J., Morin, J., and Banin, A. 1980. Scanning electron microscope observations on soil crusts and their formation. *Soil Science* 130:49–55.

Chepil, W.S. 1945a. Dynamics of wind erosion: I. Nature of movement of soil by wind. *Soil Science* 60:305–320.

Chepil, W.S. 1945b. Dynamics of wind erosion: II. Initiation of soil movement. *Soil Science* 60:397–411.

Chepil, W.S. 1945c. Dynamics of wind erosion: IV. The translocating and abrasive action of the wind. *Soil Science* 61:169–177.

Chepil, W.S. and Woodruff, N.P. 1963. The physics of wind erosion and its control. *Advances in Agronomy* 15:211–302.

Chung, A., Chang, D.P.Y., Kleeman, M.J., Perry, K.D., Cahill, T.A., Dutcher, D., McDougall, E.M., and Stroud, K. 2001. Comparison of real-time instruments used to monitor airborne particulate matter. *Journal of the Air and Waste Management Association* 51(1): 109–120.

Ciccone, A.D., Kawall, J.G., and Keffer, J.F. 1987. An experimental investigation of a wind-generated suspension of particulate matter from a tailings disposal area. *Boundary-Layer Meteorology* 38:339–368.

Countess, R., Barnard, W., Claiborn, C., Gillette, D., Latimer, D., Pace, T., and Watson, J. 2001. *Methodology for Estimating Fugitive Windblown and Mechanically Resuspended Road Dust Emissions Applicable for Regional Scale Air Quality Modelling.* Final Report for WGA Contract No. 30203-9, Countess Environmental, Westlake Village, CA.,103pp.

Draxler, R.R., Gillette, D.A., Kirkpatrick, J.S., and Heller, J. 2001. Estimating PM10 air concentrations from dust storms in Iraq, Kuwait and Saudi Arabia. *Atmospheric Environment* 35(25):4315–4330.

Etyemezian, V., Kuhns, H., Gillies, J., Green, M., Pitchford, M., and Watson, J. 2003. Vehicle-based road dust emission measurement: Part I—methods and calibration. *Atmospheric Environment* 37:4559–4571.

Etyemezian, V., Nikolich, G., Ahonen, S., Pitchford, M., Sweeney, M., Purcell, R., Gillies, J., and Kuhns, H. 2007. The Portable *In Situ* Wind Erosion Laboratory (Pi-Swerl): A new method to measure PM1-windblown dust properties and potential for emissions. *Atmospheric Environment*, 41:3789–3796.

Falkovich, A.H., Schkolnik, G., Ganor, E., and Rudich, Y. 2004. Adsorption of organic compounds pertinent to urban environments onto mineral dust particles. *Journal of Geophysical Research* 109: D02208, doi:10.1029/2003JD003919.

Fergusson, J.E. and Ryan, D.E. 1984. The elemental composition of street dust from large and small urban areas related to city type, source and particle size. *Science of the Total Environment*, 34:101–116.

Fisher, R.A. 1926. On the capillary forces in an ideal soil. Correction of formulae given by W.B. Haines. *Journal of Agricultural Science* 16:492–505.

Formenti, P., et al. 2001. Physical and chemical characteristics of aerosols over the Negev Desert (Israel) during summer 1996, *Journal of Geophysical Research* 106(D5):4871–4890.

Garrison,V.H., Shinn, E.A., Foreman, W.T., Griffin, D.W., Holmes, C.W., Kellogg, C.A., Majewski, M.S., Richardson, L.L., Ritchie, K.B., and Smith, G.W. 2003. African and Asian dust: From desert soils to coral reefs. *BioScience* 53(5):469–480.

Gillette, D.A. 1974. On the production of soil wind erosion aerosols having the potential for long-range transport. *Journal de Recherches atmosphériques* 8(3–4):735–744.

Gillette, D.A. 1977. Fine particle emissions due to wind erosion. *Transactions of the American Society of Agricultural Engineers* 20(5):890–897.

Gillette, D.A. 1978. A wind-tunnel simulation of the erosion of soil: Effect of soil moisture, sandblasting, wind speed and soil consolidation on dust production. *Atmospheric Environment* 12(8):1735–1743.

Gillette, D.A. and Goodwin, P.A. 1974. Microscale transport of sand-sized soil aggregates eroded by wind. *Journal of Geophysical Research* 79(27):4080–4084.

Gillette, D.A. and Passi, R. 1988. Modelling dust emission caused by wind erosion. *Journal of Geophysical Research* 93(D11):14233–14242.

Gillette, D.A., Blifford, I.H., and Fryrear, D.W. 1974. The influence of wind velocity on size distributions of soil wind aerosols. *Journal of Geophysical Research* 79(27): 4068–4075.

Gillette, D.A., Adams, J., Endo, A., Smith, D., and Kihl, R. 1980. Threshold velocities for input of soil particles into the air by desert soils. *Journal of Geophysical Research* 85(C10):5621–5630.

Gillette, D.A., Adams, J., Muhs, D., and Kihl, R. 1982. Threshold friction velocities and rupture moduli for crusted desert soils for the input of soil particles to the air. *Journal of Geophysical Research* 87(C10):9003–9015.

Gillies, J.A. and Berkofsky, L. 2004. Eolian suspension above the saltation layer: The concentration profile. *Journal of Sedimentary Research* 74(2):176–183.

Gillies, J.A., Nickling, W.G., and King, J. 2007. Shear stress partitioning in large patches of roughness in the atmospheric intertial sublayer. *Boundary Layer Meteorology* 122:367–396.

Gordon, M. and McKenna-Neuman, C.L. 2009. A comparison of collisions of saltating grains with looseand consolidated silt surfaces. *Journal of Geophysical Research—Earth Surface*. Vol. 114, F0415, doi: 10.1029/2009JF00130.

Greeley, R., White, B.R., Leach, R.N., Iversen, J.D. and Pollack, J.B. 1976. Mars: Wind friction speeds for particle movement. *Geophysical Research Letters* 3(8): 417–420.

Hagen, L. 1984. Soil aggregate abrasion by impacting sand and soil particles. *Transactions of the American Society of Agricultural Engineers* 27:805–808.

Haines, W.B. 1925. Studies in the physical properties of soils. II. A note on the cohesion developed by capillary forces in an ideal soil. *Journal of Agricultural Science* 15:525–535.

Heal, M.R., Tunes, T., and Beverland, I.J. 2000 Using archive data to investigate trends in the sources and composition of urban PM10 particulate matter: Application to Edinburgh (UK) between 1992 and 1997, *Environmenal Monitoring and Assessment* 62: 333–340.

Houser, C.A. and Nickling, W.G. 2001a. The emission and vertical flux of particulate matter <10 microns from a disturbed clay-crusted surface. *Sedimentology* 48(2): 255–268.

Houser, C.A. and Nickling, W.G. 2001b. The factors influencing the abrasion efficiency of saltating grains on a clay-crusted playa. *Earth Surface Processes and Landforms* 26(5):491–505.

Hunt, J.C.R. and Nalpanis, P. 1985. Saltating and suspended particles over flat and sloping surfaces. I. Modelling concepts. In: Barndorff-Nielsen, O., Møller, J.-T., Rasmussen, K.R., and Willetts, B.B. (eds), *Proceedings of the International Workshop on the Physics of Blown Sand*, May 28–31, Department of Theoretical Statistics, Institute of Mathematics, University of Aarhus, Memoir 8, pp. 9–36.

Iversen, J.D. and White, B.R. 1982. Saltation threshold on Earth, Mars and Venus. *Sedimentology* 29(1):111–119.

Iversen, J.D., Greeley, R., and Pollack, J.B. 1976. Windblown dust on Earth, Mars and Venus. *The Journal of the Atmospheric Sciences* 33(12):2425–2429.

Jaffe, D.A., Anderson, T., Covert, D., Kotchenruther, R., Trost, B., Danielson, J., Simpson, W., Berntsen, T., Karlsdottir, S., Blake, D., Harris, J., Carmichael, G., and Uno Itsushi. 1999. Transport of Asian air pollution to North America. *Geophysical Research Letters* 26(6):711–714.

Jickells, T. D., An, Z. S., Andersen, K. K., Baker A. R., Bergametti, G., Brooks, N., Cao, J. J. et al. 2005. Global iron connections between desert dust, ocean biogeochemistry, and climate. Science 308 (1):67–71.

Junge, C. E. 1971. The nature and residence times of tropospheric aerosols. In: Matthews, W. H., Kellogg, W. W., and Robinson, G. D. (eds.), *Man's Impact on the Climate*. MIT Press, Cambridge, MA, pp. 302–309.

Lancaster, N., Rasmussen, K.R., and Greeley, R. 1991. Interactions between unvegetated desert surfaces and the atmospheric boundary layer: A preliminary assessment. *Acta Mechanica Supplementum* 2:89–102.

Leenders, J.K., van Boxel, J.H., and Sterk, G. 2005. Wind forces and related saltation transport. *Geomorphology* 71(3-4):357–372.

Lettau, H. 1969. Note on aerodynamic roughness-parameter estimation on the basis of roughness element description. *Journal of Applied Meteorology* 8:828–832.

López, M.V. 1998. Wind erosion in agricultural soils: An example of limited supply of particles available for erosion. *Catena* 33(1):17–28.

Lopez, M.V., Sabre, M., Gracia, R., Arrue, J.L., and Gomes, L. 1998. Tillage effects on soil surface conditions and dust emission by wind erosion in semi-arid Aragon (NE Spain). *Soil and Tillage Research* 45:91–105.

Lu, H. and Shao, Y. 1999. A new model for dust emission by saltation bombardment. *Journal of Geophysical Research* 104(D14):16827–16841.

McKenna-Neuman, C., Boulton, J. W., and Sanderson, S. 2009. Wind tunnel simulation of environmental controls of fugitive dust emissions from mine tailings. *Atmospheric Environment* 43(3):520–529.

McKenna-Neuman, C. and Langston, G. 2006. Measurement of water content as a control of particle entrainment by wind. *Earth Surface Processes and Landforms* 31(3): 303–317.

McKenna-Neuman, C. and Maxwell, C. 2002. Temporal aspects of the abrasion of microphytic crusts under grain impact. *Earth Surface Processes and Landforms* 27(8):891–908.

McKenna-Neuman, C. and Nickling, W.G. 1989. A theoretical and wind tunnel investigation of the effect of capillary water on the entrainment of sediment by wind. *Canadian Journal of Soil Science* 69(1):79–96.

McKenna-Neuman, C., Maxwell, C.D., and Boulton, J.W. 1996. Wind transport on sand surfaces with photoautotrophic microorganisms. *Catena* 27(3–4):229–247.

McKenna-Neuman, C., Maxwell, C., and Rutledge, C. 2005. Spatial and temporal analysis of crust deterioration under particle impact. *Journal of Arid Environments* 60(2): 321–342.

Namikas, S.L., Bauer, B.O., and Sherman, D.J. 2003. Influence of averaging interval on shear velocity estimates for aeolian transport modelling. *Geomorphology* 53(3–4): 235–245.

Nickling, W.G. 1978. Eolian sediment transport during dust storms: Slims River Valley, Yukon Territory. *Canadian Journal of Earth Sciences* 15(7):1069–1084.

Nickling, W.G. and Gillies, J.A. 1989. Emission of fine-grained particulates from desert soils. In: Leinen, M. and Sarnthein, M. (eds.), *Palaeoclimatology and Palaeometeorology: Modern and Past Patterns of Global Atmospheric Transport*. Dordrecht: Kluwer Academic, pp. 133–165.

Nickling, W.G. and Gillies, J.A. 1993. Dust emission and transport in Mali, West Africa. *Sedimentology* 40(5):859–868.

Nickling, W. G. and McKenna-Neuman, C. 2009. Aeolian sediment transport. In: Parsons, A. and Abrahams, A.D. (eds.) *Desert Geomorphology*, 2nd ed. Berlin: Springer, 760pp.

Nickling, W.G., McTainsh, G.H., and Leys, J.F. 1999. Dust emissions from the Channel Country of western Queensland, Australia. *Zeitschrift für Geomorphologie Supplementbände* 116:1–17.

Oke, T. R. 1978. *Boundary Layer Climates*. New York: Methuen, 371pp.

Owen, P.R. 1964. Saltation of uniform grains in air. *Journal of Fluid Mechanics* 20(2): 225–242.

Perry's Chemical Engineering Handbook, 7th edn., 1997. McCraw-Hill, New York, pp. 20-14.

Pye, K. and Sherwin, D. 1999. Loess. In: Goudie, A.S., Livingstone, I. and Stokes, S. (eds.), *Aeolian Environments, Sediments and Landforms*. Chichester: Wiley, pp. 213–237.

Rajot, J.-L., Alfaro, S.C., Gomes, L., and Gaudichet, A. 2003. Soil crusting on sandy soils and its influence on wind erosion. *Catena* 53(1):1–16.

Rasmussen, K.R. and Sørensen, M. 1999. Aeolian mass transport near the saltation threshold. *Earth Surface Processes and Landforms* 24(5):413–422.

Roney, J.A. and White, B.R. 2006. Estimating fugitive dust emission rates using an environmental boundary layer wind tunnel. *Atmospheric Environment* 40(40):7668–7685.

Roney, J.A. and White, B.R. 2010. Comparison of a two-dimensional numerical dust transport model with experimental dust emission from soil surfaces in a wind tunnel. *Atmospheric Environment*, 44:512–522.

Sanderson, R.S., McKenna-Neuman, C., and Boulton, J.W. 2009. Quantification and modelling of fugitive dust emissions from nickel slag. AGU/CGU Joint Assembly 2009. Toronto, Ontario. May 24, 2009.

Saxton, K.E., Chandler, D., Stetler, L., Lamb, B., Claiborn, C., and Lee, B.H. 2000. Wind erosion and fugitive dust fluxes on agricultural lands in the Pacific Northwest. *Transactions of the American Society of Agricultural Engineers* 43(3):623–630.

Shao, Y. 2000. *Physics and Modelling of Wind Erosion*. Dordrecht: Kluwer Academic, 393pp.

Shao, Y. and Lu, H. 2000. A simple expression for wind erosion threshold friction velocity. *Journal of Geophysical Research* 105(D17):22437–22443.

Shao, Y., Raupach, M.R., and Findlater, P.A. 1993. The effect of saltation bombardment on the entrainment of dust by wind. *Journal of Geophysical Research* 98(D7): 12719–12726.

Stull, R.B. 1988. *Introduction to Boundary Layer Meteorology*. Dordrecht, Boston: Kluwer Academic.

Sweeney, M., Etyemazian, V., Macpherson, T., Nickling, W. G., Gillies, J., Nikolich, G., and McDonald, E. 2008. Comparison of PI-SWERL with dust emission measurements from a straight-line field wind tunnel., *Journal of Geophysical Research* 113: F01012 doi: 01010.01029/02007JF000830.

Tegen, I. and Fung, I. 1994. Modeling of mineral dust in the atmosphere: Sources, transport and optical thickness. *Journal of Geophysical Research—Atmospheres* 99 (D16):22897–22914.

Tsoar, H. and Pye, K. 1987. Dust transport and the question of desert loess formation. *Sedimentology* 34(1):139–154.

U. S. Environmental Protection Agency, AP-42 5th edn. *Compilation of Air Pollutant Emission Factors*, Vol. I, Stationary Point and Area Sources. Published online at http://www.epa.gov/ttn/chief/ap42/index.html.

van Boxel, J.H., Sterk, G., and Arens, S.M. 2004. Sonic anemometers in aeolian sediment transport research. *Geomorphology* 59(1–4):131–147.

Walker, I.J. 2000. Secondary airflow and sediment transport in the lee of transverse dunes. Doctoral thesis, University of Guelph, Ontario, Canada, 256pp.

Walker, I.J. 2005. Physical and logistical considerations of using ultrasonic anemometers in aeolian sediment transport research. *Geomorphology* 68(1–2):57–76.

Walker, I.J. and Nickling, W.G. 2003. Simulation and measurement of surface shear stress over isolated and closely spaced transverse dunes in a wind tunnel. *Earth Surface Processes and Landforms* 28(10):1111–1124.

Wiggs, G.F.S. 2001. Desert dune processes and dynamics. *Progress in Physical Geography* 25(1):53–79.

Wiggs, G.F.S., Livingstone, I., Thomas, D.S.G., and Bullard, J.E. 1996. Airflow and roughness characteristics over partially vegetated linear dunes in the southwest Kalahari, Desert. *Earth Surface Processes and Landforms* 21(1):19–34.

Wiggs, G.F.S., Bullard, J.E., Garvey, B., and Castro, I. 2002. Interactions between airflow and valley topography with implications for aeolian sediment transport. *Physical Geography* 23(5):366–380.

Zobeck, T.M. and Van Pelt, R.S. 2006. Wind-induced dust generation and transport mechanics on a bare agricultural field. *Journal of Hazardous Materials* 132:26–38.

ADDITIONAL RESOURCES

The bibliography of aeolian research is hosted at http://www.csrl.ars.usda.gov/wewc/biblio/bar.htm

The site contains over 31,000 references to papers that address aeolian research, inclusive of

- the study of the physics of blowing sand, dust, and other granular material (i.e., saltation, suspension, creep, and abrasion) and
- the study of mineral dust, dust emissions, atmospheric transport, and the effects of fine particulate matter on climate, weather, and air quality.

ISAR, the International Society for Aeolian Reseach, launched a new journal *Aeolian Research* in January 2009 which will include papers that address "fundamental studies of the physics of blowing sand and dust and the deposition of sediment. Practical applications including environmental impacts and erosion control will be covered as well. The objective of the journal is to offer a single platform for papers dealing with aeolian processes which are presently published in a wide variety of journals. Research articles, case histories, short communications, book reviews, thematic issues and review articles will be included in the journal."

17 Deposition of Dissolved and Particulate-Bound Chemicals from the Surface Ocean

Rainer Lohmann and Jordi Dachs

CONTENTS

17.1 Introduction 495
17.2 Quantifying Particulate Matter and Deep Water Formation Fluxes 498
    17.2.1 Settling Fluxes of OM 498
        17.2.1.1 Estimation of Settling Fluxes of Particulate Matter....499
    17.2.2 Formation of Deep Water Masses 499
        17.2.2.1 Upwelling of Water Masses 500
        17.2.2.2 Estimates of Oceanic Deep Water Formation Rates ...501
        17.2.2.3 Other Ocean Water Mass Rates 501
17.3 Chemical Fluxes due to the Biological and Physical Pump 502
    17.3.1 Chemical Flux and MTC for Biological Pump 502
    17.3.2 Chemical Flux and MTC for Physical Pump 503
    17.3.3 Transport to Phytoplankton and PM Sinking 503
17.4 A Guide to Users 505
17.5 Example Calculations...505
Literature Cited 510

17.1 INTRODUCTION

The deep ocean is an important reservoir and sink for chemicals, such as persistent organic pollutants (POPs) (Dachs et al., 2002). Over the last decade, several studies have attempted to parameterize the settling fluxes of particle-associated and dissolved compounds to the deep ocean (Jurado et al., 2004, 2005; Lohmann et al., 2006a). Indeed, this process, which was often overlooked in the past, is now considered one of the most important global sinks of chemicals, together with atmospheric degradation (Wania and Daly, 2002; Lohmann et al., 2006b). The overall paucity of regional and global settling estimates of chemicals is related to both the difficulties of sampling this process and to the complexity of the processes driving the sinking fluxes of organic matter and associated pollutants (Dachs et al., 1999, 2000). While our previous work

495

has focused on POPs, the processes involved will affect organic compounds in general. In this chapter we discuss and quantify the importance of deep marine waters as a sink for chemicals. Specifically, the importance of the "biological pump" (phytoplankton production) and the "physical pump" (deep water formation) as removal processes of chemicals are discussed, rates are quantified, and methods for estimating these fluxes are provided for a variety of cases. A major vector or vehicle for chemical transport to deep marine waters is organic matter (OM) or organic carbon (OC). Note that in general, OC is actually the measured quantity (by CHN analysis), but it is often used synonymously for OM. Fluxes of OM can be obtained by multiplying the F_{OC} values by 2, which is based on an averaged composition of a diatom, and implies that OM is 50% OC (Hedges et al., 2002).

The term "biological pump" usually refers to the process by which primary producers fix carbon dioxide to produce organic matter (see Figure 17.1) (e.g., Valiela, 1995). A fraction of this organic matter sinks below the marine photic zone, thus removing (pumping) carbon from the atmosphere (e.g., Sarmiento and Gruber, 2006). Similarly, scavenging of organic pollutants and other materials occurs because of their tendency to accumulate in and bind to OC. The settling flux of chemicals is estimated as the product of chemicals' concentration on settling OC (e.g., $ng\,g_{OC}^{-1}$) and the settling flux of OC (e.g., $g_{OC}\,m^{-2}\,day^{-1}$). An underlying assumption for all such calculations is that the measured chemical/OC ratio in surface waters applies to particles that actually sink.

Considerable effort has been devoted to better quantifying settling fluxes of OC over the last three decades because of the importance of the carbon cycle for understanding global climate change. However, despite numerous studies, estimations of settling fluxes of OC, and thus of the associated chemicals, still have a high uncertainty. This is due to the natural complexity and heterogeneity of the driving forces of sinking fluxes of OC, and the difficulties in obtaining accurate experimental measurements of OC and chemical fluxes. Nonetheless, a number of methods are currently used for estimating settling fluxes as discussed later. Initially, we describe chemical fluxes associated with settling fluxes as a result of the biological pump (Figure 17.2). Nonbiotic factors, however, can also influence the magnitude and variability of these fluxes in some environments as reported elsewhere (Lohmann et al., 2007).

The "physical pump," or the sinking flux of ocean surface water is part of the global "ocean conveyor belt" (e.g., Broecker, 1974) (see Figure 17.1). It is entirely driven by nonbiotic factors: The formation of deep oceanic waters occurs as part of the global thermohaline circulation. It is caused by gradients in salinity and temperature, and moves surface waters, including chemicals, directly to the deep ocean. The formation of deep oceanic water is especially important in the North Atlantic (Norwegian and Labrador Seas) and the Southern Ocean (Weddel and Ross Seas). Deep water formation is one of the few processes that efficiently transport both soluble and particle-bound compounds to deep waters. The formation of deep oceanic waters also occurs to a lesser degree in other water basins, for example, in the Gulf of Lion in the Northeastern Mediterranean Sea.

It is noteworthy that there are other oceanic processes that will remove chemicals from the surface mixed layer, but which are difficult to quantify. The reader should, however, be aware of their existence. First, numerous factors, such as small-scale

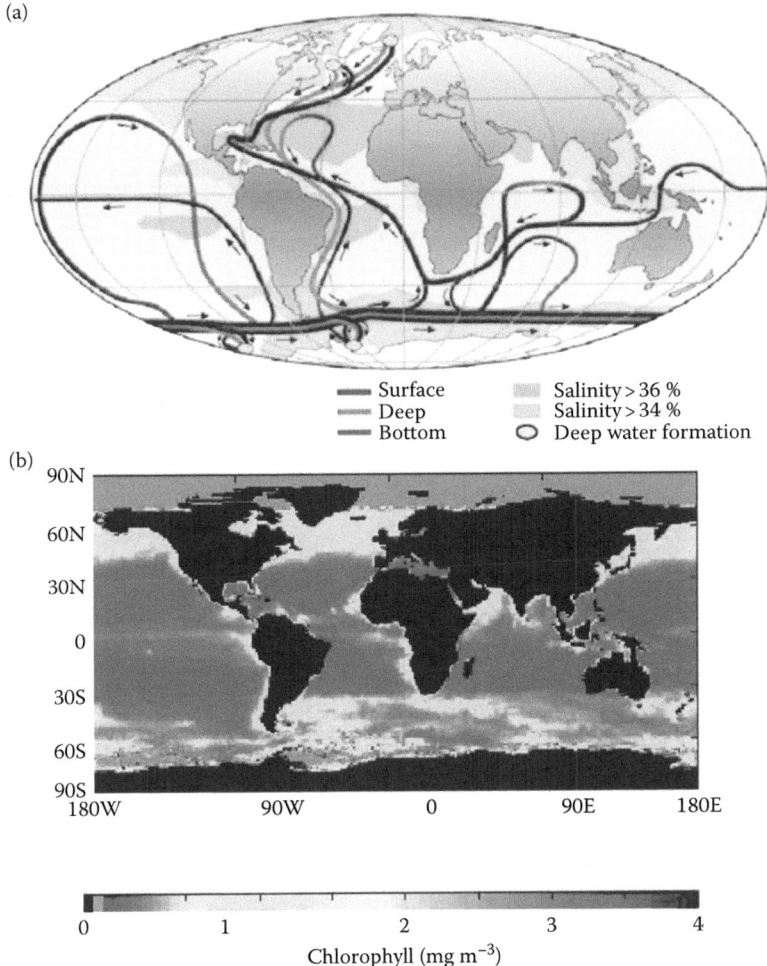

(a)

▬ Surface	▨ Salinity > 36 %	
▬ Deep	▨ Salinity > 34 %	
▬ Bottom	◯ Deep water formation	

(b)

Chlorophyll (mg m^{-3})

FIGURE 17.1 (**See color insert following page 300.**) (a) Principal deep water formation areas; (b) Global chart of chlorophyll a, as a proxy for phytoplankton biomass. (From
Rahmstorf, S. 2002. *Nature* **419**(6903): 207–214. With permission.)

eddies facilitate the mass transfer from the mixed surface layer to the water below.
Second, biologically induced mixing occurs through the vertical diurnal migration of
zooplankton, which induces a similar motion in predators of all sizes (Kunze et al.,
2006). The diurnal migration of zooplankton across the mixed layer depth also moves
organic compounds out of the mixed layer depth (Steinberg et al., 2000). This process
is not addressed by many existing carbon flux models and is very difficult to quantify.
Finally, there are several regionally important mixing processes resulting in enhanced
mass transfer. These include upwelling, smaller deep-water formation regions, and
the transport of surface waters to shallower depth during the formation of the so-called

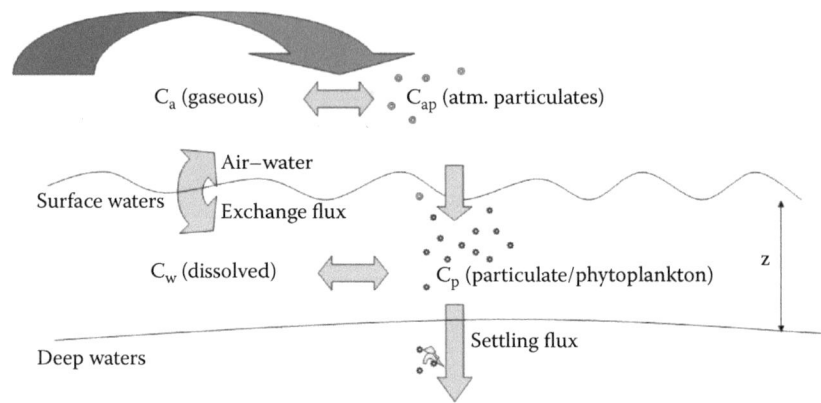

C_a (gaseous) C_{ap} (atm. particulates)

Air–water

Surface waters   Exchange flux

C_w (dissolved) C_p (particulate/phytoplankton)   z

Settling flux

Deep waters

FIGURE 17.2 Conceptual diagram of the air–water–phytoplankton–settling flux. (Modified from Dachs, J. et al. 2002. *Environmental Science & Technology* **36**(20): 4229–4237.)

mode waters (e.g., Apel, 1987). Mode waters are water masses of uniform properties over an extended depth range, formed by the sinking of surface waters.

17.2 QUANTIFYING PARTICULATE MATTER AND DEEP WATER FORMATION FLUXES

17.2.1 SETTLING FLUXES OF OM

Whenever possible, it is preferable to use measured fluxes of OC to those predicted from empirical or mechanistic models. Fluxes of particulate matter and its constituents (including OC as well as associated compounds) can be measured by deploying sediment traps or derived from measurements employing radioisotope techniques. A number of artifacts have been described associated with sediment trap measurements, mainly due to turbulence effects and so-called "swimmers" (Buesseler, 1991). Turbulence introduced by the presence of the sediment trap can modify the flux of particulate matter, and organisms such as copepods or other zooplankton can increase the measured flux significantly (Buesseler, 1991). To minimize these artifacts, traps are usually deployed well below the photic zone (200 m or below) or much deeper, reaching depths of several thousand meters. The overall trend observed with depth is a decrease in OC fluxes, due to mineralization of OC as it sinks. In addition, reported fluxes vary seasonally, caused by differences in phytoplankton communities among other factors, and geographically, due to differences in trophic status and particulate matter inputs.

During the last two decades, attention has increasingly been given to radioisotopes as tracers of organic matter settling. The most frequently used technique is based on the thorium/uranium (^{234}Th/^{238}U) disequilibrium caused by settling fluxes. ^{234}Th is a daughter product (half-life, $t_{1/2} = 24.1$ d) of ^{238}U ($t_{1/2} = 4.47 \times 10^9$ year). The basis of this technique is that a theoretical equilibrium exists between the abundance of both isotopes. However, unlike soluble ^{238}U, ^{234}Th binds to particulates and is referred to as being "particle-reactive." The net result is a disequilibrium between ^{238}U and

^{234}Th due to the settling fluxes of OC with its associated ^{234}Th; see Figure 17.3. The ratio of ^{234}Th/^{238}U allows the determination of settling fluxes on timescales of a few weeks (limited by the half life of ^{234}Th). In this case the flux uncertainties stem from the determination of the OC/^{234}Th ratio of sinking particulate matter needed to calculate the rate at which particulate biogenic carbon is transported downward (Moran et al., 2003). Measurements of settling fluxes combine different types of particles, such as phytoplankton, fecal pellets, and other aggregates of particles ("marine snow") (Fowler and Knauer, 1986) and particles of different sizes that may sink at different rates.

17.2.1.1 Estimation of Settling Fluxes of Particulate Matter

The modeling or numerical estimation of settling fluxes of OC has been a key issue in earth system sciences due to its importance in the carbon cycle and thus on global climate change. However, there is no single approach that can be used universally due to the high diversity of biological and physical factors driving the fluxes of particulate matter in the oceans and lakes. In cases where measured fluxes are not available, a number of methods exist that provide a first-order flux estimate as a function of biological variables such as primary productivity and chlorophyll content. As mentioned above, settling fluxes combine phytoplankton and other organic materials, so that primary productivity/chlorophyll are only proxies for the general carbon export.

Some of the classical empirical methods to estimate settling flux of OC (F_{OC}, mg m^{-2} d^{-1}) are based on measured time series of sediment trap data at different depths. They are as follows (please note the specific units):

$$F_{OC} = 3.52 \times z^{(-0.73 \times PP)} \quad \text{(Pace et al., 1987)}, \tag{17.1}$$

$$F_{OC} = 6.3 \times PP \times z^{-0.8} \text{ (Berger et al., 1987,}$$
$$\text{as reviewed by Bishop, 1989)}, \tag{17.2}$$

$$\log(F_{OC,100}) = 2.09 + 0.81 \times \log(\text{Chl}) \quad \text{(Baines and Pace, 1994)}, \tag{17.3}$$

where z is the depth (m) at which the flux is estimated (Equation 17.3 is for just below the photic zone, about 100 m for the open ocean), PP is the primary productivity (g m^{-2} year^{-1}), and Chl is the chlorophyll a concentration (mg m^{-3}). A comparison of the estimated global OC settling fluxes shows good agreement between the different approaches (Figure 17.4). There are other empirical and theoretical models for the prediction of settling fluxes (see Dunne et al., 2005; Sarmiento and Gruber, 2006 and references therein), but due to the complexity of the topic, there is a lack of generally accepted methods for different environments. It is emphasized that there is considerable global heterogeneity of carbon export fluxes even though primary productivity or biomass, as expressed by chlorophyll concentrations, may be one of the most used and useful predictors.

17.2.2 FORMATION OF DEEP WATER MASSES

The thermohaline circulation is the result of processes generating density gradients in ocean water (attributable to salinity and temperature), and processes acting to erase

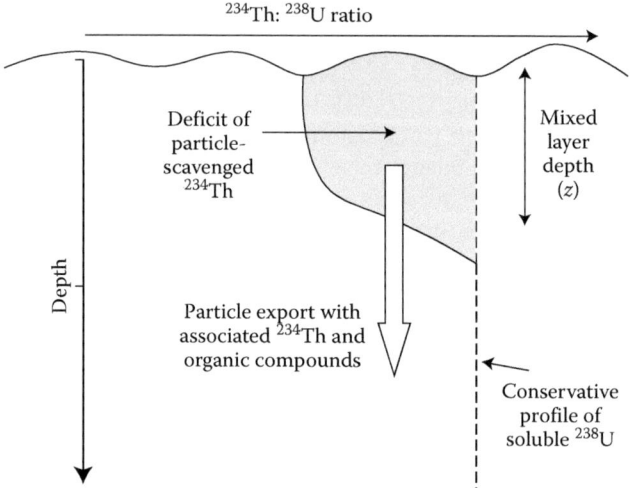

FIGURE 17.3 Systematic diagram indicating how ^{234}Th–^{238}U disequilibrium can be used to derive settling fluxes of carbon and associated organic compounds. (Adapted from Moran, S.B. et al. 2003. *Limnology and Oceanography* **48**(3): 1018–1029.)

these gradients. The net result is a global circulation and overturning of the oceans (also known as the ocean conveyor belt) on timescales of ~1000 years (Broecker, 1974). Of interest to this chapter is the formation of oceanic deep waters and the role it plays with regard to chemical mobility. The strength of the deep oceanic water formation in the North Atlantic and the Southern Ocean regions has been estimated by inverse modeling of measured heat transport across hydrographic sections (Ganachaud and Wunsch, 2000). Smaller areas of deep water formation occur in the Mediterranean Sea in the Gulf of Lion and the Adriatic Sea (Testor and Gascard, 2006).

Surface waters in the mid-latitudes are entrained to hundreds of meters of depth ("subtropical mode waters") (Marsh, 2000; Hanawa and Talley, 2001). They originate as thick winter mixed layers, which are moved to depth and advected from their original formation areas. In the North Atlantic Ocean, this water mass (also referred to as "18 degree" water due to its temperature) display elevated CO_2 concentrations, indicative of having been out of equilibrium with the atmosphere for several years. By analogy, the subtropical mode waters presumably remove organic compounds from the surface water on similar time scales (Bates et al., 2002). Finally, the turnover of deep lakes also results in mixing of surface layer with its contained compounds into deeper waters as discussed in more detail in Chapter 20.

17.2.2.1 Upwelling of Water Masses

Upwelling is the transport of deeper water masses to the surface ocean. It is driven by wind-shear, forcing colder, nutrient-rich waters to the surface, resulting in phytoplankton blooms (Broecker, 1974). Coastal upwelling occurs off western South America, California, Western South Africa, Eastern New Zealand, and the Arabian Sea (see Figure 17.1b). Its effect on most chemicals settling fluxes is currently unknown. On the one hand, deeper, older water is transported to the surface (which could liberate chemicals from deep ocean waters), on the other, the high productivity results in increased settling fluxes and a high turnover of organic carbon (which would tend to remove chemicals from the surface ocean) (Sarmiento and Gruber, 2006).

Finally, there is upwelling in cyclonic eddies. For example, enantiomer-specific analysis of α-hexachlorocyclohexane (α-HCH) showed that samples from the East Greenland Sea (75°N around the meridian) had much lower enantiomer ratios, such as typically encountered in deeper, older Arctic waters (Jantunen and Bidleman, 1996). This could be interpreted as surface waters from 75°N represented older/deeper water masses that had surfaced.

17.2.2.2 Estimates of Oceanic Deep Water Formation Rates

For the present purposes we assume the estimated rates of deep water formation to be constant for each region (Table 17.1). The following main deep water formation regions are considered: in the North Atlantic, the Norwegian Sea, and Labrador Sea; in the Southern Ocean, the Weddell and Ross Sea. These regions broadly represent the main deep-water formation sites of the world ocean (Rahmstorf, 2002). To simplify the complex situation, each region is considered to be a rectangular basin with a characteristic deep-water formation rate (Table 17.1). In the North Atlantic, the relative strength of the Labrador and Norwegian Sea is apportioned according to Marsh (Marsh, 2000). Deep water formation rates are on average around 10 Sv per region, where 1 Sv is $10^6 \, m^3 \, s^{-1}$ (Table 17.1). Only the Labrador Sea has a slightly lower rate of 5 Sv.

17.2.2.3 Other Ocean Water Mass Rates

The estimated average strength of the Atlantic subtropical Mode Water (18° water) is 4 Sv (Marsh, 2000). As discussed above, this mixing is only to a depth of a few hundred meters, with removal of chemicals on the timescale of years (Bates et al., 2002). Apart from regions of formation of deep oceanic water, the role of upwelling

---

TABLE 17.1
Oceanic Deep Water Formation Regions

Region	Boundary Windows	Area (m^2)	r_{dwf} (Sv)[a]
Norwegian Sea	68°N–80°N, 0–25°E	4.0×10^{12}	10
Labrador Sea	50°N–60°N, 60°W–40°W	2.7×10^{12}	5
Weddell Sea	60°S–85°S, 70°W–20°W	1.7×10^{13}	11
Ross Sea	60°S–80°S, 170°E–155°W	4.4×10^{13}	10

Source: Modified from Lohmann, R. et al. 2006a. *Geophysical Research Letters* **33**, ppL12607, 4pp, Doi:10.1029/2006GL025953.

Note: 1 Sv $= 10^6 \, m^3 \, s^{-1}$.

[a] Rates taken from Ganachaud and Wunsch (2000), with the relative strength of the Labrador and Norwegian Sea apportioned according to Marsh (2000).

and downwelling of water masses have not been studied in terms of chemicals. The upwelling and downwelling velocities are estimated to be ca. $1-2 \cdot 10^{-4}$ m s^{-1} for most of the ocean (Sarmiento and Gruber, 2006), except in the regions of deep water formation (see above) and upwelling regions off Namibia, Arabia, Sahara, and equatorial oceans.

17.3 CHEMICAL FLUXES DUE TO THE BIOLOGICAL AND PHYSICAL PUMP

17.3.1 CHEMICAL FLUX AND MTC FOR BIOLOGICAL PUMP

The chemical flux, $F_{\text{Chem},z}$ (in g m^{-2} s^{-1}) for particulate matter (PM) settling is

$$F_{\text{Chem},z} = k_{\text{cp}} C_{\text{Chem,aq}}, \tag{17.4}$$

where k_{cp} is the PM MTC (m s^{-1}) and $C_{\text{Chem,aq}}$ is the concentration of the chemical in aqueous solution (g m^{-3}). The definition of the PM settling MTC is

$$k_{\text{cp}} = F_{\text{OC}}/C_{\text{OC}}, \tag{17.5}$$

where F_{OC} is the flux of organic (g$_{\text{OC}}$ m^{-2} s^{-1}) and C_{OC} is the particulate surface water OC concentration (g$_{\text{OC}}$ m^{-3}).

Depending on measured concentrations, there are alternate ways to expressing the fluxes. For example, if $C_{\text{Chem,OC}}$ is known rather than $C_{\text{Chem,aq}}$, the flux is

$$F_{\text{Chem},z} = [k_{\text{cp}}/K_{\text{OC}}] C_{\text{Chem,OC}}. \tag{17.6}$$

If $C_{\text{Chem,p}}$ is known, then the flux is

$$F_{\text{Chem},z} = [k_{\text{cp}}/K_{\text{OC}} f_{\text{OC}}] C_{\text{Chem,p}}, \tag{17.7}$$

where $C_{\text{Chem,OC}}$ and $C_{\text{Chem,p}}$ are the concentrations of chemicals in OC (g g$_{\text{OC}}^{-1}$) and PM (g g$_{\text{PM}}^{-1}$), respectively, f_{OC} is the fraction of OC in the PM (in general 0.5 g$_{\text{OC}}$ g$_{\text{PM}}^{-1}$) and K_{OC} is the organic carbon–water partition coefficient (L kg$_{\text{OC}}^{-1}$). (Remember that a conversion factor might be needed if SI units are used for $K_{\text{OC}} - 10^6$ m^3 g$_{\text{OC}}^{-1}$.) It should be noted that water particle partitioning is different from water–sediment partitioning. Therefore, K_{OC} should be viewed as being similar to a bioconcentration factor from water to phytoplankton (Gerofke et al., 2005), rather than those used for the compounds' accumulation in sediment (Karickhoff et al., 1979).

The parameterizations of the carbon and water mass fluxes from Section 17.2 contain general OC flux estimates (Equations 17.1 through 17.3, which can be used in Equations 17.4 through 17.7 to estimate chemical fluxes provided their concentrations in the particulate phase are known. The fluxes of organic carbon can be taken from measurements or from estimations as shown above. It is also possible to define a transfer rate constant (k_{OC}, in s^{-1}) for settling fluxes of OC. This is given by

$$k_{\text{OC}} = F_{\text{OC}}/(z C_{\text{OC}}), \tag{17.8}$$

where C_{OC} is the OC concentration in the photic zone ($g\,m^{-3}$) at depth z (m). Knowing the value of k_{OC}, the flux of settling particulate chemicals can also be estimated as

$$F_{Chem,z} = C_{Chem,OC} C_{OC} z k_{OC}. \tag{17.9}$$

This assumes that the average concentration of chemicals in the OC, $C_{Chem,OC}$ ($g\,g_{OC}^{-1}$) in the photic zone is known, and that transfer rates of OC and particle phase chemicals are equal.

17.3.2 CHEMICAL FLUX AND MTC FOR PHYSICAL PUMP

For chemical flux due to deep water formation and movement,

$$F_{Chem,z} = k_{cdw} C_w, \tag{17.10}$$

where k_{cdw} is the deep water formation MTC ($m\,s^{-1}$) and C_w is the total aqueous concentration of the chemical (dissolved plus particulate phase, in $g\,m^{-3}$). It may be expressed as

$$C_w = C_{Chem,aq} + C_{Chem,p} = C_{Chem,aq}(1 + C_{OC} K_{OC}) \tag{17.11}$$

in terms of previously defined quantities. The rate of deep water formation of the chosen area, r_{dwf}($m^3\,s^{-1}$), divided by the area of interest, A(m^2) is the MTC k_{cdw} ($m\,s^{-1}$) as

$$k_{cdw} = r_{dwf}/A. \tag{17.12}$$

Deep water formation rates are on average around $10\,Sv$, where $1\,Sv = 10^6\,m^3\,s^{-1}$ (Table 17.1). Only the Labrador Sea has a slightly lower rate of $5\,Sv$.

17.3.3 TRANSPORT TO PHYTOPLANKTON AND PM SINKING

It is possible to estimate the marine settling flux knowing the gas phase concentration and environmental conditions such as temperature, wind speed, and chlorophyll (see Dachs et al., 2002 and Jurado et al., 2004 for details and justification). The approach used here assumes that air–water exchange is the principal mechanism delivering chemicals to the water column. This assumption is likely valid for the open ocean and chemicals which are mainly in the gas phase. For estimating this flux, we first derive the appropriate equations for air-to-deep water transfer (Figure 17.2).

We take into account air–water exchange, water–phytoplankton exchange, and settling flux. Air–water exchange may be treated in the traditional manner where the air–water flux (F_{AW}, $g\,m^{-2}\,s^{-1}$) is given by, as discussed in Chapter 9,

$$F_{AW} = k_{aw}(C_A/K_{AW} - C_w), \tag{17.13}$$

where C_A and C_W are the dissolved and gas-phase chemical concentrations ($g\,m^{-3}$), respectively, K_{AW} is the dimensionless and temperature and salinity corrected

Henry's law constant, and k_{AW} is the overall water phase air–water mass transfer coefficient (m s^{-1}).

Once pollutants enter the oceanic surface mixed layer they sorb to phytoplankton and a fraction will sink associated to phytoplankton cells, fecal pellets, and other aggregates (Figure 17.2). Fluxes of chemicals between water and plankton (F_{WP}, g m^{-2} s^{-1}) are given by Equation 17.14 (Del Vento and Dachs, 2002):

$$F_{WP} = k_{WP} \left(C_W - \frac{k_d}{k_u} C_P \right), \tag{17.14}$$

where C_P is the chemical concentration in phytoplankton (g g$_{phyto}^{-1}$), k_u (m^3 g$_{phyto}^{-1}$ s^{-1}) and k_d (s^{-1}) are the uptake and depuration rate constants, respectively. k_{WP} (m s^{-1}), the MTC between water and plankton, depends on the mixed layer depth z (m), the phytoplankton biomass (B_P, g$_{phyto}$ m^{-3}), and the uptake constant (Dachs et al., 1999).

$$k_{WP} = z B_P k_u. \tag{17.15}$$

Values of k_{WP} are estimated using Equation 17.15 with z values from vertical profiles of density as determined in the field or using climatological data for the mixed layer depth. Biomass or suspended material concentrations can be derived from chlorophyll data if available (see Figure 17.1b). Values of k_u and k_d for a number of POPs have been reported by Del Vento and Dachs (2002).

Vertical fluxes of chemicals out of the mixed layer are given by the product of chemical concentrations in sinking particles and the vertical flux of sinking organic carbon (Equation 17.4). To compare these fluxes with those of air–water and water–phytoplankton exchange, sinking fluxes can be parameterized by Equations 17.16 and 17.17:

$$F_{Sink} = k_{cp,chem} \frac{k_d}{k_u} C_P, \tag{17.16}$$

$$k_{cp,chem} = F_{OC} \frac{k_u}{k_d}, \tag{17.17}$$

where $k_{cp,chem}$ (m s^{-1}) is the MTC for the sinking of the chemicals.

The flux between air and deep water can thus be estimated by considering the air–water exchange, the water–phytoplankton exchange, and the sinking fluxes occurring in series (Figure 17.2). Water–deep water flux (F_{WDW}) associated with absorption to the phytoplankton organic matter and subsequent sinking is given by Equation 17.18, which is derived from Equations 17.14 and 17.16.

$$F_{WDW} = \frac{k_{WP} k_{Sink}}{k_{WP} + k_{Sink}} C_W. \tag{17.18}$$

F_{WDW} is supported by air–water exchange (F_{AW}), and therefore, assuming steady-state conditions:

$$F_{AW} = F_{WDW} = F_{ADW}, \tag{17.19}$$

where F_{ADW} is the air-deep water flux $(g\,m^{-2}\,s^{-1})$. From Equations 17.13 and 17.18, we obtain the expression for the air–deep water transfer rate used in the equation given the settling flux (equal here to the air–water exchange flux):

$$k_{ADW} = \frac{k_{AW}\dfrac{k_{WP}k_{Sink}}{k_{WP} + k_{Sink}}}{k_{AW} + \dfrac{k_{WP}k_{Sink}}{k_{WP} + k_{Sink}}}, \tag{17.20}$$

$$F_{ADW} = k_{ADW}C_G/K_{WA}. \tag{17.21}$$

The processes explained here for estimating fluxes of chemicals have been derived for persistent and hydrophobic chemicals. For nonhydrophobic chemicals, mineral fractions of settling particles may play an important role, and thus the parameterization would need to be modified accordingly. In addition, sinking particles and aggregates are rich in bacteria (Sarmiento and Gruber, 2006). Therefore, for nonpersistent chemicals, degradation occurs and the concentration of the chemical in particles will decrease during transport. These and other aging processes, such as repartitioning to dissolved phase due to OC mineralization, may be important, but have received little research effort so far except for some biogenic organic compounds (Minor et al., 2003).

17.4   A GUIDE TO USERS

Chemicals can be removed from the surface ocean by two main processes: sinking bound to particles, or as part of surface water masses that are entrained to deeper waters. To determine the settling flux of compounds at depth, use Equations 17.1 through 17.3 to estimate the flux of OC to depth, and multiply by the chemicals' ratio to OC (Equation 17.5) to obtain the PM MTC. The MTC can now be used with Equations 17.4, 17.6, or 17.7 to estimate the settling fluxes of chemicals, depending on whether total, particulate or OC-normalized concentrations are known. For deep water formation fluxes, only total aqueous concentrations are needed for Equation 17.10, with MTCs derived from Equation 17.12 using Table 17.1. In cases where only the gas-phase concentration is known, settling fluxes can be estimated very approximately by modeling the coupling of air–water exchange and settling fluxes (Equations 17.20 and 17.21). In general, the uncertainty in estimation of chemical vertical fluxes is associated with the uncertainty in OC flux estimation.

17.5   EXAMPLE CALCULATIONS

Please note that the units used in the case studies reflect those predominantly used in the literature, which are not necessarily SI units, as used throughout the chapter.

Case study 1: Estimate the settling flux of particle bound polychlorinated biphenyls (PCBs) in the North Atlantic, knowing their concentrations in the particle phase. Assume general chlorophyll a concentrations and fluxes of organic carbon. This case study estimates the flux and MTC using Equations 17.4 and 17.5.

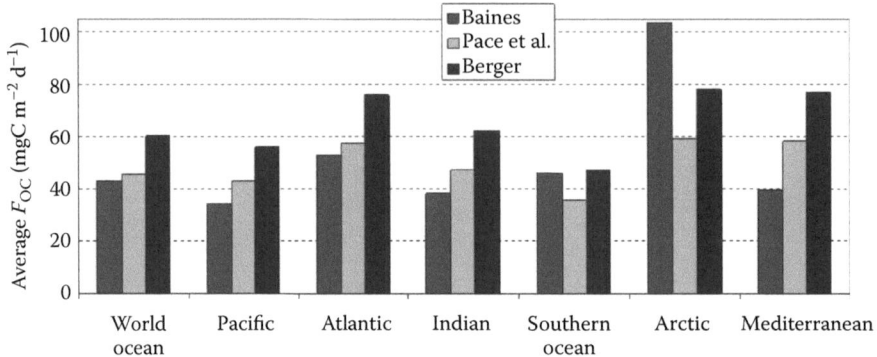

FIGURE 17.4 Comparison of predicted CO settling fluxes for the different ocean basins using the approximations from Equations 17.1 through 17.3.

For a productive region in North Atlantic, primary productivity is of the order of $20 \, \text{mol} \, C \, m^{-2} \, \text{year}^{-1}$ (Sarmiento and Gruber, 2006). Applying Equation 17.2, we can estimate the flux of OC carbon out of the photic zone ($z = 100 \, \text{m}$):

$$F_{OC,100} = 6.3 \times 20 \times 100^{-0.8} = 3.165 \, \text{mol} \, C \, m^{-2} \, \text{year}^{-1}$$
$$= 37.98 \, \text{g} \, C \, m^{-2} \, \text{year}^{-1}$$

With a particulate organic carbon concentration of $C_{OC} = 2 \, \mu \, \text{mol} \, C \, L^{-1}$, the MTC of the OC settling settling flux, k_{cp}, can be calculated using Equation 17.5:

$$k_{cp} = \frac{37.98 \, \text{g} \, C \, m^{-2} \, \text{year}^{-1}}{24 \, \text{mg} \, C \, m^{-3}} = 1.58 \times 10^3 \, \text{m} \, \text{year}^{-1} = 5.0 \times 10^{-5} \, \text{m} \, \text{s}^{-1}.$$

The settling flux of a PCB can now be estimated using Equation 17.4. For example, the concentration of particle-phase PCB 52 has been reported as $C_{Chem,aq} = 0.08 \, \text{pg} \, L^{-1}$ (Schulz-Bull et al., 1998):

$$F_{PCB,100} = 5.0 \times 10^{-5} \, \text{m} \, \text{s}^{-1} \times 80 \, \text{pg} \, m^{-3}$$
$$= 4.0 \times 10^{-3} \, \text{pg} \, m^{-2} \, \text{s}^{-1} \quad \text{or} \quad 340 \, \text{pg} \, m^{-2} \, \text{d}^{-1}$$

Gustafsson et al. (1997) derived settling fluxes of PCBs using the $^{234}\text{Th}/^{238}\text{U}$ disequilibrium approach. For samples taken from close to the New England coast to the oligotrophic Sargasso Sea, fluxes of PCB 52 ranged from ca. 2 to 40 pmol m^{-2} d^{-1}. The flux derived here compares well to the $^{234}\text{Th}/^{238}\text{U}$ derived fluxes of $\sim$600–1800 pg m^{-2} d^{-1} (Gustafsson et al., 1997).

Case study 2: Estimate the residence time of polycyclic aromatic hydrocarbons (PAHs) in the water column of the Mediterranean. In this hypothetical scenario, PAHs sorbed to settling particles are the only removal process. Average water column dissolved and particulate phase PAH concentrations and their settling fluxes are known (data from Bouloubassi, 2006; Dachs et al. 1997).

Dachs et al. (1997) reported concentrations of dissolved and particulate phenanthrene of 200 and 45 ng m^{-3} in the surface waters of the open Mediterranean Sea. For benzofluoranthenes the concentrations were much lower, at 4 and 1.6 ng m^{-3} for the particle and dissolved phases, respectively. Assuming these concentrations of phenanthrene and benzofluoranthenes to be representative of the photic zone (100 m depth), the inventory of phenanthrene and benzofluoranthenes in the photic zone is

$$\text{Inventory phenanthrene} = (200 + 45)\,\text{ng m}^{-3} \times 100\,\text{m} = 245{,}000\,\text{ng m}^{-2}$$

$$\text{Inventory benzofluoranthenes} = (4 + 1.6)\,\text{ng m}^{-3} \times 100\,\text{m} = 560\,\text{ng m}^{-2}$$

Concentrations below the photic zone are one order of magnitude lower than those in surface waters and are neglected here. Bouloubassi et al. (2006) reported settling fluxes of particle-bound phenanthrene at 250 m depth of $\sim$6 ng m^{-2} d^{-1} and of particle-bound benzofluoranthenes of $\sim$7 ng m^{-2} d^{-1}. It is thus possible to estimate the residence time (t_{res}) of phenanthrene and benzofluoranthenes as

$$t_{res} \text{ phenanthrene} = \frac{245{,}000\,\text{ng m}^{-2}}{6\,\text{ng m}^{-2}\,\text{d}^{-1}} = 4083\,\text{days},$$

$$t_{res} \text{ benzofluoranthenes} = \frac{560\,\text{ng m}^{-2}}{7\,\text{ng m}^{-2}\,\text{d}^{-1}} = 80\,\text{days}.$$

The calculated long residence time of phenanthrene in surface waters, of over 4000 days, is unlikely to be real. Instead, other processes than particle-bound settling of phenanthrene affects its fate and transport. Phenanthrene is mainly present in the dissolved phase and is thus bioavailable for bacterial degradation and for diffusive air–water exchange. On the other hand, the residence time of 80 days for benzofluoranthenes may be closer to reality.

Case study 3: Estimate the settling flux of PCB 52 for a case where only its gas phase concentration is known (e.g., 6 pg m^{-3}) and the prevailing environmental conditions: Temperature is 298 K and wind speed is such that the air–water exchange MTC is 1 m d^{-1}. The water column particle (or phytoplankton) load is 1 mg L^{-1}, with a resulting flux of organic matter of 1 g m^{-2} d^{-1}. The oceanic mixed layer depth is 100 m.

For this example, we consider the combination of air–water exchange, water–phytoplankton exchange and settling flux. The air–water mass transfer rate, k_{AW} (m d^{-1}), can otherwise be estimated from the two resistance model as discussed in Chapter 9.

Here we use a value of 100 m for z and a value of 1 mg L^{-1} for the biomass. Alternatively, biomass or suspended material concentrations can be derived from chlorophyll data if available (see Figure 17.1b).

With the values for k_{AW}, biomass, and MLD given in this case study, and k_u and k_d for PCB 52 of $400\,m^3\,kg^{-1}\,d^{-1}$ and $0.89\,d^{-1}$, respectively, as given by Del Vento and Dachs (2002), then, we can calculate the MTC between water and plankton (Equation 17.15):

$$k_{WP} = 400\,m^3\,kg^{-1}\,d^{-1} \times 0.001\,kg\,m^{-3} \times 100\,m = 40\,m\,d^{-1}.$$

Using Equation 17.17, we can also calculate the MTC for the sinking of the chemicals:

$$k_{Cp,chem} = 0.001\,kg\,m^{-2}\,d^{-1} \times 400\,m^3\,kg^{-1}\,d^{-1}/0.89\,d^{-1} = 0.45\,m\,d^{-1}.$$

Therefore, we can estimate the MTC from air to deep water according to Equation 17.20:

$$k_{ADW} = \frac{1 * \dfrac{40 * 0.45}{40 + 0.45}}{1 + \dfrac{40 * 0.45}{40 + 0.45}} = 0.64\,m\,d^{-1}.$$

The flux will be calculated using Equation 17.21, for which we need K_{AW}. Henry's law constant H (from Li et al., 2003) for PCB 52 is $1.4\,Pa\,m^3\,mol^{-1}$:

$$K_{AW} = H/RT = \frac{1.4\,Pa\,m^3\,mol^{-1}}{(8.31\,Pa\,m^3\,mol^{-1}\,K^{-1} \times 298\,K)} = 0.000565.$$

Hence, F_{ADW} is

$$F_{ADW} = \frac{0.64\,m\,d^{-1} \times 6\,pg\,m^{-3}}{0.000565} = 6792\,pg\,m^{-2}\,d^{-1}.$$

The reason that the settling flux of PCB 52 derived here is higher than the flux derived in case study 17.1 ($340\,pg\,m^{-2}\,d^{-1}$) can be traced back to the differing starting PCB concentrations. At equilibrium, an air concentration for PCB 52 of $6\,pg\,m^{-3}$ supports a water concentration of $10.6\,pg\,L^{-1}$. Based on a $2\,\mu mol$ OC in the water column ($24\,\mu g\,OC\,L^{-1}$, or $48\,\mu g\,OM\,L^{-1}$), and an average bioconcentration factor in the field of $10^{6.1}$ (Gerofke et al., 2005), we can derive a particulate concentration of $0.63\,pg\,L^{-1}$. In contrast, case study 1 started off with a lower particulate concentration of $0.08\,pg\,L^{-1}$.

Case study 4:   Compare the fluxes of PCBs 52 and 153 in the Norwegian Sea due to settling of particles and deep water formation, from known dissolved and particulate concentrations, and rates of deep water formation and settling (data from Lohmann et al., 2006a). This case study uses the chemical flux (Equation 17.10) and MTC (Equation 17.12) of the physical pump deep water formation.

In this case, we use Equation 17.10, for which we need total concentrations (C_w), the MTC of deep-water formation, k_{cdw} (Equation 17.12) and the area of the deep water formation region (Table 17.1).

C_w (dissolved and particulate) of PCBs 52 and 153 in the surface waters of the Norwegian Sea were $0.12\,\text{pg}\,\text{L}^{-1}$ for PCB # 52 ($0.12 \times 10^{-12}\,\text{kg}\,\text{m}^{-3}$) and $0.056\,\text{pg}\,\text{L}^{-1}$ for PCB # 153 ($0.056 \times 10^{-12}\,\text{kg}\,\text{m}^{-3}$) in 2001 (Sobek and Gustafsson, 2004). First, to obtain the MTC, we use Table 17.1, which gives r_{dwf} of the Norwegian Sea as 10 Sv ($10 \times 10^6\,\text{m}^3\,\text{s}^{-1}$), and the area of the Norwegian Sea as $4.0 \times 10^{12}\,\text{m}^2$:

$$k_{cdw} = \frac{10 \times 10^6\,\text{m}^3\,\text{s}^{-1}}{4.0 \times 10^{12}\,\text{m}^2} = 2.5 \times 10^{-6}\,\text{m}\,\text{s}^{-1}.$$

For PCB # 52, the resulting flux is calculated as

$$F_{Chem,z} = k_{cdw}C_w = 2.5 \times 10^{-6}\,\text{m}\,\text{s}^{-1} \times 0.12 \times 10^{-12}\,\text{kg}\,\text{m}^{-3}$$

$$= 0.3 \times 10^{-18}\,\text{kg}\,\text{m}^{-2}\,\text{s}^{-1},$$

$$F_{Chem,z} = 9.4 \times 10^{-9}\,\text{g}\,\text{m}^{-2}\,\text{year}^{-1},$$

or a flux of $9.4\,\text{ng}\,\text{m}^{-2}\,\text{year}^{-1}$. Similarly, for PCB #153, the resulting flux is $4.4\,\text{ng}\,\text{m}^{-2}\,\text{year}^{-1}$.

To calculate the flux of PCBs due to the settling of OC, we can use Equation 17.4 or 17.6. We need to know F_{OC}, the flux of OC ($\text{g}\,\text{OC}\,\text{m}^{-2}\,\text{s}^{-1}$) and $C_{Chem,aq}$, or $C_{Chem,P}$, the chemical concentration in solution ($\text{g}\,\text{m}^{-3}$) or in the particulate OC phase ($\text{g}\,\text{g}^{-1}\text{OC}$). From Seawifs (see also Figure 17.1b), we can get an average estimate of the abundance of chlorophyll in the surface of the Norwegian Sea, which is $0.90\,\mu\text{g}\,\text{L}^{-1}$ (or $\text{mg}\,\text{m}^{-3}$). According to Equation 17.3, we calculate the flux of OC at 100 m depth, well below the surface mixed ocean:

$$\log(F_{OC,100}) = 2.09 + 0.81 \times \log(0.9),$$

$$\log(F_{OC,100}) = 2.05\,\text{mg}\,\text{m}^{-2}\,\text{day}^{-1} \quad \text{or} \quad F_{OC,100} = 112.2\,\text{mg}\,\text{m}^{-2}\,\text{day}^{-1}.$$

We can estimate $C_{Chem,P}$ for the PCBs by first calculating the concentration of OC in the water and then applying a bioconcentration factor, BCF, to use Equation 17.6 (see case study 3 for example). Alternatively, as detailed in case study 1, we use aqueous concentrations of particulate PCBs and then derive the MTC of the OC settling flux, k_{cp}, using Equation 17.5.

With a particulate organic carbon concentration of $C_{OC} = 2\,\mu\text{mol}\,\text{C}\,\text{L}^{-1}$, k_{cp} is

$$k_{cp} = \frac{112.2\,\text{mg}\,\text{C}\,\text{m}^{-2}\,\text{day}^{-1}}{24\,\text{mg}\,\text{C}\,\text{m}^{-3}}$$

$$= 4.7\,\text{m}\,\text{day}^{-1} = 5.4 \times 10^{-5}\,\text{m}\,\text{s}^{-1}.$$

The settling flux of a PCB can now be estimated using Equation 17.4 and a particle phase concentration for PCB 52 of $80\,\text{pg}\,\text{m}^{-3}$ (Schulz-Bull et al., 1998):

$$F_{PCB,100} = 5.4 \times 10^{-5}\,\text{m}\,\text{s}^{-1} \times 80\,\text{pg}\,\text{m}^{-3}$$

$$= 4.3 \times 10^{-3}\,\text{pg}\,\text{m}^{-2}\,\text{s}^{-1} \quad \text{or} \quad 0.37\,\text{ng}\,\text{m}^{-2}\,\text{d}^{-1}.$$

For PCB 153, particle-bound concentrations of $420\,fg\,L^{-1}$ were reported, and we derive

$$F_{PCB,100}(PCB\ 153) = 1.97\,ng\,m^{-2}\,d^{-1}.$$

These settling fluxes of PCBs are lower than those derived for deep water formation, especially for the more soluble, less chlorinated PCB congeners (see also Lohmann et al., 2006a). This example highlights the importance of removal fluxes of dissolved compounds from surface waters via deep water formation. However, if we calculate the importance of PCB removal fluxes by the main oceanic deep water formation regions, we obtain approximately $870\,kg\,year^{-1}$ for seven PCBs (Lohmann et al., 2006a). The organic carbon settling flux for the entire Atlantic Ocean for the same PCBs can be calculated to be ca. $22{,}000\,kg\,year^{-1}$ (see Jurado et al., 2004, 2005). Hence, the total amount of PCBs removed by deep water formation is regionally important but relatively small on the scale of the entire Atlantic Ocean.

LITERATURE CITED

Apel, J.R. 1987. *Principles of Ocean Physics*. International Geophysics series 38. New York: Academic Press.

Baines, S.B. and M.L. Pace 1994. Why does the relationshop between sinking flux and planktonic primary production differ between lakes and oceans? *Limnology Oceanography* **39**: 213–226.

Bates, N.R., A.C. Pequignet, R.J. Johnson, and N. Gruber 2002. A short-term sink for atmospheric CO_2 in subtropical mode water of the North Atlantic Ocean. *Nature* **420**: 489–493.

Berger, W.H., K. Fisher, C. Lai, and G. Wu. 1987. Ocean carbon flux: Global maps of primary production and export production. In: *Biogeochemical Cycling and Fluxes between the Deep Euphotic Zone and Other Oceanic Realms*. C. Agegian, Ed. NOAA Symp. Ser. for Undersea Research, NOAA Undersea research programm vol 3, preprint in SIO ref. 87–30.

Bishop, J.K.B. 1989. Regional extremes in particulate matter composition and flux: Effects on the chemistry of the ocean interior. In *Productivity of the Ocean: Present and Past*. W.H. Berger, V.S. Smetacek, and G. Wefer, Eds. John Wiley & Sons.

Broecker, W.S. 1974. *Chemical Oceanography*. New York: Harcourt Brace Jovanovich, 214pp.

Buesseler, K.O. 1991. Do upper ocean sediment traps provide an accurate record of particle flux? *Nature* **353**: 420–423.

Dachs, J., J.M. Bayona, and J. Albaiges 1997. Spatial distribution, vertical profiles and budget of organochlorine compounds in Western Mediterranean seawater. *Marine Chemistry* **57**(3–4): 313–324.

Dachs, J., S.J. Eisenreich, J.E. Baker, F.C. Ko, and J.D. Jeremiason 1999. Coupling of phytoplankton uptake and air–water exchange of persistent organic pollutants. *Environmental Science & Technology* **33**(20): 3653–3660.

Dachs, J., S.J. Eisenreich, and R.M. Hoff 2000. Influence of eutrophication on air–water exchange, vertical fluxes, and phytoplankton concentrations of persistent organic pollutants. *Environmental Science & Technology* **34**(6): 1095–1102.

Dachs, J., R. Lohmann, W.A. Ockenden, L. Mejanelle, S.J. Eisenreich, and K.C. Jones 2002. Oceanic biogeochemical controls on global dynamics of persistent organic pollutants. *Environmental Science & Technology* **36**(20): 4229–4237.

Del Vento, S. and J. Dachs 2002. Prediction of uptake dynamics of persistent organic pollutants by bacteria and phytoplankton. *Environmental Toxicology and Chemistry* **21**(10): 2099–2107.

Dunne, J.P., R.A. Armstrong, A. Gnanadesikan, and J.L. Sarmiento 2005. Empirical and mechanistic models for the particle export ratio. *Global Biogeochemical Cycle* **19**: GB4027, Doi: 10.1029/2004GB002390.

Fowler, S.W. and G.A. Knauer 1986. Role of large particles in the transport of elements and organic compounds through the oceanic water column. *Progress in Oceanography* **16**: 147–194.

Ganachaud, A. and C. Wunsch 2000. Improved estimates of global ocean circulation, heat transport and mixing from hydrographic data. *Nature* **408**(6811): 453–457.

Gerofke, A., P. Komp, and M.S. McLachlan 2005. Bioconcentration of persistent organic pollutants in four species of marine phytoplankton. *Environmental Toxicology and Chemistry* **24**(11): 2908–2917.

Gustafsson, O., P.M. Gschwend, and K.O. Buesseler 1997. Settling removal rates of PCBs into the Northwestern Atlantic derived from U-238-Th-234 disequilibria. *Environmental Science & Technology* **31**(12): 3544–3550.

Hanawa, K. and L.D. Talley 2001. Mode waters. In *Ocean Circulation and Climate*, Vol. 77. G. Siedler, J. Church, and J. Gould, Eds. London: Academic Press, pp.373–386.

Hedges, J.I., J.A. Baldock, Y. Gelinas, C. Lee, M.L. Peterson, and S.G. Wakeham 2002. The biochemical and elemental compositions of marine plankton: A NMR perspective. *Marine Chemistry* **78**(1): 47–63.

Jantunen, L.M. and T. Bidleman 1996. Air–water gas exchange of hexachlorocyclohexanes (HCHs) and the enantiomers of alpha-HCH in arctic regions. *Journal of Geophysical Research—Atmospheres* **101**(D22): 28837–28846.

Jurado, E., F. Jaward, R. Lohmann, K.C. Jones, R. Simo, and J. Dachs 2005. Wet deposition of persistent organic pollutants to the global oceans. *Environmental Science & Technology* **39**(8): 2426–2435.

Jurado, E., F.M. Jaward, R. Lohmann, K.C. Jones, R. Simo, and J. Dachs 2004. Atmospheric dry deposition of persistent organic pollutants to the Atlantic and inferences for the global oceans. *Environmental Science & Technology* **38**(21): 5505–5513.

Karickhoff, S.W., D.S. Brown, and T.A. Scott 1979. Sorption of hydrophobic pollutants on natural sediments. *Water Research* **13**: 241–248.

Kunze, E., J.F. Dower, I. Beveridge, R. Dewey, and K.P. Bartlett 2006. Observations of biologically generated turbulence in a Coastal Inlet. *Science* **313**(5794): 1768–1770.

Lohmann, R., K. Breivik, J. Dachs, and D. Muir 2007. Global fate of POPs—current and future research directions. *Environmental Pollution* **150**: 150–165.

Lohmann, R., E. Jurado, M.E.Q. Pilson, and J. Dachs 2006a. Oceanic deep water formation as a sink of persistent organic pollutants. *Geophysical Research Letters* **33**, ppL12607, 4pp. Doi:10.1029/2006GL025953.

Lohmann, R., E. Jurado, J. Dachs, U. Lohmann, and K.C. Jones 2006b. Quantifying the importance of the atmospheric sink for polychlorinated dioxins and furans relative to other global loss processes. *Journal of Geophysical Research* **111**, D21303. Doi:10.1029/2005JD006923.

Marsh, R. 2000. Recent variability of the North Atlantic thermohaline circulation inferred from surface heat and freshwater fluxes. *Journal of Climate* **13**(18): 3239–3260.

Moran, S.B., S.E. Weinstein, H.N. Edmonds, J.N. Smith, R.P. Kelly, M.E.Q. Pilson, and W.G. Harrison 2003. Does 234Th/238U disequilibrium provide an accurate record of the export flux of particulate organic carbon from the upper ocean? *Limnology and Oceanography* **48**(3): 1018–1029.

Pace, M.L., G.A. Knauer, D.M. Karl, and J.H. Martin 1987. Primary production, new production and vertical flux in the eastern Pacific Ocean. *Nature* **325**: 803–804.

Rahmstorf, S. 2002. Ocean circulation and climate during the past 120,000 years. *Nature* **419**(6903): 207–214.

Sarmiento, J.L. and N. Gruber 2006. *Ocean Biogeochemical Dynamics.* Princeton, NJ: Princeton University Press.

Schulz-Bull, D.E., G. Petrick, R. Bruhn, and J.C. Duinker 1998. Chlorobiphenyls (PCB) and PAHs in water masses of the northern North Atlantic. *Marine Chemistry* **61**(1–2): 101–114.

Sobek, A. and O. Gustafsson 2004. Latitudinal fractionation of polychlorinated biphenyls in surface seawater along a 62 degrees N-89 degrees N transect from the southern Norwegian Sea to the North Pole area. *Environmental Science & Technology* **38**(10): 2746–2751.

Steinberg, D.K., C.A. Carlson, N.R. Bates, S.A. Goldthwait, L.P. Madin, and A.F. Michaels 2000. Zooplankton vertical migration and the active transport of dissolved organic and inorganic carbon in the Sargasso sea. *Deep-Sea Research I* **47**: 137–158.

Testor, P. and J.-C. Gascard 2006. Post-convection spreading phase in the northwestern Mediterranean Sea. *Deep Sea Research I* **53**: 869–893.

Valiela, I. 1995. *Marine Ecological Processes*, 2nd ed. New York: Springer-Verlag.

Wania, F. and G.L. Daly 2002. Estimating the contribution of degradation in air and deposition to the deep sea to the global loss of PCBs. *Atmospheric Environment* **36**(36–37): 5581–5593.

18 Chemical Exchange between Snow and the Atmosphere

Torsten Meyer and Frank Wania

CONTENTS

18.1 Introduction 513
18.2 Chemical Phase Distribution in Snow and Atmosphere–Snow Exchange
Availability 515
    18.2.1 Estimating Sorption to the Snow Surface 515
    18.2.2 Estimating the Presence of Organic Chemicals in the
Pore Space of Bulk Snow 516
18.3 Theoretical Descriptions of Atmosphere–Snow Gas Exchange 518
    18.3.1 The Two-Resistance Model of Snow Pack–Atmosphere
Exchange 518
    18.3.2 Mass Transfer Coefficient k_{bl} for Transport through the
Boundary Layer Above Snow 519
    18.3.3 Mass Transfer Coefficient k_s for Transport through Bulk Snow ..519
    18.3.4 The Role of Snow Ventilation or Wind Pumping.................520
18.4 Measurements of Atmosphere–Snow Exchange Transport Parameters 521
18.5 Factors Influencing the Rate of Atmosphere–Snow Gas Exchange 522
    18.5.1 Influence of Surface Roughness on Boundary
Layer Resistance 522
    18.5.2 Influence of Wind Speed and Surface Roughness on
Snow Ventilation 523
    18.5.3 Influence of Snow Permeability on Snow Ventilation............524
18.6 Case Study 525
18.7 Conclusion 527
Literature Cited 527

18.1 INTRODUCTION

In cold regions at high altitudes and latitudes, snow can significantly affect the environmental fate of contaminants (Wania et al., 1998; Halsall, 2004) and the chemistry of the atmosphere (Dominé and Shepson, 2002). This motivates the desire for a mechanistic understanding and quantitative description of the transport processes of chemical species between a snow pack and the atmosphere. This chapter seeks to provide an overview of the key processes involved in gaseous atmosphere–snow exchange (Figure 18.1) and of the parameterizations that have been used to describe such processes.

Snow is an efficient scavenger of particles and gaseous contaminants from the lower atmosphere. Scavenging efficiencies of snow are often much larger than those of rain (Lei and Wania, 2004). The porous nature of a snow pack also facilitates the exchange of gases with the atmosphere. Snow metamorphism, in particular a decrease of the internal snow surface area, can lead to a substantial release of volatile and semivolatile contaminants to the atmosphere. Peaks in air concentrations of organic contaminants coinciding with the snow melt period have been observed around the Laurentian Great Lakes (Hornbuckle et al., 1994; Gouin et al., 2002, 2005). The potential for snow aging and melting to cause temporary concentration maxima in the atmospheric layer above snow has also been predicted by simulation models (Daly and Wania, 2004; Gouin et al., 2005; Hansen et al., 2006). Snow also greatly influences the oxidative capacity of the lower atmosphere in polar regions and may even affect global atmospheric chemistry (Dominé and Shepson, 2002). Arctic field studies indicate a strong relationship between snow pack chemistry and elevated atmospheric concentrations of hydroxyl radicals (Yang et al., 2002; Dominé and Shepson, 2002). A change of the total snow covered area due to climate warming may notably affect the oxidative potential in the higher latitude atmosphere (Dominé and Shepson, 2002) and the global distribution of persistent organic pollutants (Macdonald et al., 2003; Stocker et al., 2007).

Chemical exchange between snow and the overlying atmosphere may take place by gas exchange (absorption and volatilization), by rain or snow scavenging, dry particle

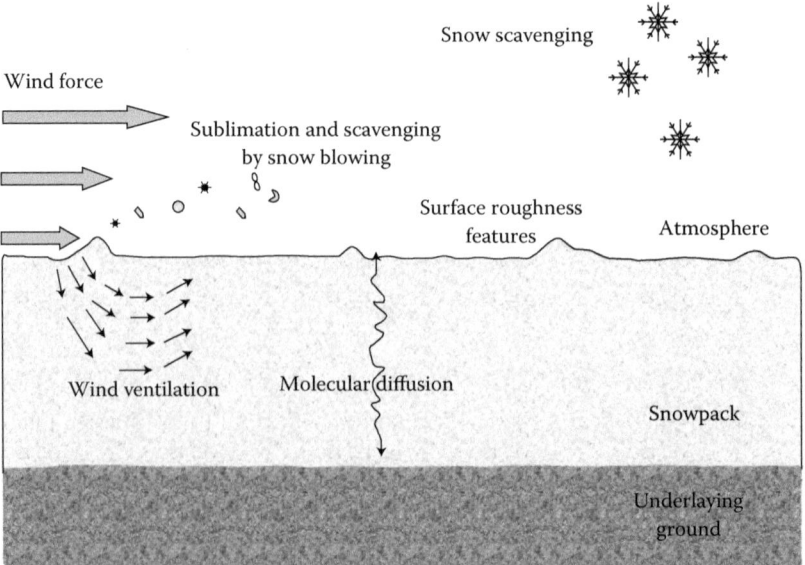

FIGURE 18.1   Chemical transport processes between a snow pack and the atmosphere.

deposition, snow blowing, and fog deposition. Because of its relative significance, only the gaseous chemical exchange is discussed in this chapter. A mechanistic description and further references of wet deposition by falling snow can be found in Lei and Wania (2004). Deposition to surfaces other than snow is covered in Chapter 8. The principles of mass transfer with blowing snow and associated chemical concentration by sublimation are described in Pomeroy et al. (1991, 1997) and Woo and Marsh (2005).

Section 18.2 of this chapter presents a method for assessing the availability of chemicals for gaseous atmosphere–snow exchange. Section 18.3 provides a summary of mechanistic descriptions of air–snow gas exchange and Section 18.4 is devoted to findings from field-based and laboratory measurements. Section 18.5 discusses key factors influencing air–snow exchange and Section 18.6 provides a sample calculation of the chemical exchange with two different types of snow pack.

18.2 CHEMICAL PHASE DISTRIBUTION IN SNOW AND ATMOSPHERE–SNOW EXCHANGE AVAILABILITY

The rate of trace gas exchange between snow and the atmosphere is, in most cases, limited by the transport within the air-filled pore space of the snow pack. Thus, it is advisable to first investigate the extent of a chemical's presence within the air-filled pore space of bulk snow.

18.2.1 ESTIMATING SORPTION TO THE SNOW SURFACE

The availability of a chemical for gas exchange between snow pack and atmosphere depends on its phase distribution within the bulk snow. The chemical can reside as gas in the pore space, be embedded within the ice lattice of the snow grains, be attached to particles present in the snow, be sorbed to the snow surface, and—during melting—be dissolved within the liquid aqueous phase (Meyer and Wania, 2008). During snow aging, chemicals accumulate near the surface of the snow grain while its interior becomes gradually purified (Colbeck, 1981).

The distribution of organic chemicals between the surface of the snow grains and the interstitial pore space can be expressed with a temperature-dependent equilibrium snow surface/air sorption coefficient $K_{IA} = C_I/C_A$. If the concentrations on the surface C_I and in the gas phase C_A are in units of $mol\, m^{-2}$ and $mol\, m^{-3}$, respectively, K_{IA} has units of length (m). Early, and only modestly successful, attempts at estimating K_{IA} were based on assuming equivalence between sorption to the water and ice surface and on empirically regressing the sorption coefficient to the water surface with other commonly encountered physical chemical properties, such as the air–water partition coefficient, water solubility, vapor pressure, or the octanol–water partition coefficient (see, e.g., comparison in Wania et al., 1999). More recently, Roth et al. (2004) measured the propensity for sorption to a snow surface for a wide variety of organic vapors and developed a linear free energy relationship based on intermolecular interactions:

$$\log K_{IA}(-6.8°C) = 3.53 \sum \alpha_2^H + 3.38 \sum \beta_2^H + 0.639 \log L^{16} - 6.85 \quad (18.1)$$

where $\log L^{16}$ is the hexadecane/air partition coefficient at 25°C, which expresses quantitatively the van der Waals interactions a chemical can undergo, and $\sum \alpha_2^H$ and $\sum \beta_2^H$ are measures of a compound's ability to act as electron acceptor and donor, respectively. Equation 18.1 allows for the estimation of K_{IA} for any organic nonelectrolyte for which $\log L^{16}$, $\sum \alpha_2^H$, and $\sum \beta_2^H$ are known. It should, however, be applied cautiously to large and complex molecules, such as organochlorine pesticides (Goss, 2004; Schüürmann et al., 2006; Burniston et al., 2007). The $\log K_{IA}$ value may be temperature corrected using

$$\log K_{IA}(T) = \log K_{IA}(T_{ref}) + \frac{\Delta_{ads}H_i}{2.303\,R}\left[\frac{1}{T_{ref}} - \frac{1}{T}\right] \tag{18.2}$$

where R is the ideal gas constant and T_{ref} is the reference temperature ($-6.8°C = 266.3$ K). Goss and Schwarzenbach (1999) provide relationships between enthalpies of adsorption on a water surface $\Delta_{ads}H_i$ and corresponding adsorption constants.

18.2.2 ESTIMATING THE PRESENCE OF ORGANIC CHEMICALS IN THE PORE SPACE OF BULK SNOW

The availability of organic substances for atmosphere–snow exchange may be studied by displaying their equilibrium phase distribution in snow using a chemical space plot as a function of the snow surface/air sorption coefficient ($\log(K_{IA}/m)$) and the organic matter/air distribution expressed by the humic acid/air partition coefficient ($\log K_{HA/A}$) (Figure 18.2, Meyer and Wania, 2008). The plot delineates regions of predominant sorption to the snow surface, presence as gas in the snow pore space, and sorption to organic particles in dry snow. Figure 18.2 represents recently fallen snow with an internal specific surface area (SSA) of 1000 cm^2 g^{-1}, a density of 0.05 g cm^{-3}, and an organic matter content of 9 ng μL^{-1} snow volume. With snow metamorphism, the phase composition and thus also the chemical phase distribution in the snow pack will change. In particular, the K_{IA} threshold value indicating the transition from air pore space to ice surface would shift to the left with increasing snow density, and shift to the right with decreasing SSA (Meyer and Wania, 2008).

The initial days after snow deposition are characterized by an increase of the snow density and a significant decrease of the SSA (Cabanes et al., 2002; Jellinek, 1967; Hanot and Dominé, 1999). Measured SSA values range from 19 to 1558 cm^2 g^{-1} (Dominé et al., 2007). A rapid decrease in SSA due to snow compaction and metamorphism leads to the loss of a snow pack's capacity to store chemicals within the interstitial air and on the ice surface. This can result in the release of substantial amounts of sorbed constituents from the snow pack (Wania et al., 1998; Hanot and Dominé, 1999; Herbert et al., 2005; Hansen et al., 2006; Taillandier et al., 2006; Burniston et al., 2007).

Selected organic contaminants covering a wide range of partitioning properties, including polycyclic aromatic hydrocarbons (PAHs), polychlorinated biphenyls (PCBs), hexachlorocyclohexanes (HCHs), fluorotelomer alcohols (FTOHs), chemicals from the BTEX group, light aldehydes (formaldehyde, acetaldehyde), and acetone

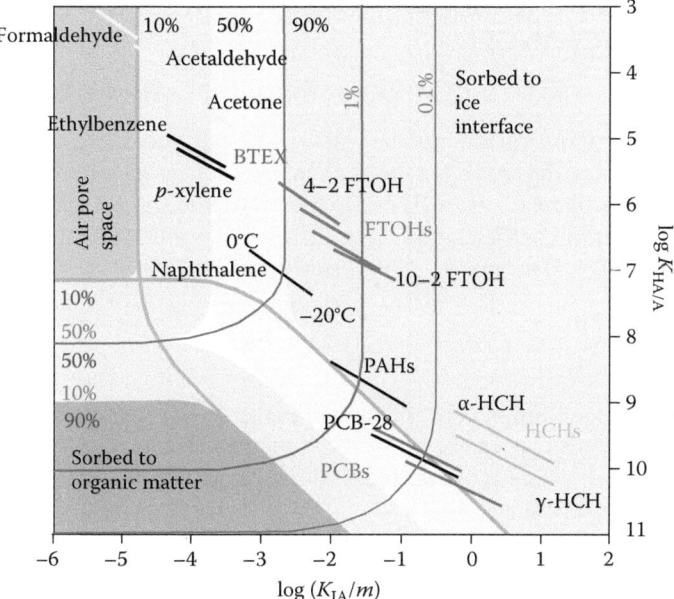

FIGURE 18.2 (See color insert following page 300.) Chemical space plot of the phase distribution of different chemicals in a dry snow pack as a function of the snow surface/air sorption coefficient (log K_{IA}/m) and the humic acid/air partition coefficient (log $K_{HA/A}$). K_{IA} and $K_{HA/A}$ values were determined based on Roth et al. (2004) and Niederer et al. (2006).

are placed on the chemical space plots as a function of temperature in the range of $-20°C$ to $0°C$. With decreasing temperature the position of chemicals in the partitioning map shifts to the lower right, implying that those in the transition areas become increasingly sorbed to the snow surface. Volatile chemicals, such as the aldehydes, acetone, ethylbenzene, and p-xylene, are predicted to partition predominantly into the air-filled pore space of the snow and should thus be highly susceptible to loss by evaporation, especially in depth hoar and at higher snow pack temperatures. Depth hoar is low-density snow consisting of large facetted crystals with sizes between 2 and 20 mm. It is formed when large temperature gradients cause strong water vapor fluxes and rapid grain growth (Dominé et al., 2008). A notable fraction of somewhat less volatile substances, such as naphthalene and short FTOHs, is also predicted to be present as gases in snow (Figure 18.2) and should be available for gas exchange with the atmosphere. Even semivolatile chemicals such as tri- and tetrachlorinated PCBs and three-ring PAHs can be susceptible to loss by evaporation if sufficient ventilation in snow provides rapid replacement of the air in the pore space. The depleted quantity of gas phase chemical is then replenished by rapid desorption from the ice surface. As a result, even small fractions in the gas phase may be important as illustrated in Figure 18.2 by the lines delineating a presence of 1% and 0.1% of a chemical within the air pore space.

18.3 THEORETICAL DESCRIPTIONS OF ATMOSPHERE–SNOW GAS EXCHANGE

18.3.1 THE TWO-RESISTANCE MODEL OF SNOW PACK–ATMOSPHERE EXCHANGE

Mechanistic descriptions of atmosphere–snow exchange processes generally calculate a flux ($g\,m^{-2}\,h^{-1}$) as the product of the gradient between the gas phase concentrations in the lower atmosphere C_A ($g\,m^{-3}$) and in the snow pore space C_{AS} ($g\,m^{-3}$), and an overall mass transfer coefficient (MTC) k_m ($m\,h^{-1}$) (Wania, 1997; Daly and Wania, 2004; Halsall, 2004; Hansen et al., 2006; Herbert et al., 2006):

$$F = k_m[C_{AS} - C_A] \tag{18.3}$$

By assuming equilibrium partitioning within the snow pack, the vapor concentration in the pore space C_{AS} can be estimated from the bulk snow concentration C_S (in $g\,m^{-3}$ melt water), which is accessible to measurements, and a dimensionless bulk snow/air partition coefficient K_{SA}:

$$C_{AS} = \frac{C_S}{K_{SA.}} \tag{18.4}$$

K_{SA} can be estimated from the snow surface sorption coefficient K_{IA} by multiplication with the specific snow surface area SSA ($m^2\,g^{-1}$) and the density ρ of melt water ($g\,m^{-3}$)

$$K_{SA} = K_{IA} \times SSA \times \rho \times 100 \tag{18.5}$$

Equation 18.3 thus becomes

$$F = k_m\left[\frac{C_S}{K_{IA} \times SSA \times \rho} - C_A\right] \tag{18.6}$$

The existing parameterizations of k_m treat the exchange across the atmosphere–snow pack interface on the basis of the Whitman two-film resistance model, which assumes that a chemical that moves from snow pack to atmosphere has to first transfer from the bulk of the snow pack to the snow pack's surface, and then through the atmospheric boundary layer to the bulk atmosphere.

$$\frac{1}{k_m} = \frac{1}{k_{bl}} + \frac{1}{k_s} \tag{18.7}$$

where k_{bl} is the snow–air boundary layer MTC and k_s is the bulk snow phase MTC. The limiting MTC determines the extent of the total mass transfer. The overall MTC k_m comprises most of the uncertainties associated with atmosphere–snow exchange. Accordingly, its parameterization varies between the studies.

18.3.2 MASS TRANSFER COEFFICIENT k_{bl} FOR TRANSPORT THROUGH THE BOUNDARY LAYER ABOVE SNOW

Similar to the widely adopted model of air–water exchange (Schwarzenbach et al., 2003), the mass transfer from the snow pack–air interface to the bulk atmosphere is usually interpreted as occurring by molecular diffusion across a thin stagnant boundary layer film. Because the thickness of this boundary layer is unknown, some approaches (Wania, 1997; Daly and Wania, 2004) simply treat k_{bl} as a constant. In particular, it is assumed that k_{bl} adopts values similar to a typical MTC for the boundary layer above soil (e.g., $0.14 \, \mathrm{cm \, s^{-1}}$ or $5 \, \mathrm{m \, h^{-1}}$; Mackay and Stiver, 1991). In a more realistic approach, Hansen et al. (2006) allow for the influence of variable wind speed on k_{bl}, which they expressed as:

$$k_{bl} = k^2 U \frac{1}{\ln\left(\dfrac{z_w}{z_0}\right) \ln\left(\dfrac{z_{ref}}{z_0}\right)} \tag{18.8}$$

where U is the wind speed, z_0 is the aerodynamic surface roughness length (a generic value of $z_0 = 0.001 \, \mathrm{m}$ was applied), z_w is the wind height, z_{ref} is the reference wind height, and $k = 0.4$ is the von Karman's constant.

18.3.3 MASS TRANSFER COEFFICIENT k_s FOR TRANSPORT THROUGH BULK SNOW

Some models of atmosphere–snow exchange assume that transport within the snow pack occurs in parallel both in the air and water-filled interstitial snow pores (Hoff et al., 1995; Wania, 1997; Daly and Wania, 2004; Stocker et al., 2007):

$$k_s = \frac{k_{sw}}{K_{AW}} + k_{sa} \tag{18.9}$$

where k_{sw} is the snow–water phase MTC, k_{sa} is the snow–air phase MTC, and K_{AW} is the air–water partition coefficient. However, the liquid water content in aged and coarse-grained snow rarely exceeds 5% of the snow pack's volume (Colbeck, 1978). Because of the relatively small pore volume filled by water and the slower diffusivity in water compared to air, k_{sw}/K_{AW} is typically much smaller than k_{sa}, even for chemicals with a very low K_{AW}. Accordingly, some studies (Hansen et al., 2006) neglect chemical transport within the water-filled pore space and assume:

$$k_s = k_{sa}. \tag{18.10}$$

Initially, most studies of air–snow exchange assumed that transport within the snow pack occurs by molecular diffusion through the pore space only. The parameter k_{sa} is then calculated from the molecular diffusivity in air D_A, the diffusion path length (dpl), and a factor pt accounting for the porosity of the snow pack and the tortuosity of the diffusion pathway:

$$k_{sa} = D_A \cdot \frac{pt}{dpl}. \tag{18.11}$$

The diffusion path length dpl has variably been assumed to be equal to the snow pack depth z (Herbert et al., 2006), z divided by 2 (Hansen et al., 2006), or $\ln(2)$ or 0.69 times z (Wania et al., 1997; Daly and Wania, 2004; Stocker et al., 2007). The porosity–tortuosity term pt is typically expressed as a function of the volume fraction of air in bulk snow, or porosity, v_{sa}, which is equal to 1 minus the ratio of the bulk snow density ρ_{snow} and the density of ice ρ_{ice} ($0.917\,\mathrm{g\,cm^{-3}}$):

$$\mathrm{pt} = v_{sa}^{n} = \left(1 - \frac{\rho_{snow}}{\rho_{ice}}\right) \qquad (18.12)$$

The exponent n in Equation 18.12 has been assumed to be either 1.5 (Albert and Shultz, 2002; Hansen et al., 2006), or 4/3 (Wania et al., 1997; Daly and Wania, 2004; Stocker et al., 2007), or 2/3 (Herbert et al., 2005). Equivalent expressions are used if diffusion in the water-filled pore space is considered (Wania et al., 1997; Daly and Wania, 2004; Stocker et al., 2007).

18.3.4 THE ROLE OF SNOW VENTILATION OR WIND PUMPING

Equations 18.11 and 18.12 state that with increasing snow pack depth z and decreasing snow porosity v_{sa} the resistance to molecular diffusive transport increases. Accordingly, a shallow snow pack exhibiting a low density has the highest predicted k_{sa}. However, because of the slow rate of molecular diffusion, this approach generally predicts negligible volatilization rates in all but extremely shallow snow packs. Such low volatilization rates are clearly not reconcilable with the observational evidence (e.g., Herbert et al., 2005; Burniston et al., 2007). In response, Daly and Wania (2004), citing studies on the importance of wind ventilation (Albert and Shultz, 2002), incorporated the process of wind pumping into their description of atmosphere–snow exchange. Modifying Equation 18.10, the mass transfer in the bulk snow was described as

$$k_{s} = \mathrm{wpf} \times k_{sa} \qquad (18.13)$$

where the wind pumping factor wpf varies with snow depth and was selected, somewhat arbitrarily, so that both resistances in the air-filled pores space and in the atmospheric boundary layer are of similar magnitude, that is, $\mathrm{wpf} \times (k_{sa}/k_{bl}) \approx 1$. This assumes moderate wind ventilation at all times, making the atmosphere–snow interface relatively permeable.

Hansen et al. (2006) developed an Arctic snow model, in which the bulk snow MTC was calculated using

$$k_{s} = k_{sa} \times K_{SA} \qquad (18.14)$$

Although the appearance of the snow–air partition coefficient K_{SA} in Equation 18.14 is theoretically not correct (Halsall et al., 2008), K_{SA} effectively adopts a role similar to that of the wind pumping factor in Equation 18.13, greatly increasing k_{s} to the point that the resistance posed by transport within the snow pack $1/k_{s}$ is no longer much larger than that posed by the boundary layer resistance $1/k_{bl}$. The rapid gas exchange predicted using Equation 18.14 is consistent with the windy conditions in polar regions (Hansen et al., 2006; Halsall et al., 2008) and the observation of volatilization loss from aging snow packs (Herbert et al., 2005) and highlights further that it is often not appropriate to assume that chemical transport in the snow pack occurs by molecular diffusion only (Halsall et al., 2008).

Although the simple approaches described in this section provide a convenient way to describe air–snow gas exchange on the scale of several days or weeks, they provide only limited quantitative insight. They can, however, serve to identify the most important processes, and thus the key parameters that need to be further investigated. In particular, it is important to accurately define the impact of permeability, snow layering, and wind on snow ventilation (Daly and Wania, 2004).

18.4 MEASUREMENTS OF ATMOSPHERE–SNOW EXCHANGE TRANSPORT PARAMETERS

Several studies have investigated empirically the flux of chemicals within snow or between snow and the atmosphere (Guimbaud et al., 2002; Albert and Shultz, 2002; Herbert et al., 2006). In particular, measured concentration gradients within the atmospheric boundary layer or within the snow pack have been used to calculate a chemical's flux into or out of the snow pack. This approach has resulted in miscellaneous parameterizations to calculate fluxes of, for example, carbonyl compounds and NO_x species from the snow pack as a result of photochemical processes in snow (Dominé and Shepson, 2002; Hutterli et al., 1999; Guimbaud et al., 2002; Grannas et al., 2002). However, flux measurements can only be used to derive kinetic transport parameters, such as diffusivities and mass transport coefficients, if the chemicals involved are reasonably persistent and do not undergo rapid conversions within the snow pack. For example, measurements of the flux of carbonyl compounds out of snow are more likely to reflect the kinetics of formation in the snow pack than the kinetics of snow–air gas exchange. As a result, there is a very limited number of experimental studies that provide quantitative information on the rate of chemical transport in snow.

Albert and Shultz (2002) used the stable and inert trace gas sulfur hexafluoride (SF_6) to study both diffusive and advective transport in Arctic snow at Summit, Greenland. Chemical diffusivity in wind-packed surface snow was determined by measuring the rate of disappearance of SF_6 from the headspace of a steel cylinder inserted into the snow pack. Wind pack is high-density snow and consists of small rounded crystals, generated by strong and cold winds (Dominé et al., 2008). The effective diffusivity D_{eff} was $0.06\ cm^2\ s^{-1}$. The reduction of the molecular diffusivity by the presence of the snow pack was well described by Equation 18.12 and the empirically derived exponent n was close to the theoretically expected value of 1.5.

Transport velocities in snow due to wind ventilation, k_s as in Equation 18.13, were determined by injecting SF_6 into a 15 cm deep hoar layer beneath a dense wind pack in a snow pack with little surface roughness. The arrival of the SF_6 pulse 1 m upwind and 1 m downwind of the injection point was measured as a function of time (Albert and Shultz, 2002). At lower wind speeds ($U_{10} = 3\ m\ s^{-1}$) arrival at upwind and downwind locations occurred at the same time, indicating that there was no ventilation effect on the chemical's transport, even though the measured transport velocities were faster than what would be expected based on molecular diffusion alone. Under higher wind speed conditions ($U_{10} = 9$ m s^{-1}) the SF$_6$ arrived much more quickly at the downwind location, indicative of strong wind ventilation. The transport velocity was 1.3 cm s^{-1} (46.8 m h^{-1}). This study revealed that strong winds allow for substantial ventilation in deeper more permeable snow layers even in the absence of notable surface roughness and in the presence of a dense and poorly permeable surface layer (Albert and Shultz, 2002).

In a modification of the approach by Albert and Schultz (2002), Herbert et al. (2006) measured the diffusion of several semivolatile organic chemicals (γ-HCH, α-HCH, HCB) into fresh and aged snow by maintaining a constant air concentration for 24 h in the head space of a cylinder inserted into the snow pack. Subsequently, the concentration gradient with depth was measured in the melted snow and used to derive an effective diffusivity. By dividing this measured diffusivity (m^2 h^{-1}) by the snow pack depth z (0.4 m), they estimated MTCs k_{sa} of 1.2×10^{-3} cm s^{-1} (0.043 m h^{-1}). Those empirically derived MTCs were of the same order of magnitude as those estimated theoretically using an equation of type (18.11).

In the only known laboratory study of atmosphere–snow gas exchange the volatilization flux of PAHs covering a wide range of partitioning properties were determined by measuring the air concentrations in an air stream passing over an homogeneous snow pack of 35 cm depth, which was artificially contaminated with those PAHs (Daly et al., 2007). The wind speed was negligible because the air velocity above the snow surface was kept at approximately 0.1 cm s^{-1}. Mass transfer coefficients k_m of the three relatively volatile PAHs naphthalene, acenaphthylene, and fluorene were found to be on the order of 6×10^{-6} to 6×10^{-5} cm s^{-1} (2×10^{-4} to 2×10^{-3} m h^{-1}). Those values are smaller than those estimated using an equation of type (18.11).

18.5 FACTORS INFLUENCING THE RATE OF ATMOSPHERE–SNOW GAS EXCHANGE

18.5.1 INFLUENCE OF SURFACE ROUGHNESS ON BOUNDARY LAYER RESISTANCE

The resistance to the transport of chemicals through the boundary layer above snow ($1/k_{bl}$ in Equation 18.7) may be neglected when the flow within the atmospheric boundary layer is sufficiently rough to prevent the formation of a laminar sublayer where chemicals only move due to molecular diffusion. If this happens above snow, only the MTC k_s becomes important. The aerodynamic roughness length z_0 in combination with the friction velocity u° and the kinematic viscosity v are commonly used to describe whether the flow at the boundary layer between two media is rough or smooth (Seinfeld and Pandis, 1998). Specifically, the shear Reynolds number, which is the product of the roughness length and friction velocity divided by the kinematic viscosity of air, is used to distinguish rough vs. smooth flow. Empirically, it was found that $v > 2.5$ implies rough or turbulent flow while $v < 0.13$ implies smooth or laminar flow.

The roughness length z_0 is defined as the height above the surface at which a wind profile assumes a zero velocity (Brock et al., 2006), and must not be confused with the surface roughness amplitude A_{sr} (Albert and Hawley, 2002). Empirical studies indicate the relationship (Seinfeld and Pandis, 1998):

$$z_0 \approx \frac{A_{sr}}{30}. \tag{18.15}$$

Typical values for friction velocities u° in the lower atmosphere fall in the range of $10-80\,\mathrm{cm\,s^{-1}}$ (Weber, 1999) and the kinematic viscosity ν of air is approximately $0.1\,\mathrm{cm^2\,s^{-1}}$. With prevalent snow roughness lengths z_0 ranging between 0.01 and 0.50 cm (see Brock et al., 2006), the atmospheric boundary flow would be considered rough under most circumstances. However, roughness lengths can vary widely from as low as 0.0004–0.0150 cm for Antarctic plateau snow to 3 cm for tropical glacier snow penitentes (Brock et al., 2006).

18.5.2 INFLUENCE OF WIND SPEED AND SURFACE ROUGHNESS ON SNOW VENTILATION

As discussed in Section 18.3.4, advective transport due to wind-induced air flow is generally much faster than diffusive transport through snow and can thus greatly enhance atmosphere–snow exchange and the snow depth involved in such processes (Albert and Shultz, 2002). Wind pumping can cause vapor movement in snow covers at depths of several meters and enhances the available internal snow surface area for uptake and release of chemical substances (Albert et al., 2002). Wind ventilation in snow is caused by "form drag" pressure variations across snow roughness features and by pressure variations due to wind turbulence (Albert and Hawley, 2002). The intensity of snow ventilation depends on the wind speed above the snow surface, the profile of the surface roughness features, and snow permeability. A common roughness feature in Arctic and Antarctic regions are sharp and irregularly shaped *sastrugi* that can move in time and may seasonally vary in size (Albert and Hawley, 2002). Surface roughness amplitudes A_{sr} for different kinds of snow packs may be estimated using Equation 18.15 and surface roughness lengths z_0 listed in Brock et al. (2006).

Albert and Hawley (2002) modeled the wind induced airflow in snow at Summit using measured snow permeability profiles within the snow pack. The amplitude of the maximum pressure difference caused by a steady flow over a sinusoidal snow surface is accordingly

$$p_0 = \frac{C\rho_{air}U_{10}^2 A_{sr}}{\lambda}, \tag{18.16}$$

where p_0 is the amplitude of the pressure (Pa), ρ_{air} is the density of air (kg m^{-3}), A_{sr} is the amplitude of the surface roughness (m), and λ is the wavelength of the surface roughness (m). The proportionality constant C was determined by Colbeck (1989) to be approximately 3. During the summer, the surface roughness profiles were smoothly curved and approximately 5 cm high, while in the winter those profiles increased to a height of roughly 20 cm, which strongly impacts the wind ventilation in snow. Under winter surface roughness conditions, the transport velocity within the dense and wind-packed surface layer is three times that during the summer assuming similar strong wind conditions ($U_{10} = 12$ m s^{-1}). A relatively short surface roughness wave length λ of 5 m induces ventilation down to 2 m depth in the snow pack, while a longer wavelength causes ventilation even in 4 m snow depth (Albert and Hawley, 2002). An increasing distance between surface roughness features causes a deeper air flow penetration within the snow pack. However, the air flow velocity in the upper layers decreases.

A similar calculation using measurements of physical snow properties and interstitial ozone concentrations further confirmed the significance of surface roughness features for a chemical's transport in snow (Albert et al., 2002). Beneath moderate winds ($U_{10} = 7$–9 m s^{-1}), surface roughness amplitudes of $\sim$10 cm and roughness wavelengths of $\sim$10–20 m almost the entire approximately 40 cm deep snow pack was subject to ventilation.

18.5.3   INFLUENCE OF SNOW PERMEABILITY ON SNOW VENTILATION

Snow permeability is crucial for the transport of chemicals by wind ventilation (Albert and Shultz, 2002) and is substantially influenced by snow layering and snow microstructure. Increasing snow grain size and homogeneity enhances the permeability (Albert and Perron Jr., 2000). The grain size increase is fast when exposed to large macroscopic temperature gradients. However, well-rounded and uniform snow grains usually develop under isothermal conditions. Figure 18.3 depicts snow crystals in different stages of metamorphism in isothermal snow. As a result of the development of larger, rounded ice crystals, the permeability is considerably enhanced. Although

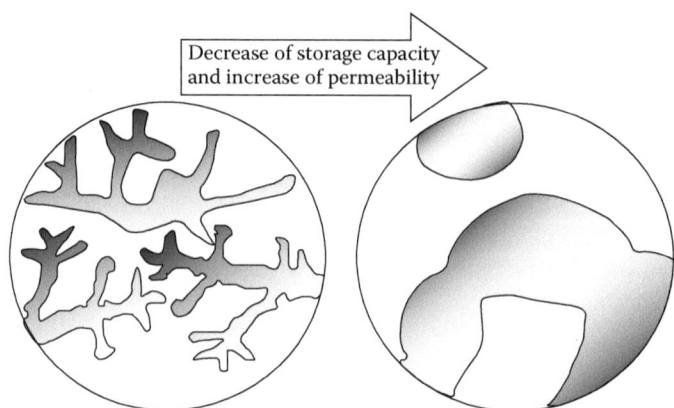

FIGURE 18.3 Snow crystal sketches describing recently deposited dendritic snow with large surface area (left) and isothermal aged snow crystals with small surface area (right); both exhibit similar porosity. Dendritic snow consists of six-branched crystals with a diameter of a few millimeters and a thickness of a few tens of micrometers. (Adapted from Dominé F. et al. 2008. *Atmospheric Chemistry and Physics* **8**, 171–208.)

commonly applied to describe permeability variations, the snow density is poorly correlated to snow permeability (Albert et al., 2000).

The snow permeability P (m^2) is defined as the proportionality factor between the pressure gradient and the flow velocity within the bulk snow, known as Darcy's law:

$$k_{sa} = -\frac{P}{\mu}\frac{\partial p_0}{\partial x} \qquad (18.17)$$

where $\partial p_0/\partial x$ is the pressure drop (Pa m^{-1}) and μ (Pa s) is the fluid viscosity.

The permeability of seasonal snow varies within two orders of magnitude depending on grain size and shape, porosity, and the type of snow (Dominé et al., 2008). Albert et al. (2000) measured variations by almost an order of magnitude in different snow layers within the top 2 m of a snow pack at Siple Dome, Antarctica. A typical permeability of newly deposited seasonal snow is 30–70×10^{-10} m^2. After months of ageing and metamorphism, the permeability of that snow may have decreased to 8–50×10^{-10} m^2 (Jordan et al., 1999). Snow containing notable fractions of refrozen melt water exhibited permeability of 1–19×10^{-10} m^2 (Albert et al., 2000; Albert and Perron Jr., 2000). While Arctic wind pack typically features permeability of 5–30×10^{-10} m^2, that of depth hoar usually is very high, between 100 and 600×10^{-10} m^2 (Dominé et al., 2008). The latter two types of snow packs characterize typical snow in the Arctic and in the Subarctic, respectively.

18.6   CASE STUDY

The influence of the various environmental parameters on atmosphere–snow exchange of organic chemicals is illustrated by a conceptual case study using the semivolatile chemical PCB-28 as an example (Table 18.1). The mass transfer of PCB-28 between two different types of snow pack and the overlaying atmosphere is estimated for wind under still and moderately windy conditions. Under wind still conditions, chemical transport is assumed to be controlled by molecular diffusion, while stronger winds cause chemical movement in snow to be controlled by advection. One snow pack, consisting mostly of depth hoar, is typical for sub-Arctic regions and the other, a high-density wind pack, can often be found in the Arctic. Snow packs in sub-Arctic and Arctic regions are the most extended snow covers on Earth in terms of area and duration. At the same time, they are relatively easily distinguishable from each other (Taillandier et al., 2006). The applied parameters and estimation results are shown in Table 18.1.

Surface roughness features are highly variable and it is difficult to assign typical values for the two types of snow covers. Therefore, roughness amplitude A_{sr} and wavelength λ were assumed to be the same for sub-Arctic and Arctic snow packs. With a λ between 0.05 and 10 m, the ratio A_{sr}/λ is approximately 0.15 and 0.025 for smaller and larger surface roughness features, respectively (see Colbeck, 1989). Here, we assume a relatively high A_{sr} (Albert et al., 2002). The calculation does not consider the influence of λ on the depth of the snow pack that will be affected by ventilation.

TABLE 18.1
Parameters Applied to Calculate the Atmosphere–Snow Mass Transfer of PCB-28 in the Sub-Arctic and Arctic

Parameter	Subarctic Snow	Arctic Snow	Reference/Equation
Average snow temperature (°C)	−10	−30	Taillandier et al., 2006; Sturm et al., 1995
K_{IA} (m)	0.58	14.45	Equations 18.1 and 18.2
SSA (cm^2 g^{-1})	150	300	Dominé et al., 2007
Snow density (g cm^{-3})	0.22	0.40	Dominé et al., 2007
K_{SA} (dimensionless)	536	39,372	Equation 18.5
C_S (g cm^{-3})	4×10^{-14}	4×10^{-14}	Gregor et al., 1995; Herbert et al., 2005
C_A (g cm^{-3})	4×10^{-18}	4×10^{-18}	Hung et al., 2005
C_{AS} (g cm^{-3})	7.5×10^{-17}	1.0×10^{-18}	Equation 18.4
$C_{AS} - C_A$ (g cm^{-3})	7.1×10^{-17}	-3.0×10^{-18}	
Air density (kg m^{-3})	1.34	1.44	
Wind speed U_{10} (m s^{-1}) (in the case of advection controlled chemical movement in snow)	7	10	Canadian Wind Energy Atlas (2003)
Surface roughness amplitude A_{sr} (m)	0.1	0.1	Albert et al., 2002; Brock et al., 2006; Equation 18.15
Surface roughness wavelength λ (m)	5	5	Albert et al., 2002
Pressure amplitude p_0 (Pa)	3.9	8.6	Equation 18.16
Snow permeability P (m^2)	2×10^{-8}	2×10^{-9}	Albert and Shultz, 2002
Air viscosity μ (Pa·s)	1.68×10^{-5}	1.58×10^{-5}	
Snow depth z (m)	0.7	0.4	Sturm et al., 1995
Diffusion coefficient D_{eff} (cm^2 s^{-1}) (molecular diffusion controlled)	0.06	0.04	Dominé et al., 2008
MTC in snow k_s (cm s^{-1})			
Advection controlled	0.97	0.39	Equation 18.18
Molecular diffusion controlled	0.0017	0.0020	$k_s = 2(D_{eff}/z)$ (Hansen et al., 2006)
Flux from snow (ng m^{-2} day^{-1})			Equation 18.3
Advection controlled	59.0	−1.02	
Molecular diffusion controlled	0.105	−0.005	

The depth-averaged MTC k_s was estimated using Equations 18.16 and 18.17 whereby ln2 z is used for the mean diffusion path length.

$$k_{sa} = -\frac{p}{\mu} \frac{p_0}{\ln 2\ z} \tag{18.18}$$

The dimensionless snow-pore space partition coefficient K_{SA} is notably lower in the sub-Arctic snow pack largely because of the influence of temperature on K_{IA} (see Equation 18.2). The bulk of the snow pack in sub-Arctic regions consists of depth hoar that exhibits both low density and SSA, which further contributes to this discrepancy. The differences in partitioning of PCB-28 within the sub-Arctic and Arctic snow pack cause significantly different concentration gradients between snow cover and lower atmosphere and thus, different exchange behavior. While PCB-28 is released from the sub-Arctic snow pack, the comparably low pore space concentration in the Arctic snow leads to a net deposition of this chemical (Table 18.1).

The sub-Arctic snow pack exhibits permeability that is approximately one order of magnitude larger than that in Arctic snow, which significantly increases the chemical exchange in the former. By combining the effects of chemical partitioning and snow properties, the MTCs are two to three orders of magnitudes larger under the influence of moderately strong winds than under still wind conditions.

18.7 CONCLUSION

A variety of models of air–snow exchange have been developed in recent years that have been modified along with, and on the basis of, progressing field measurements. There is still need to perform field and laboratory measurements to explore the influence of physical snow properties, wind and a chemical's partitioning properties on the exchange dynamics. The models differ most in terms of their description of chemical movement within the bulk snow, which is in most cases also the limiting factor of atmosphere–snow exchange. Whether snow ventilation due to wind pumping needs to be more mechanistically described depends on the particular model application. For some scenarios, the application of a generic wind pumping factor may be sufficient, for example, when long-term processes such as the release of chemicals from snow in the course of weeks or months are the scope of the investigation. When the atmospheric boundary conditions are extremely static, mere molecular diffusion may represent such processes well, and an effective molecular diffusion coefficient is sufficient. However, the description of diurnal exchange processes within the upper snow layers of substances, such as those generated by photolytic reactions, will require a more specific parameterization of wind ventilation in snow.

LITERATURE CITED

Albert, M.R., Perron Jr., F.E. 2000. Ice layer and surface crust permeability in a seasonal snow pack. *Hydrological Processes* **14**, 3207–3214.

Albert, M.R., Shultz, E.F., Perron Jr., F.E. 2000. Snow and firn permeability at Siple Dome, Antarctica. *Annals of Glaciology* **31**, 353–356.

Albert, M.R., Hawley, R.L. 2002. Seasonal changes in snow surface roughness characteristics at Summit, Greenland: Implications for snow and firn ventilation. *Annals of Glaciology* **35**, 510–514.

Albert, M.R., Shultz, E.F. 2002. Snow and firn properties and transport processes at Summit, Greenland. *Atmospheric Environment* **36**, 2789–2797.

Albert, M.R., Grannas, A.M., Bottenheim, J., Shepson, P.B., Perron, F.E. 2002. Processes and properties of snow–air transfer in the high Arctic with application to interstitial ozone at Alert, Canada. *Atmospheric Environment* **36**, 2779–2787.

Brock, B.W., Willis, I.C., Sharp, M.J. 2006. Measurement and parameterization of aerodynamic roughness length variations at Haut Glacier d'Arolla, Switzerland. *Journal of Glaciology* **52**(177), 281–297.

Burniston, D.A., Strachan, W.J.M., Hoff, J.T., Wania, F. 2007. Changes in surface area and concentrations of semivolatile organic contaminants in ageing snow. *Environmental Science & Technology* **41**, 4932–4937.

Cabanes, A., Legagneux, L., Dominé, F. 2002. Evolution of specific surface area and of crystal morphology of arctic fresh snow during the ALERT 2000 campaign. *Atmospheric Environment* **36**, 2767–2777.

Canadian Wind Energy Atlas 2003. EOLE Wind Energy Project, Environment Canada, Quebec, Canada.

Colbeck, S.C. 1978. The physical aspects of water flow through snow. *Advances in Hydroscience* **11**, 165–206.

Colbeck, S.C. 1981. A simulation of the enrichment of atmospheric pollutants in snow cover runoff. *Water Resources Research* **17**, 1383–1388.

Colbeck, S.C. 1989. Air movement in snow due to windpumping. *Journal of Glaciology* **35**, 209–213.

Daly, G.L., Wania, F. 2004. Simulating the influence of snow on the fate of organic compounds. *Environmental Science & Technology* **38**, 4176–4186.

Daly, G.L., Westgate, J.N., Lei, Y.D., Wania, F. 2007. Development of a laboratory method for investigations of the air–snow exchange of PAHs. Ph.D. thesis, Chapter Four, Department of Chemistry, University of Toronto.

Dominé, F., Shepson, P.B. 2002. Air–snow interactions and atmospheric chemistry. *Science* **297**, 1506–1510.

Dominé, F., Taillandier, A.-S., Simpson, W.R. 2007. A parameterization of the specific surface area of seasonal snow for field use and for models of snow pack evolution. *Journal of Geophysical Research* **112**, F2, F02031.

Dominé, F., Albert, M., Huthwelker, T., Jacobi, H.-W., Kokhanovsky, A.A., Lehning, M., Picard, G., Simpson, W.R. 2008. Snow physics as relevant to snow photochemistry. *Atmospheric Chemistry and Physics* **8**, 171–208.

Goss, K.-U., Schwarzenbach, R.P. 1999. Empirical prediction of heats of vaporization and heats of adsorption of organic compounds. *Environmental Science & Technology* **33**, 3390–3393.

Goss, K.U. 2004. The air/surface adsorption equilibrium of organic compounds under ambient conditions. *Critical Reviews in Environmental Science and Technology* **34**, 339–389.

Gouin, T.W., Thomas, G.O., Cousins, I., Barber, J., Mackay, D., Jones, K.C. 2002. Air–surface exchange of polybrominated diphenyl ethers and polychlorinated biphenyls. *Environmental Science & Technology* **36**, 1426–1434.

Gouin, T.W., Harner, T., Daly, G.L., Wania, F., Mackay, D., Jones, K.C. (2005) Variability of concentrations of polychlorinated biphenyls and polybrominated diphenyl ethers in air: Implications for monitoring, modeling and control. *Atmospheric Environment* **39**, 151–166.

Grannas, A.M., Shepson, P.B., Guimbaud, C., Sumner, A.L., Albert, M., Simpson, W., Dominé, F., Boudries, H., Bottenheim, J., Beine, H.J., Honrath, R., Zhou, X. 2002. A study of photochemical and physical processes affecting carbonyl compounds in the Arctic atmospheric boundary layer. *Atmospheric Environment* **36**, 2733–2742.

Gregor, D.J., Peters, A.J., Teixeira, C., Jones, N., Spencer, C. 1995. The historical residue trend of PCBs in the Agassiz Ice Cap, Ellesmere Island, Canada. *The Science of the Total Environment* **160/161**, 117–126.

Guimbaud, C., Grannas, A.M., Shepson, P.B., Fuentes, J.D., Boudries, H., Bottenheim, J.W., Dominé, F., Houdier, S., Perrier, S., Biesenthal, T.B., Splawn, B.G. 2002. Snow pack processing of acetaldehyde and acetone in the Arctic atmospheric boundary layer. *Atmospheric Environment* **36**, 2743–2752.

Halsall, C.J. 2004. Investigating the occurrence of persistent organic pollutants (POPs) in the arctic: Their atmospheric behaviour and interaction with the seasonal snow pack. *Environmental Pollution* **128**, 163–175.

Halsall, C.J., Hansen, K., Christensen, J. 2008. Correction to: "A dynamic model to study the exchange of gas-phase persistent organic pollutants between the air and a seasonal snow pack". *Environmental Science & Technology* **42**, 2205–2206.

Hanot, L., Dominé, F. 1999. Evolution of the surface area of a snow layer. *Environmental Science & Technology* **33**, 4250–4255.

Hansen, K.M., Halsall, C.J., Christensen, J.H. 2006. A dynamic model to study the exchange of gas-phase persistent organic pollutants between air and a seasonal snow pack. *Environmental Science & Technology* **40**, 2640–2652.

Herbert, B.M.J., Halsall, C.J., Villa, S., Jones, K.C., Kallenborn, R. 2005. Rapid changes in PCB and OC pesticide concentrations in Arctic snow. *Environmental Science & Technology* **39**, 2998–3005.

Herbert, B.M.J., Halsall, C.J., Jones, K.C., Kallenborn, R. 2006. Field investigation into the diffusion of semi-volatile organic compounds into fresh and aged snow. *Atmospheric Environment* **40**, 1385–1393.

Hoff, J.T., Wania, F., Mackay, D., Gillham, R. 1995. Sorption of non-polar organic vapors by ice and snow. *Environmental Science & Technology* **29**, 1982–1989.

Hornbuckle, K.C., Jeremiason, J.D., Sweet, C.W., Eisenreich, S.J. 1994. Seasonal variation in air–water exchange of polychlorinated biphenyls in Lake Superior. *Environmental Science & Technology* **28**, 1491–1501.

Hung, H., Lee, S.C., Wania, F., Blanchard, P., Brice, K. 2005. Measuring and simulating atmospheric concentration trends of polychlorinated biphenyls in the Northern Hemisphere. *Atmospheric Environment* **39**, 6502–6512.

Hutterli, M.A., Röthlisberger, R., Bales, R.C. 1999. Atmosphere-to-snow-to-firn transfer studies of HCHO at Summit, Greenland. *Geophysical Research Letters* **26**, 1691–1694.

Jellinek, H.H.G. 1967. Liquid-like (transition) layer on ice. *Journal of Colloid and Interface Science* **25**, 192–205.

Jordan, R.E., Hardy, J.P., Perron, F.E., Fisk, D.J. 1999. Air permeability and capillary rise as measures of the pore structure of snow: An experimental and theoretical study. *Hydrological Processes* **13**, 1733–1753.

Lei, Y.D., Wania, F. 2004. Is rain or snow a more efficient scavenger of organic chemicals? *Atmospheric Environment* **38**, 3557–3571.

Macdonald, R. W., Mackay, D., Li, Y.-F., Hickie, B. 2003. How will global change affect risks from long-range transport of persistent organic pollutants? *Human and Ecological Risk Assessment* **9**, 643–660.

Mackay, D., Stiver, W. 1991. Predictability and environmental chemistry. In *Environmental Chemistry of Herbicides*. Grover, R. and Cessna, A.J. eds., CRC Press, Boca Raton, FL, pp. 281–297.

Meyer, T., Wania, F. 2008. Organic contaminant amplification during snow melt—a review. *Water Research* **42**, 1847–1865.

Niederer, C., Goss, K.-U., Schwarzenbach, R.P. 2006. Sorption equilibrium of a wide spectrum of organic vapors in leonardite humic acid: Modeling of experimental data. *Environmental Science & Technology* **40**, 5374–5379.

Pomeroy, J.W., Davies, T.D., Tranter, M. 1991. The impact of blowing snow on snow chemistry. *NATO ASI Series* **G28**, 71–113.

Pomeroy, J.W., Marsh, P., Gray, D.M. 1997. Application of a distributed blowing snow model to the Arctic. *Hydrological Processes* **11**, 1451–1464.

Roth, C.M., Goss, K.-U., Schwarzenbach, R.P. 2004. Sorption of diverse organic vapors to snow. *Environmental Science & Technology* **38**, 4078–4084.

Schüürmann, G., Ebert, R.U., Kühne, R. 2006. Prediction of physicochemical properties of organic compounds from 2D molecular structure-fragment methods vs. LFER models. *Chimia* **60**, 691–698.

Schwarzenbach, R.P., Gschwend, P.M., Imboden, D.M. 2003. *Environmental Organic Chemistry*. Wiley, New York.

Seinfeld, J.H., Pandis, S.N. 1998. *Atmospheric Chemistry and Physics: From Air Pollution to Climate Change*. Wiley, New York.

Stocker, J., Scheringer, M., Wegmann, F., Hungerbühler, K. 2007. Modeling the effect of snow and ice on the global environmental fate and long-range transport potential of semi-volatile organic compounds. *Environmental Science & Technology* **41**, 6192–6198.

Sturm, M., Holmgren, J., Liston, G.E. 1995. A seasonal snow cover classification system for local to global applications. *Journal of Climate* **8**, 1261–1283.

Taillandier, A.-S., Dominé, F., Simpson, W.R., Sturm, M., Douglas, T.A., Severin, K. 2006. Evolution of the snow area index of the subarctic snow pack in Central Alaska over a whole season. Consequences for the air to snow transfer of pollutants. *Environmental Science & Technology* **40**, 7521–7527.

Wania, F. 1997. Modelling the behaviour of non-polar organic chemicals in an ageing snow pack. *Chemosphere* **35**, 2345–2363.

Wania, F., Hoff, J.T., Jia, C.Q., Mackay, D. 1998. The effects of snow and ice on the environmental behaviour of hydrophobic organic chemicals. *Environmental Pollution* **102**, 25–41.

Wania F., Semkin, R., Hoff, J.T., Mackay, D. 1999. Modelling the fate of non-polar organic chemicals during the melting of an Arctic snow pack. *Hydrological Processes* **13**, 2245–2256.

Weber, R.O. 1999. Remarks on the definition and estimation of friction velocity. *Boundary-Layer Meteorology* **93**, 197–209.

Woo M.-K., Marsh, P. 2005. Snow, frozen soils and permafrost hydrology in Canada, 1999–2002. *Hydrological Processes* **19**, 215–229.

Yang, J., Honrath, R.E., Peterson, M.C., Dibb, J.E., Sumner, A.L., Shepson, P.B., Frey, M., Jacobi, H.-W., Swanson, A., Blake, N. 2002. Impacts of snow pack emissions on deduced levels of OH and peroxy radicals at Summit, Greenland. *Atmospheric Environment* **36**, 2523–2534.

19 Chemical Dynamics in Urban Areas

Miriam L. Diamond and Louis J. Thibodeaux

CONTENTS

19.1 Introduction to Urban Surfaces..531
    19.1.1 Urban Areas 532
    19.1.2 Overview of Urban Chemical Dynamics..........................533
    19.1.3 Urban Surfaces and Surface Films 534
19.2 Chemical Transport to and from Urban Surfaces..........................536
    19.2.1 Film Growth 537
    19.2.2 Chemical Partitioning into the Film 539
    19.2.3 Film Wash-Off 542
    19.2.4 Chemical Reactions within Surface Films 544
19.3 The Dynamic Model of Chemical Transport to and
    from Urban Surfaces..544
    19.3.1 The Kinetics of Transport to the Surface Film....................544
    19.3.2 Linear Film Growth Rate Model 547
    19.3.3 The Film Wash-Off Model 548
19.4 Parameter Estimation Procedures 549
    19.4.1 Film Growth Times and Precipitation
          Wash-Off Time Periods.......................................549
    19.4.2 The Air-Side MTC for Urban Surfaces 550
19.5 A Guide for Users..550
19.6 Case Studies and Example Calculations 552
    19.6.1 Multimedia Model (MUM): Toronto, Canada....................552
    19.6.2 Chemical Loading Levels on an Urban Surface:
          Toronto, Canada...556
    19.6.3 Film Chemical Mass and Runoff Concentration:
          Toronto, Canada...557
Literature Cited 559

19.1 INTRODUCTION TO URBAN SURFACES

Critical interactions occur at interfaces with air–water, soil, vegetation, and biota. We add to this list the interface between air and urban surfaces. Its importance comes from the role of impervious surfaces in mediating chemical fate in the urban environment and in facilitating exposure of urban air pollutants to humans and biota. In contrast to exchanges between air and natural media, which (we hope) are dominated by natural biogeochemical cycling, exchanges in urban areas reflect our anthropogenic activities including the myriad chemicals which we import and/or discharge (Diamond and Hodge, 2007; Hodge and Diamond, 2009).

Cities are chemical "hotspots" because of their tremendous drawing power of human inhabitants. With nearly two-thirds of the world's population of 6.7 billion, cities require enormous resources. Cities continue to expand due to population growth and increasing urbanization, which come at the expense of rural populations and agricultural and natural land. The unique physical feature of our cities is human-built impervious surfaces that form roadways, sidewalks, building walls, and roofs and other built infrastructure. These surfaces interact directly with the adjoining air and water compartments (and by connection sediment), and indirectly alter air and water interactions with soil. The first goal of this chapter is to focus on the chemical transport processes between the atmosphere and urban surfaces where these processes are critical to the understanding and modeling urban chemical dynamics. The second goal is to provide the user with guidance and tools for applying the chemical dynamics of urban surfaces. To this end, multimedia models, procedures, and algorithms will be presented to account for chemical fate in an urban context and for estimating chemical uptake and release rates from surfaces.

19.1.1 URBAN AREAS

Urban areas are global focal points of people, resources, and activities (Diamond and Hodge, 2007; Hodge and Diamond, 2009). To support large populations of over 10 million in each of the world's 25 mega-cities and one million or more in each of the 411 "small" cities (United Nations, 1995), we have built impressive infrastruc-tures including those for housing, work space, transportation, communication, food distribution, water and waste treatment and distribution, solid waste collection and dis-posal, hospitals and schools, and cultural institutions. The draw on resources to cities in order to support urban populations, infrastructure, and activities is enormous. Brun-ner and Rechberger (2001) have commented that "modern cities are material hotspots, containing more hazardous materials than most hazardous landfills." Whereas most of the materials drawn into the urban technosphere are embedded in the built land-scape, releases of far less than 1% of hazardous chemicals from this vast reservoir are sufficient to be problematic (e.g., Diamond et al., 2009; Hodge and Diamond, 2009).

There is a long list of hazardous chemicals used in, and/or emanating from cities (Diamond and Hodge, 2007; Hodge and Diamond, 2009). Their release comes from the various stages involved with chemical aggregation and use, as well as intentional and unintentional releases by industry and people. Although emissions on an individ-ual basis seem trivial, in aggregation they are considerable and urban surfaces play a key role in promoting their mobility.

The chemicals to which we are referring include trace organics such as persistent organic pollutants (POPs), trace metals, and "pseudopersistent" compounds that are ubiquitous in the urban environment because of their volume of usage. Examples of the latter include bisphenol A used to make polycarbonate plastics and various phthalate compounds that are used as plasticizers and in personal care products (e.g., diethyl phthalate [DEP], diethyl hexyl phthalate [DEHP]) (e.g., Wilson et al., 2003; Focazio et al., 2008; Rudel and Perovich, 2009). We can also list chemicals discharged in cities in terms of their functions, such as the flame retardants that are used in products and building materials (e.g., polybrominated diphenyl ethers [PBDEs], hexabromo-cyclododecane [HBCD]), solvents used in commercial (e.g., use of perchlorethylene or PCE in dry cleaning) and industrial applications (e.g., polychlorinated paraffins), surfactants used in detergents and industrial applications (e.g., alkyl sulfonates, alkyl phenol ethoxylates), and inert "fillers" used in commercial products such as the polysiloxanes (Hodge and Diamond, 2009). We consume and emit myriad pharmaceuticals, illicit drugs and biocides contained in products (e.g., ibuprofen, cocaine, caffeine, triclosan), all of which are found in our waste water discharges (e.g., Glassmeyer et al., 2008; Van Nuijs et al., 2009). Because of our enormous reliance on fossil fuels, we emit a cocktail of combustion products from the tail pipes of our cars in addition to lubricants and oils emitted from crank cases and other noncombustion sources. Polycyclic aromatic hydrocarbons (PAH) are perhaps the most infamous of the hazardous trace organics emitted from the transportation sector. The list of chemicals goes on and on.

19.1.2 Overview of Urban Chemical Dynamics

The physical urban environment is overlain on this image of cities as vast concentrators and hence geographic point sources of hazardous and nonhazardous chemicals. We replace natural ecosystems with their complex architecture and multiple biogeochemical pathways with relatively simple physical structures and flows (Figure 19.1) (Diamond et al., 2001; Kaye et al., 2006; Pouyat et al., 2007). Cities are complex in terms of human activities, but have relatively few and simple biogeochemical pathways. Cities are (understandably) designed to rapidly remove water and wastes. Sewer systems connect to impervious surfaces accomplish this rapid transfer, a consequence of which is a highly disturbed hydrologic regime. For example, Schueler (1994) estimated that urban runoff was 16 times greater from a watershed dominated by impervious surfaces in comparison to a forested watershed of the same area. Because of the multiple chemical emissions to the city and the conveyance efficiency of impervious surfaces, it is not surprising that virtually all surface waters nearby cities are polluted (e.g., Van Metre and Mahler, 2005; Kimborough and Dickhut, 2006; Zarnadze and Rodenburg, 2008). Soils not covered by impervious surfaces become polluted as well due to atmospheric deposition of urban-emitted chemicals (Covaci et al., 2001; Motely-Massei et al., 2004; Jamshidi et al., 2007; Wong et al., 2009). Soil contamination arises because of the high sorptive capacity of soils (urban or nonurban), aided by higher soil organic carbon in urban than forested soils (Pouyat et al., 2009).

Whereas cities are efficient geographic "concentrators" of chemicals, they are not efficient at retaining these emissions. As noted above, the key reason that urban environments are efficient at exporting rather than retaining chemical emissions is the abundance of impervious surfaces that act as transient, but not permanent chemical

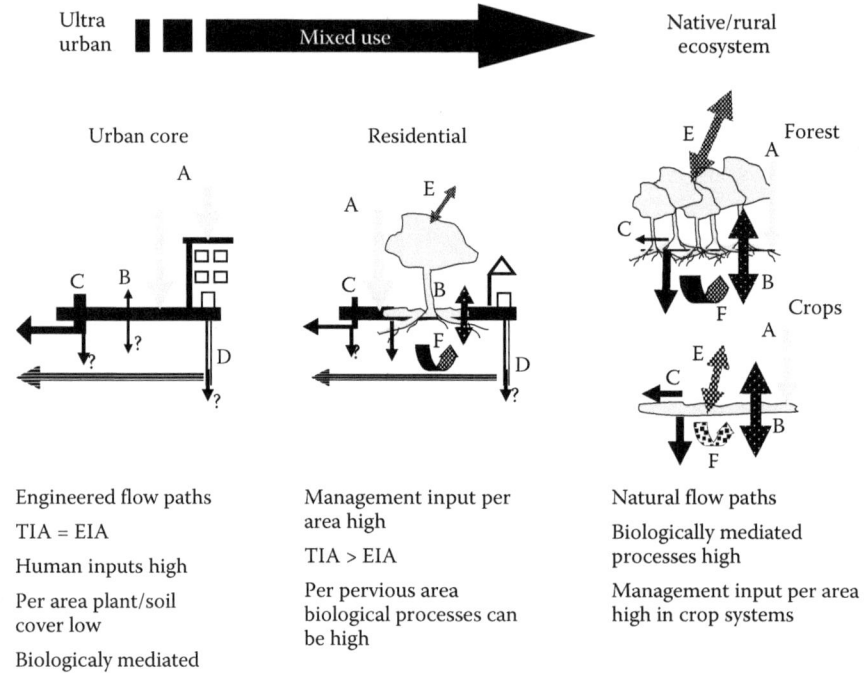

FIGURE 19.1 **(See color insert following page 300.)** Conceptual model of ecosystems along an urban to native or rural continuum. Material inputs and flows (A) to an urban system are disconnected from natural processing by impervious surfaces and other features of the built environment, resulting in low connectivity (B) between the processes taking place in the built and natural systems. In residential areas, connectivity between the built and natural system processes can be relatively high resulting in rapid cycling rates (F) for processing and storing often high material and natural inputs and flows (A and E). (Reprinted from Pouyat, R.V., et al. 2007. In: *Terrestrial Ecosystems in a Changing World.* Canadell, J.G., D.E. Pataki, and L.F. Pitelka, eds. Berlin: Springer, pp. 45–58. With permission.)

sinks of emitted chemicals. In addition, impervious surfaces themselves can be sources of chemicals (e.g., metals and PAHs in asphalt, pesticides in roofing materials, Bucheli et al., 1998a,b; Sulaiman et al., 2009; Van Metre et al., 2009). Overall, these surfaces dramatically alter chemical dynamics relative to undisturbed systems.

19.1.3 URBAN SURFACES AND SURFACE FILMS

Impervious surfaces are our cities' roadways, building walls and roofs, walkways, and parking lots. Impervious surfaces comprise from 5 (low-density suburbs) to 98% (dense downtown areas) of urban land area (Boyd et al., 1993 from Diamond et al., 2001). Esch et al. (2009) estimated that from 40% to 45% of three German states were covered with impervious surfaces. Schueler (1994) reported that nearly 110,000 km^2 of mainland United States is covered in impervious surfaces with 1000 km^2 added annually. An example of the rapid pace of the expansion of impervious surface coverage comes from the Seattle area in Washington State, where Powell et al. (2008) quantified a 225% increase in impervious surface coverage between 1972 and 2006. Over this period, the population grew by 79%. On an aerial basis, Schueler (1994) approximated that two-thirds of impervious surfaces is pavement and the remaining third is building roofs. To this we need to add building walls, which can significantly increase the total impervious depending on building height.

As a result of gas-phase condensation and particle deposition, thin chemical films develop on impervious surfaces (Law and Diamond, 1998; Diamond et al., 2000; Weschler, 2003). The accumulation of the films varies from negligible on highways where high-speed traffic prevents accumulation, to maximal for horizontal surfaces that are shielded from precipitation and activity. The soiling of buildings and monuments is the visual evidence of surface films (e.g., Grossi and Brimblecombe, 2002).

The chemical profiles of semivolatile organic compounds (SOCs) in films and atmospheric particulate matter (PM) can be indistinguishable, which is not surprising since both have the same atmospheric source and the films accumulate by gravitational settling or impaction of PM on surfaces (Gingrich et al., 2001). Film mass is comprised mainly of inorganic, soil-related compounds such as mineral silicates and carbonates, but the films are enriched relative to soils with sulfates, nitrates, and trace metals (Lam et al., 2005). The organic phase, which comprises only 4–10% of ground-level films, consists of aliphatics, carbohydrates, aromatics and to a lesser extent, carbonyl compounds. Most abundant among the organic compounds that have been identified are biogenic monocarboxylic acids (fatty acids from plant and animal sources) and alkanes. Trace organic pollutants such as polycyclic aromatic hydrocarbons (PAHs), polychlorinated biphenyls (PCBs), and polychlorinated dibenzodioxins and furans (PCDD/Fs) are found at well less than 0.01% of the total film mass (Lam et al., 2005). It is important to note that of the organic compounds analysed, many are amphiphiles and as such, can act as surfactants. The organic fraction of the film allows partitioning of nonpolar organic compounds, as well as other amphiphilic compounds that are common in the oxidizing air of urban pollution. The film is hygroscopic which creates an organic, gel-like layer with its solid particles. Overall, the film serves as an efficient absorber-concentrator for air-borne substances in particle- as well as vapor-phase forms.

The other significant aspect of the films is that they change the functionality of impervious surfaces vis-à-vis urban chemical dynamics. An atmospheric chemical, whether it is particulate or gas phase, is thought to interact chemically with the surface film constituents on an impervious surface, rather than the underlying impervious surface itself. This holds for most surfaces with weak electronegative charges that cannot extend this charge beyond the film. For example, the surface hydroxyl groups of glass, and hydroxyl and carbonate groups of brick and stone have weak electronegative charges. Concrete, which is at least two-thirds calcium silicates with lesser amounts of calcium sulfate, also has a relatively weak surface charge. Asphalt pavement is more complex: its chemical influence likely extends beyond an accumulated film. Asphalt pavement is stone aggregate coated with bitumen. The "stickiness" of the bitumen holds the aggregate together to create a hard, durable surface. The bitumen is comprised of asphaltenes, which are very high molecular homo- and heteropolycyclic

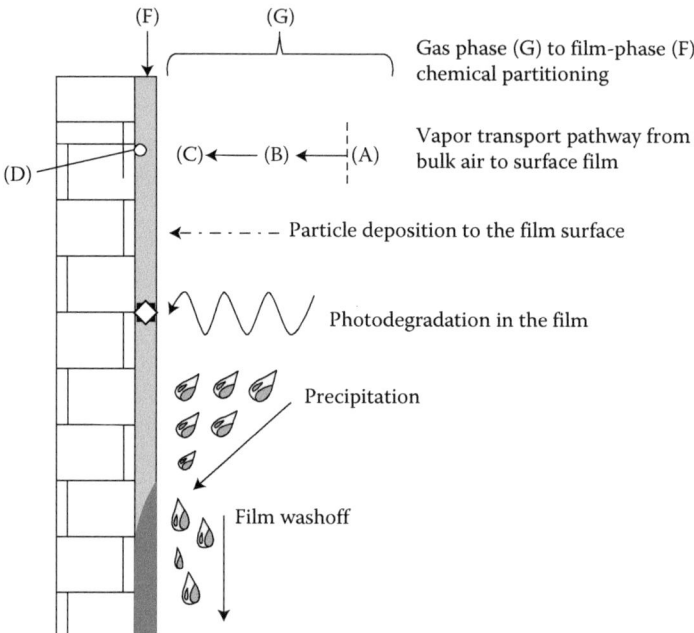

FIGURE 19.2 Conceptual model of film dynamics and chemodynamics.

aromatic hydrocarbons with alkyl chains. The complex bitumous mixture affords chemical reactivity and a high sorptive capacity able to attract an array of compounds that come in contact with its surface.

19.2 CHEMICAL TRANSPORT TO AND FROM URBAN SURFACES

As a compartment, films on impervious surfaces are highly dynamic, which is central to their impact on SOC fate in urban areas. Figure 19.2 illustrates the processes involved in film dynamics and chemodynamics. Depending on the location, the nature of the impervious surface, season, films grow and are rapidly lost through wash-off (Figure 19.3).

SOC dynamics are superimposed on the background of surface film dynamics. SOCs are subject to constant accumulation as the films grow; here the film is functioning as a transient sink. Loss to surface waters occurs as the films are removed; the film is now a source by acting as the intermediary between the atmosphere and surface waters. Particulate SOCs can accumulate and be lost as a function of particle dynamics and precipitation events. Diffusive exchange occurs between the film and gas-phase air constituents as a function of the fugacity gradient between these phases. The very high surface area to volume ratio of $\sim 1.4 \times 10^7$ promotes rapid rates of these transport processes (Diamond et al., 2001). In addition, SOCs may undergo transformation processes in the film, again promoted by the large surface area with its simple architecture in comparison to a vegetation canopy (Lam et al.,

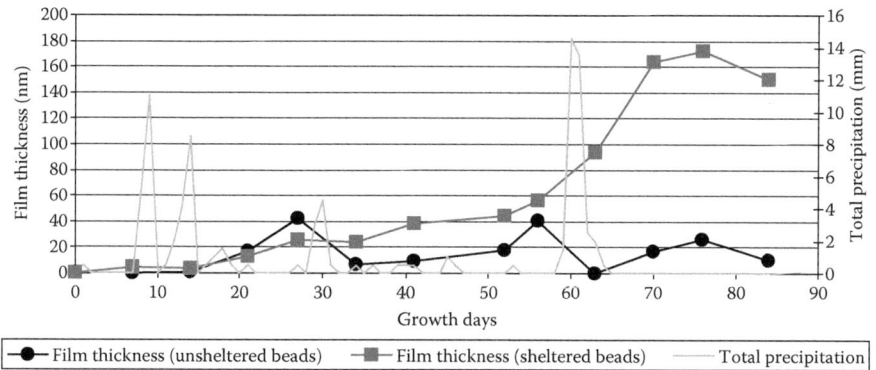

FIGURE 19.3 Changes in film thickness over time as the film grows during dry periods and is removed by precipitation. (Reprinted from Wu, R. W., Harner, T., and Diamond, M. L. 2008a. *Atmos. Environ.* 42(24): 6131–6143. With permission.)

2005). This character of the film as a large surface collector has key consequences for chemical fate and ecosystem exposure in urban areas. Simply put, these films facilitate rapid chemical-film dynamics and promote SOC mobility in urban areas with high impervious surface coverage.

Practically, these rapid dynamics demand a transport model that accounts for (1) film growth and chemical removal, (2) film washoff by precipitation, and (3) chemical transformation. In addition, we must understand the partitioning behavior of a chemical between air and the film. A unique aspect of the transient nature of films is that the dynamic model is an integral part of the MTC algorithms. Film growth time, $\tau_f(s)$, and the duration of the precipitation or washoff event, $\tau_w(s)$, are parameters in addition to the usual MTCs.

There are characteristics of this new environmental medium, which set it apart from the others and require an alternative approach to its mass-transport process. This section contains a vignette chemodynamic model from which estimates of the chemical concentration in the surface film and the chemical mass removed during rain events that is delivered to the urban water bodies can be made.

19.2.1 FILM GROWTH

Films on impervious surfaces grow as a result of the deposition of air-borne particles, condensation of gas phase compounds, and the accumulation of moisture (Law and Diamond, 1998; Liu et al., 2003b; Lam et al., 2005). Downtown, they accumulate at an unexpectedly linear rate of $\sim$1.6–3.1 nm d^{-1} (measured gravimetrically and not including moisture), as seen in Figure 19.4 (Wu et al., 2008a). The implication of this constant growth is that you do need to clean your windows! These growth rates were derived from replicate film accumulation experiments using horizontally arranged glass beads from which the bulk material was removed up to 287 days. In comparison, film growth rates in rural locations were $\sim$10 times less or 0.1 nm d^{-1}

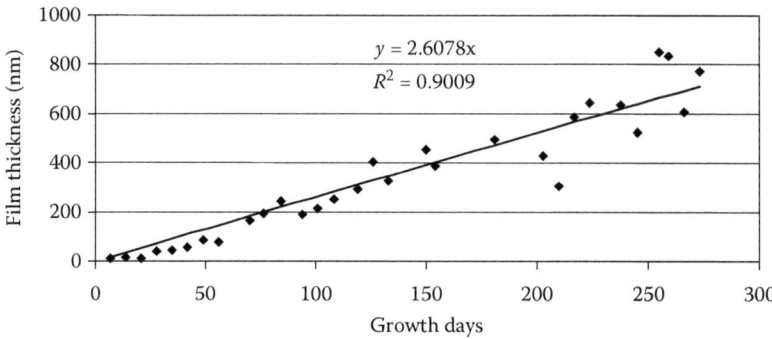

FIGURE 19.4 Surface film growth on covered glass beads deployed in downtown Toronto, Canada, as a function of time. (Reprinted from Wu, R. W., Harner, T., and Diamond, M. L. 2008a. *Atmos. Environ.* 42(24): 6131–6143. With permission.)

and were also linear. The downtown growth rate is comparable to that expected strictly on the basis of particle deposition, which would yield a rate of 2.7 nm d^{-1}. Growth rates are higher in summer than winter, which is presumably related to the abundance of biogenic emissions and humidity (e.g., Liu et al., 2003a).

Film growth also depends on aspect (north–south–east–west), angle (vertical–horizontal), and the underlying material. Eckley et al. (2008a), investigating mercury accumulation in surface films, found the greatest accumulation on glass windows presented toward the prevailing wind direction and those that were near horizontal in comparison to vertical windows facing other directions. The chemistry of the impervious surface also influences film growth. Labencki et al. (in preparation) found the greatest film accumulation rates on PTFE plastic or Teflon, followed by glass and PVC (Figure 19.5). Supported by other observations in the literature, the authors suggested that particle attraction and hence film growth rate was positively related to the electrostatic charge of the surface, which is greatest for PTFE plastic, followed by PVC. Glass, brick, and concrete also have a negative surface charge as a result of terminal hydroxyl and oxygen species. However, as discussed in Section 19.1.3, the attractive forces of glass, brick, and concrete are much less than that from the terminal fluorines in PTFE.

The presence of the film itself is also thought to enhance film growth by providing a "sticky" surface that minimizes particle rebound and maximizes particle capture as part of the dry deposition process. The reasoning behind this supposition is that greased surfaces have a higher dry deposition velocity and flux than clean surfaces (Turner and Hering, 1987; Wu et al., 1992). The "stickiness" is afforded by the polar terminal groups (e.g., O$^-$, –OH, NO$_4^-$) at the film's surface, which is consistent with the abundance of polar rather than neutral compounds in film constituents (Liu et al., 2003a; Lam et al., 2005). For example, Liu et al. (2003b) found that a dirty window (e.g., with a surface film) accumulated 1.4–94 times more trace metals than an immediately adjacent clean window.

In contrast to windows, the accumulation of surface pollutant loads on roadways has been found to reach an asymptote after initial rapid accumulation following a storm

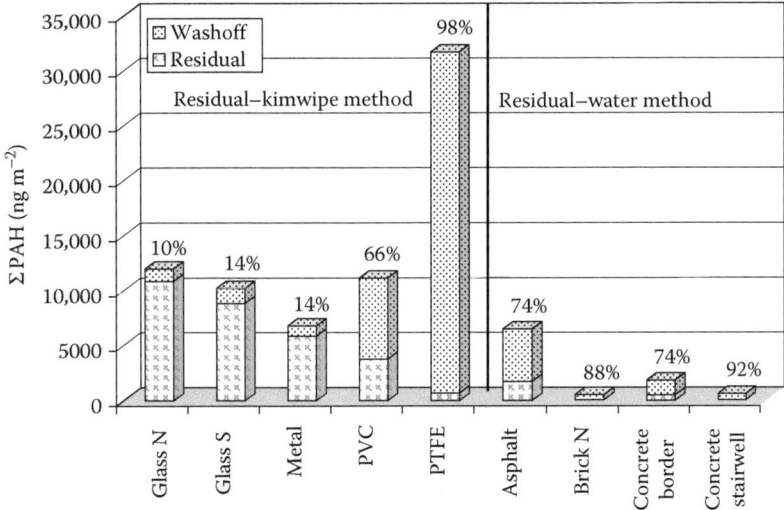

FIGURE 19.5 Differential accumulation and washoff of Σ14 PAH in surface film for various types of surfaces. The washoff fraction was removed using simulated water sprayed at $0.16\,L\,min^{-1}$ after 3 months accumulation during which there was intermittent rain. The residual fraction was removed by either wiping the surface (Kimwipe method) or by spraying additional water (water method). (Adapted from Labencki, T., et al. Variability in and mechanisms of PAH washoff from urban impervious surfaces. In preparation.)

event (Vaze and Chiew, 2002; Kim et al., 2006). The difference between window and roadway accumulation could be due to the effect of traffic causing film abrasion.

19.2.2 CHEMICAL PARTITIONING INTO THE FILM

As surface films grow by accumulating material from the atmosphere, the film contains a SOC burden as well (Wu et al., 2008a). The process governing chemical partitioning involves three separate phenomena: (1) the transport of SOC in both vapor and particle-bound fractions from the atmosphere to the film surface, (2) the equilibrium solvation of the SOC into the organic matrix of the film, and (3) the growth of the film due to the deposition of other vapor-phase substances and particles from the atmosphere. First we discuss the empirical results and then relate the results to the equilibration solvation, the relevance of which is the choice of the appropriate physical–chemical predictor and equation to express chemical partitioning. The discussion of the transport and coupled film growth is found in Section 19.3.2.

Wu et al. (2008a,b) and Butt (2003) examined the uptake of PCBs and PAH, respectively, into ambient window films. The results of Wu et al. (Figure 19.6) show an initial pattern where PCB concentrations appear to be high relative to film mass. There after a 1:1 relationship between chemical and film accumulation occurs. The initial nonlinearity is likely due to analytical error because of very low analyte mass,

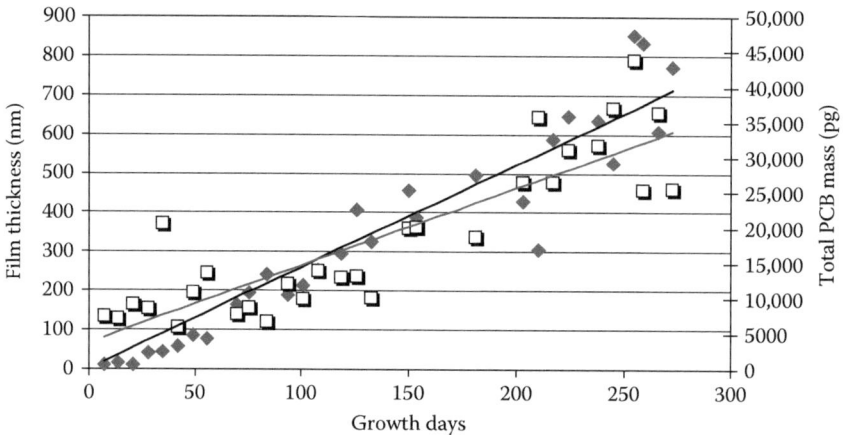

FIGURE 19.6 Linear accumulation of PCBs (square symbols) at a rate comparable to film accumulation (diamond symbols) on glass beads deployed in downtown Toronto, Canada, over a 300 day period (Wu, R., unpublished data).

changes in the chemical composition of the film, and/or rapid changes in air chemistry that are not captured in time-integrated measurements.

The presumed process leading to the linear accumulation of chemical mass as a function of film growth over time is a consequence of all the material, both vapor-phase and particles, arriving and sticking to the film and remaining. The theory suggests that the time to equilibrium is a function of film growth rate, transport parameters, and the SOC equilibrium partition coefficient (see Equation 19.10). Consequently, high molecular weight compounds may not reach equilibrium, as discussed in Section 19.3.2. These assumptions are consistent with that regarding equilibrium between gas- and particle-phases in air (e.g., Pankow and Bidleman, 1992). However, empirical result show that the nature of the partitioning can vary event-to-event, at urban versus rural sites and in different climatic zones (Cotham and Bidleman, 1995; Tasdemir et al., 2004; Tasdemir and Esen, 2007; He and Balasubramanian, 2009).

Next we discuss the equilibrium solvation of SOCs in the film and its consequences for developing expressions to quantify partitioning. If we assume that equilibrium partitioning of gas-phase SOCs is with the organic phase of the film, then we can estimate the film–air partition coefficient, K_{FA} as $f_{OC}K_{OA}$ where f_{OC} is the octanol-equivalent content of the surface film. This provides a convenient equivalence with expressions for soil–air and leaf–air partitioning. However, the slopes of regression of $\log K_{FA-E}$ versus $\log K_{OA}$ are consistently less than unity (Figure 19.7). Consequently, we require a proportionality constant to translate between 1-octanol and the matrix since 1-octanol is not a perfect surrogate for the organic matrix of the film. Donaldson et al. (2005) found experimentally that 1-octanol is a more effective solvent for the nonreactive uptake of PAH than the longer chain organic compounds oleic acid (C-18) and squalene ($c = 30$). These authors inferred that the solvation process of PAH in 1-octanol was closer to ambient films than the long chain organics. Wu et al. (2008b)

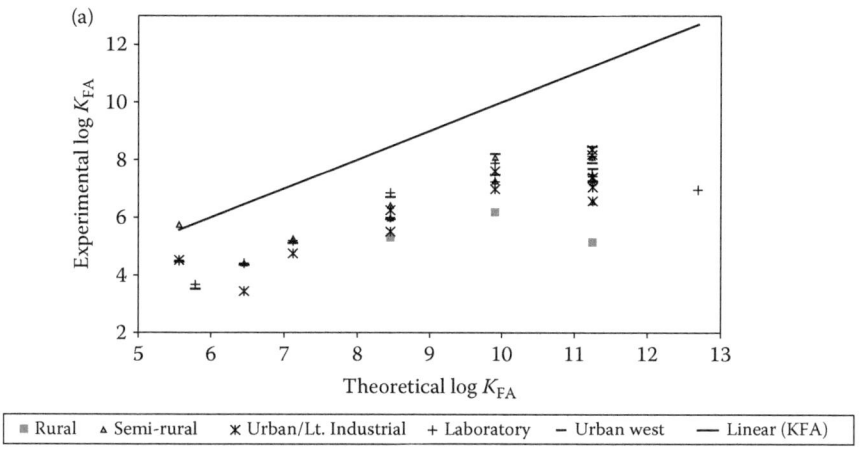

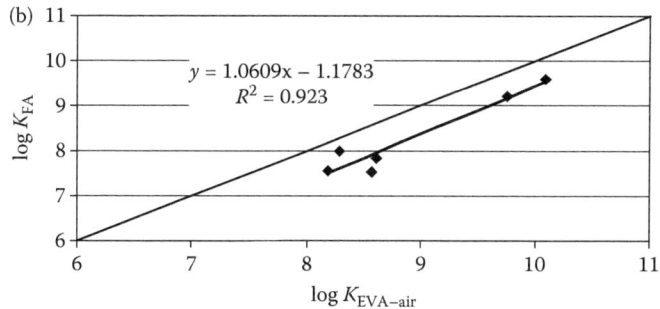

FIGURE 19.7 Log–log relationships between empirically derived air–film partition coefficients, K_{FA-E}, and surrogate partition coefficients. (a) Values of K_{FA-E} for PAH plotted against theoretical values of K_{FA} derived from $K_{OA}f_{OC}$ (Butt, 2003), and (b) values of K_{FA-E} for PCBs plotted against $K_{EVA-air}$ (Wu et al., 2008b). All surrogate partition cofficients were temperature corrected. Empirical partition coefficients were obtained from film measurements of (a) covered glass beads and (b) window films. Gas-phase air concentrations were obtained from passive sampling with PUF disks. (From Wu, R.W., et al. 2008b. *Atmos. Environ.* 42(22): 5696–5705. With permission.)

found that empirically derived values of K_{FA} were better described by ethylene vinyl acetate or EVA rather than 1-octanol; see Figure 19.7b. They surmised that the polarity of the carbonyl ester functional group of EVA was closer to the polar film constituents (e.g., carboxyl groups) rather than the hydroxyl functional group of 1-octanol. Another explanation for the improved fit between K_{FA-E} and K_{EVA} rather than K_{OA} is the similarity in physical structure of the EVA polymer: in contrast to the relatively disordered nature of bulk 1-octanol, the polar carbonyl ester groups of EVA are exposed in the well-ordered polymer. In the film, there is some evidence suggesting that the carboxyl and hydroxyl groups are oriented outwards in hydrated films (Lam et al., 2005). The major drawback to using EVA as a well-defined surrogate for the organic phase of the surface film is that EVA–air partition coefficients are available for very few compounds and hence out of necessity, we use K_{OA}. Thus, we proceed with using K_{OA} to calculate K_{FA}, noting the need for further empirical studies that will allow developing a robust empirical relationship between K_{FA} and K_{OA}.

The other aspect of chemical partitioning into film is that for chemicals with log K_{OA} of ~ 11 and greater, the relationship with a surrogate partition coefficient reaches an asymptote. This is illustrated in Figure 19.7A for PAH (Butt et al., unpublished data). We posit that the leveling off of the slope is due to a kinetic limitation between film growth and chemical accumulation, as discussed in Section 19.3.2.

19.2.3 FILM WASH-OFF

Chemical loads from precipitation events are related to (1) the loads contained in the precipitation itself, (2) the chemical loads on impervious surfaces, and (3) the efficiency with which these loads are removed to become runoff. The latter is influenced by film removal processes and the generation of runoff from rain events. The first component, loadings from precipitation, is attributable to precipitation in urban locations "scrubbing" the usually polluted air mass. After precipitation, film regrowth commences immediately with growth time periods controlled by antecedent dry days that separate the rain events. The washed-off film mass becomes a component of urban runoff; it enters the aquatic conveyances that deliver the water to the receiving bodies, such as rivers, lakes, reservoirs, and temporary impoundments. In this chapter, we turn our main attention to the removal of chemical loads from surface films.

Since surface films are efficiently removed by precipitation, their net mass accumulation is a direct function of antecedent dry days (Vaze and Chiew, 2002, 2003). This was clearly demonstrated in experiments with adjacent sheltered and unsheltered windows or glass surfaces and experiments conducted on pavement (Eckley et al., 2008; Eckley and Branfireun, 2008; Wu et al., 2008a).

The overall efficiency with which the films and their constituents are mobilized into runoff is attributable to the efficiency of film removal from surfaces and the efficiency with which the mobilized films contribute to urban runoff. Several studies have shown that surface films and their chemical constituents are efficiently, but not entirely removed by precipitation (Vaze and Chiew, 2002; Eckley and Branfireun, 2008, 2009; Labencki et al., in preparation). This is fortunate since linear growth rates with no apparent asymptote would lead to insufferably dirty windows and walls throughout our cities. The "flip side" of efficient film removal is the degradation of water quality in urban receiving waters! (Diamond and Hodge, 2007; Hodge and Diamond, 2009).

The removal efficiency of the films from the surface is a function of the energy provided by precipitation and the chemistry of the impervious surface. Labencki et al. (in preparation) found that 98% of the PAHs in films were washed off of PTFE or Teflon in a controlled spray experiment versus 66–92% for PVC, asphalt, brick, and concrete surfaces (Figure 19.5). Diamond et al. (2000) and Eckley et al. (2008) found that 70–80% of several groups of SOCs and mercury in films, respectively, washed off of window glass. In contrast, Labencki et al. (in preparation) measured on 10–14% removal efficiency of the PAHs from window glass. Eckley et al. (2008) found that removal efficiency is related to angle of the impervious surface, and thus, presumably the energy with which the precipitation can mobilize the film. Chemical profiles in the washoff and the surface residual are nearly identical, suggesting that chemical-specific solubilization is not occurring. These findings are consistent with the relationship between the impact energy of rain and the liberation of particles on surfaces (Vaze and Chiew, 2002, 2003). In addition, it is possible that physical removal may be promoted by the abundance in the film of highly soluble and amphilic compounds that can act as surfactants.

An important difference between window film washoff and pavement washoff is the supply of particles from the underlying substrate. Eckley and coworkers and Vaze and Chiew (2003) found that the washoff load is overwhelmed by particulate and not dissolved-phase chemicals. Whereas films removed from windows and other smooth surfaces will have a finite supply of particles mostly attributable to atmospheric deposition, washoff from pavement mobilizes a fraction of the considerable particle-dust load.

The runoff coefficient or RC has been used to quantify the percentage of precipitation that is collected as runoff. RC is defined as the ratio of runoff to rainfall for a given time period. The ratio has been parameterized at a small or "patch" scale (e.g., a parking lot) where the values vary according to surface type, to a citywide scale where values of RC are derived for different land uses. At the patch scale, values of RC vary according to rates of precipitation, evaporation, and infiltration of water through cracks and crevices, and surface storage. Losses to these factors depend on surface temperature (hence surface albedo), roughness, and the age of the impervious surface. For example, Ragab et al. (2003) estimated that low values of RC were due to pavement infiltration (6–9% of annual rainfall) and evaporation (21–24% with higher values in summer) in runoff experiments conducted in the United Kingdom. Ramier et al. (2006) found similar losses of 20–30% of precipitation to infiltration plus evaporation in an instrumented site at Nantes, France. Values of RC typically vary between 0.50 and 0.75 (Ragab et al., 2003; Ramier et al., 2006; Eckley and Branfireun, 2009), with lower values during short rain events and higher values with longer rain events as the pavement infiltration capacity becomes saturated. Impervious surfaces with greater infiltration capacity such as bricks and brick paving, and with greater surface microtopography, have lower values of RC (Mansell and Rollet, 2009). Ragab et al. (2003) reported values of RC from an experiment in Wallingford, UK, of 0.7, 0.9, and 0.5 for annual, winter, and summer periods, respectively.

At a citywide scale, RC approaches 1 where impervious surface coverage ranges from 40% to 100% of total urban land use, and diminishes in a stepwise fashion for locations with 20–40% and less than 20% impervious surface area (Goldshleger et al., 2009).

Temporally, the efficiency of film washoff and the initial contribution from scrubbing polluted urban air causes a "first flush" in water, particles, and chemical loads in urban stormwater (e.g., Saget et al., 1996). Bertrand-Krajewski et al. (1998) define "first flush" as the removal of 80% of the chemical loads in the initial 30% of runoff. In detailed experiments, Eckley et al. (2008) showed that 87% of mercury accumulated on exposed window surfaces was mobilized by the first few millimeters of rain, reinforcing the highly mobile nature of constituents in the film on glass. In pavement experiments, Eckley and Branfireun (2009) measured 70 times higher Hg concentrations in the first minutes of a simulated rain experiments than later in the hydrograph. This load was closely correlated with the particle load. The "first flush" measured in pavement runoff is then seen in urban streams, which is not surprising considering that urban drainage is engineered to promote efficient water removal (e.g., Sansalone and Buchberger, 1997; Sansalone and Cristina, 2004; Robson et al., 2008).

19.2.4   CHEMICAL REACTIONS WITHIN SURFACE FILMS

Chemical transformation and reaction in films could be a significant, but not dominant loss process, for low-volatility chemicals that are reactive (e.g., photoreactive) and significantly partition into or deposit onto the film (Donaldson et al., 2005; Kwamena et al., 2007). Recently, reactions at air–surface interfaces, particularly organic-coated surfaces, have garnered much attention (e.g., Nissenson et al., 2006, Finlayson-Pitts, 2009). SOCs in films have the potential to undergo direct and indirect photolysis, and oxidation reactions that can be catalyzed by metals in the films.

The experimental evidence suggests that oxidative reactions of PAH in films are slower than in those that coat particles, the reason for which is not known of (Poschl et al., 2001; Kwamena et al., 2004; Kahan et al., 2006). Despite potentially slower reaction rates, more loss of SOC could occur in surface films than in films coating atmospheric particles because of the $\sim$10-fold greater surface area to volume ratios of films on impervious surfaces compared with atmospheric particles (Kwamena et al., 2007). This holds for lower volatility, higher molecular weight compounds that partition into condensed phases such as surface films since higher volatility chemicals are more likely to remain in the gas phase.

Mmereki et al. (2004) have suggested that oxidation of aliphatics by reaction with atmospheric oxidants such as hydroxyl or nitrate radicals may be more efficient in surface films than in the atmospheric because of higher probabilities of collision with the stationary film phase. With the finding in surface films of polymeric structures consistent with polybutadiene, Lam et al. (2005) postulated that the presence of these structures could arise from deposition of abraded tires made of polybutadiene and/or *in situ* polymerization processes. They commented that polymerization was possible given the abundance of precursor compounds in films together with acid and metal catalysts. Clearly, additional research is necessary to assess the extent to which organic compounds in surface films react and if so, at what rate.

19.3   THE DYNAMIC MODEL OF CHEMICAL TRANSPORT TO AND FROM URBAN SURFACES

19.3.1   THE KINETICS OF TRANSPORT TO THE SURFACE FILM

The chemodynamic pathways of organic compounds moving from the bulk gas-phase air to the surface film on an impervious material are illustrated in Figure 19.2. The compound moves as vapor and particulate bound from the turbulent well mixed atmospheric air at *point A* then through the boundary layer/air-side film, *point B*. It moves close to the interface with both vapor and particles entering the film contacting the surface at *point C*. Once within the film, the chemical molecules diffuse within the porous, semisolid, gel-like material, and become "sorbed" at an internal site, *point D*. The pathway within the surface film is visualized as a pore.

The chemical flux n (ng m^{-2} s^{-1}) from air to film by both vapor and particle transport processes can be approximated by

$$n = K_A [C_A - C_F/K_{FA}] + v_P P_P C_A K_{PA}, \tag{19.1}$$

where K_A is the overall MTC (m s^{-1}), C_A (ng m^{-3}) concentration of the chemical in the air phase, C_F (ng m^{-3}) the concentration in the film, K_{FA} (m^3 kg^{-1}) is the thermo-dynamic film-to-air equilibrium partition coefficient, v_P (m s^{-1}) is the atmospheric particle deposition velocity onto the film, P_P (kg m^{-3}) is the particulate matter concentration in the air and K_{PA} (m^3 kg^{-1}) is the particle-to-air partition coefficient. The general equation for the overall MTC that combines the individual air- and film-side MTCs is

$$\frac{1}{K_A} = \frac{1}{k_A} + \frac{1}{k_F K_{FA}}. \tag{19.2}$$

The terms k_A and k_F are the individual air- and film-side MTCs, respectively (m s^{-1}).

Table 19.1 contains a list of chemicals and their properties that are likely to be found in both urban air and urban surface films. At this point we must consider the situation of a gas-phase chemical in relation to a bulk surface film containing a fraction of organic material, f_{oc}, into which the gas-phase chemical will partition. K_{FA} as discussed in Section 19.2.2, can be approximated by $f_{oc} K_{OA}$. In this case, f_{oc} becomes a proportionality constant translating between the sorptive capacity of octanol and the film's sorptive capacity. We retain the use f_{oc} as a proportionality constant, which is consistent with its use to describe organic carbon in other matrices such as soil and vegetation. As seen in Table 19.1, K_{FA} is a highly variable parameter and its numerical value will impact the value of K_A. From Equation 19.2, we see that this resistance-in-series expression combines the individual MTCs. (See Chapter 4, Section 4.4.3 for details on its development.) Thibodeaux and Diamond (in preparation) present an analysis and options for the transport of organic molecules within the film by assuming it is composed of various material compositions. Details on the composition and physical structures of surface films are still uncertain at this time. Three film types were assumed: air-filled porous material, water-filled porous material, and an organic matter, gel-like material. Calculations were performed using Equation 19.2 and in all three cases, they indicate that the air-side resistances are greater than those for the film-side except for the volatile chemicals such as benzene.

Table 19.2 summarizes air-side MTCs obtained from various literature sources. The MTCs increase as a function of wind speed: windy-daytime represented by 5–10 mile h^{-1} (8–16 km h^{-1}) winds and calm nighttime represented by 0.5–3.0 mile h^{-1} (0.8–5 km h^{-1}) winds with average MTCs of 500 and 100 cm h^{-1}, respectively. This is equivalent to air-side resistances (i.e., $1/k_A$) of 0.002–0.01 h cm^{-1}, respectively. Therefore, as with other interfacial exchanges, the air-side mass transfer controls the kinetics of chemical sorption to the surface film for lower vapor pressure SOCs whereas the reverse is true for higher vapor pressure volatile organic compounds

TABLE 19.1
Physical and Chemical Properties of Some Common Urban Chemicals

Chemical	Molecular Weight (g mol^{-1})	log K_{OC} (K_{OC} in L kg^{-1})	Henry's Constant (H, dimensionless)	Vapor Pressure (atm, at 25°C)	log K_{FA} (K_{FA} in L kg^{-1})
Benzene (BZ)	78	1.9	0.224	0.125	1.6
Naphthalene (NA)	128	3.1	0.017	1.10×10^{-4}	3.9
Anthracene (AN)	178	4.3	0.014	1.40×10^{-7}	5.2
Pyrene (PY)	202	4.8	8.81×10^{-3}	1.60×10^{-7}	5.9
Benzo(a)pyrene (BAP)	252	6.0	1.80×10^{-5}	7.00×10^{-12}	9.7
2,3,7,8-TCCD	322	6.7	1.23×10^{-6}	1.80×10^{-12}	11.6

Note: $K_{FA} = f_{OC} K_{OC}/H$ was assumed with $f_{OC} = 0.1$ (i.e., 10%).

TABLE 19.2
Values of the Air-Side Mass Transfer Coefficients for SOC for Wind Speeds <20 mph[a]

| | | Air-Side Mass Transfer Coefficient (m h^{-1}) | | | | | |
| | | Wind Speed (miles h^{-1}) | | | | | |
Chemical	Diffusivity in Air x 10^4 (m^2 s^{-1})	20	10	5	3	1	0.5
Benzene (BZ)	0.088	1.500	0.870	0.520	0.310	0.140	0.083
Naphthalene (NA)	0.0513	1.030	0.605	0.360	0.218	0.100	0.058
Anthracene (AN)	0.0421	0.900	0.530	0.320	0.190	0.088	0.051
Pyrene (PY)	0.0395	0.870	0.510	0.300	0.180	0.084	0.049
Benzo(a)pyrene (BAP)	NA	0.800	0.470	0.280	.0170	0.077	0.045
2,3,7,8-TCCD	0.0524	1.040	0.610	0.370	0.220	0.100	0.059

[a] All values for 25°C and 1 atm.

(VOCs). In other words, for SOCs the $1/k_A$ term in Equation 19.2 determines the magnitude of K_A and can be used to approximate the flux in Equation 19.1.

19.3.2 LINEAR FILM GROWTH RATE MODEL

The following theoretical model describes the qualitative and quantitative aspects of gas-phase SOCs in air transported to, and accumulating in a growing surface film (Thibodeaux and Diamond, in preparation). The film depicted in Figure 19.2 is assumed to grow at the constant rate of a (m s^{-1}) by the relationship

$$h = h_0 + at, \tag{19.3}$$

where h_0 (m) is the initial thickness and t is the time (s). It is further assumed that the concentration of chemical within the film is uniform at C_F (ng m^{-3}) and it is initially equal to C_{F0}, no degradation or chemical reactions other than sorption are occurring within the film, and the background concentration in air, C_A (ng m^{-3}), is assumed to be constant.

A transient chemical species Lavoisier mass balance is performed, which incorporates Equations 19.2 and 19.3. A simple, linear, ordinary differential equation results:

$$K_A [C_A - C_F/K_{FA}] + v_P \rho_p C_A K_{PA} = (h_0 + at) \frac{dC_F}{dt} + aC_F. \tag{19.4}$$

Upon integration over the growth time t (s) the solution is arranged to yield the chemical concentration in the growing film:

$$C_F = C_A K_{FA-E} + (C_{F0} - C_A K_{FA-E}) \left(\frac{h_0}{h_0 + at} \right)^{1 + \frac{K_A}{aK_{FA}}}, \tag{19.5}$$

where the constant K_{FA-E} is defined as

$$K_{FA-E} = K_{FA} \left(\frac{\frac{K_A}{a} + v_P P_P \frac{K_{PA}}{a}}{\frac{K_A}{a} + K_{FA}} \right) \qquad (19.6)$$

Equation 19.5 shows film concentration increasing with increasing growth time, t (s^{-1}). For a large growth time, the right side of Equation 19.5 approaches the asymptotic relationship:

$$C_F^* = C_A K_{FA-E}, \qquad (19.7)$$

where C_F^* is defined as the combined equilibrium and steady-state (ESS) concentration in the film and K_{FA-E} is the effective film-to-air partition coefficient (m^3 kg^{-1}).

This is an unusual theoretical result for a chemical partition coefficient. Equation 19.6 contains both kinetic parameters and traditional equilibrium partition parameters for characterizing the expected or field-observed film-to-air partition coefficient. The equation suggests that for higher volatility compounds (e.g., small values of K_{FA} such that $K_{FA} \ll K_A/a$) the experimental or measured and theoretical film-to-air coefficients equals $K_{FA-E} = K_{FA}(1 + v_P P_P C_A K_{PA}/K_A)$. For low volatility compounds (e.g., large values of K_{FA} such that $K_{FA} \gg K_A/a$) Equation 19.5 suggests that $K_{FA-E} = K_A/a + v_P P_P C_A K_{PA}/a$, which is also a constant. The latter contains the ratio of the air-side MTC to the film growth rate plus the particle deposition contribution to film growth. Such unusual partitioning characteristics have been observed for selected PAHs on urban glass windows (Thibodeaux and Diamond, in preparation). The data presented on measured K_{FA-E} values do not increase linear with increasing K_{FA} but approach asymptotic values of K_{FA-E} that are significantly lower than the corresponding K_{FA} values; see Figure 19.7a as an example of this behavior. In other words, thermodynamic equilibrium is not achieved for the PAHs with very large values of K_{FA}. The partitioning to surface films becomes mass transfer controlled. The lower values result when $K_{FA} \gg K_A/a$.

The film growth Equations 19.5 through 19.7 can be used to estimate the chemical loadings on urban surfaces for expected chemical concentrations in air. See Example Problem 19.6.2 for details. Such calculations yield surface concentrations which allow estimates of stormwater runoff. It also allows chemical concentration estimates for dust particles generated from cleanings, decommissioning, or otherwise handling the solid exteriors of urban structures.

19.3.3 The Film Wash-Off Model

In Section 19.2.3, data were presented that shows the surface films can be partially removed quickly by washing with water, leaving in place residual film. Removal efficiencies, E_w, were shown to vary from 0.10 to 0.98 depending on the surface composition/texture. This section provides estimation methods for the mass quantities M_a (kg) and the storm-water concentrations, C_W (mg m^{-3}) due to precipitation wash-off of surface films.

We define the average precipitation rate as v_w (m h^{-1} or m^3 m^{-2} h^{-1}) and its duration is τ_{ppt} (h). For an urban area of footprint A (km^2), the volume of water V_w

(m^3) involved is equal to $v_w \tau_{ppt} A$. Based on the concentration in air, C_A (μg m^{-3}), the maximum possible mass of chemical on the impervious surfaces is:

$$M_A = C_A K_{FA-E} A \cdot ISI(h_0 + a\tau)\rho_F E_W, \tag{19.8}$$

where ISI, the impervious surface index, is the ratio of area occupied by film to the footprint of the urban area, τ (h) is the time-period of film growth, ρ_F (kg L^{-1}) is film density, and E_W is the efficiency of film wash-off by precipitation. The average concentration of chemical in the run-off water departing the urban area is

$$C_W = \frac{M_A}{v_w \tau_{ppt} A}, \tag{19.9}$$

See Example Problem 19.6.2.

These final results, Equations 19.8 and 19.9 are the maximum values of M_A and C_W, provided the film concentration has reached its ESS partition coefficient, K_{FA-E} value. The time period for the growing surface film to achieve 95% of its final steady-state chemical loading, τ_{95}(h), can be estimated by

$$\tau_{95} = \frac{h_0}{a}\left[\exp\left(\frac{3K_{FA}}{K_{FA} + K_A/a}\right) - 1\right]. \tag{19.10}$$

This result is obtained from Equation 19.5 for $C_{FO} = 0$. If the time-period for film growth, τ, is less than τ_{95}, Equation 19.5 with $t = \tau$ should be used to estimate C_F rather than the product $C_A K_{FA-E}$ in Equation 19.7. Also, C_F rather than the $C_A K_{FA-E}$ product must be used in Equation 19.8.

Example Problem 19.6.2 illustrates the use of the above wash-off model for estimating urban run-off concentrations due to atmospherically derived chemicals accumulating in films that grow on urban impervious surfaces.

19.4   PARAMETER ESTIMATION PROCEDURES

The following section recommends values of parameters needed to estimate chemical mass transfer between air and surface films, film growth rates, and hence chemical washoff, and reaction rates.

19.4.1   FILM GROWTH TIMES AND PRECIPITATION WASH-OFF TIME PERIODS

Section 19.2 summarizes experimental values found for film growth, washoff efficiencies, and runoff coefficients. Film growth rates, between precipitation events, ranges from ~1.6 to 3.1 nm d^{-1} in an urban midlatitude city and are about 10 times less in a rural location. Anecdotal evidence suggests that film accumulation on windows can be negligible for surfaces receiving direct and intense sunlight (Diamond, M., unpublished data). Film accumulation is also negligible for heavily travelled roadways.

Data on antecedent dry days or conversely the frequency of precipitation events is highly site-specific. Fortunately considerable data are available through, for example, the National Oceanographic and Atmospheric Administration (NOAA) for the United States and all regions of the world, Environment Canada, and EUMETNET or the Network of European Meterological Services that links the meteorological services of 26 countries.

Film washoff efficiencies were listed in Section 19.2.3. The efficiencies vary widely from 10% in an experiment with sum-total PAHs on glass, to over 90% for accumulation on PTFE or Teflon. A mid-range of efficiencies is 60–85% for most surfaces. As discussed in Section 19.2.3, values of RC are published for patch up to city-wide scales. At the scale of a patch, values of RC are typically 0.5–0.9 where lower values are reasonable for warm or hot weather and/or more pervious surfaces and vice versa. Goldshleger et al. (2009) developed generalities for values of RC based on percentage of imperviousness, where RC approaches 1 for areas with 40–100% imperviousness. Park et al. (2009) summarized values of RC reported in the literature for various urban land uses (Table 19.4).

19.4.2   THE AIR-SIDE MTC FOR URBAN SURFACES

The artificial geomorphology of urban surfaces consists of a few simple shapes. These include predominately flat surfaces that extend upward from the ground into the atmospheric boundary layer. Approaching horizontal winds will flow over and around single structures. It arrives and is directed perpendicular and parallel to the plane surfaces as well as all angles in between with velocities of various magnitudes. As a result the forced convection MTCs on these surfaces vary considerably with circumferential position as well as height above the ground. A group of several buildings nestled together will dampen to force of the wind and change the shape of the vertical velocity profile. However, high speeds will occur in the so called "street canyons" running parallel to the wind. In addition, direct sunlight and shading on surfaces produce very different solid surface temperatures, which drive complex natural convection fluid dynamic flow patterns. The effective natural convection air-side MTCs can be significant during light and no wind conditions. Clearly, attempts at estimating representative MTCs for urban structures is a challenge.

To our knowledge, there are no data published on these air-side MTCs. Numerous studies with plane and extended solid surfaces aimed at improving heat transfer efficiencies to and from flowing fluids are available. Correlations exists for estimating heat transfer coefficients (HTCs) around solid shapes that do, to a degree, mimic urban surface geometric forms. These are for vertical cylinders, spheres, and flat planes and are presented in Tables 2.3 and 2.4 of Chapter 2. Using the appropriate transport analogy, these HTCs computations can be converted to MTCs. See Section 2.6 in Chapter 2 for further details. A few MTC derived correlations do exist; see Section 2.5.5 in Chapter 2. Table 19.2 contains some numerical values of MTC representative of urban structures estimated from heat transfer correlations.

19.5   A GUIDE FOR USERS

One group of users will be those desiring to learn about this unique environmental media compartment. Clearly it is very different from the others in the natural

TABLE 19.3

Urban Surface Parameters for Examples 19.6.2 and 19.6.3

Chemical	K_A (cm h^{-1})	C_A (μg m^{-3})	$\dfrac{K_A}{a}$	$\dfrac{v_\rho \rho_\rho K_{FA}}{a}$	K_{FA-E} (L kg^{-1})	C_f (μg^{-1})	τ_{95} (day)	M_A (kg)	C_W (μg L^{-1})
NAPH	61	5.8E5	2.44E9	1.27E2	7.9E3	4600	≪1	350	87.5
PYR	88	1.3E3	2.04E9	1.27E4	7.94E5	1050	≪1	80.3	20.1
TCDD	61	2.4E–2	2.49E9	6.42E9	3.76E9	89.1	3.12	0.19	0.048

environment. Its distinguishing compartmental characteristics are: (1) it has a very small mass in comparison to other environmental compartments, (2) it exists as an extremely thin layer on all solids, and (3) it is continually being formed, washed away, and reformed. Details about its discovery, origin, formation processes, measurement, growth rate, composition, and so on are presented in Sections 19.1 through 19.3.

The other group of users will be the environmental modelers. This group will be interested in the chemical dynamics of these urban films and quantifying their role accumulating and transporting chemicals between other compartments in urban environments. Three media phases are involved in the overall transport process. Chemicals originating in the atmosphere accumulate within the growing films that form on solid surfaces. They are then washed off by rain events. As part of the rain-generated urban runoff the film material and chemicals are delivered to the aquatic environment. With regard to its chemical dynamics, two modeling approaches are available for use. One is the comprehensive multimedia urban model (MUM) described in the case study presented as Example 19.6.1. The other is a vignette chemodynamic model describing the mass transport between the three phases and is presented in Examples 19.6.2 and 19.6.3. The use of a comprehensive model is necessary to estimate overall chemical fate if emissions are known, and to back-calculate emissions from measured ambient concentrations. The three phase model is useful for the detailed exploration of site-specific chemodynamics.

19.6   CASE STUDIES AND EXAMPLE CALCULATIONS

19.6.1   Multimedia Model (MUM): Toronto, Canada

In this chapter, we have explored the mechanism behind the high mobility of chemicals emitted into urban areas. Films on impervious surfaces, and of course the surfaces themselves, promote high export rates of emitted chemicals by air advection or transfer of atmospherically deposited chemicals into surface waters via storm water runoff, rather than sequestration in the urban environment itself. We illustrate this quantitatively using the fugacity-based multimedia urban model or MUM (Diamond et al., 2001; Priemer and Diamond, 2002; Jones-Otazo et al., 2005; Diamond et al., 2010). MUM is a seven-compartment fugacity Level III model. The model is coded in Visual Basic and is run on a PC in a Windows© environment.

MUM considers SOC dynamics among air (0–50 and 50–500 m height), water, sediment, vegetation, soil, and imperivous surface films. The reader is referred to the publications listed above for full details of the model. Let us consider the fate of emissions of a persistent SOC such as (the famous or infamous) PCBs in the city of Toronto (see Diamond et al., 2010 for further details). PCBs are a convenient tracer for chemical fate because of their relatively low reactivity and unfortunate abundance in urban areas.

To simulate year-round conditions, we assume a temperature of $9.3°C$ and a wind speed of 15 mph or 23 km h^{-1}. We use an illustrative emission to the lower air compartment of 1000 g d^{-1}. The 636 km^2 area of Toronto has 58% impervious surfaces, 42% soil and vegetation, and 0.4% water. This simulation does not include Lake Ontario on which Toronto is situated. Rather, surface water and the underlying sediment are assumed to represent a relatively small urban river. The three-dimensional area of these surfaces is about twice that of the planar area, but can be much higher for cities such as Manhattan or Hong Kong. As noted above, the ratio between the three- and two-dimensional surfaces areas is the impervious surface index (ISI) (Diamond et al., 2001) and so in this simulation we assumed a value of 2. We assume that impervious surfaces convey precipitation to nearby surface waters but not to underlying soils through infiltration (e.g., via cracks in the pavement). We ignore the complexities of stormwater storage in the sewer system and leakage from the sewer system to recharge groundwater. Stormwater storage in and evaporation from impervious surfaces is captured by the use of the runoff coefficient (RC) as described in Section 19.2.3.

We assume that soil is covered by vegetation. In contrast to a temperate forest that has canopy and shrub layers giving a leaf area index (LAI) of ~4, the vegetation in most cities is highly simplified by grassy lawns, which have a LAI of 1.2 (Preimer and Diamond, 2002). The vegetation accumulates atmospherically deposited constituents and intercepts precipitation, which then flows, along with its chemical burden, to the underlying soil. Lower values of LAI result in atmospheric deposition to a smaller surface area and hence less leaf washoff to soils.

Model results show that from 97% to 86% of di- to hepta-CBs (CB-28 and -180) entering Toronto from direct emissions to air are advected downwind (Figure 19.8). The export rate is highly sensitive to the windspeed. Kwamena et al. (2007) showed that at lower windspeeds a higher proportion of PAH partitions into the condensed phases of film, vegetation, and soil than at lower wind speeds. The very high rate of advective loss contrasts with a forest where $\sim <1$–15% (a lower value for chemicals with higher vapor pressures and vice versa) less chemical is advected downwind due to the increased capture efficient of the forest versus impervious surfaces—the "forest filter" effect explained by McLachlan and Horstmann (1998).

Of the total mass not advected downwind, 66–80% of di and hepta are transferred from air to film to surface water, with most of the remaining 40–20% transferred from air to vegetation to soil. The net direction of transport from air–film–surface water and air–vegetation–soil is maintained by washoff for chemicals with log K_{OA} greater than ~9 such as the penta- to nona-CBs. For chemicals with log K_{OA} less than ~9, the rate of film-to-air transport (volatilization) is predicted to exceed that of washoff, particularly at higher temperatures. Volatilization back to air promotes the cycling and recycling of chemicals between air and films and can increase air concentrations. This reciprocal relationship of loss mechanisms from the film of volatilization versus washoff, as a function of K_{OA}, is discussed by Priemer and Diamond (2002) and Hodge and Diamond (2009).

For a photoreactive reactive chemical such as benzo[a]pyrene, the main loss mechanism is downwind advection. However, reaction losses in the film can be of secondary importance and can exceed losses due to reaction in air (Kwamena et al., 2007).

Although air plays a key role in terms of export, air holds <1% of chemical mass and (by order of magnitude) the lowest concentrations of CB-28 and -180 (Figures 19.9 and 19.10). In contrast, the film with its minuscule volume of 76 m^3 holds 1% and 14% of mass of CB-28 and -180, respectively, and has concentrations ~10^8 and 10^9 greater than air. In particular, the film holds a mass of CB-180 of 14 g that is greater than the ~6 g held in the lower air compartment with its volume of 3.2×10^{10} m^3.

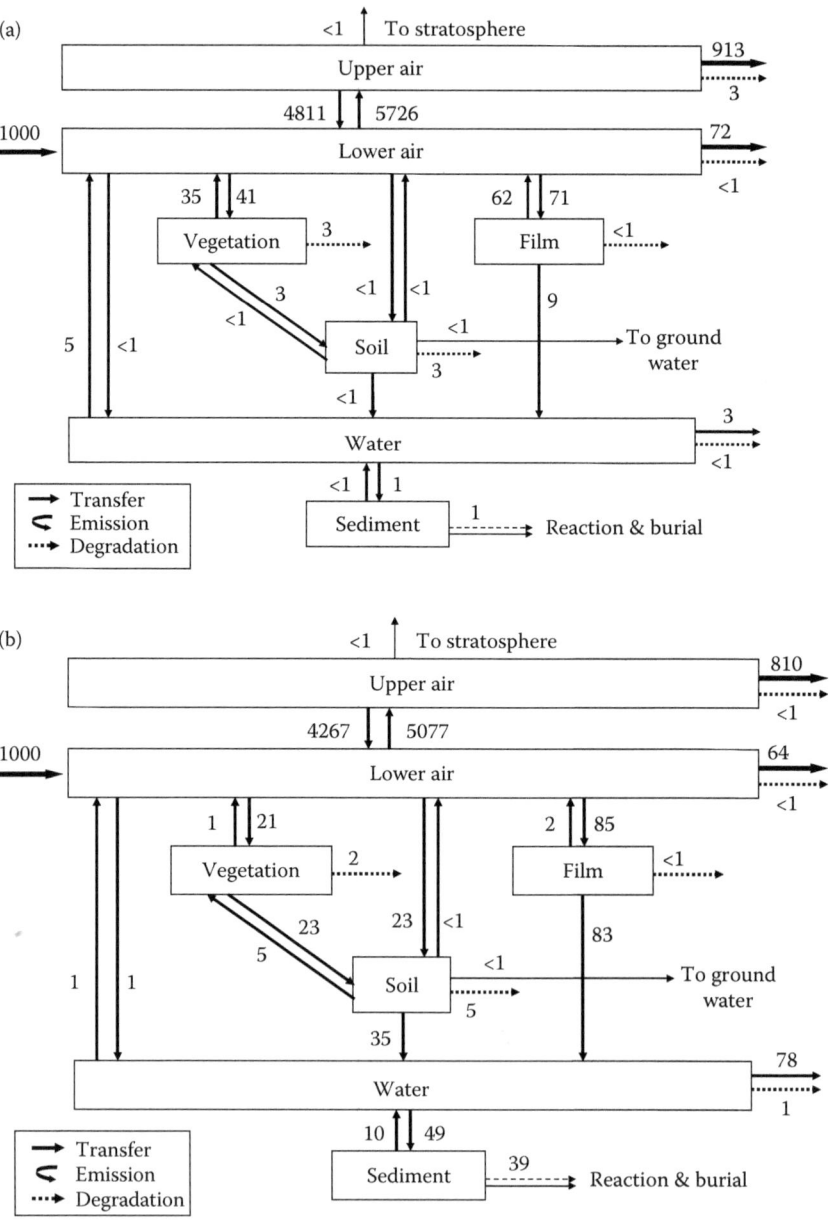

FIGURE 19.8 Fate of a low reactivity SOC in Toronto, Canada, estimated using the Multimedia Urban Model, MUM (S. Csiszar, unpublished data). Model is run for 1000 g d^{-1} emission into the lower air compartment (0–50 m) assuming an average annual temperature of 9.3°C. (a) PCB congener 28 (2,4,4′-trichlorobiphenyl), and (b) PCB congener 180 (2,2′,3,4,4′,5,5′-heptachlorobiphenyl). Arrows indicate intercompartmental transfer rates and rates of chemical transformation (g d^{-1}). (Adapted from Diamond, M. L., et al. 2010. *Environ. Sci. Technol.* 44(8): 2777–2783.)

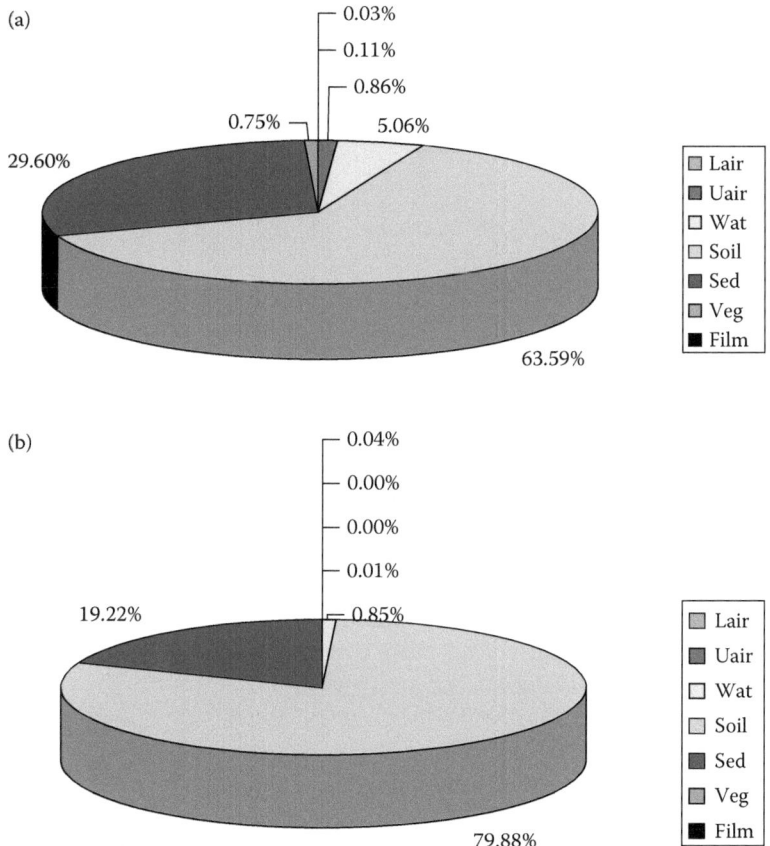

FIGURE 19.9 Mass of PCB congeners CB-28 (a) and CB-180 (b) estimated in compartments of the Multimedia Urban Model of Toronto, Canada.

Soil (depth 0–2 cm depth, $5 \times 10^6 \, \mathrm{m}^3$ volume) is the largest sink for chemical in the system, holding 3700 and 63,400 g or 63 and 80% of total CB-28 and -180, respectively. The next most important chemical sink is sediment (also 0–2 cm depth, $10^5 \, \mathrm{m}^3$) and water (5 m depth, $2.7 \times 10^7 \, \mathrm{m}^3$). Here too, the film plays a key role as the main conduit delivering to water and sediment the emissions to air.

The inefficiency of the surface film as a sink or source or conversely, its efficiency as an intermediary between air and surface water, is conferred by the fact that it has an equal or even higher temperature than air and it is very thin—a few nanometers up to a few micrometers after 300 days (Wu et al., 2008a). In contrast, soil is an efficient sink of deposited/condensed chemical because of its depth and high surface area available for sorption. Deep lakes and the ocean act as sinks by virtue of volume and usually cold temperatures.

The film is critical to the persistence of chemical in the urban system. The persistence of CB-28 and -180 is very short in film at 30 min and <4 h, respectively. The

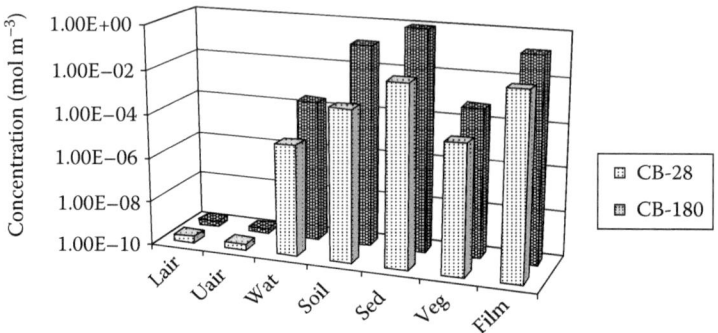

FIGURE 19.10 Concentrations ($mol\ m^{-3}$) of PCB congeners CB-28 and CB-180 estimated in the compartments of the multimedia urban model of Toronto, Canada.

short persistence is attributable to the return of CB-28 to air and delivery of CB-180 to water and sediment. This compares with the persistence of <2 h in the combined air compartments, which is largely controlled by wind speed followed by the tendency of the chemical to partition into the condensed phases of film, soil, vegetation, and water. The longest persistence is 2.5 and 38 years in soil for the two congeners, respectively. The greatest loss mechanism from soil is particle erosion into water. These fate processes result in an overall persistence of 6 days and ~2 years for CB-28 and -180, which is short owing to the role of surface films, relative to a natural system (Priemer and Diamond, 2002).

19.6.2 CHEMICAL LOADING LEVELS ON AN URBAN SURFACE: TORONTO, CANADA

Naphthalene (NAPH), pyrene (PYR), and 2,3,7,8-tetrachlorodibenzodioxin (TCDD) were chosen as typical air contaminants. For purposes of human exposure and risk assessments, the concentrations in films at 25°C and 1 atm total pressure assuming air concentrations are at the saturated vapor levels. Choose any realistic parameters needed and justify any key assumptions.

Solution

Based on the data available for Toronto an initial thickness and growth rate of 1 nm and 6 nm d^{-1} seem appropriate. Values of 4 cm h^{-1} and 100 µg m^{-3} are generally appropriate for aerosol particle deposition velocity and concentrations in air at street level where a 10 mile h^{-1} (4.6 m s^{-1}) wind speed is selected. The MTCs for each chemical taken from Table 19.2 appear in Table 19.3; other properties are in Table 19.1. It is assumed that $K_{PA} = K_{FA}$ for particles and film alike. The ideal gas law, $C_A = MWp°/RT$, yields the air concentration values in column 3 of Table 19.3.

The effective film-to-air partition coefficient (Equation 19.6) is used to estimate the maximum film concentration in Equation 19.7. For NAPH the first step calculations are as follows:

$$K_A/a = 2.44E9 \quad \text{and} \quad v_\rho \rho_p K_{PA}/a = 127.$$

In the calculations the particle-to-air K_{PA} was set to equal to K_{FA}, the latter being 7940 L kg^{-1} for NAPH. The effective film-to-air partition coefficient is estimated using Equation 19.6:

$$K_{FA-E} = 7940 \left[\frac{2.44E9 + 127}{2.44E9 + 7940} \right] = 7940 \text{ L/kg}$$

and is unchanged from the K_{FA} value. The calculated results for PYR and TCDD appear in column 6 of Table 19.3. For TCDD the log K_{FA-E} is 9.58, which is a lower value than the log K_{FA} of 11.6. This result suggests partitioning for TCDD is also limited by its mass transfer rate to the film. The particle deposition contribution is also a significant factor in TCDD accumulation within the film.

The concentration of each chemical in the film appears in column 7; Equation 19.5 was used to obtain these estimates. Such large film loading ratios result because the air is assumed to be saturated with the chemical vapor; the air concentrations appear in column 3. Nevertheless, the results indicate that impervious surface films do have the ability to be an effective medium for concentrating air-borne pollutants to high levels in urban setting where human exposure is sure to occur.

19.6.3 FILM CHEMICAL MASS AND RUNOFF CONCENTRATION: TORONTO, CANADA

For the chemicals used in the previous example, estimate the mass of each present on the impervious surfaces and the urban runoff concentration resulting from its solubilization by a typical rain event. Assume the urban footprint size is 50×10^7 m^2 with an ISI of 2. Choose realistic parameters needed and justify key assumptions.

Solution

Film chemical uptake time can be estimated by Equation 19.10. For TCDD the numbers are:

$$\tau_{95} = \frac{1}{6}[\exp(3 \cdot 0.909) - 1] = 3.1 \text{ days.}$$

The results for all three appear in Table 19.3, column 8. The time period for the film to achieve the maximum concentration as given by Equation 19.10 is short. Only for rain events spaced less than 3.1 days should Equation 19.5 be used to estimate C_f for TCDD. The C_f values appearing in column 7 will be used in the calculation of the chemical content in the film mass. Equation 19.8 is the appropriate one for estimating the mass of chemical accumulating in the film after growth time t. It is necessary to use the time periods between rain events that cause significant film build-up for estimating this film growth time. As shown in Figure 19.3, the values for Toronto vary between 1 and 20 days, approximately; a value 14 days will be used in the following calculations. The film thickness is equal $h = 1 + 6(14) = 85$ nm. The computation for NAPH is

$$M_A = 4.6 \frac{\text{g}}{\text{kg}} \left| 50 \times 10^7 \text{ m}^3 \right| 2 \left| 85E - 9 \text{ m} \right| \frac{1000 \text{ kg}}{\text{m}^3} \left| 0.9 = 350{,}000 \text{ g} \right.$$

TABLE 19.4
Runoff Coefficients for Urban Land Uses

	Location	SFR	MFR	C	P	I	T	O	Other
Ackerman and Schiff, 2003	S. CA	0.39	0.39	0.61		0.64		0.06	0.41
Line et al., 2002	NC	0.57	0.57					0.47	0.52
									0.70
Charbeneau and Barrett, 1998	TX	0.04	0.62	0.83			0.56/0.85	0.007	
		0.17		0.75					
Imperviousness									
Pitt et al., 2004	USA	0.37	0.85	0.45	0.75	0.8	0.02		
Goonetilleke et al., 2005	Australia	0.7					0.02		
		0.5							

Source: From Park, M.H., Swamikannu, X., and Stenstrom. M. K. 2009. *Water Res.* 43(11): 2773–2786. With permission.

Note: SFR is single-family residential, MFR is multiple-family residential, C is commercial, P is public, I is industrial, T is transportation, and O is open land use and Pitt et al.'s values are median imperviousness.

In this calculation a film density of $1\,kg\,L^{-1}$ and film washoff efficiency of 0.90 were used. The solubilized mass of each chemical appears in Table 19.3, column 9. Equation 19.9 can be used for estimating the runoff concentration, a 8 mm rain event will be assumed. For NAPH the calculation is:

$$C_W = 352E6\,mg/(50 \times 10^7 m^2)(8E - 3m) = 87.5\,mg/m^3.$$

All three water concentrations appear in the last column of the table. These estimates indicate that the washoff of urban surface films by rain can deliver short duration but significant chemical mass to the aquatic receiving bodies. In addition the concentrations in the runoff water may be significant.

LITERATURE CITED

Ackerman, D. and K. Schiff. 2003. Modeling storm water mass emissions to the Southern California bight. *J. Environ. Eng., ASCE* 129(4): 308–317.

Bertrand Krajewski, J.-L., G. Chebbo, and A. Saget. 1998. Distribution of pollutant mass vs volume in stormwater discharges and the first flush phenomenon. *Water Res.* 8: 2341–2356.

Brunner, P. and H. H. Rechberger. 2001. Anthropogenic metabolism and environmental legacies. In *Encyclopedia of Global Environmental Change.* T. Munn, ed. Chichester, UK: Wiley.

Bucheli, T. D., S. R. Muller, S. Heberle, and R. P. Schwarzenbach. 1998a. Occurrence and behaviour of pesticides in rainwater, roof runoff, and artificial stormwater infiltration, *Environ. Sci. Technol.* 32: 3457–3464.

Bucheli, T. D., S. R. Muller, A. Voegelin, and R. P. Schwarzenbach. 1998b. Bituminous roof sealing membranes as major sources of the herbicide (*R,S*)-mecoprop in roof runoff waters: Potential contamination of groundwater and surface waters. *Environ. Sci. Technol.* 32: 3465–3471.

Butt, C. M. 2003. Chemical and physical characterization of organic films on an impervious surface. MS thesis, University of Toronto, National Library of Canada, Ontario, Canada, 168 pp.

Charbeneau, R. J. and M. E. Barrett. 1998, Evaluation of methods for estimating stormwater pollutant loads. *Water Environ. Res.* 70(7): 1295–1302.

Cotham, W. E. and T. F. Bidleman. 1995. Polycyclic aromatic hydrocarbons and polychlorinated biphenyls in air at an urban and a rural site near Lake Michigan, *Environ. Sci. Technol.* 29(11): 2782–2789.

Covaci, A., C. Hura, and P. Schepens. 2001. Selected persistent organochlorine pollutants in Romania. *Sci. Total Environ.* 280(1–3): 143–152.

Diamond, M. L., S. E. Gingrich, G. A. Stern, and B. E. McCarry. 2000. Wash-off of SOC's from organic films on an urban impervious surface. *Organohal. Compds.* 45: 272–275.

Diamond, M. and S. Harrad. 2009. The chemicals that will not go away: Implications for human exposure to reservoirs of POPs. In*Persistent Organic Pollutants: Current Issues and Future Challenges,* S.J. Harrad, ed. Chichester, UK: Wiley.

Diamond, M. L. and E. Hodge. 2007. Urban contaminant dynamics: From source to effect. *Environ. Sci. Technol.* 41(11): 3796–3800.

Diamond, M. L., D. A. Priemer, and N. L. Law. 2001. Developing a multimedia model of chemical dynamics in an urban area. *Chemosphere* 44(7): 1655–1667.

Diamond, M. L., L. Melymuk, S. Csiszar, and M. Robson. 2009. PCB stocks, emissions and urban fate: Will our policies be effective? *Environ. Sci. Technol.* 44(8): 2777–2783.

Donaldson, D. J., B. T. Mmereki, S. R. Chaudhuri, S. Handley, and M. Oh. 2005. Uptake and reaction of atmospheric organic vapours on organic films. *Faraday Dis.* 130: 227–239.

Eckley, C. S., B. Branfireun, M. Diamond, P. C. van Metre, and F. Heitmuller. 2008. Atmospheric mercury accumulation and washoff processes on impervious urban surfaces. *Atmos. Environ.* 42(32): 7429–7438.

Eckley, C. S. and B. Branfireun. 2008. Mercury mobilization in urban stormwater runoff. *Sci. Total Environ.* 403(1–3): 164–177.

Eckley, C. S. and B. Branfireun. 2009. Simulated rain events on an urban roadway to understand the dynamics of mercury mobilization in stormwater runoff. *Water Res.* 43(15): 3635–3646.

Esch, T., V. Himmler, G. Schorcht, et al. 2009. Large-area assessment of impervious surface based on integrated analysis of single-date Landsat-7 images and geospatial vector data. *Remote Sens. Environ.* 113(8): 1678–1690.

Finlayson-Pitts, B. J. 2009. Reactions at surfaces in the atmosphere: Integration of experiments and theory as necessary (but not necessarily sufficient) for predicting the physical chemistry of aerosols. *Phys. Chem. Chem. Phys.* 11(36): 7760–7779.

Focazio, M. J., D. W. Kolpin, K. K. Barnes, et al. 2008. A national reconnaissance for pharmaceuticals and other organic wastewater contaminants in the United States—II: Untreated drinking water sources. *Sci. Total Environ.* 402(2–3): 201–216.

Gingrich, S. E., M. L. Diamond, G. A. Stern, and B. E. McCarry. 2001. Atmospherically derived organic surface films along an urban–rural gradient. *Environ. Sci. Technol.* 35(20): 4031–4037.

Glassmeyer, S. T., D. W. Kolpin, E. T. Furlong, and M. J. Focazio. 2008. Environmental presence and persistence of pharmaceuticals—An overview. *Fate of Pharmaceuticals in the Environment and in Water Treatment Systems.* Boca Raton, FL: CRC Press, Taylor & Francis Group, pp. 3–51.

Goldshleger, N., M. Shoshany, L. Karnibad, S. Arbel, and M. Getker. 2009. Generalising relationships between runoff-rainfall coefficients and impervious areas: An integration of data from case studies in Israel with data sets from Australia and the USA. *Urban Water J.* 6(3): 201–208.

Goonetilleke, A., E. Thomas, S. Ginn, and D. Gilbert. 2005. Understanding the role of land use in urban stormwater quality management, *J. Environ. Manage.* 74(1): 31–42.

Grossi, C. M. and P. Brimblecombe. 2002. The effect of atmospheric pollution on building materials. *J. Phys. IV* 12(PR10): 197–210.

He, J. and R. Balasubramanian. 2009. A study of gas/particle partitioning of SVOCs in the tropical atmosphere of Southeast Asia. *Atmos. Environ.* 43(29): 4375–4383.

Hodge, E. and M. Diamond. 2009. Sources, fate and effects of contaminant emissions in urban areas. In *Persistent Organic Pollutants: Current Issues and Future Challenges,* S.J. Harrad, ed. Chichester, UK: Wiley.

Jamshidi, A., S. Hunter, S. Hazrati, and S. Harrad. 2007. Concentrations and chiral signatures of polychlorinated biphenyls in outdoor and indoor air and soil in a major U.K. conurbation. *Environ. Sci. Technol.* 41(7): 2153–2158.

Jones-Otazo, H. A., J. P. Clarke, M. L. Diamond, et al. 2005. Is house dust the missing exposure pathway for PBDEs? An analysis of the urban fate and human exposure to PBDEs. *Environ. Sci. Technol.* 39(14): 5121–5130.

Kahan, T. F., N. O. A. Kwamena, and D. J. Donaldson. 2006. Heterogeneous ozonation kinetics of polycyclic aromatic hydrocarbons on organic films. *Atmos. Environ.* 40(19): 3448–3459.

Kaye, J. P., P. M. Groffman, N. B. Grimm, L. A. Baker, and R. V. Pouyat. 2006. A distinct urban biogeochemistry? *Trends Ecol. Evol.* 21(4): 192–199.

Kim, L-H., K.-D. Zoh, S.-M. Jeong, M. Kayhanian, and M. K. Stenstrom. 2006. Estimating pollutant mass accumulation on highways during dry periods. *J. Environ. Eng.* 132: 985–993.

Kimbrough, K. L. and R. M. Dickhut. 2006. Assessment of polycyclic aromatic hydrocarbon input to urban wetlands in relation to adjacent land use. *Marine Pollut. Bull.* 52(11): 1355–1363.

Kwamena, N. O. A., J. A. Thornton, and J. P. D. Abbatt. 2004. Kinetics of surface-bound benzo[a]pyrene and ozone on solid organic and salt aerosols. *J. Phys. Chem. A* 108(52): 11626–11634.

Kwamena, N. O. A., J. P. Clarke, T. F. Kahan, M. L. Diamond, and D. J. Donaldson. 2007. Assessing the importance of heterogeneous reactions of polycyclic aromatic hydrocarbons in the urban atmosphere using the Multimedia Urban Model. *Atmos. Environ.* 41(1): 37–50.

Labencki, T., M. L. Diamond, A. Motelay-Massei, J. Truong, B. Branfireun, and T. Dann. Variability in and mechanisms of PAH washoff from urban impervious surfaces. In preparation.

Lam, B., M. L. Diamond, A. J. Simpson, et al. 2005. Chemical composition of surface films on glass windows and implications for atmospheric chemistry. *Atmos. Environ.* 39(35): 6578–6586.

Law, N. L. and M. L. Diamond. 1998. The role of organic films and the effect on hydrophobic organic compounds in urban areas: An hypothesis. *Chemosphere* 36(12): 2607–2620.

Line, D. E., N. M. White, D. L. Osmond, G. D. Jennings, and C. B. Mojonnier. 2002. Pollutant export from various land uses in the Upper Neuse River Basin, *Water Environ. Res.* 74(1): 100–108.

Liu, Q. T., R. Chen, B. E. McCarry, M. L. Diamond, and B. Bahavar. 2003a. Characterization of polar organic compounds in the organic film on indoor and outdoor glass windows. *Environ. Sci. Technol.* 37(11): 2340–2349.

Liu, Q. T., M. L. Diamond, S. E. Gingrich, et al. 2003b. Accumulation of metals, trace elements and semi-volatile organic compounds on exterior window surfaces in Baltimore. *Environ. Pollution* 122(1): 51–61.

McLachlan, M. S. and M. Horstmann. 1998. Forests as filters of airborne organic pollutants: A model. *Environ. Sci. Technol.* 32(3): 413–420.

Mansell, M. and F. Rollet. 2009. The effect of surface texture on evaporation, infiltration and storage properties of paved surfaces. *Water Sci. Technol.* 60(1): 71–76.

Mmereki, B. T., D. J. Donaldson, J. B. Gilman, T. L. Eliason, and V. Vaida. 2004. Kinetics and products of the reaction of gas-phase ozone with anthracene adsorbed at the air-aqueous interface. *Atmos. Environ.* 38(36): 6091–6103.

Motelay-Massei, A., D. Ollivon, B. Garban, et al. 2004. Distribution and spatial trends of PAHs and PCBs in soils in the Seine River basin, France. *Chemosphere* 55(4): 555–565.

Nissenson, P., C. J. H. Knox, B. J. Finlayson-Pitts, L. F. Phillips, and D. Dabdub. 2006. Enhanced photolysis in aerosols: Evidence for important surface effects. *Phys. Chem. Chem. Phys.* 8(40): 4700–4710.

Pankow, J. F. and T. F. Bidleman. 1992. Interdependence of the slopes and intercepts from log–log correlations of measured gas–particle partitioning and vapor pressure—I. Theory and analysis of available data. *Atmos. Environ.* 26A: 1071–1080.

Park, M. H., X. Swamikannu, and M. K. Stenstrom. 2009. Accuracy and precision of the volume–concentration method for urban stormwater modeling. *Water Res.* 43(11): 2773–2786.

Pitt, R., A. Maestre, and R. Morquecho. 2004. *The National Stormwater Quality Database (NSQD, version 1.1)*. http://unix.eng.ua.edu/ isn't in document rpitt/Research/ms4/mainms4.shtml.

Platford, R.F., J.H. Carey, and E.J. Hale. 1982. The environmental significance of surface films: Part 1—Octanol–water partition coefficients for DDT and hexachlorobenzene. *Environ. Pollut.* 3: 125–128.

Poschl, U., T. Letzel, C. Schauer, and R. Niessner. 2001. Interaction of ozone and water vapor with spark discharge soot aerosol particles coated with benzo[*a*]pyrene: O-3 and H_2O adsorption, benzo[a]pyrene degradation, and atmospheric implications. *J. Phys. Chem. A* 105(16): 4029–4041.

Powell, S. L., W. B. Cohen, Z. Yang, J. D. Pierce, and M. Alberti. 2008. Quantification of impervious surface in the Snohomish Water Resources Inventory Area of Western Washington from 1972–2006. *Remote Sensing Environ.* 112(4): 1895–1908.

Pouyat, R.V., K. Belt, D. Pataki, P.M. Groffman, J. Hom, and L. Band. 2007. Urban land-use change effects on biogeochemical cycles. In:, *Terrestrial Ecosystems in a Changing World. Global Change*, The IGBP Series, Canadell, J.G., D.E. Pataki, and L.F. Pitelka, eds. Berlin: Springer, pp. 45–58.

Pouyat, R. V., I. D. Yesilonis, and N. E. Golubiewski. 2009. A comparison of soil organic carbon stocks between residential turf grass and native soil. *Urban Ecosyst.* 12(1): 45–62.

Priemer, D.A. and M.L. Diamond. 2002. Application of the multimedia urban model to compare the fate of SOCs in an urban and forested watershed. *Environ. Sci. Technol.* 36: 1004–1013.

Ragab, R., P. Rosier, A. Dixon, J. Bromley, and J. D. Cooper. 2003. Experimental study of water fluxes in a residential area: 2. Road infiltration, runoff and evaporation. *Hydrol. Process.* 17(12): 2423–2437.

Ramier, D., E. Berthier, P. Dangla, and H. Andrieu. 2006. Study of the water budget of streets: Experimentation and modelling. *Water Sci. Technol.* 54(6–7): 41–48.

Robson, M., L. Melymuk, B. Gilbert, et al. 2008. Comparison of concentrations and loadings of PCBs and PAHs in urban and rural streams during base flow and storm events. *Organohalogen Compds.* 70: 685–688.

Rudel, R. A. and L. J. Perovich. 2009. Endocrine disrupting chemicals in indoor and outdoor air. *Atmos. Environ.* 43(1): 170–181.

Saget, A., G. Chebbo, and J.I. Bertrand Krajewski 1996. The first flush in sewer systems. *Water Sci. Technol.* 33:101–108.

Sansalone, J. J. and S. G. Buchberger. 1997. Partitioning and first flush of metals in urban roadway storm water. *J. Environ. Engng ASCE* 123(2): 134–143.

Sansalone, J. J. and C. M. Cristina. 2004. First flush concepts for suspended and dissolved solids in small impervious watersheds. *J. Environ. Eng. ASCE* 130(11): 1301–1314.

Schueler, T.R. 1994. The importance of imperviousness. *Watershed Protect. Techn.* 1: 100–111.

Sulaiman, F. R., P. Brimblecombe, and C. M. Grossi. 2009. Mobilization and loss of elements from roofing tiles. *Environ. Geol. (Berlin)* 58(4): 795–801.

Tasdemir, Y., N. Vardar, M. Odabasi, and T. M. Holsen. 2004. Concentrations and gas/particle partitioning of PCBs in Chicago. *Environ. Pollution* 131(1): 35–44.

Tasdemir, Y. and F. Esen. 2007. Urban air PAHs: Concentrations, temporal changes and gas/particle partitioning at a traffic site in Turkey. *Atmos. Res.* 84(1): 1–12.

Thibodeaux, L. and M.L. Diamond. Chemodynamics of films on impervious surfaces: Data and modelling. In preparation.

Turner, V.R. and S.V. Hering. 1987. Greased and oiled substrates as bounce-free impaction surfaces. *J. Aerosol Sci.* 18: 215–224.

United Nations. 1995. *World Urbanization Prospects. The 1994 Revision—Estimates and Projections of Urban and Rural Populations and of Urban Agglomerations.* New York: United Nations, pp. 4–5.

Van Metre, P. C. and B. J. Mahler. 2005. Trends in hydrophobic organic contaminants in urban and reference lake sediments across the United States, 1970–2001. *Environ. Sci. Technol.* 39(15): 5567–5574.

Van Metre, P. C., B. J. Mahler, and J. T. Wilson. 2009. PAHs underfoot: Contaminated dust from coal-tar sealcoated pavement is widespread in the United States. *Environ. Sci. Technol.* 43(1): 20–25.

van Nuijs, A. L. N., B. Pecceu, L. Theunis, et al. 2009. Cocaine and metabolites in waste and surface water across Belgium. *Environ. Pollut.* 157(1): 123–129.

Vaze, J. and F.H.S. Chiew. 2002. Experimental study of pollutant accumulation on an urban road surface. *Urban Water* 4: 379–389.

Vaze, J. and F.H.S. Chiew. 2003. Study of pollutant washoff from small impervious experimental plots. *Water Resources Res.* 39: 1160–1170.

Weschler, C. J. 2003. Indoor/outdoor connections exemplified by processes that depend on an organic compound's saturation vapor pressure. *Atmos. Environ.* 37(39–40): 5455–5465.

Wilson, N. K., J. C. Chuang, C. Lyu, R. Menton, and M. K. Morgan. 2003. Aggregate exposures of nine preschool children to persistent organic pollutants at day care and at home. *J. Exposure Anal. Environ. Epidemiol.* 13(3): 187–202.

Wong, F., M. Robson, M. L. Diamond, S. Harrad, and J. Truong. 2009. Concentrations and chiral signatures of POPs in soils and sediments: A comparative urban versus rural study in Canada and UK. *Chemosphere* 74(3): 404–411.

Wu, Y. L., C. I. Davidson, D. A. Dolske, and S. I. Sherwood. 1992. Dry deposition of atmospheric contaminants—the relative importance of aerodynamic, boundary-layer, and surface resistances. *Aerosol Sci. Technol.* 16(1): 65–81.

Wu, R. W., T. Harner, and M. L. Diamond. 2008a. Evolution rates and PCB content of surface films that develop on impervious urban surfaces. *Atmos. Environ.* 42(24): 6131–6143.

Wu, R. W., T. Harner, M. L. Diamond, and B. Wilford. 2008b. Partitioning characteristics of PCBs in urban surface films. *Atmos. Environ.* 42(22): 5696–5705.

Zarnadze, A. and L. A. Rodenburg. 2008. Water-column concentrations and partitioning of polybrominated diphenyl ethers in the New York/New Jersey Harbor, USA. *Environ. Toxicol. Chem.* 27(8): 1636–1642.

20 Mixing in the Atmosphere and Surface Waters with Application to Compartmental Box Models

Ellen Bentzen, Matthew MacLeod, Brendan Hickie, Bojan Gasic, Konrad Hungerbühler, and Donald Mackay

CONTENTS

20.1 Introduction 565
20.2 Mixing in the Atmosphere..567
    20.2.1 Introduction..567
    20.2.2 Vertical Mixing in the Atmosphere..............................567
    20.2.3 Horizontal Mixing in the Atmosphere...........................570
    20.2.4 Summary of Mixing Processes in the Atmosphere 571
    20.2.5 Example Calculation 572
20.3 Mixing in Surface Waters..573
    20.3.1 Introduction..573
    20.3.2 Thermal Stratification of Lakes 576
    20.3.3 Biological Implications and Particle Deposition 580
    20.3.4 Summary of Mixing and Diffusive Processes within
             Lake Strata...581
    20.3.5 Modeling and Parameterizing Mass Transport in
             Surface Waters..582
            20.3.5.1 Lakes 582
            20.3.5.2 Rivers 582
    20.3.6 Case Study of a Lake Mass Balance with and without
             Stratification 584
References...586

20.1 INTRODUCTION

When modeling chemical distribution and fate in environmental systems, adopting the "well-mixed box" assumption to describe a unit world has proven to be immensely convenient. Only one concentration need be defined in the well-mixed compartment, thus the contained mass is the product of that concentration and the compartment volume. In many environmental applications, compartments describing the atmosphere and water bodies are "open," that is, air or water is continuously exchanged between the system under consideration and the background outside of the system. When setting up models of this type, it is necessary to specify mass transfer coefficients for advective exchange of air and water to describe the key input and output processes. This chapter illustrates generic methods of doing this in the absence of site-specific information about wind speed, water current velocity, or related diffusive or dispersive parameters.

Volumes of air and water are often sufficiently mixed that the "well-mixed box" assumption is acceptable. Even if it is known that concentration gradients exist, the benefits of applying the assumption of homogeneity within compartments may outweigh the loss in accuracy. There are, however, exceptions, and these are also the subject of this chapter. The issue of errors possibly incurred by applying this assumption have been discussed by Warren et al. (2009).

Mixing in the atmosphere is described in Section 20.2. The potential for inhomogeneous mixing within a modeled region is addressed, primarily as a function of vertical height. Understandably, most measurements of contaminants in air are made close to ground level since this is the region of the atmosphere that is of most concern for exposure of humans and the ecosystem. If contaminant sources (or sinks) are at ground level, it is likely that concentrations near the ground will be greater (or less) than those aloft. Ground level concentrations may be more variable since they respond more rapidly to local changes in source or sink fluxes. Observations of diel (24-h) concentration changes may be driven by changes in mixing height or by temperature-controlled source or sink terms. When interpreting or modeling such situations, it is essential that the role of vertical mixing be fully appreciated.

In Section 20.3 the analogous situation in surface water bodies is discussed. Rivers usually have longitudinal (upstream–downstream) concentration gradients, but are usually well mixed vertically because of relatively high current velocities and shallow depths. Wide rivers may have horizontal (shore to shore) gradients. Ponds, lakes, estuaries, and oceans may be less well mixed or even stratified vertically, thus the well mixed box assumption can be inapplicable. Stratification can be enhanced by stable density gradients derived from temperature or salinity differences in the water column. Inputs or outputs from the atmosphere may then primarily affect near-surface layers and effects on deeper layers can be damped or delayed. Likewise interactions with bottom sediments may fail to penetrate to surface layers. In temperate regions, these effects will vary seasonally.

This chapter aims to sensitize and inform the reader about these processes by describing the nature of the horizontal and vertical mixing processes using relatively simple terms, with a particular focus on situations where rates of mixing are insufficient for the compartment to be considered well mixed. No attempt is made to treat these phenomena in the detail they deserve, but references are provided as an entry to the large literature on these topics. Suggestions are offered to assist the environmental scientist or engineer to make reasonable assumptions as to well-mixed heights and depths, and the corresponding residence times for air and water in typical model regions. Data sources that can be exploited for more detailed and site-specific parameterization are described.

20.2 MIXING IN THE ATMOSPHERE

20.2.1 INTRODUCTION

The Earth's atmosphere is a layer of gases that surround the planet and is retained by gravity. More than 99.999% of the atmosphere is held within 100 km of sea level, and this height is commonly used to define a boundary between the atmosphere and outer space. The atmosphere is therefore very thin relative to the size of the Earth, which has a mean radius of 6371 km. On a typical desk globe model of the Earth with diameter 30 cm, the atmosphere would be only 2.4 mm thick, and the lower atmosphere or troposphere to a height of some 10 km would be only 0.24 mm thick (NASA, 2008).

20.2.2 VERTICAL MIXING IN THE ATMOSPHERE

Vertical transport in the atmosphere at local scales is driven by the buoyancy of air masses. Buoyancy is the net force on an air mass caused by a local difference between the pressure-gradient force from surrounding air masses and the force of gravity (Jacob, 1999). In the atmosphere, buoyancy is determined by the decrease in temperature with altitude ($-dT/dz$), namely the *lapse rate*. In a column of atmosphere where the pressure-gradient force and the gravitational force are perfectly balanced (i.e., one in which there is no buoyancy) the lapse rate exactly corresponds to the rate that a rising air mass would cool as a result of the expansion of the air mass as pressure falls with height. A dry air mass that (1) does not exchange energy with its surroundings (i.e., is adiabatic), and (2) behaves as an ideal gas, cools at a rate of 9.8°C/km as it rises in the atmosphere (Jacob, 1999). Therefore, -9.8°C/km is a good estimate of the *adiabatic lapse rate*, the vertical temperature gradient in a column of atmosphere that has no buoyant force.

If the lapse rate of the atmosphere is less than the adiabatic lapse rate, for example, -15°C/km, a rising air parcel will be accelerated upward by buoyancy, and likewise, a sinking air parcel will be accelerated downward. In this case the atmosphere is called unstable, since any initial vertical motion results in positive feedback and acceleration in the same direction. Conversely, if the lapse rate of the atmosphere is greater than the adiabatic lapse rate, for example -5°C/km, a rising air parcel will be propelled back downward, and a sinking air parcel will be propelled upward. In such a case the atmosphere is called stable, since any initial vertical motion is dampened. Stable conditions with very little vertical mixing are especially encountered during a temperature inversion, when temperature increases with altitude and thus the lapse rate is positive.

The transport and dispersion of chemical contaminants near ground level is dictated by the part of the atmosphere that is in close contact with the Earth's surface; the troposphere. The troposphere extends from the ground to a typical height of 11 km, depending on latitude. Within the bulk of the troposphere, temperature decreases with increasing height (i.e., the lapse rate is negative), and turbulent vertical mixing and unstable conditions are therefore an important characteristic of this atmospheric layer. The tropopause is the upper boundary of the troposphere. Here, temperature remains constant with increasing height, producing a stable atmospheric layer. Therefore there is only very slow mixing across the tropopause between the troposphere and the atmospheric layer above, the stratosphere.

The Earth's surface is at the lower boundary of the troposphere, and exerts a strong influence over a layer that is 100 to 3000 m high, depending on local conditions. The part of the troposphere that is directly influenced by the Earth's surface over a timescale of hours or less is called the boundary layer, and the remainder is referred to as the free troposphere (Stull, 1988). Figure 20.1 summarizes the vertical structure of the troposphere up to the stratosphere, including generic estimates of the timescale for vertical transport between the surface and the different layers recommended in the text by Jacob (1999).

Atmospheric processes in the boundary layer are of particular interest and importance since they directly impact contaminant concentrations in air near the surface. The text by Stull (1988) provides an excellent introduction to the meteorology of the boundary layer. Within the boundary layer, strong mechanical and thermal turbulence

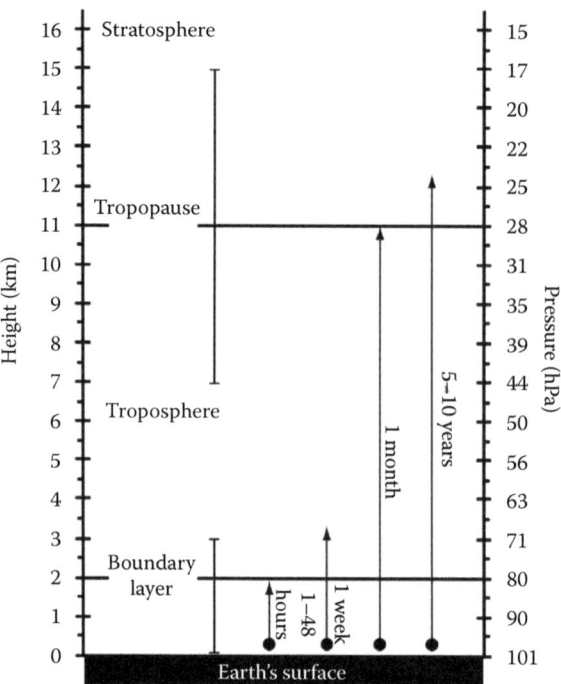

FIGURE 20.1 Vertical structure of the atmosphere near the Earth's surface. Error bars indicate the typical range of variability in the height of the boundary layer and the tropopause. Vertical arrows indicate timescales for transport (Jacob, 1999).

mixes the atmosphere, while the free troposphere is characterized by advective flow with only sporadic occurrence of turbulence. The height of the boundary layer is usually lower during a high pressure system due to subsidence of high-level air and low-level horizontal advection towards areas of lower pressure. Conversely, during a low pressure system, the height of the boundary layer is usually larger due to energetic turbulent mixing, and air may be transported to high altitudes of the troposphere within an hour or less.

When the temperature of the surface is lower than the overlying air, a local temperature inversion can form and the boundary layer height can be very low. For example, a stable low-level boundary layer regularly forms at night during high pressure conditions over land, when the surface cools more rapidly than the atmosphere after sunset. In such cases the boundary layer is referred to as a nocturnal boundary layer, since it forms shortly after sunset and breaks up at sunrise when solar heating of the surface and turbulent vertical mixing by large thermal eddies resumes. A more persistent low-level boundary layer that lasts for up to several days can be formed when warm air is advected into an area where the surface temperature is cooler. Within a stable boundary layer, horizontal winds are typically calm near the surface. It is difficult to define the height of a stable boundary layer because at the top it blends gradually into the turbulent air mass above, but a typical effective height of 100 m for the nocturnal boundary layer has been proposed by Stull (1988). In contrast to the situation over land, the boundary layer height over the ocean is relatively constant because the high heat capacity of water and mixing of the surface water layer prevents strong heating and cooling of the surface relative to the overlying atmosphere.

The formation of a stable low-level or nocturnal boundary layer can have a significant influence on concentrations of contaminants in air near the surface. Where there are sources of contaminants at ground level, concentrations can build up during the night because of inefficient mixing to higher altitudes. This is particularly evident during the early morning hours in major cities, when volatile air pollutants from motor vehicles accumulate in the nocturnal boundary layer before it is dispersed by convective mixing as the sun rises higher in the sky and heats the surface. The nocturnal boundary layer can also play an important role in determining the short-term variability in concentrations of semivolatile organic chemicals in air. MacLeod et al. (2007) have used a compartment-based multimedia contaminant fate model to interpret patterns of diel (24-h) variability of polychlorinated biphenyls (PCBs) at several different locations in terms of competing influences of atmospheric mixing height, temperature, windspeed, and hydroxyl radical concentration. They found that, at background sites, the formation of the nocturnal boundary layer drives lower concentrations during the night relative to the day due to (1) nighttime depletion of PCBs in the shallow nocturnal boundary layer, followed by (2) downward mixing of PCBs from higher attitudes in combination with temperature controlled volatilization from vegetation during the day. In contrast, in the city of Chicago, concentrations were higher at night, consistent with the urban area acting as a volatilization source of PCBs during both the day and night.

It should be noted that the description of vertical mixing processes in this section is appropriate for modeling movement of pollutants that are in the gas phase in the atmosphere, and pollutants attached to small, accumulation mode aerosols with diameters less than 1 μm that are mixed by turbulent eddies just like gasses. Pollutants attached to larger aerosols with diameters greater than 1 μm settle under the influence of gravity and may be transported vertically downward through a stable boundary layer (see Chapter 6 for information about settling velocities of aerosol particles). For a case study, the reader is referred to Moeckel et al. (2008), who developed a mass balance model that differentiates between the fine and a coarse particle size fraction in the atmosphere, and applied it to interpret the diel variability of concentrations of polybrominated diphenyl ethers at a rural site in northwest England during a five-day period characterized by formation of a nocturnal boundary layer.

20.2.3   HORIZONTAL MIXING IN THE ATMOSPHERE

At the global scale, the horizontal movement of air in the troposphere is driven by two major influences: pressure gradients caused by uneven heating of the Earth's surface and the Coriolis force produced by the rotation of the planet. Mathematical descriptions of the global circulation of the atmosphere can be formulated by balancing the Coriolis force and the pressure gradient (Jacob, 1999; Seinfeld and Pandis, 1998). The net effect of these two forces is that transport in the troposphere is most efficient along the east–west axis, parallel to the equator. Windspeeds along this axis in the troposphere are typically 10 m/s. Winds are slower along the north–south axis in the northern and southern hemispheres, typically 1 m/s, and are characterized more by turbulent mixing than by advection. Mixing between the northern and southern hemispheres is very inefficient because large-scale temperature differences favor vertical transport near the equator followed by migration of air masses aloft toward the respective poles. Figure 20.2 summarizes the typical timescales for large-scale horizontal transport in the troposphere recommended by Jacob (1999).

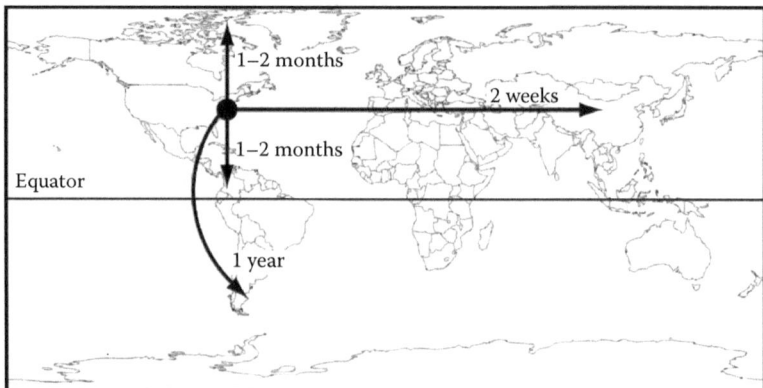

FIGURE 20.2 Typical timescales for global horizontal transport in the troposphere. (Adapted from Jacob, D.J. 1999. *Introduction to Atmospheric Chemistry.* Princeton University Press, Princeton, NJ.)

At regional scales, the mean daily horizontal windspeed is determined by a number of variable factors including local geographic features, proximity to oceans or seas, and, most importantly, the local weather conditions. Under stable high-pressure conditions over land the horizontal windspeed at the surface can display diel variability with a low at night concurrent with the formation of the nocturnal boundary layer. Wherever possible, site-specific data should be used to parameterize windspeeds in regional-scale contaminant fate models. However, a useful generic estimate of the mean tropospheric windspeed for a typical region is 4 m/s (14.4 km/h). This value has been used to represent the typical average horizontal windspeed in the troposphere in many model applications (e.g., in Beyer and Matthies, 2008; Wegmann et al., 2008; Beyer et al., 2000).

20.2.4 Summary of Mixing Processes in the Atmosphere

Order-of-magnitude estimates of mass transfer coefficients for vertical mixing across the different types of boundary layers shown in Figure 20.1 can be estimated by dividing the vertical height traveled by the characteristic time required for transport. Thus, the mass transfer coefficient across the tropopause can be estimated as (10,000 m/10 years) or approximately 0.1 m/h. This value can also be used to estimate mass transfer through a stable low-level or nocturnal boundary layer. The mass transfer coefficient across the planetary boundary layer can be estimated as (2000 m/1 week) or approximately 10 m/h. This value is higher because the planetary boundary layer is not always defined by a thermal inversion layer like the tropopause or the nocturnal boundary layer. In the bulk of the troposphere, typical vertical mass transfer coefficients of 60–360 m/h are associated with turbulent mixing (Seinfeld and Pandis, 1998). Therefore stable boundary layers and the planetary boundary layer are logical choices for defining the vertical height of well-mixed compartments when setting up a mass balance model.

Table 20.1 summarizes atmospheric mixing processes that control the transport and dispersion of chemical contaminants in the troposphere at various temporal and spatial scales. Some global-scale contaminant fate models use high spatial and temporal resolution and high computational effort to attempt to describe all of these processes simultaneously (Lammel et al., 2007; Hansen et al., 2004; Malanichev et al., 2004). We describe here an alternative and simpler approach in which the modeler considers the temporal and spatial scale of the phenomena that are of interest and then defines model characteristics accordingly. For example, many contaminant fate models are intended to be applied to describe the long-term mass balance of contaminants at the "regional scale," that is, in an area that is typically 100,000 km^2, or 316 km on a side if the region is square. These are the dimensions of the regional environment in the generic EQuilibrium Criterion (EQC) model (Mackay et al., 1996; Mackay, 2001), and similarly sized regions are used in other regional contaminant fate models. Assuming a windspeed of 14.4 km/h, the residence time of air in such a region is 22 h, or approximately one day. Over this timescale, contaminants can be expected to be mixed fairly efficiently within the planetary boundary layer, but not into the free troposphere. Thus, a single atmospheric compartment with a height of 1–2 km representing the planetary boundary layer has been adopted in such models. Considered together,

TABLE 20.1

Temporal and Spatial Scales of Atmospheric Mixing Processes that Control Chemical Concentrations near the Earth's Surface

Temporal Scale	Spatial Scale	Relevant Atmospheric Mixing Processes	
		Vertical	Horizontal
Hours	10s to 100s of kilometers	Turbulent mixing in the planetary boundary layer; formation and dissipation of a nocturnal boundary layer or a local temperature inversion boundary layer.	Local weather conditions, thunderstorms, sea breezes, diel variability in windspeed
1–7 days	100s to 1000s of kilometers	Turbulent mixing in the planetary boundary layer and exchange between the planetary boundary layer and the free troposphere	Local weather conditions, frontal weather systems
1–4 weeks	1000s of kilometers	Mixing in the troposphere	Intercontinental transport along the east–west axis and episodic north–south transfer due to pressure systems
1–6 months	>10,000 km	Mixing in the troposphere	Hemispheric mixing along east–west and north–south axes
>6 months	Global scale	Mixing in the troposphere; transfer to the stratosphere	Interhemispheric transfer, hemispheric mixing

Figures 20.1 and 20.2, and Table 20.1 provide guidance for setting up the atmosphere compartment (or compartments) in contaminant fate models based on the well-mixed box assumption.

When setting up a mass balance model based on the well-mixed compartment assumption, the modeler must be aware of the spatial and temporal scale of atmospheric mixing processes and formulate the model accordingly. The mass transfer coefficients suggested in this section can be used as generic first estimates to characterize atmospheric transport processes, however site-specific data should be used whenever possible to parameterize these processes. Depending on the spatial and temporal timescales of interest the model might consider only a single atmospheric compartment or two or more connected compartments. A simple example of model formulation for different atmospheric mixing conditions is provided as an example calculation in the following section.

20.2.5 Example Calculation

Case Study: Effect of the Nocturnal Boundary Layer on Ground-Level Concentration of Air Pollutants

The urban area of Indianapolis, Indiana, is approximately square with a total area of 1000 km^2. Sulfur hexafluoride (SF$_6$), a volatile, nondegrading substance is being released at ground level into the atmosphere in Indianapolis at a rate of 1 kg/h. Use a

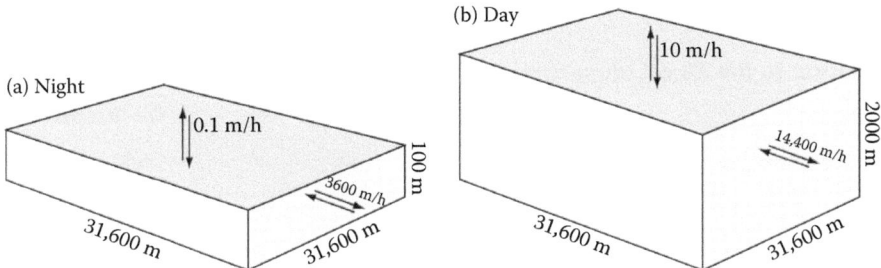

FIGURE 20.3 One-compartment box models of the air over Indianapolis, Indiana during; (a) night during a nocturnal temperature inversion, and (b) day with mixing to a planetary boundary layer. Dimensions are not drawn to scale.

one-compartment box model to estimate the steady-state concentration of SF_6 in air at ground level: (A) at night assuming a stable nocturnal boundary layer is in place and (B) during the day when there is turbulent mixing up to the planetary boundary layer. Assume that the horizontal windspeed is 1 m/s during the night and 4 m/s during the day.

In the nighttime scenario, assume that the height of the nocturnal boundary layer is 100 m and that the vertical mass transfer coefficient across the nocturnal boundary layer is 0.1 m/h. In the daytime scenario, assume efficient mixing up to the planetary boundary layer at 2000 m, and a vertical mass transfer coefficient across the planetary boundary layer of 10 m/h. Using these assumptions, we can represent the two scenarios using the one-compartment box models illustrated in Figure 20.3. Note that in a model of this type we must treat the emission as a "volume source" that is instantaneously mixed throughout the available volume of air. Characteristics of these two models systems and calculated steady-state concentrations under the two scenarios are presented in Table 20.2.

Since advection in air is the only significant removal process for SF_6 from the air compartment, the residence time of air in the region provides an estimate of the response time of the system under the two scenarios. In both cases the residence time of air is less than 12 h, so it is likely that the steady-state concentration will be approached during each day (or night). Because of a combination of smaller dilution volume and less efficient advection, the calculated steady-state concentration of SF_6 during the night-time scenario is 87 times higher than during the daytime scenario. In both scenarios, horizontal winds are much more efficient at removing the chemical from the air compartment than vertical mixing.

20.3 MIXING IN SURFACE WATERS

20.3.1 INTRODUCTION

The principal focus here is on lakes with only a brief discussion of mixing in rivers. The first and simplest generation of contaminant fate models treat lake stratification as simple homogenous or well-mixed units. Contaminant concentrations in the

TABLE 20.2
Solution to the One-Compartment Model Problem

	Nighttime Scenario	Daytime Scenario
Cross-sectional area	$3.16 \times 10^6 \, \text{m}^2$	$6.32 \times 10^7 \, \text{m}^2$
Area of region	$1 \times 10^9 \, \text{m}^2$	$1 \times 10^9 \, \text{m}^2$
Air volume	$1 \times 10^{11} \, \text{m}^3$	$2 \times 10^{12} \, \text{m}^3$
Horizontal volumetric airflow rate	$1.14 \times 10^{10} \, \text{m}^3/\text{h}$	$9.10 \times 10^{11} \, \text{m}^3/\text{h}$
(= cross sectional area × horizontal wind velocity)		
Vertical volumetric airflow rate	$1 \times 10^8 \, \text{m}^3/\text{h}$	$1 \times 10^{10} \, \text{m}^3/\text{h}$
(= area of region × vertical mass transfer coefficient)		
Total volumetric airflow rate	$1.15 \times 10^{10} \, \text{m}^3/\text{h}$	$9.20 \times 10^{11} \, \text{m}^3/\text{h}$
Residence time of air in compartment	8.7 h	2.2 h
(= air volume/total volumetric airflow rate)		
Chemical inventory at steady-state	8.7 kg	2.2 kg
(= emission rate × residence time of air)		
Chemical concentration in air	$87 \, \text{ng/m}^3$	$1.1 \, \text{ng/m}^3$
(= chemical inventory/air volume)		

lake water are functions of the rates of both inputs and outputs via rivers but are also modified by degradation processes and exchange at the sediment–water and water–atmosphere interfaces (Figure 20.4a). Mass transport through these interfaces is discussed in Chapters 6, 9, and 10. The presence of organic particles in the water is also an important factor because these particles provide absorptive sites for organic contaminants and are the vector for contaminant flux to and from sediments. They also influence the bioavailability of chemicals present in the water column to biota. However, this "well-mixed" approach is often too simplistic because many lakes of concern are not homogeneously mixed units for significant periods of time during the year. This is particularly true during the biologically active period of summer warming, which results in distinct zones in the lake that have minimal exchange, as shown in Figure 20.4b. The second and more realistic generation of contaminant fate models and corresponding estimates of the mass balance of contaminants in lakes require taking into account internal lake processes, particularly lake thermal stratification. This may become of increasing importance because the impact of climate change on lakes includes changes in the hydrology of the watershed and the physical features of stratification processes (e.g., Dillon et al., 2003; Schindler, 1997; Fee et al., 1996; Schindler et al., 1996). These climate induced changes may influence contaminant fate over time and change long-term predictions of impacts (e.g., French et al., 2006).

This chapter provides a review of the physical dynamics of typical temperate lakes with a view to evaluating the likely effects of these and related processes on the fate and mass balance of contaminants in lakes. We focus only on those processes that may have the largest contributions to measurable differences in contaminant concentrations relative to analytical accuracy or precision of measurement in environmental media such as water, sediments, and biota.

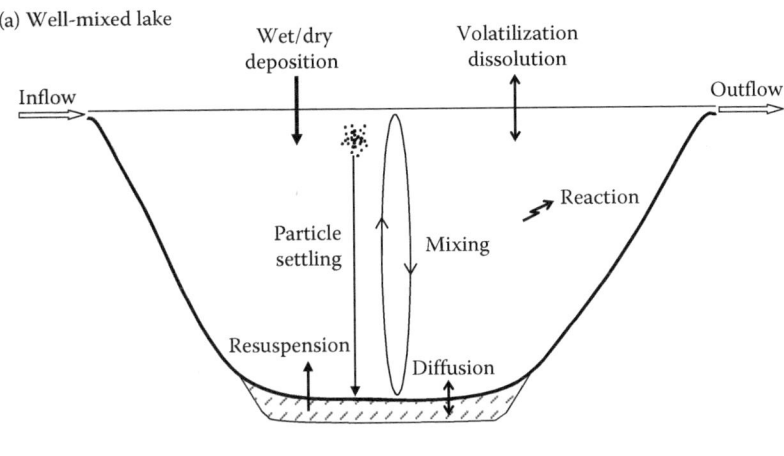

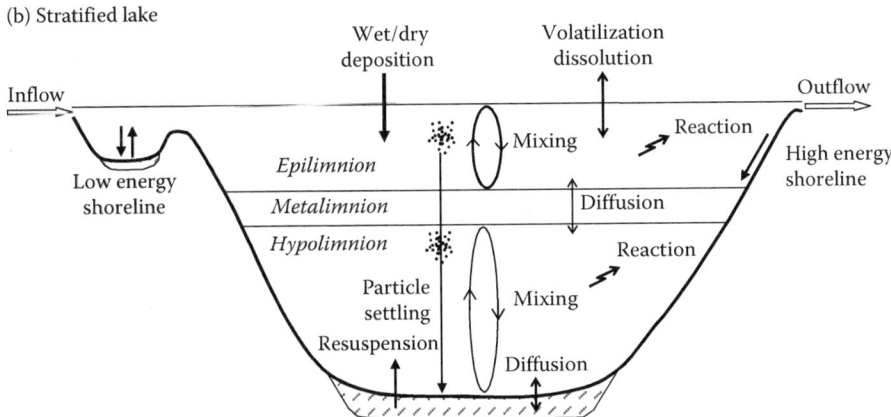

FIGURE 20.4 Cross-sectional representation of a lake showing the contaminant fate processes for (a) the "well-mixed" scenario and (b) the scenario that includes thermal stratification. The latter figure also includes representation of variable shorelines.

The focus is on two major physical processes in the lake and water column that affect contaminant fate and the overall mass balance as well as influencing air/water and sediment/water exchange processes. The first processes is thermal stratification, which is characteristic of temperate lakes, and affect the extent of mixing of water masses. This is a function of the heat budget in the lake as it controls temperature and density gradients along with the role of wind. Stratification phenomena may invalidate the application of the well-mixed, whole-lake assumption, necessitating a more complex treatment in which the water column is divided into thermally distinct regions with different rates of mixing with adjacent regions. Temperature also plays a role in determining physical–chemical properties, degradability, and rates of transfer. For example, a recent study by French et al. (2006) considered the role of air temperature on the persistence of organic contaminants and mercury in Lake Ontario salmonids. The second process is the movement and mixing of organic-rich particles and their fate and ultimate delivery to the sediments. Features of lake morphology as well as characteristics of the surrounding watershed need to be considered for both processes. Lake morphology is critical because it essentially defines the lake (e.g., Schindler, 1971). Geographic position (latitude and altitude) of the lake and regional climatic impacts are also important considerations in the heat budgets of lakes.

Temperate freshwater lakes in the Northern Hemisphere have been the focus of most limnological studies due to the historic prevalence of research institutions in temperate regions with high population densities adjacent to important water bodies such as the North American Great Lakes. These lakes have generally experienced the greatest anthropogenic disturbances over long time periods, including exposure to contaminants as well as problems with water quality attributed to nutrient enrichment (eutrophication) and acidification. More recent studies have included Southern Hemispheric lakes such as those in South America and Africa also subject to degradation as well as a broader variety of water bodies, including tropical and polar lakes and the highly abundant smaller systems such as ponds, which globally contribute to a significant fraction of aquatic productivity (Downing et al., 2006). Ponds, large but shallow tropical lakes, and lakes in colder climates differ in their physical and chemical properties relative to temperate lakes, and many of these may, or may not, be sufficiently characterized by the well-mixed models.

The aim of this chapter is to provide the reader who is assessing the fate of organic contaminants in a lake with guidance on the factors that render the "well-mixed" assumption invalid and on how parameters such as stratified layer depths, duration of stratification, and organic carbon fluxes may be obtained. In some cases, a complete limnological monitoring program may be justified and necessary. In other less critical cases, simplifying assumptions may be made and approximate parameter values selected. The sensitivity of chemical fate to these assumptions and parameters can then be explored. The reader is urged to consult other sources for more detailed explanations of the mechanisms (Kalff, 2002; Hutchinson, 1957; Wetzel, 2001; Imboden and Wüest, 1995).

20.3.2 THERMAL STRATIFICATION OF LAKES

This section describes the temporal changes in a typical temperate freshwater lake, but includes brief discussions of divergences from these patterns by lakes in other environments, including polar and tropical regions. Fortunately, despite a huge range in lake size and productivity, northern temperate lakes share common features in terms of their heat budgets. A fairly simple descriptive approach is taken here and those readers wishing to understand the underlying physics of these physical processes should refer to Imboden and Wüest (1995).

A fundamental determinant of physical mixing in a lake is the relationship between density of water and temperature, as modified by the lake's size and shape (Kalff, 2002; Wetzel, 2001; Hutchinson, 1957). The density of water varies over the range of temperatures characteristic of temperate lakes (0 to about 24°C; Figure 20.5). Water is unique in that it has higher density as a liquid relative to its solid state (ice) with the greatest density at 3.94°C. Lower densities are observed both below and

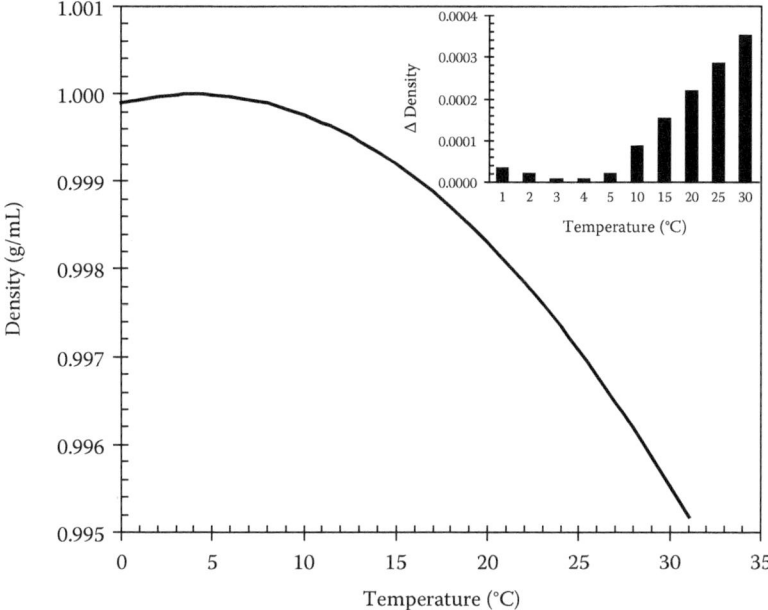

FIGURE 20.5 The relationship between density (D, g/mL) of average freshwater and temperature (T, °C) is nonlinear and can be estimated from the equation $D = 1 - 6.63 \times 10^{-6}$ (T-4)^2 (Kalff, 2002). The insert illustrates the changes in water density are greatest at the higher end of typical lake water temperatures.

above 3.94°C, with the difference in density substantially increasing for each degree increase in temperature as the water warms (Figure 20.5 insert). Thus, at 0°C the density of ice is 920 kg/m^3 and that of water is 999.8 kg/m^3. As temperature rises the density of water increases to its maximum 1000 kg/m^3 at 3.94°C. Density falls to 999.7 kg/m^3 at 10°C and further to 998.2 kg/m^3 at 20°C. Water has a high heat capacity of approximately 1 cal/(g K) or 4.2 J/(g K). These density differences may appear to be small but they have profound effects on the degree of mixing, as discussed below. Waters in temperate lakes in spring following ice out are generally close to 4°C and are isothermal from surface to bottom. Incoming solar radiation warms surface waters, thereby reducing the density near the surface, continuing throughout the spring to summer season. Since light intensity attenuates with depth as a result of absorption by water and particles, heating is not uniform and the warmer, less dense waters "float" over colder, higher density waters. A density gradient is formed with measurable differences in temperature producing the thermal stratification that is a fundamentally important feature driving many lake processes. It should be noted that salinity also affects water density (Kalff, 2002; Wetzel, 2001; Imboden and Wüest, 1995) and the stratification profiles of saline lakes may thus differ. There are convenient online density calculators available that include salinity; for example from the CSGNetwork (www.csgnetwork.com).

Temperate lakes may be completely or partially ice covered during winter. As spring progresses with increasing inputs of solar radiation and warm rains, the ice slowly degrades and becomes fractured by air and water currents, to a point where the loss of ice can occur quite suddenly over some hours. The water column temperature is around 4°C from top to bottom and with constant density. This period of cold isothermal waters that have little resistance to mixing by small amounts of wind energy is referred to as the spring turnover. Smaller and more protected lakes may have a relatively short period of spring turnover, responding rapidly to solar heating inputs while lakes with larger surface area and greater volume may extend this period for days to weeks. For example, the water of large, deep lakes such as the Laurentian Great Lakes or Lake Baikal take longer to warm from solar radiation and demonstrate a later onset of thermal stratification relative to nearby smaller lakes, which may stratify within days of ice-out. Overall, morphological features including mean depth (related to lake volume), surface area, and also fetch (the linear distance over which wind can pass) are modifying parameters on the type and duration of thermal stratification (Schindler, 1971; Hanna, 1990; Demers and Kalff, 1993; Edmundson and Mazumder, 2002).

Surface waters heat rapidly as the solar intensities increase during spring, but in deeper lakes (generally more than about 5–6 m depth) heat distribution by mixing is slower. Spring often includes periods of warm, calm weather persisting over several days. Lower density-warmer surface waters overlie deep, cool, higher-density waters and vertical stratification increases to a point that a difference of a few degrees Celsius prevents complete water circulation from top to bottom. This establishes a period of thermal stratification in which the open water column divides into three distinct regions. The top warm layer, the epilimnion, is relatively uniform in temperature; the water freely mixes and the extent of this layer is often referred to as the mixing depth (Figures 20.4b and 20.6). The reader is advised that there is some discrepancy in terms of defining mixing depth, which may be estimated to be at the bottom of the epilimnion or measured to the thermocline (e.g., Fee et al., 1996; Håkanson, 2004). With higher temperatures and solar inputs, the epilimnion is generally the most productive region of open waters (the pelagic zone). The lower region is the hypolimnion and is also relatively uniform in temperature, often from about 8°C down to 4°C near the sediments, with actual hypolimnetic temperature a function of the water temperature at onset of thermal stratification, lake size, and water transparency. Lakes with maximum depths in the range of 5–10 m generally have hypolimnia several degrees above 4°C during the summer whereas the hypolimnia of deeper lakes (i.e., over 10 m) more likely remain closer to 4°C (Schindler, 1971). The stratum that separates epilimnion from hypolimnion is the metalimnion and is characterized by a steep decline in temperature. The thermocline is defined as the plane of maximum rate of temperature decrease. The temperature change in the metalimnion may exceed 1°C/m. The metalimnetic depth deepens over the course of the season; as thermal inputs continue during the summer, the epilimnetic stratum becomes larger while the metalimnetic stratum shrinks. The hypolimnion will marginally gain heat over the course of the warm season and this compresses the metalimnetic stratum from below. Lake morphology influences metalimnetic depth, and thus thermocline profiles vary among lakes. It should also be noted that these

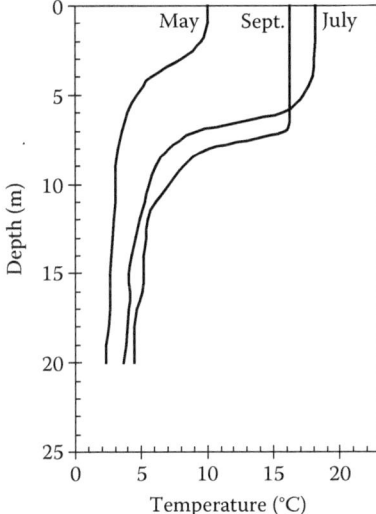

FIGURE 20.6 A temperature–depth profile for a temperate lake in Ontario, Canada (Eels Lake, 1992 data). Stratification occurs rapidly following loss of ice in late April and deepens over the summer; this was a unseasonably cool summer and surface temperatures never exceeded 20°C, about 2–3° cooler than most summers.

layers apply to the open waters of the lake. Depending upon the slope of the lake bottom from the shoreline, the inshore region (littoral zone) may exchange with the eplimnetic waters only (Figure 20.4b); this is a higher energy zone subject to wave turbulence.

A reversal of the above process occurs with progression towards autumn, with declining air temperatures and reduced solar inputs. Heat is gradually lost from the lake as air temperatures fall, with resulting increasing density of epilimnetic waters. Resistance to mixing is slowly reduced until a point is reached when energy from wind is sufficient to cause complete mixing during the fall turnover. Timing of fall turnover is a function of both air temperatures and average hypolimnetic temperature (Nurenberg, 1988). However, larger lakes are slower to cool and typically experience later turnover relative to nearby smaller lakes. Eventually, cold winter temperatures may result in ice cover over the lake. Water immediately below the ice nears the freezing point but water overlying sediments may be several degrees warmer. This bottom water is denser, and an inverse stratification may become established during the winter, albeit the range of temperatures is reduced relative to the summer stratification period.

Lakes with two turnover events per year are dimictic. Monomictic lakes may have a single turnover event, while other lakes described as polymictic experience multiple episodes of weaker stratification and turnover events. Some permanently ice covered lakes never turn over, are amictic and maintain isothermal temperatures due to a lack of wind induced mixing. At times of calm, stable weather during the period of summer stratification, near-surface thermoclines may become established but these are typically less than 1 m in depth (Xenopoulos and Schindler, 2001); these temporary thermoclines may influence mass balances but are not expected to be important within the error of most mass balances.

Several key factors modify the timing, extent, and stability of thermal stratification, notably climate together with lake-specific morphological features. Regionally, air temperature and solar radiation are important factors. Other parameters of importance are the water temperatures, the thermocline depth (which controls the volume of the epilimnion) and the duration of stratification. Direct measurement of these parameters is preferable but correlations also exist to estimate these quantities as reported in the literature (e.g., Cahill et al., 2005; Futter, 2003; Gao and Stefan, 1999).

20.3.3 BIOLOGICAL IMPLICATIONS AND PARTICLE DEPOSITION

The presence of stratification results in significant changes to the biological/ecological processes within the lake. The more intensely illuminated epilimnion is inevitably the zone of greatest biological (autotrophic) production, namely the conversion of inorganic carbon (OC) into organic carbon (C). For mass balances, this can be considered the zone of organic carbon "solvent" production. The production and amount of particulate organic matter (POM) is related to concentrations of inorganic nutrients, notably phosphorus (P) and nitrogen (N) as the major limiting forces of autotrophic production in freshwaters (Sterner, 2008). However, it should be noted that there is substantial recycling of materials within the epilimnion due to grazing by heterotrophic species (e.g., zooplankton) resulting in remineralization of the nutrients followed by rapid reutilization by both autotrophs and heterotrophic bacteria (Bentzen et al., 1992). Overall, particle-mediated mass transfer is likely to be largely a one-way process associated with settling particles in the form of fecal pellets and occurs at a rate corresponding to some fraction of total epilimnetic primary production (Figure 20.4). This issue and the importance of the organic carbon budget and fluxes are discussed more fully in Chapter 10.

The metalimnion, which may be a few meters thick by midsummer, provides a real barrier to chemical movement by advection and diffusion due to the density changes of the water. Diffusion through the metalimnion is likely slow given the relatively long path length and is a function of the concentration/fugacity gradient. The movements of dissolved gases such as oxygen and carbon dioxide have been well studied and provide the basis for modeling the mass transfer of contaminants using Fick's law of diffusion. As a consequence, lakes with high production often experience hypolimnia, which may go anoxic over the summer due to the net flux of organic rich particles settling down from the high-production epilimnia. The lower light environment of the hypolimnion typically results in production dominated by bacteria (heterotrophic production) who consume oxygen while degrading the suspended material (Kalff, 2002). The decline of oxygen concentrations in the hypolimnion again may alter chemical partitioning and reactions both for organic and metallic substances and especially at and below the sediment–water interface. There may be relatively little or no mixing in the hypolimnion during summer stratification, thus there is a tendency toward net loss to the sediments at least until the lake turns over. The deepest lakes such as Lake Baikal may have bottom hypolimnia that rarely, if ever, experience mixing even during turnover (Kalff, 2002). As a result, some models treat the hypolimnion as a series of strata. Limited mixing is also evident by the establishment of the nepheloid layer just above sediments. This layer has been studied more extensively in marine systems but has been the subject of recent studies in large lakes. Some deep water mixing may be the result of either internal currents or internal waves (seiches). For our example, we restrict the hypolimnion to a single compartment but this is an obvious simplification.

20.3.4 SUMMARY OF MIXING AND DIFFUSIVE PROCESSES WITHIN LAKE STRATA

In terms of mass balances and contaminant fate, stratification is important for a number of reasons. It reduces the active volume of water that exchanges between the atmosphere during the warm seasons that dominate chemical exchange processes. This is a seasonally ongoing process beginning in spring continuing to the maximum extent before fall turnover (Figure 20.7). It is represents a smaller volume receiving contaminant inputs from inflows and nonpoint sources (e.g., runoff from land). For a given loading of chemical, predicted concentrations in epilimnia will be higher than levels predicted from the well-mixed models based upon the whole lake volume. Volatilization may then become more important. The atmosphere likely plays a smaller role in transfer of contaminants to deeper waters while stratification persists, and this may increase the importance of particle-mediated transport to deeper waters and the sediments. Additionally, the hypolimnion is not well-mixed since the

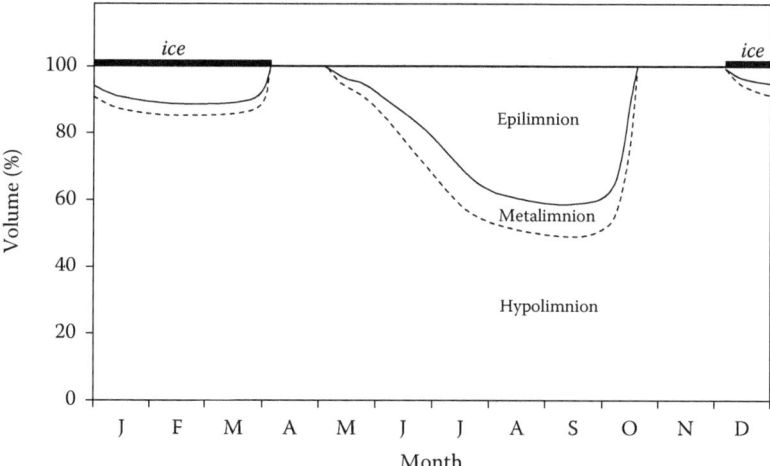

FIGURE 20.7 A theoretical profile of volume of the thermal regions due to changes in lake temperature over the course of a calendar year. The period of ice cover is indicated by the black bar.

effect of wind-driven mixing does not extend through the thermocline when the lake is stratified. On the other hand, sediments in shallow water are subject to the high energy of wind and wave action, and thus the particle-mediated processes of sediment deposition, resuspension, and composition (particle size, density, and organic carbon content) will have more influence on contaminant concentrations and fate. A net effect of this is the process known as sediment focusing, which results in particle transport and preferential deposition in the deepest (low energy) areas of the lake; near-shore sediments may not retain contaminants as readily as deep water sediments (Kalff, 2002). Deeper sediments where most of the deposition occurs are also typically characterized by fine inorganic silts and clay and small grained organic material. The location and depth of these regions of deposition is a function of prevailing currents the maximum fetch (Kalff, 2002).

20.3.5 MODELING AND PARAMETERIZING MASS TRANSPORT IN SURFACE WATERS

20.3.5.1 Lakes

The effect of mixing on mass transport is most serious in lakes, especially as a result of stratification as discussed earlier. When compiling and parameterizing any mass balance model it is essential to address the existence, nature, and duration of stratification. A critical parameter is the vertical diffusion coefficient, which controls transport across the thermocline. This quantity can be calculated from vertical profiles of temperature or from radiotracer measurements.

Schnoor (1987) has compiled these coefficients for a number of lakes. In most cases, the thermocline depth is 10 ± 5 m. The vertical dispersion coefficients across the thermocline are approximately 0.02 cm^2/s plus or minus a factor of 5, that is, 0.004–0.10 cm^2/s whereas whole lake coefficients are considerably larger and in the range 0.10–4 cm^2/s.

A mass transfer coefficient can be defined as the vertical dispersion coefficient (cm^2/s or m^2/h) divided by the mean path length (cm or m) between the water compartments, yielding a transport velocity (cm/s or m/h). This in turn can be multiplied by the area of transfer (cm^2 or m^2) to give a hypothetical volumetric exchange flow rate (cm^3/s or m^3/h). In the absence of data for a specific system, an initial estimate of 0.02 cm^2/s or 0.0072 m^2/h can be assumed and the sensitivity of the model results to this parameter evaluated. If the value proves to be critical, measurements of temperature profiles may be desirable as a basis for estimating the mass transfer rate.

20.3.5.2 Rivers

Rivers present a number of challenges when developing and parameterising mass balance models. In addition to the continuous flow of water down the channel, water velocities generally are not consistent across the width of the river. Rivers rarely flow in a straight path, and the curves can result in relative calm embayments to the sides of the main flow. Water depth is also variable both across the width as well as along the length of the river and this influences the velocity of water along the sediments.

Furthermore, rivers often have multiple contaminant point source inputs from either streams or discharge as well as variable runoff from land as a function of land use. Dams along the river provide further complications.

The general approach is to segment the river or the river network into a number of reaches and write mass balance equations for each reach including all relevant input and output processes. The advective–dispersive equation can be applied to each reach as a control volume, the key processes and parameters being:

Contaminant mass inputs from sources such as runoff and effluents
Exchange with the atmosphere and bed sediments
Advective inputs and outputs associated with the river flow and tributaries
Transformation or degradation reactions
Mixing, diffusion or dispersion in all three dimensions

These mixing processes are generally characterized by diffusion or dispersion coefficients, all with units such as m^2/h. Molecular diffusion coefficients in water are isotropic and typically $0.4\,m^2/h$ or $10^{-5}\,cm^2/s$. They are usually negligible in comparison to eddy diffusion or dispersion. Eddy or turbulent diffusion coefficients are controlled by water flow, wind, biotic, and buoyancy effects and may be anisotropic. Dispersion coefficients are controlled by velocity gradients in the water and are anisotropic with vertical, lateral, and longitudinal components. These individual coefficients can be added to give a net coefficient that depends on flow conditions and the system geometry.

Vertical turbulent diffusion rates are usually sufficiently fast that vertical homogeneity can be assumed. A typical vertical dispersion coefficient is $100\,cm^2/s$ or $0.01\,m^2/s$ thus a characteristic time for vertical mixing in a 1 m deep river is about 100 s or a few minutes. If the water residence time in the reach is 2000 s (e.g., a 1000 m reach with a velocity of 0.5 m/s) vertical mixing is essentially complete and vertical homogeneity can be assumed.

Lateral dispersion coefficients can be important in wide, slow flowing rivers. These coefficients are typically $0.005\,m^2/s$, thus the characteristic time for lateral mixing in a river 10 m wide can be 20,000 s or 6 h. Lateral homogeneity is unlikely to be achieved in a short reach, but dispersion rates are very site-specific and are best determined empirically using, for example, dye tracer tests.

Longitudinal dispersion coefficients are influenced by current velocity, depth, river bends, and back-waters. Longitudinal dispersion coefficients range typically from 5 to $100\,m^2/s$. The modeler thus faces the problem of selecting a plug flow model of the river (implying zero longitudinal dispersion) or a well-mixed box model (implying fast longitudinal and lateral dispersion over each reach) or a more complex dispersion model (implying limited dispersion). These models can give very different results if other loss processes are relatively rapid.

The reader is referred to the report and text by Schnoor (1987, 1996) to citations therein, to texts on river and lake modeling such as those by Chapra and Reckhow (1983), Shanahan and Harleman (1984), and the texts cited in Chapter 1, Table 1.1 for more detailed accounts of models and the relevant parameters.

TABLE 20.3
Parameters Used for the Lake Case Study
Described in Section 20.3.6

Parameter	Value
Lake area	$40 \times 10^6 \, m^2$
Lake depth	$25 \, m$
Lake volume	$10^9 \, m^3$
Thermocline depth	$10 \, m$
Water outflow	$10^5 \, m^3/h$
Reaction rate constant	$10^{-3}/h$ (half life 693 h)
Volatilization MTC	$0.01 \, m/h$
Net MTC to sediment	$0.001 \, m/h$
Thermocline diffusivity	$0.0072 \, m^2/h$
Path length	$12.5 \, m$
Thermocline MTC	$5.76 \times 10^{-4} \, m/h$
Chemical input	$100 \, g/h$

Note: MTC, mass transfer coefficient.

20.3.6 CASE STUDY OF A LAKE MASS BALANCE WITH AND WITHOUT STRATIFICATION

We consider a lake 10 km × 4 km and depth 25 m subject to input of a volatile organic chemical. Typical values for mass transfer and degradation parameters are assumed as shown in Table 20.3. Mass balances are compiled for the well mixed lake under spring and fall conditions and for summer stratified conditions with a thermocline at 10 m depth. For simplicity, a relatively small net rate constant for loss to bottom sediments is assumed.

Figure 20.8a gives the mass balance diagram for the well mixed lake. Degrading reactions and volatilization largely control the concentration in the water. This figure also gives the corresponding diagram for the stratified lake (Figure 20.8b). The concentration in the epilimnion increases by a factor of 1.6 because of a reduced reaction rate caused by the lower reaction volume. There is faster volatilization caused by the increase in surface concentration. The most significant difference is the concentration in the hypolimnion, which is reduced by a factor of 17. This is caused by the thermocline restricting the rate of transport of chemical into and out of this region. Reaction is the principal loss process in the hypolimnion.

It is relatively straightforward to test the sensitivity of the results to changes in the diffusion coefficient at the thermocline. For example, reducing the coefficient by a factor of 5 virtually isolates the hypolimnion from the epilimnion resulting in low concentrations in the hypolimnion. Increasing it by a factor of 5 brings the two concentrations closer, the ratio falling from 17 to 6.5. A large increase in diffusion rate by a factor of 10 or more results in near well-mixed conditions.

The concentrations and fluxes are thus profoundly influenced by lake thermal stratification, however, the nature of the effect is specific to chemical properties such

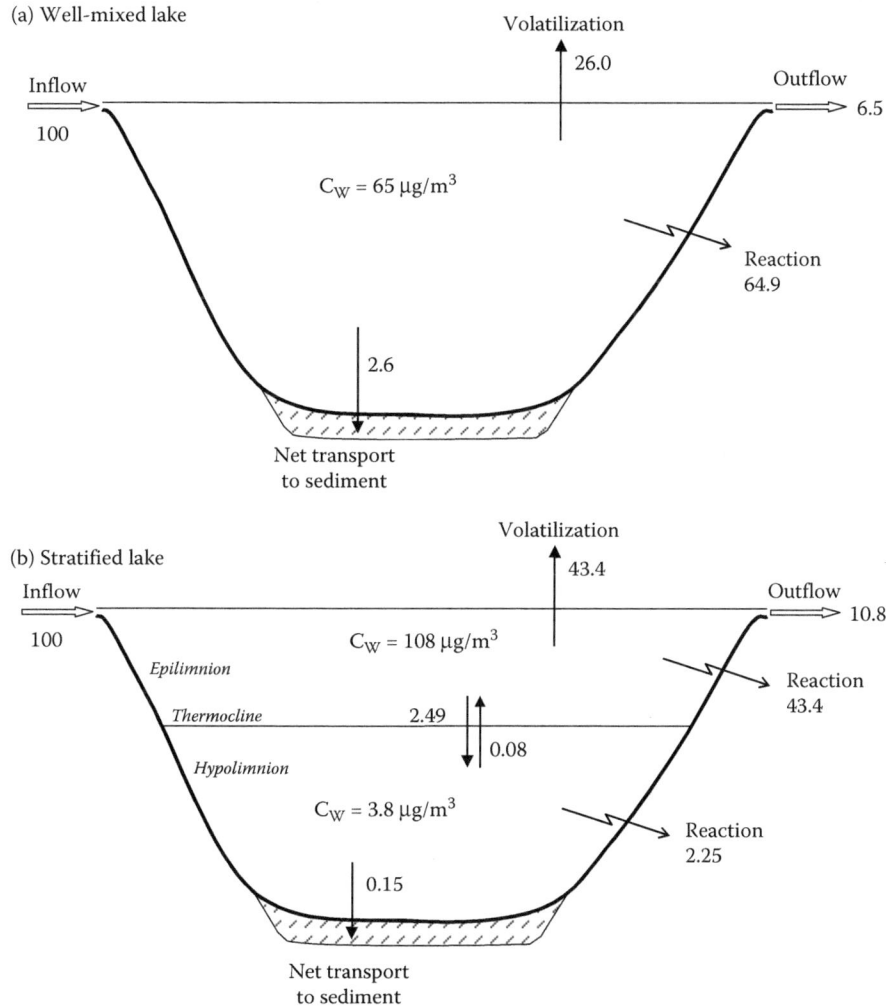

FIGURE 20.8   Mass balance diagram showing the effect of thermal stratification on chemical fate and transport for the (a) "well-mixed" model and (b) the stratified lake. Refer to Table 20.2 and the text for details of the equations and morphometric parameters used.

as partition coefficients from water to air and solid phases and especially to potential degradation rate constants, which are likely to differ between the layers. Transport of chemical by particle deposition will also influence concentrations and bioavailability, especially during periods of high productivity in the summer. Although the case study presented here is simplistic and hypothetical, it does demonstrate that restrictions to mixing because of stratification processes can have a profound effect on contaminant fate and that these effects can be, and in some cases must be, addressed in fate simulations.

REFERENCES

Bentzen, E., W.D. Taylor, and E.S. Millard. 1992. The importance of dissolved organic phosphorus to phosphorus uptake by limnetic plankton. *Limnol. Oceanogr.* 37: 217–231.

Beyer, A., D. Mackay, M. Matthies, F. Wania, and E. Webster. 2000. Assessing long-range transport potential of persistent organic pollutants. *Environ. Sci. Technol.* 34: 699–703.

Beyer, A. and M. Matthies. 2008. ELPOS—Criteria for persistence and long-range transport potential of pesticides and industrial chemicals. http://www.usf.uni-osnabrueck.de/projects/elpos/#model (accessed January 15, 2008).

Cahill, K.L., J.M. Gunn, and M.W. Futter. 2005. Modelling ice cover, timing of spring stratification, and end-of-season mixing depth in small Precambrian Shield lakes. *Can. J. Fish. Aquat. Sci.* 62: 2134–2142.

Chapra, S.C. and K.H. Reckhow. 1983. *Engineering Approaches for Lake Management*, Vol. 2: *Mechanistic Modeling*. Butterworth Publishers/Ann Arbor Science, Woburn, MA.

Demers, E. and J. Kalff. 1993. A simple model for predicting the date of spring stratification in temperate and subtropical lakes. *Limnol. Oceanogr.* 38: 1077–1081.

Dillon, P.J., B.J. Clark, L.A. Molot, and H.E. Evans. 2003. Predicting the location of optimal boundaries for lake trout (*Salvelinus namaycush*) in Canadian Shield lakes. *Can. J. Fish. Aquat. Sci.* 60: 959–970.

Downing, J.A., Y.T. Prairie, J.J. Cole, C.M. Duarte, L.J. Tranvik, R.G. Striegl, W.H. McDowell, P. Kortelainen, N.F. Caraco, J.M. Melack, and J.J. Middelburg. 2006. The global abundance and size distribution of lakes, ponds, and impoundments. *Limnol. Oceanogr.* 51: 2388–2397.

Edmundson, J.A. and A. Mazumder. 2002. Regional and hierarchical perspectives of thermal regimes in subarctic, Alaskan lakes. *Freshwater Biol.* 47: 1–17.

Fee, E.J., R.E. Hecky, S.E.M. Kasian, and D.R. Cruikshank. 1996. Effects of lake size, water clarity, and climatic variability on mixing depths in Canadian Shield lakes. *Limnol. Oceanogr.* 41: 912–920.

French, T.D., L.M. Campbell, D.A. Jackson, J.M. Casselman, W.A. Scheider, and A. Hayton. 2006. Long-term changes in legacy trace organic contaminants and mercury in Lake Ontario salmon in relation to source controls, trophodynamics, and climatic variability. *Limnol. Oceanogr.* 51: 2794–2807.

Futter, M.N. 2003. Patterns and trends in Southern Ontario ice phenology. *Environ. Monitor. Assess.* 88: 431–444.

Gao, S. and H.G. Stefan. 1999. Multiple linear regression for lake ice and lake temperature characteristics. *J. Cold Regimes Eng.* 13: 59–77.

Håkanson, L. 2004. *Lakes: Form and Function*. Blackburn Press, New Jersey. 201pp.

Hanna, M. 1990. Evaluation of models predicting mixing depth. *Can. J. Fish. Aquat. Sci.* 47: 940–947.

Hansen, K.M., Christensen, J.H., Brandt, J., Frohn, L.M., and Geels, C. 2004. Modelling atmospheric transport of alpha-hexachlorocyclohexane in the Northern Hemisphere with a 3-D dynamical model: DEHM-POP. *Atmos. Chem. Phys.* 4: 1125–1137.

Hutchinson, G.E. 1957. *A Treatise on Limnology*, Vol. 1: *Geography, Physics, and Chemistry*. Wiley, New York, 1015pp.

Imboden, D.M. and A. Wüest. 1995. Mixing mechanisms in lakes, in A. Lerman, D. Imboden and J. Gat (eds) *Physics and Chemistry of Lakes*, 2nd ed. Springer-Verlag, Berlin, pp. 83–138.

Jacob, D.J. 1999. *Introduction to Atmospheric Chemistry*. Princeton University Press, Princeton, NJ.

Kalff, J. 2002. *Limnology*. Prentice-Hall, Englewood Cliffs, NJ, 592pp.

Lammel, G., Klopffer, W., Semeena, V.S., Schmidt, E., and Leip, A. 2007. Multicompartmental fate of persistent substances: Comparison of predictions from mulimedia box models and a multicompartment chemistry—atmospheric transport model. *Environ. Sci. Pollution Res.* 14: 153–165.

Mackay, D. 2001. *Multimedia Environmental Models: the Fugacity Approach*, 2nd ed. Lewis Publishers, Boca Raton, FL.

Mackay, D., A. Di Guardo, S. Paterson, and C.E. Cowan. 1996. Evaluating the environmental fate of a variety of types of chemicals using the EQC model. *Environ. Toxicol. Chem.* 15: 1627–1637.

MacLeod, M., M. Scheringer, H. Podey, and K. Hungerbühler. 2007. The origin and significance of short-term variability of semi-volatile contaminants in air. *Environ. Sci. Technol.* 41: 3249–3253.

Malanichev, A., E. Mantseva, V. Shatalov, B. Strukov, and N. Vulykh. 2004. Numerical evaluation of the PCBs transport over the Northern Hemisphere. *Environ. Pollution* 128: 279–289.

Moeckel, C., M. MacLeod, K. Hungerbühler, and K.C. Jones. 2008. Measurement and modeling of diel variability of polybrominated diphenyl ethers and chlordanes in air. *Environ. Sci. Technol.* 42(9): 3219–3225.

NASA. 2008. National Aeronautics and Space Administration. "Earth Fact Sheet" http://nssdc.gsfc.nasa.gov/planctary/factsheet/earthfact.html, Last accessed 10 May 2010.

Nurenberg, G.K. 1988. A simple model for predicting the date of fall turnover in thermally stratified lakes. *Limnol. Oceanogr.* 33: 1190–1195.

Schindler, D.W. 1971. Light, temperature, and oxygen regimes of selected lakes in the Experimental Lakes Area, Northwestern Ontario. *J. Fish. Res. Bd. Canada* 28: 157–169.

Schindler, D.W. 1997. Widespread effects of climatic warming on freshwater ecosystems in North America. *Hydrol. Process.* 11: 1043–1067.

Schindler, D.W., S.E. Bailey, B.R. Parker, K.G. Beaty, D.R. Cruikshank, E.J. Fee, E.U. Schindler, and M.P. Stainton. 1996. The effects of climatic warming on the properties of boreal lakes and streams at the Experimental Lakes Area, northwestern Ontario. *Limnol. Oceanogr.* 41: 1004–1017.

Schnoor, J.L. 1987 Processes, coefficients and models for simulating toxic organics and heavy metals in surface waters. US EPA report EPA/600/3-87/015.

Schnoor, J.L. 1996 *Environmental Modeling: Fate and Transport of Pollutants in Water, Air, and Soil*. Wiley, New York.

Sterner, R. 2008. On the phosphorus limitation paradigm for lakes. *Int. Rev. Hydrobiol.* 93: 433–445.

Stull, R.B. 1988. *An Introduction to Boundary Layer Meteorology*. Kluwer Academic, Boston, MA.

Seinfeld, J. H. and S.N. Pandis. 1988. *Atmospheric Chemistry and Physics: From Air Pollution to Climate Change*. Wiley, New York.

Shanahan, P. and D.R.F. Harleman. 1984. Transport in lake water quality modeling. *J. Environ, Eng. ASCE* 110: 42–57.

Warren, C.S., D. Mackay, E. Webster, and J.A. Arnot. 2009. A cautionary note on implications of the well-mixed compartment assumption as applied to mass balance models of chemical fate in flowing systems. *Environ. Toxicol. Chem.* 28(9): 1858–1865.

Wegmann, F., L. Cavin, M. MacLeod, M. Scheringer, and K. Hungerbühler. 2008. A software tool for screening chemicals of environmental concern for persistence and long-range transport potential. *Environ, Modeling Software* 24(3): 228–237.

Wetzel, R.G. 2001. *Limnology: Lake and River Ecosystems*, 3rd ed. Academic Press, San Diego, CA, 1006pp.

Wikipedia, "Karman Line" http://en.wikipedia.org/wiki/K%C3%A1rm%C3%A1n_line (accessed January 15, 2008).

Xenopoulos, M.A. and D.W. Schindler. 2001. The environmental control of near-surface thermoclines in Boreal lakes. *Ecosystems* 4: 699–707.

21 Environmental Mass Transfer State-of-the-Art, Deficiencies, and Future Directions

Justin E. Birdwell, Louis J. Thibodeaux, and Donald Mackay

CONTENTS

21.1 Introduction 589
21.2 The Current State of Estimating Environmental Mass Transport
 Coefficients 589
21.3 Deficiencies in the Field of Environmental Mass Transfer 591
21.4 A Glimpse at the Way Forward...593
Literature Cited 594

21.1 INTRODUCTION

In the Preface to this handbook we argued that the subject of mass transfer is a "poor relation" compared to the sciences of chemical partitioning and reactivity in the environment. The science of mass transfer is less mature in its development. Uncertainties about transport rates are, we believe, a significant cause of error in simulations of chemical fate in the environment. In this final chapter we present a personal perspective assessing the state-of-the-art for chemical transport coefficient estimation and we offer a cursory review of the deficiencies in the field of environmental mass transfer that should be the focus for future research endeavors.

21.2 THE CURRENT STATE OF ESTIMATING ENVIRONMENTAL MASS TRANSPORT COEFFICIENTS

The transport rates of chemical substances and particles are relatively rapid when compared to the scale of individual human life times. These fast rates have caused all environmental media on the surface of the Earth to be contaminated with both naturally occurring and anthropogenic substances. Luckily, the levels of contamination are generally low and most are below levels of concern. The levels at some locales within nearly all media are high and are properly termed polluted and therefore hazardous to selected organisms. Multimedia chemical fate models of all genres will continue to play a significant role in assessing contamination levels over a wide range of scales from local to global so as to protect human and ecological health.

This handbook presents the accumulated knowledge for predicting environmental transport rates as needed in these chemical fate models. Therefore it is appropriate to assess the state of development by performing a judging and ranking activity involving the individual processes. The outcome will, we hope, be helpful in identifying the knowledge gaps and directions for future studies and research.

The fundamental principles of chemical transport are well advanced particularly in the fields of chemistry, chemical engineering, and physics. This handbook extends these transport principles to addressing the many applications required in predicting chemical fate in nature using multimedia models. Thirty-eight experts have contributed, producing 20 chapters of technical information, summarizing theoretical material and offering practical data, equations, and guidance for estimating these parameters. All together, it is a broad subject area. It provides, among other things, the technical information on the current state of science and engineering for estimating environmental mass transport coefficients (MTCs). In Table 4.1 (Chapter 4), 41 individual processes were identified as being significant transport processes. Many of these processes, though not all, have been studied sufficiently so that it is possible to make meaningful predictions of MTCs to forecast mass transfer rates.

The evaluation will focus only on the individual transport processes listed in the table. They are conveniently divided into four groups based primarily on the four environmental media compartments: air, soil, water, and plants and involve interphase as well as intraphase processes. The levels of good, average, and poor are used as the grade scales. They indicate the state-of-the-engineering correlating measurements and empirical algorithms for making reliable estimates of the MTCs. Judgments are made considering (1) the complexity and accessibility of the media interface, (2) the combination of biological, chemical, and physical processes occurring, (3) degree of development of theoretical models describing and laboratory studies performed simulating processes, and (4) the quantity and quality of field measurements and validation studies performed.

The results of the evaluation made by the authors of this chapter follows; brief comments explaining our assessment accompany the rankings.

1. *The air–water interface.* It is readily accessible and dominated by physical processes. Both the air-side and water-side boundary layer coefficients estimation methods are ranked as good.

2. *The water–sediment interface.* It is not readily accessible and may be occupied by macro and micro fauna as well as macro flora. Except for sites where the presence of subaquatic vegetation dominates the transport behavior at the benthic boundary layer, the grade for the water-side is good to average, though this depends to a certain extent on the type of aquatic system (fresh,

estuarine, marine, etc.). The sediment side is judged to be average to poor due to the general absence of chemical flux data.

3. *The soil–air interface.* It is readily accessible but may be occupied by diverse biological plant and animal species. For the case of the air-side MTC adjacent to generally flat and bare soil the rank is good. It falls to average to poor if the plants canopies are present and the same ranking applies to the soil resuspension MTCs. The soil-side usually contains solid, liquid, and gas phases. Not only does the presence or absence of liquid water directly impact chemical transport by its effect of porosity and the availability of an aqueous diffusive route, but it is a significant determinant of the nature and extent of biological activity as in deserts vs. wetlands. Therefore, it is ranked poor. Plant uptake from soil is ranked poor and from air is ranked average.

4. *Intraphase transport.* It is generally termed an environmental dispersion process and applies to the atmosphere, water bodies, and groundwater. Those in the atmosphere and water bodies in both horizontal and vertical dimensions are ranged good to average. Groundwater chemical mass transport is ranked average mainly due to limited direct accessibility.

Although tempered by the information available as presented in the handbook, the rankings also reflect the experiences and opinions of the authors. Nevertheless, the rankings provide an indication of the remaining gaps in knowledge on the subject of chemical transport in the environment. Because most transport processes occur in parallel and series simultaneously and "a chain is only as strong as its weakest link," it is fair to conclude that the technology ranking of estimating methods for overall environmental MTCs is well below average and inadequate in many aspects so that much work remains to be done in the field.

21.3 DEFICIENCIES IN THE FIELD OF ENVIRONMENTAL MASS TRANSFER

In the realm of environmental mass transport processes, there are a wide range of issues that have not been addressed sufficiently in order to facilitate robust, reliable forecasts of the distribution and fate of many important, longstanding pollutants as well as a whole host of emerging contaminants. For some hazardous materials, their behavior in the environment is poorly documented and virtually unknown. For many substances that have been examined, too little information is available to assess the reliability of and uncertainty in modeling forecasts of concentrations and fluxes. There are even new or previously undefined environmental compartments that have been found to be significant sources and sinks for a range of pollutants and pathogens for which little if any information is available describing the transport rates and mechanisms relevant to these natural (animal feces) and human-made (urban organic films) phases. Another example is the indoor environment in which human exposure occurs by ingestion, inhalation of gases and dust, and dermal contact.

The vast majority of research on the transport behavior of anthropogenic chemicals in the environment has involved materials often described as "Persistent Organic Pollutants (POPs)," which includes specific compound classes such as chlorinated solvents (trichloroethylene, perchloroethylene, chloroform, etc.), polycyclic aromatic hydrocarbons (PAHs) (naphthalene, phenanthrene, benzo(*a*)pyrene, etc.), and various chlorinated aromatic compounds such as polychlorinated biphenyls (PCBs), dichlorodiphenyltrichloroethane (DDT), and chlorobenzenes. Other materials that have been extensively studied include simple gases, aerosols, particles, radionuclides, and certain metals. Recently, many additional products and by-products produced by human activities have become the focus for concern. These materials can be broadly described as emerging contaminants, though many have been used for decades. More specifically, human-made nanoparticles, personal care products and pharmaceuticals, new agricultural and industrial chemicals, and various pathogens (microbial, bacterial, and viral) are being examined by a number of regulatory agencies and researchers to assess their environmental impact.

The same transport mechanisms that affect the movement of well studied environmental pollutants can be expected to affect emerging contaminants. Diffusion and advection processes are expected to affect carbon nanotubes, antibiotics, endocrine disruptors, biocides, fire retardants, surfactants, ionizing substances, bacteria, viruses, or other solutes very similar to the molecular and particulate pollutants mentioned above. However, the interactions of these "new" pollutants with environmental interfaces and their potential for transformation and alteration concomitant with transport through the environment will, in many cases, be much more important to understanding the impact they will have on environmental quality and human health. The persistent pollutants have been found to be fairly stable in many cases in the environment; PCBs found in Arctic and Antarctic regions retain much of the same character they possess when they were originally released to the air or water despite moving thousands of kilometers through various environmental media. This is an issue that has been explored in some specific cases, such as the extensive study of particular pollutants or classes of pollutants that produce toxic metabolites; one example is the explosive 2,4,6-trinitrotoluene (TNT), whose metabolites have been examined in soil and water collected from military bases and former munitions facilities in several places around the world.

Much of the concern for emerging pollutants has to do with uncertainty regarding how the differences in their molecular structure and properties relative to better understood pollutants will manifest in terms of transport and fate in the environment. Based on the available information on the behavior of well-studied pollutants, some generalities for transport behavior of various emerging contaminants can be proposed. Aggregation of nanoparticles, biologicals, and other amphiphilic compounds will cause these materials to behave more like particles than discrete molecules in air and water and will change the way they interact with the surfaces of particulate phases (i.e., soil and sediment). Nanoparticles have much larger molar masses and volumes than conventional contaminants. Their behavior is often dominated by coagulation kinetics. Another high molar mass group is the dyes and pigments that may have unmeasurably low solubilities in water. Ionizing substances present a challenge because both the protonated and deprotonated species can diffuse in parallel in water but the ion cannot enter and diffuse in the air phase. The extent of ionization is conventionally deduced using the Henderson–Hasselbalch equation but this may not always apply, especially in the presence of other ions and when the ion has surface-active properties and may associate in solution to form micelles. Polar compounds exhibit higher water solubilities and will interact differently with mineral surfaces and natural organic matter (NOM) than the more extensively examined nonpolar chemicals. Uptake of some pollutants by organisms can lead to excretion at a new location and may include the release of metabolites that have different transport properties and could be more toxic than the parent material. Considerable caution must be exercised when modeling substances with extreme properties, especially those with low solubilities in water (e.g., siloxanes) and experience speciation changes (e.g., ionizing substances and metals).

Processes that change the apparent molecular weight and size (via aggregation, coagulation, micelle formation, etc.), surface properties (fouling of nanoparticles surfaces with NOM), and chemical structure (via reaction, degradation, etc.) will be particularly challenging to deal with in transport models. Mass transfer coefficients describing diffusional and advective processes as well as the retardation factors describing the environmental interactions of the original and "transformed" materials will change as functions of time and with changes in the environmental composition (mineralogy, biota, soil quality, etc.) as they move within a single ecosystem and across different geographical regions.

21.4   A GLIMPSE AT THE WAY FORWARD

Many of the issues related to the behavior of emerging pollutants have been with us since the advent of mass transport assessment and modeling. Much progress has been made in the study of coupled reactive transport processes and modeling, and the tools that have been developed can be modified and improved to deal with emerging contaminant issues. The uncertainty in environmental mass transfer coefficients and other parameters that affect the rate of chemical transport is an issue that has received little attention, but is vital to the accurate forecasting of chemical distribution in the natural environment. For all the processes described in this handbook, the observed chemical transport rates will vary even within highly localized regions, reflecting small but potentially important variability in the transport mechanisms (wind speed and direction, bio- and phyto-turbation, etc.) and environmental media (temperature, composition, biota, etc.) as well as "noise" in the measured data. This indicates that a stochastic approach to describing the values of transport parameters is warranted in order to obtain a reasonable range of probable values for chemical concentrations within and fluxes between the phases of interest.

There is clearly a need to better measure and document mass transfer coefficients (MTCs) or velocities in the 41 individual processes described in this handbook. It would be very useful to have available compilations of MTCs for specific chemicals in specific environmental settings thus providing guidance as to the likely range of values. It is understandably difficult to measure MTCs in the highly variable natural environment, even for example, in a relatively simple system such as the water–air interface of a small lake. Complications result from thermal and meteorlogical variability and the presence of particles. These variables can be better controlled in the laboratory in wind-wave tanks but this introduces problems of scale, wind-driven currents, and fetch. What is needed are careful experimental measurements in both settings with appropriate interpretation of differences caused by scale. This is the best foundation for correlating MTCs as a function of environmental variables such as fluid velocity, geometric shape, surface roughness, particle concentration, stratification, temperature, and so on. This principle of complementary laboratory and field measurements can be applied to all intermedia transport processes. This body of information could form the foundation of new and improved estimation methods. We hope this handbook will encourage further measurements and theoretical developments in this important field of environmental science.

This handbook contains state-of-the-art estimation methods for environmental MTCs. The Internet has websites for estimating various environmentally relevant chemical properties, the U.S. Environmental Protection Agency's (EPI) Suite (2009) being an example. To our knowledge none exist for estimating environmental MTCs however, the availability of this handbook now provides the opportunity to develop and disseminate the much-needed information, data, and correlations.

LITERATURE CITED

Battaglin, W.A. and Kolpin, D.W. 2009. Contaminants of emerging concern: Introduction to a featured collection, *Journal of the American Water Resources Association* **45**, 1–3.

Blanford, W.J., Brusseau, M.L., Jim Yeh, T.C., Gerba, C.P., and Harvey, R. 2005. Influence of water chemistry and travel distance on bacteriophage PRD-1 transport in a sandy aquifer, *Water Research* **39**, 2345–2357.

Cao, H., Tao, S., Xu, F., Coveney, R.M., Jr., Cao, J., Li, B., Liu, W., Wang, X., Hu, J., Shen, W., Qin, B., and Sun, R. 2004. Multimedia fate model for hexachlorocyclohexane in Tianjin, China, *Environmental Science and Technology* **39**, 2126–2132.

Christian, P., Von der Kammer, F., Baalousha, M., and Hofmann, T. 2008. Nanoparticles: Structure, properties, preparation and behaviour in environmental media, *Ecotoxicology* **17**, 326–343.

Estimating Programs Interface (EPI) Suite v. 4.00. 2009. U.S. Environmental Protection Agency. http://www.epa.gov/oppt/exposure/pubs/episuitedl.htm

Hursthouse, A. and Kowalczyk, G. 2009. Transport and dynamics of toxic pollutants in the natural environment and their effect on human health: Research gaps and challenge, *Environmental Geochemistry and Health* **31**, 165–187.

Klaine, S.J., Alvarez, P.J., Batley, G.E., Fernandes, T.F., Handy, R.D., et al. 2008. Nanomaterials in the environment: Behavior, fate, bioavailability, and effects, *Environmental Toxicology and Chemistry* **27**, 1825–1851.

Meyer, T., Wania, F., and Breivik, K. 2005. Illustrating sensitivity and uncertainty in environmental fate models using partitioning maps, *Environmental Science and Technology* **39**, 3186–3196.

Snow, D.D., Bartelt-Hunt, S.L., Devivo, S., Saunders, S., and Cassada, D.A. 2007. Detection, occurrence, and fate of emerging contaminants in agricultural environments, *Water Environment Research* **81**, 941–958.

Walker, F.R. Jr. and Stedinger, J.R. 1999. Fate and transport model of cryptosporidium, *Journal of Environmental Engineering* **125**, 325–333.

Wania, F. and Mackay, D. 1999. The evolution of mass balance models of persistent organic pollutant fate in the environment, *Environmental Pollution* **100**, 223–240.

Index

A

ADE. *See* Advection–dispersion equation (ADE)
AD equation. *See* Advective–diffusive (AD)
    equation
Adiabatic lapse rate, 567
Advection, 178, 317, 573
    darcian flow, by, 306
    differential, 415–416
    downwind, 553
    porewater, 307
    process, 414
Advection–dispersion equation (ADE), 221, 418
    analytical solution, 425, 427
    breakthrough curves, 430
    Graetz–Nusselt problem solution, 442
    three-dimensional solution, 427
Advective
    conveyor-belt feeding, 362
    fluxes, 58–59, 306, 340
    flux formulation uses, 21
    processes, 138
    transport process, 302
    water flow, 188
Advective–diffusive equation (AD)
Advective flow modifier, 308. *See also* Peclet
    number
Advective-type fluxes, 58. *See also* Chemical flux
    air-to-soil deposition velocity, 59
    dust resuspension, 59
    flux equations, 58
    total deposition velocity, 58
Aeolian mass transport estimation, 469
    dust emission rate computation, 469, 472
    resuspension empirical studies, 472
Aeolian transport modes, 458
    modified saltation, 459
    particle suspension, 458
    particle tracking velocimetry, 459
Aerodynamic roughness length, 476, 522
    typical values, 462–464
Aerosol property descriptors, 110
Air-deep water flux, 505
Air-side mass transfer coefficients, 176
Air–water exchange, 503, 504
    atmospheric vapor pressure, 235
    buoyancy flux, 236
    convection types, 234, 235
    evaporative heat flux, 237
    gas film coefficient, 234, 235
    mass flux equation, 237
    momentum-flux stability function, 236
    momentum/unit volume difference, 236
    saturation vapor pressure, 235
    values for SOC, 547
    virtual temperature difference, 235
Air–water interface, 58, 213–214, 590
    analogous transport equations, 217–218
    diffusive flux, 216
    Henry's law constant, 215
    mass flux, 214, 215
    phenol concentration substitution, 216–217
    phenol flux rate, 215
Alabaster. *See* Gypsum
α-HCH. *See* α-hexachlorocyclohexane (α-HCH)
α-Hexachlorocyclohexane (α-HCH), 501
American Society of Civil Engineers
    (ASCE), 402
Amictic lakes, 579
Analogy theories, 27, 40, 312
    Chilton–Coburn analogy, 28
    Manning's roughness coefficient, 31
    MTC, 28, 29
    resistance equation, 31
    skin friction coefficients, 28, 29, 30
    velocity profile law, 28
Anemometers, 461
Annular flumes, 270
Apoplastic transport, 391
Aquaporins, 391
Aquifer zone. *See* Saturated zone
Archie's law, 340
    diffusion coefficient reduction, 344
    molecular diffusivity correction, 350
Arctic snow model, 520
    low pore space concentration, 527
    parameters, 526
ASCE. *See* American Society of Civil Engineers
    (ASCE)
Atmosphere, 567
    buoyancy, 567
    horizontal mixing, 570–571
    one-compartment box model, 573, 574
    snow exchange analysis, 521–522
    stable, 567
    unstable, 567
    vertical mixing, 567–570
    vertical structure, 568
    washoff, 543
Atmospheric mixing
    chemical availability, 515

595

Atmospheric mixing (*Continued*)
 chemical's flux calculation, 521
 horizontal winds impact, 573
 organic chemicals distribution, 515
 PCB-28 parameters influence, 525–527
 semivolatile chemicals, 522
 snow permeability influence, 524–525
 sorption coefficient, 515–516
 surface roughness influence, 522–523
 temporal and spatial scales, 572
 two-film resistance model, 518
 vapor concentration estimation, 518
 wind speed influence, 523–524
 wind ventilation, 520–521
Atmospheric residence time, 113–114
Average porewater velocity, 302. *See also*
  Advective—transport process; Darcian
  velocity
AW2 interface. *See* Air–water interface

B

Bank-water exchange process, 310
 fluxes, 308–309
 MTC estimation, 313
Barometric pumping, 193, 194–195
Batch technique, 228
 equation, 228
 Sherwood vs. Schmidt number, 229
BCFs. *See* Bioconcentration factors (BCFs)
Bedload, 263, 268, 293. *See also* Sediment
 deposition fluxes calculation, 288
 erosion flux, 286
Bed sediment diffusive transport, 348
 chemical flux, 348–349
 colloidal particle flux MTC, 348
 solute flux MTC, 348
Bed surface properties, 373
Benzoic acid MTC correlation, 335–336
Bioconcentration factors (BCFs), 392
 plant tissue water sorption, 395
 RCF, 394
 transpiration stream concentration factors, 396
Biological pump, 496
 chemical fluxes and MTC, 502–503
 phytoplankton biomass, 497
Biomantle, 363
Bioturbation, 274, 360, 361, 362
 agriculture and tillage mixing, 365
 aquatic sediments, 361
 conveyer belt, 362
 earthworm role, 364
 process, 4
 sediment MTC, 383–384
 soil bioturbation, 363, 384–385
 termites role, 364
 vertebrate species, 365

Bioturbation transport parameters estimation, 370
 chemical flux, 381
 sediment bioturbation, 370
 soil bioturbation, 377, 380–382
 time-period, 281
 transport processes, 382
Bitumen, 535, 536
Bottom-water MTC, 334
 natural convection density, 337–338
 non stratified water body, 335
 stratified lake, 336
 wind-generated bottom water currents, 336
Boundary layer, 568, 569
 aerodynamic roughness length, 522
 planetary, 114, 124, 571
 shear Reynolds number, 522
 surface roughness, 522, 523
 theory, 26–27, 335, 338
Breakthrough curve, 415, 419, 430
Brownian diffusion, 20, 346
 coefficients, 345
 deposition, 148
Buildings, underpressurization, 195
 relative factors, 195–196
 values, 195
Bulk
 advection velocity, 173
 aerodynamic resistance, 144, 145
 snow/air partition coefficient, 518
 soil diffusivity, 172
Bulk density, 280
 critical shear stress, 265
 displacement height, 146
 erosion rate, 267
 friction velocity, 145
 roughness heights, 147
Buoyancy, 567
 flux, 236
Burial velocity. *See*
  Sedimentation—velocity

C

C. *See* Inorganic carbon (C)
Canopy resistance, 144, 145, 146–147, 148
Capillary fringe, 189, 439
Capillary pressure, 423. *See also* Porous medium
Casparian strip, 391. *See also* Suberin
CE. *See* Continuity equation (CE)
Cementation factor, 342
CF&T models. *See* Chemical fate and transport
  (CF&T) models
CFD. *See* Computational fluid dynamic (CFD)
Chapman–Enskog equation, 74, 76
Chemical
 biotic ingestion, 7
 exposure assessment, 2
 mass balances, 19

mass transport, 16
mobility, 1–2, 16
potential, 72
solubilization, 61–62
vapor transport, 64–65
volatilization, 63–64
Chemical fate and transport (CF&T) models, 5
Chemical flux, 140–141, 307, 348–349, 375, 376, 475, 502
    aeolian process mass transport coefficient, 475
    biological pump, 502
    categories, 19
    dust resuspension MTC, 475
    emissions from slag field estimation, 478–483
    emissions from tailings as slurry estimation, 483–487
    one-dimension expression, 19
    physical pump, 503
    phytoplankton and PM sinking, 503
    soil materials and specific gravity, 476
Chemical transport, 1–2, 138
    advective processes, 7
    canopy, 137, 138
    diffusive processes, 6–7
    dry sedimentation, 141
    flowing media velocity, 7
    flux, 5–6
    soil particle resuspension, 141
    transport processes, 138, 139
    vegetation/air partition coefficient, 143–144
    volatilization, 143
Chilton–Coburn analogy, 28
Cohesive sediments, 254
    deposition, 262, 279
    flux, 275
    resuspension, 255, 256, 279
    turbulence impact, 257
Compartment box model, 63, 65
    advective-diffusive combination, 65–66
    chemical vapor transport, 64
    chemical volatilization, 63
    flux expression, 63, 65, 66
    interface compartment execution, 63, 64, 66
    interface concentration, 66
    Ohm's law application, 65
Computational fluid dynamic (CFD), 23
Concentration gradient, 22, 32, 54, 56, 179
    chemical, 348
    dispersion/diffusion processes, 168
    flux equation, 54, 56, 68
    multiple soil layers, 182–184
    porewater, 62, 307
    two-compartment model, 179–180
    water, 348
Concrete, 535
Contaminant, 160, 163, 165
    burial, 256

concentrations, 569, 573, 574
penetration in soil, 168
sources, 566
transport, 459
Contaminant fate process, 575
Contaminant fluxes, 457–458. See also Dust
Continuity equation (CE), 187, 189
    mass accumulation, 187
    species, 3, 8, 16, 187
Continuous stirred tank reactor (CSTR), 9
Controlled flux technique, 233
Convection–dispersion equation. See Advection–dispersion equation (ADE)
Convection mass transfer, mixed, 35–36
Convective mass flux equation, 22
Conveyor-belt feeding, 362
Coriolis force, 332, 570
CPD. See Cumulative probability distribution (CPD)
Critical shear stress, 263, 264–265, 266, 271
    motion initiation, for, 264
    small particles, for, 266, 267
Critical shear velocity, 269
    motion initiation, for, 281
    suspension initiation, for, 285
Crusting, 468
Cryoturbation, 360, 362–363. See also Bioturbation
CSTR. See Continuous stirred tank reactor (CSTR)
Cumulative probability distribution (CPD), 309, 371, 378
Cup anemometry, 461

D

Damkoehler number, 170
Darcian advection, 306
    bed-sediment MTCs, 306
    chemical flux, 307
    Darcian velocity, 306
    porewater advection, 307
Darcian velocity, 7, 21, 303, 305, 306, 308, 313
    bank exchange water fluxes, 308–309
    bank-water exchange process, 310
    Darcian water velocities, 309
    estimation equation, 312, 313
    superficial porewater, 312
Darcian velocity estimation, 312
Darcy's law, 188, 305, 525
DDT. See Dichlorodiphenyltrichloroethane (DDT)
Deep water formation, 496, 497, 499. See also Physical pump
    heat transport inverse modeling, 500
    rate estimation, 501
    regions, 501
DEHP. See Diethyl hexyl phthalate (DEHP)
Dense nonaqueous phase liquids (DNAPLs), 190, 438, 439

DEP. *See* Diethyl phthalate (DEP)
Deposition, 261
   biological implications, 580–581
   contaminant burial, 256
   dry, 124–125
   dry aerosol-bound, 147–149
   dry gaseous, 144–146
   dry particle velocities, 120
   estimation, 275
   factors in, 257
   function probability, 285
   probability of deposition, 261, 262
   shear stress impact, 293
   wet, 125–126
Depth hoar, 517
Des. *See* Differential equations (DEs)
De-topped plant, 396
Dichlorodiphenyltrichloroethane (DDT), 592
Diethyl hexyl phthalate (DEHP), 533
Diethyl phthalate (DEP), 533
Differential advection, 415
   breakthrough curve, 415
   mechanical dispersion, 415
   one-dimensional column experiment, 416
Differential equations (DEs), 56
   chemical solubilization, 61, 62
   interface compartment concept, 62
   re-suspension, 62
Diffusion, 71–72, 89, 197, 198, 199, 580, 592
   coefficients, 198
   depth, 165
   diffusivities, 90
   mineral lattices distance, 91
   parameters, 90
   path length (dpl), 146, 520
   processes, 138
   reactive decay, with, 37
   solid minerals, through, 89–90
Diffusion coefficient, 72, 74, 77, 86, 198
   APLs, in, 89
   Chapman–Enskog equation, 74, 76
   chemical interactions, 87
   chemical resistance factor, 87
   collision diameter calculation, 74–75
   collision integral calculation, 75–76
   comparison studies, 201
   correction factor, 84
   diffusion volumes, 78, 82
   dynamic viscosity, 85, 86
   effective, 199, 200
   empirical correlation, 88
   empirical equations, 199
   estimation, 73, 80
   Eyring theory, 81
   FSG method, 199
   gypsum in water, 96–97
   Hayduk–Laudie method, 81

   hydrodynamic theory, 80–81
   ice and snow, in, 88
   influenced factors, 199
   interaction energy calculation, 75
   limiting ionic conductance, 83
   mean free path, 74
   Millington's expressions, 200, 201
   molecular weight correlation, 78–79, 84
      in air, 92–94
      in hydraulic oil, 98–100
      in saturated sediment, 97–98
      in unsaturated soil, 98
      in water, 95–96
   Nernst equation, 83
   physical effects, 86–87
   porous media, in, 86
   Scheibel method, 81, 82
   semiempirical correlations, 77
   snow porosity estimation, 88–89
   solid–water partition coefficients, 87
   sorption, 88
   temperature correction, 79–80
   temperature dependence, 85
   weighted average approach, 80
   Wilke–Chang correlation, 81
Diffusive chemical transport, 197, 339
   aqueous phase, in, 197
   bed sediment, 348
   dispersion process, 201
   in-bed naphthalene MTC, 356
   molecular diffusion process, 339
   natural convection MTC for gypsum, 355–356
   particle bio-diffusion, by, 197–198
   smooth and rough surface MTCs, 351–352
   upper sediment layers, 339
   vapor phase, in, 198–201
   water-side MTCs, 350–351, 352–355
Diffusive flux, 6
   air–water interface, 216
   direction, 7
   porous media expression, 340
Diffusive gaseous exchange, 140
Diffusive mixing models, 367
   advective–diffusive transport equation, 367
   application, 368–369
   flux equation, 368
Diffusive-type fluxes, 54. *See also* Chemical flux
   compartment box model, 63–65
   concentration gradient, 54, 56
   DEs, 56
   media concentrations, 57
   MTC, 56
   transfer rate constants, 57
   transport parameters, 57–58
   WS interface, 54, 56
Diffusivity. *See* Diffusion coefficient
Dimictic lakes, 579

Dispersion, 419
Dispersivity, 414, 418, 433
    coefficients estimation, 422–425, 427, 428
    longitudinal, 418, 423
    pore-scale, 423, 425
    transverse, 426
Dispersivity coefficients, Forced gradient
        test, 428
    advantage, 428
    single-well injection test, 428
    single-well push–pull test, 428–429
    two-well test, 429
Dispersivity determination, 425
    device for transverse, 426–427
    methods for column tests, 425–426
    single-well withdrawal tests, 431–433
    unsaturated medium sand, 429–431
Dissolved organic carbon (DOC), 345, 356
    bed sediment DOC diffusivities, 346
    production, 345
DNAPLs. See Dense nonaqueous phase liquids
        (DNAPLs)
DOC. See Dissolved organic carbon (DOC)
dpl. See diffusion—path length (dpl)
Dry aerosol-bound deposition, 147
    Brownian diffusion deposition, 148
    chemical's partition ability, 147
    mass transfer coefficients, 152
    particle impaction/interception, 148–149
    sedimentation, 148
Dry deposition, 104, 163. See also Wet deposition
    aerosols, 148
    associated factors, 105
    chemical, 140, 147
    conceptual resistance model, 111, 112
    flux calculation, 111
    mass transfer coefficient, 105, 152
    parameter calculations, 124, 125
    pathways, 105
    values for volume fractions, 118
    velocity calculations, 111, 116–117, 118,
        119–121
Dry gaseous deposition, 144
    bulk aerodynamic resistance, 144–146, 153
    canopy resistance, 144, 145, 146–147
    mass transfer coefficient, 144, 147, 151
    quasilaminar sublayer resistance, 144,
        145, 153
Dryoturbation, 360, 363. See also Bioturbation
Dry particle deposition, 105, 111, 116
    conceptual resistance model, 112
    MTC, 127
    pathways, 105
Dry sedimentation, 141
Dual tracer technique, 228
    analysis with gas tracers, 228, 229
    first-order-second moment analysis, 232

mass balance influence, 229–230
    nonvolatile tracer analysis, 230–231
    precision uncertainty association, 231, 232
    Sherwood vs. Schmidt number, 233
    tracers transport relation, 229
    wind velocity importance, 232
Dust, 454, 455
    classification, 455
    concentration measurement, 468
    fugitive, 455
    monitoring instruments, 470
    particles, 459
    plume emission, 455
    respiratory illnesses, 455
    resuspension process, 475
    samplers, 469
    storm, 454
Dust emission rate, 469, 472
    diffusive component, 471
    flux, 470, 472
    power law, 471

E

Earth surface processes, 2
Eddy
    accumulation, 150
    correlation technique, 238, 239
    diffusivity, 22, 471
EF. See Emission factor (EF)
EFDC. See Environmental Fluid Dynamics Code
        (EFDC)
Effective diffusion coefficient, 54, 199, 340, 348,
        374. See also Diffusion coefficient
    literature relations, 200
    moisture content effect, 199, 206
Effective diffusivities, 86, 315. See also Diffusion
        coefficient
    estimation, 88
    snow and ice, in, 88–89
"18 degree" water, 500
Emission
    estimation, 456, 476–477, 478–483, 483–487
    factor (EF), 456
    rate, 456, 469, 472
Emission factor (EF), 456
Empirical correlations, 34–35
    molecular weight, 77, 78
    snow, for, 88
Endogeic species, 364
Endophytes, 400
Environmental
    chemodynamics, 5
    solvents, 71
Environmental Fluid Dynamics Code (EFDC),
        277, 278
    cohesive sediments, 279–280
    noncohesive sediments, 280–281, 281–282

Environmental Fluid Dynamics Code (EFDC)
    (*Continued*)
    sediment bed representation and armoring,
        283–284
    and SEDZLJ comparison, 287–288
    SNLEFDC, 288
EO. *See* Oxidation potential (EO)
Epifauna, 323, 361. *See also* Macrofauna
Epigeic species, 364
Epilimnion, 332, 578, 580
EQC. *See* EQuilibrium Criterion (EQC)
EQuilibrium Criterion (EQC), 571
Erosion, 263, 270. *See also* Resuspension;
        Scour
    Annular flumes, 270
    Sedflume, 270–271
    Straight flumes, 271
Erosional/depositional process, 274, 275, 595
    in estuary, 275
    factors affecting, 263
    noncohesive sediments, 281
    preferential, 283
    rate, 263, 265–268
    sediment in SEDZLJ, 284–285
    sediment transport models, 276–277
Ethylene vinyl acetate (EVA), 541
ETo. *See* Evapotranspiration (ETo)
Eutrophication, 576
EVA. *See* Ethylene vinyl acetate (EVA)
Evapotranspiration (ETo), 403
Exposure period, 396
Eyring theory, 81

F

Fickian-type diffusion coefficient, 198
Fick's first law, 8
    molecular diffusion, 20
    turbulent diffusion analogy, 21–22
Film theory, 23, 24
Film wash-off, 542, 548. *See also* Surface film
    efficiencies variation, 550
    films removal efficiency, 542–543
    first flush, 543–544
    ISI, 549
    maximum chemical mass, 549
    RC, 543
    steady-state chemical loading, 549
    surface films removal, 542, 543
Fine particulate (FP), 455
Fine-grained sediments. *See* cohesive sediments
First flush, 543–544
Flocculation, 259. *See also* Aggregation
    factors in, 257
    fluid shear, 260
Floralturbulation, 364. *See also* Bioturbation
Flow water exchange process, 313

Darcian velocity, 313
    MTC estimation, 316
Fluid mixing, 374
Fluorotelomer alcohols (FTOHs), 516
Flux, 5–6
    advective-type, 58–59
    air–water–phytoplankton–settling, 498
    chemical, 19–20, 475
    diffusive-type, 54–58
    dimensions, 6
    in estuary, 275
    global atmospheric, 272, 273
    impact of, 256
    mechanics, 258
    organic carbon (OC), 496, 498
    resuspension, 280
    sinking, 496
Forced convection, 34, 36, 234, 235, 236
    Chilton–Colburn analogy, 35
    fluid-to-bed surface friction velocity, 328
    MTC, 324, 550
    water-side boundary layer, 323
Forest filter
    effect, 142, 553
    factor, 142, 143
FP. *See* Fine particulate (FP)
Free convection, 32, 235
    creation, 234
    density decrement, 32
    heat transfer, 32–33, 36
    mass transfer, 33–34
Friction velocity, 30, 144, 328, 464
    quadratic drag coefficient, 328–331
    quadratic drag law, 328
FSG method. *See* Fuller–Schettler–Giddings
        (FSG) method
FTOHs. *See* Fluorotelomer alcohols (FTOHs)
Fugacity approach, 46
    air–water partition coefficient, 44
    characteristic time calculation, 47–48
    diffusion process, 46–47
    fugacity capacity, 44
    ideal gas law, 44
    mass balance model, 48–50
    partition coefficients list, 45
    rate equation, 45
    rate parameter definitions, 46
    two-resistance theory relationship, 47
    volatilization rate, 47
Fugacity mass balance model, 48, 49
    characteristic time, 49, 50
    input data, 49
    sediment fugacity level, 48
    steady-state equations, 48
    unsteady-state differential equations, 50
Fuller–Schettler–Giddings (FSG) method,
        77, 199

G

Gaseous deposition, 27
Gaseous transport, 142
Gas film coefficient, 219, 234, 238
　air–water transfer, 238
　characteristic relations application, 246, 247,
　　248
　compound A concentration, 238
　computation over water body, 245–246, 247
　developed relationships, 240
　eddy correlation technique, 239
　estimation, 237
　pan evaporation data, 238
　prediction, 239, 243
　water evaporation, 238
Gas/particle partitioning, 106
　chemicals distribution, 106
　one-parameter correlations, 107–108
　parameter calculations, 122, 123–124
　particle-associated chemical fraction, 107
　partition coefficient, 106–107
　polyparameter energy relationships, 109–110
　sorption capacity, 106
　surface area-to-volume ratio, 107
Gilland–Sherwood
　correlations, 443, 444, 445
　models, 443. *See also* Graetz–Nusselt problem
　　solution
Graetz–Nusselt problem solution, 442
Grain size, 524
　critical shear stress, 265
　erosion rate, 267
　suspension initiation, 269
Grashoff number, 33
　additivity, 36
Ground-surface-soil layer, 165
Groundwater (GW), 302
GW. *See* Groundwater (GW)
Gypsum, 324
　accumulation, 325
　diffusivity, 96–97
　MTC measurement, 325
　natural convection MTC, 355–356

H

Hanratty's approach, 26
Harmful compounds, 194
Hayduk–Laudie method, 81
Hayduk-Minhas method, 89
HBCD. *See* Hexabromocyclododecane (HBCD)
HCB. *See* Hexachlorobenzene (HCB)
HCHs. *See* Hexachlorocyclohexanes (HCHs)
Heat transfer coefficients (HTCs), 33, 550
Henry's law constant, 13, 215
Hexabromocyclododecane (HBCD), 533
Hexachlorobenzene (HCB), 123

Hexachlorocyclohexanes (HCHs), 516
High-volume (HiVol), 455
HiVol. *See* High-volume (HiVol)
HOCs. *See* Hydrophobic organic compounds
　　(HOCs)
Hot-wire anemometry, 461
HTCs. *See* Heat transfer coefficients
　　(HTCs)
Hydraulic conductivity, 189, 304,
　　305, 420
　estimation, 310
　partially saturated, 189
　ranges for earthen materials, 311
Hydrodynamic dispersion. *See also* Mechanical
　　dispersion
　coefficient, 417
　dispersive solute mass flux, 416
　macroscale, 419
　microscopic pore scale, 416
　Peclet number, 417
Hydrodynamic theory, 80
Hydrolytic transformation, 166
Hydrophobic
　contaminants, 163
　organic compounds (HOCs), 87
　substances, 7
Hydrophobic organic compounds (HOCs), 87
Hypolimnion, 333, 578, 580–581
　concentration effect, 584
　maximum average current velocity, 336
　separation, 578
Hyporheic flow, 302, 316. *See also*
　　Advective—transport process
　noncircular, 306
　pathway, 303
　types, 304

I

Ideal gas law, 44, 567
Impervious surface, 532, 534. *See also* Urban
　　areas
　air-side MTC, 550
　chemical films development, 535
　chemical films effect, 535–536
　chemical loading levels, 556–557
　chemodynamic pathways, 544–547
　coverage, 534–535
　diffusive exchange, 536
　effect, 533, 534
　film growth, 537–539
　film wash-off model, 548–549
　index (ISI), 549, 553
　infiltration capacity, with, 543
　linear film growth rate model, 547–548
　possible mass of chemical, 549
Impervious surface index (ISI), 549, 553
Infauna, 323, 361. *See also* Macrofauna

Inhalable particulate (IP), 455
Inorganic carbon (C), 580
Instantaneous uniform mixing model,
    366–367
Interface compartment, 60–61, 64, 67
Interfacial flux, 218–219
Interstitial water. *See* Porewater
Intraphase transport, 591
IP. *See* Inhalable particulate (IP)
ISI. *See* Impervious surface index (ISI)

L

Lake
    mass balance case study, 584–585
    mixing and diffusive processes, 581
Landfill gas generation, 196
    affecting factors, 196
    anaerobic gas production phase, 196
    migration, 196–197
Langmuir circulations, 332–333
Lapse rate, 567
Lavoisier mass balance, 6, 17, 65
Leaking underground storage tanks (LUSTs),
    187, 205
Light nonaqueous phase liquids (LNAPLs),
    190, 438
Linear driving model, 440
    mass flux rate, 440
    NAPL pool dissolution, 441
Liquid film coefficient, 219, 226
    batch technique, 228
    C^{14} and Rn^{222} measurements, 227
    controlled flux technique, 233
    dual tracer technique, 228–233
    measurements, 242
    predictive relationships, 233–234
    wind velocity adjustment, 227
Litter fall, 141
Littoral zone, 579
LNAPLs. *See* Light nonaqueous phase liquids
    (LNAPLs)
LUSTs. *See* Leaking underground storage tanks
    (LUSTs)

M

Macrofauna, 323, 364
    biomixing by, 198
    surface mixing depths, 380
Magnus–Tetons formula, 235
Manning's roughness coefficient, 31, 330
Marine deposit feeders, 362
Mass balance model, 570
    mechanistic, 404
    setting up, 572
    surface area-to-volume ratio, 107

Mass flux, 214
    air–water interface, 214–217
    dust, 477
    horizontal and vertical, 470
Mass rate
    flux equation, 202–203
    ocean water, 501–502
Mass transfer
    eddies, 497
    immobile to mobile phase, 202
    mode waters, 497–498
    parameters, 180
    vertical diurnal migration, 497
Mass transfer coefficient (MTC), 6, 23, 56, 111,
    162, 176, 313, 324, 370, 375–377, 518,
    582, 590
    air-side estimation, 40, 41
    analogy theories, 27–32
    bank-water exchange, 313
    bioturbation, 376
    bottom roughness correction, 327–328
    boundary layer theory, 26–27
    chemical flux, 375, 376
    data correlations, 326
    differential surface water elevation
        exchange, 316
    diffusive, 306
    dry particle deposition, 127, 131
    film theory, 23–24
    Hanratty's approach, 26
    measurements, 67–69, 324–326
    natural convection theory, 32–34
    Nusselt correlations, 34, 35
    penetration theory, 24–25
    roughness induced water exchange,
        314–316
    Sherwood correlations, 34
    surface renewal theory, 25–26
    theoretical correction factors, 338
    transport through boundary layer, 519
    transport through bulk snow, 519–520
    water-side MTCs estimation, 39, 40–41,
        327, 332
    wet gaseous deposition, 129–130
    wet particle deposition, 128, 131
Mass transfer conceptual models, 439
    linear driving model, 440
    one 1D vertical dispersion model, 439
    stagnant film model, 441
Mass transfer rate coefficients, 442
    dissolution using spherical blobs
        model, 447
    dissolution using tubular model, 447–449
    Gilland–Sherwood models, 443
    Graetz–Nusselt problem, 442
    numerical experiments, 449–450
    2D complete dissolution problem, 446

upscaled mass transfer correlation, 445
wall dissolution, 442, 443
Mass transfer rates calculation, 175, 176
multiple soil layers, 175, 182–185
single soil compartment, 175
two-compartment model, 176–180
Mass transport, 16
categories, 52
current status, 589–591
deficiencies, 591–593
emerging pollutants behavior, 593–594
Fick's first law, 8
flux categories, 19
MTC, 9
plant BCF—chemical properties relationship, 403–404
processes, 53–54, 55
quantification, 403
rate estimation, 404–405
roughness correction factors, 327
species CE, 8
sulfolane removal from wetland, 405
two-resistance theory, 9
uptake plans, 405–406
Mathematical simulations, 5
Mean free path, 74
Mean multiplicative error (MME), 222
advantages, 223
Mechanical dispersion, 201, 415
one-dimensional column experiment, 415, 416
relative contribution, 417
Metalimnion, 333, 578, 580
Microbial transformations, 166
Millington's expressions, 200, 201
Mixing in surface waters
density effect, 577–578
thermal stratification, 578
wind-driven mixing, 582
MME. *See* Mean multiplicative error (MME)
Mobile–immobile model, 418–419
Mode waters, 498
subtropical, 500
Moisture, 86, 196, 199
capacity, 189
effect, 206
Molecular diffusion, 24, 71, 339
Archie's law, 340
Brownian diffusion coefficient, 345
chemical mobility, 339
coefficients, 91, 368, 583
colloidal matters, 343, 345
diffusion coefficients, 341
initial porosity, 343, 344
kinetic theory indication, 73
porous media, 342
sediment bed pore sizes, 342, 346

sedimentary material porosity, 339
solute diffusive transport, 340
Momentum diffusion, 217
Momentum-flux stability function, 236
Monomictic lakes, 579
MTC. *See* Mass transfer coefficient (MTC)
Multimedia urban model (MUM), 167, 552–556
conceptual issue, 164
requirement, 167
Multiple soil layers, 173, 182, 183
concentration vs. normalized depth, 185
gradients in concentration, 175
mass-balance equations, 182
transfer factors, 183, 184
MUM. *See* Multimedia urban model (MUM)

N

NAAQS. *See* National Ambient Air Quality Standards (NAAQS)
NAPH. *See* Naphthalene (NAPH)
Naphthalene (NAPH), 556
in air, 92–94
in hydraulic oil, 98–100
in saturated sediment, 97–98
in unsaturated soil, 98
in water, 95–96
NAPL. *See* Nonaqueous phase liquid (NAPL)
National Ambient Air Quality Standards (NAAQS), 455
National Oceanographic and Atmospheric Administration (NOAA), 550
Natural
capping, 373
colloids, 345
convection models, 32–34
porous media, 342
sedimentary materials, 343
solid surfaces, 22
Natural organic matter (NOM), 593
Near-surface thermoclines, 580
Nernst equation, 83
Net deposition. *See* Sedimentation
Net erosion. *See* Scour
Net sedimentation, 255, 271–276
NMOCs. *See* Nonmethane organic compounds (NMOCs)
NOAA. *See* National Oceanographic and Atmospheric Administration (NOAA)
Nocturnal boundary layer, 569, 570
impact on air pollutants, 572–573
NOM. *See* Natural organic matter (NOM)
Nonaqueous phase liquid (NAPL), 82, 89, 187, 190, 202, 408
capillary fringe, 439
DNAPLs, 438
expressions, 191
hydraulic gradient, 193

Nonaqueous phase liquid (NAPL) (*Continued*)
   LNAPLs, 438
   migration in saturated soil, 191
   mobilization, 192
   pool dissolution, 441
   properties, 190
   relative permeability, 192
   residual saturation, 191
   steady-state saturated flow, 191–192
   transport dependence, 190
Noncohesive sediments, 254, 255, 270
   particle settling, 254
   probability of deposition, 261, 262
   resuspension, 255, 281–282
   scour, 255
   settling and deposition, 280–281
Nonlocal mixing, 362. *See also* Conveyor-belt
        feeding
   conveyor-belt species, by, 369
Nonmethane organic compounds (NMOCs), 196
Non-stratified lake, 333. *See also* Stratified lake
   wind–water interaction current, 335
Nonvolatile compounds, 217
   air–water transfer, 238
   transportation, 406

O

OC. *See* Organic carbon (OC)
Ocean conveyor belt, 496, 500
1-octanol, 540
Octanol–water partition coefficient, 394
OM. *See* Organic matter (OM)
One-compartment box model, 573
One-dimensional column experiment, 415, 416
   breakthrough curve, 415
One-dimensional vertical dispersion model, 439
   governing equation, 440
   time to complete dissolution, 440
   vertical concentration profile, 440
One-parameter correlations, 107–108, 109
Organic carbon (OC), 12, 88, 496, 580
   dissolved, 345, 346
   fluxes, 502
   normalized sorption coefficient, 400–401
   settling flux, 510
   water partition coefficient, 88, 502
Organic chemicals estimation, 516
   chemical space plot, 517
   depth hoar, 517
   organic contaminants prediction, 516, 517
   SSA values, 516
Organic matter (OM), 12, 496
   absorption, 109, 110
   aerosol, 118
   chemical transport vehicle, 496
   DOC, 345

   environmental solvents, 71
   fulvic acid, 164
   litter fall, 163
   NOM, 593
   partition coefficients, 12, 45
   sediment, 88
   snow, 516
   soil profile, 161
   SOM, 401
Organic phases, 7
Orographic fog, 139
Oscillations, 334
Overlying buildings effects, 195
Oxidation potential (EO), 166

P

PAHs. *See* Polycyclic aromatic hydrocarbons
        (PAHs)
Pankow adsorption model, 107–108
Particle aggregation, 259. *See also*
        Flocculation
   mechanism, 259–260
   settling speed, 260–261
Particle biodiffusion, 198
   bed-side, 382
   coefficient, 371–372, 378–379
   soil side flux, 66
Particle entrainment, 464, 467
   constraints, 467
   crusting, 468
   friction velocity, 464
   Kelvin equation, 467
   physical dust emission model, 466
   saltation, 467
   threshold friction velocity variation, 465
Particle mixing depth, 372–373
Particle-reactive, 498
Particle resuspension empirical studies, 472
   vertical dust flux, 472–473
   windblown dust flux prediction, 473, 474
Particulate matter (PM), 454, 535
   deep water formation fluxes, 498
   deep water masses formation, 499
   ocean water mass rates, 501
   oceanic deep water formation rates, 501
   settling fluxes estimation, 499
   water masses upwelling, 500
Particulate organic matter (POM), 580
Patch scale, 543
Pavement
   advection, 573
   asphalt, 535
PBDEs. *See* Polybrominated diphenyl ethers
        (PBDEs)
PCBs. *See* Polychlorinated biphenyls (PCBs)
PCDD/Fs. *See* Polychlorinated dibenzodioxins
        and furans (PCDD/Fs)

PCE. *See* Perchlorethylene (PCE)
Peclet number, 307, 417
　advective flow modifier, 308
　sedimentation and bioturbation rates, 376
　settling rates vs. mixing rates, 367
Pedbioturbation, 362. *See also* Bioturbation
Penetration depth, 171
Penetration theory, 24–25
Penman–Monteith (PM) equation, 403
Perchlorethylene (PCE), 533
Persistent organic pollutants (POPs), 163, 495,
　532, 591
PFR. *See* Plug flow reactor (PFR)
Photic zone, 498
Phototransformation, 166
Physical dust emission model, 466
Physical pump, 496
　chemical fluxes and MTC, 503
Phytoplankton and PM sinking, 503–505
Phytoremediation, 138
Phytovolatilization, 398–399
Planetary boundary layer, 571
Plant biotransformation, 399
　endophytes, 400
　phases, 400
Plant canopy, 137, 138
　bulk aerodynamic resistance, 144–146
　chemical flux, 140–141
　cuticle erosion, 141
　diffusive gaseous exchange, 140
　dry aerosol-bound deposition, 147–149
　dry gaseous deposition, 144–147
　dry sedimentation, 141
　eddy accumulation, 150
　mass transfer coefficients, 151, 152, 153–156
　organic chemicals accumulation, 149
　quasilaminar sublayer resistance, 144, 145
　soil particle resuspension, 141
　vertical gradient measurement, 149–150
　wet deposition, 139
Plant tissue water sorption coefficients, 395
　K_{lignin} and K_{OW}, 395–396
　K_{wood} and K_{OW}, 396
　SCF, 395
Plant uptake of chemicals, 396
Plug flow reactor (PFR), 9
PM. *See* Particulate matter (PM)
PM equation. *See* Penman–Monteith (PM)
　equation
Polybrominated diphenyl ethers (PBDEs), 533
Polychlorinated biphenyls (PCBs), 273, 274, 516,
　535, 569, 592
　linear accumulation, 539–540
　probabilistic processes, 254–255
Polychlorinated dibenzodioxins and furans
　(PCDD/Fs), 535

Polycyclic aromatic hydrocarbons (PAHs),
　516, 592
　residence time estimation, 506–507,
　533, 535
Polymictic lakes, 579
Polyparameter linear free energy relationships
　(pp-LFERs), 109, 110
　Richards' equation, 189
　soil system, 188
　solute-to-solid phase transport, 203
　specific discharge of water, 418
　tortuosity, 199
　van Genuchten, 423
　vapor-to-solid phase transport, 204
POM. *See* particulate organic matter (POM)
POPs. *See* Persistent organic pollutants (POPs)
Porewater, 342
　biodiffusion coefficient, 374–375
　flow paths, 303
　substantial inflow, 304
Porosity, 86
Porosity–tortuosity, 520
Porous medium, 346, 423
　capillary rise, 189
　diffusional flux, 199
　longitudinal dispersivity coefficients,
　423, 424
　NAPL, 190, 202
　physical transport, 187
pp-LFERs. *See* polyparameter linear free energy
　relationships (pp-LFERs)
Prandtl–von Karman equation, 460
Precipitation, 58, 105, 176
　average rate, 548
　and canopy, 140
　chemical removal, 104
　and film thickness, 537
　film washoff, 536, 537, 542–544,
　549–550
　global average, 121
　impervious surface, 553
　mass transfer, 202
　quantification, 543
　wet deposition, 139
Pressure differentials, 304
Probability of deposition, 261, 262
　Drag forces, 261
　Floc porosity, 261
　individual particles, 258
　salinity effect, 260
　speed, 260
　velocity, 259
Proportionality constant, 8, 545
Pumping, 312
PYR. *See* Pyrene (PYR)
Pyrene (PYR), 556

Q

QSS flux. *See* Quasisteady state (QSS) flux
Quadratic drag
    coefficient equivalent, 329
    coefficient estimation, 328–331
    law, 328, 335
Quantitative water air sediment interaction
    (QWASI), 48
Quasilaminar sublayer resistance, 144, 145
Quasisteady state (QSS) flux, 348
QWASI. *See* Quantitative water air sediment
    interaction (QWASI)

R

Radioactive decay, 47, 167
RC. *See* Runoff coefficient (RC)
RCF. *See* Root concentration factor (RCF)
Reaeration coefficient, 221
    estimation, 243–245
    measurements, 221–222, 223
    MME, 222
    predictions, 222, 224–225
    ramifications, 225, 226
    surface renewal rate, 224
    surface vortices, 225
Redox potential, 166
Remineralization, 580
Representative elemental volume (REV),
    438, 442
Reptation, 459
Residence time estimation, 506–507
Residual saturation, 191
Resuspension
    cohesive sediments, 255
    contaminants, 255
    impact of, 256
    particle transport, 269
    river, 255
Retention function, 423. *See also* Capillary
    pressure
REV. *See* Representative elemental
    volume (REV)
Richards' equation, 189
River parameterizing, 582–583
Root concentration factor (RCF),
    394–395
Root uptake, 390
    apoplastic transport, 391
    aquaporins, 391
    organic chemical transport, 393
    phloem channels, 392
    symplastic transport, 391
    transpiration, 390–391
    xylem channels, 392
Runoff coefficient (RC), 543, 558

S

SA2 interface. *See* Soil–air (SA2) interface
Saltation, modified, 459
Sandia National Laboratories (SNL), 288
Saturated zone, 162, 414
Saturation degree, 423
SAV. *See* Subaquatic vegetation (SAV)
Scavenging ratio, 112
SCF. *See* Stem concentration factors (SCF)
Scheibel method, 81, 82
Scour, 255
Sedflume, 270–271
Sediment
    erosion, 293
    focusing, 582
    properties, 288–289
    water exchange processes, 254
Sedimentation, 148
    dry, 141
    net, 255, 256, 271–276
    rate, 373
    velocity, 21, 373
Sediment bed, 272
    armoring, 283
    consolidation, 284
    hiding factor, 283
    vertical distributions, 272
Sediment bioturbation transport
    parameters, 370
    bed surface properties, 373
    biodiffusion coefficient, 371, 374
    flux, 371
    mass-transfer coefficient, 375
    mixing depth, 372
    sedimentation velocity, 373
Sediment transport
    calibration, 288–289
    case study, 289–292
    Delft 2D, 277
    Delft 3D, 277
    ECOMSED, 277
    EFDC, 277
    GSTARS, 276
    HEC-6, 276
    MIKE21, 277
    MIKE-3, 277
    MOBED, 276
    modes, 269
    research need, 294–296
    RMA-10, 277
    SEDZL, 277
    SEDZLJ, 277
    USTARS, 277
    WASP, 276
Sediment-water interface, 339
    chemical solubilization, 62
    porosity adjustment, 344

sediment erosion, 263
turbulence, 261
SEDZLJ
and EFDC comparison, 287–288
sediment bed, 286–287
sediment erosion simulation, 284–285
settling and deposition, 285–286
Seepage
rate distribution, 309
rates determination, 307–308
Seiches, 333
deep water mixing, 581
surface layer, 334
time-period, 336
and wind, 350
Semivolatile organic compounds (SOCs),
535, 536
air-side MTCs, 545, 547
dynamics, 536
equilibrium solvation, 540, 541
film dynamics model, 536
log–log relationships, 541
Settling
flux comparison of PCBs, 508–510
flux disequilibrium to derive, 500
flux estimation, 499–500, 503, 505–506,
507–508
nonhydrophobic chemicals, 505
OC flux, 498–499
speed, 258
transport, 284
SF$_6$. *See* Sulfur hexafluoride (SF$_6$)
Shear
critical shear stress, 263
effects, 260
stress, 255
velocity and sediment transport modes, 269
Sherwood number, 33–34
Shields curve, 264
Shrink–swell behavior. *See* dryoturbation
Sinking flux
ocean surface water, 496
parameterization, 504
Sinks. *See also* Contaminant—sources
cuticle, 146
deep ocean, 495, 555
Skin friction coefficients, 28, 29, 30
SNL. *See* Sandia National Laboratories (SNL)
Snow, 514
chemical transport, 514–515
influences, 514
metamorphism, 514
permeability, 524–525
surface/air sorption coefficient, 515–516
temperature correction, 516
Snow ventilation
arctic snow model, 520–521

mass transfer, 520
snow permeability, 524–525
surface roughness, 524
wind speed, 523–524
SOCs. *See* Semivolatile organic compounds
(SOCs)
Soil, 160, 161
advective water flow, 188
bioturbation, 165
chemicals partitioning, 164, 165
composition, 163–164
contaminants, 162–163, 164
ground-surface-soil layer, 165
horizons types, 161
oxidation/reduction reactions, 166
radioactive decay, 167
saturated zone, 162
tortuosity, 199
transformations, 166–167
Soil organic matter (SOM), 401
Soil–air (SA2) interface, 58, 591
Soil biodiffusion coefficients, 378, 379
Soil bioturbation transport parameters, 377
flux, 377
soil particle biodiffusion coefficient, 378
soil particle "mixing" depth, 379
soil-side MTC, 380
vapor-phase diffusive transport, 378
Soil concentration gradient,
170–171, 174
characteristic time, 171, 174
concentration curves, 171
Damkoehler number, 170
penetration depth, 171, 174
Soil contamination, 162
contaminant releases, 163
dry deposition, 163
transfer processes, 164
Soil gas movement, 193, 204
barometric pumping, 193, 194–195
landfill gas generation, 196–197
soil vapor intrusion, 194, 195–196
TCE flux measurements, 204–206
Soil gas pressure gradients, 193, 194
Soil mass transfer, 167
bulk advection velocity, 173
bulk-soil effective diffusivity, 172
compartment property terms, 172
conceptual model, 168
fugacity capacity definitions, 173
mass balance equations, 168–170
partitioning models, 167
partition properties, 168
reaction rate constant, 171–172
soil concentration gradient, 170–171, 174
transfer rates calculation, 175–176
vertical transport, 168

Soil particle resuspension, 141
Soil pore water concentrations, 400–401
Soil vapor intrusion, 194, 195–196
Solid–water partition coefficients, 87
SOM. *See* Soil organic matter (SOM)
Sorbed-phase transport, 360. *See also* Bioturbation
    bioturbation and, 361–365
    processes, 362
Sorption, 88
    adsorption model, 107
    air-borne particles, 454
    bulk aerosols, 108
    coefficients, 394, 395–396, 400
    dry soils, 204
    estimation, 106, 515–516
    ionic effect, 401
    OM absorption, 109
    snow, 115
SP. *See* Suspended particulate (SP)
Species continuity equation (CE), 8, 17–18, 187
    causative terms, 18
    chemical fate, 17–18
    chemical flux, 3
    chemical mobility, 16
    Lavosier mass balance, 17
    media box models, 18
Specific surface area (SSA), 516
    flocs, 259
    particle size and, 196
    snowflakes, 115, 516, 517
    wet deposition, 139
Specimen chemicals, 9, 174
    partition coefficients, 12, 13
    properties, 12
Spring turnover, 578
SSA. *See* Specific surface area (SSA)
Stable boundary layer, 569
Stable low-level boundary layer. *See* Nocturnal
        boundary layer
Stack effect, 195
Stagnant film model, 441
    mass rate equation, 442
    mass transfer coefficient, 442
    steady mass flux, 441, 442
Standard correction, 377
Steady-state flux equations, 20
    advective transport, 21
    convective mass transfer, 22–23
    diffusion across films, 20, 22
    eddy diffusivity, 22
    Fick's first law, 20
    turbulent flux, 21
Stem concentration factors (SCF), 395
Stokes–Einstein's equation, 346. *See also*
        Brownian diffusion—coefficients
Stokes' law, 254, 257–258
Straight flumes, 271

Stratified lake, 332–333, 575
    bottom-water MTC model, 336–337
    internal circulating water current, 333–334
    water-side MTC estimate, 354–355
Subaquatic vegetation (SAV), 323
Suberin, 391
Subtropical mode waters, 500
Sulfur hexafluoride (SF_6), 521
Surface
    mixing depths, 379–380
    renewal theory, 25–26
    water movements, 332
Surface film. *See also* Film wash-off
    accumulation, 535, 539–540, 542
    chemical mass, 557
    chemical partitioning, 539–542
    chemodynamic pathways, 544–545
    film dynamics model, 536
    flux approximation, 545
    impervious surfaces, 535
    linear growth rate model, 547–548
    MTC equation, 545
    organic fraction, 535
    oxidative reactions, 544
    removal efficiency, 542
    runoff concentration, 557, 559
    thickness, 537
    types, 545
    wash-off, 542, 548–549
Surface film growth, 537
    covered glass beads, on, 538
    differential accumulation, 539
    growth rates, 537, 538, 549
    mercury accumulation, 538
    washoff, 539
Surface roughness
    bed, 323
    canopy, 144, 146
    feature, 523, 525
    friction coefficients, 28
    low, 111
    MTC, 29
    natural surfaces, 146
    snow pack, 514
    snow ventilation, 523–524
    water, 238
Surface wind, 460
    aerodynamic roughness length, 462–464
    bottom water currents, 350
    impact on MTC, 32
    Prandtl–von Karman equation, 460
    speed, 476
    wind profile equation, 461
Suspended load, 268
    bedload, 268–269
    factors affecting, 268
    sorbed contaminants, 263

Suspended particulate (SP), 455, 471
Symplastic transport, 391

T

TBA. See *tert*-butyl alcohol (TBA)
TCDD. *See* 2,8-tetrachlorodibenzodioxin
          (TCDD), 3, 7
TCE. *See* Trichloroethylene (TCE)
Tension-saturated zone. *See* Capillary fringe
tert-butyl alcohol (TBA), 398
2,8-tetrachlorodibenzodioxin (TCDD), 3, 7, 556
Thermal gaseous contaminants, 33, 36, 41, 163,
          187, 355
      two-parameter correlations, 108
      volumetric partition coefficient, 106, 107
Thermal stratification, 575
      in autumn, 579
      density-temperature impact, 576
      depth impact, 578
      factors affecting, 580
      temperature–depth profile, 579
      in temperate water, 578
      volume profile, 581
Thermocline, 332, 578
      near-surface thermoclines, 580
      vertical dispersion coefficients, 582
Thermodine, 333
Thompson Island Pool (TIP), 352
Tillage mixing, 360, 363
      agriculture and, 365
Timescales
      and diffusion, 72, 90
      global horizontal transport, 570
      regional scales, 571
      settling flux determination, 499
TIP. *See* Thompson Island Pool (TIP)
TMDL calculations. *See* Total daily maximum
          load (TMDL)
TNT. *See* 2,6-trinitrotoluene (TNT), 4
Tortuosity, 87
      dispersivity, 418
      factor, 199, 443
      porosity–tortuosity, 520
      soil, 199
Total daily maximum load (TMDL), 222
Total suspended particle (TSP), 106, 455
Trace organics, 532
      PAH, 533
      pollutants, 535
Tracers, radioisotopes, 498
Transpiration, 390–391, 401
      controlling factors, 401
      estimation, 402
      evapotranspiration (ETo), 403
      Penman–Monteith (PM) equation, 403
      rates, 401–403

Transpiration stream concentration factor (TSCF),
          393, 396–398
Transport parameter, 150–153
      bioturbation, 370–382
      and correlations, 116–122, 308–316
      Darcian velocity, 308
      dispersivity determination, 424, 425, 427
      estimation, 421–429
      hydraulic conductivity, 304
      mass-transfer coefficient, 313
      measurement, 149–150, 307–308
      molecular diffusion, 71
      MTC, 56
      regression laws, 422
      soil, 171–174
Transport processes, 138–143
      bioturbation and sorbed-phase, 361–365
      diffusive and advective, 6–7
      hyporheic exchange, 302
      individual mass, 53–54
      simultaneous parallel and series, 382
      in soil column, 164–165
Transport stage parameter, 281
      grain shear stress, 282
Transport theory, 305, 366, 416, 458
      covariance function, 420
      Darcian advection, 306
      Darcy's law theory, 305
      diffusive mixing models, 367
      dispersion coefficients, 420
      dust concentration, 468
      equations, 418
      hydrodynamic dispersion, 416, 419
      nonlocal mixing, 369
      particle entrainment, 464, 467
      surface wind, 460
      transport modes, 458
      transport parameters, 307
      transverse macrodispersivity, 421
      uniform mixing model, 366
      wind speed measurement, 461
Trichloroethylene (TCE), 204, 396
2,6-trinitrotoluene (TNT), 4, 592
Tropopause, 568
      air movement, 570
      mass transfer coefficient, 571
      transport in, 570
Troposphere, 567
      atmosphere Vertical structure, 568
      atmospheric mixing, 572
      Earth's surface, 568
      free troposphere, 568, 569
      mixing, 569, 570
      windspeeds, 570
TSCF. *See* Transpiration stream concentration
          factor (TSCF)
TSP. *See* Total suspended particle (TSP)

Turbidity maximum, 255
Two-compartment model, 176
    advection/dispersion losses, 179
    chemical persistence variation, 180–181
    concentration gradient, with, 179–180
    mass balance, 177–178
    mass loss, 178–179
    mass transfer factor expressions, 178
    MTC, 176–177
    net flow, 176, 177
    soil–air interface, 177
    specimen chemicals properties, 174
Two-film theory. *See* Two-resistance theory
Two-parameter correlations, 109, 110
Two-region model. *See* Mobile–immobile model
Two-resistance theory, 9

U

Unsaturated zone
    atmospheric pressure impact, 193
    barometric pumping, 194–195, 204
    capillary fringe, 189
    in soil column, 161
    soil pores, 189, 438
    vapor transport, 198
    water infiltration, 439
    water table, 414
Upper sediment layers, 339
    bioturbation, 367
    contaminants, 360
Upwelling, 500–501
    mass transfer, 497
    point seepage, 304
    velocity, 502
Urban areas, 532. *See also* Impervious surface
    conceptual model of ecosystems, 534
    discharged chemicals list, 532–533
    impervious surfaces effect, 533, 534
    physical and chemical properties, 546
Urban runoff, 542, 552, 553, 557

V

Vadose zone, 414
    contaminants advection, 198
    soil-layer compartments, 182
    solute spreading, 421
Vapor–particle partitioning, 106
Vapor transport, 198
Vegetation, 137
Vignette chemodynamic models, 36–37
    advection–diffusion, 38
    concentration profile, 38
    diffusion with reactive decay, 37
VOCs. *See* Volatile organic compounds (VOCs)
Void ratio. *See* Bulk density

Volatile chemicals, 517
Volatile organic compounds (VOCs), 195,
        202, 547
Volatilization, 581
    chemical transfer, 63, 139, 399
    contaminant transfer, 164, 575
    diffusive gaseous exchange, 140
    phytovolatilization, 398
    rate, 47
    surface concentration, 584
    temperature effect, 553
    velocity, 7, 19
Volumetric flux 21
Volumetric partition coefficient, 106, 107
von Karman's constant, 328, 460, 479, 519

W

WASP. *See* Water Quality Simulation Program
        (WASP)
Water advection processes
    bottom roughness, 317–318
    effect on flux, 307
    ground water, 55
    river bed MTC, 317
Water currents, 332
    internal water movements, 333
    surface layer seiches, 334
    surface water movements, 332
Water currents, 332, 353
Water partition coefficient, 142
    air, 7, 13, 44, 87, 125, 127, 168
    lipid, 12
    octanol, 127, 168, 173, 394, 395, 515
    organic carbon, 12, 88, 502
    solid, 87
Water quality models, 57
Water Quality Simulation Program (WASP), 276
    sediment transport simulation, 277–278
Water–sediment (WS) interface, 54, 590–591
    chemical transport, 322
    in mass transfer chemodynamics, 52
Water-side boundary layer, 323
    chemodynamic process, 382
Water-side mass transfer coefficients, 220
    convection–dispersion equation, 221
    gas tracer pulses, 222
    reaeration coefficient, 221–222, 223
    reaeration coefficient, 222–226
    theoretical relationships, 220
Water, surface
    anthropogenic disturbances, 576
    contaminant fate, 575
    density-temperature, 577
    organic contaminants, 574, 576
    parameterizing mass transport, 582–583
    thermal stratification, 575
Water table, 414

Weighted average approach, 80
Well-mixed box, 565, 566
    in atmosphere compartment, 572
    in river, 583
Wet deposition, 104, 105. *See also* Dry deposition
    associated factors, 105
    chemical, 139
    chemical flux calculation, 112
    chemical flux with snowflakes, 115
    long-term average flux, 114
    long-term precipitation rate, 121–122
    mass transfer coefficient, 105, 152
    parameter calculations, 125–126
    partitioning, 115, 139
    pathways, 105
    residence time, 113–114
    scavenging ratio, 122, 123
    snow, 114
Wet gaseous deposition, 105, 112, 121
    equations, 125
    flux, 113
    Lindane and B (*a*)P removal, 126

MTC, 113, 115, 125, 129, 130
    temperature effect, 115
Wet particle deposition, 105, 125
    calculation, 115
    chemical flux, 112
    MTC, 128
    nickel removal, 126
Wilke–Chang correlation, 81
Wilke–Lee model, 77
Wind-generated
    bottom water currents, 336
    bottom-water MTC, 334
    convective mass transport, 332
Wind pack, 521, 525
Wind profile equation, modified, 461
Wind pumping, 523
wind pumping factor (wpf), 520
Wind speed measurement, 461
Wind–water interaction model, 335
wpf. *See* wind pumping factor (wpf)
WS interface. *See* Water–sediment (WS) interface